Shpol'skii Spectroscopy and Other Site-Selection Methods

CHEMICAL ANALYSIS

A SERIES OF MONOGRAPHS ON ANALYTICAL CHEMISTRY
AND ITS APPLICATIONS

Editor
J. D. WINEFORDNER

VOLUME 156

A JOHN WILEY & SONS, INC., PUBLICATION

New York • Chichester • Weinheim • Brisbane • Singapore • Toronto

Shpol'skii Spectroscopy and Other Site-Selection Methods

Applications in Environmental Analysis, Bioanalytical Chemistry, and Chemical Physics

Edited by

Cees Gooijer, Freek Ariese

Department of Analytical Chemistry and Applied Spectroscopy
Vrije Universiteit Amsterdam
The Netherlands

and

Johannes W. Hofstraat

Molecular Photonics Group, University of Amsterdam &
Department of Polymers and Organic Chemistry
Philips Research, Eindhoven
The Netherlands

A JOHN WILEY & SONS, INC., PUBLICATION

New York • Chichester • Weinheim • Brisbane • Singapore • Toronto

This book is printed on acid-free paper. 

Published simultaneously in Canada.

For ordering and customer service, call 1-800-CALL-WILEY

Library of Congress Cataloging-in-Publication Data:

Gooijer, C.
Shpol'skii spectroscopy and other site selection methods applications in environmental analysis, bioanalytical chemistry, and chemical physics / C. Gooijer, F. Ariese and J.W. Hofstraat.
p. cm. – (Chemical analysis a series of monographs on analytical chemistry and its applications ; v. 156)
Includes index.
ISBN 0-471-24508-9 (alk. paper)
1. Molecular spectroscopy. I. Ariese, F. II. Hofstraat, J. W. III. Title. IV. Chemical analysis ; v. *156*.

QD96.M65 G66 2000
543'.08584–dc21 99-055651

Printed in the United States of America.

10 9 8 7 6 5 4 3 2 1

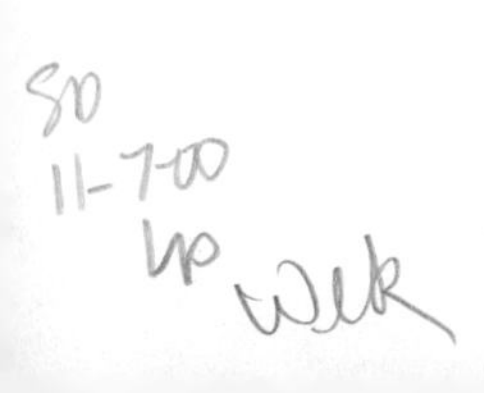

To Professor Nel Velthorst (Vrije Universiteit, Amsterdam), one of the pioneers in high-resolution molecular luminescence spectroscopy, on the occasion of her sixty-fifth birthday and (formal) retirement in 1997.

CONTENTS

PREFACE

It is a great pleasure for the editors to dedicate this volume to our stimulating teacher, colleague in science, and close friend Nel Velthorst, professor in General and Analytical Chemistry at the Vrije Universiteit in Amsterdam. For three decades, Nel has been fascinated by the high-resolution phenomena in cryogenic molecular luminescence. As early as the mid-1970s, she started research projects on the Shpol'skii technique and in later years also on fluorescence line-narrowing spectroscopy (FLNS), both in analytical and physical chemistry. Nel's group has reported impressive results in the latter discipline, a highlight being the study on the Jahn–Teller effects observed for the photochemically generated phenalenyl radical. Nevertheless, we feel that the main part of her efforts has been directed on the improvement of the robustness of these techniques and on the demonstration of their potential and relevance in analytical practice. The significance of her work is illustrated by the references quoted by the various authors contributing to this volume.

The editors fully agree with Nel Velthorst that recording high-resolution luminescence spectra is truly fascinating. Spectra are obtained with an extremely high information content compared to the well-known fluorescence and phosphorescence spectra as measured for organic molecules under conventional conditions. Simply stated, high-resolution luminescence spectroscopy combines the selectivity inherent to luminescence with the selectivity (or even specificity) characteristic of infrared and Raman spectroscopy.

This book aims to demonstrate that high-resolution spectroscopy, not only in the luminescence but also in the absorption mode, deserves attention from a variety of disciplines. It concentrates on solid, low-temperature matrices; supersonic jet approaches to achieve high resolution will not be considered in this monograph. It will be shown that recent technological breakthroughs, notably in the field of lasers, detection, and data processing, and also in the construction of sample holders, have significantly improved the userfriendliness of the techniques concerned; they will undoubtedly stimulate new applications in disciplines like environmental analysis, bioanalytical chemistry, as well as chemical physics. The book is a multiauthor monograph, and we greatly appreciate the willingness of internationally recognized leaders in their respective speciality areas to write the individual chapters.

A general introduction in the field, following the historical development, is presented by Personov, the founder of FLNS who has been active in cryogenic molecular luminescence for over 25 years. The order of the subsequent chapters reflects a fundamental distinction between high-resolution spectra obtained for molecules in crystalline and amorphous matrices. Basically, there are two approaches to obtain highly resolved spectra in molecular electronic spectroscopy. For analytes in crystalline matrices, including *n*-alkane and mixed crystals, it is based on analyte–matrix compatibility. The analyte molecules should fit into—ideally—identical sites of the crystal structure for the inhomogeneous broadening of the spectral lines to be reduced. In contrast, for amorphous matrices, laser-based selection within a broad inhomogeneous band is performed. Only those analyte molecules that are resonant with the laser wavelength are observed.

The section on crystalline matrices focuses on cryogenic *n*-alkane solutions, usually denoted—after their inventor—as Shpol'skii matrices. It starts with a highly informative chapter on the principles of matrix-induced high-resolution spectroscopy, written by Renge and Wild, followed by a comprehensive overview by Lamotte discussing the trapping mechanisms of polycyclic aromatic hydrocarbons (PAHs) in Shpol'skii matrices.

The three subsequent chapters deal with the applicability of the Shpol'skii technique to modern, relevant analytical problems. First of all, attention is focused on large PAHs, that is, polycyclic aromatic hydrocarbons consisting of six or more aromatic rings—a class of analytes very difficult to handle with current techniques—in a thorough chapter written by Zander. The next chapters show that the applicability of the Shpol'skii technique is not limited to PAHs. It has been successfully invoked to study both in-ring and at-ring substituted heteroaromatics, as outlined by Kozin and Matsuzawa. The same holds for PAH metabolites, a topic evaluated by Ariese. This is an important finding, since modern environmental scientists attempt to follow the complete and detailed pathway of toxic compounds through the environment and within organisms. The identification of PAH metabolites in complex environmental samples, present at (extremely) low levels, is a tremendous challenge. The crystalline section is completed with a chapter from the area of chemical physics, written by Migirdicyan, Parisel, and Berthier. They show that Shpol'skii matrices are very appropriate to obtain detailed spectroscopic information on photochemical intermediates, such as organic biradicals.

In the second part of this book, dealing with site- or energy-selection methods in cryogenic amorphous matrices, the high-resolution techniques are studied not only in luminescence, but also in absorption. Both organic and inorganic systems are discussed.

The section starts with an in-depth treatment of the fundamental aspects of FLNS in a chapter written by Jankowiak. It includes a thorough discussion of zero-phonon lines and phonon side bands, evaluates model calculations of fluorescence spectral line shapes, gives a brief description of instrumental aspects of FLNS, and ends with a sampling of representative applications. In the next chapter by Creemers and Völker, the fundamental aspects of spectral hole burning are explained. Like in FLNS, in this technique the inhomogenously broadened absorption band is irradiated with a narrow-band laser and only the molecules resonant with the laser wavelength are excited. In hole-burning spectroscopy it is crucial that the excited molecules undergo a phototransformation (photochemical or photophysical) so that they can no longer be excited at the laser wavelength. Thus a hole is created in the original absorption band at the laser frequency. The chapter focuses on the possibility to extract information about the dynamics of the amorphous matrix from the shape of such holes. Interestingly, both "normal" organic glasses and proteins can be employed as amorphous matrices. In other words, the technique can be used to study the highly relevant relaxation processes in proteins.

The subsequent four chapters describe recent applications of FLNS and hole-burning spectroscopy in environmental and bioanalytical chemistry. Chapter 10, by Gooijer and Kok, is directed at the coupling of FLNS with liquid chromatography (LC) and capillary electrophoresis (CE). The objective of this coupling is to exploit fully the identification potential of FLNS in analytical practice. Although, obviously, the cryogenic sample conditions inherent to FLNS prohibit a direct coupling, both for LC and for CE interesting interface devices have been developed and convincing applications are shown. Chapter 11 by Ariese and Jankowiak further illustrates the identification power of the FLNS technique. It has been successfully applied to elucidate the (stereo)isomeric structures of PAH–DNA and PAH–protein adducts. Furthermore, the FLN spectra of these compounds provide information on the immediate environment of the fluorophore, for example, the degree of intercalation inside a DNA helix. Next, Vanderkooi evaluates the possibility to study protein conformations and relaxation processes by means of FLNS. The information obtained is more or less complementary to that achieved by spectral hole burning, outlined in Chapter 9. Chapter 13, by Rätsep and Small, discusses the use of spectral hole burning to investigate photosynthetic complexes. The authors review how the combination with high pressure and external electric (Stark) fields leads to new insights into the excited-state electronic structures and excitation energy transfer processes of light harvesting complexes. Recent results, obtained for various bacteriophyll $\underline{a}$ antenna complexes, are presented.

The last four chapters are devoted to chemical physics studies in amorphous matrices. First, Zilker and Haarer give an in-depth, clear overview of site-selection spectroscopy of amorphous polymers, which serves to illustrate the state of the art, both theoretically and experimentally. Possible applications in the field of data storage are evaluated as well.

Chapter 15, written by Tamarat, Jelezko, Lounis, and Orrit, describes the evolution of the fascinating field of single molecule spectroscopy during the past decade. Spectroscopy of single guest molecules in host solids under cryogenic conditions has been performed for various host–guest systems. Single molecules can be used for various purposes, for example, as ultrasensitive probes of the dynamical degrees of freedom active in solids at cryogenic temperatures.

In the next chapter, Hofstraat and Wild present appealing experimental results showing the development of high-resolution excitation–emission approaches. The three-dimensional plots, combining highly resolved excitation with highly resolved emission spectra, provide a direct insight into the structural and dynamic aspects of the environment of the luminescent molecules. Of course, they also provide an extremely high identification power.

This monograph ends with a chapter written by Murdoch and Wright on site-selection spectroscopy of defects in inorganic materials. It gives a comprehensive overview, focusing on rare-earth doped systems, in particular fluorite crystals. There is wide interest in these systems, since they may be useful for energy upconversion processes—optica1 pumping at one wavelength and emission at a shorter one—as well as for data storage purposes. The editors feel that this book provides a balanced overview of the possibilities of high-resolution molecular spectroscopy. We express the sincere hope that it will inspire readers from various disciplines to explore and implement these fascinating techniques.

Cees Gooijer, Freek Ariese, and Johannes W. Hofstraat

CONTRIBUTORS

FREEK ARIESE Dept. of Analytical Chemistry and Applied Spectroscopy, Vrije Universiteit Amsterdam, Amsterdam, the Netherlands

GASTON BERTHIER Laboratoire d'Etude Théorique des Milieux Extrêmes, Ecole Normale Supérieure, Paris, France

TIJSBERT M. H. CREEMERS Center for the Study of Excited States of Molecules, Huygens and Gorlaeus Laboratories, University of Leiden, Leiden, the Netherlands

CEES GOOIJER Dept. of Analytical Chemistry and Applied Spectroscopy, Vrije Universiteit Amsterdam, Amsterdam, the Netherlands

DIETRICH HAARER Physics Department and Bayreuther Institut für Makromolekülforschung, University of Bayreuth, Bayreuth, Germany

JOHANNES W. HOFSTRAAT Molecular Photonics Group, University of Amsterdam, & Dept. of Polymers and Organic Chemistry, Philips Research, Eindhoven, the Netherlands

RYSZARD JANKOWIAK Ames Laboratory-USDOE, Iowa State University, Ames, Iowa, USA

FEDOR JELEZKO Centre de Physique Moléculaire Optique et Hertzienne, CNRS, Université de Bordeaux I, Talence, France

STEVEN J. KOK NV Organon, div. Pharmaceutics, Oss, the Netherlands

IGOR S. KOZIN Department of Chemistry, Queen's University, Kingston, Ontario, Canada

MICHEL LAMOTTE Laboratoire de Physico-Toxico-Chimie des Milieux Naturels, Université de Bordeaux I, Talence, France

BRAHIM LOUNIS Centre de Physique Moléculaire Optique et Hertzienne, CNRS, Universit de Bordeaux I, Talence, France

SADAO MATSUZAWA Atmospheric Environmental Department, National Institute for Resources and Environment (NIRE), Tsukuba, Japan

EVA MIGIRDICYAN Laboratoire de Photophysique Moléculaire du CNRS, Université Paris-Sud, Orsay, France

KEITH M. MURDOCH Department of Chemistry, University of Wisconsin, Madison, Wisconsin, USA

MICHEL ORRIT Centre de Physique Moléculaire Optique et Hertzienne, CNRS, Université de Bordeaux I, Talence, France

OLIVIER PARISEL Laboratoire d'Etude Théorique des Milieux Extrêmes, Ecole Normale Supérieure, Paris, France

ROMAN I. PERSONOV Institute of Spectroscopy, Academy of Sciences of Russia, Troitsk, Moscow Region, Russia

MARGUS RATSEP Ames Laboratory-USDOE and Dept of Chemistry, Iowa State University, Ames, Iowa, USA

INDREK RENGE Institute of Physics, Tartu University, Tartu, Estonia

GERALD J. SMALL Ames Laboratory-USDOE and Dept of Chemistry, Iowa State University, Ames, Iowa, USA

PHILIPPE TAMARAT Centre de Physique Moléculaire Optique et Hertzienne, CNRS, Université de Bordeaux I, Talence, France

JANE M. VANDERKOOI Department of Biochemistry and Biophysics, School of Medicine, University of Pennsylvania, Philadelphia, Pennsylvania, USA

SILVIA VOLKER Center for the Study of Excited States of Molecules, Huygens and Gorlaeus Laboratories, University of Leiden, Leiden, the Netherlands

URS P. WILD Physical Chemistry Laboratory, Swiss Federal Institute of Technology/ETH, Zürich, Switzerland

JOHN C. WRIGHT Department of Chemistry, University of Wisconsin, Madison, Wisconsin, USA

MAXIMILIAN ZANDER Castrop-Rauxel, Germany

STEPHAN J. ZILKER Physics Department and Bayreuther Institut für Makromolekülforschung, University of Bayreuth, Bayreuth, Germany

CHEMICAL ANALYSIS

A SERIES OF MONOGRAPHS ON
ANALYTICAL CHEMISTRY AND ITS APPLICATIONS

J. D. Winefordner, *Series Editor*

Vol. 39. **Activation Analysis with Neutron Generators**. By Sam S. Nargolwalla and Edwin P. Przybylowicz
Vol. 40. **Determination of Gaseous Elements in Metals**. Edited by Lynn L. Lewis, Laben M. Melnick, and Ben D. Holt
Vol. 41. **Analysis of Silicones**. Edited by A. Lee Smith
Vol. 42. **Foundations of Ultracentrifugal Analysis**. By H. Fujita
Vol. 43. **Chemical Infrared Fourier Transform Spectroscopy**. By Peter R. Griffiths
Vol. 44. **Microscale Manipulations in Chemistry**. By T. S. Ma and V. Horak
Vol. 45. **Thermometric Titrations.** By J. Barthel
Vol. 46. **Trace Analysis: Spectroscopic Methods for Elements**. Edited by J. D. Winefordner
Vol. 47. **Contamination Control in Trace Element Analysis**. By Morris Zief and James W. Mitchell
Vol. 48. **Analytical Applications of NMR**. By D. E. Leyden and R. H. Cox
Vol. 49. **Measurement of Dissolved Oxygen**. By Michael L. Hitchman
Vol. 50. **Analytical Laser Spectroscopy**. Edited by Nicolo Omenetto
Vol. 51. **Trace Element Analysis of Geological Materials**. By Roger D. Reeves and Robert R. Brooks
Vol. 52. **Chemical Analysis by Microwave Rotational Spectroscopy**. By Ravi Varma and Lawrence W. Hrubesh
Vol. 53. **Information Theory As Applied to Chemical Analysis**. By Karl Eckschlager and Vladimir Stepanek
Vol. 54. **Applied Infrared Spectroscopy: Fundamentals, Techniques, and Analytical Problem-solving**. By A. Lee Smith
Vol. 55. **Archaeological Chemistry.** By Zvi Goffer
Vol. 56. **Immobilized Enzymes in Analytical and Clinical Chemistry**. By P. W. Can and L. D. Bowers
Vol. 57. **Photoacoustics and Photoacoustic Spectroscopy.** By Allan Rosenewaig
Vol. 58. **Analysis of Pesticide Residues**. Edited by H. Anson Moye
Vol. 59. **Affinity Chromatography**. By William H. Scouten
Vol. 60. **Quality Control in Analytical Chemistry.** *Second Edition.* By G. Kateman and L. Buydens
Vol. 61. **Direct Characterization of Fineparticles**. By Brian H. Kaye
Vol. 62. **Flow Injection Analysis**. By J. Ruzicka and E. H. Hansen
Vol. 63. **Applied Electron Spectroscopy for Chemical Analysis**. Edited by Hassan Windawi and Floyd Ho
Vol. 64. **Analytical Aspects of Environmental Chemistry**. Edited by David F. S. Natusch and Philip K. Hopke
Vol. 65. **The Interpretation of Analytical Chemical Data by the Use of Cluster Analysis.** By D. Luc Massart and Leonard Kaufman

Vol. 90. **Inductively Coupled Plasma Emission Spectroscopy: Part 1: Methodology, Instrumentation, and Performance; Part II: Applications and Fundamentals.** Edited by J. M. Boumans

Vol. 91. **Applications of New Mass Spectrometry Techniques in Pesticide Chemistry**. Edited by Joseph Rosen

Vol. 92. **X-Ray Absorption: Principles, Applications, Techniques of EXAFS, SEXAFS, and XANES.** Edited by D. C. Konnigsberger

Vol. 93. **Quantitative Structure-Chromatographic Retention Relationships**. By Roman Kaliszan

Vol. 94. **Laser Remote Chemical Analysis**. Edited by Raymond M. Measures

Vol. 95. **Inorganic Mass Spectrometry.** Edited by F. Adams, R. Gijbels, and R. Van Grieken

Vol. 96. **Kinetic Aspects of Analytical Chemistry.** By Horacio A. Mottola

Vol. 97. **Two-Dimensional NMR Spectroscopy.** By Jan Schraml and Jon M. Bellama

Vol. 98. **High Performance Liquid Chromatography.** Edited by Phyllis R. Brown and Richard A. Hartwick

Vol. 99. **X-Ray Fluorescence Spectrometry.** By Ron Jenkins

Vol. 100. **Analytical Aspects of Drug Testing**. Edited by Dale G. Deustch

Vol. 101. **Chemical Analysis of Polycylclic Aromatic Compounds**. Edited by Tuan Vo-Dinh

Vol. 102. **Quadrupole Storage Mass Spectrometry**. By Raymond E. March and Richard J. Hughes

Vol. 103. **Determination of Molecular Weight**. Edited by Anthony R. Cooper

Vol. 104. **Selectivity and Detectability Optimization in IIPLC.** By Satinder Ahuja

Vol. 105. **Laser Microanalysis**. By Lieselotte Moenke-Blankenburg

Vol. 106. **Clinical Chemistry**. Edited by E. Howard Taylor

Vol. 107. **Multielement Detection Systems for Spectrochemical Analysis.** By Kenneth W. Busch and Marianna A. Busch

Vol. 108. **Planar Chromatography in the Life Sciences.** Edited by Joseph C. Touchstone

Vol. 109. **Fluorometric Analysis in Biomedical Chemistry: Trends and Techniques Including HPLC Applications.** By Norio Ichinose, George Schwedt, Frank Michael Schnepel, and Kyoko Adochi

Vol. 110. **An Introduction to Laboratory Automation.** By Victor Cerd and Guillermo Ramis

Vol. 111. **Gas Chromatography: Biochemical, Biomedical, and Clinical Applications.** Edited by Ray E. Clement

Vol. 112. **The Analytical Chemistry of Silicones.** Edited by A. Lee Smith

Vol. 113. **Modern Methods of Polymer Characterization.** Edited by Howard G. Barth and Jimmy W. Mays

Vol. 114. **Analytical Raman Spectroscopy.** Edited by Jeanette Graselli and Bernard J. Bulkin
Vol. 115. **Trace and Ultratrace Analysis by HPLC.** By Satinder Ahuja
Vol. 116. **Radiochemistry and Nuclear Methods of Analysis.** By William D. Ehmann and Diane E. Vance
Vol. 117. **Applications of Fluorescence in Immunoassays.** By Ilkka Hemmila
Vol. 118. **Principles and Practice of Spectroscopic Calibration.** By Howard Mark
Vol. 119. **Activation Spectrometry in Chemical Analysis.** By S. J. Parry
Vol. 120. **Remote Sensing by Fourier Transform Spectrometry.** By Reinhard Beer
Vol. 121. **Detectors for Capillary Chromatography.** Edited by Herbert H. Hill and Dennis McMinn
Vol. 122. **Photochemical Vapor Deposition.** By J. G. Eden
Vol. 123. **Statistical Methods in Analytical Chemistry.** By Peter C. Meier and Richard Zünd
Vol. 124. **Laser Ionization Mass Analysis**. Edited by Akos Vertes, Renaat Gijbels, and Fred Adams
Vol. 125. **Physics and Chemistry of Solid State Sensor Devices**. By Andreas Mandelis and Constantinos Christofides
Vol. 126. **Electroanalytical Stripping Methods**. By Khjena Z. Brainina and E. Neyman
Vol. 127. **Air Monitoring by Spectroscopic Techniques**. Edited by Markus W. Sigrist
Vol. 128. **Information Theory in Analytical Chemistry**. By Karel Eckschlager and Klaus Danzer
Vol. 129. **Flame Chemiluminescence Analysis by Molecular Emission Cavity Detection**. Edited by David Stiles, Anthony Calokerinos, and Alan Townshend
Vol. 130. **Hydride Generation Atomic Absorption Spectrometry**. By Jiri Dedina and Dimiter L. Tsalev
Vol. 131. **Selective Detectors: Environmental, Industrial, and Biomedical Applications**. Edited by Robert E. Sievers
Vol. 132. **High Speed Countercurrent Chromatography**. Edited by Yoichiro Ito and Walter D. Conway
Vol. 133. **Particle-Induced X-Ray Emission Spectrometry**. By Sven A. E. Johansson, John L. Campbell, and Klas G. Malmqvist
Vol. 134. **Photothermal Spectroscopy Methods for Chemical Analysis.** By Stephen E. Bialkowski
Vol. 135. **Element Speciation in Bioinorganic Chemistry.** Edited by Sergio Caroli
Vol. 136. **Laser-Enhanced Ionization Spectrometry.** Edited by John C. Travis and Gregory C. Turk

Vol. 137. **Fluorescence Imaging Spectroscopy and Microscopy.** Edited by Xue Feng Wang and Brian Herman
Vol. 138. **Introduction to X-Ray Powder Diffractometry.** By Ron Jenkins and Robert L. Snyder
Vol. 139. **Modern Techniques in Electroanalysis.** Edited by Petr Vanýsek.
Vol. 140. **Total Reflection X-Ray Fluorescence Analysis.** By Reinhold Klockenkamper.
Vol. 141. **Spot Test Analysis: Clinical, Environmental, Forensic, and Geochemical Applications,** *Second Edition* By Ervin Jungreis
Vol. 142. **The Impact of Stereochemistry on Drug Development and Use**. Edited by Hassan Y. Aboul-Enein and Irving W. Wainer
Vol. 143. **Macrocyclic Compounds in Analytical Chemistry**. Edited by Yury A. Zolotov
Vol. 144. **Surface-Launched Acoustic Wave Sensors: Chemical Sensing and Thin-Film Characterization.** By Michael Thompson and David Stone
Vol. 145. **Modern Isotope Ratio Mass Spectrometry.** Edited by T. J. Platzner.
Vol. 146. **High Performance Capillary Electrophoresis: Theory, Techniques, and Applications.** Edited by Morteza G. Khaledi
Vol. 147. **Solid Phase Extraction: Principles and Practice.** By E. M. Thurman
Vol. 148. **Commercial Biosensors: Applications to Clinical, Bioprocess and Environmental Samples.** Edited by Graham Ramsay
Vol. 149. **A Practical Guide to Graphite Furnace Atomic Absorption Spectrometry.** By David J. Butcher and Joseph Sneddon
Vol. 150. **Principles of Chemical and Biological Sensors.** Edited by Dermot Diamond
Vol. 151. **Pesticide Residue in Foods: Methods, Technologies, and Regulations.** By W. George Fong, H. Anson Moyc, James N. Seiber, and John P. Toth
Vol. 152. **X-Ray Fluorescence Spectrometry.** *Second Edition.* By Ron Jenkins
Vol. 153. **Statistical Methods in Analytical Chemistry.** *Second Edition.* By Peter C. Meier and Richard E. Zünd
Vol. 154. **Modern Analytical Methodologies in Fat- and Water-Soluble Vitamins.** Edited by Won O. Song, Gary R. Beecher, and Ronald R. Eitenmiller
Vol. 155. **Modern Analytical Methods in Art and Archaeology.** Edited by Enrico Ciliberto and Guiseppe Spoto
Vol. 156. **Shpol'skii Spectroscopy and Other Site-Selection Methods: Applications in Environmental Analysis, Bioanalytical Chemistry, and Chemical Physics.** Edited by Cees Gooijer, Freek Ariese, and Johannes W. Hofstraat

Shpol'skii Spectroscopy and Other Site-Selection Methods

CHAPTER

1

THE HISTORICAL DEVELOPMENT OF HIGH-RESOLUTION SELECTIVE SPECTROSCOPY OF ORGANIC MOLECULES IN SOLIDS

R. I. PERSONOV

Institute of Spectroscopy of the Academy of Sciences of Russia, 142190 Troitsk, Moscow Region, Russia

1.1. INTRODUCTION

The optical spectra of organic molecules in solutions and in solid matrices are extremely important sources of information, not only on their molecular structure but also on peculiarities and dynamics of their environment. However, in contrast to the spectra of atoms and simple molecules, electronic absorption and emission (fluorescence and phosphorescence) spectra of polyatomic organic molecules in solutions at room temperature are diffuse as a rule and consist of one or several broad bands of a width ranging from about a few hundred to a few thousand wavenumbers. The diffuse character of such spectra reduces their information content and drastically limits their power for sciences and applications. Therefore, the problems of the origin of broad bands in the spectra of complex organic molecules and the search for conditions that would enable one to observe finer details in spectra have always been a matter of attention for spectroscopists. By now various new spectroscopic methods (based on the use of low-temperature and laser techniques) have been developed: fluorescence line narrowing (FLN), persistent spectral hole burning (HB), single-molecule spectroscopy (SMS), and photon echo (PE) spectroscopy. These methods enable one to obtain very informative fine-structured spectra and are collectively known as "selective spectroscopy." Owing to the large gain in optical resolution (3–5 orders of magnitude), the field of applications of selective spectroscopy in physics and chemistry is now growing rapidly. It is extending from organic molecular systems (crystals, polymers, and glasses) to very complex biological objects (proteins and nucleic acids containing chromophores, reaction-center complexes of photosynthetic units, etc.). Selective spectroscopy techniques and their applications are discussed in detail in the forthcoming chapters. In this introductory chapter I will

Shpol'skii Spectroscopy and Other Site-Selection Methods, Edited by Cees Gooijer, Freek Ariese, and Johannes W. Hofstraat.
ISBN 0-471-24508-9

present a short historical overview of the main steps taken in the development of high-resolution selective spectroscopy.

1.2. QUASI-LINE SPECTRA OF ORGANIC MOLECULES IN *n*-PARAFFIN MATRICES (SHPOL'SKII METHOD)

In the 1930s several scientists [1, 2] observed very sharp vibronic structure exhibited by electronic spectra of crystals of benzene, naphthalene, and some of their derivatives at the temperature of liquid hydrogen. Later similarly structured spectra were also obtained for some mixed aromatic crystals, for example, naphthalene in durene and some others (for a review of the pioneer works in this field see Ref. 3). But only a few systems with sharp electronic spectra (the simplest aromatic crystals and some mixed crystals) were obtained and investigated until the 1950s.

At the beginning of the 1960s Shpol'skii and co-workers [4] made an important experimental discovery. They found that at low temperatures and with a definite type of solvent (crystallized short-chain *n*-paraffins) fluorescence and absorption spectra of some aromatic polyatomic molecules consist of dozens of comparatively narrow bands or "quasi-lines," and not of broad bands as in most solvents. The first experiments of this kind were performed at 77 K and demonstrated spectra with linewidths of about 10–30 cm^{-1}. At present, it is known that at liquid helium temperatures the linewidth in such spectra is usually about 1–10 cm^{-1}. One example of such a spectrum, the organic molecule perylene, in a Shpol'skii (*n*-alkane) matrix at helium temperature is shown in Figure 1.1.

Quasi-line spectra immediately started to be utilized in spectroscopy of polyatomic organic molecules and for selective and sensitive luminescence analysis of complex organic products [5–7], but at the same time spectroscopists (who usually worked with diffuse broadband spectra) were puzzled by the fact that such large organic molecules (in spite of their very high density of vibronic states and, in addition, interactions with many intermolecular vibrations of the solvent) could show such narrow spectral lines. Understanding of the nature of Shpol'skii spectra came later, after theoretical works devoted to the so-called "optical analog of the Mössbauer effect" introduced in spectroscopy the concept of zero-phonon lines.

1.3. OPTICAL ANALOG OF THE MÖSSBAUER EFFECT

In the systems under consideration (molecules in transparent solid matrices) the electronic excitation of the matrix requires a much higher energy than that of the impurity molecules, the molecules to be studied. This means that only the

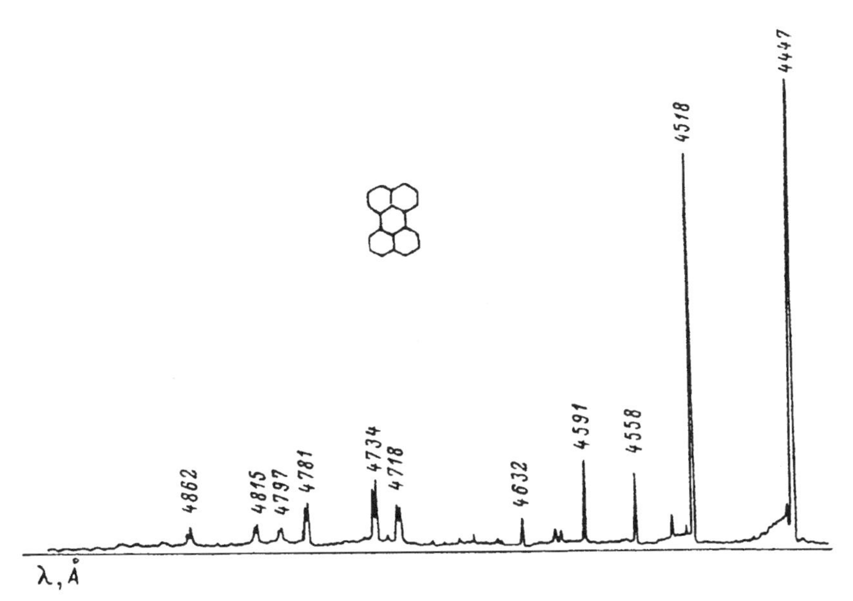

Figure 1.1. Fluorescence spectrum of perylene in *n*-octane at 4.2 K.

electrons of the impurity interact with light. At low concentrations of impurity molecules we can also neglect interactions between them. Under such conditions the spectrum of the impurity molecule is determined by electron-vibration interactions of two types. The first one is the interaction of the molecular electrons with intramolecular vibrations (vibronic coupling), and the second one is the interaction with intermolecular vibrations of the matrix (electron–phonon coupling). Vibronic coupling leads to the presence in the molecular spectrum of a series of vibronic bands, along with a band in the region of the purely electronic transition. The shape of each vibronic band is determined by electron–phonon coupling.

After the discovery of the Mössbauer effect and the development of its theory, it was found that there is a close analogy between γ transitions in nuclei and optical transitions in impurity centers embedded in crystals. This analogy is based on the symmetry of the Hamiltonians, which describe both types of transitions, replacing momentum for coordinate changes; in the case of γ transitions in nuclei, the momentum (recoil energy) plays an important role, but in the case of an optical transition in a molecular impurity, the coordinate displacement is important. As a consequence, if in the Mössbauer effect γ transitions without recoil (without participation of any phonons) are possible, in the optical region transitions without the appearance and disappearance of matrix phonons are also possible [8–10].

The main conclusions of the theory in the case of optical transitions are the following. Each vibronic band may consist, in principle, of two parts (Fig. 1.2). The first is a narrow zero-phonon line (ZPL), corresponding to transitions in the impurity taking place without a change in the number of the matrix phonons (an optical analogy of the resonance γ line in the Mössbauer effect). The second is a relatively broad phonon wing (PW) due to phototransitions in the impurity molecule accompanied by the creation or annihilation of matrix phonons (for details see, for example, Refs. 11 and 12). What, in particular, is actually observed in the spectrum (ZPL, PW, or both ZPL and PW) depends on the strength of the electron–phonon coupling and on the temperature.

In the framework of the adiabatic and Franck–Condon approximations, the ZPL profile can be represented as Lorentzian:

$$I(\omega, T) = \frac{1}{\pi} \frac{e^{-f(T)}\Gamma(T)}{[\omega - \Omega(T)]^2 + \Gamma^2(T)} \tag{1.1}$$

where $\Gamma(T)$ and $\Omega(T)$ express the temperature broadening and the shift of the ZPL, respectively, as determined by the electron–phonon coupling. In the case of $T \to 0$ the quantities $\Gamma(T)$ and $\Omega(T)$ approach zero, and the ZPL turns into a δ-shaped peak (or, strictly speaking, into a narrow line with the natural, lifetime-determined, width). The temperature-dependent function $f(T)$ is determined by the phonon spectrum of the crystal and by the electron–phonon coupling.

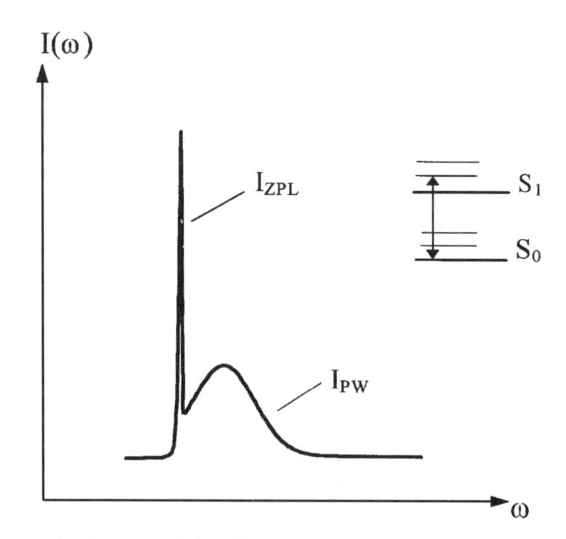

Figure 1.2. Diagram showing zero-phonon line and phonon wing in the spectral absorption band of an impurity center (one vibronic transition).

To characterize the share of the ZPL intensity in the integrated intensity of the hole spectral band the following quantity is usually used:

$$\alpha = \frac{I_{\mathrm{ZPL}}}{I_{\mathrm{ZPL}} + I_{\mathrm{PW}}} = e^{-f(T)} \quad (1.2)$$

where I_{ZPL} and I_{PW} are the integrated intensity of the ZPL and PW, respectively, and $f(T)$ is the same function as in Eq. (1.1). The parameter α may serve as a characteristic of the strength of the electron–phonon coupling and (in analogy to the theory of neutron and X-ray scattering) is usually called the Debye–Waller factor.

In a sufficiently good approximation the function $f(T)$ can be written in the form:

$$f(T) = \int_0^{\infty} f_0(\omega)\left(\frac{2}{\exp(\hbar\omega/kT) - 1} + 1\right) d\omega \quad (1.3)$$

where $f_0(\omega) = C\xi^2(\omega)\rho(\omega)$ is the so-called weighted density of phonon states; $\rho(\omega)$ is the phonon state density; $\xi^2(\omega)$ is the electron–phonon coupling function.

It is seen from Eqs. 1.1–1.3 that the stronger the electron–phonon coupling, the lower the ZPL intensity. At a given strength of electron–phonon coupling, the ZPL intensity and the α factor decrease very rapidly with increasing temperature. After development of the ZPL theory many papers have been devoted to systems that display optical bands with narrow ZPLs and broad PWs, to the temperature dependence of the Debye–Waller factor, and to the measurement and analysis of the temperature broadening and shift of the ZPLs. Such investigations of the nature of quasi-line spectra led to the conclusion that the lines in these spectra really correspond to optical zero-phonon transitions and that they possess all the theoretically expected features of such transitions (see, for example, Refs. 13–16). By now, several hundreds of compounds are known to give quasi-line spectra in n-paraffin matrices. However, powerful as it is, the Shpol'skii method does not succeed in all cases. Many molecules give broadband spectra in n-paraffin matrices even at low temperature. Taking into account the great variety of organic compounds and the wide choice of organic solvents, one can say that the majority of solutions of organic molecules (and practically always in glass and polymer matrices) possess broadband spectra even at helium temperatures. The chapters by Renge and Wild and by Lamotte in this volume discuss the principles of the Shpol'skii effect and the matrix-related limitations of the method in great detail.

1.4. FLUORESCENCE AND PHOSPHORESCENCE LINE-NARROWING EFFECTS

The next breakthrough came in 1972 when Personov and co-workers discovered that the low-temperature spectra of many organic solutions are broadened mainly inhomogeneously and possess hidden fine structure. This fine structure can be revealed by selective laser excitation [17–19]. This important finding was connected to previous investigations of the quasi-line spectra of molecules in *n*-paraffin matrices. When the main features of the nature of the quasi-line spectra are understood (the spectra consist of narrow ZPLs and broad PWs with their characteristic temperature behavior), the question arises: "Why in most cases do the spectra of organic molecules in solution consist of broad bands even at very low temperature?"

From the viewpoint of the concepts mentioned about spectral bands of impurity centers in crystals, there are two possible extreme cases that concern the origin of the broad bands in the spectra of molecules in solutions.

1. Strong electron–phonon coupling takes place between the impurity molecules and the solvent. In this case the ZPL intensity can be very low even at low temperatures (see Eqs. 1.1–1.3). In this case the broad spectral bands are wide PWs. The broadening of the spectral bands of a molecular ensemble is connected to the spectral broadening of each molecule and is called homogeneous.
2. The electron–phonon coupling is weak, and the spectra of all molecules consist of narrow ZPLs, but the molecules under investigation experience different local conditions (local fields) in the matrix. These conditions can be, in principle, distinguished as electrostatic and dispersion interactions, specific interactions like hydrogen bonds, etc. This leads to a statistical distribution of the positions of their electronic levels and to a relative displacement of their spectra on the frequency scale. In this case each broad band observed in the spectrum is the envelope of a large number of ZPLs and the spectral broadening is inhomogeneous.

In order to establish which of the above-mentioned cases actually occurs, experiments were conducted in which the fluorescence was excited by means of monochromatic laser radiation. It was expected that in the case of homogeneous broadening the change from broadband to laser excitation would not lead to a significant modification of the character of the fluorescence spectrum. On the contrary, in the case of inhomogeneous broadening the situation should change considerably. In the latter case, monochromatic excitation will affect mainly the molecules that have a ZPL in absorption at the frequency of the laser. Then the emission spectrum should display narrow lines that belong only to these molecules.

As a result of these experiments, performed in our laboratory in the years 1972–1973 on a number of different organic systems, it was found that at sufficiently low temperatures, the decisive role, in many cases, is played by inhomogeneous broadening. Under monochromatic laser excitation (in the region of the purely electronic or of the lowest vibronic transitions) this inhomogeneous broadening is eliminated to a great extent, and many narrow ZPLs are displayed in the fluorescence spectra [17–20]. This was subsequently confirmed in the works of many other authors [21–25]. Two examples obtained in our laboratory illustrating the above are given in Figure 1.3. One can see that under selective laser excitation narrow ZPLs are observed, instead of a few broad bands. These ZPLs are accompanied on the long-wavelength side by relatively wide wings. Such line spectra can be obtained for various compounds, both in neutral and in ionic form, and in a great variety of solvents (glassy and crystalline, polar and nonpolar, etc.). During the next 8–10 years many features of the fine-structured spectra, such as the dependence on wavelength of laser excitation, on temperature, and on the nature of the matrix, etc., have also been investigated (see, for example, the review by Personov [26], and references therein). The emergence of fine structure in fluorescence spectra under selective excitation is now referred to as "fluorescence line narrowing" (FLN). (One can note that laser excitation has been used earlier to eliminate the inhomogeneous broadening of the zero-phonon *R*-line in the inorganic ruby crystal [27]. In the case of the broadband spectra of organic systems, however, the principal reasoning behind such experiments is very different. This is related to the fact that the existence of the narrow zero-phonon lines inside the broad bands in molecular spectra is not evident, and the results of the experiments on selective laser excitation have to prove or to refute that!)

Many organic compounds, along with fluorescence, also show rather intensive phosphorescence due to electronic T_1–S_0 transitions from the metastable T_1 state to the ground state. The triplet state is usually populated due to the nonradiative intersystem S_1–T_1 transition after S_1–S_0 absorption. It was reasonable to believe that the broad bands in phosphorescence spectra are also broadened inhomogeneously and that this broadening can be eliminated upon selective excitation as well. But experiments showed that the application of laser excitation in the region of the S_1-S_0 transition (at 4 K), leading to the appearance of fine structure in the fluorescence spectrum, does not change the character of the broadband phosphorescence spectrum [23, 28]. Various hypotheses were proposed and various causes analyzed to explain this specific feature of phosphorescence spectra. Here we will only indicate the actual cause for this effect.

In inhomogeneous systems the scatter of the energies of the S_1–S_0 and T_1–S_0 transitions of molecules in matrix depends, in general, on different parameters that characterize the mutual arrangement of the impurity molecule and its surroundings. This implies that molecules that have the same energy of the S_1–S_0 transition can have different energies of the T_1–S_0 transitions and vice versa.

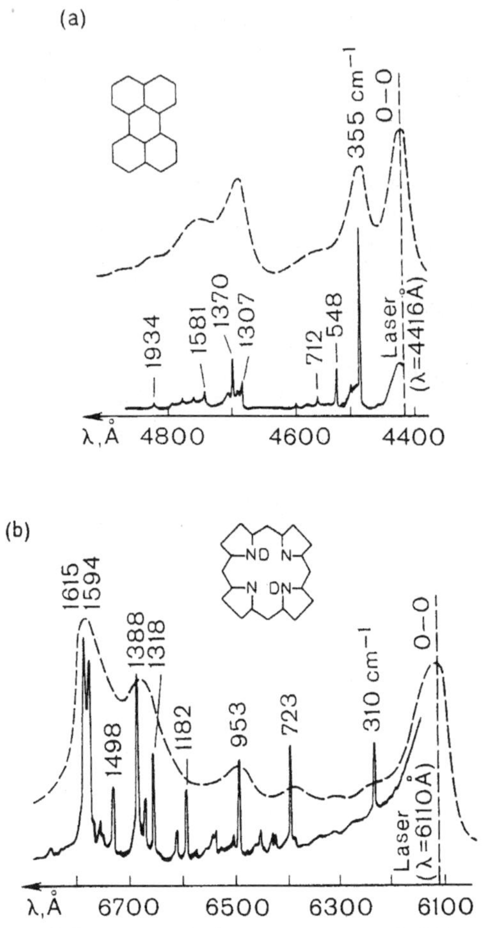

Figure 1.3. Emergence of fine structure in fluorescence spectra (full lines) of solid solutions of organic compounds upon selective laser excitation at $T = 4.2\,K$: (a) perylene in ethanol, (b) D_2-porphin in deuterated ethanol. Dashed lines indicate the broadband spectra upon ordinary Hg-lamp excitation.

Under laser excitation in the region of the S_1–S_0 transition, one selects molecules with the same energies of this transition. But the energies of the T_1–S_0 transitions of these molecules may be much different, and therefore their phosphorescence spectra under such conditions of excitation are diffused. Hence, to eliminate inhomogeneous broadening in the phosphorescence spectra, it is necessary to provide selective laser excitation directly in the region of the T_1–S_0 transition. Such experiments are difficult due to the very low absorption coefficient in the region of the spin-forbidden T_1–S_0 transition (this coefficient is lower than that of the S_1–S_0 transition by a factor of 10^6–10^7). But these difficulties for T_1–S_0 excitation were overcome by using sufficiently intensive laser lines and very sensitive registration systems and, eventually, very fine line structure has been obtained in the phosphorescence spectra as well [29–31]. Figure 1.4 presents one such example obtained in our laboratory.

It should be noted that (due to the long lifetime of the T_1 state) in a phosphorescence spectrum under selective T_1–S_0 excitation it is easy to record (along with many vibronic lines) the resonance 0–0 line at the laser frequency. This line is so narrow that in a number of cases it is possible to observe so-called "zero-field splitting" of this line, resulting from the spin–spin magnetic coupling between two electrons in the triplet state [32]. This splitting, which was usually investigated by means of the ESR method, for the first time has been recorded directly in the optical spectrum of large organic molecules upon selective T_1–S_0 excitation.

The literature on selective spectroscopy discussed so far has been dominated by studies on organic molecules. However, inorganic systems have also been studied, mainly by luminescence line-narrowing methods. The publication by Szabo on ruby, mentioned above, is the first publication in this field [27]. The body of literature reported since has mainly been devoted to the study of crystals and glasses containing rare-earth ions. Since the valence 4*f* electrons of lanthanides are well shielded by the filled 5*s* and 5*p* orbitals, they only weakly interact with their environment. Therefore, the 4*f* orbitals are only weakly split by local crystal fields, and thus the lanthanide luminescence spectra display sharp

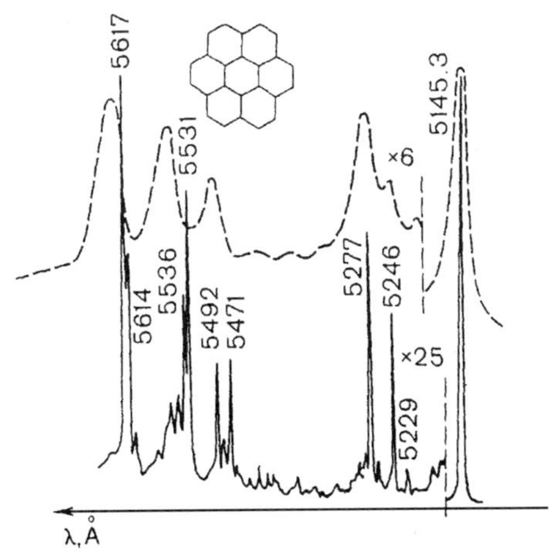

Figure 1.4. Emergence of fine structure in phosphorescence spectrum of coronene in butyl bromide (full line) upon selective laser T_1–S_0 excitation ($T = 4.2$ K). The dashed line indicates the spectrum upon ordinary S_1–S_0 excitation.

line transitions and are very amenable to selective techniques. Selective techniques appear to be very useful for the determination of the sometimes very complex local structure of rare-earth ions in doped crystals and in glasses, due to the specificity of the spectra. The chapter by Murdoch and Wright in this volume describes the state of the art in selective spectroscopy of defects in inorganic materials, mainly considering studies on systems containing lanthanide ions.

The chapter by Jankowiak in this volume deals with fundamental aspects of fluorescence line-narrowing spectroscopy.

1.5. PERSISTENT SPECTRAL HOLE BURNING

In the first experiments on fluorescence line narrowing [17–19] it was also shown that the zero-phonon line intensities drop during laser irradiation; that is some photochemical or photophysical "burning out" of the selectively excited centers takes place. This observation suggests that due to such burning processes in the wide inhomogeneously broadened absorption band some narrow "gap," "dip," or "hole" should appear at the laser frequency. Actually, in 1974 we recorded such a hole in the broadband spectra of perylene and 9-aminoacridine in ethanol glass [33, 34]. At the same time another experimental group obtained similar holes in the zero-phonon line contour of the absorption spectrum of phthalocyanine in an *n*-octane matrix [35]. The method was called "persistent hole burning" or simply "hole burning" (HB). It is evident that the holes should appear not only at the laser (burning) frequency, but also at all vibronic transition frequencies of the "burnt out" molecules. The difference between absorption spectra before and after burning presents a set of these holes and was called by us the "hole-burning spectrum" [36]. Hole-burning spectra contain information on vibrations in the excited electronic state and on the homogeneous linewidths. As an example, the hole-burning spectrum of perylene is shown in Figure 1.5.

The lifetime of the holes is determined by the rate of the back reaction (if it is reversible). At low temperature and in the dark persistent holes may last many hours, days, and even months. Photoinduced burning processes can have either a photochemical or a photophysical cause. Photochemical hole-burning mechanisms lead to a relatively large frequency shift of the photoproduct, but in the case of the photophysical process the photoproduct absorbs in the region of the inhomogeneously broadened band. Photochemical hole-burning mechanisms include, in particular, proton and electron phototransfer, photoionization, photodissociation, and reorganization of hydrogen bonds. For photostable molecules only photophysical transformations involving the matrix can occur. This photophysical (or nonphotochemical) hole burning takes place in many polar glassy and polymer matrices. To explain the nonphotochemical hole-burning mechanism and temperature line broadening in spectra of disordered systems, a so-called

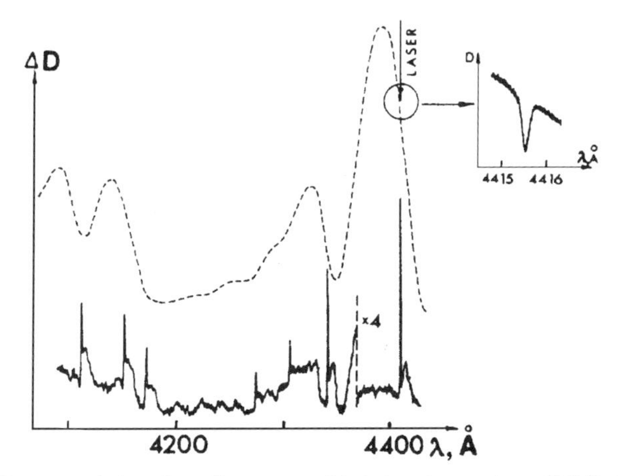

Figure 1.5. Resonance hole at laser frequency and hole-burning spectrum (full line) of perylene in ethanol (burning time $t = 5$ min, $P = 5$ mW/cm^2). The dashed line indicates the broadband absorption spectrum of the same sample.

two-level system (TLS) model of glass [37, 38] was also introduced in selective spectroscopy [39, 40]. The coupling of the impurity to TLSs in glassy matrices may lead to structural rearrangements.

The persistent hole-burning method allows very precise measurement of the homogeneous linewidth, and is a very powerful technique for the detailed spectroscopic investigation of solid-state dynamics and for many other applications. We should note here that, along with high-resolution frequency-domain spectroscopy (FLN and HB), high-resolution time-domain methods that permit one to measure homogeneous linewidths have been developed as well. Such techniques as photon echo spectroscopy [41] are now widely used together with HB and FLN to study the dynamics of solids. Some steps in the development of this field of research (applications of the frequency-domain technique mainly) will be briefly considered at the end of this chapter.

Creemers and Völker have devoted a chapter in this book to fundamental aspects of time-resolved HB as a method to investigate the dynamics of impurity molecules in glasses.

1.6. SINGLE-MOLECULE SPECTROSCOPY

The FLN and HB methods of selective spectroscopy have improved the real spectral resolution by a few orders of magnitude and have thus opened up many new, important, possibilities for science and applications. But at the same time it is important to draw attention to the fact that these selective methods cannot completely remove all inhomogeneity in optical spectra. Actually, in FLN and HB spectroscopies the laser excitation selects a large set of molecules with the same energy of their electronic transitions (the more correct name for such spectroscopic approaches in fact should be "energy-selection spectroscopy" instead of the often-used name "site-selection spectroscopy"). But all these molecules with the same electronic transition energies may differ from each other in their host–guest interaction, in their homogeneous linewidths, in the lifetime of the excited state, etc. It means that even energy-selective spectroscopy at very high resolution provides average information about large ensembles of impurity molecules. Therefore, after the development of energy-selective techniques a few scientific groups started to work on the challenging problem: "What is the way to obtain spectra of a single molecule?"

At first sight it looks almost impossible. But this is not the case. In reality, molecules in solids at low temperature are quietly sitting in the matrix and hence we can collect their emission for a relatively long time. A molecule may, in principle, emit about 10^7–10^8 photon/s and it is not a very serious problem to record such emission intensities. But the main aim was not to detect single-molecules (which has been done in several earlier works), but to perform single-molecule spectroscopy. For this purpose one should remember that the considered spectrum is inhomogeneously broadened and that the ratio between the inhomogeneous and the homogeneous width may be in the order of 10^4–10^5. Therefore, if we are able to measure the spectrum of a sample with only 10^4–10^5 impurity molecules, we can assume that the lines of the different molecules will not overlap. This means that we should work with very small and dilute samples.

An important advancement in this direction was the discovery in 1989–1990 by two research groups of a new branch of high-resolution selective spectroscopy of the solid state. In 1989, Moerner and Kador [42] reported results of the spectral line detection of single pentacene molecules in a *p*-terphenyl crystal via the sophisticated technique of double-modulated absorption spectroscopy. In 1990, Orrit and Bernard [43] presented very clear results on the fluorescence excitation spectra of single molecules in the same system. They also showed that this simpler method provides dramatic improvement of the signal-to-noise ratio. These pioneer papers served as a starting point for the development of a novel branch of high-resolution selective spectroscopy: "single-molecule spectroscopy of doped solids," in which averaging over a molecular ensemble is completely suppressed. One example of a single molecule spectrum obtained by Orrit and Bernard is presented in Figure 1.6. Single-molecule spectroscopy is now a fast-growing field of solid-state spectroscopy. After the first experiments in this field many fascinating results were obtained. The chapter by Tamarat et al. describes fundamental aspects of the spectroscopy of single molecules in solid matrices at cryogenic temperatures.

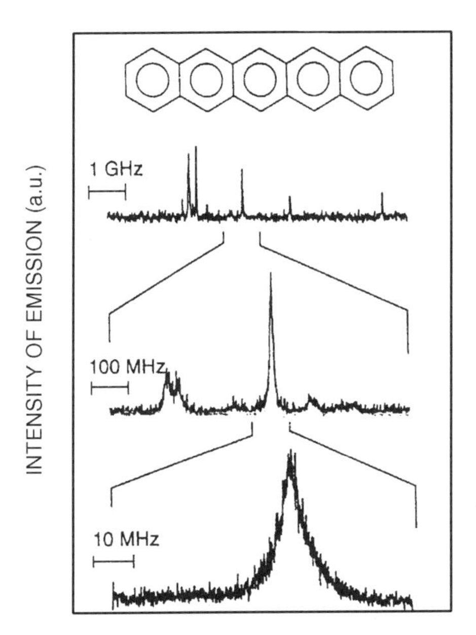

Figure 1.6. Example of a section of the excitation profile of a small crystal of *p*-terphenyl doped with pentacene at 1.8 K. The spectra were recorded in the red wing of the O_1 line. The width of the line of a single pentacene molecule is approximately the natural width of 8 MHz. (Adapted from Ref. 62)

1.7. CERTAIN APPLICATIONS OF SELECTIVE SPECTROSCOPY

Selective spectroscopy allows precise measurements within inhomogeneously broadened bands. It permits to increase the spectral resolution by a factor of 10^3–10^5. This is now a rapidly growing field of molecular and solid-state physics. FLN and HB methods are used in research of both organic and inorganic systems. But here, taking into account the character of this volume, I will mainly focus on organic materials. Even in this case only a few important points can be briefly touched upon. In a number of chapters in this book, applications of selective spectroscopy are described in more detail.

One of the very important applications of selective spectroscopy is in the research of solid-state dynamics. The homogeneous linewidth of impurity centers contains a lot of information, which is directly related to these dynamics. Immediately after the discovery of the HB technique this method started to be used for precise linewidth measurements [44–46]. A large number of subsequent works were devoted to the investigation of the homogeneous linewidth and its temperature dependence, both in crystalline and in amorphous materials. In this field a few intriguing facts were soon discovered. It was established that in amorphous matrices (at liquid helium temperatures) the ZPL width is 1–2 orders of magnitude larger than in crystals. The temperature dependence is also very different. At low temperature the linewidth in amorphous solids follows the law: $\Gamma \propto T^n$, with $1 \leq n \leq 2$ (whereas in crystals the low-temperature line broadening due to coupling with acoustic phonons should obey this power law with $n = 7$, and in the case of coupling with a local phonon temperature-dependent line broadening should be exponential). It was also discovered that relaxation processes in glasses and polymers produce so-called "spectral diffusion." This process results in the drift of the electronic transition frequency with time and, consequently, leads to line broadening, depending on the duration of the measurement [47–49]. In SMS measurements the spectral diffusion manifests itself as sudden intensity changes attributed to spectral jumps. Such jumps were found in one of the first papers on SMS [43]. All these peculiarities of amorphous solids result from the existence in such systems of particular low-energy excitations in the so-called two-level systems (TLSs) [37, 38], which may be used to describe the relaxation processes of guest molecules in amorphous materials. The main features of TLS dynamics are related to the very broad distribution of energies and relaxation times (ranging from picoseconds to many hours and days) observed in amorphous matrices.

After the development of the FLN and HB techniques, it was realized almost immediately that such methods provide an excellent possibility for the investigation of the effects of external fields: both the Stark (electric field) effect [50–52] and the Zeeman (magnetic field) effect [53, 54] have been investigated. The Stark effect on single-molecule lines has also been studied [55, 56]. Among the most important results obtained in Stark experiments is the observation that centrosymmetric molecules embedded in polymer and glass show a linear Stark effect, instead of a quadratic one. This effect arises because the molecular symmetry is broken by the interaction with the matrix and opens up a new possibility for investigation of the internal electric field in complex molecular systems (see, for example, the review of Kohler et al. [57] and references therein).

At present selective spectroscopy is a fast-growing field of research, and in this short chapter it is impossible to discuss all possible applications. In this volume a number of applications will be discussed in more detail. For instance, Zander, Kozin and Matsuzawa cover in two chapters the use of Shpol'skii spectroscopy for the identification of polynuclear aromatic hydrocarbons. Bioanalytical applications of Shpol'skii spectroscopy are discussed by Ariese. The application of selective methods in photochemistry is described by Migirdicyan et al. In particular in the chapters by Creemers and Völker and by Rätsep and Small, the use of selective spectroscopy to probe ultrafast dynamics in proteins and in photosynthetic complexes is discussed in detail. Analytical applications of fluorescence line-narrowing spectroscopy are described in the chapter by Kok and Gooijer. Vanderkooi and Ariese and Jankowiak in their chapters cover bioanalytical applications of selective spectroscopy. Polymeric systems are discussed in the chapter by Zilker and Haarer. The use of excitation–emission matrices in selective spectroscopy is the subject of the chapter by Hofstraat and Wild. Finally, in the chapter by Murdoch and Wright the application of selective techniques for the study of inorganic materials is described. Other subjects, which are not touched upon in this volume, are several other aspects of selective spectroscopy, such as adsorbates and thin films, and the application of high-resolution spectroscopy to optical memories and molecular computing. Much important and interesting information concerning the present state of the art in high-resolution selective spectroscopy can be found in reviews published during recent years, such as Jankowiak et al. [58]; Haarer and Kador [59]; Personov [60]; Reddy et al. [61]; Orrit et al. [62]; Moerner and Basche [63]; Jankowiak et al. [64]; Schellenberg and Friedrich [65]; Osad'ko [66]; Kohler et al. [57]; Kador [67]; Orrit et al. [68]; Skinner and Moerner [69], and Jankowiak and Small [70]. In two earlier contributions to the Wiley series on Chemical Analysis, chapters have been devoted to high-resolution spectroscopy, both by Hofstraat, Gooijer and Veltharst [71, 72].

REFERENCES

1. A. Kronenberger, *Z. Phys.* **63**, 494 (1930).
2. I. Obreimov and A. F. Prikhotjko, *Physik. Z. Sowietunion* **9**, 48 (1936).
3. D. S. McClure in F. Seitz and D. Turnbull, eds., *Solid State Physics*, Academic Press, New York, London, Vol. 8, 1959, p. 1.
4. E. V. Shpol'skii, A. A. Il'ina, and L. A. Klimova, *Dokl. Akad. Nauk SSSR* **87**, 935 (1952).
5. E. V. Shpol'skii, *Soviet Physics Uspekhi* **3**, 372 (1960).
6. E. V. Shpol'skii, *Soviet Physics Uspekhi* **5**, 522 (1962).
7. E. V. Shpol'skii, *Soviet Physics Uspekhi* **6**, 411 (1963).
8. R. N. Silsbee, *Phys. Rev.* **28**, 1726 (1962).
9. E. D. Trifonov, *Soviet Physics-Doklady* **7**, 1105 (1963).
10. K. K. Rebane and V. V. Khizhnyakov, *Opt. Spektrosk.* **14**, 362, 491 (1963) (in Russian).
11. A. A. Maradudin, *Theoretical and Experimental Aspects of the Effects of Point Defects and Disorder on the Vibrations of Crystals*, Academic Press, New York, London, 1966.
12. K. K. Rebane, *Impurity Spectra of Solids,* Plenum Press, New York, 1970.
13. R. I. Personov, E. D. Godyaev, and O. N. Korotaev, *Sov. Phys.—Solid State* **13**, 88 (1971).
14. J. L. Richards and S. A. Rice, *J. Chem. Phys.* **54**, 2017 (1971).
15. E. I. Al'shits, E. D. Godyaev, and R. I. Personov, *Sov. Phys.—Solid State* **14**, 1385 (1972).

16. I. S. Osad'ko, R. I. Personov and E. V. Shpol'skii, *J. Luminescence* **6**, 369 (1973).
17. R. I. Personov, E. I. Al'shits and L. A. Bykovskaya, *Opt.Commun.* **6**, 169 (1972).
18. R. I. Personov, E. I. Al'shits and L. A. Bykovskaya, *JETP Lett.* **15**, 431 (1972).
19. R. I. Personov, E. I. Al'shits, L. A. Bykovskaya, and B. M. Kharlamov, *Zh. Eksp. Teor. Fiz.* **65**, 1825 (1973) (in Russian); Engl. transl.: *JETP* **38**, 912 (1974).
20. L. A. Bykovskaya, R. I. Personov, and B. M. Kharlamov, *Chem. Phys. Lett.* **27**, 80 (1974).
21. R. A. Avarmaa, *Izv. Akad. Nauk ESSR, Fiz.-Mat. Ser.* **23**, 93 (1974) (in Russian).
22. J. H. Eberly, W. C. McColgin, K. Kawaoka, and A. P. Marchetti, *Nature (London)* **251**, 214 (1974).
23. K. Cunningham, J. M. Morris, J. Funfschilling, and D. F. Williams, *Chem. Phys. Lett.* **32**, 581 (1975).
24. I. I. Abram, R. A. Auerbach, R. R. Birge, B. E. Kohler, and J. M. Stevenson, *J. Chem. Phys.* **61**, 3875 (1974).
25. I. I. Abram, R. A. Auerbach, R. R. Birge, B. E. Kohler, and J. M. Stevenson, *J. Chem. Phys.* **63**, 2473 (1975).
26. R. I. Personov, in V. M. Agranovich and R. M. Hochstrasser, eds., *Spectroscopy and Excitation Dynamics of Condensed Molecular Systems*, North-Holland, Amsterdam, 1983, Ch. 10.
27. A. Szabo, *Phys. Rev. Lett.* **25**, 924 (1970).
28. T. B. Tamm and P. M. Saari, *Chem. Phys. Lett.* **30**, 219 (1975).
29. E. I. Al'shits, R. I. Personov, and B. M. Kharlamov, *Chem. Phys. Lett.* **40**, 116 (1976).
30. E. I. Al'shits, R. I. Personov, and B. M. Kharlamov, *Opt. Spectr.* **41**, 803 (1976) (in Russian).
31. K. Brenner, Z. Ruzievich, G. Suter, and U. P. Wild, *Chem. Phys.* **59**, 157 (1981).
32. E. I. Al'shits, R. I. Personov, and B. M. Kharlamov, *JETP Lett.* **26**, 586 (1977).
33. B. M. Kharlamov, R. I. Personov, and L. A. Bykovskaya, *Opt. Commun.* **12**, 191 (1974).
34. B. M. Kharlamov, R. I. Personov, and L. A. Bykovskaya, *Opt.Spectrosc.* **39**, 137 (1975).
35. A. A. Gorokhovskii, R. K. Kaarli, and L. A. Rebane, *JETP Lett.* **20**, 216 (1974).
36. B. M. Kharlamov, L. A. Bykovskaya, and R. I. Personov, *Chem. Phys. Lett.* **50**, 407 (1977).
37. P. W. Anderson, B. I. Halperin, and C. M. Varma, *Philos. Mag.,* **25**, 1 (1972).
38. W. A. Philips, *J. Low Temp. Phys.* **7**, 351 (1972).
39. J. M. Hayes, K. P. Stout, and G. J. Small, *J. Chem. Phys.* **74**, 4266 (1981).
40. G. J. Small, in V. M. Agranovich and R. M. Hochstrasser, eds., *Spectroscopy and Excitation Dynamics of Condensed Molecular Systems*, North-Holland, Amsterdam, 1983, Ch. 9.
41. N. A. Kurnit, I. D. Abella, and S. R. Hartmann, *Phys. Rev. Lett.* **13**, 567 (1964).
42. W. E. Moerner and L. Kador, *Phys. Rev. Lett.* **62**, 2535 (1989).

43. M. Orrit and J. Bernard, *Phys. Rev. Lett.* **65**, 2716 (1990).
44. A. A. Gorokhovskii, R. K. Kaarli, and L. A. Rebane, *Opt. Commun.* **16**, 282 (1976).
45. H. De Vries and D. A.Wiersma, *Phys. Rev. Lett.* **36**, 91 (1976).
46. S. Völker, R. M. Macfarlane, A. Z. Genak, H. P. Trommsdorff, and J. H. Van der Waals, *Chem. Phys.* **67**, 1759 (1977).
47. W. Breinl, J. Friedrich, and D. Haarer, *J. Chem. Phys.* **81**, 3915 (1984).
48. L. W. Molenkamp and D. A. Wiersma, *J. Chem. Phys.* **83**, 1 (1985).
49. M. Berg, C. A. Walsh, L. R. Narasimhan, K. A. Littau, and M. D. Fayer, *Chem. Phys. Lett.* **139**, 66 (1987).
50. A. P. Marchetti, M. Scozzafava, and R. H. Young, *Chem. Phys. Lett.,* **51**, 424 (1977).
51. V. D. Samoilenko, N. V. Rasumova, and R. I. Personov, *Opt. Spectrosc. (USSR)* **52**, 346 (1982).
52. F. A. Burkhalter, G. W. Suter, U. P. Wild, V. D. Samoilenko, N. V. Rasumova, and R. I. Personov, *Chem. Phys. Lett.,* **94**, 483 (1983).
53. B. M. Kharlamov, E. I. Al'shits, R. I. Personov, N. I. Nizhankovsky, and V. G. Nazin, *Opt. Commun.* **24**, 199 (1978).
54. A. I. Dicker, M. Noort, S. Völker, and D. H. Van der Waals, *Chem. Phys. Lett.* **73**, 1 (1980).
55. M. Orrit, J. Bernard, A. Zumbush, and R. I. Personov, *Chem. Phys. Lett.* **196**, 595 (1992); **199**, 408 (1992).
56. U. P. Wild, F. Guttler, M. Pirotta, and A. Renn, *Chem. Phys. Lett,* **193**, 451 (1992).
57. B. Kohler, R. I. Personov, and J. C. Woehl in A. B. Meyers, and T. R. Tizzo, eds., *Laser Techniques in Chemistry,* John Wiley & Sons, Inc., New York, 1995, Ch. 8.
58. R. Jankowiak and G. J. Small, *Chem. Res. Toxicol.* **4**, 256 (1991).
59. D. Haarer and L. Kador, *Makromol. Chem., Macromol. Symp.* **44**, 139 (1991).
60. R. I. Personov, *J. Photochem. Photobiol. A: Chem.* **62**, 321 (1992).
61. N. R. S. Reddy, P. A. Lyle, and G. J. Small, *Photosynth. Res.* **31**, 167 (1992).
62. M. Orrit, J. Bernard, and R. I. Personov, *J.Phys. Chem.* **97**, 10256 (1993).
63. W. E. Moerner and Th. Basche, *Angew. Chem. Int. Ed. Engl.* **32**, 457 (1993).
64. R. Jankowiak, J. M. Hayes, and G. J. Small, *Chem. Rev.* **93**, 1471 (1993).
65. P. Schellenberg and J. Friedrich, in Richert/Blumen, eds., *Disordered Effects on Relaxation Processes*, Springer-Verlag, Berlin, Heidelberg, 1994.
66. I. S. Osad'ko, *Advances in Polymer Sciences* **114**, 123 (1994).
67. L. Kador, *Phys. Stat. Sol.(b)* **189**, 11 (1995).
68. M. Orrit, J. Bernard, R. Brown, and B. Lounis, in *Progress in Optics,* Elsevier, Amsterdam, Vol. 35, 1996.
69. J. L. Skinner and W. E. Moerner, *J. Phys. Chem.* **100**, 13251 (1996).
70. R. Jankowiak and G. J. Small, in A. H. Neilson, ed., *Handbook of Environmental Chemistry,* Springer, Berlin, 1998, p. 119.

71. J. W. Hofstraat, C. Gooijer, and N. H. Velthorst, in S. G. Schulman, ed., *Molecular Luminescence Spectroscopy, Methods and Applications: Part 2*, John Wiley & Sons, Inc., New York, 1988, p. 283.

72. J. W. Hofstraat, C. Gooijer, and N. H. Velthorst, in S. G. Schulman, ed., *Molecular Luminescence Spectroscopy, Methods and Applications: Part 3*, John Wiley & Sons, Inc., New York, 1993, p. 323.

CHAPTER

2

PRINCIPLES OF MATRIX-INDUCED HIGH-RESOLUTION OPTICAL SPECTROSCOPY AND ELECTRON–PHONON COUPLING IN DOPED ORGANIC CRYSTALS

INDREK RENGE AND URS P. WILD*

Institute of Physics, University of Tartu, EE51014 Tartu, Estonia
**Physical Chemistry Laboratory, Swiss Federal Institute of Technology, CH-8092 Zürich, Switzerland*

2.1. INTRODUCTION

In the first half of last century the prevailing point of view was that the optical spectra of polyatomic molecules are congested as a result of large number of intramolecular vibrations. In solids at low temperatures no improvement of spectral resolution was expected because of the huge density of phonon states in the condensed phase. In his first publication in 1952, Shpol'skii et al. [1] expressed surprise about the observation of luminescence fine structure in frozen solutions of coronene and benzo[*a*]pyrene in *n*-alkanes at 77 K. Independently, the narrow lines in doped aromatic crystals were measured very soon afterwards (Ref. 2 and references therein). Sharp excitonic transitions in neat aromatic crystals were known even before (for reviews of older work, see Refs. 2 and 3). It became clear that the occurrence of vibrationless (zero-phonon) transitions in the optical spectra of organic crystals is a rule, rather than an exception.

The nature of broad spectra in frozen solvent glasses remained obscure for another 20 years until the laser excitation was applied, leading to the observation of the site-selection (fluorescence line narrowing) effect [4] and spectral hole burning [5, 6]. Fluorescence line-narrowing experiments could have been done earlier by using a lamp and a monochromator for excitation and photographic detection. However, optical bands in dyes and strong p(^1L_a) transitions in polycyclic arenes are subject to severe broadening at 77 K. In tetrapyrrolic pigments (e.g., phthalocyanine) and arenes with α (^1L_b) type S_1–S_0 transitions (coronene, pyrene) the thermal broadening is less, but the low solubility of these

Shpol'skii Spectroscopy and Other Site-Selection Methods, Edited by Cees Gooijer, Freek Ariese, and Johannes W. Hofstraat.
ISBN 0-471-24508-9

compounds in glass-forming solvents may pose a problem of getting enough optical density in the 0–0 band. Similarly, small absorptivity in the 0–0 bands of suitable samples and the generally low efficiency of photochemical processes at low temperatures obviously precluded earlier discovery of hole burning. Therefore, the revelation of extremely narrow zero-phonon lines in disordered environments had to await until lasers and liquid He came in use, and, last but not least, the inhomogeneous nature of broadband spectra was recognized.

Narrow optical transitions proved highly useful for analytical and structural studies of organic molecules. The fluorescence analysis is known as one of the most sensitive methods. The large cross section of zero-phonon transitions opens the way for dramatic enhancement of the sensitivity of detection as well as unprecedented selectivity [7, 8]. The combination of narrow inhomogeneous bandwidth in crystals with selective laser excitation appears particularly promising [9]. The rich vibrational structure in emission and excitation spectra encodes practically complete information about the molecular geometry in the ground and the excited states, respectively. However, for larger molecules the structure-vibration frequency problem is not yet resolved exactly enough to derive molecular structure on the basis of spectra. The vibronic frequencies and intensities are used mostly as characteristic fingerprints for identification of large molecules. Recently, the luminescence analysis has reached its ultimate sensitivity limit as single-molecule spectroscopy [10]. Informative spectra of single molecules of certain compounds can be relatively easily recorded in doped *n*-alkanes and aromatic crystals cooled to liquid helium temperatures [10]. Again, it is of interest to note that both the “organic ruby” (pentacene solution in *p*-terphenyl crystal) [11] and the tunable single-mode lasers [12] have been available in spectroscopy laboratories a decade before the single molecules were actually detected in this system in 1989 [10].

In this review we deal with the optical spectra of mainly crystalline solutions displaying small inhomogeneous bandwidths as compared to those in typically amorphous media, such as glass-forming solvents and polymers. The host matrices will be divided into four groups: solid rare gases, crystalline *n*-alkanes, other solvents forming crystals upon freezing, and crystals of aromatic compounds that are solid at room temperature.

Organic molecules as dopants (absorbers, analytes, chromophores, guests, impurities, solutes) absorbing in the visible or near-ultraviolet region will be grouped according to the dimensionality of their π-electronic system as follows:

1. Linear chromophores: polyenes, polymethine ions
2. Planar chromophores: polycyclic aromatic hydrocarbons, tetrapyrrolic pigments
3. Three-dimensional chromophores: fullerenes

Table 2.1. Classification of Organic Crystalline Solutions[a]

	Matrices			
Dopants	Solid Rare Gases	n-Alkanes	Other Frozen Solvents	Aromatic Crystals
Polyenes		[41–43]		[124, 125]
Polymethine ions			[98]	[126]
Arenes	[13–26]	[1, 44–74]	[53, 62, 99–110]	[11, 127–169]
Tetrapyrroles	[27–33]	[5, 12, 75–88]	[111]	[170–174]
Others[b]	[34, 35]	[89–95]	[107, 112–120]	[175–184]
Fullerenes	[36–40]	[96, 97]	[97, 121–123]	

[a]The numbers refer to publications describing the optical properties of the pertinent systems.
[b]Mainly planar chromophores, including heteroaromatic and polar ones, molecules with charge-transfer and n–π^* transitions, etc.

Table 2.1 summarizes the guest–host combinations treated in this chapter. We describe linear spectroscopic properties of these systems basing on a representative selection of publications as indicated in Table 2.1. In addition, original spectra will be presented with the aim to provide new solute–solvent combinations, in particular, for single-molecule studies. In several cases the reproducibility of published data has been checked experimentally.

2.2. OPTICAL SPECTRA OF SOLID SOLUTIONS

2.2.1. Solvent Shift

The optical transition frequencies shift if the molecule is transferred from the isolated state to the matrix. The solvent shift can be calculated as a difference between the stabilization energies of the ground- and the excited-state levels in the condensed phase. Mainly the intermolecular dispersive interaction contributes to the optical band shifts in nonpolar and weakly polar molecules. Bakhshiev has proposed a simple expression for dispersive solvent shift in liquids [185] based on the London formula and the Lorentz–Lorenz relation between the molecular polarizability and the refractive index:

$$\begin{aligned} \nu &= \nu_0 + p\phi(n^2) \\ p &= -3II'(\alpha_e - \alpha_g)/[2(I + I')a^3] \end{aligned} \tag{2.1}$$

where ν and ν_0 are the maximum wavenumbers in the condensed and gas phase, respectively, $\phi(n^2) = (n^2 - 1)/(n^2 + 2)$, n is the refractive index of the solvent for

Na D line, I and I' are the ionization energies of the solute and solvent molecules, $\alpha_e - \alpha_g = \Delta\alpha$ is the average static polarizability difference between the excited (e) and the ground (g) state, and a is the Onsager cavity radius.

Equation (2.1) is perfectly applicable in liquid n-alkanes at room temperature [186–190]. The slope can vary in a broad range from $-130 \pm 20\,\text{cm}^{-1}$ for the S_1 band of octaethylporphine [188] to $-19200 \pm 2400\,\text{cm}^{-1}$ for the 211-nm band in fullerene C_{70} [189]. The α (^1L_b) and p (^1L_a) transitions in polyarenes possess p values of -1500 ± 500 and $-5000 \pm 1000\,\text{cm}^{-1}$, respectively. It follows from Eq. (2.1) that p is closely related to the change of average polarizability density upon electronic excitation. For nonspherical solutes the interaction distance cannot be approximated by a single value of the cavity radius. To estimate the polarizability changes in flat chromophores, a calibration plot has been proposed by us earlier [187]. It will be demonstrated that besides vacuum-to-matrix and pressure shifts, the inhomogeneous broadening and thermal broadening depend on the magnitude of p. In the following the slope of Eq. (2.1) p for liquid n-alkanes will be referred to as the Bakhshiev number. The intercept of Eq. (2.1) yields a transition frequency of the nonsolvated molecule [186] that is very close to the value obtained in supersonic jets [25, 191–214]. The applicability of Eq. (2.1) will be discussed in Section 2.3.1 and 2.3.2.

2.2.2. Inhomogeneous Broadening

The inhomogeneous or statistical broadening of spectra is the result of differences in microscopic solvent shifts for chromophore molecules. The inhomogeneous width in glassy environments is well correlated with the average solvent shift [216]. In crystalline solutions the inhomogeneous width is typically two orders of magnitude smaller. The broadband contours can often be regarded as a convolution of the spectrum of a single molecule (the homogeneous spectrum) and the distribution function of the zero-phonon transition frequencies (Fig. 2.1). The homogeneous spectrum can be presented as a superposition of a narrow zero-phonon line (ZPL) and a broad sideband of vibronic origin. In amorphous matrices the inhomogeneous site-distribution function (IDF) has a shape of a broad, smooth (usually Gaussian) curve (Fig. 2.1). In crystalline solutions the width of IDF is usually smaller than the phonon sideband, and the phonon wings are more or less clearly separated from the 0–0 band.

The presence of multiple sites in crystalline hosts had been noticed already in early publications. Sometimes the multiplet structure is very complicated, containing numerous 0–0 lines. The peak positions are well defined (usually within $1\,\text{cm}^{-1}$) and serve as excellent markers of the compounds to be analyzed. The intensity distribution between different sites can depend on the crystallization conditions. The crystallization rate also influences the inhomogeneous widths of the 0–0 line. The narrowest IDF in molecular systems is observed in sublimed

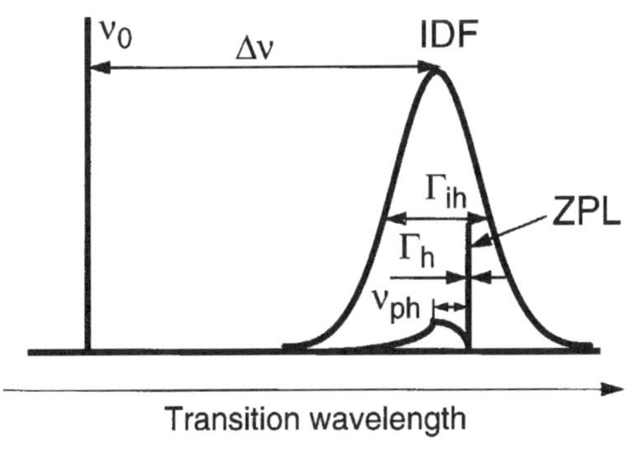

Figure 2.1. Transformation of the molecular absorption line in the disordered environment at low temperature. The purely electronic transition ν_0 of nonsolvated molecules undergoes different solvent shifts characterized by the inhomogeneous site distribution function (IDF) and the average vacuum-to-matrix shift $\Delta\nu$. The spectrum of a single absorber in the solid consists of a zero-phonon line (ZPL) and a phonon sideband with the maximum at ν_{ph}; Γ_{ih}, inhomogeneous line (band) width; Γ_h, homogeneous linewidth.

crystal flakes, for example, pentacene in *p*-terphenyl ($\Gamma_{ih} = 0.07\ cm^{-1}$) [127]. The multiplet structure in single crystals is simpler since at slow cooling only the most stable sites are populated [44–48, 75, 76]. It has been shown by two-dimensional fluorescence spectroscopy [49] and spectral hole-burning that in polycrystalline hosts the 0–0 transitions also occur between the lines ("glassy" regions), forming, together with the phonon wings, a broad background.

Several instrumental techniques have been elaborated for the determination of IDF contour [50, 51, 170]. The first of these methods is based on double scanning of both excitation and recording frequencies in such a way that their distance remains constant [50]. The intensity distribution of a selected vibronic zero-phonon line yields the IDF. Similarly, the IDF can be simply extracted from a two-dimensional spectrum by plotting the intensities of selected vibronic ZPLs on excitation wavelength that is tuned over the 0–0 region [216]. Another group of methods makes use of the largely different homogeneous widths of ZPLs and phonon wings. Fluorescence spectra measured at high excitation intensities show less resolution, since the zero-phonon absorption becomes saturated. The difference between the properly normalized excitation spectra measured under linear and saturated conditions gives the contribution of ZPLs to the absorption spectrum or, in other words, the IDF [51, 170]. The IDF of polycrystalline solutions can have a complicated shape consisting of sharp features superimposed on a broad background [50, 170]. In crystalline solutions the IDF approximately coincides with the shape of the narrow 0–0 line at 5 K. The IDF of broadband spectra in glasses and polymers consisting of at least in 50% of purely electronic ZPLs resemble closely the measured contour, since the phonon wing is relatively narrow. Besides the solvent shift, the character of low-frequency motions will change when an absorber is transferred from vacuum to the condensed phase. First of all, the rovibronic contour with a width of several wavenumbers will either disappear or be replaced by librational modes of 5–50 cm^{-1}. The low-frequency torsional modes in flexible molecules may be modified or disappear altogether. The internal vibrations of rigid chromophores of 500–3500 cm^{-1} usually undergo only a small shift of several cm^{-1}.

2.2.3. Electron–Phonon Coupling

In contrast to amorphous solids, the phonon sideband is more or less well distinguishable from the zero-phonon line in crystalline environment. The strength of linear electron–phonon coupling is characterized by the Debye–Waller factor α, which shows the intensity of the ZPL relative to the total intensity of the transition (line plus wing). Occasionally another parameter, the Huang–Rhys factor, is used, equal to $-\ln \alpha$.

The quadratic electron–phonon coupling responsible for the line broadening and shift vanishes at $T = 0$ K. The zero-phonon linewidth reaches the excited-state lifetime-limited value at $T = 0$ K, even in disordered host–guest systems [217, 218]. The lines broaden at $T > 0$ K, as the rate of phase relaxation increases. Dephasing times are directly measured in photon-echo experiments [128–132]. Alternatively, the dephasing rates may be calculated from the widths of spectral holes that are burned and recorded at the same temperature. The shift and broadening can be directly measured in crystalline matrices beginning at fairly low temperatures ($T > 10$–20 K) since the inhomogeneous lines are narrow. In principle, both the temperature and the pressure effects on narrow lines can be applicable for analytical identification purposes.

Generally speaking, the solvent shift and the electron–phonon coupling phenomena ought to be closely related [219]. On the one hand, the matrix-induced shift of the transition frequency depends on the relative stabilization of respective energy levels upon transfer of the free molecule to the condensed phase. For instance, a stronger interaction with the environment in the S_1 state results in the red shift of the S_1–S_0 transition frequency and vice versa. On the other hand, the change in the intermolecular forces leads to the coupling of the electronic excitation with nuclear motions, including low-frequency modes and two-level systems. According to the Franck–Condon principle, the probability of exciting intermolecular vibrations in the course of an electronic transition is higher when the minima of the intermolecular interaction potentials are displaced (Fig. 2.2B). Also, the ZPL shifts and broadens as a result of differences in the energies of corresponding vibronic levels in the initial and the final states. The change in intermolecular force constants means that the curvature of the respective potential functions are different, and so are the vibrational quanta (Fig. 2.2C). Therefore, both the zero-phonon transition probability or the Debye–

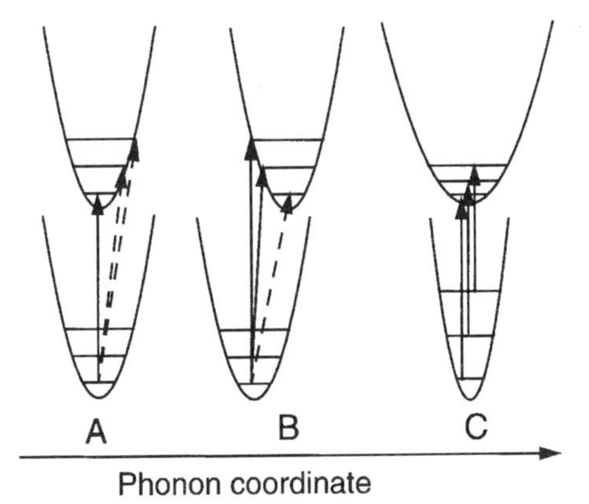

Figure 2.2. Franck–Condon principle for the vibronic transitions in the case of weak (A) and strong (B) linear electron-phonon coupling to vibrations at $T = 0$ K. The vertical transitions have higher probability than nonvertical ones (dashed arrows). Quadratic electron–phonon coupling manifests itself at $T > 0$ K as the broadening and the shift of the zero-phonon line (ZPL). Thermally populated phonon levels have shorter lifetimes than the purely electronic upper level resulting in broadening. No-phonon transitions between the thermally populated sublevels can have different frequencies leading to the internal structure, shift, and broadening of ZPL, (C).

Waller factor (linear electron–phonon coupling) and the linewidth (quadratic electron–phonon coupling) are expected to show correlations with the change in van der Waals forces upon excitation. For a general introduction into the vibronic coupling theory of the impurity spectra, see Ref. 220.

2.3. SPECTRAL NARROWING IN CRYSTALLINE MATRICES

2.3.1. Solid Noble Gases

Fast condensation of mixed vapors and gases on a cold surface constitutes the essence of the matrix isolation method [221]. Atomically or molecularly dispersed guest–host mixtures of totally insoluble compounds can be prepared in this way. Matrix isolation studies in solid inert gases have been focusing mainly on creation or trapping of highly reactive species, such as atoms, atomic or molecular ions, free radicals, etc. The large transparency window of atomic solids ranging from the far infrared to the vacuum ultraviolet facilitates studies of molecular vibrations and highly excited electronic states [221].

Reviews [221] and bibliographies [222] covering all aspects of matrix isolation spectroscopies are available. Spectroscopic data on large stable molecules are less extensive. Absorption or emission of benzene [13], its methyl-substituted derivatives [14], biphenyl [15], naphthalene [16, 17], and anthracene [18, 19] have been studied in some detail. Hot emission from vibrationally nonrelaxed states of anthracene and perylene has been discovered in solid Ne [20, 21]. The recognition of the importance of organic molecules in astrophysics has stimulated the interest in the spectroscopy of unperturbed large aromatic hydrocarbons [22–24] and fullerenes [36–40] in their neutral and ionic states. Magnetooptical studies on porphyrins [27, 28] and phthalocyanines [29] have been performed over the spectral range of 200–1000 nm. A detailed hole-burning investigation of guest–host interactions was reported for phthalocyanine in solid Ar, Kr, and Xe [30].

A number of polycyclic hydrocarbons were trapped in solid N_2 and their quantitative fluorescence analysis has been carried out at 15 K [26]. The fluorescence intensity was found to be linear over five decades or more in concentration, in favorable contrast to frozen *n*-alkane matrices. Multiple peaks corresponding to a mixture of six four-ring arenes could be easily separated [26].

Band widths. The inhomogeneous bandwidth of matrix isolated species can be fairly narrow, for example, 0.1–0.7 cm^{-1} for porphine in Ne, Ar, Kr, and Xe [31], 0.7 cm^{-1} for *s*-tetrazine in annealed Ar [34], $\sim$10 cm^{-1} for anthracene in Ar [18, 19]. In many cases multiplet structure is observed (e.g., in naphthalene [17], anthracene [18, 19], porphine [31], and its Mg and Zn complexes [32], Zn tetrabenzoporphine [33], etc.). Therefore, the guest molecules can occupy distinct sites. A narrow bandwidth and the presence of site structure points to the crystallinity of the host. On the other hand, rather smooth and relatively broad absorption is a consequence of substantially disordered environment. Phthalocyanine is an example of broadband spectra in Ar, Kr, and Xe with Γ_{ih} equal to 44, 58, and 89 cm^{-1}, respectively [30]. Sometimes the Γ_{ih} in a rare gas host is as broad as that observed in a hydrocarbon glass, for example, naphthalene in Ne ($\Gamma_{ih} \sim 160\ cm^{-1}$) and Ar (200 cm^{-1}) [22], naphthalene-D_8 in Ar, Kr, and Xe ($\sim$100 cm^{-1}) [17], or Zn phthalocyanine in Ar (350 cm^{-1}) [29]. Surprisingly, in the latter case the annealing of the Ar matrix up at 40 K leads to a band broadening from 140 to 360 cm^{-1} [29]. The Γ_{ih} of C_{60} ranges from $< 2\ cm^{-1}$ in Ar and $\sim$20 cm^{-1} in Ne [36] (fluorescence) $\sim$5 cm^{-1} in Xe (phosphorescence) to "severely broadened" in Kr, where both types of emission have been recorded [37]. Multiplets occur only in Xe where both the $T_1 \rightarrow S_0$ and $S_1 \rightarrow S_0$ doublets are split by 100 cm^{-1} [37]. Broad inhomogeneous site distribution reflects the disorder of the system.

Solvent shifts. Noble gas atoms are devoid of electrical moments. Therefore, the shifts originate solely from the dispersion forces and the exchange repulsion [13]. The dispersive stabilization of the excited state gives rise to a bathochromic vacuum-to-matrix shift according to Eq. (2.1). The magnitudes of the Lorentz–Lorenz function $[\phi(n^2)]$ for solid Ne, Ar, Kr, and Xe at 5 K were calculated and collected in Table 2.2. The average matrix-induced shift $\Delta\nu$ for polyarenes,

Table 2.2. Lorentz-Lorenz Function $\phi(n^2)$ Values of Matrices at 5 K[a]

Matrix	$\phi(n^2)$
Neon	0.074[b]
Argon	0.182[c]
Krypton	0.232[c]
Xenon	0.295[c]
n-Heptane, *n*-octane	0.30[b]
Durene	0.35[d,e]
Benzene	0.367[f]
Biphenyl	0.411[d] 0.419[f]
Fluorene	0.411[d,g]
Naphthalene	0.414[d] 0.418[f]
Phenanthrene	0.415[d]
p-Terphenyl	0.426[d]
Anthracene	0.43[d]

[a] $\phi(n^2) = (n^2 - 1)/(n^2 + 2)$, n, refractive index for Na D line.
[b] Ref. 189, Table 7.
[c] Calculated from $\phi(n^2)$ at higher T (Ref. 223) and the density change between T and 5 K (Refs. 224, 225).
[d] $\phi(n^2)$ was calculated for each n tensor component (Ref. 226), averaged, and multiplied by 1.05 to take into account the density change of the matrix between 293 and 5 K.
[e] For pentamethylbenzene (Ref. 227).
[f] Tensor components of n at 0 K were taken from Ref. 228.

porphine, phthalocyanine, *s*-tetrazine, C_{60}, and C_{70} can very well be estimated from the Bakhshiev number and the $\phi(n^2)$ of the matrix (Fig. 2.3):

$$\Delta\nu = (45 \pm 23) + (0.99 \pm 0.04)p\phi(n^2), \qquad N = 40, \ r = 0.971 \tag{2.2}$$

where N is the number of data points and r is the linear regression coefficient.

The polarizability increases fast in the series of Ne $\ll$ Ar $<$ Kr $<$ Xe (Table 2.2). The spectral red shifts increase in the same order, confirming that the dispersive interaction plays the predominant role. Solid Ne is the least polarizable material, with $\phi(n^2) = 0.074$. Accordingly, the transition energy shifts are also small: $-53\,\text{cm}^{-1}$ for naphthalene [22], $-16\,\text{cm}^{-1}$ for phenanthrene [23], $-200\,\text{cm}^{-1}$ for anthracene [20, 23], $-181\,\text{cm}^{-1}$ for perylene [21], -47–$-68\,\text{cm}^{-1}$ for C_{60} [36, 40] and $-38\,\text{cm}^{-1}$ for C_{70} [38, 189]. As a remarkable exception, hypsochromic shifts with respect to the S_1 origin ν_0 in the cold jet [196] are observed for the blue sites of free-base porphine [31]. The average energy of the $S_1 \leftarrow S_0$ transition is practically not shifted in all rare gas

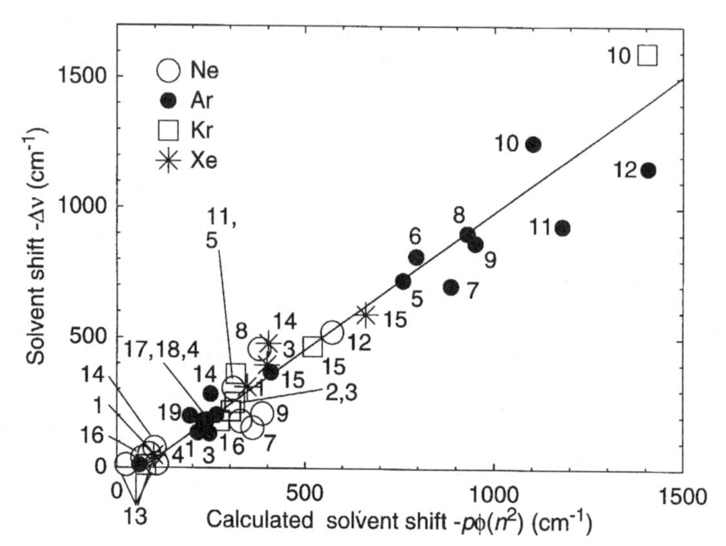

Figure 2.3. Vacuum-to-matrix shifts (Δv) of absorption energies of aromatic hydrocarbons, porphyrins, fullerenes, and *s*-tetrazine embedded in solid rare gas matrices. Vacuum frequencies (v_0) are either from supersonic jet spectra or room-temperature solvent shift measurements. Δv is usually very close to the Bakhshiev number p (from Refs. 186–189, 229) multiplied by the Lorentz–Lorentz function of matrix $\phi(n^2)$ [Eq. (2.1)]. The transitions are arranged in the order of increasing $-p$. The transition/source of maximum/source of v_0 are indicated: 1, naphthalene S_{11} in Ne/[22]/[191], in Ar, Kr, and Xe red site/[17]/[191]; 2, toluene S_1/[14]/[215]; 3, benzene S_{11}/[13]/[-]; 4, phenanthrene S_1/[23]/[192, 193]; 5, anthracene S_1 red site/[19, 20]/[194]; 6, perylene S_1/[21, 24]/[195]; 7, pyrene S_3/[22]/[229]; 8, pyrene S_2/[22]/[229]; 9, phenanthrene S_2/[23]/[187]; 10, diphenylacetylene S_1/[25]/[25]; 11, pyrene S_4/[22]/[229]; 12, phenanthrene S_3/[21]/[187]; 13, porphine S_1 red site/[31]/[196]; 14, porphine S_2/[31]/[196]; 15, phthalocyanine S_1/[30]/[197]; 16, C_{70} S_{11}/[38]/[189]; 17, C_{60} A_0/[39]/[189]; 18, C_{60} S_{11} (γ_0)/[39]/[189]; 19, *s*-tetrazine S_1/[34]/[198].

solids [31]. Therefore, the first singlet transition in porphine should be characterized by a negligible polarizability change [187].

In most cases the average matrix shifts in solid rare gases can be estimated on the basis of the Bakshiev formula [Eq. (2.1)], treating the matrix as a continuous dielectric. A microscopic approach to the solvent shifts has been applied by Najbar et al. [17] in an attempt to calculate the site structure for the S_1–S_0 and $T_1 \rightarrow S_0$ bands based on a model of substitutional sites created by replacing 6 or 7 rare gas atoms in the hexagonal plane with a naphthalene molecule.

2.3.2. *n*-Alkanes

Spectral band narrowing in frozen *n*-alkane solutions is commonly referred to as the Shpol'skii effect in the narrow sense. A handson coverage of earlier investigations in the prelaser era has been given by Nurmukhametov in 1969 [230]. Note that the actual marriage of the Shpol'skii effect with lasers dates back to later work on site selection in 1973 [52] and hole burning in the next year [5]. A large number of luminescence (fluorescence and/or phosphorescence) spectra of heteroaromatic and substituted aromatic compounds is depicted in spectral atlases [231, 232]. The positions of lines are collected in extensive tables [231, 232]. In many papers the main emphasis has been placed on vibrational analysis of large molecules, since the band narrowing in n-alkanes makes the exact determination of vibrational frequencies possible in both the ground and the excited states. The establishing relationships between the structure of π-electronic molecules and vibronic patterns turned out to be a far more complicated task than structure elucidation by using characteristic group frequencies in the IR spectra. Reviews of the applications of Shpol'skii spectra in chemical analysis are presented in Refs. 7–9.

A more recent handbook on low-temperature spectra of polycyclic aromatic compounds by Nakhimovsky, Lamotte, and Joussot-Dubien shows both fluorescence and absorption spectra at 5 K [233]. Overview spectra recorded in the broad wavelength range at 77 K are also displayed. The absorption spectra are of particular value as a unique characteristic of a system, since, strictly speaking, for inhomogeneously broadened samples the nonselective excitation of luminescence is hard to realize. Alternatively, the total luminescence method can be used in order to obtain a unique signature of a guest–host system [53–55, 97]. The handbook [233] also reviews the experimental and theoretical work on transition energies, intensities and polarizations. References are given to the studies on supersonic jets, equilibrium vapors, alkane and aromatic hosts, as well as neat crystals.

Structure of solid solutions. Perhaps it is no exaggeration in the statement that every flat molecule embedded in a n-alkane crystal would have a dramatically narrowed spectrum. Experience shows that the most favorable conditions for the narrowing are created when the length of the hydrocarbon chain and the size of the guest molecule match [56]. The suitability of n-alkane crystal matrices as hosts for planar chromophores relies on the fact that the removal of two or three adjacent hydrocarbon chains leaves behind a cavity that has the form of a parallelepiped. The orientation of polyarenes [44–48] and porphyrins [75–77] in n-alkane single crystals was established by dichroic absorption and polarized emission measurements [44, 45], as well as triplet state ESR [46], ODMR [47], and the Zeeman effect [75, 76]. In particular, rotation of a coronene molecule in the same cavity formed by removing three n-heptane (C_7) molecules is the reason for three multiplet components in the spectrum [47]. Possible ways of incorporating porphine in C_8 were treated by atom–atom potential method and Monte Carlo modeling and compared with experimental observations [77]. Here two different cavities are formed, one by replacement of three matrix molecules that can host the porphine macrocycle in two different positions, and the other one of two alkane chains [77]. An interesting conclusion of Koehler is that the crevices do not collapse [77]. It appears that the empty volume in aliphatic hydrocarbon crystals is slightly thicker than the planar molecules of nonsubstituted polycyclic hydrocarbons and porphine. The nonideality of the fit may give rise to two phenomena, both unfavorable from the analytical point of view: the multiplet structure and the creation of low-frequency modes. The latter show up as phonon wings in most of the spectra at 5 K and give rise to the line broadening and shift at higher temperatures. At 77 K the spectra are largely thermally (homogeneously) broadened (see below).

Solvent shifts. The absolute solvent shifts ($\Delta\nu$) can be determined with high precision, since the 0–0 origins of large molecules in both cold supersonic expansions [25, 191–214] and in crystals are available. The bathochromic shifts range from $-300 - -400\,\mathrm{cm}^{-1}$ for α transitions in naphthalene, pyrene, and coronene to $-2500 - -3500\,\mathrm{cm}^{-1}$ for the S_2 (β) band in tetracene and the S_2 band in 1,8-diphenyl-1,3,5,7-octatetraene. The maximum wavenumbers of the strongest sites in polyarenes and two diphenylpolyenes at 5 K are taken from the graphs in Ref. 233. On the other hand, one can calculate the solvent shift from the Bakshiev number (solvent shift per unit Lorentz–Lorenz function in liquid n-alkanes at 293 K) and the average refractive index of host crystals [Eq. (2.1)]. As the latter is poorly known, an estimate of $\phi(n^2) = 0.30$ has been obtained for C_6 and C_7 at 5 K [189]. A linear relationship between the observed and calculated solvent shifts holds, with a slope close to unity (Fig. 2.4):

$$\Delta\nu_{\mathrm{calc}} = 0.3p = -(74 \pm 48) + (1.06 \pm 0.03)\,\Delta\nu, \qquad N = 34,\ r = 0.985 \quad (2.3)$$

The calculated shift for diphenylpolyenes is underestimated, since the polarizability function for C_{13} and C_{14} matrices is probably larger than 0.30. Therefore, the expected line position in the crystal can be estimated from the broadband measurements at ambient conditions with an error of 5–10% relative to the absolute solvent shift. Evidently, effective averaging of highly anisotropic dispersive interaction in the doped crystal takes place owing to the large size of the π-electronic system of the chromophore. The deviations from Eq. (2.1) can be accounted for in terms of free space in the cavity housing the guest molecule.

Site splitting. It is noteworthy that the maximum site splitting also depends on the solvent shift or the magnitude of the Bakhshiev number (Table 2.3). The best matrices for every solute have been chosen in Ref. 233. In such a crystal, most of the polyarenes display either a single 0–0 line or a doublet. For relatively weak α-type transitions with $-p = 1000$–$2000\,\mathrm{cm}^{-1}$ the spread of zero-phonon lines rarely exceeds $100\,\mathrm{cm}^{-1}$. In case of p bands in anthracene and tetracene the distance between the multiplet components can reach

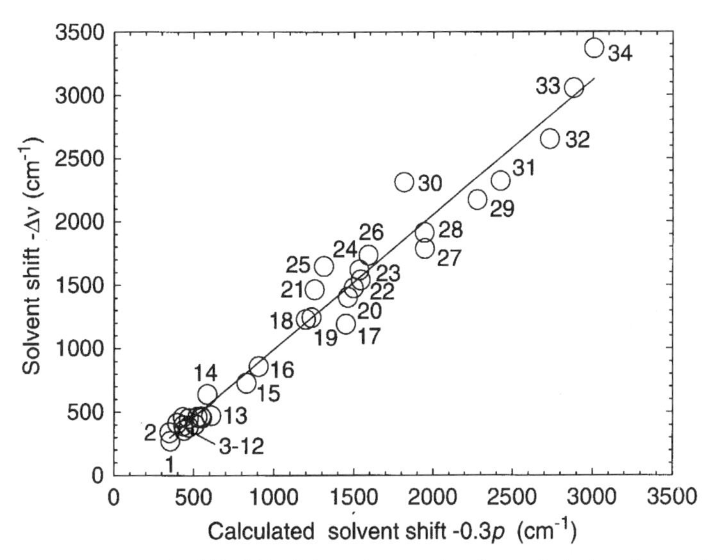

Figure 2.4. Vacuum-to-matrix shifts ($\Delta\nu$) of absorption energies of aromatic hydrocarbons embedded in n-alkane crystals at 5 K. The position of the strongest site or the average value for several sites with similar intensity were taken from the graphs in Ref. 233. Vacuum frequencies (ν_0) are either from supersonic jet spectra or room-temperature solvent shift measurements (Refs. 187, 229). The values of $\Delta\nu$ are very close to the Bakhshiev number p multiplied by 0.3, where 0.3 stands for the average Lorentz–Lorenz function of solid C_7 and C_8 (Ref. 189). The transition/source of ν_0 (if not Ref. 229) are indicated (S_{11}, first singlet transition to a vibronic level, etc.): 1, naphthalene S_{11}/[191]; 2, pyrene S_1/[199, 200]; 3, benzo[*e*]pyrene S_{11}; 4, benzo[*g,h,i*]perylene S_1; 5, phenanthrene S_1/[193]; 6, chrysene S_1/[187]; 7, coronene S_1/[201]; 8, benzo[*a*]pyrene S_1/[202]; 9, benz[*a*]anthracene S_1; 10, acenaphthylene S_1; 11, dibenzo[*a,e*]pyrene S_1; 12, dibenz[*a,c*]anthracene S_1; 13, picene S_1; 14, azulene S_2/[203]; 15, acenaphthylene S_2 (77 K); 16, benzo[*k*]fluoranthene S_1; 17, dibenzo[*a,h*]pyrene S_1; 18, coronene S_{21} (77 K); 19, benzo[*g,h,i*]perylene S_2; 20, pyrene S_3 (77 K); 21, anthracene S_1/[194]; 22, coronene S_3 (77 K); 23, tetracene S_1, red site/[204]; 24, pyrene S_2 (77 K); 25, perylene S_1/[195]; 26, terrylene S_1 (Ref. 229); 27, dibenz[*a,h*]anthracene S_3 (77 K); 28, pyrene S_4 (77 K); 29, benz[*a*]anthracene S_2; 30, diphenylacetylene S_1/[25]; 31, anthracene S_2 (77 K); 32, tetracene S_3 (77 K); 33, 1,6-diphenyl-1,3,5-hexatriene S_2; 34, 1,8-diphenyl-1,3,5,7-octatetraene S_2.

400 cm^{-1}. This value is similar to the inhomogeneous bandwidth in solvent glasses [216]. In many cases the site splitting is only smaller by a factor of 3 than the absolute solvent shift (Fig. 2.4). Therefore, if the red-shifted sites belong to the molecules fully surrounded by alkane chains at a van der Waals distance, the hypsochromically shifted molecules should be residing in the cavities containing 20–25% of free volume. The matrix can no longer be regarded as a continuous dielectric, as far as the site splitting is concerned, and the calculation of vacuum-to-matrix shifts should take into account the microscopic structure or at least anisotropy of guest–host interactions. Such theoretical models are being worked out [234].

Table 2.3. Splitting of the 0–0 Multiplet in the S_1–S_0 Absorption/Fluorescence Spectra of PAHs[a]

Dopant	Matrix	Transition[b]	Site Splitting (cm^{-1})	$-p$ (cm^{-1})[c]
Benz[*a*]anthracene	C_8	α	18	1841 ± 60
Phenanthrene	C_6	α	27	1451 ± 52
Benzo[*a*]pyrene	C_8	α	34	1567 ± 60
Dibenzo[*a*,*l*]pyrene	C_8	α	41	
Chrysene	C_7	α	45	1697 ± 32
Dibenzo[*a*,*i*]pyrene	C_6	α	65	
Pyrene	C_6	α	71	1158 ± 34
Dibenz[*a*,*c*]anthracene	C_6	α	87	1822 ± 64
Dibenz[*a*,*h*]anthracene	C_6	α	88	1340 ± 70
Dibenzo[*e*,*l*]pyrene	C_9	α	90	
1-Methylphenanthrene	C_6	α	103	
Dibenzo[*a*,*e*]pyrene	C_8	α	113	1745 ± 63fl
Dibenzo[*a*,*h*]pyrene	C_6	p	114	4828 ± 255fl
Anthanthrene	C_8	p	116	
Perylene	C_6	p	131	4370 ± 133
	C_7	p	224[d]	
Picene	C_9	α	140	2034 ± 40
Tetracene	C_9	p	337	5137 ± 207
Anthracene	C_7	p	398	4175 ± 85

[a]Taken from the graphs in Ref. 233, error $\pm 5\,cm^{-1}$.
[b]Weak α (^1L_b) or medium strong p (^1L_a) transition.
[c]Bakhshiev number, i.e., the shift of absorption or fluorescence (fl) band maxima in *n*-alkanes at 293 K per unit Lorentz–Lorenz function $\phi(n^2)$, from Refs. 187 and 229.
[d]Ref. 57.

Correlation between the multiplets. Another aspect of microscopic solvent shifts that should be mentioned is the correspondence between the multiplet structure of different electronic transitions. By means of dispersive interaction, the molecules in tightly packed surroundings should have all the transitions bathochromically shifted [Eq. (2.1)]. Accordingly, the red component of N-ethylcarbazole phosphorescence in C_7 ($24192\,cm^{-1}$) appears at the excitation to the red line of the $S_1 \leftarrow S_0$ doublet ($28830\,cm^{-1}$), [89]. The multiplet splittings for $T_1 \rightarrow T_0$ and $S_1 \leftarrow S_0$ are also similar, 163 and $150\,cm^{-1}$, respectively [89]. Similarly, both the fluorescence and phosphorescence multiplets consisting of 4–5 lines at 631–634 and 790–794 nm are correlated for protoporphyrin IX dimethyl ester in C_8 [78]. The $S_2 \leftarrow S_0$, $S_1 \rightarrow S_0$, and $T_1 \rightarrow S_0$ transitions of benzo[*g*,*h*,*i*]perylene in C_6 have two spectral components [58, 59]. Here the site with low-frequency fluorescence corresponds to the high-frequency component in phosphorescence and $S_2 \leftarrow S_0$ absorption. Evidently, the directions of polarizability tensor components can change in different electronic states and so does the strength of dispersive forces and the solvent shift.

The most extensively studied and important group of planar π-systems, the polyarenes have been treated in previous reviews [233] and addressed in several chapters in this book. Therefore, only the peculiarities of less common linear (polyenes), planar (tetrapyrroles), and three-dimensional (fullerenes) π-conjugated molecules embedded in alkane crystals will be mentioned below.

Polyenes. Conjugated linear polyenes in an all-*trans* configuration, such as 1,3,5,7-octatetraene are nicely incorporated in *n*-alkane crystals [41, 42]. The S_1 band emerges as a tiny sharp feature. Its intensity arises as a result of symmetry breaking in the matrix cage and is strongly dependent on the alkane chain length [41]. α,ω-Diphenylpolyenes are chemically more stable and have much higher fluorescence quantum yields than the parent compounds [41]. They have sharp S_1 features in fluorescence and absorption spectra (as tiny peaks) [233]. For 1,8-diphenyl-1,3,5,7-octatetraene in C_{14} two sites split by 96 cm^{-1} are observed [43]. Interestingly, the blue site at 442.2 nm is "blind" in one-photon excitation and appears only under 2-photon excitation [43].

Tetrapyrroles. The second large group of planar nonpolar compounds besides polyarenes, the porphyrins, are very suitable as dopants of *n*-alkanes. The geometry of substitution has been considered in detail, both experimentally [76] and theoretically [77]. Porphine can replace two or three alkane chains, forming the so-called A- and B-type sites. Each of them is split due to the inequivalence of two proton tautomers in the crystal field. The simplest Shpol'skii multiplet is obtained for nonsubstituted compounds: porphine [12, 76, 79], chlorin [80], bacteriochlorin [81], isobacteriochlorin [82], tetrabenzoporphine [83], and phthalocyanine [84]. The insertion of the central metal and, in particular, the peripheral substituents lead to both the decrease of the line-to-background ratio and a complex multiplet, as in pheophytin *a* [85], *meso*-cyano-octamethyl-isobacteriochlorin [86], tetra-*tert*-butylphthalocyanine [87, 229], etc. A large background can occur as a result of the larger disorder or/and the reduction of the Debye–Waller factor. However, the inhomogeneous width of the individual multiplet components can be still as narrow as 1 cm^{-1} in rapidly frozen solutions of octaethylporphine and tetra-*tert*-butylphthalocyanine in C_8 at 8 K [229].

Other compounds. Extensive redistribution of charges upon optical excitation inevitably leads to different interatomic and intermolecular distances in the ground and the excited states. Therefore, the coupling strength to both the internal high-frequency modes and the low-frequency "phonons" is tremendously enhanced for transitions with considerable charge-transfer character (Fig. 2.2B). However, in crystalline solutions narrow 0–0 bands can be observed for typical charge-transfer transitions, such as for coumarin 7 (in C_6) [90]. Observation of the Shpol'skii effect for biologically relevant polar chromophores, 3-hydroxyflavon [91], and several flavins [92] is worth noting. A number of coordination and metallo-organic compounds produce narrow-lined phosphorescence and excitation spectra in *n*-alkanes [93–95]. The common feature of bis(8-quinolinato–O,N)platinum [93] and Pd and Pt complexes with two molecules of deprotonated 2-(2-thienyl)pyridine [94, 95] is their planarity and nearly rectangular shape enabling them to fit snugly into the *n*-paraffin host.

Fullerenes. The three-dimensional π-systems, the fullerenes, are subject to a nice line narrowing effect in normal paraffins as hosts. In full accordance with the size-matching rule, *n*-pentane is the best solvent for C_{70} [96]. Sharp peaks are observed both in fluorescence and phosphorescence. On account of the small solvent shift of the S_1–S_0 transition in C_{70} ($-p = 920\,\mathrm{cm}^{-1}$ [189]), the width of individual peaks ($3\,\mathrm{cm}^{-1}$), and the spread of the S_1–S_0 multiplet ($70\,\mathrm{cm}^{-1}$) are quite small [96]. The matrix-induced line narrowing in C_{70}/C_6 at 77 K [97] seems to be unsubstantial, because the observed width of the 0–0 band ($\sim 80\,\mathrm{cm}^{-1}$) is only slightly smaller than the width of inhomogeneous site distribution function for pyrene ($145\,\mathrm{cm}^{-1}$, $-p = 1160\,\mathrm{cm}^{-1}$) in 1,3-dimethylcyclohexane glass [216].

Nonplanar dopants. From the point of view of luminescence analysis, the worsening of Shpol'skii spectra equals the increase of the complexity of site structure and the decrease of the line-to-background ratio. The latter may originate from enhanced linear electron–phonon coupling and the creation of disordered (glassy, liquid-like) regions. The quality of the fine-line fluorescence spectra decreases in the following series of sterically crowded polycyclic arenes with α-type S_0–S_1 transitions: tetrabenzonaphthalene [60] > tetrahelicene [53] > dinaphtho[1,2-a;1′,2′-h]anthracene (a propeller-like double tetrahelicene shearing a common benzene ring) [61] > pentahelicene [54] > hexahelicene [53]. With increasing number of angularly annelated benzo rings the hydrogen–hydrogen contacts force the molecule more and more extensively out of plane. Finally, for hexahelicene fluorescence no suitable alkane host could be found [53]. Still quite structured phosphorescence and phosphorescence excitation spectra of hexahelicene could be observed in *n*-heptane [53].

From the given examples it is obvious that neither nonplanarity of polyarenes nor bulky alkyl substituents (e.g., *tert*-butyl) attached to the π-electronic systems of arenes and tetrapyrroles are capable of destroying the matrix-induced line narrowing completely. Far more deleterious is the influence of phenyl and other π-electronic substituents of aromatic molecules. The spectra for cold molecules seeded in supersonic expansions show extensive progressions of torsional modes in *meso*-tetraphenylporphine [205], and, in particular, in phenyl-substituted anthracenes [206] and rubrene [207]. In the latter molecule the 0–0 lines vanish. The alkane crystal is hardly able to clamp the pending phenyl groups and flatten the dopant molecule. Instead, the crystalline ordering is perturbed, and the low-frequency modes are retained. By contrast, in nitrobenzene host crystal the torsional modes disappear and a well-ordered environment is created for tetraphenylporphine [111] (see the next chapter).

2.3.3. Other Frozen Solvents

In the following chapters the matrices capable of giving rise to spectral narrowing, except for solid gases and *n*-alkanes, will be divided into two groups: frozen solvents and aromatic crystals. The spectroscopic behavior is in principle very similar in these two cases, although the sample preparation procedure depends on the melting point of the host. Solutions that are liquid at ambient conditions are usually frozen in a cell or tubing, whereas the solid crystals can be handled without a container. The first category of materials is quite often termed Shpol'skii systems in a broad sense.

Polymethine ions. The oxazine and xanthene dyes behave as linear polymethine ions as far as the optical transition energies and oscillator strength are concerned. An unprecedented narrowing of fluorescence excitation spectra of resorufin and cresyl violet in solid formamide at 4.2 K has been reported [98]. The inhomogeneous width of 6 cm^{-1} for resorufin and 22 cm^{-1} for cresyl violet would mean diminishing of spectral disorder by a factor of 10 to 20 as compared to glassy alcohol solutions. However, it seems that a laser- rather than matrix-induced narrowing effect has occurred, because we were not able to reproduce these data and observe anything peculiar at the spectra of snowy formamide solutions [229]. Thus the suitable solvent for a large number of cyanine cations, oxonole anions, and related dyes still remains to be discovered.

Polyarenes. A dramatic reduction of inhomogeneous bandwidth is expected if the size and shape of the solute and solvent molecules are similar. This effect has been documented for small, planar π-electron systems such as benzene in cyclohexane [99], toluene in benzene [100], cyclohexane [101] and methylcyclohexane [102], etc. (Table 2.4). As expected, the normal benzene doped in C_6D_6 host undergoes a dramatic sharpening of optical transitions [103, 105]. Solutions of larger polycyclic aromatic hydrocarbons display line narrowing in several frozen solvents, both aliphatic (cyclohexane [106], methylcyclohexane [107], methylcyclopentane [109], CCl_4 [120]) or aromatic (benzene [62]). For reasons that are not at all obvious, tetrahydrofuran (THF) appears to be a particularly favorable matrix for many polycyclic hydrocarbons [108]. It has been observed that in THF 12 compounds out of 32 have a narrower fluorescence bandwidth than 100 cm^{-1}. Except for perylene they belong to α-type transitions. Because of the high temperature (77 K), the bands are homogeneously broadened and may conceal a multiplet structure.

After an unexpected observation of a rather pronounced multiplet structure in perylene/acetone at 10 K, consisting of two components at 436.1 and 440.85 nm,

Table 2.4. Transition Energies of Solutes in Frozen Nonalkane Solvent Crystals at 2–10 K

Dopant	Matrix	Transition[a]	$-\Delta\nu$ (cm^{-1})[b]	Ref.	ν_0 (cm^{-1})	Ref. (ν_0)[c]
		Aromatic hydrocarbons				
Benzene	Benzene-D_6	$T_1 \rightarrow S_0$	29658.1	[103]		
Benzene	cyclohexane	$T_1 \rightarrow S_0$	29503,[d,e] 29444[e,f]	[99]		
Benzene	borazine	$T_1 \rightarrow S_0$	29568.0	[104]		
Benzene	benzene-D_6	$S_1 \rightarrow S_0$	224	[105]	38078[g]	
Toluene	benzene	$T_1 \rightarrow S_0$	28994	[100]		
Toluene	methylcyclohexane	$S_1 \rightarrow S_0$	247.5[h]	[102]	37477.5	[215]
Toluene	cyclohexane	$S_1 \rightarrow S_0$	316.5[h]	[101]		
Biphenyl	cyclohexane	S_1–S_0	33660, 33588	[106]		
Azulene	methylcyclohexane	$S_2 \leftarrow S_0$	350,[h,i] 440[h,j]	[107]	28757	[203]
Many arenes	tetrahydrofuran	$S_1 \rightarrow S_0$, $T_1 \rightarrow S_0$	*h*	[108]		
Pyrene	methylcyclopentane	$S_1 \rightarrow S_0$	268–383[k]	[109]	27208	[199]
Tetracene	benzene	S_1–S_0	1701[l]	[62]	22360[e]	[204]
Hexahelicene	chlorobenzene	$S_1 \rightarrow S_0$	24110	[53]		
		Porphyrins				
Tetraphenylporphine	nitrobenzene	$S_1 \rightarrow S_0$	–200,–143, 247, 303	[111]	15617[e]	[205]
		Other compounds				
s-Tetrazine	benzene	S_1–S_0	894.1	[112]	18128	[198]
s-Tetrazine	pyridazine	S_1–S_0	100–452[k]	[113]		
Pyrazine	cyclohexane	$T_1 \rightarrow S_0$	26436	[114]		
Benzonitrile	cyclohexane	$S_1 \rightarrow S_0$	426[e,h]	[115]	36501	[209]
Aniline-D_2	*p*-xylene	$T_1 \rightarrow S_0$	27659[e]	[116]		
Benzaldehyde	methylcyclohexane	$T_1 \rightarrow S_0$	25070	[117]		

Benzaldehyde	methylcyclohexane-D_{14}	$T_1 \rightarrow S_0$	25068, 24943	[118]		
Benzaldehyde	acetophenone	$T_1 \rightarrow S_0$	25613	[119]		
p-Chlorobenzaldehyde	methylcyclohexane	$T_1 \rightarrow S_0$	24937, 24893	[118]		
p-Chlorobenzaldehyde	p-xylene	$T_1 \rightarrow S_0$	24896	[118]		
Dibenzothiophene	CCl_4	$T_1 \rightarrow S_0$	24514^{h}	[120]		
1,2,5,6-Dibenzoxalene	methylcyclohexane	$S_2 \leftarrow S_0$	$27952^{h,j}$	[107]		
3-Phenyl-1,2,5,6-dibenzoxalene	methylcyclohexane	$S_2 \leftarrow S_0$	$27498^{h,i}$	[107]		
		Fullerenes				
C_{60}	toluene	S_1–S_0	605–1015^{m}	[123]	15695	[36]
C_{70}	toluene	$T_1 \rightarrow S_0$	12400, 12350	[122]		
C_{70}	toluene	S_1–S_0	440–570^{m}	[122]	15630 ± 8	[189]
$C_{60}[Cr(\pi\text{-allyl})_2](allyl)$	toluene	S_1–S_0	~ 14200	[123]		

[a] $T_1 \rightarrow S_0$, $S_1 \rightarrow S_0$, 0–0 phosphorescence and fluorescence; $S_1 \leftarrow S_0$, $S_2 \leftarrow S_0$, 0–0 absorption; S_1–S_0, absorption and fluorescence in resonance.
[b] Absolute solvent shift $\Delta v = v - v_0$. The transition frequency v is given when 0–0 wavenumber in nonsolvated molecule v_0 was not available.
[c] Source of v_0.
[d] Metastable phase III.
[e] Both the wavenumber and wavelength λ of the transition have been reported, whereas $v = \lambda^{-1}$. It was not clearly indicated that λ has been corrected for the refractive index of the air.
[f] Stable phase II.
[g] 6^1 frequency from Ref. 208 minus vibrational quantum 528 cm^{-1} (Ref. 2).
[h] 77 K.
[i] β form.
[j] α form.
[k] Multiplet of 5 lines.
[l] 25 K.
[m] Multiplet of 3–5 lines.

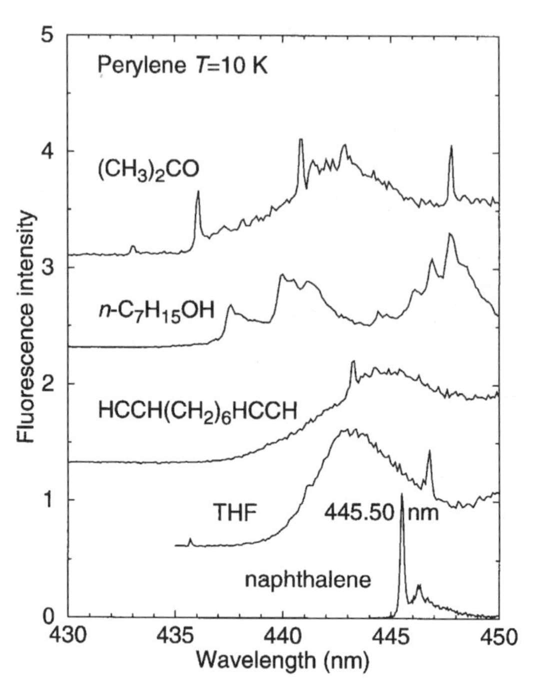

Figure 2.5. Fluorescence spectra of perylene (5×10^{-6} M) in fine crystalline acetone, *n*-heptyl alcohol, 1,7-octadiyne, and tetrahydrofuran (2×10^{-7} M) and in a melt with naphthalene at 10 K. Excitation was with 1.5 nm bandwidth between 380 and 400 nm to the higher S_1 vibronic transitions or at 420 nm (naphthalene), recording bandwidth was 0.1 nm. The site selection effect was negligible.

further spectra were recorded for 5×10^{-6} M solutions in a number of crystallizing liquids (Fig. 2.5) [229]. A multiplet structure of decreasing line-to-background contrast was observed in acetone > THF > *n*-heptanol > 1,7-octadiyne > cyclohexane (Fig. 2.5). In CH_2Cl_2, CCl_4, CH_3CN, CH_3NO_2, cyclopentanone, and benzene host matrix only broadband fluorescence was detected [229]. The line narrowing in *n*-alkane derivatives is expected and has been well known for a long time [230]. The most puzzling are the matrices of compact solvent molecules hosting large chromophores. A nonplanar chomophore, hexahelicene, yields well-resolved spectra in chlorobenzene [53]. As an extreme case, terrylene in THF shows several well-defined 0–0 sites spreading over a considerable wavelength interval between 554.9 and 572.4 nm (Fig. 2.6).

Vapor-phase codeposition of coronene, benzo[*g*,*h*,*i*]perylene, and benzo[*e*]-pyrene with benzene and cyclohexane onto a glass surface cooled with liquid nitrogen produced samples with comparatively narrow fluorescence lines at 77 K (10–20 cm^{-1}), although the spectra of frozen solutions were broad [110].

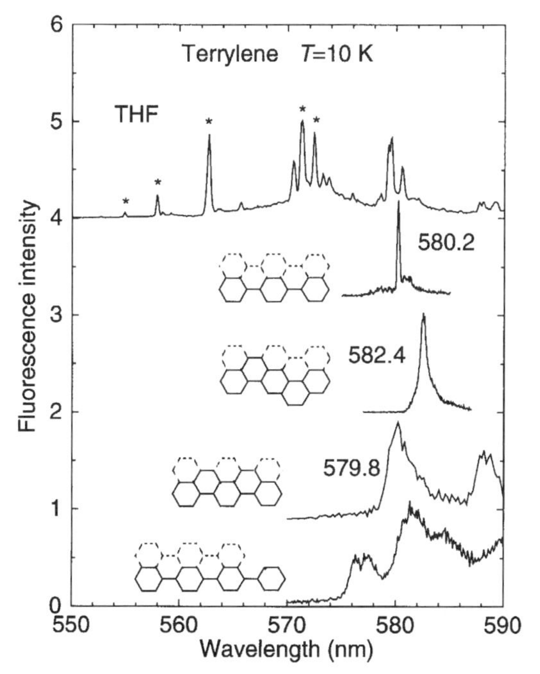

Figure 2.6. Fluorescence spectra of terrylene in fine crystalline tetrahydrofuran (THF) (5×10^{-7} M) and in sublimed layers of *p*-terphenyl, dibenz[*a*,*h*]anthracene, picene and *p*-quaterphenyl. Fluorescence was excited with 1.5-nm bandwidth at 480 nm (THF) or 540 nm, recording bandwidth was 0.2 nm. The 0–0 transitions in THF occur at 554.9, 557.9, 562.7, 571.3, and 572.4 nm (marked with asterisks) and coincide both in fluorescence and excitation spectra within 0.05 nm. The maximum wavelengths of other peaks are indicated in nm. The structure of terrylene (dashed line) is overlaid with that of the matrix molecules in order to show the structural similarities. Although no obvious isomorphism between THF and terrylene exists, at least five very well-defined sites are formed.

Evidently, the fast crystallization prevented the dopant from being expelled from the solvent matrix.

Tetrapyrroles. Line narrowing for another 'hopeless' guest, tetraphenylporphine (TPP) has been discovered by chance in nitrobenzene in the course of an attempt to observe solid-state fluorescence quenching in an electron-accepting matrix [111]. Because of nonplanarity of the molecule and strong vibrational activity of torsional modes of *meso*-phenyl substituents [205], the TPP fails to give Shpol'skii effect in *n*-alkanes. In rapidly frozen nitrobenzene the TPP shows two pairs of sites at 628.3/630.5 and 646.4/648.8 nm. The intensity of low-frequency wings is considerably less than that of the torsional vibrations in nonsolvated TPP [205]. Therefore, the torsional vibrations should be clamped to a large extent. The absolute solvent shifts are 303/247 and $-143/-200\ \mathrm{cm}^{-1}$ for the site pairs under discussion (Table 2.4). Based on the large hypsochromic shift, the large change in vibrational frequencies in the S_1 state and the decay times of the S_1 and T_1 levels, it was proposed that the blue-shifted centers correspond to the molecules with phenyls switched off the resonance and, as a result, resembling the parent porphine [111]. It was found recently that structurally similar tetraphenylchlorin and tetrakis-(4-pyridyl)porphine also have good fine-line spectra in nitrobenzene (Fig. 2.7).

Fullerenes. The matrix-induced reduction of inhomogeneous bandwidth has also been reported for 3-dimensional π-systems, fullerenes [97, 121–123]. Two 0–0 sites of C_{70} fluorescence in polycrystalline toluene (15180 and 15080 cm^{-1}) are barely resolved at 77 K [97]. At 2 K numerous 0–0 lines are observed in both fluorescence and phosphorescence spectra in toluene [121–123]. The line-to-

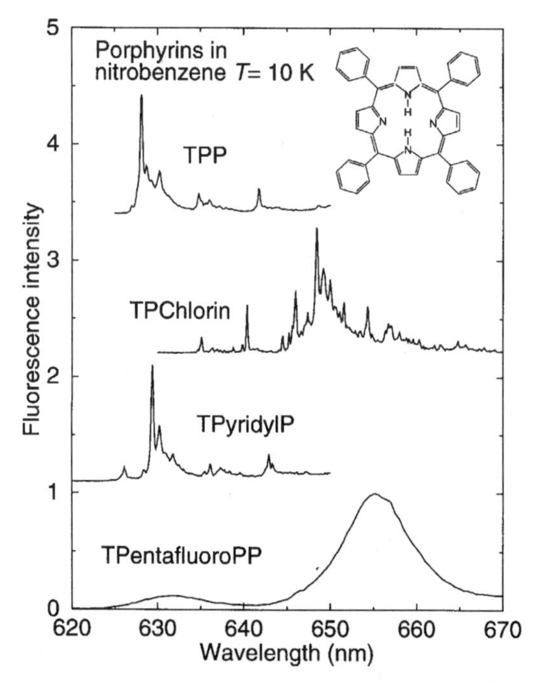

Figure 2.7. Fluorescence spectra of *meso*-substituted porphyrins (4×10^{-5} M) in rapidly frozen crystalline nitrobenzene: TPP, tetraphenylporphine. Excitation was with 1.5-nm bandwidth at 410, 520, 510, and 510 nm for TPP, tetraphenylchlorin, 4-tetrapyridylporphine and *meso*-tetrakis(pentafluorophenyl)porphine, respectively; recording bandwidth was 0.1–0.2 nm. Under present rapid cooling conditions the less perturbed sites of TPP at 646.4 and 648.87 nm (see Ref. 111) are practically not formed and the porphinelike sites with maxima at 628.3 and 630.5 nm predominate. It has been shown on the basis of vibronic structure and the singlet- and triplet-state lifetimes that these centers correspond to the molecule with phenyl rings turned out of conjugation with the tetrapyrrolic macrocycle.

background ratio reaches 1:1, and the smallest linewidth is in the order of 4 cm^{-1}. The Shpol'skii effect is even stronger in *o*-xylene as solvent [121]. Crystalline toluene was also a good matrix for C_{60} and its derivatives, $C_{60}C_3H_7(H)$ and C_{60}[chromium(π-allyl)$_2$](allyl) [123].

The selection of a proper matrix remains largely an empirical task to be solved with the help of rather intuitive considerations. In order to provide the researcher with useful ideas, a number of successful guest–host combinations is listed in Table 2.4.

2.3.4. Aromatic Crystals

Mixed crystalline solutions that are solid at ambient conditions have been the favorite systems in studies of vibronic coupling mechanisms and electron–phonon interactions. In particular, the "organic ruby," pentacene/*p*-terphenyl, has been extensively used in optical coherence experiments [128–133]. Subsequently, single-molecule absorption (1989) and fluorescence (1990) [10] have been discovered in this crystal. For several years organic ruby remained a unique system for experiments with single molecules [10].

The absolute solvent shifts or the peak positions of a number of dopant–aromatic host matrix combinations are collected from the literature and displayed in Table 2.5.

Polyenes. The list of linear polyenes embedded in a host matrix that is solid at room temperature is restricted to a single example: all-*trans*-1,8-diphenyl-1,3,5,7-octatetraene in bibenzyl [41, 124, 125]. At 4 K the forbidden S_1 transition has three 0–0 peaks at ~453 nm with multiplet splitting 80–100 cm^{-1} and the shift relative to that in C_{14} −450 cm^{-1} [233]. In the aromatic crystal the very strong S_2 origin is located at 426 nm [125], shifted by −720 cm^{-1} relative to C_{14} [233] and by −4200 cm^{-1} relative to the nonsolvated molecule [229].

Polymethine ions. Isomorphous substitution of 1,1′-diethyl-2,2′-cyanine (pseudoisocyanine) iodide into the crystal lattice of 1,1′-diethyl-9-aza-2,2′-cyanine serves perhaps as the only example of an ionic polymethine dye showing the elimination of inhomogeneous broadening in the well-defined local environment [126]. Three weak but sharp origins at ~550 nm (Table 2.5) spread over 230 cm^{-1} and have an average solvent shift of −1600 cm^{-1} [235]. The vacuum-to-matrix shift corresponds to an average refractive index of the host ~2 [$\phi(n^2)=0.5$]. It is worth noting that the average wavelength of two excitonically split levels of *J* aggregates of pseudoisocyanine at 540 and 570 nm is just about the same [235].

Polyarenes. The most dramatic narrowing of inhomogeneous bandwidths can be achieved in mixed crystals grown from the vapor phase. Upon sublimation many aromatic hydrocarbons form very thin two-dimensional crystals that are essentially free from internal strains. The inhomogeneous width of the

Table 2.5. Transition Energies of Solutes in Aromatic Crystals at 2–10 K

Dopant	Matrix	Transition[a]	$-\Delta\nu$ (cm^{-1})[b]	Ref.[c]	ν_0 (cm^{-1})	Ref. (ν_0)[d]
		Linear polyenes				
1,8-Diphenyloctatetraene	bibenzyl	S_1–S_0	22118, 22108, 22038	[125, 41, 124]		
1,8-Diphenyloctatetraene		$S_2 \leftarrow S_0$	4230	[125]	27710±40	[229]
		Polymethine dyes				
1,1′-Diethyl–2,2′-cyanine iodide	1,1′-diethyl-9-aza-2,2′-cyanine iodide	$S_1 \leftarrow S_0$	1493, 1620, 1719	[126]	19710±40	[229]
		Aromatic hydrocarbons				
Naphthalene	naphthalene-D_8	$T_1 \rightarrow S_0$	21209.1[e]	[134]		
Naphthalene	naphthalene-D_8	$S_1 \leftarrow S_0$	487	[128, 134[f]]	32019	[191]
Naphthalene	durene	$S_1 \leftarrow S_0$	471.3	[128]		
Biphenyl	biphenyl-D_{10}	$T_1 \rightarrow S_0$	23163	[135]		
Azulene	naphthalene	$S_1 \leftarrow S_0$	–366.6	[136–138]	14284	[203]
Azulene	naphthalene	$S_2 \leftarrow S_0$	709	[136]	28757	[203]
Phenanthrene	durene	$S_1 \leftarrow S_0$	596.7	[139]	29328	[192, 193, 210[f]]
Phenanthrene	biphenyl	$T_1 \rightarrow S_0$	21378	[140]		
Phenanthrene	biphenyl	$S_1 \leftarrow S_0$	779	[140]		
Phenanthrene	fluorene	$S_1 \leftarrow S_0$	604	[140]		
Phenanthrene-D_{10}	durene	$S_1 \leftarrow S_0$	599.5	[139]	29421	[192, 193]
Phenanthrene-D_{10}	biphenyl	$S_1 \leftarrow S_0$	787	[140]		
Anthracene	naphthalene	$S_1 \rightarrow S_0$	1839	[141, 142]	27695	[194][g]
Anthracene	biphenyl	$S_1 \leftarrow S_0$	1639	[143, 144]		
Anthracene	fluorene	$S_1 \leftarrow S_0$	1720	[143]		
Anthracene	carbazol	$S_1 \leftarrow S_0$	2178	[145]		

Anthracene	phenanthrene	$S_1 \leftarrow S_0$	1615[h]	[142]		
Anthracene	p-terphenyl	$S_1 \leftarrow S_0$	1765, 1820, 1985, 1999, 2014	[146]		
Anthracene-D_{10}	fluorene	$S_1 \leftarrow S_0$	1733	[143]	27770	[194][g]
Pyrene	biphenyl	$S_1 \rightarrow S_0$	474	[147]	27208	[199, 200[f,g]]
Pyrene	fluorene	$S_1 \rightarrow S_0$	518	[147]		
Tetracene	benzoic acid	$S_1 \leftarrow S_0$				
		$S_1 \rightarrow S_0$	1675, 2505	[148]	22360	[204][g]
Tetracene	naphthalene	S_1–S_0	2108[h]	[149]		
Tetracene	anthracene	$S_1 \leftarrow S_0$	2128	[128, 149]		
Tetracene	p-terphenyl	$S_1 \rightarrow S_0$	2083, 2086, 2220, 2271	[150–152, 129, 131]		
Pentacene	benzoic acid	$S_1 \leftarrow S_0$	1623	[154, 153]		
			1762	[155]	18625	[211][g]
Pentacene	benzoic acid-D_1	$S_1 \leftarrow S_0$	1603	[154]		
Pentacene	naphthalene	$S_1 \leftarrow S_0$	2037.04	[156, 128, 130, 157]		
Pentacene	stilbene	$S_1 \leftarrow S_0$	2038	[157]		
Pentacene	anthracene	S_1–S_0	2224, 2230, 2242	[158, 157]		
Pentacene	p-terphenyl	$S_1 \leftarrow S_0$	1560.1, 1619.2, 1738, 1742.2	[132, 11, 131, 133[e], 159[e]]		
Pentacene	p-terphenyl-D_{14}	$S_1 \leftarrow S_0$	1534, 1593.3, 1712.5	[132]		
Coronene	triphenylene	$S_1 \rightarrow S_0$	736	[160][g]	23872	[201]
2,3,8,9-Dibenzanthanthrene	naphthalene	$S_1 \rightarrow S_0$	16159	[161][g]		
Terrylene	p-terphenyl	$S_1 \leftarrow S_0$	1920, 1931, 1938, 1994	[162, 163]	19229 ± 47	[229]
7,8,15,16-Dibenzoterrylene	naphthalene	$S_1 \leftarrow S_0$	1724	[165]	14918 ± 37	[229]
Kekulene	1,2,4,5-tetrachlorobenzene	$S_1 \rightarrow S_0$	22082.4, 21978	[166]		
		Tetrapyrroles				
Porphine	biphenyl	$S_1 \rightarrow S_0$	35–55[i]	[171]	16320	[196][g]

(continued)

Table 2.5. *(Continued)*

Dopant	Matrix	Transition[a]	$-\Delta\nu$ (cm^{-1})[b]	Ref.[c]	ν_0 (cm^{-1})	Ref. (ν_0)[d]
Porphine	benzophenone	$S_1 \rightarrow S_0$	20–229[j]	[171]		
Porphine	anthracene	$S_1 \leftarrow S_0$	303, 326	[171, 172]		
Porphine	triphenylene	$S_1 \rightarrow S_0$	106–268[j]	[171]		
Porphine	anthracene	$S_2 \leftarrow S_0$	1066	[172]	19884	[196][g]
Mg porphine	triphenylene	$S_1 \rightarrow S_0$	17285	[173]		
Zinc porphine	biphenyl	$S_1 \rightarrow S_0$	17937–17919[i]	[171]		
Zinc porphine	anthracene	$S_1 \rightarrow S_0$	17613–17478[i]	[171]		
Chlorin	benzophenone	$S_1 \leftarrow S_0$	50–430[j]	[174]	15912	[196][g]
Zinc tetrabenzoporphine	benzophenone	S_1–S_0	700–930[j]	[170]	16579	[212][g]
		Other compounds				
Dimethyl-*s*-tetrazine	durene	S_1–S_0	483	[128]	17500	[213]
Quinoxaline	durene	$T_1 \rightarrow S_0$	21639	[175]		
Quinoxaline	naphthalene	$S_1 \leftarrow S_0$	25825	[176]		
Carbazole	fluorene	$S_1 \rightarrow S_0$	1003	[177]	30694	[214]
4,4′-Dibromobenzophenone	4,4′-dibromodiphenyl ether	$T_1 \rightarrow S_0$	23542.2	[178]		

[a] $T_1 \rightarrow S_0$, $S_1 \rightarrow S_0$, 0–0 phosphorescence and fluorescence, respectively; $S_1 \leftarrow S_0$, $S_2 \leftarrow S_0$, 0–0 absorption; S_1–S_0, absorption and fluorescence in resonance.

[b] Absolute solvent shift $\Delta\nu = \nu - \nu_0$. The transition frequency ν is given when the 0–0 wavenumber in nonsolvated molecule ν_0 was not available.

[c] Data are taken from the first reference.

[d] Source of ν_0.

[e] Plausibly not corrected for the refractive index of the air.

[f] The ν value is higher and plausibly not corrected for the refractive index of the air.

[g] Both the wavenumber and wavelength λ of the transition have been reported, whereas $\nu = \lambda^{-1}$. At the same time it was not clearly indicated that λ has been corrected for the refractive index of the air.

[h] 20 K.

[i] Several sites.

[j] ~ ten-line multiplet.

602.847 nm pentacene line in naphthalene sublimation flakes is as small as $0.11\,\mathrm{cm}^{-1}$ [156]. The even sharper line of $\Gamma_{\mathrm{ih}} = 0.025\,\mathrm{cm}^{-1}$ in the *p*-terphenyl host crystal allowed the authors of Ref. 167 to resolve the satellite peaks in the natural isotopic mixture of pentacene containing ^{13}C atoms at different positions.

Following its discovery by Marchetti et al. [11] in 1975, pentacene/*p*-terphenyl has turned out to be extremely useful in advanced spectroscopic studies due to the fact that the four site origins are in resonance with rhodamine 6G dye lasers [10, 128–133]. The distance between the O_1 and O_4 lines in five published data sets [11, 131–133, 159] equals to $182 \pm 0.5\,\mathrm{cm}^{-1}$. Very similar multiplet structure is observed for tetracene with a splitting of $188\,\mathrm{cm}^{-1}$ [150]. For a larger dopant molecule, terrylene, the splitting is less ($74\,\mathrm{cm}^{-1}$) [162–164]. The smaller anthracene molecule gives rise to a multiplet structure spreading over $249\,\mathrm{cm}^{-1}$ [146]. Note that the closely spaced sites in anthracene (denoted as O_1–O_3, separated by $29\,\mathrm{cm}^{-1}$ [146]) and pentacene (O_1 and O_2, $4\,\mathrm{cm}^{-1}$ [11, 131–133, 159]) are the most red-shifted ones, whereas the close line pair in tetracene (O_1, O_2, $3\,\mathrm{cm}^{-1}$) [129, 131, 150–152] and a group of three close lines in terrylene (X_2–X_4, $18\,\mathrm{cm}^{-1}$ [162, 163]) are the least shifted centers, that is, there is no one-to-one correspondence between the multiplets of different compounds. The absolute solvent shifts $-\Delta\nu$ are rather similar for these *p*-type transitions: 1765–2014, 2083–2271, 1560.1–1742.2, and $1920–1994\,\mathrm{cm}^{-1}$ for anthracene, tetracene, pentacene, and terrylene, respectively. From the average Lorentz–Lorenz function of *p*-terphenyl, $\phi(n^2) = 0.426$ (Table 2.2), and the Bakhshiev number $-p$ of 4175 (anthracene), 5137 (tetracene) [187], and 5301 ± 195 (terrylene) [229], one obtains the solvent shifts of -1779, -2188, and $-2258\,\mathrm{cm}^{-1}$, respectively. For pentacene the solvent shifts in liquid alkanes have not been measured because of its low solubility. For the same reason the value of p for terrylene was determined for fluorescence. The fluorescence bands tend to have slightly larger solvent sensitivity than the absorption maxima. Thus the respective $-p$ values for perylene are 5122 ± 213 (fluorescence) [229] and $4370\,\mathrm{cm}^{-1}$ (absorption) [187]. Therefore, despite the strong polarizability anisotropy of the *p*-terphenyl crystal [226], the average solvent shift can be very well predicted by treating the matrix as a continuous dielectric.

Average polarizabilities of polyarene matrices vary in a relatively narrow range from 0.411 (fluorene, diphenyl) to 0.43 (anthracene) (Table 2.2). The absolute shift $-\Delta\nu$ of the anthracene S_1 transition in a series of matrices increases symbatically with polarizability: 1639 [biphenyl, $\phi(n^2) = 0.411$], 1720 (fluorene, 0.411) [143], 1839 (naphthalene, 0.414) [141], 1765–2014 (*p*-terphenyl, 0.426) [146], and $2178\,\mathrm{cm}^{-1}$ (carbazol) [145].

In accordance with the smaller Bakhshiev numbers (Table 2.3, [187]), the α-type transitions in naphthalene [128, 134], phenanthrene [139, 140], pyrene [147], and coronene [160] have roughly three times smaller dispersive solvent shifts (-500–$-800\,\mathrm{cm}^{-1}$) in aromatic crystals than the p type transitions

($-1600 - -2200\,cm^{-1}$). A very large hypsochromic shift of the S_1 band of azulene in naphthalene ($366.6\,cm^{-1}$) [136–138] probably cannot be understood in terms of either the change of dipole moment direction (from 0.796 [236] to −0.42 D [136]) or the possible diminishing of the polarizability in the excited state [187]. This interesting example can serve as a test case of the validity of any microscopic approach to the generalized solvent-shift problem in crystals.

Similarly to the *n*-alkane hosts, a geometric similarity principle should hold in aromatic crystals. A single site structure of azulene [136–138] and quinoxaline [176] in naphthalene illustrates the validity of the key-and-lock principle. The crystal should remain essentially unperturbed if the protonated guest is introduced into a fully deuterated matrix whose excitonic band is somewhat higher than the monomeric transition of the impurity [103, 105, 128, 134, 135, 169]. In the case of large solute molecules the best results are obtained in crystals consisting of the structural fragments of impurity molecules, such as *p*-terphenyl embedded in terrylene [162–164]. Rendering *p*-terphenyl rigid by annelating two benzo rings to its opposite sides (resulting in dibenz[*a*,*h*]anthracene) or the same side (picene) counteracts somewhat the ordered fashion of terrylene incorporation, yielding inhomogeneous bandwidths of 15 and $60\,cm^{-1}$, respectively (Fig. 2.6). The spectra in *p*-quaterphenyl display an extremely broad double band as the 0–0 origin, obviously as a result of the length mismatch (Fig. 2.6). It follows from the "fragments rule" that for perylene, the biphenyl and naphthalene ought to be the best matrices. Indeed, the best resolution is obtained in naphthalene, displaying a single site at $445.490 \pm 0.05\,nm$ with an inhomogeneous width of $4\,cm^{-1}$ (melt-grown polycrystalline sample) (Fig. 2.5). In diphenyl the line is somewhat broader [229].

Tetrapyrroles. Site structures have been reported for porphine and Zn porphine in biphenyl, benzophenone, anthracene, and triphenylene crystals [171]. A single narrow S_1 peak has been documented for free-base porphine in anthracene [172] and Mg porphine in triphenylene [173]. The phosphorescence of Cu and Pd porphine also reveal fine structure in the latter medium [173]. We have found that chlorin possesses narrow sites in phenanthrene and 1,3-dihydroxybenzene crystals grown from the melt [229]. On the other hand, the nonplanar and low-symmetry benzophenone in the crystalline state represents a matrix where planar tetrapyrroles, chlorin [174], and Zn tetrabenzoporphine [170] have a multiplet of more than 10 sharp 0–0 lines spread over $250–400\,cm^{-1}$.

Other compounds. The potential of the isomorphous substitution approach has been convincingly demonstrated in the case of *tris*-bipyridinium complexes of transition metals that are characterized with broad spectra. Highly resolved emission and absorption spectra of $[Os(bpy)_3]^{2+}$ in a single crystal $[Ru(bpy)_3]$ $(ClO_4)_2$ have been reported by Yersin et al. at 2 K [179]. Further, metal-to-ligand charge-transfer transitions of $[Ru(bpy)_3](ClO_4)_2$ display zero-phonon components at 17684, 17695, and $17747\,cm^{-1}$ with the full width at half-maximum of $7\,cm^{-1}$

at 1.4 K when diluted in the colorless closed-shell Zn complex $[Zn(bpy)_3](ClO_4)_2$ [180, 181].

2.4. ELECTRON–PHONON COUPLING IN DOPED CRYSTALS

The presence of phonon wings and the thermal broadening of zero-phonon lines influence directly the resolution of spectral luminescence analysis. With optical transitions having Debye–Waller factor of unity and the lifetime-limited linewidth, one could have an analytical method of unprecedented selectivity and sensitivity. In order to understand how far one can approach this ideal limit and what are the ways of improving the conditions of chemical analysis the temperature and matrix dependence of zero-phonon lines will be discussed. Moreover, when reasonably well known and predictable, the thermal and pressure-induced line shifts and broadenings can serve as additional characteristic parameters of luminescent compounds.

2.4.1. Theory

Accoustic modes. Simple equations describing the shift and broadening of optical transitions of Cr^{3+} in ruby and other inorganic crystals as a result of the interaction with Debye phonons were derived by McCumber and Sturge [237, 238]. Later, these expressions were applied to the *n*-alkane [63–65] and aromatic host crystals [173, 182, 183]. Despite the satisfactory fitting of the experimental data, serious inconsistencies became obvious. Thus the Debye temperature (T_D) required to accommodate the line broadening between 80 and 200 K was much higher ($T_D = 300$ K for benzo[*a*]pyrene/C_8) [63] than expected for *n*-alkane crystals ($T_D = 80$ K for C_6 and 50–70 K for C_7) [65]. The broadening of $S_1(S_{11}) \leftarrow S_0$ absorption lines for naphthalene/C_5, benzo[*g,h,i*]perylene/C_6 and coronene/C_7 between 4 and 120–140 K yielded $T_D = 144$ K [64]. Both the broadening and shift of the fluorescence of a weakly coupled system, benzo[*g,h,i*]perylene/C_6 between 10 and 80 K could be treated correctly with $T_D = 80$ K [65]. On the other hand, the analysis of broadening for a number of lines of azulene and 1,3-diazaazulene in naphthalene at $T = 10$–30 K gave T_D values varying in the range of 50–100 K [182, 183]. However, the T_D as a matrix parameter should remain constant and have a higher value of 113–131 K [239].

Objections were also raised from the theoretical point of view, since the formulas derived for the very weak-coupling limit should not be applied to organic materials, where the electron–phonon interaction can be strong [65, 239]. Alternatively, simple two-parameter coth formulas containing an effective temperature and a coupling constant were used to describe the Debye–Waller factor, width, and shift of several Shpol'skii lines between 10 (80) and 200 K with moderate success [66]. More recently, the data of Ref. 183 were reconsidered in terms of a more general nonperturbative expression for coupling to acoustic phonons [239].

(Pseudo)local modes. Both the dephasing rate and line broadening below 5–20 K often obeys the Arrhenius law. The activated behavior of spectral hole shift and broadening in porphine doped in C_8 and C_{10} matrices have been treated in terms of the exchange model, elaborated first for microwave and Raman spectroscopy [12, 79, 88]. The respective expressions contain the activation energy (ΔE) that can be equal to the ground- or excited-state phonon frequency (ω_g or ω_e), the difference (Δ) between ω_g and ω_e, as well as the phonon lifetime τ as parameters. The ω_g and ω_e were accessible from the fluorescence and absorption spectra [79, 88], and τ was determined from the width of a hole burned in the phonon wing [79]. As a result, both the shift (for porphine/C_8) and broadening could be nicely described.

The interaction of a couple of phonon levels with a zero-phonon transition is qualitatively illustrated with the aid of a four-level system (see Fig. 2.2C, where six levels are drawn). At $T = 0$ only the 0–0 and the one-phonon transitions occur. Since the phonon levels can have different energies, another transition with no change in phonon numbers becomes possible at $T > 0$ having a different frequency, and the average transition energy will shift. The phonon levels are usually short-lived and lead to the broadening of the averaged 0–0 line at $T > 0$ even if the ω_g and ω_e are the same. The physical meaning of the averaging mechanisms has been discussed by Wiersma et al. [130, 155, 240]. The dephasing rate is expected to obey a double Arrhenius law, if $\omega_g \neq \omega_e$ [239–241]. The advanced single-mode models require both the phonon frequencies and lifetimes in the ground and excited states as input parameters [239–241]. For dibenzo[*a*,*e*]pyrene/C_8 it has been possible to extract the phonon lifetimes from the widths of phonon sidebands in the fluorescence and excitation spectra, and to compare the broadening behavior of the 0–0 line (396.2 nm) between 10 and 40 K with theoretical prediction without adjustment of parameters [67]. However, in doped organic crystals the occurrence of single-peaked narrow phonon sidebands [67, 79, 88] is not a common phenomenon, and for that reason the theoretical treatment of data cannot always be unambiguous.

2.4.2. Debye–Waller Factors and Phonon Wings

Debye-Waller factors (DWF). In contrast to the quadratic electron–phonon coupling (EPC) processes, which freeze out at low temperatures, the linear EPC strength remains finite at $T = 0$, and one cannot get rid of phonon wings by lowering the temperature. The determination of DWF as a relative 0–0 line intensity is straightforward in the case of a narrow single-site spectrum displaying the well-separated 0–0 line and the phonon wing. The DWFs at the low-temperature limit are collected in Table 2.6.

Because of their "inertness," the rare gas solids are sometimes believed to assure weak EPC. Indeed, the purely electronic component of the n-π^* transition in s-tetrazine embedded in Ar belongs to the most weakly coupled ones among the molecular transitions in the solid state (DWF $= 0.99$) [34]. The DWFs in solid Ne are moderately high: ~ 0.3 for perylene [21] and ~ 0.5 naphthazarine in Ne [35]. Unfortunately, the DWFs of several arenes in Ne [22, 23], phthalocyanine [30], and fullerenes [36, 38, 40] cannot be estimated from the published spectra because of large bandwidth or vibronic congestion. The S_1–S_0 transition in porphine [31] and perhaps its Mg and Zn complexes [32] in inert gas matrices show strong purely electronic components (DWF > 0.5).

The DWFs reported for the α type transition of benzo[g,h,i]perylene/C_6 (0.85) and the p-type transition of perylene/C_7 (0.23) constitute essentially the limiting cases for nonsubstituted polyarenes n-alkane matrices [57, 68]. In spite of large Bakhshiev numbers, the EPC of some p-type transitions is surprisingly weak, with DWFs of 0.72–0.78 for anthracene in C_6 and C_7 and 0.61 for perylene in C_6 [68] (Table 2.6). However, the inspection of the spectra in the handbook [233] reveals that the α-type transitions in general show particularly intense lines and weaker wings as compared to the p-type transitions. The overall correlation between the DWF and the dispersive solvent shifts is most obvious when comparing porphyrins ($-p = 300$–$600\,\mathrm{cm}^{-1}$) and arenes with the lowest bands of α ($-p = 1000$–$2000\,\mathrm{cm}^{-1}$) and p type ($-p = 4000$–$5000\,\mathrm{cm}^{-1}$) [219]. As already mentioned, it follows from the Franck–Condon principle that the large change in intermolecular forces upon electronic excitation will produce both large solvent shifts and a high probability to excite intermolecular vibrations. The interaction with a local phonon is a microscopic one, and therefore, large differences in DWF (and thermal shift and broadening) between various sites can be observed in the same system [69, 229].

Even larger variations in DWFs occur for different sites in aromatic crystals. Thus, for anthracene [146] and tetracene [152] in p-terphenyl the sites have very dissimilar phonon spectra and DWFs. The phonon wings are more intense for the lines that have larger solvent shifts, such as O_1–O_3 in anthracene [146], O_3 and O_4 in tetracene [152], and O_1 and O_2 in pentacene [11]. In a polar benzoic acid matrix the DWF for tetracene does not exceed 0.05 [148], but surprisingly, for a pentacene photosite in benzoic acid the DWF can be as high as 0.51 ± 0.03 in fluorescence and 0.49 ± 0.03 in absorption [155]. In general, the DWFs in aromatic crystals are lower than those in n-alkanes, on account of the large polarizability and quadrupolar moments of aromatic molecules.

Phonon wings. The phonon structure accompanying optical transitions in the impurity or excitonic spectra have been amply discussed from the point of view of the similarity to the frequencies measured by Raman and far infrared spectroscopy

Table 2.6. Debye–Waller Factors of Electronic Transitions in Crystalline Systems at 2–10 K

Dopant	Matrix	Transition[a]	ν (cm^{-1})[b]	DWF	Ref.
		Solid rare gases			
Perylene	Ne	$S_1 \rightarrow S_0$	23877	$\sim 0.3^c$	[21]
s-Tetrazine	Ar	$S_1 \leftarrow S_0$	17928	~ 0.99	[34]
Naphthazarine	Ne	$S_1 \leftarrow S_0$	18119	$\sim 0.5^c$	[35]
		n-Alkanes			
Anthracene	C_6	$S_1 \rightarrow S_{01}$	25694	0.72	[68]
Anthracene	C_7	$S_1 \rightarrow S_{01}$	25427	0.78	[68]
Tetracene	C_6	$S_1 \rightarrow S_0$	21033	0.31	[68]
Perylene	C_6	$S_1 \rightarrow S_0$	22410	0.61	[68]
Perylene	C_7	$S_1 \rightarrow S_0$	22680	0.23	[57]
			22455	$0.3–0.5^c$	
Benzo[*g*,*h*,*i*]perylene	C_6	$S_1 \rightarrow S_0$	24624	0.85	[57]
Dibenz[*a*,*h*]anthracene	C_6	$S_1 \rightarrow S_0$	25451	$0.8–0.9^c$	[233]
Dibenzo[*a*,*e*]pyrene	C_8	$S_1–S_0$	25233	0.71–0.73	[67]
		Frozen solvent crystals			
Tetracene	benzene	$S_1–S_0$	20659	$0.1–0.2^{c,d}$	[62]
Hexahelicene	chlorobenzene	$S_1 \rightarrow S_0$	24110	$0.1–0.2^c$	[53]
Tetraphenylporphine	nitrobenzene	$S_1 \rightarrow S_0$	15864, 15920	$\sim 0.5^c$	[111]

		Aromatic crystals			
1,1′-Diethyl–2,2′-cyanine iodide	1,1′-diethyl–9-aza–2,2′-cyanine iodide	$S_1 \leftarrow S_0$	18217, 18090	~ 0.01	[126]
Naphthalene	*p*-dibromobenzene	$T_1 \rightarrow S_0$	20721	0.23	[168]
Naphthalene	*p*-dichlorobenzene, α-phase	$T_1 \rightarrow S_0$	20712	0.3	[168]
Azulene	naphthalene	$S_1 \leftarrow S_0$	14651	0.1–0.2[c]	[138]
Anthracene	naphthalene	$S_1 \rightarrow S_0$	25856	~ 0.02[c]	[141]
Anthracene	carbazol	$S_1 \leftarrow S_0$	25517	0.02–0.05[c]	[145]
Pentacene	benzoic acid	$S_1 \rightarrow S_0$	16863	0.51 ± 0.03	[155]
		$S_1 \leftarrow S_0$		0.49 ± 0.03	
Perylene	naphthalene	$S_1 \rightarrow S_0$	22441	0.3–0.4	[229]
2,3,8,9-Dibenzanthanthrene	naphthalene	$S_1 \rightarrow S_0$	16159	0.14	[161]
Kekulene	1,2,4,5-tetrachlorobenzene	$S_1 \rightarrow S_0$	21978	~ 0.4[c]	[166]
Quinoxaline	durene	$T_1 \rightarrow S_0$	21639	0.1–0.2[c]	[175]
1-Hydronaphthyl radical	naphthalene		18557	0.2	[184]
			18527	0.18	

[a] $T_1 \rightarrow S_0$, $S_1 \rightarrow S_0$, 0–0 phosphorescence and fluorescence; $S_1 \rightarrow S_{01}$, vibronic fluorescence; $S_1 \leftarrow S_0$, $S_2 \leftarrow S_0$, 0–0 absorption; S_1–S_0, absorption and fluorescence in resonance.

[b] Transition frequency.

[c] Estimated from the figure.

Table 2.7. Phonon Sideband Maxima and Activation Energies of Optical Dephasing

Dopant	Matrix	Transition[a]	ν (cm^{-1})[b]	$\omega/\Delta E$[c]	Value (cm^{-1})	Ref.
			n-Alkanes			
Porphine, B_1 site	C_{10}	S_1–S_0	16291	$\omega_g/\omega_e/\Delta E_b$	7/5.6/7	[79]
Dibenzo[*a*,*e*]pyrene	C_8	S_1–S_0	25233	ω_g/ω_e	13.1/15.6	[67]
Porphine, B sites	C_8	S_1–S_0	16294, 16268	ΔE_b	13.5–15	[12]
Mg porphine	C_8	S_1–S_0	17434	$\omega_g/\omega_e/\Delta E_b$	14/15/18.5	[88]
2,3,8,9-Dibenzanthanthrene	C_{16}	$S_1 \rightarrow S_0$	16970	ω_g	15	[161]
Perylene	C_7	$S_1 \rightarrow S_0$	22679	ω_g	16, 29, 124	[57]
Anthracene	C_7	S_1–S_0	26217	ω_g/ω_e	18/25, 46	[70]
Tetracene	C_6	$S_1 \rightarrow S_0$	21033	ω_g	21, 35, 52, 76, 87, 94, 105	[68]
Anthracene	C_7	$S_1 \rightarrow S_{01}$	25427	ω_g	21, 55, 76	[68]
Anthracene	C_6	$S_1 \rightarrow S_{01}$	25694	ω_g	27, 39, 77	[68]
Perylene	C_6	$S_1 \rightarrow S_0$	22410	ω_g	29, 45, 65, 87, 107, 125, 140	[68]
Porphine, A sites	C_8	S_1–S_0	16327, 16261	ΔE_b	~ 30	[12]
Benzo[*g*,*h*,*i*]perylene	C_6	$S_1 \rightarrow S_0$	24624	ω_g	31	[68]
1,3,5,7-Octatetraene	C_6	$S_1 \leftarrow S_0$	28744	$\omega_g/\Delta E_b$	32/17	[42]
INS (9 K)[d]	C_5	25sh, 40sh, 58, 73, 95, 102, 136, 159, 190				
	C_6	35, 55, 71, 93, 105, 124, 142, 182				[244]
Raman (20 K)	C_6	53, 74, 87, 174				[245]
	C_7	31, 37.5, 52, 62.5, 77.5, 116, 131, 169.5, 178				
	C_8	47.5, 65, 72, ~ 140, ~ 170				
Heat uptake[e,f]	C_8	$Q = -113 + 0.425T^{1.968}$				[246]
			Naphthalene matrix			
Naphthalene[g]		$T_1 \rightarrow S_0$	21209.1	ω_g	8, ~29, 47, 56, 70, 78, 89	[134]
Naphthalene[g]		$S_1 \rightarrow S_0$	31542.4	ω_g	10, ~33, 47, 58, 66, 90	[134]

Anthracene	$S_1 \rightarrow S_0$	25856	ω_g	14sh, 31, 45, 59, 75, 82, 92, 103	[141]
Tetracene[h]	$S_1 \rightarrow S_0$	20252	ω_g	19, 40, 49, 65, 121, 176	[149]
	$S_1 \leftarrow S_0$	20252	ω_e	24, 115	
Naphthalene[g]	$T_1 \rightarrow S_0$	21208.2	ω_g	27, 45.9, 55.2, 76.8, 90.9	[169]
Anthracene[h]	$S_1 \rightarrow S_0$	25857	ω_g	30, 50, 81, 102	[142]
	$S_1 \leftarrow S_0$	25855	ω_e	26	
Pentacene	$S_1 \rightarrow S_0$	16580	ω_g	33, 61, 139	[157]
	$S_1 \leftarrow S_0$	16584	ω_e	19, 69, 125	
Pentacene	$S_1 – S_0$	16587.7	$\omega_g/\omega_e/\Delta E_e$	~36/27.5/16.3	[130]
Naphthalene[g]	$S_1 \leftarrow S_0$	31532	ΔE_e	36	[128]
Azulene	$S_1 \leftarrow S_0$	14651	$\omega_e^{\,i}$	36, 50, 64, 74, 86, 95	[137]
Perylene	$S_1 \rightarrow S_0$	22441	ω_g	40	[229]
2,3,8,9-Dibenzanthanthrene	$S_1 – S_0$	16159	$\omega_g/\Delta E_b$	~40/40	[161]
1-Hydronaphthyl radical		18527	ω_g	43	[184]
		18557	ω_g	48	
Azulene	$S_1 \leftarrow S_0$	14651	$\omega_e^{\,i}$	~45	[138]
INS (5 K)[d,i]	20(onset), 47sh, 60, 67, 94				[247]
Raman (4 K)	56, 69, 81, 88, 120, 141				[243]
IR (293 K)	44, 49, 57, 63, 80, 99, 192				[248]
Heat uptake[e,j]	$Q = -75 + 0.357T^{1.886}$				[249]

[a] $T_1 \rightarrow S_0$, $S_1 \rightarrow S_0$, 0–0 phosphorescence and fluorescence, respectively; $S_1 \rightarrow S_{01}$, vibronic fluorescence; $S_1 \leftarrow S_0$, 0–0 absorption; $S_1 – S_0$, both absorption and fluorescence.

[b] Transition frequency.

[c] ω_g, ground state phonon frequency; ω_e, excited state phonon frequency; ΔE_b, activation energy of line broadening or shift; ΔE_e, activation energy of photon echo decay.

[d] Inelastic neutron scattering spectra; sh, shoulder.

[e] Increase of heat content, cm^{-1}/molecule.

[f] Between 13 and 100 K.

[g] In naphthalene-D_{10}.

[h] 20 K.

[i] Taken from figure.

[j] Between 16 and 100 K.

[146, 242]. The low-frequency vibronic modes different from optical phonons were regarded as anomalous [242]. It was also evident from theoretical considerations that incorporation of guest molecules with different masses into the crystal lattice would lead to (quasi)localization of phonons [220]. Although direct proofs of the nature of localized phonons are scarce [75, 130], it is widely accepted that they have librational character [12, 130]. This is in accordance with the Raman spectra of pure crystals where several rotational modes have the lowest frequencies among optically active lattice vibrations (although a translational mode is the lowest one) [243].

In many cases, the wings are structured and display multiple peaks and humps. In Table 2.7 the wing frequencies are collected for *n*-alkane and naphthalene host matrices together with the Arrhenius activation energies of line broadening and dephasing rates (see also Ref. 128). For comparison, the peak maxima in inelastic neutron scattering [244, 247], Raman [243, 245], and far-infrared spectra [248] are listed. The heat content as a function of T was calculated (in cm^{-1}/molecule units) from the specific heat data for *n*-octane [246] and naphthalene [249], and approximated to a power-law dependence.

In general, the wing modes have lower frequencies than the lattice modes of neat host crystals, particularly in *n*-alkanes. Since the dopant molecules are usually heavier than those of the matrix, the wing modes have been assigned to the pseudolocal vibrations [220]. However, the highest wing frequencies are close either to the maxima in the Raman spectra [243, 245] or the shoulders in rather broad neutron scattering profiles [244, 247] of matrix crystals. The phonon structure in the phosphorescence and fluorescence spectra of isotopically substituted naphthalene that have ideal mutual solubility and no extra-free space left upon doping [134, 169] has some lines in common with the Raman spectra (56, 81, 88 cm^{-1}), albeit with different intensities [243]. A very low-frequency component appears in the wing of the $S_1 \rightarrow S_0$ and $T_1 \rightarrow S_0$ lines of the naphthalene in perdeuterated host crystal (8–10 cm^{-1}) [134, 169] that can hardly belong to an optical vibration. The quite characteristic 40–45 cm^{-1} mode of naphthalene [138, 161, 184, 229] corresponds well to the average frequency of acoustic vibrations estimated from thermodynamic functions in Ref. 243 (40 cm^{-1}).

The heat content of crystals increases as T^2. A large amount of kinetic energy per molecule is already present at 40 K: 500 and 300 cm^{-1} for C_8 and naphthalene, respectively. At 80 K the corresponding values are as large as 2300 and 1300 cm^{-1} (Table 2.7). One can speculate that the electron–phonon coupling function can form the wing profiles by selecting a narrow interval from the smooth acoustic phonon density of states.

2.4.3. Temperature Broadening

The homogeneous widths (Γ_h) of singlet transitions in crystals approach the lifetime-limited value at 1–2 K because the frequencies of (pseudo)local modes are rarely less than 5 cm^{-1} [12, 79]. Perhaps the smallest Γ_h has been measured by photon echo for the long-lived T_1–S_0 transition of 4,4′-dibromobenzophenone incorporated in a 4,4′-dibromodiphenyl ether matrix at 1.2 K (80 kHz) [178].

The smallest Γ_h at 80 K, estimated as the broadening between 5–10 and 80 K, is observed for Mg porphine/triphenylene (1.6 cm^{-1} [173]) and α-type S_1 transitions in polyarenes: coronene/C_7 (2.7 cm^{-1}), benzo[*g,h,i*]perylene/C_6 (2.9 [64] or 4 cm^{-1} [65]). The p bands show much stronger broadening: perylene/C_8 23 cm^{-1} and terrylene/C_{10} 37 cm^{-1} [229]. The Γ_h for the main components of the complicated multiplets of octaethylporphine and tetra-*tert*-butylphthalocyanine in C_8 is close to 3–4 cm^{-1} [229]. A surprisingly small broadening at 80 K has been reported for p bands of anthracene/C_7 (9 cm^{-1}) [64] and tetracene/C_9 (5–7 cm^{-1}) [62]. In these systems the line intensity decreases dramatically at 80 K, resulting in a spectrum of a peculiar shape: a small, relatively narrow peak on top of a huge triangular pedestal. It remains to be clarified whether we have here a case of strong linear EPC but weak quadratic EPC or simply a line-wing separability problem leading to an artefact. For most of the p-type transitions the matrix-induced narrowing is essentially lost at 80 K, and the cooling of the sample down to 5–10 K becomes a must. On the other hand, the α-type transitions retain their analytical significance at 77 K.

Concerning the aromatic host crystals, a slightly different broadening at 25 K has been observed for various multiplet components of 2-phenyl-1-azaazulene embedded in *p*-terphenyl ranging between 2.1 and 2.6 cm^{-1} [182]. Interestingly, even larger variations have been seen for different vibronic lines for a single site of 1,3-diazaazulene/naphthalene extending from 1.2 to 3.6 cm^{-1} [183]. By contrast, between 80 and 160 (200) K rather similar broadening behavior has been demonstrated for 0–0 and vibronic fluorescence lines for benzo[*g,h,i*]perylene/C_6, benzo[*e*]pyrene/C_6, benzo[*a*]pyrene/C_8, and perylene/C_6 [71]. Unexpectedly, the broadening of absorption and fluorescence lines was different in benzo[*a*]pyrene and perylene [71]. It appears that the line (band) broadening above liquid N_2 temperatures may be regarded as a modulation of the solvent shift by a vibrating matrix, since the extent of broadening is roughly proportional to the absolute solvent shift or Bakhshiev number.

Most of the broadening curves reported until now are featureless, in accordance with the McCumber–Sturge formula [237, 238], a coth [66], a power law or an exponential function. A sigmoid dependence resembling the full Arrhenius curve has been found on several occasions only, for single pentacene molecules in *p*-terphenyl between 1.5 and 10 K [127] and perylene/C_8 under the pressure on 3 kbar (5–30 K) [72]. Figure 2.8 shows that the T dependencies having a more complicated shape may occur more commonly than generally believed. Such curves with inflections can be approximated to a double Arrhenius law with largely different ΔE or, alternatively, to a combination of an activated process due to a local mode and the acoustic-phonon effect coming into play at higher T.

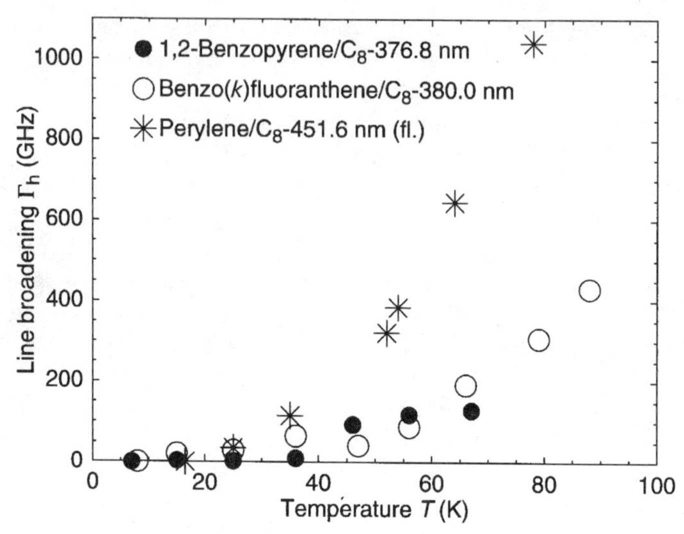

Figure 2.8. Temperature broadening of absorption (0–0 origin of benzo[*e*]pyrene, 1540 cm^{-1} vibronic line of benzo[*k*]fluoranthene) and fluorescence spectra (350 cm^{-1} vibronic line of perylene) in crystalline *n*-octane matrix. A smooth dependence (perylene) can be fitted to a power law, the McCumber–Sturge expression for Debye phonons (Refs. 237, 238), the coth formula of Sapozhnikov (Ref. 66), or the initial portion of the Arrhenius law (exponential function). Quite often the broadening reaches a plateau in some *T* range showing Arrhenius behavior for a single activation energy (phonon frequency) (benzo[*e*]pyrene, benzo[*k*]fluoranthene). In many cases the distinguishing of the broadening of the 0–0 line from the growing and broadening phonon pedestal requires a careful deconvolution procedure and may prove difficult at higher *T*.

2.4.4. Temperature and Pressure Shift

Observed temperature shift. According to the McCumber–Sturge theory the line shift is proportional to the heat content of the material [238], which in turn follows the quadratic dependence [Table 2.7]. On the other hand, coupling to a single promoting mode should lead to an Arrhenius dependence, as in the case of broadening [12]. Neither of these models takes into account the shift of transition frequency caused by the thermal expansion of the matrix. It has been verified that in inorganic crystals the matrix expansion effect is small [238], but this is definitely not true for van der Waals solids [73]. The solvent shift has been neglected in the theoretical analysis of spectral shifts for arenes in *n*-alkanes [63–66], porphyrins in *n*-alkanes [12, 79, 88], and triphenylene [173] as well as 2-phenyl-1-azaazulene [182] and pentacene [130] in *p*-terphenyl. Later, very few papers mention the thermal shift [130, 161], and the main attention is focused on the electronic dephasing processes that broaden the zero-phonon lines [132, 183]. It has been established that the lines of 2-phenyl-1-azaazulene [182] and benzo[*e*]pyrene as

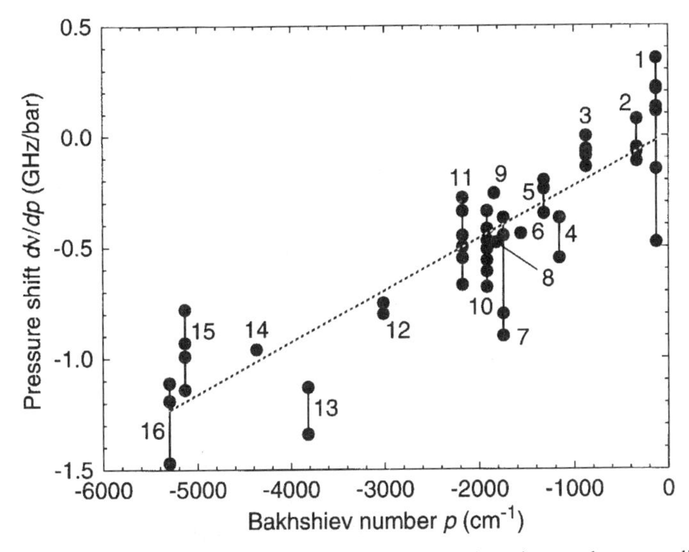

Figure 2.9. Pressure shifts of fluorescence lines in aromatic hydrocarbons and tetrapyrrolic pigments doped in *n*-octane (C_8) plotted versus the Bakshiev number, *p* (the solvent shift in liquid *n*-alkanes per unit Lorentz–Lorenz function). The positions of lines were measured under the He gas pressure between 1 and 200 bar at 10 K. The variation range of pressure shifts in a given system for different multiplet components is indicated with black dots connected with a vertical line. In most cases the pressure shift $-d\nu/dP$ increases with increasing solvent shift. A linear regression line is drawn for all points showing that the expected correlation between the pressure shift and absolute solvent shift ($\Delta\nu \sim 0.3p$) is observed for the central multiplet components. 1, octaethylporphine; 2, porphine; 3, *cis*-isobacteriochlorin; 4, pyrene (in C_6); 5, coronene (in C_7); 6, benzo[*g,h,i*]perylene; 7, dibenzo[*a,e*]pyrene; 8, dibenz[*a,c*]anthracene; 9, benz[*a*]anthracene; 10, tetra-*tert*-butylphthalocyanine; 11, *trans*-isobacteriochlorin; 12, benzo[*k*]fluoranthene; 13, 2-methylanthracene; 14, perylene; 15, tetracene (in C_9); 16, terrylene (in C_8 and C_{10}).

well as benzo[*a*]pyrene shift bathochromically, whereas those of perylene and benzo[*g,h,i*]perylene undergo a hypsochromic shift with the increase of temperature [63, 65, 66, 71]. In porphyrins both red and blue shifts have been reported [12]. Careful determination of the ratio between the broadening and shift is of considerable interest from the point of view of EPC mechanisms [12, 132, 241].

Pressure shift. The simplest way to estimate the matrix expansion effect is to assume that the absolute solvent shift is proportional to the density of the matrix [Eq. (2.1)] [250]. Alternatively, the matrix polarizability change and the Bakhshiev number may be used [190]. Perhaps the most correct method is to utilize the pressure shift and the crystal compressibility [69, 73]. All three methods are based on the use of thermal expansion coefficients that have been investigated for rare gas crystals [224, 225], but are poorly known for solid *n*-alkanes and for such well-studied materials as naphthalene [251] and anthracene.

We have recently started pressure-shift studies on matrix-narrowed spectral bands by using a simple cell that can be filled with gaseous He up to 200 bar. In polyarenes the bathochromic shifts range from −0.15 (coronene phosphorescence in C_8) to −1.1 – −1.4 GHz/bar for terrylene in C_8 and C_{10}. Blue shifts are observed for porphine and octaethylporphine sites that are already hypsochromically shifted with respect to the vacuum frequency [188, 196]. The pressure shifts show considerable variation for different components of the multiplet. However, the average dv/dP correlates quite well with the Bakhshiev number of the transition (Fig. 2.9):

$$\frac{dv}{dP} = (0.008 \pm 0.043) + (23.4 \pm 1.7)p, \qquad N = 55,\ r = 0.883 \qquad (2.4)$$

In its turn, the Bakhshiev number is proportional to the vacuum-to-matrix shift in *n*-alkane hosts [Fig. 2.4, Eq. (2.3)].

Pure thermal shift. The shift resulting from the matrix expansion is calculated as $-(\alpha_p/\beta_T)(dv/dP)$ (α_p, isobaric thermal expansion coefficient; β_T, isothermal compressibility) [73]. Figure 2.10 shows the integrated ratio of α_p/β_T (in bar

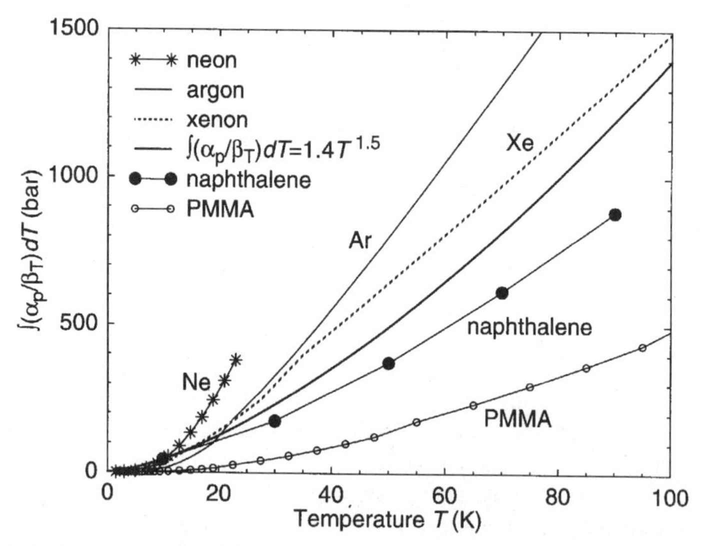

Figure 2.10. Integrated value of the volume (V) thermal expansion coefficient [$\alpha_p = (dV/dT)_p/V$] divided by isothermal compressibility [$\beta_T = -(dV/dP)_T/V$] for solid rare gases (data from Ref. 225), naphthalene (Refs. 251, 252) and poly(methyl methacrylate) (Refs. 253, 254). It will be assumed that the behavior of solid *n*-alkanes is intermediate between Xe and naphthalene, and the integral is approximated with a power-law dependence $1.4T^{1.5}$. Multiplication of the curves by spectral pressure shift $-dv/dP$ yields the thermal line shift due to the expansion of the matrix which is subtracted from the observed dependence in order to find out the pure thermal, phonon-induced change of transition frequency.

units) for Ne, Ar, Xe, naphthalene, and poly(methyl methacrylate) (PMMA) as a function of T. Remarkably, at a given T (e.g., 20 K) the highest pressure (~280 bar) should be exerted on solid Ne in order to reduce its volume to the extent that is equivalent to cooling down to 0 K. For PMMA the required pressure is less by more than an order of magnitude, since the volume expansion of the polymer is small [253]. The magnitude of α_p/β_T in solid alkanes is probably intermediate to that in Xe and naphthalene, so an empirical power-law dependence was applied: $\int \alpha_p/\beta_T dT = 1.4T^{1.5}$ (Fig. 2.10). To obtain the phonon-induced or pure thermal component, the solvent shift should be subtracted from the observed temperature dependence.

The observed thermal line shifts of anthracene, benzo[k]fluoranthene, and perylene in C_8 and terrylene in C_{10} are shown in Figure 2.11. The shift component owing to the thermal expansion of the matrix was calculated as $-1.4T^{1.5}(dv/dP)$ and subtracted from the measured one to obtain the pure phonon-induced frequency shift (Fig. 2.12). Up to 100 K the pure thermal shift is bathochromic for these and other Shpol'skii systems studied. The largest pure thermal red shift occurs for α-type transitions and a nonalternant polyarene, benzo[k]fluoranthene. The p-type chromophores with the largest broadening and solvent shifts have small pure thermal shifts (perylene, terrylene). This

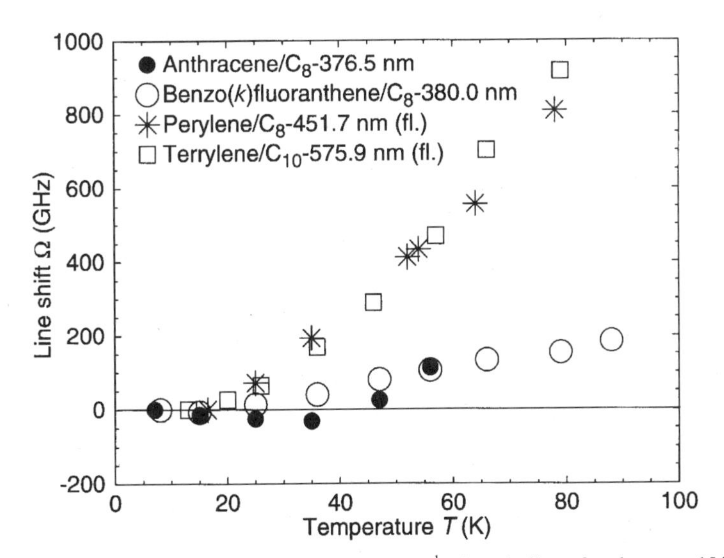

Figure 2.11. Temperature shift of absorption (395 cm^{-1} vibronic line of anthracene, 1540 cm^{-1} vibronic line of benzo[k]fluoranthene) and fluorescence spectra [350 cm^{-1} vibronic line of perylene, 0–0 origin of terrylene (in C_{10})] in crystalline n-octane matrix. Quite often the initial small bathochromic shift is changed to a hypsochromic one at higher T. The hypsochromic shift occurs as a result of the decrease of the solvent shift due to the expansion of the matrix.

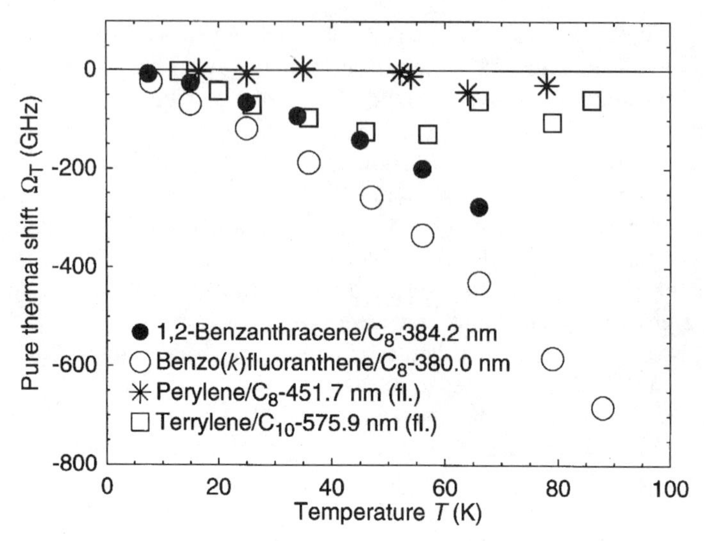

Figure 2.12. Pure thermal, phonon-induced shift of absorption (0–0 origin of anthracene, 1540 cm^{-1} vibronic line of benzo[*k*]fluoranthene) and fluorescence spectra [350 cm^{-1} vibronic line of perylene, 0–0 origin of terrylene (in C_{10})] in crystalline *n*-octane matrix. The pure thermal shift appears to be exclusively bathochromic. However, at higher T the red shift of transitions with large solvent shifts (perylene, terrylene) decreases, because the intermolecular vibrations that have higher frequencies in the excited state come into play.

unexpected effect raises a question about the nature of phonons that are involved in the shift. In the case of bathochromic shifts the phonon frequency ought to be lower in the excited state (Fig. 2.2C). On the contrary, in systems with large solvent shifts the intermolecular forces strengthen in the excited state and the corresponding intermolecular modes should have higher frequencies. For perylene/C_8 and terrylene/C_{10} the shifts caused by different kinds of phonons seem to be roughly equal resulting in a small pure thermal shift. The observed large blue shift of p transitions originates almost entirely from the matrix expansion.

In anthracene/C_8 at first a bathochromic shift is observed that changes to a blue one at higher T (Fig. 2.11). A similar dependence has been reported for pentacene in naphthalene [130] and a reverse one for coronene fluorescence and phosphorescence in C_7 [69]. Therefore, the interplay between the matrix shift and the EPC can lead to fairly complicated behavior.

The components of a multiplet can show net shifts of different sign. As a rule, the lines with larger absolute solvent shifts undergo a hypsochromic shift. Such an effect occurs in octaethylporphine and tetra-*tert*-butylphthalocyanine in C_8 [229] as well as 3,4-benzopyrene [74] and coronene [69] in C_7. However, the pure thermal shift is bathochromic for all sites, if the influence of the matrix expansion is corrected for by means of pressure shifts.

Both the temperature- and pressure-induced shifts were first considered for benzo[*e*]pyrene, benzo[*g*,*h*,*i*]perylene, and perylene doped in *n*-hexane matrix between 80–150 K [73]. The pressure range was extended up to 7 kbar. In the present work the $-d\nu/dP$ values obtained at low pressures were several times larger: 0.44 GHz/bar in benzo[*g*,*h*,*i*]perylene and 0.96 GHz/bar for perylene (both in C_8), whereas the respective values in Ref. 73 are 0.20 and 0.36 GHz/bar (both in C_6). The application of high pressures is not at all necessary, bearing in mind the small magnitude of thermal expansion of the matrix (Fig. 2.10). Therefore, the solvent shift components calculated from the pressure effect in Ref. 73 are too small, since at such high pressures the compressibility of molecular crystals is less and so is the average pressure shift. It was assumed earlier that α_p/β_T is constant [69, 73]. However, this simplification is invalid over the broad *T* range since the thermal expansion coefficient is strongly *T* dependent, while the compressibility change is small [225]. The dearth of precise thermal expansion data is a serious obstacle, complicating the accurate analysis of thermal shifts of optical transition energies in doped molecular crystals, especially below 10–20 K.

2.5. SUMMARY

The selectivity of luminescence analysis depends largely on both the width of spectral components and the line-to-background ratio in the spectra. Neither the cooling of the sample nor the monochromatic laser excitation enables one to get rid of background emission caused by the excitation of low-frequency matrix modes. However, trapping of analyte molecules in well-ordered environments creates a very favorable situation from the point of view of spectral analysis when the zero-phonon lines can be well separated from the phonon wings. Addressing such lines by means of laser excitation or narrow-band detection allows one to measure relatively easily the signals from a small number of molecules or even a single molecule, and to resolve the vibronic structure of selected components in complicated mixtures. Therefore, the selection of a proper matrix that shows a spectral narrowing effect for a single compound, a group of compounds with similar structures, or, alternatively, a large set of molecules with different structures remains a practical task.

Without any doubt, the normal alkanes constitute the most universal host matrix capable of producing line narrowing. Solid noble gases and N_2 [26], frozen tetrahydrofuran [108], cyclohexane, methylcyclohexane, benzene, and naphthalene (Table 2.4) can serve as good matrices for a fairly wide group of compounds. On the other hand, very special combinations, such as tetraphenylporphine/nitrobenzene [111] or hexahelicene/ chlorobenzene [53], can be of interest. Of course, it would be premature to state that the inhomogeneous width of every optical transition can be drastically reduced by a proper choice of environment. However, the sharp lines of the charge-transfer transition in $Ru(bpy)_3^{2+}$ in the colorless matrix of the respective Zn complex [180, 181] demonstrate that the isomorphous substitution principle is far from being exhausted. Several "nonobvious" cases of good guest–host fits leading to surprising band narrowing have been discovered by chance. It seems plausible that extensive computer-aided modeling of guest–host compatibility based on atom–atom potentials [255] would provide us with numerous combinations suitable for spectroscopic and analytical applications.

In the present chapter we attempted to elucidate the applicability ranges of the simplest continuum models of the host matrix for understanding the solvent shift and dephasing phenomena. The merits of the Lorentz–Lorenz expression for polarizability [185] and the Debye formula for the phonon density of states [237, 238] are their generality and simple use for practical calculations of absolute solvent shifts and T-induced line broadening, respectively. Particularly, the Bakhshiev model for average dispersive solvent shifts is applicable in all kinds of environments. Deviations from Eq. (2.1) point to the presence of a loose fit of the dopant or the importance of dipolar or quadrupolar solute–solvent interactions. Similarly, the failure of the Debye model, or more generally speaking, the lack of correlation between bulk heat capacity and the line broadening, points to the existence of localized low-frequency modes that are created by doping.

In general, both the solvent shift phenomena and the electron–phonon coupling processes depend on the change of intermolecular interactions upon electronic excitation. The transitions with small solvent shifts, such as the α bands in polycyclic hydrocarbons and the S_1 bands in porphyrins and fullerenes are less strongly coupled to low-frequency vibrations. The literature as well as our own experiments seem to confirm that the Debye–Waller factor of molecular transitions in van der Waals solids can be as high as 0.85 or ultimately 0.9. Crystalline solutions are more favorable for spectroscopy and analysis owing to the dramatic reduction of inhomogeneous broadening. The zero-phonon lines in pyrene (DWF = 0.75) or anthracene (0.5) in 1,3-dimethylcyclohexane glass [216] are by no means weaker than in crystalline n-alkanes. The usefulness of crystalline samples stems mainly from the separation of zero-phonon lines and phonon sidebands in the frequency scale, rather than from the decrease of electron–phonon coupling strength.

REFERENCES

1. E. V. Shpol'skii, A. A. Il'ina, and L. A. Klimova, *Dokl. Akad. Nauk SSSR* **87**, 935 (1952).
2. D. S. McClure, in F. Seitz and D. Turnbull, eds., *Solid State Physics*, Vol. 8, Academic Press, New York, 1959, pp. 1–47.

3. H. C. Wolf, in F. Seitz and D. Turnbull, eds., *Solid State Physics*, Vol. 9, Academic Press, New York, 1959, pp. 1–81.
4. R. I. Personov, E. I. Al'shits, and L. A. Bykovskaya, *Zh. Eksp. Teor. Fiz., Pis'ma Red.* **15**, 609 (1972); *Opt. Commun.* **6**, 169 (1972).
5. A. A. Gorokhovskii, R. K. Kaarli, and L. A. Rebane, *Zh. Eksp. Teor. Fiz., Pis'ma Red.* **20**, 474 (1974) [*JETP Lett.* **20**, 216 (1974)].
6. B. M. Kharlamov, R. I. Personov, and L. A. Bykovskaya, *Opt. Commun.* **12**, 191 (1974).
7. C. G. de Lima, in W. L. Zielinski, Jr., ed., *Critical Reviews in Analytical Chemistry*, Vol. 16, CRC Press, Boca Raton, Florida, 1986, pp. 177–221.
8. J. W. Hofstraat, C. Gooijer, and N. H. Velthorst, in S. G. Schulman, ed., *Molecular Luminescence Spectroscopy: Methods and Applications, Part 2*, John Wiley, New York, 1988, pp. 283–400.
9. J. W. Hofstraat, W. J. M. van Zeijl, F. Ariese, J. W. G. Mastenbroek, C. Gooijer, and N. H. Velthorst, *Mar. Chem.* **33**, 301 (1991).
10. T. Basché, W. E. Moerner, M. Orrit, and U. P. Wild, eds., *Single-Molecule Optical Detection, Imaging and Spectroscopy*, VCH, Weinheim, 1997.
11. A. P. Marchetti, W. C. McColgin, and J. H. Eberly, *Phys. Rev. Lett.* **35**, 387 (1975).
12. S. Voelker, R. M. MacFarlane, and J. H. van der Waals, *Chem. Phys. Lett.* **53**, 8 (1978).
13. G. W. Robinson, *J. Mol. Spectr.* **6**, 58 (1961).
14. B. Katz, M. Brith, B. Sharf, and J. Jortner, *J. Chem. Phys.* **54**, 3924 (1971).
15. A. Baca, R. Rossetti, and L. E. Brus, *J. Chem. Phys.* **70**, 5575 (1979).
16. M. Gutmann, P.-F. Schönzart, and G. Hohlneicher, *Chem. Phys.* **140**, 107 (1990).
17. J. Najbar, A. M. Turek, and T. D. S. Hamilton, *J. Lumin.* **26**, 281 (1982).
18. J. Wolf and G. Hohlneicher, *Chem. Phys.* **181**, 185 (1994).
19. R. Fraenkel, U. Samuni, Y. Haas, and B. Dick, *Chem. Phys. Lett.* **203**, 523 (1993).
20. P. M. Saari and T. B. Tamm, *Izv. Akad. Nauk SSSR, Ser. Fiz.* **42**, 562 (1978) [*Bull. Acad. Sci. USSR, Phys. Ser.* **42**, 96 (1978)].
21. T. Tamm and P. Saari, *Chem. Phys.* **40**, 311 (1979).
22. F. Salama and L. J. Allamandola, *J. Chem. Soc. Faraday Trans.* **89**, 2277 (1993).
23. F. Salama, C. Joblin, and L. J. Allamandola, *J. Chem. Phys.* **101**, 10252 (1994).
24. J. Szczepanski, C. Chapo, and M. Vala, *Chem. Phys. Lett.* **205**, 434 (1993).
25. M. Gutmann, M. Gudipati, P.-F. Schönzart, and G. Hohlneicher, *J. Phys. Chem.* **96**, 2433 (1992).
26. R. C. Stroupe, P. Tokousbalides, R. B. Dickinson, Jr., E. L. Wehry, and G. Mamantov, *Anal. Chem.* **49**, 701 (1977).
27. Z. Gasyna, D. H. Metcalf, P. N. Schatz, C. L. McConnell, and B. E. Williamson, *J. Phys. Chem.* **99**, 5865 (1995).
28. T. C. VanCott, M. Koralewski, D. H. Metcalf, P. N. Schatz, and B. E. Williamson, *J. Phys. Chem.* 97, 7417 (1993).

29. T. C. VanCott, J. L. Rose, G. C. Misener, B. E. Williamson, A. E. Schrimpf, M. E. Boyle, and P. N. Schatz, *J. Phys. Chem.* **93**, 2999 (1989).
30. P. Geissinger, L. Kador, and D. Haarer, *Phys. Rev. B* **53**, 4356 (1996).
31. J. Radziszewski, J. Waluk, and J. Michl, *J. Mol. Spectr.* **140**, 373 (1990).
32. A. Starukhin, A. Shulga, and J. Waluk, *Chem. Phys. Lett.* **272**, 405 (1997).
33. L. Bajema, M. Gouterman, and C. B. Rose, *J. Mol. Spectr.* **39**, 421 (1971).
34. B. Dellinger, D. S. King, R. M. Hochstrasser, and A. B. Smith III, *J. Am. Chem. Soc.* **99**, 7138 (1977).
35. V. E. Bondybey, S. V. Milton, J. H. English, and P. M. Rentzepis, *Chem. Phys. Lett.* **97**, 130 (1983).
36. A. Sassara, G. Zerza, and M. Chergui, *J. Phys. B: At. Mol. Opt. Phys.* **29**, 4997 (1996).
37. A. Sassara, G. Zerza, and M. Chergui, *Chem. Phys. Lett.* **261**, 213 (1996).
38. J. Fulara, M. Jakobi, and J. P. Maier, *Chem. Phys. Lett.* **206**, 203 (1993).
39. Z. Gasyna, P. N. Schatz, J. P. Hare, T. J. Dennis, H. W. Kroto, R. Taylor, and D. R. M. Walton, *Chem. Phys. Lett.* **183**, 283 (1991).
40. W.-C. Hung, C.-D. Ho, C.-P. Liu, and Y.-P. Lee, *J. Phys. Chem.* **100**, 3927 (1996).
41. B. S. Hudson, B. E. Kohler, and K. Schulten, in E. C. Lim, ed., *Excited States*, Vol. 6, Academic Press, New York, 1982, pp. 1–95.
42. G. Adamson, G. Gradl, and B. E. Kohler, *J. Chem. Phys.* **90**, 3038 (1989).
43. T. Plakhotnik, D. Walser, A. Renn, and U. P. Wild, *Chem. Phys. Lett.* **262**, 379 (1996).
44. M. Lamotte and J. Joussot-Dubien, *J. Chem. Phys.* **61**, 1892 (1974).
45. M. Lamotte, S. Risemberg, A.-M. Merle, and J. Joussot-Dubien, *J. Chem. Phys.* **69**, 3639 (1978).
46. A. M. Merle, M. Lamotte, S. Risemberg, C. Hauw, J. Gaultier, and J. Ph. Grivet, *Chem. Phys.* **22**, 207 (1977).
47. A. M. Merle, W. M. Pitts, and M. A. El-Sayed, *Chem. Phys. Lett.* **54**, 211 (1978).
48. A. M. Merle, M. F. Nicol, and M. A. El-Sayed, *Chem. Phys. Lett.* **59**, 386 (1978).
49. K. Palewska, E. C. Meister, and U. P. Wild, *J. Lumin.* **50**, 47 (1991).
50. T. B. Tamm, Ya. V. Kikas, and A. É. Sirk, *Zh. Prikl. Spektr.* **24**, 315 (1976) [*J. Appl. Spectr.* **24**, 218 (1976)].
51. J. V. Kikas and A. B. Treshchalov, *Chem. Phys. Lett.* **98**, 295 (1983).
52. T. Vo-Dinh and U. P. Wild, *J. Lumin.* **6**, 296 (1973).
53. K. Palewska, Z. Ruziewicz, H. Choinacki, and E. C. Meister, *Chem. Phys.* **161**, 437 (1992).
54. K. Palewska, E. C. Meister, and U. P. Wild, *J. Photochem. Photobiol. A: Chem.* **50**, 239 (1989).
55. K. Palewska, J. Lipinski, J. Sworakowski, J. Sepiol, H. Gygax, E. C. Meister, and U. P. Wild, *J. Phys. Chem.* **99**, 16835 (1995).
56. E. P. Lai, E. L. Inman, Jr., and J. D. Winefordner, *Talanta* **29**, 601 (1982).

57. R. I. Personov, I. S. Osad'ko, É. D. Godyaev, and Al'shits, *Fiz. Tverd. Tela* **13**, 2653 (1971) [*Soviet Phys. Solid State* **13**, 2224 (1972)].
58. T. B. Tamm and P. M. Saari, *Chem. Phys. Lett.* **30**, 219 (1975).
59. P. M. Saari and T. B. Tamm, *Izv. Akad. Nauk SSSR, Ser. Fiz.* **39**, 2321 (1975).
60. K. Palewska, E. C. Meister, and U. P. Wild, *Chem. Phys.* **138**, 115 (1989).
61. K. Palewska, Z. Ruziewicz, and H. Choinacki, *J. Lumin.* **39**, 75 (1987).
62. J. W. Hofstraat, A. J. Schenkeveld, M. Engelsma, C. Gooijer, and N. H. Velthorst, *Spectrochim. Acta* **44A**, 1019 (1988).
63. V. A. Kizel and M. N. Sapozhnikov, *Phys. Stat. Sol.* **41**, 207 (1970).
64. J. L. Richards and S. A. Rice, *J. Chem. Phys.* **54**, 2014 (1971).
65. E. I. Al'shits, É. D. Godyaev, and R. I. Personov, *Fiz. Tverd. Tela* **14**, 1605 (1972).
66. M. N. Sapozhnikov, *Phys. Stat. Sol.* **56b**, 391 (1973).
67. O. N. Korotaev and M. Yu. Kaliteyevskii, *Zh. Eksp. Teor. Fiz.* **79**, 439 (1980).
68. I. S. Osad'ko, E. I. Al'shits, and R. I. Personov, *Fiz. Tverd. Tela* **16**, 1974 (1974) [*Soviet Phys. Solid State* **16**, 1286 (1975)].
69. V. V. Nizhnikov, I. E. Zalesskii, and A. M. Sarzhevskii, *Opt. Spektrosk.* **55**, 863 (1983) [*Opt. Spectrosc.* **55**, 519 (1983)].
70. R. M. Macnab and K. Sauer, *J. Chem. Phys.* **53**, 2805 (1970).
71. R. I. Personov and V. V. Solodunov, *Fiz. Tverd. Tela* **10**, 1848 (1968) [*Soviet Phys. Solid State* **10**, 1454 (1968)].
72. O. N. Korotaev, I. P. Kolmakov, M. F. Shchanov, V. P. Karpov, and É. D. Godyaev, *Zh. Eksp. Teor. Fiz., Pis'ma Red.* **55**, 417 (1992) [*JETP Lett.* **55**, 424 (1992)].
73. A. I. Laisaar, A. K.-I. Mugra, and M. N. Sapozhnikov, *Fiz. Tverd. Tela* **16**, 1155 (1974) [*Soviet Phys. Solid State* **16**, 741 (1974)].
74. I. E. Zalesskii and S. M. Gorbachov, *Fiz. Tverd. Tela* **27**, 493 (1985) [*Soviet Phys. Solid State* **27**, 301 (1985)].
75. G. Jansen, M. Noort, G. W. Canters, and J. H. van der Waals, *Mol. Phys.* **35**, 283 (1978).
76. G. Jansen, M. Noort, N. van Dijk, and J. H. van der Waals, *Mol. Phys.* **39**, 865 (1980).
77. T. R. Koehler, *J. Chem. Phys.* **72**, 3389 (1980).
78. A. Suisalu and R. Avarmaa, *Chem. Phys. Lett.* **101**, 182 (1983).
79. A. I. M. Dicker, J. Dobkowski, and S. Völker, *Chem. Phys. Lett.* **84**, 415 (1981).
80. S. Völker and R. M. Macfarlane, *J. Chem. Phys.* **73**, 4476 (1980).
81. H. P. H. Thijssen and S. Völker, *Chem. Phys. Lett.* **82**, 478 (1981).
82. W-Y. Huang, U. P. Wild, and L. W. Johnson, *J. Phys. Chem.* **96**, 6189 (1992).
83. I. E. Zalesskii, V. N. Kotlo, K. N. Solovyov, and S. F. Shkirman, *Opt. Spektrosk.* **38**, 917 (1975) [*Opt. Spectrosc.* **38**, 527 (1975)].
84. A. A. Gorokhovskii, *Opt. Spektrosk.* **40**, 477 (1976) [*Opt. Spectrosc.* **40**, 272 (1976)].
85. K. H. Mauring, I. V. Renge, and R. A. Avarmaa, *Zh. Prikl. Spektr.* **48**, 429 (1988) [*J. Appl. Spectr.* **48**, 298 (1988)].

86. F. A. Burkhalter and U. P. Wild, *Chem. Phys.* **66**, 327 (1982); S. M. Arabei, S. F. Shkirman, and K. N. Solovyov, *Spectrochim. Acta* **48A**, 155 (1992).
87. J. Zollfrank and J. Friedrich, *J. Chem. Phys.* **93**, 8586 (1990).
88. A. I. M. Dicker, L. W. Johnson, S. Völker, and J. H. van der Waals, *Chem. Phys. Lett.* **100**, 8 (1983).
89. T. Vo-Dinh, U. T. Kreibich, and U. P. Wild, *Chem. Phys. Lett.* **24**, 352 (1974).
90. V. T. Koyava and V. V. Sakovich, *Zh. Prikl. Spektrosk.* **43**, 210 (1985).
91. D. McMorrow and M. Kasha, *Proc. Natl. Acad. Sci. USA* **81**, 3375 (1984).
92. R. J. Platenkamp, A. J. W. G. Visser, and J. Koziol, in V. Massey and C. H. Williams, eds., *Flavins and Flavoproteins*, Elsevier, Amsterdam, 1982, pp. 561–67.
93. D. Donges, J. K. Nagle, and H. Yersin, *Inorg. Chem.* **36**, 3040 (1997).
94. H. Wiedenhofer, S. Schützenmeier, A. von Zelewsky, and H. Yersin, *J. Phys. Chem.* **99**, 13385 (1995).
95. H. Yersin, S. Schützenmeier, H. Wiedenhofer, and A. von Zelewsky, *J. Phys. Chem.* **97**, 13496 (1993).
96. J. B. M. Warntjes, I. Holleman, G. Meijer, and E. J. J. Groenen, *Chem. Phys. Lett.* **261**, 495 (1996).
97. K. Palewska, J. Sworakowski, H. Chojnacki, E. C. Meister, and U. P. Wild, *J. Phys. Chem.* **97**, 12167 (1993).
98. R. van den Berg and S. Völker, *Chem. Phys.* **128**, 257 (1988).
99. P. J. Vergragt and J. H. van der Waals, *Chem. Phys. Lett.* **36**, 283 (1975).
100. D. M. Haaland and G. C. Nieman, *J. Chem. Phys.* **59**, 4435 (1973).
101. L. Watmann-Grajcar, *J. Chim. Physique* **66**, 1023 (1969).
102. V. J. Morrison and J. D. Laposa, *Spectrochim. Acta* **32A**, 443 (1976).
103. G. C. Nieman and D. S. Tinti, *J. Chem. Phys.* **46**, 1432 (1967).
104. J. M. van Pruyssen and S. D. Colson, *Chem. Phys.* **6**, 382 (1974).
105. J. M. van Pruyssen, F. B. Tudron, and S. D. Colson, *Mol. Phys.* **31**, 699 (1976).
106. G. D. Gillispie, M. H. van Benthem, and M. A. Connolly, *Chem. Phys.* **106**, 459 (1986).
107. A. Olszowski, *Chem. Phys. Lett.* **78**, 520 (1981).
108. G. F. Kirkbright and C. G. De Lima, *Chem. Phys. Lett.* **37**, 165 (1976).
109. A. Pellois and J. Ripoche, *Chem. Phys. Lett.* **3**, 280 (1969).
110. V. V. Padalka, N. A. Kovrizhnykh, and V. A. Butlar, *Opt. Spektr.* **41**, 1078 (1976) [*Opt. Spectr.* **41**, 635 (1976)].
111. R. Tamkivi, I. Renge, and R. Avarmaa, *Chem. Phys. Lett.* **103**, 103 (1983).
112. R. M. Hochstrasser and D. S. King, *J. Am. Chem. Soc.* **98**, 5443 (1976).
113. R. M. Hochstrasser and D. S. King, *Chem. Phys.* **5**, 439 (1974).
114. M. A. El-Sayed and W. R. Moomaw, in A. B. Zahlan, ed., *Excitons, Magnons and Phonons in Molecular Crystals*, Cambridge University Press, 1968, pp. 103–23.
115. G. L. LeBel and J. D. Laposa, *J. Mol. Spectr.* **41**, 249 (1972).

116. H. M. van Noort, Ph. J. Vergragt, J. Herbich, and J. H. van der Waals, *Chem. Phys. Lett.* **71**, 5 (1980).
117. J. Olmsted III and M. A. El-Sayed, *J. Mol. Spectr.* **40**, 71 (1971).
118. O. S. Khalil and L. Goodman, *J. Chem. Phys.* **65**, 4061 (1976).
119. M. Koyanagi and L. Goodman, *J. Chem. Phys.* **55**, 2959 (1971).
120. D. M. Grebenshchikov, N. A. Kovrizhnykh, and S. A. Kozlov, *Opt. Spektr.* **31**, 733 (1971) [*Opt. Spectr.* **31**, 392 (1971)].
121. B. S. Razbirin, A. N. Starukhin, A. V. Chugreev, Yu. S. Grushko, and S. N. Kolesnik, *Zh. Eksp. Teor. Fiz., Pis'ma Red.* **60**, 435 (1994) [*JETP Lett.* **60**, 451 (1994)].
122. A. N. Starukhin, B. S. Razbirin, A. V. Chugreev, D. K. Nelson, Yu. S. Grushko, S. N. Kolesnik, J. M. Hvam, D. Birkedal, K. Litvinenko, C. Spiegelberg, J. Zeman, and G. Martinez, *J. Lumin.* **72–74**, 457 (1997).
123. B. S. Razbirin, A. N. Starukhin, A. V. Chugreev, D. K. Nelson, Yu. S. Grushko, S. N. Kolesnik, V. N. Zgonnik, L. V. Vinogradova, and L. A. Fedorova, *Fiz. Tverd. Tela* **38**, 943 (1996) [*Phys. Solid State* **38**, 522 (1996)].
124. B. S. Hudson and B. E. Kohler, *J. Chem. Phys.* **59**, 4984 (1973).
125. S. R. Hawi and J. C. Wright, *J. Chem. Phys.* **103**, 1274 (1995).
126. A. P. Marchetti and M. Scozzafava, *Chem. Phys. Lett.* **41**, 87 (1976).
127. W. P. Ambrose, Th. Basché, and W. E. Moerner, *J. Chem. Phys.* **95**, 7150 (1991).
128. D. A. Wiersma, in J. Jortner, R. D. Levine, and S. A. Rice, eds., *Advances in Chemical Physics*, Vol. 47, Part 2, John Wiley & Sons, New York, 1981, pp. 441–85.
129. T. J. Aartsma and D. A. Wiersma, *Chem. Phys. Lett.* **42**, 520 (1976).
130. W. H. Hesselink and D. A. Wiersma, *J. Chem. Phys.* **73**, 648 (1980).
131. F. G. Patterson, W. L. Wilson, H. W. H. Lee, and M. D. Fayer, *Chem. Phys. Lett.* **110**, 7 (1984).
132. H. de Vries and D. A. Wiersma, *J. Chem. Phys.* **70**, 5807 (1979).
133. T. E. Orlowski and A. H. Zewail, *J. Chem. Phys.* **70**, 1390 (1979).
134. R. Kopelman, F. W. Ochs, and P. N. Prasad, *J. Chem. Phys.* **57**, 5409 (1972).
135. P. S. Friedman, P. N. Prasad, and R. Kopelman, *Chem. Phys.* **13**, 121 (1976).
136. R. M. Hochstrasser and L. J. Noe, *J. Chem. Phys.* **50**, 1684 (1969).
137. R. M. Hochstrasser and C. A. Nyi, *J. Chem. Phys*, **70**, 1112 (1979).
138. G. J. Small and S. Kusserow, *J. Chem. Phys.* **60**, 1558 (1974).
139. G. Fischer, *Chem. Phys. Lett.* **4**, 62 (1974).
140. R. M. Hochstrasser and G. J. Small, *J. Chem. Phys.* **45**, 2270 (1966).
141. R. Ostertag and H. C. Wolf, *Phys. Stat. Sol.* **31**, 139 (1969).
142. J. W. Sidman, *J. Chem. Phys.* **25**, 115 (1956).
143. A. Bree, A. Leyderman, and C. Taliani, *Chem. Phys. Lett.* **118**, 468 (1985).
144. R. M. Hochstrasser and G. J. Small, *J. Chem. Phys.* **48**, 3612 (1968).
145. A. Bree and R. Zwarich, *Chem. Phys.* **170**, 185 (1993).
146. G. J. Small, *J. Chem. Phys.* **52**, 656 (1970).

147. A. Bree and V. V. B. Vilkos, *Spectrochim. Acta* **27A**, 2333 (1971).
148. H. B. Levinsky and D. A. Wiersma, *J. Chem. Phys.* **79**, 2677 (1983).
149. J. W. Sidman, *J. Chem. Phys.* **25**, 122 (1956).
150. C. de La Riva, C. Kryschi, and H. P. Trommsdorff, *Chem. Phys. Lett.* **227**, 13 (1994).
151. G. J. Small, *J. Chem. Phys.* **58**, 2015 (1973).
152. A. Krüger, C. Kryschi, and D. Schmid, *J. Lumin.* **45**, 447 (1990).
153. K. Duppen, L. W. Molenkamp, J. B. W. Morsink, D. A. Wiersma, and H. P. Trommsdorff, *Chem. Phys. Lett.* **84**, 421 (1981).
154. R. W. Olson, H. W. H. Lee, F. G. Patterson, M. D. Fayer, R. M. Shelby, D. P. Burum, and R. M. MacFarlane, *J. Chem. Phys.* **77**, 2283 (1982).
155. L. W. Molenkamp and D. A. Wiersma, *J. Chem. Phys.* **80**, 3054 (1984).
156. S. Kummer, C. Bräuchle, and T. Basché, *Mol. Cryst. Liq. Cryst.* **283**, 255 (1996).
157. A. F. Prikhotko, A. F. Skorobogatko, and L. I. Tsikora, *Opt. Spektr.* **26**, 214 (1969) [*Opt. Spectr.* **26**, 115 (1969)].
158. A. Brillante and D. P. Craig, *J. Chem. Soc. Faraday Trans. II* **71**, 1457 (1975).
159. H.-C. Fleischhauer, C. Kryschi, B. Wagner, and H. Kupka, *J. Chem. Phys.* **97**, 1742 (1992).
160. K. Ohno and H. Inokuchi, *Chem. Phys. Lett.* **23**, 561 (1973).
161. F. Jelezko, B. Lounis, and M. Orrit, *J. Chem. Phys.* **107**, 1692 (1997).
162. S. Kummer, F. Kulzer, R. Kettner, Th. Basché, C. Tietz, C. Glowatz, and C. Kryschi, *J. Chem. Phys.* **107**, 7673 (1997).
163. S. Kummer, Th. Basché, and C. Bräuchle, *Chem. Phys. Lett.* **229**, 309 (1994).
164. S. Kummer, S. Mais, and Th. Basché, *J. Phys. Chem.* **99**, 17078 (1995).
165. F. Jelezko, Ph. Tamarat, B. Lounis, and M. Orrit, *J. Phys. Chem.* **100**, 13892 (1996).
166. D. Schweitzer, K. H. Hausser, H. Vogler, F. Diederich, and H. A. Staab, *Mol. Phys.* **46**, 1141 (1982).
167. J. Köhler, A. C. J. Brouwer, E. J. J. Groenen, and J. Schmidt, *Chem. Phys. Lett.* **228**, 47 (1994).
168. S. D. Colson and B. W. Gash, *Chem. Phys.* **1**, 182 (1973).
169. D. M. Hanson, *J. Chem. Phys.* **51**, 5063 (1969).
170. R. Jaaniso and J. Kikas, *Chem. Phys. Lett.* **123**, 169 (1986).
171. B. F. Kim, J. Bohandy, and C. K. Jen, *Spectrochim. Acta* **30A**, 2031 (1974).
172. B. F. Kim and J. Bohandy, *J. Mol. Spectr.* **73**, 332 (1978).
173. J. Bohandy and B. F. Kim, *J. Chem. Phys.* **73**, 5477 (1980).
174. P. Schellenberg, J. Friedrich, and J. Kikas, *J. Chem. Phys.* **100**, 5501 (1994).
175. S. Yamauchi and T. Azumi, *Chem. Phys. Lett.* **21**, 603 (1973).
176. R. H. Clarke, R. M. Hochstrasser, and C. J. Marzzacco, *J. Chem. Phys.* **51**, 5015 (1969).
177. A. Bree and R. Zwarich, *J. Chem. Phys.* **49**, 3344 (1968).
178. G. Wäckerle, H. Zimmermann, and K. P. Dinse, *Chem. Phys. Lett.* **110**, 107 (1984).

179. H. Yersin, E. Gallhuber, and G. Hensler, *Chem. Phys. Lett.* **140**, 157 (1987).
180. M. Kato, S. Yamauchi, and N. Hirota, *Chem. Phys. Lett.* **157**, 543 (1989).
181. H. Riesen, L. Wallace, and E. Krausz, *J. Phys. Chem.* **99**, 16807 (1995).
182. F. P. Burke and G. J. Small, *Chem. Phys.* **5**, 198 (1974).
183. F. P. Burke and G. J. Small, *J. Chem. Phys.* **61**, 4588 (1974).
184. T. Nakayama, *J. Lumin.* **28**, 313 (1983).
185. N. G. Bakshiev, O. P. Girin, and I. V. Piterskaya, *Opt. Spektrosk.* **24**, 901 (1968) [*Opt. Spectrosc.* **24**, 483 (1968)].
186. I. Renge, *J. Photochem. Photobiol.* **A69**, 135 (1992).
187. I. Renge, *Chem. Phys.* **167**, 173 (1992).
188. I. Renge, *J. Phys. Chem.* **97**, 6582 (1993).
189. I. Renge, *J. Phys. Chem.* **99**, 15955 (1995).
190. I. Renge, R. van Grondelle, and J. P. Dekker, *J. Photochem. Photobiol.* **A96**, 109 (1996).
191. J. A. Warren, J. M. Hayes, and G. J. Small, *J. Chem. Phys.* **80**, 1786 (1984).
192. J. A. Warren, J. M. Hayes, and G. J. Small, *Chem. Phys.* **102**, 323 (1986).
193. N. Ohta and H. Baba, *Mol. Phys*, **59**, 921 (1986).
194. W. R. Lambert, P. M. Felker, J. A. Syage, and A. H. Zewail, *J. Chem. Phys.* **81**, 2195 (1984).
195. S. A. Wittmeyerand and M. R. Topp, *Chem. Phys. Lett.* **171**, 29 (1990).
196. U. Even and J. Jortner, *J. Chem. Phys.* **77**, 4391 (1982).
197. P. S. H. Fitch, C. A. Haynam, and D. H. Levy, *J. Chem. Phys.* **73**, 1064 (1980).
198. C. A. Haynam, D. V. Brumbaugh, and D. H. Levy, *J. Chem. Phys.* **79**, 1581 (1983).
199. N. Ohta, H. Baba, and G. Marconi, *Chem. Phys. Lett.* **133**, 222 (1987).
200. E. A. Mangle and M. R. Topp, *J. Phys. Chem.* **90**, 802 (1986).
201. G. Bermudez and I. Y. Chan, *J. Phys. Chem.* **91**, 4710 (1987).
202. G. D. Greenblatt, E. Nissani, E. Zaroura, and Y. Haas, *J. Phys. Chem.* **91**, 570 (1987).
203. T. Suzuki and M. Ito, *J. Phys. Chem.* **91**, 3537 (1987).
204. A. Amirav, U. Even, and J. Jortner, *J. Chem. Phys.* **75**, 3770 (1981).
205. U. Even, J. Magen, J. Jortner, J. Friedman, and H. Levanon, *J. Chem. Phys.* **77**, 4374 (1982).
206. D. W. Werst, A. M. Brearley, W. R. Gentry, and P. F. Barbara, *J. Am. Chem. Soc.* **109**, 32 (1987).
207. H.-G. Löhmannsröben, D. Bahatt, and U. Even, *J. Phys. Chem.* **94**, 4025 (1990).
208. T. A. Stephenson and S. A. Rice, *J. Chem. Phys.* **81**, 1083 (1984).
209. T. Kobayashi and O. Kajimoto, *J. Chem. Phys*, **86**, 1118 (1987).
210. A. Amirav, M. Sonnenschein, and J. Jortner, *J. Phys. Chem.* **88**, 5593 (1984).
211. A. Amirav, U. Even, and J. Jortner, *Chem. Phys. Lett.* **72**, 21 (1980).
212. U. Even, J. Magen, J. Jortner, and J. Friedman, *J. Chem. Phys.* **77**, 4384 (1982).

213. C. A. Haynam, L. Young, C. Morter, and D. H. Levy, *J. Chem. Phys.* **81**, 5216 (1984).
214. A. R. Auty, A. C. Jones, and D. Phillips, *Chem. Phys.* **103**, 163 (1986).
215. N. Ginsburg, W. W. Robertson, and F. A. Matsen, *J. Chem. Phys.* **14**, 511 (1946).
216. I. Renge and U. P. Wild, *J. Lumin.* **66&67**, 305 (1996).
217. F. A. Burkhalter, G. W. Suter, U. P. Wild, V. D. Samoilenko, N. V. Rasumova, and R. I. Personov, *Chem. Phys. Lett.* **94**, 483 (1983).
218. H. P. H. Thijssen, R. van den Berg, and S. Völker, *Chem. Phys. Lett.* **97**, 295 (1983).
219. I. Renge, *J. Opt. Soc. Am.* **B9**, 719 (1992).
220. K. K. Rebane, *Impurity Spectra of Solids*, Plenum Press, New York, 1970.
221. M. J. Almond and A. J. Downs, in R. J. H. Clark and R. E. Hester, eds., *Advances in Spectroscopy*, Vol. 17, John Wiley & Sons, New York, 1989.
222. D. W. Ball, Z. H. Kafafi, L. Fredin, R. H. Hauge, and J. L. Margrave, *A Bibliography of Matrix Isolation Spectroscopy, 1954–1985*, Rice University Press, Houston, Texas, 1987.
223. B. Sonntag, in M. L. Klein and J. A. Venables, eds., *Rare Gas Solids*, Vol. 2, Academic Press, London, 1977, pp. 1021–117.
224. P. Korpiun and E. Lüscher, in M. L. Klein and J. A. Venables, eds., *Rare Gas Solids*, Vol. 2, Academic Press, London, 1977, pp. 743–822.
225. V. A. Rabinovich, A. A. Vasserman, V. I. Nedostup, and L. S. Veksler, *Thermophysical Properties of Neon, Argon, Krypton and Xenon*, Hemisphere Publishing Corp., Washington, 1988.
226. K. S. Sundararajan, *Z. Kristallogr.* **93**, 238 (1937).
227. Landolt-Börnstein, *Zahlenwerte und Funktionen*, 6 Aufl., II/8, Springer, Berlin, 1962, p. 2–287.
228. R. M. Hochstrasser, G. R. Meredith, and H. P. Trommsdorff, *J. Chem. Phys.* **73**, 1009 (1980).
229. I. Renge, unpublished data.
230. R. N. Nurmukhametov, *Usp. Khim.* **38**, 351 (1969) [*Russ. Chem. Rev.* 38, 180 (1969)].
231. N. S. Strokach, D. N. Shigorin, and N. A. Sheglova, *Atlas of Quasiline Spectra of Aromatic Molecules*, Nauka, Moscow, 1982.
232. Ya. M. Kolotyrkin and D. N. Shigorin, eds., *Vibronic Spectra of Aromatic Compounds with Heteroatoms*, Nauka, Moscow, 1984.
233. L. A. Nakhimovsky, M. Lamotte, and J. Joussot-Dubien, *Handbook of Low Temperature Electronic Spectra of Polycyclic Aromatic Hydrocarbons*, Elsevier, Amsterdam, 1989.
234. E.-U. Wallenborn, U. P. Wild, and R. Brown, *J. Chem. Phys.* **107**, 8338 (1997).
235. I. Renge and U. P. Wild, *J. Phys. Chem.* **101**, 7977 (1997).
236. H. J. Tobler, A. Bauder, and Hs. H. Günthard, *J. Mol. Spectr.* **18**, 239 (1965).
237. D. E. McCumber and M. D. Sturge, *J. Appl. Phys.* **34**, 1682 (1963).
238. G. F. Imbusch, W. M. Yen, A. L. Schawlow, D. E. McCumber, and M. D. Sturge, *Phys. Rev.* **133**, A1029 (1964).

239. J. L. Skinner and D. Hsu, in I. Prigogine and S. A. Rice, eds., *Advances in Chemical Physics*, Vol. 65, Wiley, New York, 1986, pp. 1–44.

240. P. de Bree and D. A. Wiersma, *J. Chem. Phys.* **70**, 790 (1979).

241. I. S. Osad'ko, *Phys. Rep.* **206**, 43 (1991).

242. A. B. Zahlan, in A. B. Zahlan, ed., *Excitons, Magnons and Phonons in Molecular Crystals*, Cambridge University Press ,1968, pp. 153–63.

243. M. Suzuki, T. Yokoyama, and M. Ito, *Spectrochim. Acta* **24A**, 1091 (1968).

244. W. B. Nelligan, D. J. LePoire, T. O. Brun, and R. Kleb, *J. Chem. Phys.* **87**, 2447 (1987).

245. L.-C. Brunel and D. A. Dows, *Spectrochim. Acta* **30A**, 929 (1974).

246. H. L. Finke, M. E. Gross, G. Waddington, and H. M. Huffman, *J. Am. Chem. Soc.* **76**, 333 (1954).

247. A. V. Belushkin, É. L. Bokhenkov, A. I. Kolesnikov, I. Natkaniec, R. Righini, and E. F. Sheka, *Fiz. Tverd. Tela* **23**, 2607 (1981) [*Soviet Phys. Solid State* **23**, 1529 (1981)].

248. A. Hadni, in A. B. Zahlan, ed., *Excitons, Magnons and Phonons in Molecular Crystals*, Cambridge University Press, 1968, pp. 31–41.

249. J. Timmermans, *Physico-Chemical Constants of Pure Organic Compounds*, Elsevier, New York, 1950, p. 179.

250. I. Renge, *J. Chem. Phys*, **106**, 5835 (1997).

251. Yu. A. Kovalevskaya, *Fiz. Tverd. Tela* **9**, 344 (1967) [*Soviet Phys. Solid State* **9**, 258 (1967)].

252. S. N. Vaidya and G. C. Kennedy, *J. Chem. Phys.* **55**, 987 (1971).

253. K. G. Lyon, G. L. Salinger, and C. A. Swenson, *Phys. Rev.* **B19**, 4231 (1979).

254. I. Perepechko, *Low Temperature Properties of Polymers*, Mir Publishers, Moscow, 1980, p. 272.

255. A. J. Pertsin and A. I. Kitaigorodsky, *The Atom–Atom Potential Method*, Springer Series in Chemical Physics, Vol. 43, Springer, Berlin, 1987.

CHAPTER

3

TRAPPING OF PAHs IN SHPOL'SKII MATRICES: ORIENTATION AND DISTORTION

MICHEL LAMOTTE

Laboratoire de Physico-Toxico-Chimie des Milieux Naturels,
Université de Bordeaux I 351, Cours de la Libération, F33405 Talence, France

3.1. INTRODUCTION

Progress in our understanding of the electronic spectra and electronic properties of polycyclic aromatic hydrocarbons (PAH) has been largely dependent on the possibilities to record highly vibronically resolved emission or absorption spectra. This goal can be reached by at least five different methods: supersonic molecular beam spectroscopy, laser line-narrowing spectroscopy in glassy matrices, matrix isolation spectroscopy, spectroscopy with mixed single crystals, and Shpol'skii spectroscopy.

The first method is the most recent one and, contrary to the others, does not require low temperatures but is technically the most elaborate to operate.

Matrix isolation spectroscopy, although of interest for analytical purposes, most often requires some thermal treatment for optimizing the quality of the spectra, which complicates somehow the application of the technique. Such a difficulty can be efficiently overcome by using the fluorescence line-narrowing technique, which is based on the selective excitation of a class of molecules whose absorption matches a very narrow excitation wavelength domain. Besides it selectivity, the interest of the method is based on the fact that it can be applied to a large variety of matrices, as glassy matrices, but it requires a wavelength tunable laser as excitation source.

Spectroscopy in mixed single crystals can lead to very sharp line spectra but requires very stringent conditions of isomorphism between guest and host molecules in order to form molecularly dispersed substitutional solid solutions, a widely accepted requirement for minimizing inhomogeneous line broadening. These conditions were first defined by Grimm and Wolff [1] and developed later by Kitaigorodsky [2]. They are referred to as isomorphism conditions in the broad sense by Shpol'skii [3]. They include close molecular packing density

Shpol'skii Spectroscopy and Other Site-Selection Methods, Edited by Cees Gooijer, Freek Ariese, and Johannes W. Hofstraat.
ISBN 0-471-24508-9

values as well as geometrical similarity of the components and minimal disturbance of the host lattice.

This method has its limitation due to the difficulty in finding an appropriate host compound. However, there are in the literature many examples of studies of mixed crystals whose components do no strictly satisfy isomorphism criteria but still lead to sharp line spectra provided low enough guest concentrations are used. Nevertheless, two major drawbacks of the method arise from the tedious task of growing single crystals and from the fact that the suitable host molecule is often a parent compound of the guest molecule with the consequence of absorbing in a range close to that of the guest compound, thus limiting the absorption or the excitation emission study of the guest molecule to a narrow wavelength domain.

Shpol'skii spectroscopy makes use of ordinary n-alkanes as host molecules. These matrices do not present the above-mentioned inconveniences since alkanes have absorption bands in the far UV well below 200 nm [4] and lead to very large sharp line spectra of PAHs. They constitute solid matrices, which apparently can accommodate a very large variety of PAHs.

In the case of PAH/n-alkane systems it is clear that the isomorphism criteria required for the formation of real solid solutions are not fulfilled. While the thickness and largest dimension of aromatics and alkane molecules can be matched within 20%, the width of many PAHs is about, or more than, 1.5 times larger than that of normal alkane molecules. This fact, added to their quite different electronic configurations, leads one to question the real nature of the Shpol'skii solutions. In fact, this problem has been the subject of a large number of studies since the very first report by Shpol'skii himself and his co-workers and is still under debate. An important aspect of the Shpol'skii effect concerns the fact that depending on the PAH solute molecule, its observation can be loosely or tightly dependent on experimental conditions like the choice of the n-alkane, the cooling rate, or the concentration.

In the first section, we shall illustrate the fundamental differences between solid solutions made of isomorphic or quasi-isomorphic components and Shpol'skii matrices, by comparing their geometric and crystallographic properties as well as the effects of concentration and cooling rate on the spectra.

In the second section we shall propose a classification of the different types of PAH/n-alkane systems based on the effect of cooling rate and concentration on the occurrence of quasiline spectra.

The examples considered deal with polycrystallized solutions whose formation is inherent to the fast cooling process usually required for obtaining quasiline (QL) spectra. However, it has been found that in some PAH/n-alkane systems, QL spectra of PAHs trapped in monocrystallized Shpol'skii solutions have been observed in spite of the very slow cooling required to grow n-alkane single crystals. This fact has opened the possibility to verify whether the PAH molecules are oriented in the n-alkane lattice as could be expected from an insertion of the solute molecules in substitutional sites, a hypothesis that was proposed by some authors but without experimental proof.

In the third section, we shall describe results obtained from the studies of doped single crystals of *n*-alkanes. These results, which were obtained from various techniques, have contributed to give a precise interpretation of the multiplet structure of some solute molecules. They also provide valuable information on the deformation of some highly symmetric PAH and porphin molecules trapped in Shpol'skii solution, as well as on the properties of their singlet and triplet states.

3.2. THE SPECIFICITY OF PAH/*n*-ALKANE SYSTEMS

3.2.1. Guest–Host Matching

Most useful details on the spectra of guest molecules in a solid medium are obtained when both homogeneous and inhomogeneous broadening are reduced to a minimum. Lowering the temperature partly allows one to attain the first objective while the second one can be attained by using mixed single crystals.

Mixed crystals are stable solid solutions of guest molecules, occupying well-defined substitutional sites in the lattice of a matching host. To obtain such an arrangement, the host and guest molecules must satisfy stringent conditions of isomorphism in order to form a molecularly dispersed solid solution. Since the pioneering work of McClure [5], these types of material have been the subject of a tremendous amount of work that considerably contributed to our knowledge of the spectroscopic properties of a number of molecules as well as of excitation migration and energy transfer in the solid state.

The conditions required for the formation of mixed crystals is closely related to the problem of isomorphism, which was already under debate at the beginning of this century [6, 7]. Isomorphism has been clearly defined by Grimm and Wolff [1] in the case of inorganic compounds. They stated that the following three conditions must be fulfilled by the host and guest molecules:

- They must crystallize in the same crystalline group;
- They must have close chemical structures (homeomorphism);
- They must have identical interatomic distances in the crystal (crystalline isomorphism).

These conditions were extended to organic compounds by Kitaigorodsky [2] and expressed in terms of geometrical similarity of the components, close molecular packing density values, and minimal disturbance of the host lattice.

He introduced the parameters Θ and ε expressed as:

$$\Theta = (V_A - V_B)/V_a \qquad \text{and} \qquad \varepsilon = 1 - G/g$$

where V_A and V_B are the specific unit volume for compounds A and B, with V_A larger than V_B; G and g are, respectively, the volumes of the nonoverlapping (protruding) and overlapping parts of the molecules.

It was established that for the formation of a mixed crystal with the guest molecules substitutionnally dispersed in the host lattice, ε must not be lower than 0.8 [8] and Θ must be as low as possible. Such conditions are, for instance, ideally fulfilled in a crystal doped with a deutero analogue [9, 10]. In practice, however, this rule does not appear very critical. In the literature, we can find many examples of mixed crystals giving sharp line spectra at least for low guest concentration, in spite of ε values that may be slightly less than 0.8. In the cases of aromatic hydrocarbons, well-known examples of widely investigated systems are naphthalene [5] and its aza analogues [11, 12] in durene, carbazol in fluorene [13], phenanthrene in biphenyl [14, 15], tetracene in anthracene [16, 17], anthracene in naphthalene [18], stilbene in dibenzyl or tolan [19], and also methylbenzene radicals in durene (Migirdicyan, this volume, Chapter 7, and references therein). It should be noticed, however, that in all cases the components are molecules of the same chemical nature and with relatively similar shapes, thus satisfying to a good degree the isomorphism criteria.

If we refer now to the PAH/*n*-alkane systems, it is obvious that these two components do not satisfy the isomorphism conditions.

In the crystalline state, *n*-alkane molecules are planar with all-*s*-*trans* configuration, the carbon atoms forming the so-called zigzag chain. Pentane crystals are orthorombic, whereas the crystals of *n*-alkanes from *n*-hexane to *n*-nonane are triclinic [20–22]. However, while *n*-hexane and *n*-octane crystals have one molecule per unit cell, in *n*-heptane [20, 23] and *n*-nonane [24] there are two molecules per unit cell. For larger *n*-alkanes, some crystallographic data are available in work published by Norman and Mathisen [24]. Figure 3.1 is a schematic view of the *n*-heptane crystal in a direction perpendicular to the molecular plane [23]. Its space group is $P\bar{1}$. The molecules are lined up along the *c* axis and in the *bc* plane; a C atom of one molecule fits into the hollow formed by the three C atoms of the adjacent molecules. Along the *a* axis, the molecules are stacked in parallel planes with a spacing of 3.92 Å.

Contrary to *n*-alkanes, for which molecular arrangements in the crystal are very similar, aromatic compounds crystallize in quite different systems with different arrangements depending on their molecular structure [25]. After Stevens [26], aromatic crystals can be classified into three types, as illustrated in Figure 3.2. Examples of type A are fluorene, naphthalene, anthracene, phenanthrene, or chrysene. Type B1 is characteristic of pericondensed aromatics like pyrene,

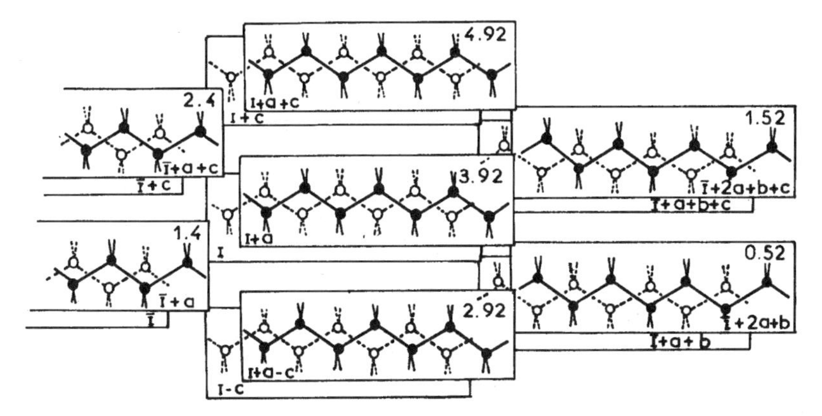

Figure 3.1. Packing of the n-heptane molecules along the a axis. **I** and **Ī** denote the two symmetry-related molecules of the unit cell. The rectangles represent the mean plane through the zigzag carbon chains. The numbers indicate the distance in Å along the growth a axis to the mean plane of the molecule **I**. (After Merle et al. [23] with permission.)

perylene, or 1,12-benzoperylene, while type B2 concerns large pericondensed aromatics like coronene or ovalene. Depending on the temperature, some PAHs may crystallize in different systems. This seems to be the case for perylene, whose low-temperature crystalline structure would be of type A, while that at high temperature would be of type B1 [27]. Table 3.1 gives the main crystallographic parameters for some common PAHs. Some examples of molecular arrangements in the unit cells of PAH crystals are given in Figure 3.3.

From these crystallographic data it follows that n-alkanes and PAHs present very dissimilar crystallographic structures. Their shapes and molecular dimensions also differ substantially (Fig. 3.4). In n-alkanes, the C–C bond lengths range between 1.516 and 1.537 Å and the average distance, d, between two adjacent C atoms along the chain axis is about 1.27 Å (Fig. 3.4). This value is the

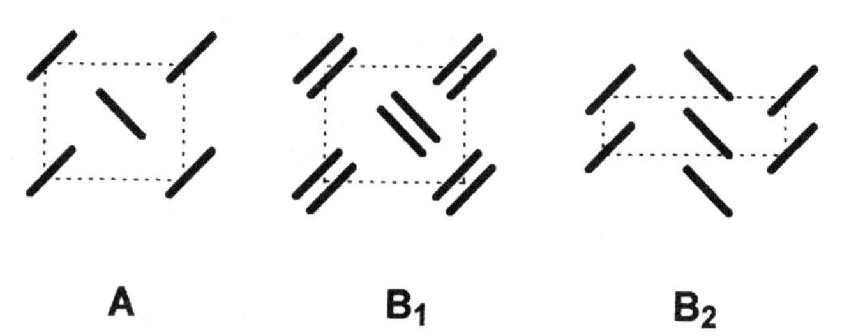

Figure 3.2. The three types of molecular packing in PAH crystals as seen along the molecular planes. (After Stevens [26] with permission.)

TABLE 3.1. Crystallograhic Data and Unit Cell Parameters for Some PAH Crystals; *Z*, Number of Molecules per Unit Cell

PAH	Type	Z	a (Å)	b (Å)	c (Å)	α (°)	β (°)	γ (°)	Ref.
Phenanthrene	$M:P2_1$	2	8.46	6.16	9.47	90	97.7	90	[123]
Naphthalene	$M:P2_{1/a}$	2	8.24	6.00	8.66	90	122.7	90	[124,125]
Anthracene	$M:P2_{1/a}$	2	8.,56	6.04	11.16	90	124.7	90	[124]
Tetracene	$T:P1$	2	7.90	6.03	13.53	100.3	112.2	86.3	[124]
Chrysene	$M:I2/c$	4	8.39	6.20	25.20	90	116.2	90	[126]
Dibenzo[*ah*]	$M:P2_1$	2	6.59	7.884	14.17	90	103.5	90	[127]
anthracene	$O:Pcab$	4	8.26	11.47	15.24	90	90	90	[128]
Fluorene	$O:Pnam$	4	8.48	5.72	18.92	90	90	90	[129]
Pyrene	$M:P2_{1/a}$	4	13.64	9.25	8.47	90	100.3	90	[130]
Perylene	$M:P2_{1/a}$	4	11.35	10.87	10.31	90	100.8	90	[27,131, 132]
	$M:P2_{1/a}$	2	11.27	5.88	9.65	90	92.1	90	[27]
Benzo[*ghi*] perylene	$M:P2_{1/a}$	2	11.7	11.9	9.90	90	98.5	90	[131]

approximate difference in length between two successives homologues. In the crystal the shortest intermolecular distance, D, is about 3.65 Å. The overall molecular length, L, can be estimated using the following expression:

$$L = d(n - 1) + D \tag{3.1}$$

where n is the number of carbon atoms. The largest dimensions thus calculated are approximately 8.7 Å for n-pentane, 10.0 Å for n-hexane, and 11.3 Å for n-heptane. The two other dimensions (width and thickness) are practically the same for all n-alkanes and amount approximately to 4.0 and 4.5 Å.

The dimensions of the PAH molecules can be estimated by using the following mean values [2]: 1.4 Å for the length of C–C bonds (l_{CC}); 1.0 Å for that of the C–H bonds (l_{CH}); 1.8 Å for the intermolecular radius of carbon (R_C), and 1.2 Å for that of hydrogen (R_H). The geometic model for anthracene is shown in Figure 3.4.

The widths (W) and lengths (L) of all polyacenes (naphthalene, anthracene, tetracene, ...) can be estimated from the relations:

$$W = 2(l_{CC} + R_H) = 7.2\text{Å} \tag{3.2}$$

$$L = 2(Nl_{CC} + l_{CH}) \cos 30^\circ + 2R_H \tag{3.3}$$

where N is the number of benzene rings.

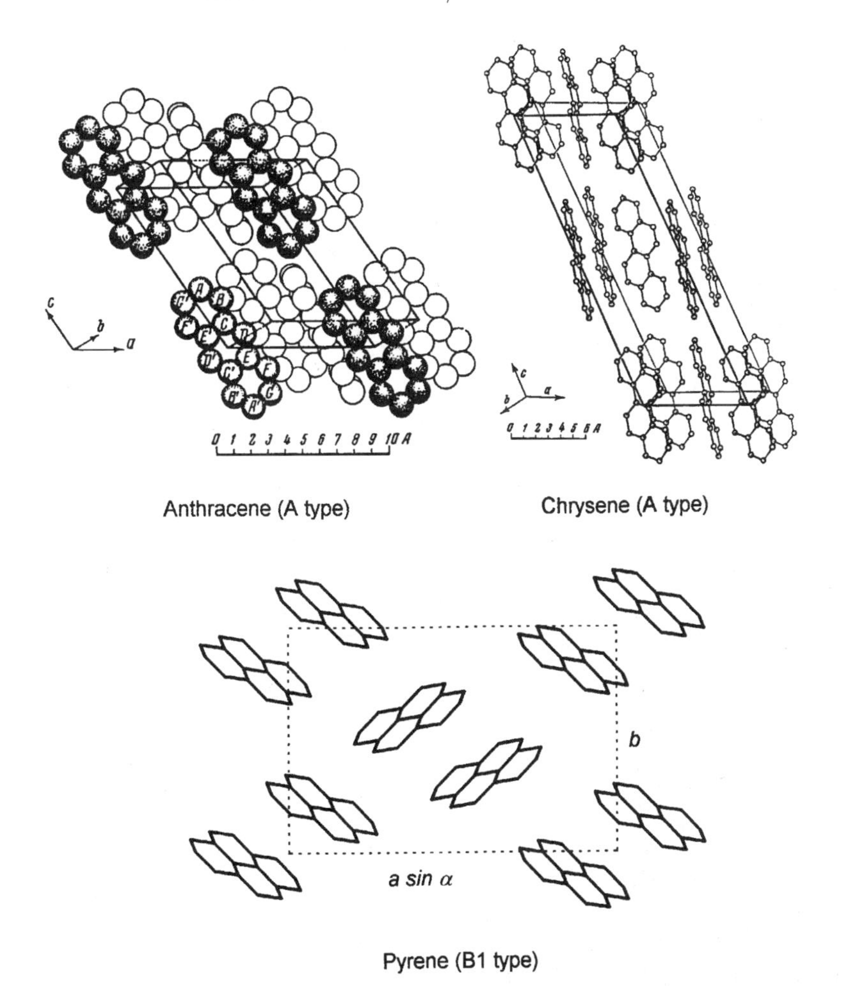

Figure 3.3. Examples of molecular arrangements in the unit cell of some PAHs. (After Kitaigorodskii [25] with permission.)

The thickness of all planar PAHs is around 3.6 Å. The estimated length of naphthalene is around 9.0 Å, that of anthracene 11.3 Å, and that of tetracene (naphthacene) 13.6 Å. The dimensions of all other PAHs can be found in a similar way. The overall dimensions of orthocondensed aromatics, like phenanthrene, chrysene, etc., are very close to those of the corresponding acenes. So, except for the largest dimensions, which can be set very closely to the length of an n-alkane by a suitable choice of the host, the other PAH dimensions and particularly the intermediate one (the width) are all quite different. At best, the coefficient of geometrical similarity ε is hardly larger than 0.5; so neither the shape nor the

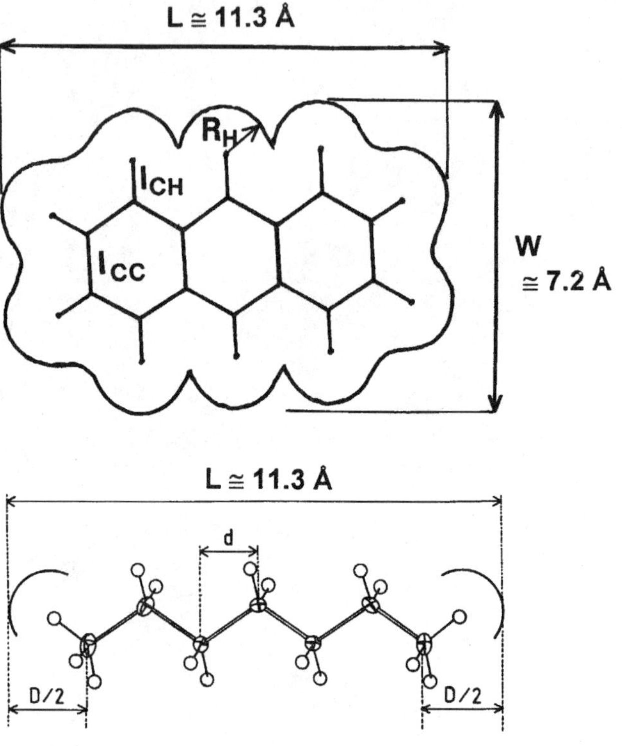

Figure 3.4. Comparison of shapes and dimensions of anthracene (top) and *n*-heptane (bottom) molecules. The molecular model for *n*-heptane is derived from the Ortep plot. (After Merle et al. [23] with permission.)

crystalline structure appears to be similar enough to satisfy any of the isomorphism criteria. Accordingly, PAH/*n*-alkane systems are not expected to form neatly substitutional solid solutions. The fact that sharp line spectra of PAHs in frozen *n*-alkane solutions are nonetheless observed has lead Shpol'skii and his co-workers to conclude that these types of solutions can be formed under conditions much broader than those imposed by the isomorphism conditions even if these are used in a broad sense [3]. They noticed, however, that QL spectra occur only in specific combinations of PAH and *n*-alkane molecules. Bolotnikova [28] was the first to report that the sharpest fluorescence spectra of naphthalene, anthracene, and naphthacene (tetracene) are obtained, respectively, in *n*-pentane (n-C_5), *n*-hexane (n-C_6), and *n*-heptane (n-C_7), that is, in cases where the largest dimensions of the guest and host molecules are the closest (Fig. 3.4). A similar correlation has also been noted in the case of other types of PAH like orthocondensed phenanthrene in n-C_6 and chrysene in n-C_8. Besides the good matching of the lengths of the molecules, Shpol'skii also noticed that in both types of molecules there is a close geometric similarity of the zigzag structure of the carbon atom chains; in both types of molecules the mean distance between two equivalent carbons atoms along the largest dimensions are close to 2.5 Å (Fig. 3.4). It was deduced by Shpol'skii that such features of form analogy or *synmorphism*, after Bruni [6], may play an essential role in the occurrence of QL spectra. It was proposed to formulate this phenomenon as the *lock and key* rule [29] or *key and hole* rule [30, 31].

In fact, since the very first investigations of these solutions by Shpol'skii and co-workers, the results obtained on a large number of PAH compounds investigated in n-alkane frozen solutions by many authors show that this rule suffers many exceptions [32–38], and accordingly it cannot be considered to be valid in all cases.

In agreement with the key and hole rule, very sharp lines are obtained for naphthalene in n-pentane, but sharp line spectra were also reported by Shpol'skii et al. [39] in n-C_7. Further convincing evidence for this conclusion was provided by Dekkers [33], who showed that although QL spectra cannot be observed in every n-alkane, they can be obtained in several consecutive n-alkanes. In particular, they showed that QL spectra are not restricted to anthracene in n-C_7 and naphthacene in n-C_9 but that sharp line spectra can also be obtained in alkanes from n-C_5 to n-C_9 with anthracene and from n-C_8 to n-C_{11} with naphthacene. Moreover, in the case of anthracene, using their cooling conditions, they obtained sharper lines in n-C_5 and n-C_6 than in n-C_7, which is supposed to be the best solvent according to the key and hole rule.

As a conclusion, in the case of PAH/n-alkane systems, although the key and hole rule can be useful as a first approach to choose an appropriate n-alkane, a good matching of the host and guest lengths does not constitute the determining factor for obtaining QL spectra. As we shall see in the next section, cooling and concentration conditions also play an important role. Other parameters, such as the planarity of the host molecule, are also important. Non-totally planar molecules, which have attracted particular attention, are the helicenes [41, 41]. Other interesting cases are those dealing with molecules that may exist under different conformations, each with various deviation from planarity. Examples are diarylethenes [42–44], dinaphthonaphthalene [45], dinaphthoanthracene [40], tetrabenzoperylene [47], or terrylene [48]. However, this aspect will not be discussed here.

3.2.2. Cooling Rate Effect on Spectra

In the case of mixed crystals made of isomorphic components, an ideal arrangement is obtained when one guest molecule substitutes one host molecule, without perturbing too much the host crystal lattice. A minimum inhomogeneous broadening would be obtained if each guest molecule experiences identical interaction, a situation that can be achieved if the host crystal lattice does not present too large a concentration of defects. This is usually the case if mixed single crystals are grown using slow cooling conditions.

An example of a solid solution with isomorphic components is carbazole incorporated in a fluorene crystal [13]. These two components have similar molecular and crystalline structures and identical molecular (C_{2v}) and crystal (D_{2h}) symmetries [49, 50] and so fulfill the major conditions for formation of true equilibrium solid solutions. The effect of the cooling rate on the absorption spectra of carbazole in fluorene is shown in Figure 3.5. In this experiment, slow

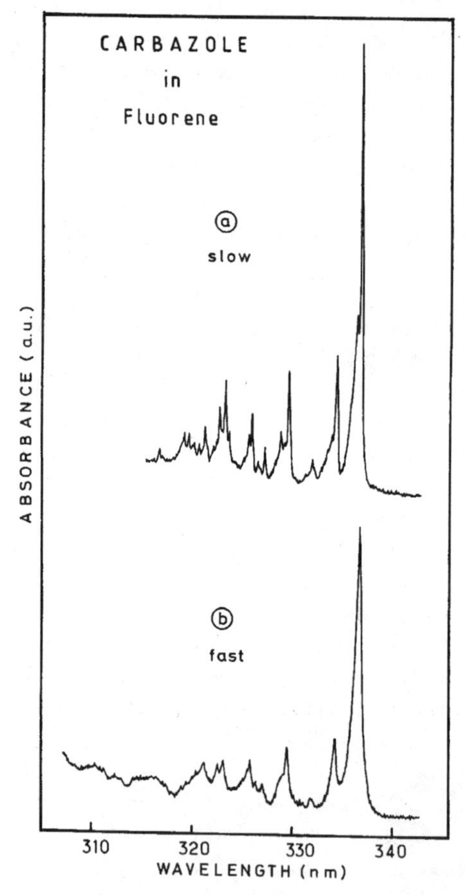

Figure 3.5. Cooling rate dependence of the absorption spectrum of carbazole in fluorene crystal. $T = 5\,\text{K}$, $c = 0.01\%$ (mole/mole), sample thickness $20\,\mu\text{m}$. (a) Slow crystallization rate; (b) fast cooling by immersion into liquid nitrogen. (After Nakhimovsky et al. [36] with permission.)

cooling refers to a cooling rate of about 2.0 K per minute, while fast cooling is performed by immersing the sample directly into liquid nitrogen [36]. As expected for a binary system formed with isomorphic components, the best resolved spectrum is obtained under slow cooling, which is the most favorable procedure to build up good crystalline arrangements with a minimum concentration of defects. This is a quite general phenomenon, which accounts for the fact that organic mixed solid solutions have been almost exclusively investigated as mixed single crystals [5].

A quite different behavior is observed in the case of PAHs in n-alkane matrices. Whatever the system under study, provided it is suitable for the appearance of the Shpol'skii effect, sharp line spectra are always observed when the solution is rapidly frozen. Contrary to mixed single crystals of isomorphic components, slow cooling leads most generally to the occurrence of broadband spectra. An example is given in Figure 3.6 for dibenzofuran in

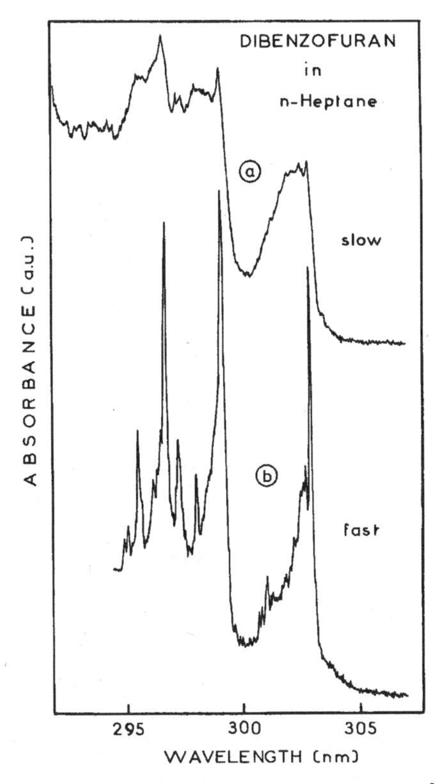

Figure 3.6. Cooling rate dependence of the absorption spectrum of a 10^{-5} M solution of dibenzofuran in n-heptane. $T = 5$ K; sample thickness 2.0 mm. (a) "Slow crystallization" rate; (b) fast cooling into liquid nitrogen. (After Nakhimovsky et al. [36] with permission.)

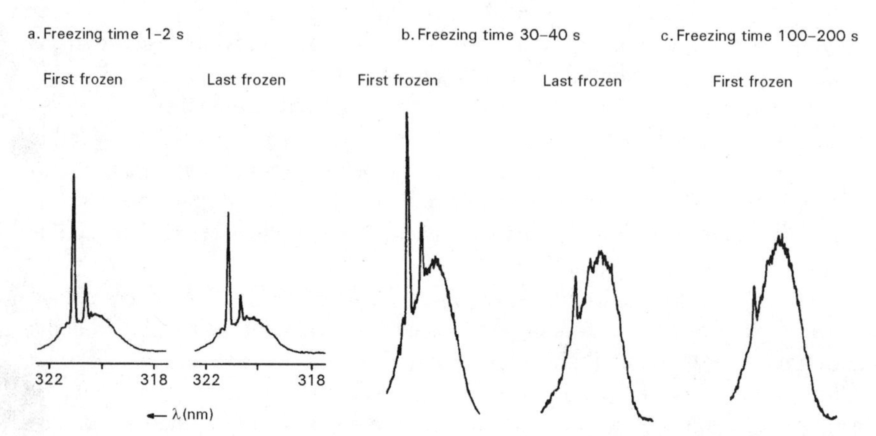

Figure 3.7. Influence of freezing time and distance from the freezing starting point in the sample, on the 0–0 fluorescence region of acenaphthene in *n*-hexane for various types of windows/spacer sample holder combinations. $c = 10^{-5}$ M. (a) Sapphire/indium; (b) quartz/indium; (c) quartz/Teflon. (After Hofstraat [34] with permission.)

n-C_7. Upon slow cooling, the absorption spectrum loses its sharp line character and becomes diffuse. A similar result was also reported for the fluorescence spectra of benzo[*a*]pyrene in *n*-C_8 [51] or acenaphthene in *n*-hexane [34, 52], as shown in Figure 3.7. In all these cases, the broad bands manifest practically at the same wavelength range as the quasilines and, as a consequence, have been thought to be molecular in nature. In addition, depending on the concentration, a second broadband spectrum assigned to aggregates or crystallites is frequently observed [53].

3.2.3. Concentration Dependence of Spectra

As stated above, isomorphic component systems form molecularly dispersed equilibrium substitutional solid solutions. However, in most of the cases of interest, the components are not perfectly isomorphic, but still suitably resolved spectra are usually observed at low concentrations. The concentration dependence of the absorption spectra of a slowly crystallized solution of carbazole in fluorene is illustrated in Figure 3.8 [36]. Very narrow absorption lines of carbazole are observed for a guest–host molecular ratio below 10^{-4}%. Upon increasing this ratio, the spectrum resolution decreases gradually, and for a ratio around 25% only a broadband absorption spectrum is observed. For such a high guest–host ratio, the band broadening is supposed to arise both from guest–guest interactions and from an increase of the inhomogeneous broadening due to a higher degree of disorder induced by the presence of the guest molecules. This is a quite general behavior for this type of crystal.

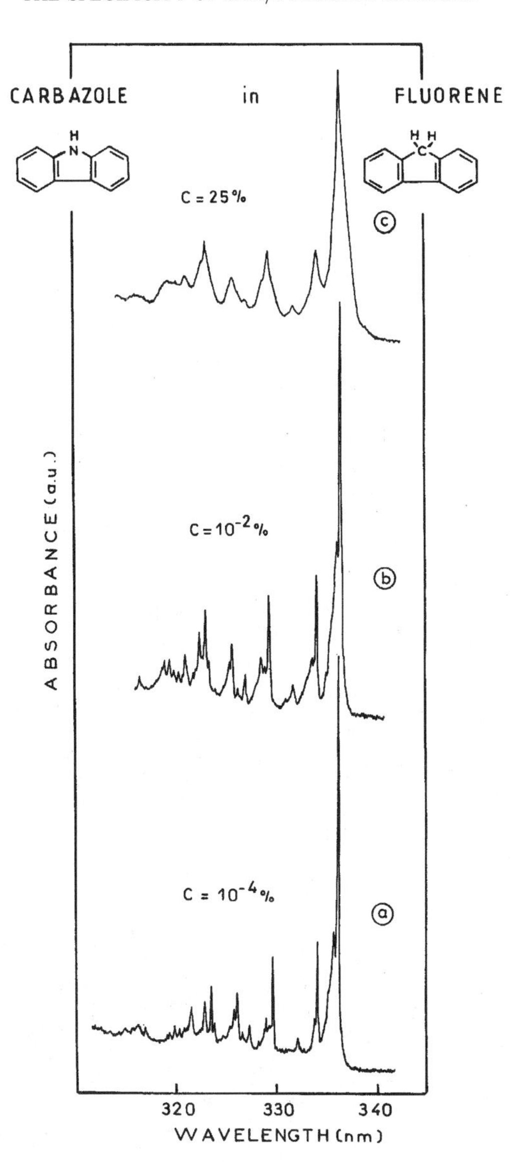

Figure 3.8. Concentration dependence of the absorption spectra of slowly crystallized solutions of carbazole in fluorene. $T = 5\,\mathrm{K}$. (a) $c = 10^{-4}\%$ (mole/mole), sample thickness, $l = 500\,\mu\mathrm{m}$. (b) $c = 10^{-2}\%$, $l = 20\,\mu\mathrm{m}$; (c) $c = 25\%$, $l < 1\,\mu\mathrm{m}$. (After Nakhimovsky et al. [36] with permission.)

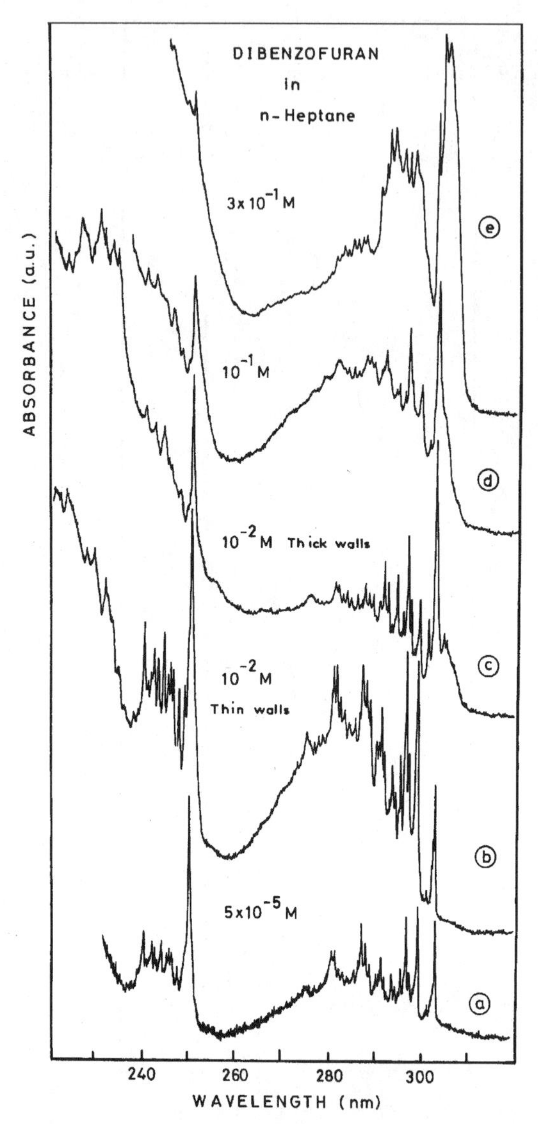

Figure 3.9. Concentration dependence of the absorption spectra of fast frozen solutions of dibenzofuran in *n*-heptane. $T = 5\,\text{K}$. (a) $c = 5 \times 10^{-5}$ M, cell thickness, $l = 70\,\mu\text{m}$,(b) $c = 10^{-2}$ M, $l = 20\,\mu\text{m}$; (c) $c = 10^{-2}$ M, $l = 20\,\mu\text{m}$, cell walls $\approx$1.5 times thicker than in (b); (d) $c = 10^{-1}$ M, $l < 20\,\mu\text{m}$; (e) $c = 3 \times 10^{-1}$ M, $l < 1\,\mu\text{m}$. (After Nakhimovsky et al. [36] with permission.)

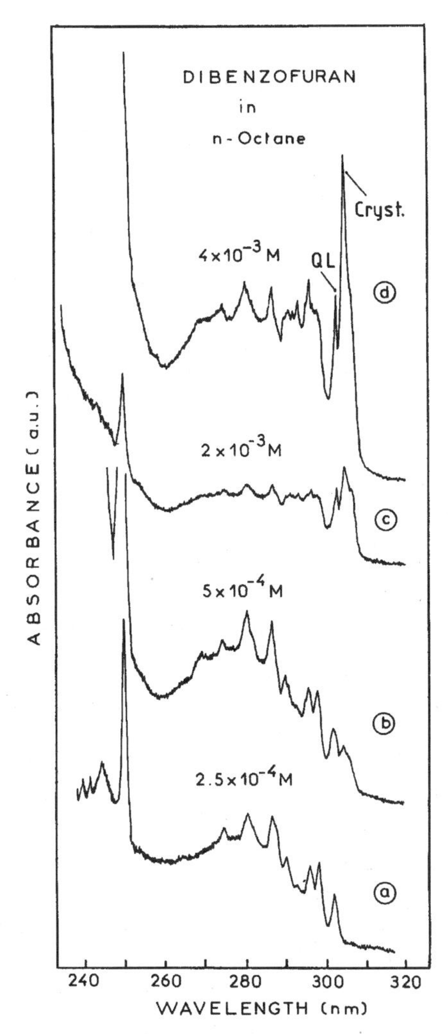

Figure 3.10. Concentration dependence of the absorption spectra of fast frozen solutions of dibenzofuran in *n*-octane. $T = 77\,\text{K}$; sample thickness $l = 150\,\mu\text{m}$. (a) $c = 2.5 \times 10^{-4}\,\text{M}$; (b) $c = 5 \times 10^{-4}\,\text{M}$; (c) $c = 2 \times 10^{-3}\,\text{M}$; (e) $c = 4 \times 10^{-3}\,\text{M}$. (After Nakhimovsky et al. [36] with permission.)

In contrast to this behavior, the concentration effect on the occurrence of QL spectra in Shpol'skii matrices has been observed to be much more complex and to manifest itself in different ways, depending on the PAH/n-alkane system. To illustrate the different behavior of mixed isomorphic component crystals and Shpol'skii solutions, we show, as an example, the case of dibenzofuran in n-heptane, which is illustrated in Figure 3.9. In this case, quasilines are observed in the absorption spectra in a very large concentration range from low concentration up to 3×10^{-1} M without a substantial loss of resolution as in isomorphic component mixed crystals. Starting from about 10^{-2} M, a broadband spectrum, slightly shifted to the red, is observed superimposed on the QL spectra. This type of band has been assigned to the presence of aggregates or pseudocrystallites [39, 54, 55]. Although quasilines are present in all spectra, it must be noticed that there are some modifications in the vibronic structure in changing the cell walls from thin to thick ones at a concentration of 10^{-2} M, as evidenced by comparison of spectra (b) and (c) in Figure 3.9. In spectrum (c), this point has been interpreted as resulting from weak solute–solute interactions [36, 56].

Another peculiar behavior, which is not yet fully understood, manifests itself in the case of systems composed of a PAH whose largest dimension does not match suitably the length of the n-alkane molecule. In these cases, it was found that at low concentrations, approximately below 10^{-4} M, the fluorescence spectra are diffuse and similar to the spectra recorded normally in glassy matrices. As in the previous case, however, at high concentrations the spectra acquire a quasiline structure while at the same time some bands due to the occurrence of aggregates or crystallites are also observed. These cases have been widely investigated. They include particularly azulene [57], phenanthrene [58], or dibenzofuran [59] in n-C_8 (Fig. 3.10). Examples have also been given by Shpol'skii et al. [39] and Klimova et al. [60, 61], who investigated the concentration dependence of the luminescence and absorption spectra of naphthalene, anthracene, pyrene, and benzo[*a*]-pyrene in different n-alkanes (from pentane to undecane) and in cyclohexane. The predominance of quasilines in the fluorescence spectra of acenaphthene at high concentration conditions in n-C_6, n-C_7, and n-C_8 matrices was also reported by Dekkers et al. [32].

So, although the above examples of PAH/n-alkane systems are particular cases of Shpol'skii solutions, they exhibit a concentration dependence that has no equivalence in isomorphic mixed crystal systems and as such, illustrate the special character of the frozen solutions of PAHs in n-alkanes.

3.3. CLASSIFICATION OF PAH/n-ALKANE SYSTEMS

The contrasting crystallization rate and concentration dependencies of the absorption and fluorescence spectra of isomorphic mixed crystals and Shpol'skii solutions described above indicate that these two types of solutions must differ substantially in their structure and thermodynamic properties. While the first type of system is accepted to form equilibrium substitutional solid solutions, the second type can lead to a variety of dispersion and trapping modes of the solute molecules, which correspond to various equilibrium and nonequilibrium situations. The phases associated with quasiline spectra are widely considered to correspond to nonequilibrium, supersaturated solid solutions. It has been proposed that the solute molecules are incorporated in the lattice in substitutional sites [32, 33, 39, 58, 61]. In contrast, when molecular broadband spectra are observed, the solute molecules are considered to be trapped in intergrain sites or in holes between microcrystals and are designated as nonincorporated molecules. It was proposed by Dekkers et al. [33] that the occurrence of such broadband spectra is indicative of a thermodynamic equilibrium. Hence it appears from previous work, and following Nakhimovsky [36], that "the observation of quasiline spectra in a binary solid system in condition of nonequilibrium crystallization can be considered to be a manifestation of the Shpol'skii effect" and from a thermodynamic point of view, binary systems that manifest this effect can be described as "metastable supersaturated substitutional solid solutions."

Whether a nonequilibrium or an equilibrium situation, or both, are reached in the frozen solutions depends on the solubility of the PAH molecule in a given n-alkane or the possible existence of an eutectic, but this is also strongly influenced by the solute concentration of the starting liquid solution and the crystallization rate. From the abundant published works dealing with the cooling rate effect on the occurrence of QL spectra, it results that among the three stages of cooling, namely (1) the cooling to the crystallization point; (2) the crystallization itself; and (3) the cooling to the temperature of measurement, the second stage plays the most decisive role. This explains why in some cases, such parameters as the thickness of the sample or of the cell walls, the nature of the material and the size of the cell holder can greatly influence the occurrence of QL spectra for similar concentrations. It must be pointed out, however, that this behavior is limited to some particular cases and in fact PAH/n-alkane systems show a large range of sensitivities toward cooling rate and concentration conditions, which confers considerable complexity to these systems.

In an attempt to rationalize the different behavior of the PAH/n-alkane systems with respect to the occurrence of QL according to cooling rate and concentration, Nakhimovsky and co-workers [36, 56] proposed to classify the PAH/n-alkane solutions in four types of systems: the α-, β-, γ-, and δ-type systems.

3.3.1. The Two Extreme Cases: α- and δ-Type Systems

Two extreme modes of concentration dependence for polycrystallized solutions can be observed, depending on whether complete or negligible formation of metastable (nonequilibrium) supersaturated substitutional solid solutions can be achieved during the crystallization stage. The first one corresponds the α-type system, the second one to the δ-type system.

3.3.1.1. α-Type Systems

In these systems, quasilinear absorption and fluorescence spectra are observed only from low concentrations up to concentrations close to the room-temperature solubility limit. However, at high concentration, solute aggregation may take place, giving rise to a red-shifted broadband spectrum similar to the spectrum of the solute crystal [31, 61, 62]. Moreover, for these systems, the Shpol'skii effect appears over a wide range of cooling rates and even, in some cases, under the extremely slow cooling rate conditions required to grow single crystals. Most large PAHs and particularly pericondensed aromatics (pyrene, perylene, benzo[*a*]pyrene, coronene) and also porphins, form α-type systems. Besides large PAHs, it must be pointed out that this case concerns molecules for which not only the largest dimension is close to the length of the suitable n-alkane but also the width is quite close to the dimension of a pair (or more) of the host molecules, in such a way that the solute may replace two (or more) host molecules. For instance, the largest and smaller dimension of pyrene and perylene are 11.0 and 8.6 Å, respectively, values that are very close to the overall dimensions of a pair of n-C_7 molecules (ca. 11.3 Å × 9.0 Å).

Although in these systems QL spectra can be obtained with several consecutive n-alkane homologues, in each matrix the spectra exhibit different multiplet structures, which result from the simultaneous presence of identical spectra, displaced with respect to one another by some tens of wavenumbers. In spite of the fact that the occurrence and description of the multiplet structure in these systems is now well documented, there are still some points, particularly those dealing with the sensitivity of the multiplet structure to the thermal history of the sample, which are still not fully interpreted [52].

Generally, provided that not too much different crystallization rates are used, the number and intensity distribution of the multiplet components are quite reproducible from one sample to another. However, when the sample is subjected to an annealing procedure (warming the sample to a temperature around its melting point and maintening it at this temperature for some time) or upon a very slow crystallization, the number of the multiplet components can be reduced [63]. The best-known cases have been investigated by Pfister, who observed the disappearance of the highest-energy component in the spectra of coronene [30, 64], pyrene in n-C_7 and benzo[*a*]pyrene in n-C_6 [30]. The same multiplet component of coronene in n-C_7 was also found to be missing in a monocrystallized matrix prepared in very slow cooling conditions [65]. A similar result was reported by these authors for perylene in n-C_7 single crystal.

The multiplet structure was initially ascribed to the presence of rotational isomers of the n-paraffin [3, 66] or related to polymorphism in n-paraffins [67, 68], but in the large majority of cases this structure appears to be assigned, instead, to different ways of solute incorporation, or sites, in the host lattice. The components that are observed irrespective of the cooling rate seem to be assigned with good confidence to molecules trapped in substitutional sites with a slightly different surrounding configuration. In contrast, the interpretation of the components that disappear upon annealing or slow cooling has not yet been firmly established. It was tentatively described by Pfister [30] as an intermediate state between the liquid and the perfect solid solution, in which the arrangement of the molecules can be reminiscent of pre-existing orientation of the solute and the solvent molecules in the liquid state [69, 70]. Following this hypothesis, Pfister proposed to designate this site as "pseudoliquid site." However, up to now, this interpretation has not received any confirmation. The only point that gives some indication on this site is the evidence for a loose packing for the solute molecule as demonstrated by the negligible wavelength shift noticed upon applying high pressure [71], a conclusion that is also consistent with its high-energy location [30] which is indicative of molecules weakly interacting with the host ones.

3.3.1.2. δ-Type Systems

The other extreme mode of the concentration dependence of spectra in PAH/n-alkane systems is observed when the overall size (e.g., nonplanar molecule) or dimensions are unfavorable for the formation of a nonequilibrium solid solution. This is the case of small PAHs like naphthalene (and naphthalene derivatives) in n-C_8 or longer n-alkanes, acenaphthene in n-C_8 [32] or dibenzofuran in n-C_9 [36]. In these cases, no QL spectra are observed, or in other words, no supersaturated solutions are formed even upon very fast cooling. During the crystallization process, all the solute molecules are efficiently rejected from the lattice, giving rise either to nonincorporated molecules probably trapped in intergrains displaying broad, glasslike spectra, or to associated forms (aggregates, crystallites) with spectra close to solute crystal spectra. It is very likely that the evolution of such solutions during cooling is partly controlled by the diffusion rate of both solute and solvent molecules and consequently by the size and shape of the molecules. This is in line with the description of Gurov and Nersesova [72], who proposed that the molecules giving rise to QL spectra are molecules that have been "captured" during the crystallization process. In δ type systems, apparently, this capturing process cannot take place.

3.3.2. The Intermediate β- and γ-Type Systems

These types correspond to the majority of cases. They are much more complex than the previous systems because they correspond to systems in which only a partial degradation of the solid solution takes place during the crystallization process [36]. It follows that all kinds of solute dispersion states can be obtained ranging from substitutionally trapped solute molecules to microcrystals, with nonincorporated molecules, preaggregates, and aggregates as intermediate states. The formation of these species, or in other words, the degree of deviation from equilibrium and the stability of the intermediate dispersion states in these systems, is of course highly dependent on the specific binary systems, the solute concentration, and the crystallization rate. Nakhimovsky et al. [36] discerned two main intermediate systems: the β- and γ-type systems.

3.3.2.1. β-Type Systems

In this type of system, one in-plane dimension of the aromatic molecule is close to the length of the normal alkane, while the other dimension is too large to substitute for one n-alkane molecule and too small to substitute for two. Most of the orthocondensed aromatics (phenanthrene, chrysene) and also polyacenes form β-type systems with one or several n-alkanes.

In these systems, the nonequilibrium phase associated with QL spectra exists over a concentration range, almost as large as that for the α-type systems, but the extent of its occurrence with respect to other possible phases is strongly dependent on the concentration and the crystallization rate. An example of the concentration dependence of the absorption spectra of such a system is given in Figure 3.11, which illustrates the case of phenanthrene in n-C_6.

A peculiar feature of this system is the presence, at concentrations lower than 10^{-4} M, of a broadband absorption spectrum underlying the QL spectrum but similar to the absorption spectrum in a glassy solvent [54, 62]. In this case, the fluorescence counterpart of the broadband absorption spectrum could also be recorded using narrow-band selective excitation tuned to the side of the QL line [62]. Both spectra are observed up to a concentration of about 10^{-4} M, that is, just below the concentration at which aggregates or crystallites begin to be formed. The fact that an energy transfer to a low-energy impurity was observed indicates that the centers responsible for these broadband spectra are disordered groups of solutes molecules [54] or preaggregates [62]. They are described as centers in which the solute concentration is locally much higher than the average value and in which the solute molecules are only weakly mutually interacting so that the molecular character of the spectra is preserved. As such, they may be considered as a particular case of nonincorporated molecules. Another example of such preaggregates is provided by acenaphthene in n-C_6 [52].

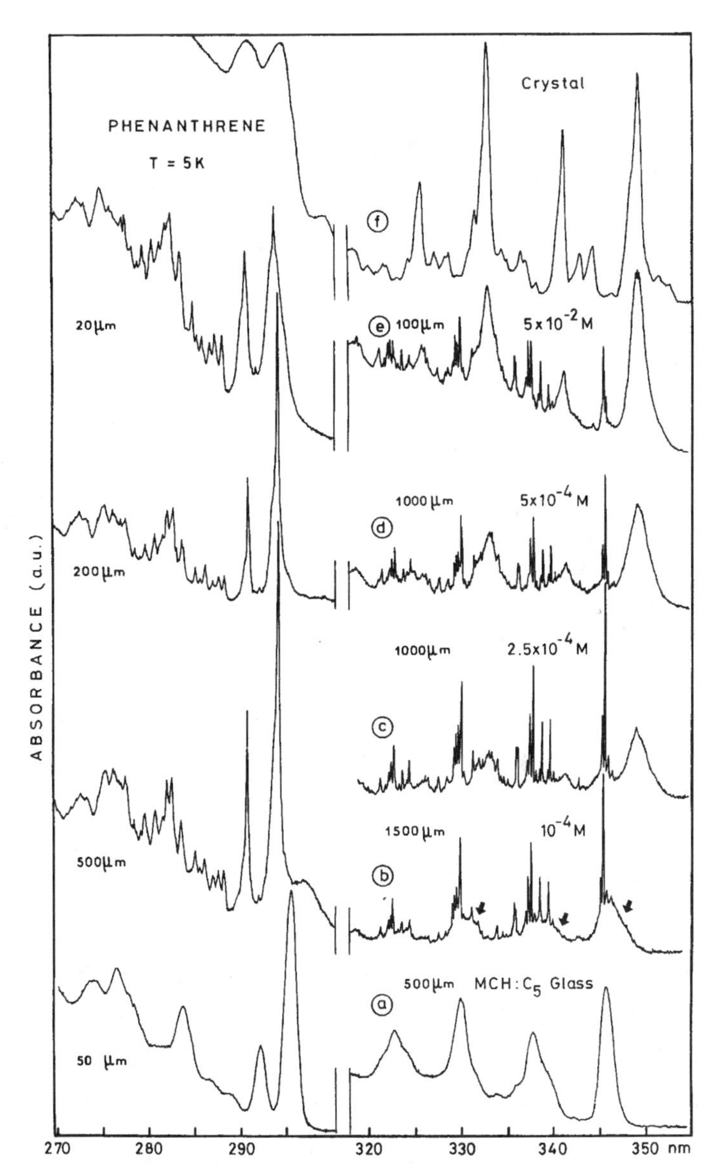

Figure 3.11. Concentration dependence of the absorption spectra of fast frozen solutions of phenanthrene in *n*-hexane. $T = 5\,\text{K}$. Comparison with the spectra of a glassy (a) and a neat (f) crystal. The arrows indicate the occurrence of molecular-type broad bands assigned to preaggregates. (After Nakhimovsky et al. [36] with permission.)

Another feature of the β-type systems is the constant intensity ratio of the crystallites to QL absorption spectra, observed over a large concentration range. This has been observed for naphthalene in n-C_5, 2-methylnaphthalene, and phenanthrene in n-C_6, dibenzofuran in n-C_5, and in several other systems [54–56].

This observation has been tentatively interpreted by Nakhimovsky et al. by the formation of one-dimensional aggregates (segregates) grouping a fixed number of solute molecules substitutionnally incorporated in the n-alkane lattice [36, 55, 73, 74].

3.3.2.2. γ-Type Systems

In these systems, broadband absorption and fluorescence spectra are recorded at low concentrations, while quasiline absorption spectra along with crystalline-type spectra are obtained at high concentrations well above the solubility limit [39, 57, 73]. An example of such a system is dibenzofuran in n-C_8; the concentration dependence of the absorption spectra as a function of concentration is given in Figure 3.10. The QL spectra are observed only for concentrations above ca. 10^{-3} M and always simultaneously with a crystalline-type spectrum. At concentrations below 10^{-3} M, only broad molecular-type spectra are recorded. This type of system was already described in the previous section to illustrate how Shpol'skii matrices are different from isomorphic mixed crystals. All PAHs that form α- or β-type systems with some alkanes are expected to form γ-type systems with adjacent n-alkane homologues.

3.4. ORIENTATION OF GUEST MOLECULES IN *n*-ALKANE SINGLE CRYSTALS

Compared to the extensive research devoted to the study of polycrystalline Shpol'skii solutions, only a small number of publications has addressed doped single crystals of n-paraffins. Interest in using such monocrystallized matrices was threefold. First, provided the guest molecules are oriented in the host lattice, it gives an opportunity to investigate the polarization of emission and absorption spectra of aromatics and other unsaturated systems, yielding valuable information about the symmetry of discrete electronic and vibronic states of these molecules. Second, it allows decisive explanation of the origin of the multiplet structure for a number of solute molecules. Third, it gives information on the orientation of the guest molecules with respect to the host molecules.

Interestingly, although the Shpol'skii effect was first observed for polycyclic aromatic hydrocarbons, it was also found to apply to porphyrin compounds. Litvin and Personov [75] were the first to obtain Shpol'skii spectra of free base and magnesium phthalocyanine and of free base protoporphyrins. Similarly to PAHs, the spectra of this type of molecules exhibit a multiplet structure and the incorporation of these molecules in single crystals of *n*-alkane turned out to be very fruitful for the interpretation of the multiplet structure of their spectra but also for a better understanding of their electronic singlet- and triplet-state properties.

Generally, when observed, the QL spectra recorded in *n*-alkane monocrystalline solutions, and their multiplet structure in particular, appear to be identical to those recorded from fast-cooled polycrystalline ones. This result indicates clearly that the same kind of trapping does occur in both matrices. Compared to fast-cooled polycrystalline solutions, the only difference is the absence in monocrystalline solutions of some components of which the most prominent ones have been referred to as "pseudoliquid sites" [30].

The presence of several spectral multiplet components in the absorption, fluorescence, and phosphorescence spectra was formerly assigned to the possible existence of *n*-alkane rotational isomers [3, 66] or to polymorphism in the host crystal [67, 68]. However, the observed small homogeneous broadening of lines in QL spectra together with numerous experimental results lead some authors to propose that multiplet components are related to molecules trapped in spatially separated centers [76–80] which, as discussed in detail below, have been shown to differ more or less from each other by their orientations [81–84].

These conclusions were drawn from various spectroscopic techniques, including dichroism measurements on emission or absorption spectra, EPR spectroscopy, and microwave-induced delayed phosphorescence (MIDP). While the first technique was mainly applied to cata- and pericondensed polycyclic aromatics (anthracene, pyrene, perylene, etc.) in which the transition moments are linearly polarized, the two last ones lead to more precise results for highly symmetric molecules such as triphenylene, coronene, or porphins.

3.4.1. Orientation of Cata- and Pericondensed PAH from Dichroism Measurements of Absorption and Emission Spectra

To our knowledge, the very first measurements with polarized light were reported by Malykhina and Shpak [85] for stilbene embedded in a monocrystallized solution of *n*-octane.

In these experiments, *n*-paraffin single crystals were directly grown in quartz cells of various thicknesses from 50 to 200 μm, or between two quartz plates. In most cases, the solution prepared at a concentration lower than 10^{-3} M was slowly cooled by lowering the cell into liquid nitrogen [78]. By selecting suitable monocrystalline domains, Malykhina and Shpak [85] observed a pronounced dichroic effect on the QL spectra, which clearly showed that the stilbene molecules are oriented in the *n*-paraffin lattice, but they did not notice any difference in the degree of polarization from one multiplet component to another.

Personov and Bykovskaya [81] used the same sample preparation method to investigate the absorption and fluorescence spectra of phthalocyanine in a single crystal of *n*-octane and the absorption spectrum of benzo[*a*]pyrene in a single crystal of *n*-heptane. Contrary to the stilbene case, they reported a small variation in the multiplet polarization. For anthracene in *n*-heptane single crystal, Bolotnikova et al. [84] established that the molecules corresponding to the two main multiplet components have noticeably different spatial orientations. However, in all these works, because of the method of preparation of the crystals, the dichroism of the spectra could be measured only through one face of the crystal, and the orientation within the crystal lattice of the solute molecules related to each multiplet component could not be deduced.

The first attempt to prepare a suitably doped *n*-alkane single crystal with well-developed crystal faces was reported by Pfister and Kahane-Paillous [64]. These authors prepared a coronene-doped monocrystallized crystal of *n*-heptane by slowly cooling a previously degassed solution, contained in a Pyrex tube ending in a capillary, which was slowly immersed into liquid nitrogen. Using a similar method, Lamotte et al. [86] obtained optically suitable doped single crystals of *n*-heptane doped with coronene, pyrene, and perylene. The *n*-heptane crystals were prepared from the bulk by cleavage while immersed in liquid nitrogen. They exhibited two well-defined cleavage planes nearly perpendicular (92°) to each other and practically parallel to the direction of the growth *a* axis with a typical size of about $3 \times 5 \times 10\,\mathrm{mm}^3$. Strongly polarized absorption, fluorescence, and phosphorescence spectra were recorded normal to each face, with light polarized, respectively, along the dielectric axes. For pyrene and perylene (Fig. 3.12), whose molecular electronic transitions are associated with in-plane linear oscillators, the dichroic ratios of the absorption spectra measured through each face allowed determination of the projection of both long and short axis polarized transitions onto each faces. This leads to an estimate of the orientation of the molecules with respect to the cleavage planes as shown in Figure 3.13. In both cases, the long axes (L) are found to point out from face I, making a small angle with respect to the normal. In the case of perylene, a similar orientation was deduced by Zalesskii et al. [87], who found an angle of about 22° between the projection of the long axis polarized dipole moment onto the *ab* face (edge II) and the surface normal to the growth axis. According to the measured dichroic ratios [86], it appeared that the perylene and pyrene molecules do not have rigorously the same orientation (Fig. 3.13). However, because of the uncertainty in the determination of the dichroic ratios and in the possibly incomplete linear polarization of the transition dipole moments, it was difficult to conclude whether this difference in orientation is real. Moreover, in these very first experiments the spectra were recorded under low resolution at 77 K; therefore, the multiplet structure was not resolved, and it could not be concluded firmly whether the spectra were quasiline in nature.

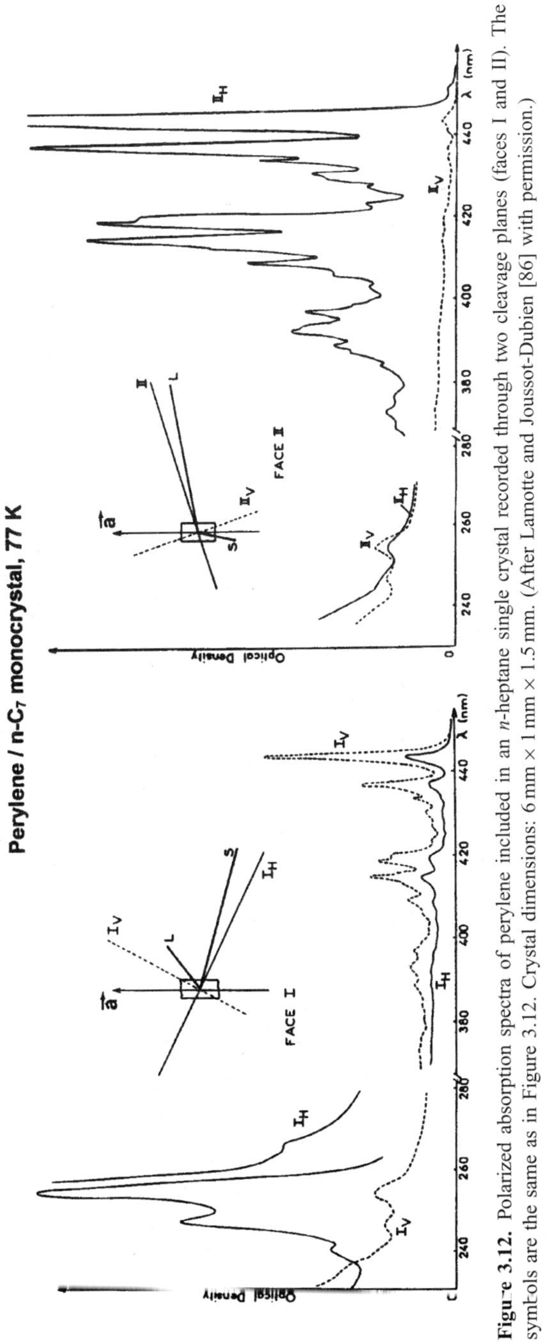

Figure 3.12. Polarized absorption spectra of perylene included in an *n*-heptane single crystal recorded through two cleavage planes (faces I and II). The symbols are the same as in Figure 3.12. Crystal dimensions: 6 mm × 1 mm × 1.5 mm. (After Lamotte and Joussot-Dubien [86] with permission.)

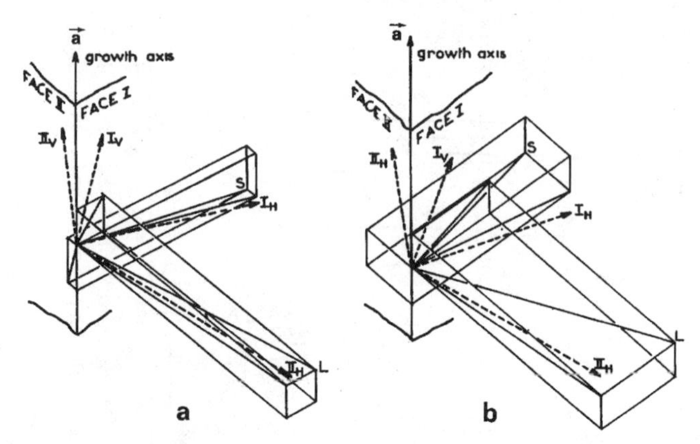

Figure 3.13. Schematic orientation of the short (S) and long (L) axes of perylene (a) and pyrene (b) molecules in the *n*-heptane crystal, as deduced from the absorption dichroic ratios through each face of the guest crystal. (After Lamotte and Joussot-Dubien [86] with permission.)

More precise and definitive results were drawn from dichroism measurements conducted at 4.2 K under high resolution [23, 65, 88–90]. Moreover, the cleavage planes were identified by X-ray measurements at 77 K [23] with the crystallographic *ac* (face I) and *ab* (face II) planes, while the growth axis was found to correspond to the *a* axis. In practice, the distinction between the *ac* and *ab* planes was made first during cleaving by noting that the *ac* plane corresponds to the easier cleavage plane (thus identified as the face I) and more precisely by the orientation of the principal dielectric axes within each face. In the face I (*ac* plane), the I_V dielectric axis makes an angle of about $25 \pm 3°$ with the *a* axis, while in the face II (*ab* plane), the angle between II_V and the *a* axis is $18 \pm 3°$. A schematic view of a typical crystal is shown in Figure 3.14, which also shows the most probable arrangement of the perylene guest and host molecules as deduced from dichroic measurements and from the complete determination of the crystal structure of *n*-heptane [23]. Due to the fact that only absolute values of the angles with respect to the dielectric axes can be determined, the orientations that have been proposed are those that are the most consistent with the molecular arrangement of the *n*-alkane chains and with the whole set of data as presented and discussed below. In the notation used in the figures, the axes *b* and *c* have been exchanged with regard to Norman and Mathisen [20] so that the *b* axis points out of the *ac* face at an angle of 85° and the chain axis makes an angle of 2° with respect to the *ab* plane.

From the dichroic data it was deduced that the most probable orientation for the plane of the aromatic molecules is parallel to the alignment plane of the *n*-alkane chains defined by the alkane zigzag axis and the crystal *c* axis. Within this hypothesis, the long axis of the perylene molecule makes an angle of about 10°

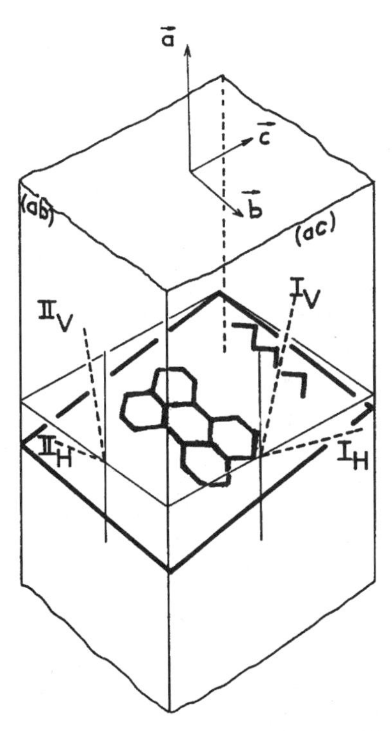

Figure 3.14. Schematic representation of a perylene molecule trapped in *n*-heptane single crystal as deduced from absorption and fluorescence dichroisms. Dashed lines are the dielectric axes in *ab* and *ac* planes, which were identified, respectively, with the faces I and II of Figure 3.14. The *a* axis is the growth axis. (After Lamotte et al. [65] with permission.)

with respect to the normal of the *ac* plane, that is, practically parallel to the direction of the zigzag axis of the alkane molecules (Fig. 3.14).

For perylene and coronene [65], as well as for pyrene [88], triphenylene [89], and anthracene [84], very sharp quasiline absorption, fluorescence, and phosphorescence spectra similar to the spectra recorded from fast-polycrystallized solutions with the same multiplet structure were obtained, thus showing definitive evidence for a similar trapping of the guest aromatic molecules responsible of QL spectra, in both fast- and slow-cooled *n*-heptane matrix. Similar conclusions were reported by Kozlov and Grebenshchikov [91] for 1,12-benzoperylene, benzo[*a*]-, and benzo[*e*]pyrene in several monocrystallized solutions of *n*-paraffins they studied at 77 K. In some cases however, a noticeable difference is the absence of a short-wavelength component in monocrystallized solutions, which was previously identified with the so-called pseudoliquid site in the case of coronene (Fig. 3.15).

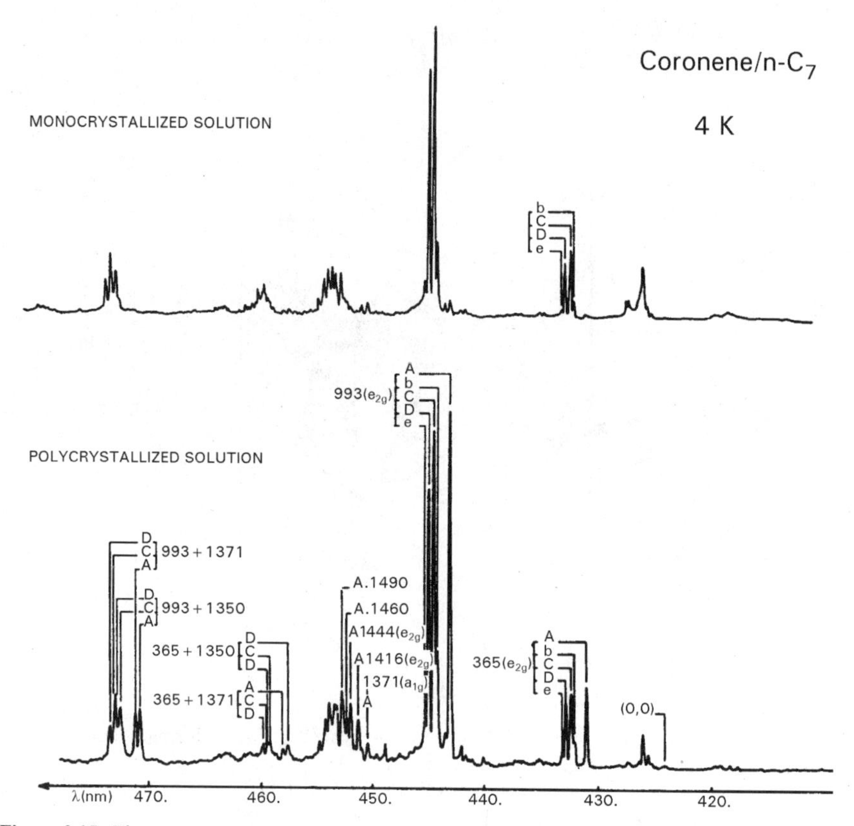

Figure 3.15. Fluorescence spectra of coronene in monocrystalline (top) and polycrystalline (bottom) solutions of *n*-heptane at 4.2 K. (After Lamotte et al. [65] with permission.)

In connection with the formerly proposed hypothesis [76, 77, 92] and previous observations [81] that the multiplet components belong to molecules spatially separated and differently trapped in the *n*-alkane lattice, attempts has been made [65] to verify whether the multiplet components of coronene and perylene in *n*-C_7 correspond to molecules having different orientations. No significant difference in the polarization of the two components observed for perylene in *n*-C_7 was noticed, in contrast with the results of Zalesskii et al. [87], who reported a change in the degree of polarization in the perylene fluorescence spectra. However, in the latter experiments, the spectra were recorded at 77 K and the resolution was too low for a precise measurement. At 4.2 K, the doublet is clearly resolved, and the absence of difference in the polarization of the two components can be explained by the fact that the weakest component could be assigned to a localized phonon band and not to a site.

3.4.2. Distortion and Orientation of Coronene and Triphenylene Molecules in the n-C_7 Lattice

In the case of coronene, and also for triphenylene, the transition dipole moments correspond to in-plane degenerate oscillators, a situation that is less favorable than linear oscillators for a precise determination of the orientation of the molecules. Coronene and triphenylene are highly symmetric molecules, and like benzene [93, 94] they were expected to be very sensitive to the crystal field anisotropy experienced in the distinct sites of the host lattice. Under such circumstances, a distortion and a consecutive lowering of their symmetry were expected. Evidence for a distortion of these molecules is given by the presence in the phosphorescence spectra of coronene (Fig. 3.16) [65, 95, 98] and in the fluorescence spectrum of triphenylene [90, 96] of the forbidden 0–0 band with a relatively high intensity. Such transitions are expected to be linearly polarized, thus offering an opportunity to estimate the orientation of the molecules.

3.4.2.1. Coronene

As in its fluorescence spectrum, the most active vibrations in the phosphorescence spectrum of coronene have e_{2g} symmetry. These modes are degenerate and

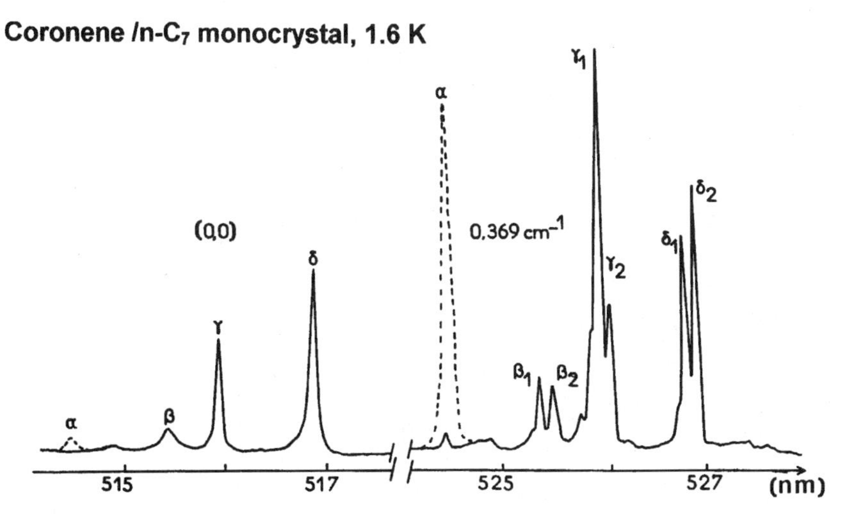

Figure 3.16. Multiplet structures of the (0,0) and first split vibronic bands of the phosphorescence spectrum of coronene in an n-heptane single crystal. The dashed line bands correspond to the normalized intensity of the α component as recorded in a polycrystalline sample. All spectra show similar site structure and site splitting of the degenerate e_{2g} vibration. (After Merle et al. [99] with permission.)

arise through vibronic spin-orbit coupling between the $T_1(^3B_{1u})$ state and an excited singlet $(^1A_{2u}, \sigma\pi^*)$ state, in agreement with an out-of-plane polarization for the phosphorescence emission [97]. By combining the dichroism measurement of the in-plane polarized fluorescence and phosphorescence and taking into account the orientation of the dielectric axes, it was found that the molecular plane is most likely oriented in such a way that it is parallel to the substitutional plane defined by the zigzag heptane chain axis and the crystal c axis, as was also deduced for perylene and pyrene [86]. No change, however, [65] of the dichroic ratios among the four multiplet components was observed for coronene.

More precise information about the respective orientation of the molecules in the main sites was obtained at very low temperature (1.6 K) and under high resolution by Merle et al. [82, 83] and Pitts et. al. [98], by taking advantage of the splitting of the e_{2g} mode into a_g and b_{3g} vibrational components, whose separations were found to in the order $\beta > \delta > \gamma$ (Fig. 3.16). Very similar multiplet structures and site-splitting frequencies were obtained in e_{2g} vibronic bands in phosphorescence, absorption, and fluorescence spectra, proving that within each site, the distortion of the molecules in its ground and first excited states are very similar. Moreover, the zero-field triplet-splitting parameters, D and E, were determined with the MIDP technique by Merle et al. [82, 83, 99] and Pitts et al. [98], through the intensity change of the 0–0 phosphorescence band (originating from the τ_B level), after the $\tau_B \leftarrow \tau_N$ microwave transition.

For the three sites, β, γ, δ, an increasingly negative value of E is found and a small positive E value is found for site α. From these results it was concluded that the coronene molecules in the α, β, γ, and δ sites, present an increasing distortion from a slightly quinoïdal distortion to a rather strong antiquinoïdal distortion, as illustrated in Figure 3.17.

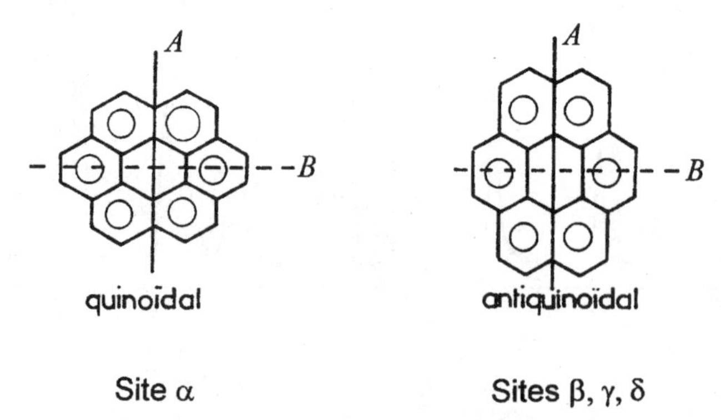

Figure 3.17. Schematic deformation of coronene in its lowest triplet state for the different trapping sites in the n-heptane lattice deduced from MIDP experiments. (After Merle [82] with permission.)

The orientations of the molecules in the three main β, γ, and δ sites were also determined by Merle et al. [82, 83, 99] by taking advantage of the fact that the $\tau_A \leftarrow \tau_N$ transition is polarized along the *B* axis, while the $\tau_B \leftarrow \tau_N$ transition is polarized along the *A* axis. For that purpose, they combined the MIDP technique with linearly polarized microwaves [100]. The polarizations of these transitions were measured along the three crystallographic faces of a doped n-C_7 single crystal. In agreement with results from fluorescence and phosphorescence polarization studies, the molecules were found to be inserted in the same substitutional plane as that found for perylene (Fig. 3.14). The results are shown in Figure 3.18, where both the different orientations and relative deformations of the coronene molecules within this plane have been schematized. It was further concluded [99] that the distortion of the molecules in the *n*-heptane lattice do not result from a strong pseudo-Jahn–Teller effect but is essentially induced by the crystal-field anisotropy.

3.4.2.2. *Triphenylene*

Contrary to coronene, of which the spectra exhibit a multiplet structure in n-C_7, the triphenylene fluorescence spectrum in this host alkane shows only one component, which indicates that the molecule is trapped in a unique type of site. The orientation of the molecule was deduced from conventional fluorescence dichroism, EPR measurements [89, 90] and MIDP experiments [82, 83].

The presence of the (0,0) transition and others of totally symmetric character in the fluorescence spectra implies a lowering of the D_{3h} symmetry to a C_{2v} symmetry by the loss of the C_3 axis [89, 101].

In Figure 3.19, two possible orientations of triphenylene molecule are schematized with respect to the *n*-alkane chain, consistent with the symmetry elements of the C_{2v} point group. Assuming an in-plane deformation, the one-

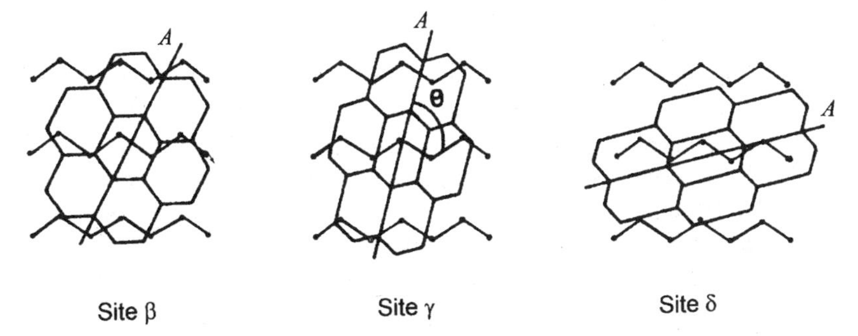

Figure 3.18. Schematized in plane distortion and orientation of the magnetic *A* axis of the coronene molecules trapped in the β, γ, and δ sites within the substitutional plane of the *n*-heptane lattice shown in Figure 3.21. (After Merle et al. [83, 99] with permission.)

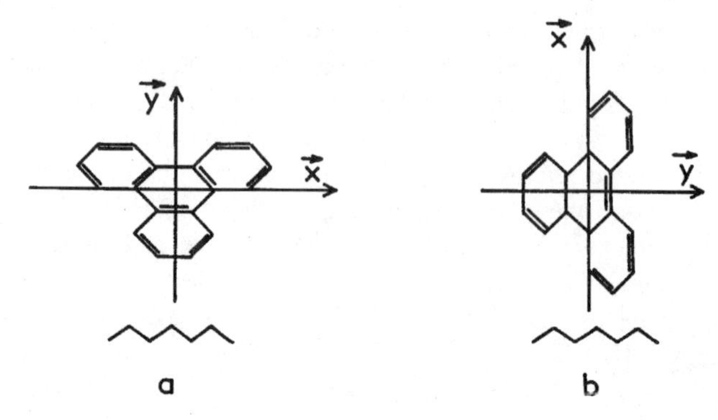

Figure 3.19. The two possible extreme cases for the orientations of the triphenylene molecule with respect to the *n*-heptane chain axis. The axis labeling is consistent with an in-plane antiquinoidal distortion lowering the symmetry from D_{2h} to C_{2v}. (After Lamotte et al. [89] with permission.)

photon-forbidden but two-photon-allowed 0–0 ($A_1' \rightarrow A_1'$) transition [82, 83] is transformed into an allowed *y*-axis-polarized $A_1 \rightarrow A_1$ transition. The larger intensity of the 0–0 band together with a larger dichroic ratio for the spectra measured through face *I* of the *n*-C_7 crystal indicate that the triphenylene *y*-axis is almost parallel to the *ab* plane and almost perpendicular to the growth axis, a result that appears consistent with orientation (b) in Figure 3.19.

A precise and definitive determination of the molecular deformation and plane orientation was provided by simulating the angular variation of the triplet state EPR signal upon a rotation of the crystal around the growth *a* axis [90]. The calculations were performed by assuming an antiquinoïdal deformation [101, 102]. The results confirmed that the triphenylene molecular plane lies in the same substitutional plane as coronene. The orientation of the *y* axis, as deduced from fluorescence dichroism (Fig. 3.19b), was confirmed by MIDP experiments conducted with polarized microwaves [82, 83].

3.4.3. Distortion and Orientation of Porphin Molecules in *n*-Alkanes

Supplementing the pioneer works of Litvin and Personov [75] and Gradyushko et al. [103] on the spectroscopy of porphyrins in polycrystalline Shpol'skii matrices, Chan et al. [104] and Canters et al. [105] were the first to show that single crystals of *n*-octane containing a low concentration of porphins can be prepared. Very sharp line spectra were obtained at 4.2 K, opening new possibilities to investigate the electronic properties of porphin molecules but also to determine their deformation and orientation in *n*-alkane matrices. As for PAHs, the spectra exhibit a simpler multiplet structure than fast-cooled solutions, with one or two major sites being populated.

Porphin-doped *n*-alkane single crystals were grown in a very similar way to PAH-doped single crystals. The degassed solution, contained in a narrow tube about 20 cm long, was slowly lowered into liquid nitrogen. The cooling rates, as controlled by the rate at which the solution was immersed into liquid nitrogen, ranged from 0.5 to 5 cm min^{-1} [106, 107]. Interestingly, it was noticed [106, 108] that when porphin is first dissolved in benzene or ethanol and the resulting solution is added to a twentyfold excess of *n*-octane, the S/N ratio of the ESR signal was improved by a factor of ten without affecting the other parameters substantially. Slight changes in the zero-field splitting and optical behavior were, however, noticed with ethanol.

Former investigations with doped *n*-paraffin single crystals at 4.2 K concerned zinc porphin (ZnP), of which the lowest triplet state was studied by optical double microwave resonance (ODMR) [104] and Zeeman effect [105] and the first excited singlet state by high-resolution optical spectroscopy [105]. Similar investigations were also made with palladiumporphin (PdP) [106, 110] and free base porphin (H_2P) [111] in *n*-octane single crystals. High-resolution spectra of palladium, platinum, and copper porphins in *n*-octane crystals were reported and analyzed by Noort et al. [112]. The spectra and the Zeeman effects on the $S_1 \leftarrow S_0$ and $T_0 \rightarrow S_0$ transitions of H_2P, ZnP, and PdP in several *n*-alkane hosts were also investigated [107]. Other studies on the implantation and spectroscopic properties of porphin derivatives in *n*-hexane single crystals have been carried on with chlorin and porphin [113], tetrapropylchlorin [114], tetrabenzoporphin [115], and tetramethylporphin [116].

In agreement with Simpson [109], it was clearly established that—when incorporated into an *n*-octane host single crystal—both the singlet (S_1) and the triplet (T_1) states of ZnP and PdP are unequivocally nondegenerate. A splitting of the S_1 state does occur and many of the absorption lines in the first $S_1 \leftarrow S_0$ forbidden transition (the Q band) appear as doublets with zero point vibronic components, separated, respectively, by 109 cm^{-1} for ZnP (Fig. 3.20) and 30 cm^{-1} for PdP, with their in-plane transition moments mutually perpendicular. As expected because of fast relaxation to the lowest-energy component of the vibrationless S_1 level, the high-energy components were only very weakly observed (as a hot band) in the fluorescence spectra (Fig. 3.20). This behavior appeared to be common to most metalloporphins [117]. It was proposed that the state splitting occurred through the crystal-field effect as the result of the Jahn–Teller instability of the ${}^1E_u(S_1)$ and ${}^3E_u(T_0)$ states.

It was concluded that in an *n*-octane host, ZnP and PdP molecules are trapped in one dominent type of site, referred to in the following as the *A* site, in which the molecules are slightly distorted. The deviation from the fourfold symmetry of the spin-density distribution in the T_1 state was derived from the observed zero-field splitting using ODMR experiments at 1.3 K [104]. Because of the Jahn–Teller instability, an in-plane deformation results [108], conferring a D_{2h} symmetry to the molecule.

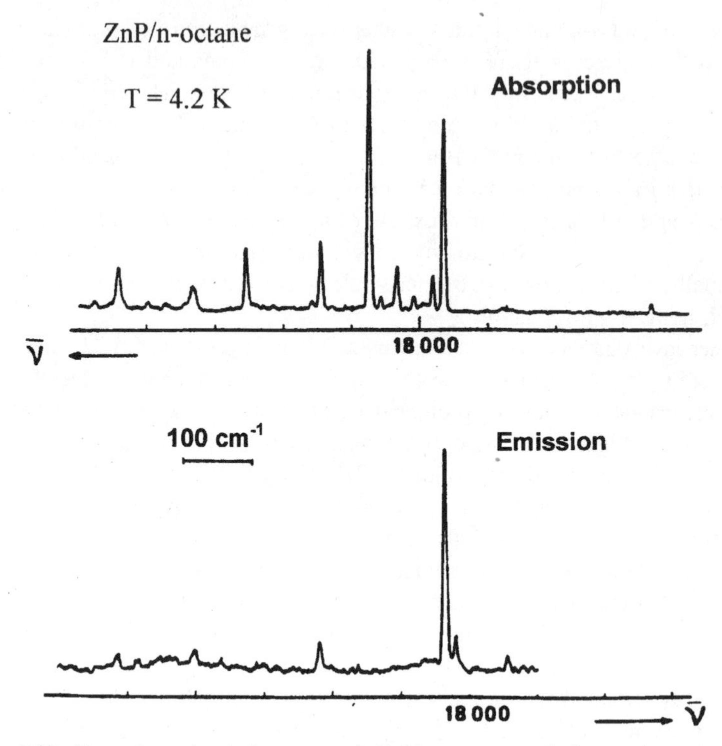

Figure 3.20. Absorption and emission spectra of a ZnP in an *n*-octane single crystal near the origin of the *Q* band. $T = 4.2\,\mathrm{K}$. (After Canters et al. [105] with permission.)

While for ZnP and magnesium porphins the principal magnetic axes were found to coincide with the axes through the N atoms [108], for PdP, results obtained from Zeeman experiments and from comparison with ESR results on H_2P and ZnP [108, 111] led the authors to propose that the PdP magnetic axes lie, instead, close to the twofold axis through the methine bridge [107].

The determination of the orientation of the guest molecules in the main sites was carried out using either the Zeeman effect on the $S_1 \leftarrow S_0$ and $T_0 \rightarrow S_0$ transitions, or based on ESR experiments on the T_0 state.

Zeeman experiments on the $S_1 \leftarrow S_0$ transition of ZnP and PdP in *n*-heptane first revealed that for the guest molecules in the predominant *A* site, the angle θ_A between the out-of-plane molecular axis *z* and the crystal *a* axis is about 25° [105, 110]. Referring to the known crystal structure of *n*-C_7 [21, 24], it was tentatively concluded that the zinc porphin molecules replace at least two octane molecules, aligned along the axis of alignment of the alkane molecules (the *n*-heptane *b* axis after Norman and Mathisen [21, 24]), which is roughly perpendicular to the carbon chains. A schematic representation of the substitutional plane characterizing the *A* site is shown in Figure 3.21. This conclusion was confirmed by ESR studies on the triplet state of ZnP [108].

A similar orientation of the molecular plane was found for free base porphin (H_2P) [108, 118]. However, in this case, the majority of guest molecules occur in two different sites (type 1 and 2 after van Dorp et al., [111]), where the molecular planes are parallel but their in-plane x and y axes are interchanged. Moreover, H_2P molecules of type 2 are found to have the same orientation as ZnP molecules in *A* sites. A qualitatively quite similar orientation of the molecular plane was determined by Arabei et al. [116] in the case of 2,3,7,8-tetramethylporphin in a single crystal of hexane.

In agreement with the direction determined from the Zeeman effect for ZnP, the axis normal to the molecular plane (the z axis) was found to make an angle of $27 \pm 4^\circ$ with respect to the crystal a axis. This mode of incorporation was found not to differ substantially in other *n*-alkane hosts, as investigated by ESR and Zeeman experiments [107]. It is in fact quite similar to that determined for

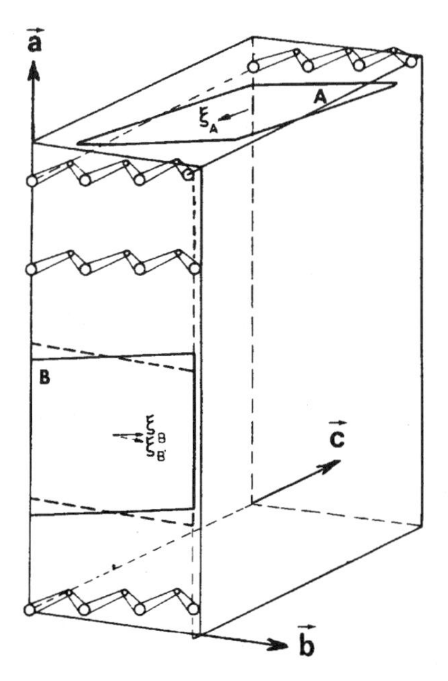

Figure 3.21. Schematic representation showing the planes (*A* and *B*) of incorporation of the porphin molecules trapped in *A* and *B* sites in the *n*-heptane crystal determined from Zeeman experiments. The arrows with labels ξ_A and ξ_B indicate the orientations of the ξ spin axes in each site. The B' site was found to arise through a transformation of the *B* site upon lowering the temperature. (After Jansen et al. [107] with permission.) For coherence with Figure 3.14, the b and c axes have been interchanged with respect to the original figure.

coronene in *n*-heptane and appears to prevail in slowly grown crystals. Upon strong illumination it was noticed that the two types of H_2P sites may be transformed into one another. It was shown that the displacement of the protons at the center of the molecule from one pair of nitrogen atoms to another is responsible for this transformation [113, 119, 120].

Another type of site (*B* site) whose occupancy, in contrast with the *A* site, is favored by increasing the rate of freezing, was not only noticed in *n*-octane but also in other *n*-alkanes ranging from n-C_7 to n-C_{10} and in n-C_{12} [121]. This site appears to be thermodynamically less stable than the *A* sites and presents some similarity with the so-called pseudoliquid sites mentioned for PAHs [30]. Interestingly it was noticed that, upon very slow cooling, the *B* sites were almost absent in even *n*-alkane solutions, while they can be still observed in odd ones [107].

The orientation of molecules in *B* sites was precisely determined in the case of PdP in n-C_7 single crystals by means of Zeeman experiments [107]. It was established that the porphin planes must lie parallel to the *a* axis with the molecules stacked perpendicular to the alkane chains between their terminal methyl groups. Such a plane of insertion, which corresponds approximately to the easiest cleavage plane of the n-C_7 and n-C_9 single crystals is the same as the plane proposed for the incorporation of thin platelike crystallites of naphthalene formed during the cooling process in these host crystals [122].

According to the characteristics of the *B* sites when compared to *A* sites, namely, weaker thermodynamic stability, smaller inhomogeneous width of the fluorescence and absorption lines, spectral location at higher energy, it was tempting to identify this site with the pseudoliquid sites postulated for some PAHs, like the α site of coronene in n-C_7 [30]. However, in the absence of definitive proof, the substitutional trapping hypothesis for the *B* site is not totally rejected by some authors [107].

3.5. CONCLUSIONS

The contrasting crystallization rate and host concentration dependencies of the host absorption and fluorescence spectra of Shpol'skii solutions, as compared to isomorphic mixed crystals, indicate clearly that these two types of solid solutions differ substantially from each other and cannot be described using similar concepts. While isomorphic mixed crystals are best described as equilibrium solid solutions, Shpol'skii solutions appear to be more complex and in fact correspond to a large variety of equilibrium, quasiequilibrium, and nonequilibrium solutions, giving rise to at least four different types of systems, exhibiting different concentration and cooling rate dependences.

It is only in a few cases that QL spectra are obtained under quasiequilibrium conditions prevailing for growing doped single crystals. In those cases, a precise determination of the orientation of the guest molecules could be made. In the majority of cases, the results have given strong evidence for a substitutional insertion of the host molecules in the *n*-alkane lattice, in the plane containing the long axis of the *n*-alkane chains and with the axis pointing to the direction of their side-by-side alignment.

The fact that in the majority of cases QL spectra are only obtained upon fast cooling is consistent with the basically nonequilibrium nature of the phase associated with the molecules responsible for these spectra. This partly explains why QL spectra could be obtained for a very large variety of planar and even nonplanar PAHs and other types of molecules as cyclic polyenes, provided a suitable host *n*-alkane is chosen. It is this unique feature of the Shpol'skii solutions that really makes them a providential tool not only for the study of spectroscopic and photophysical properties of PAHs but also for the development of very sensitive and selective methods of PAH analysis in natural media of which we may find many applications in the literature and in this volume.

ACKNOWLEDGMENT

The author thanks Professor J. Joussot-Dubien for reading the manuscript and for his support during the studies we have made on the Shpol'skii systems and Ross Brown, who kindly revised the text. Special thanks are also due to A. M. Merle, J. Rima, L. Nakhimovsky, Ph. Garrigues, and all other persons who contributed to elucidate some of the mysteries of Shpol'skii solutions.

REFERENCES

1. H. G. Grimm and H. Wolff, *Handbuch der Physik*, H. Geiger and K. Scheel, (eds.), Vol. 24, Part 2, *Atombau und Chemie, Sec. IIIf, Atombau und Kristallchemie*, Springer, Berlin (1933).
2. A. I. Kitaigorodskii, *Molecular Crystals and Molecules*, Academic Press (1973); *Mixed Crystals*, Springer Series in Solid-State Sciences 33, Springer-Verlag, Berlin (1984).
3. E. V. Shpolskii, *Soviet Physics Uspekhi Transl.* **5**, 522–31 (1962).
4. B. A. Lombos, P. Sauvageau, and C. J. Sandorfy, *Molecular Spectrosc.* **24**, 253–69 (1967).
5. D. S. McCLure, "Electronic Spectra of Molecules and Ions in Crystals, I: Molecular Crystals" in *Solid State Physics, Advances in Research and Applications*, F. Seitz and D. Turnbull, (eds.), Academic Press, N. Y., Vol. 8, 1–47 (1959).
6. G. Bruni, *Feste Lösungen und Isomorphismus*, Leipzig (1908).
7. B. Gossner, *Z. Kryst. Min.* **43**, 130–47; ibid. (1909); **44**, 417–518 (1907).

8. J. Timmermans *Les Solutions Concentrées*, Masson editor, Paris (1936).
9. D. M. Hanson, *J. Chem. Phys.* **52**, 3409–18 (1970).
10. C. L. Braun and H. C. Wolf, *Chem. Phys. Letters* **9**, 260–62 (1971).
11. S. M. Ziegler and M. A. El-Sayed, *J. Chem. Phys.* **52**, 3257–68 (1970).
12. D. Owens and M. A. El-Sayed, *J. Chem. Phys.* **52**, 4315–16 (1970).
13. A. Bree and R. Zwarich, *J. Chem. Phys.* **49**, 3355–58 (1968).
14. R. W. Brandon, R. E. Gerkin, and C. A. Hutchinson Jr., *J. Chem. Phys.* **41**, 3717–26 (1964).
15. R. M. Hochstrasser and G. J. Small, *J. Chem. Phys.* **48**, 3612–24 (1968).
16. C. D. Akon and D. P. Craig, *J. Chem. Phys.* **41**, 4000–01 (1964).
17. L. E. Lyoks and L. J. Warren, *Aust. J. Chem.* **25**, 132–37 (1972).
18. D. P. Craig and T. Thirunamachandran, *Proc. Chem. Soc.* **A271**, 207–17 (1963).
19. A. F. Prikhot'ko and M. T. Shpak, *Opt. i Spektrosk.* **4**, 17–29 (1958).
20. N. Norman and H. Mathisen, *U.S. Dept. Comm. Office, Tech. Serv. PB Depart., Report 171*, 180 pp. (1960).
21. N. Norman and H. Mathisen, *Acta Chem. Scand.* **15**, 1747–60 (1961).
22. N. Norman and H. Mathisen, *Acta Chem. Scand.* **21**, 127–35 (1967).
23. A. M. Merle, M. Lamotte, S. Risemberg, C. Hauw, J. Gaultier, and J. Ph. Grivet *Chem. Phys.* **22**, 207–14 (1977).
24. N. Norman and H. Mathisen, *Acta Chem. Scand.* **26**, 3913–16 (1972).
25. A. I. Kitaigorodskii, *Organic Chemical Crystallography*, Consultants Bureau, New York (1961).
26. B. Stevens, *Spectrochim. Acta* **18**, 439–48 (1962).
27. J. Tanaka, *Bull. Chem. Soc. Japan* **36**, 1237–49 (1963).
28. T. N. Bolotnikova, *Opt. and Spectrosc.* **7**, 138–40 (1959).
29. E. V. Shpolskii, *Soviet Physics Uspekhi Transl.* **6**, 411–27 (1963).
30. C. Pfister, *Chem. Phys.* **2**, 171–80 (1973).
31. C. Pfister, *Chem. Phys.* **2**, 181–90 (1973).
32. J. J. Dekkers, G. Ph. Hoornweg, C. MacLean, and N. H. Velthorst, *J. Molecular Spectrosc.* **68**, 56–66 (1977).
33. J. J. Dekkers, G. Ph. Hoornweg, G. Visser, C. MacLean, and N. H. Velthorst, *Chem. Phys. Letters* **47**, 357–61 (1977).
34. J. W. Hofstraat, Thesis, Vrije Universiteit Amsterdam, Free University Press, Amsterdam, 245 pp. (1988).
35. J. W. Hofstraat, C. Gooijer, and N. H. Velthorst, in *Methods in Molecular Spectroscopy*, S. G. Schulman, (ed.), Wiley, New York, Vol. 2, pp. 283–400 (1988).
36. L. A. Nakhimovsky, M. Lamotte, and J. Joussot-Dubien, *Handbook of Low Temperature Electronic Spectra of Polycyclic Aromatic Hydrocarbons*, Physical Sciences Data 40, Elsevier Science Publishers, Amsterdam, 507 pp. (1989).
37. W. Karcher, R. J. Fordham, J. J. Dubois, P. G. J. M. Glaude, and J. A. M. Ligthart,

(eds.), *Spectral Atlas of Polycyclic Aromatic Compounds*, D. Reidel Publishing Company, Vol. 1, 818 pp (1983).

38. W. Karcher, S. Ellison, M. Ewald, P. Garrigues, E. Gevers, and J. Jacob (eds.), *Spectral Atlas of Polycyclic Aromatic Compounds*, Kluwer Academic Publishers, Vol 2, 864 pp (1988).
39. E. V. Shpolskii, L. A. Klimova, G. N. Nersesova, and V. I. Glyadkovskii, *Opt. and Spectrosc.* **24**, 25–29 (1968).
40. K. Palewska and Z. Ruziewicz, *Chem. Phys. Letters* **64**, 378–82 (1979).
41. K. Palewska, Z. Ruziewicz, H. Chojnacki, and E. Meister, *Chem. Phys.* **161**, 437–45 (1992).
42. K. A. Muszkat and T. Wismontski-Knittel, *Chem. Phys. Letters* **83**, 87–90 (1981).
43. M. Lamotte, F. J. Morgan, K. A. Muszkat, and T. Wismontski-Knittel, *J. Phys. Chem.* **94**, 1302–09 (1990).
44. M. Tachon, E. Davies, M. Lamotte, K. A. Muszkat, and T. Wismontski-Knittel, *J. Phys. Chem.* **98**, 11870–77 (1994).
45. K. Palewska, E. C. Meister, and U. P. Wild, *Chem. Phys.* **138**, 115–22 (1989).
46. K. Palewska, Z. Ruziewicz, and H. Chojnacki, *J. Luminesc.* **39**, 75–85 (1987).
47. F. Morgan, Ph. Garrigues, M. Lamotte, and J. C. Fetzer, *Polycycl. Arom. Comp.* **2**, 141–53 (1991).
48. K. Palewska, J. Lipinski; J. Sworakowski, J. Sepiol, H. Gygax, E. Meister, and U. P. Wild, *J. Chem. Phys.* **99**, 16835–41 (1995).
49. M. Kurahashi, M. Fukuyo, A. Shimada, A. Furusaki, and J. Nitta, *Bull. Chem. Soc. Japan* **42**, 2174–79 (1969).
50. D. M. Burns and J. Iball, *Proc. Roy. Soc.* **A227**, 200–14 (1955).
51. N. S. Dokunikhin, V. A. Kizel, M. N. Sapozhnikov, and S. L. Solodar, *Opt. and Spectrosc.* **25**, 42–46 (1968).
52. J. W. Hofstraat, I. L. Freriks, M. E. J. de Vreeze, C. Gooijer, N. H. Velthorst, *J. Phys. Chem.* **93**, 184–90 (1989).
53. L. N. Ustyugova and L. A. Nakhimovskaya, *Zhurnal Prikladnoi Spektrosk.* **9**, 1053–56 (1968).
54. L. A. Nakhimovskaya, L. N. Ustyugova, and N. S. Proskuryakova, *Ukr. Fiz. Zh.* **16**, 268–75 (1971).
55. G. V. Kleshchev, A. I. Lyamaev, L. A. Mashina, and L. A. Nakhimovskaya *Opt. and Spectrosc.* **36**, 49–51 (1974).
56. J. Rima, L. A. Nakhimovsky, M. Lamotte, and J. Joussot-Dubien, *J. Phys. Chem.* **88**, 4302–08 (1984).
57. Z. S. Ruzevich, *Opt. and Spectrosc.* **15**, 191–93 (1963).
58. Yu. R. Red'kin and V. I. Mikhailenko, *Bull. Acad. Sci. USSR* **34**, 1205–09 (1970).
59. G. V. Gobov and L. A. Nakhimovskaya, *Opt. and Spectrosc.* **24**, 389–91 (1968).
60. L. A. Klimova, G. N. Nersesova, T. M. Naumova, A. I. Ogloblina, and V. I. Glyadkovskii, *Bull. Acad. Sci. USSR* **32**, 1361–65 (1968).

61. L. A. Klimova, A. I. Ogloblina, and E. V. Shpolskii, *Bull. Acad. Sci. USSR* **34**, 1210–14 (1970).
62. J. Rima, M. Lamotte, and A. M. Merle, *Nouveau J. de Chimie* **5**, 605–09 (1981).
63. E. V. Shpolskii and L. A. Klimova, *Opt. and Spectrosc.* **7**, 499–500 (1959).
64. C. Pfister and J. Kahane-Paillous, *J. Chim. Phys.* **65**, 876–82 (1968).
65. M. Lamotte, A. M. Merle, J. Joussot-Dubien, and F. Dupuy, *Chem. Phys. Letters* **35**, 410–16 (1975).
66. E. V. Shpolskii, *Soviet Physics Uspekhi Transl.* **6**, 411–27 (1963).
67. E. G. Moisya, *Opt. and Spectrosc.* **23**, 119–23 (1967).
68. D. M. Grebenschikov, N. A. Kovrizhnykh and S. A. Kozlov, *Opt. and Spektrosc.* **31**, 214–16 (1971).
69. P. Bothorel, C. Such, and C. Clement, *J. Chim. Phys.* **69**, 1453–61 (1972).
70. M. Lamotte and J. Joussot-Dubien, *Chem. Phys. Letters* **2**, 245–48 (1973).
71. M. Lamotte, S. Risemberg, A. M. Merle, and J. Joussot-Dubien, *J. Chem. Phys.* **66**, 875–76 (1977).
72. F. I. Gurov and G. N. Nersesova, *Bull. Acad. Sciences USSR (Engl. transl.)* **11**, 1135–38 (1970).
73. L. A. Nakhimovsky, *Bull. Acad. Sciences USSR, Phys. Ser. (Engl. transl.)* **32**, 1408–12 (1968).
74. C. Amine, L. Nakhimovsky, F. Morgan, and M. Lamotte *J. Phys. Chem.* **94**, 3931–37 (1990).
75. F. F. Litvin and R. I. Personov, *Soviet Phys. Dokl.* **6**, 134–36 (1962); *Fiz. Probl. Spektroskopii, Akad. Nauk. SSSR*, 1960, **1**, 229–30 (1961).
76. G. M. Svishchyov, *Opt. and Spectrosc.* **18**, 350–53 (1965).
77. V. A. Butlar, D. M. Grebenshchikov, and V. V. Solodunov, *Opt. and Spectrosc.* **18**, 606–07 (1965).
78. T. N. Bolotnikova and Yu. I. Glushkov *Opt. and Spectrosc.* **40**, 612–13 (1970).
79. R. I. Personov and O. N. Korotaev, *Dokl. Akad. Nauk. SSSR* **182**, 815–18 (1968).
80. R. I. Personov, V. V. Solodunov, O. N. Korotaev, and E. D. Godyaev, *Izv. Akad. Nauk SSSR, Ser. Fiz.* **34**, 1272–76 (1970).
81. R. I. Personov and L. A. Bykovskaya. *Sov. Phys. Dokl.* **16**, 556–59 (1972).
82. A. M. Merle, *Thèse d'Etat*, Université de Bordeaux I, 156 pp. (1978).
83. A. M. Merle, W. M. Pitts, and M. A. El-Sayed, *Chem. Phys. Letters* **54**, 211–16 (1978).
84. T. N. Bolotnikova, V. A. Zhukov, L. F. Utkina, and V. I. Shaposhnikov, *Opt. and Spectrosc.* **53**, 491–95 (1982).
85. N. N. Malykhina and M. T. Shpak, *Opt. and Spectrosc.* **14**, 442–43 (1963).
86. M. Lamotte and J. Joussot-Dubien, *Chem. Phys.* **61**, 1892–98 (1974).
87. E. Zalesskii, A. N. Sevchenko, V. V. Nizhnikov, and S. M. Gorbachev, *Opt. and Spectrosc.* **46**, 268–71 (1979).

88. T. Vo-Dinh, U. P. Wild, M. Lamotte, and A. M. Merle *Chem. Phys. Letters* **39**, 118–22 (1976).
89. M. Lamotte, S. Risemberg, A. M. Merle, and J. Joussot-Dubien, *J. Chem. Phys.* **69**, 3639–46 (1978).
90. M. Lamotte, A. M. Merle, and S. Risemberg, *J. Luminesc.* **18/19**, 505–07 (1979).
91. S. A. Kozlov and D. M. Grebenshikov, *Opt. and Spectrosc.* **44**, 75–78 (1978).
92. R. I. Personov, V. V. Soludunov, and O. N. Korotaev, *Izv. Akad. Nauk SSSR, Ser. Fiz.* **34**, 1272–76 (1970).
93. D. M. Burland, G. Castro and G. W. Robinson, *J. Chem. Phys.* **52**, 4100–08 (1970).
94. J. Van Egmond and J. H. Van der Waals, *Molec. Phys.* **26**, 1147–67 (1973).
95. K. Ohno, N. Nishi, M. Kinoshita, and H. Inokuchi, *Chem. Phys. Letters* **33**, 293–97 (1975).
96. Z. Ruziewicz, *Acta Physica Polonica* **28**, 389–406 (1965).
97. F. Doerr and H. Gropper, *Ber. Bunsenges. Physik. Chem.* **67**, 193–201 (1963).
98. W. M. Pitts, A. M. Merle, and M. A. El-Sayed, *Chem. Phys.* **36**, 437–46 (1979).
99. A. M. Merle, W. M. Pitts, and M. A. El-Sayed, *J. Luminesc.* **18/19**, 111–14 (1979).
100. M. A. El-Sayed, E. Gossett, and M. Leung, *Chem. Phys. Letters* **21**, 20–27 (1973).
101. J. B. Chodak, Ph. D., University of California, Los Angeles (1974).
102. M. A. El-Sayed, W. R. Moomaw, and J. B. Chodak, *J. Chem. Phys.* **57**, 4061–62 (1972).
103. A. T. Gradyushko, V. A. Mashenkov, and K. N. Solov'ev, *Biofizika* **14**, 827–35 (1969).
104. I. Y. Chan, W. G. van Dorp, T. J. Shaafsma, and J. H. Van der Waals *Molec. Phys.* **22**, 741–51; Ibid. **22**, 753–60 (1971).
105. G. W. Canters, J. van Egmond, and T. J. Schaafsma, *Molec. Phys.* **24**, 1203–15 (1972).
106. G. Jansen and M. Noort, *Spectrochim. Acta* **32A**, 747–53 (1976).
107. G. Jansen, M. Noort, N. van Dijk, and J. H. Van der Waals, *Molec. Phys.* **39**, 865–80 (1980).
108. J. A. Kooter and J. H. Van der Waals, *Molec. Phys.* **37**, 997–1013 (1979).
109. W. T. Simpson, *J. Chem. Phys.* **17**, 1218–21 (1949).
110. G. W. Canters, M. Noort, J. H. Van der Waals, *Chem. Phys. Letters* **30**, 1–4 (1975).
111. W. G. Van Dorp, M. Soma, J. A. Kooter, and J. H. Van der Waals, *Molec. Phys.* **28**, 1551–68 (1974).
112. M. Noort, G. Jansen, G. W. Canters, and Van J. H. der Waals, *Spectrochim. Acta* **32A**, 1371–75 (1974).
113. S. F. Shkirman and S. M. Arabei, *Zh. Prikl. Spektrosk.* **32**, 793–98 (1980).
114. S. M. Arabei, G. D. Egorova, and M. A. Katibnikov, *Teor. Eksp. Khim.* **25**, 306–11 (1989).
115. S. M. Arabei, K. N. Solov'ev, and Yu. I. Tatul'chenkov, *Opt. i Spektrosk.* **73**, 686–93 (1992).
116. S. M. Arabei, K. N. Solov'ev, and A. E. Turkova, *Zh. Prikl. Spektrosk.* **61**, 215–25 (1994).

117. G. W. Canters, G. Jansen, M. Noort, and J. H. Van der Waals, *J. Phys. Chem.* **20**, 2253–59 (1976).
118. J. A. Koehler, *Molec. Cryst. Liq Cryst.* **50**, 93–97 (1979).
119. S. Voelker and R. M. Macfarlane, *Mol. Cryst. Liq. Cryst.* **50**, 213–16. (1979).
120. K. Mauring and E. Avarmaa *Chem. Phys. Letters* **81**, 446–49 (1981).
121. G. Jansen, Thesis, University of Leiden (1977).
122. M. Lamotte and J. Joussot-Dubien, *J. Chim. Physique* **69**, 1687–704 (1972).
123. J. Trotter, *Acta Cryst.* **16**, 605–09 (1963).
124. R. B. Campbell, J. M. Robertson, and J. Trotter, *Acta Cryst.* **15**, 289–90 (1962).
125. D. W. J. Cruickshank, *Acta Cryst.* **10**, 504–08 (1957).
126. D. M. Burns and J. Iball, *Proc. Roy. Soc.* **A2**57, 491–514 (1960).
127. J. M. Robertson and J. G. White, *J. Chem. Soc.* 925–31 (1956).
128. J. Iball, C. H. Morgan, and D. E. Zacharias, *J.C.S. Perkin II*, 1271–72 (1975).
129. V. K. Bel'skii, V. E. Zavodnik, and V. M. Vozzhennikov, *Acta Cryst., Sect. C: Cryst. Struct. Commun.* **C40**, 1210–11 (1984).
130. A. Camerman and J. Trotter, *Acta Cryst.* **18**, 636–43 (1965).
131. D. M. Donalson, J. M. Robertson, and J. G. White, *Proc. Roy. Soc.* **A220**, 311–21 (1953).
132. A. Camerman and J. Trotter, *Proc. Roy. Soc.* **A279**, 129–46 (1964).

CHAPTER

4

LUMINESCENCE SPECTROSCOPY OF LARGE POLYCYCLIC AROMATIC HYDROCARBONS

MAXIMILIAN ZANDER

Castrop-Rauxel, Federal Republic of Germany

4.1. INTRODUCTION

The polycyclic aromatic hydrocarbons (PAHs) are widely distributed in the human environment (air, water, soil, food etc.), and it is well established that some members of this class of compounds exhibit carcinogenic activity.[1] The continuous high interest that the PAHs find in environmental chemistry and cancer research is predominantly due to this fact.

However, the PAHs are also of importance in other fields of science (e.g., theoretical and physical chemistry, cosmology and astrophysics, and in materials science). This is particularly true of what is usually called the *large polycyclic aromatic hydrocarbons*, (LPAHs) which are the subject of this chapter. It deals with a special aspect of this kind of PAHs, namely their luminescence spectroscopic properties in rigid matrices.

4.2. LARGE POLYCYCLIC AROMATIC HYDROCARBONS

4.2.1. Definition and Structures

The entire class of polycyclic aromatic hydrocarbons[2–4] can be divided into noncondensed and condensed systems. In the noncondensed PAHs, benzene rings (or other aryl units) are interlinked by C–C single bonds (biphenyl type), while in the condensed PAHs joined benzene rings have carbon atoms in common (naphthalene type). The class of condensed PAHs can be further subdivided into kata- and pericondensed hydrocarbons. In the katacondensed PAHs the quaternary carbon atoms present are centers of two interlinked rings (e.g., naphthacene), whereas in pericondensed PAHs some of the quaternary carbon atoms are centers of three interlinked rings (e.g., pyrene). Finally, pericondensed PAHs that consist exclusively of six-membered rings of sp^2 carbon atoms are

Shpol'skii Spectroscopy and Other Site-Selection Methods, Edited by Cees Gooijer, Freek Ariese, and Johannes W. Hofstraat.
ISBN 0-471-24508-9

alternant hydrocarbons (e.g., pyrene), while nonalternant pericondensed PAHs contain in addition five-membered rings of sp^2 carbon atoms (e.g., fluoranthene). (Note that PAHs with five-membered rings containing sp^3 carbon atoms, e.g., fluorenes, will according to MO theory not be classified as nonalternant.)

Large polycyclic aromatic hydrocarbons are usually defined as PAHs having seven (e.g., coronene) or more benzene rings. The currently known largest PAHs with unambiguously proven structures are depicted in Scheme 4.1. [14]Helicene (**1**) is a katacondensed PAH consisting of 14 benzene units that are angularly annelated throughout, thus forming a perfect helical structure with three face-to-face arranged decks of benzenoid layers.[5] **2** is the largest pericondensed alternant PAH known.[6] Its molecular formula is $C_{132}H_{42}$. The twelvefold alkylated hydrocarbon **2a** has been found to be sufficiently soluble in tetrachloroethane for obtaining a UV/vis absorption spectrum of the compound (for more about solubility properties of LPAHs, see Section 4.2.4). The largest nonalternant

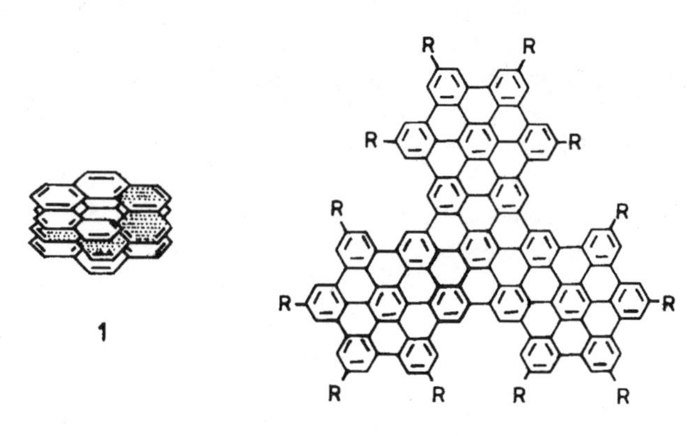

3: R = n-dodecyl

Scheme 4.1

polycyclic aromatic compound that has hitherto been synthesized is **3**.[7] The UV/vis absorption and fluorescence spectrum as well as the fluorescence quantum yield and lifetime of **3** have been measured. Some further selected examples of LPAHs are shown in Scheme 4.2.

PAH **4**[8] is a tetra-alkyl derivative of the 11-ring homologue of the series phenanthrene, chrysene, picene, benzo[*c*]picene. The pericondensed LPAHs **5**[9] and **6**[10] have the same number of six-membered rings (17) but different molecular formulae and hence different molecular weights (**5**: $C_{52}H_{20}$, MW 644; **6**: $C_{54}H_{22}$, MW 670). It is generally true that PAHs having the same number of six-membered rings must not necessarily have the same molecular formulae because the latter depends not only on the number of six-membered rings present but also on their pattern of connection (i.e., molecular topology). For example, all katacondensed four-ring PAHs have molecular formula $C_{18}H_{12}$, but that of the pericondensed four-ring PAH pyrene is $C_{16}H_{10}$. Hydrocarbon **7** is another example of a nonalternant LPAH.[11]

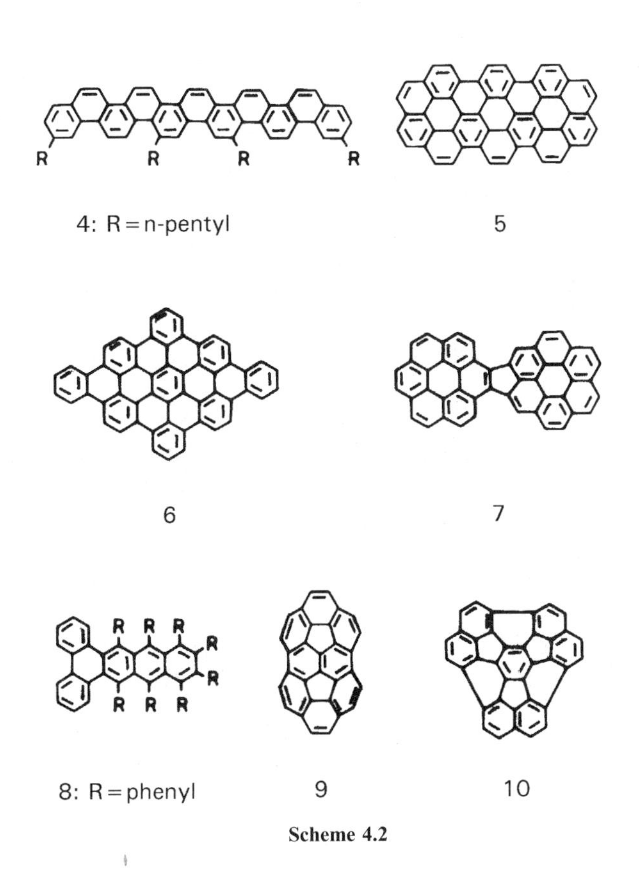

Scheme 4.2

Due to steric hindrance between neighboring (predominantly hydrogen) atoms, many LPAHs have a nonplanar molecular structure. In fact, it follows from X-ray analytical examination and theoretical work[12] that above a certain number of rings the majority of LPAHs exhibit more or less pronounced deviations from an ideal planar molecular structure. For example, only about 12% of the theoretically possible kata- and pericondensed alternant PAHs with 10 rings are assumed to be perfectly planar.[12] A recent particularly striking example of a nonplanar PAH (with 14 benzene rings) is compound **8**[13] (Scheme 4.2). The naphthacene subunit is strongly twisted, resulting in an end-to-end twist angle of the molecule of 105°.

A special class of current interest of nonplanar hydrocarbons are bowl-shaped LPAHs, which are topologically related to the C_{60} fullerene. Examples of bowl-shaped LPAHs are hydrocarbons **9**[14] and **10**[15] (Scheme 4.2). The depth of the bowl in **9** was estimated by ab initio calculation to 2.70 Å. Interestingly the UV/vis absorption spectrum of **10** bears a striking resemblance to that of C_{60}.[15]

4.2.2. Significance of LPAHs

Current interest in LPAHs with regard to practical applications originates from the search for molecules (and devices) suitable for molecular electronics.[16] Although practical solutions have not been achieved, there are some results from basic research that indicate promising tools for future work.

For example, using the technique of raster channel spectroscopy, the electric properties (current–voltage characteristics) of single molecules of the hexaalkyl derivative **11** of hexabenzocoronene (Scheme 4.3) could be measured.[17] The experiments were performed on monolayers of **11** on graphite. The observed current–voltage characteristics that originate from a single molecule being in contact with two electrodes resembles that of an electric diode: that is, the molecule exhibits the behavior of an electrical rectifier.

Hydrocarbons like **11** are discotic mesogens.[18] The corresponding liquid crystal systems exhibit electric conductivity along the column axis: that is, they can, in principle, be used as molecular wires.

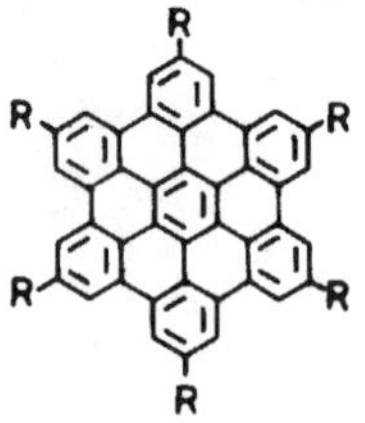

R R R R R R

11: R = n-dodecyl

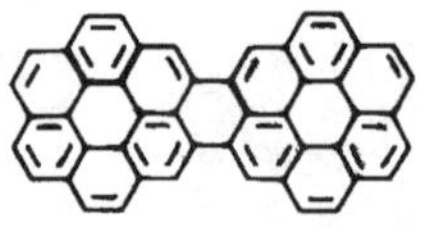

12

Scheme 4.3

It may also be mentioned that dicoronylene **12** (Scheme 4.3) has been proposed as a stationary phase in microcolumn liquid chromatography of PAHs.[19] Similarities between polymeric octadecyl silica phases and dicoronylene have been found in their planarity recognition capability for PAHs. This has been explained with the perfect planar structure of dicoronylene[20] being able to recognize planar solutes in the liquid chromatographic environment.

Studies on LPAHs with regard to carcinogenic or mutagenic properties have obviously not been carried out yet because it is assumed that due to their very low solubility in water and biological fluids, PAHs with more than seven rings will be unable to exhibit any biological activity. However, it is not clear whether this also holds for higher alkylated LPAHs, the solubility of which is normally significantly better than that of the parent hydrocarbons.

4.2.3. Occurrence

LPAHs have been detected in high-temperature coal tar (coken oven tar),[21,22] carbon black,[21] diesel engine particulates,[23] oils and deposits produced during catalytic hydrocracking of processed petroleum,[24,25] hydrothermal petroleums,[26,27] and coal-liquefaction products.[28] Examples of LPAHs found in products of anthropogenic or natural origin are compounds **5** (Scheme 4.2.), **12** (Scheme 4.3) and the hydrocarbons **13–19** shown in Scheme 4.4.

The largest PAH so far detected in coal tar is dicoronylene **12**[12](Scheme 4.3). **12** has also been identified in deposits from petroleum hydrocracker units[24] besides even larger PAHs[25] viz. **5** (Scheme 4.2) and **13** (Scheme 4.4). The LPAHs found in the high-molecular-weight portion of aromatic products are exclusively pericondensed systems, while large katacondensed PAHs have not been observed yet. The obvious preponderance of pericondensed PAHs is possibly due to the formation pathways of LPAHs under thermal or catalytical conditions.

Of particular current interest is the theory that part of the infrared fluorescence emission coming from the Universe, with intense maxima at about 3.3, 6.2, 7.7, 8.6, and 11.3 μm, originates from LPAHs present in interstellar clouds.[29–31] It is assumed that the emission comes from a collection of LPAHs having up to about 50 carbon atoms. Based on stability arguments and by comparing the observed spectra with that of known LPAHs, it has been concluded that large pericondensed hydrocarbons predominate. The interstellar infrared fluorescence emission stems from hot (i.e., vibrationally highly excited) LPAH radical cations. Excitation is assumed to occur predominantly in the UV region, and the mechanism of infrared fluorescence emission of an electronically excited LPAH has been explained in detail.[32]

It appears that approximately 1–10% of interstellar carbon may be in LPAH structures. Bearing in mind the enormous size of interstellar clouds, we are

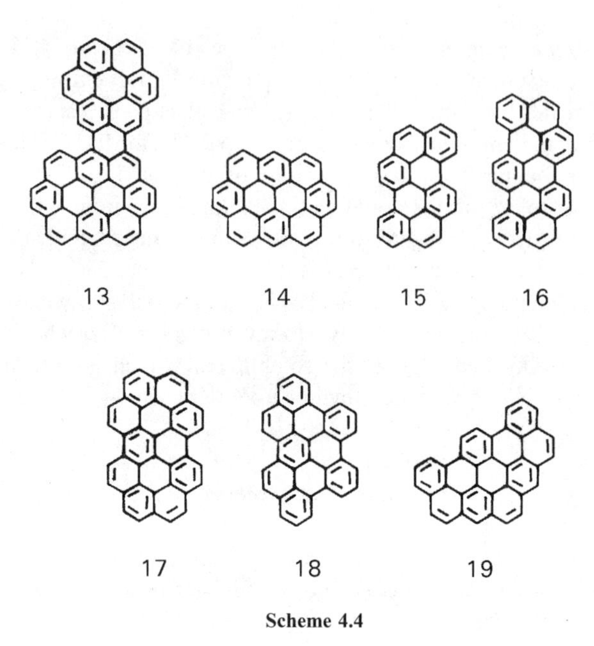

Scheme 4.4

tempted to conclude that LPAHs are among the most frequently occurring organic molecules in the Universe.

4.2.4. Properties

With regard to spectroscopic studies of LPAHs (in fluid or rigid solution), the most relevant property of the compounds is its solubility in organic media.

It is well known that the solubility of (unsubstituted, planar) PAHs depends strongly on their molecular topology. For example, the solubility of naphthacene in dichloromethane (at room temperature) amounts to $0.4\,g\,L^{-1}$, while that of the isomeric benz[*a*]anthracene is $19.5\,g\,L^{-1}$.[33]

Nonplanar PAHs are much more soluble in organic solvents than similar planar, more condensed structures. Thus the nonplanar hydrocarbon **20** ($C_{42}H_{22}$) (Scheme 4.5) is soluble in the very weak solvents methanol or *n*-hexane, while the perfectly planar hexabenzocoronene **21**[20] ($C_{42}H_{18}$) is insoluble in dichloromethane or tetrahydrofuran, and is only very slightly soluble in the extremely strong solvent 1,2,4-trichlorobenzene.[34] Very few quantitative solubility data of LPAHs are available, but dicoronylene **12** (Scheme 4.3) is a representative example; the solubility of **12** in 1,2,4-trichlorobenzene (at room temperature) has been determined to be $9\,\mu g\,L^{-1}$.[35]

Alkyl substituents (e.g., *tert*-butyl or *n*-dodecyl groups) significantly enhance the solubility of LPAHs in organic solvents. For example, the room temperature

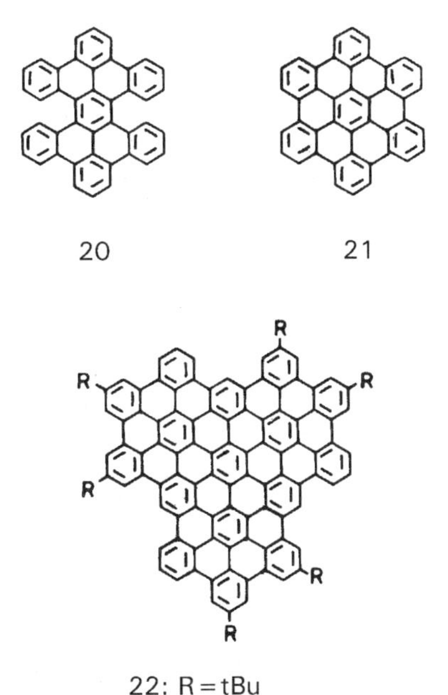

Scheme 4.5

solubility in dichloromethane of the very large aromatic compound **3** with eight *n*-dodecyl groups (Scheme 4.1) is 4.5 g L^{-1}.[7] Hydrocarbon **2** (Scheme 4.1) proved to be completely insoluble in high-boiling organic solvents or ionic liquids (e.g., 1-ethyl-3-methyl-imidazolium tetrachloroaluminate), while the twelvefold *tert*-butylated derivative **2a** (Scheme 4.1) was found to be slightly soluble in solvents like tetrachloroethane or *o*-dichlorobenzene.[6] However, a sufficient number of alkyl substituents is always necessary to improve significantly the solubility of LPAHs. Thus, hydrocarbon **22** (Scheme 4.5), which is structurally very similar to **2a** but has in contrast to **2a** only six *tert*-butyl groups, proved to be completely insoluble in organic solvents.[6]

Even an extensively purified LPAH can contain minor amounts of organic by-products, which in some cases may be significantly more soluble than the main product. Therefore, extracting the product with a spectroscopic solvent and measuring absorption or luminescence spectra of the extract obtained is a very risky procedure because the spectra observed can be considerably falsified by signals belonging to the impurity. A rather laborious but reliable procedure for measuring quantitative room temperature absorption spectra of LPAHs with extremely low solubility has been described.[35]

Another characteristic property of LPAHs (particularly of higher-annelated acenes or pericondensed PAHs with extended acene branches) that has to be taken into account when considering spectroscopic studies, is the ability of the compounds to undergo photooxidation, yielding endocyclic peroxides and/or quinones. When preparing spectroscopic solutions of highly photoreactive LPAHs, the samples must be protected against daylight. Some incorrect results reported in the literature are due to photooxidation of the compounds studied. For example, a phosphorescence spectrum of naphthacene has been reported[36] that was later shown to be in fact the spectrum of naphthacene quinone,[37] which was obviously present in the sample studied in minor amounts.

4.2.5. Chemical Analysis of LPAHs

4.2.5.1. Separation

While *capillary gas chromatography* (GC) is the method of choice for separating mixtures of PAHs with molecular weights up to 300, its applicability to larger PAHs is restricted due to the low vapor pressure of the compounds. Very high oven temperatures are required for elution, introducing temperature stability issues for both the capillary material and the stationary phase employed for the separation.[38] More recently new high-temperature GC columns have been shown to be sufficiently stable for chromatographing selected kata- and pericondensed seven- to nine-ring PAHs with molecular weights of 378–400.[39]

Although *supercritical fluid chromatography* (SFC) may be expected to be more suitable than GC for analyzing LPAHs, until now only few results have been reported.[40] For example, by using a fused silica column coated with *n*-nonyl polysiloxane and supercritical ammonia as mobile phase (temperature: 145°C) ovalene **14** (Scheme 4.4) and some even larger (unidentified) PAHs were separated in a carbon black extract.[41]

The best suited technique for separating LPAH mixtures, however, is *high-performance liquid chromatography* (HPLC). In the first report on the application of HPLC in LPAH analysis the separation of test mixtures consisting of PAHs with 7 to 14 rings (molecular weights: 306–598) was described.[42,43] Since then, many fundamental aspects of the method and a variety of applications have been studied.[34,38,44] In addition to the nature of the stationary phase used, much of the retention behavior observed for LPAHs has its origins in the three-dimensional shape assumed by these molecules.[38] Three distinct retention behaviors can be linked to three classes of molecular shape behavior. The first class of retention behavior includes LPAHs that remain planar in all solvent systems. Molecules belonging to this class are retained, under normal- or reversed-phase conditions, in a manner correlated to the number of π electrons or length-to-breadth ratio of the molecule. The second class are the highly nonplanar LPAHs while the third class of LPAHs are those molecules whose degree of planarity, hence elution behavior, is solvent composition dependent. As the mobile-phase concentration of strong solvent increases, these PAHs become more nonplanar, and when a shape-selective phase is used, their relative retention times decrease proportionally faster than those of planar PAHs.[38,45]

4.2.5.2. *Identification*

In principle, all kinds of molecular spectra (e.g., UV/vis absorption, fluorescence, phosphorescence, or infrared spectra) can be used for identification purposes in LPAH analysis, but UV/vis absorption and fluorescence spectroscopy are the best-suited techniques for on-line coupling with HPLC. First scanning spectrometers were used for this purpose[42] but have now been replaced by the much faster multichannel photodiode array detectors.[34]

There are some special fluorescence phenomena that can be used in PAH analysis (without requiring additional instrumentation) for improving the selectivity of fluorescence analysis. For example, it is well known that the vibrational structure of the room temperature fluorescence spectra of some LPAHs changes with the polarity of the solvent used. In Figure 4.1 the fluorescence spectrum of dibenzo[*h*,rst]pentaphene in the nonpolar solvent *n*-hexadecane (curve a) and in the very polar solvent dimethyl sulfoxide (curve b) are shown.[46] A characteristic

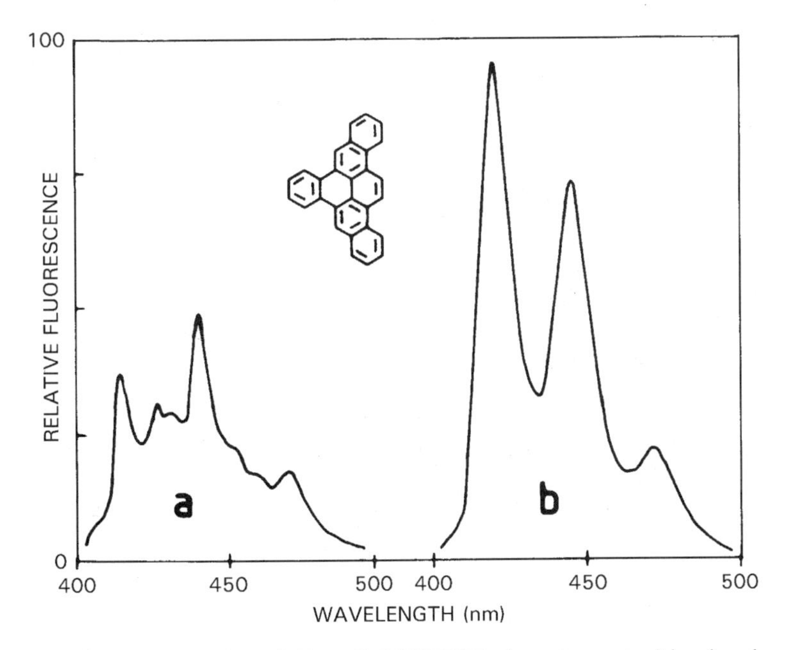

Figure 4.1. Fluorescence spectrum of dibenzo[*h*,rst]pentaphene (room temperature) in *n*-hexadecane (a) and dimethyl sulfoxide (b) (based on Ref. 46).

change of the relative band intensities when passing from the nonpolar to the polar solvent is observed.

Similar solvent-polarity-dependent spectral changes have been observed in 17 pericondensed alternant PAHs with 4–11 benzene rings from a total of 45 PAHs studied.[47] Although it is not yet clear which structural characteristics of a PAH are responsible for the solvent effect, the effect may in some cases be useful for improving the reliability of PAH identification based on fluorescence measurement.

4.2.5.3. Structure Elucidation

The most reliable method for determining the constitution (and geometry, i.e., bond lengths and bond angles as well as deviations from planarity) of an unknown PAH is X-ray analysis. The largest PAH whose crystal structure has hitherto been determined by X-ray analysis is dicoronylene **12** (Scheme 4.3).[20]

However, the rather large amounts of hydrocarbon that are normally needed for preparing suitable single crystals are in many cases not available. Thus methods requiring only milligrams of material for structure elucidation are more important, in particular photoelectron spectroscopy[48] (PES), UV, and infrared spectroscopy.[49,50] Among these methods PES is the most powerful one.[51] For each of the possible structures of the unknown PAH, the energies of the occupied π orbitals are calculated by semiempirical quantum-chemical methods. Because of Koopmans theorem the orbital energies are directly related to the corresponding ionization potentials (IPs). The theoretical results obtained are then compared with the observed PE spectrum of the PAH. The structure for which calculated and observed IPs fit best is the most likely structure of the unknown PAH. This approach of structure determination of PAHs is exceptionally successful because (1) IPs can be measured by PES with high accuracy (the PE signals can usually be located within ± 0.01–0.02 eV); (2) IPs of PAHs can be calculated in very good agreement with the experimental data (the deviations between observed and calculated IPs are normally not larger than ± 0.1 eV); (3) PE spectra, unlike UV spectra, are clear and simple even in the case of nonplanar PAHs (all ionizations have the same transition probability and the energy levels of PAHs spread over a considerable energy range); and (4) a large data set is always available (of the $N/2$ π IPs expected for a hydrocarbon with N carbon atoms, $N/3$ can be observed in the 6–10 eV range, the remaining π IPs being obscured by σ ionization).

An important step in structure determination of PAHs is to clarify whether the compound is an alternant (AH) or nonalternant PAH (NAH). A simple fluorescence quenching experiment provides a reliable means for discriminating between AHs and NAHs.[52,53] The fluorescence of AHs (in liquid solution at room temperature) is readily quenched by electron acceptor molecules, preferentially nitromethane, while that of NAHs is not. The fluorescence quenching mechanism has been identified as electron transfer and the different quenching behavior of AHs and NAHs explained within the framework of the Weller–Marcus theory.[52,53] The so-called "nitromethane quenching rule" has been extensively studied, and only few exceptions have been observed.[54] The fluorescence quenching test has also been used in the HPLC of PAH mixtures.[55–57]

4.3. RELATIONSHIPS BETWEEN STRUCTURE, STATE ENERGY DIAGRAM AND LUMINESCENCE PROPERTIES OF LPAHs

4.3.1. Some Basic Concepts and Observations

In the perimeter free-electron-orbital (PFEO) model of PAHs[58] the (singlet) electronic ground state is denoted as ^1A, while the electronically excited states are L_a, L_b, and B_b. All energy levels except the lowest (^1A) are doubly degenerate, that is, may be either singlet or triplet. In this notation L and B refer to a total ring quantum number, while the suffixes a and b refer to the two alternative positions for the nodes of the electronic wavefunction that either bisects the C–C bonds (a) or pass through the C atoms (b). According to the PFEO model, Clar's[2,59] α, para and β UV/vis absorption bands of PAHs correspond to the following electronic transitions: α, $^1A \rightarrow {}^1L_b$; para, $^1A \rightarrow {}^1L_a$; β, $^1A \rightarrow {}^1B_b$. In terms of an MO energy diagram the ^1L_a state results from the promotion of an electron from the HOMO to the LUMO; that is, the para absorption band corresponds to the HOMO–LUMO transition. (For a discussion of the relationships between the optical and photoelectron spectra of PAHs, see Ref. 48.)

While the ^1B_b state is always higher in energy than the ^1L_a and ^1L_b state (energies relative to the ground state) it depends strongly on the structure of the respective PAH whether the lowest excited singlet state is L_a or L_b.

Linear annelation of benzene rings leads to a continuous significant decrease of the energy of the ^1L_a state, and correspondingly the para band shifts to longer wavelengths, while the effect on the ^1L_b state (α absorption) is much smaller. As a result, the order of ^1L_b and ^1L_a states (α and para absorption) in the series benzene, naphthalene, anthracene, naphthacene, ... inverts: benzene, $^1L_a > {}^1L_b$; naphthalene, $^1L_a > {}^1L_b$; anthracene (naphthacene and higher acenes), $^1L_b > {}^1L_a$.[2,59] On the other hand, the influence of angular annelation of benzene rings on the energies of the ^1L_a state is less pronounced. Representative examples are shown in Scheme 4.6a: It gives the structures of two pairs of isomeric PAHs (where in both cases one is linearly and the other is angularly annelated) and the position of their absorption bands. The structures are written according to Clar's π-sextet model of PAHs,[59] that is, with the maximum number of unshared π sextets, and it can easily be recognized that the different position of bands is (in a

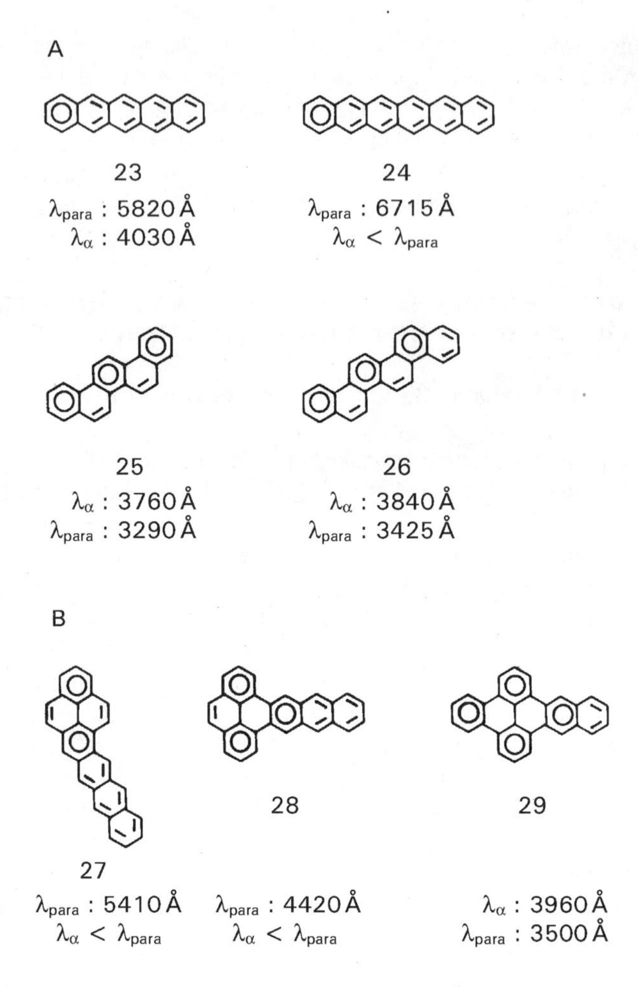

Scheme 4.6

qualitative sense) directly related to the number of Clar sextets. This relation also holds for pericondensed PAHs. Some examples are given in Scheme 4.6b.

A particular class of PAHs are the "all-benzenoid" hydrocarbons,[2,3,59,60] whose Clar structures do not possess double bonds but consist exclusively of hexagons with Clar sextets and "empty" rings; that is, the PAHs can be regarded as condensed polyphenyls. The Clar structures of some all-benzenoid PAHs are depicted in Scheme 4.7. [Further examples of all-benzenoid PAHs are hydrocarbons **2** (Scheme 4.1), **6** (Scheme 4.2), and **20–22** (Scheme 4.5).] Independent of their size and structure, the lowest singlet excited state of the all-benzenoid PAHs is always the ^1L_b state.

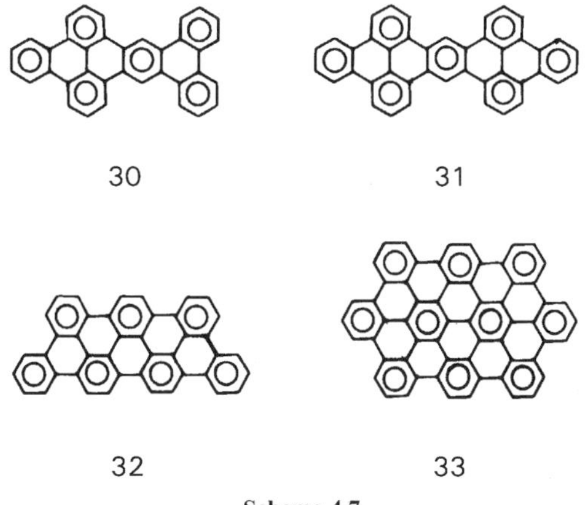

Scheme 4.7

As, according to Vavilov's rule,[61] fluorescence always occurs from the lowest singlet excited state of a PAH, it can be either ^1L_a or ^1L_b fluorescence depending on the term scheme and hence the structure of the hydrocarbon. Since the transition dipole moment of the $^1A \rightarrow {}^1L_b$ transition is smaller than that of the $^1A \rightarrow {}^1L_a$ transition, ^1L_b fluorescence has lower rate constants and hence is less competitive against radiationless intermolecular or bimolecular deactivation processes.

The L_a and L_b electronic configuration of PAHs differ markedly with regard to the size of the singlet–triplet splitting. The S–T split of the L_b configuration is about 2000 cm^{-1},[62] while that of the L_a configuration is about 10,000 cm^{-1} [63,64] (for PAHs with more than three benzene rings). Since the virtually structure-independent S–T split of the L_a configuration is much larger than that of the L_b configuration, the lowest triplet state of PAHs (which at the same time is always the lowest electronically excited state of the hydrocarbons) has the configuration ^3L_a.[63,64] No exceptions to this rule have been found yet. According to Kasha's rule,[61] the phosphorescence of PAHs always occurs from this lowest triplet state.

The most important radiationless deactivation processes of an electronically excited PAH, which compete with the radiative processes fluorescence and phosphorescence, are internal conversion (IC) and intersystem crossing (ISC). According to Siebrand's theory,[65] the rate constants k_{nr} of intramolecular radiationless transitions in electronically excited PAHs are given by

$$k_{nr} = a \exp(-\Delta E/b) \quad [\sec^{-1}] \tag{4.1}$$

where ΔE is the energy gap between the zero-point vibrational levels of the states undergoing the radiationless transition and a and b are constants for a class of structurally related compounds (e.g., PAHs). b is related to the frequency of the dominant nuclear vibration in the final state, while a is an electronic factor, which is characteristic for the respective electronic properties of the states involved; it is markedly different for IC or ISC processes and depends strongly on the extent to which spin-orbit coupling prevails in the molecule (for more about the theory of radiationless transitions, see, e.g., Refs. 61, 66, and 67).

Siebrand's energy gap rule is a powerful concept for describing (and predicting) the competition between radiationless and radiative deactivation processes in an electronically excited PAH molecule; it is this competition that predominantly determines the observables luminescence quantum yield and lifetime.

4.3.2. Fluorescence and Phosphorescence

4.3.2.1. Transition Energy and Spectral Structure

The fluorescence excitation spectrum of a PAH is normally identical with its UV/vis absorption spectrum. However, the nonplanar nine-ring PAH **34** (Scheme 4.8) has been found to exhibit quite unusual behavior inasmuch as its UV/vis absorption and fluorescence excitation spectra (in fluid solution at room temperature) are very different.[68,69] Based on extensive experimental and theoretical study, this has been explained by the presence of two ground-state conformers of **34**, which differ with regard to their UV/vis spectra and their fluorescence ability. While one of the conformers is highly fluorescent, the other is either nonfluorescent or only weakly fluorescent. The conformers were computed to have nearly identical thermodynamic stability, and it is reasonable that both should be present in a room-temperature solution. Since many LPAHs have a more or less nonplanar structure (see Section 4.2.1), the fluorescence behavior observed with **34** (and **35**[70]) may be quite common among LPAHs and deserves further attention.

The fluorescence spectra of planar LPAHs (in fluid or solid solution) exhibit vibrational structure (mirror-symmetric to that of the first absorption transition), and the Stokes shift is rather small (approx. 200–500 cm^{-1}). Since the solubility of some LPAHs is insufficient in common spectroscopic solvents for measuring fluorescence spectra (see Section 4.2.4), different approaches have been proposed to overcome this difficulty. These include measurement at elevated temperatures,

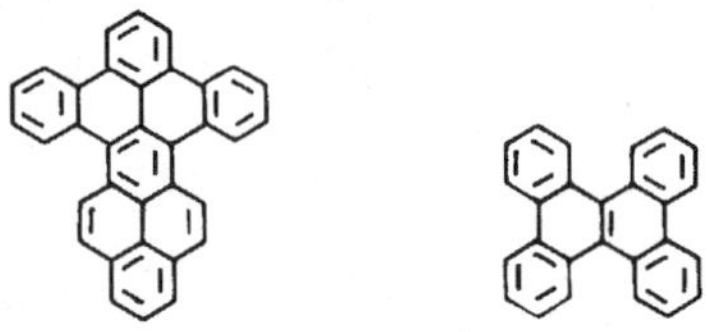

34 35

Scheme 4.8

the use of oversaturated solutions, and new solvents (e.g., 2,6-di-*tert*-butyl pyridine[71] or 1-ethyl-3-methyl-imidazolium tetrachloroaluminate[6,71,72]). A very efficient approach is to study the fluorescence properties of higher alkylated LPAHs instead of the parent hydrocarbons (see Section 4.2.4.). The introduction of alkyl groups significantly improves the solubility of the compounds without changing their electronic (spectroscopic) properties. For example, the UV/vis absorption and fluorescence spectra of the tetrafold *tert*-butylated rylenes **36–39** with 5 to 14 benzene rings could be measured in 1,4-dioxane at room temperature.[73] The spectra are shown in Figure 4.2. With increasing molecular

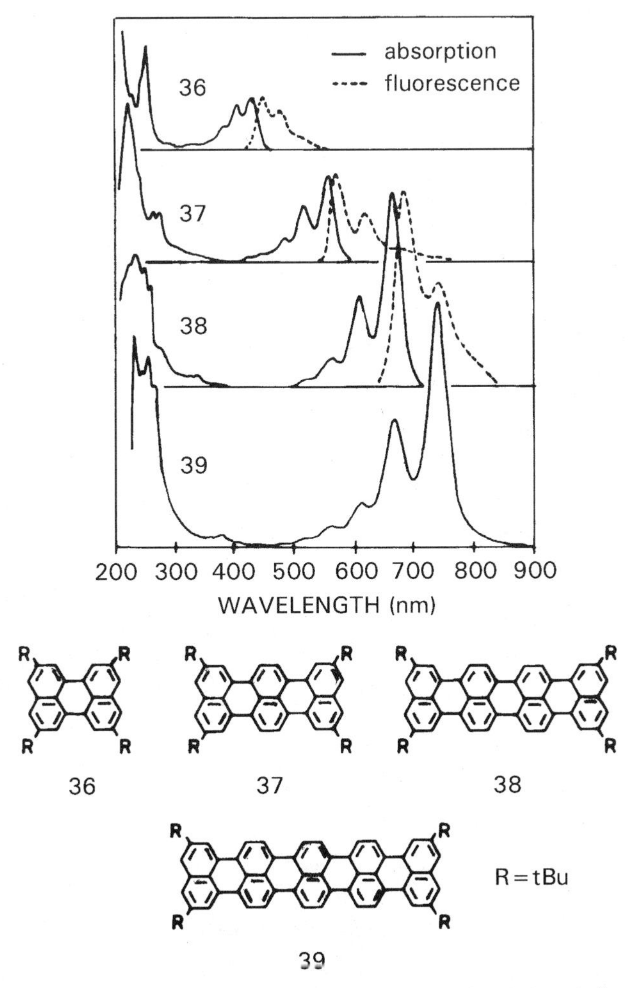

Figure 4.2. Fluorescence spectra of the rylenes **36–39** (R = *tert*-butyl) in 1,4-dioxane at room temperature (replotted with permission from Ref. 73).

size the fluorescence quantum yield decreases dramatically. **39** has been found to be nonfluorescent. It is at present not clear whether an efficient IC or ISC is responsible for the lack of fluorescence in **39**.

Interesting fluorescence behavior is shown by the nonalternant LPAH **40** (Fig. 4.3).[74] In Figure 4.3. the UV/vis absorption and fluorescence spectra (in benzene) at room temperature and the fluorescence spectrum (in toluene) at 80 K are given. It can be easily seen that the short-wavelength shoulder of the room temperature fluorescence spectrum disappears on cooling and the vibrational structure becomes much better resolved. (It should also be mentioned that neither a temperature effect on the absorption spectrum nor a solvent effect on the fluorescence spectrum has been observed.) Based on additional experimental and theoretical results, the room temperature fluorescence of **40** has been identified as a "dual fluorescence" originating from the two lowest-lying singlet excited states (^1L_b and ^1L_a) which in this particular case are separated by approximately 700 cm^{-1} only and hence are thermally equilibrated.

The phenomenon of twisted intramolecular charge transfer (TICT) fluorescence, which was first discovered during studies of the fluorescence behavior of 4-cyano-N,N-dimethylaniline,[75–77] has also been found to be a characteristic fluorescence property of certain unsubstituted noncondensed LPAHs. The

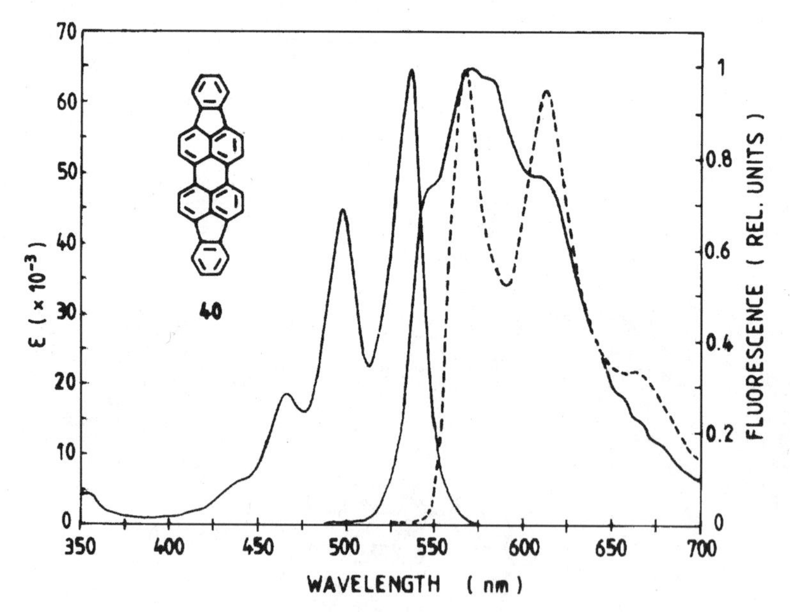

Figure 4.3. UV/vis absorption and fluorescence spectra of periflanthene **40** (for details see text) (replotted with permission from Ref. 74).

TICT fluorescence of 5,5′-bi-benzo[*a*]pyrenyl,[78] 8,8′-bi-naphtho[1,2,3,4-def] chrysenyl,[79] and 3,3′-biperylenyl[80] has been studied in detail, and the conditions that a large biaryl molecule must fulfill (regarding energy of the lowest singlet excited state, ionization potential, electron affinity, and Coulomb stabilization energy) to be capable of emitting TICT fluorescence have been derived.[78]

Phosphorescence of PAHs has predominantly been studied in solid solutions at low temperatures. However, while using a rigid medium in these experiments in order to suppress bimolecular quenching processes of the long-lived triplet state of the PAHs is essential, application of low temperature is not. In fact, under certain conditions phosphorescence of PAHs in solid solutions has been observed at temperatures up to 160°C,[81] and there is no indication that this temperature marks an upper limit. On the other hand, increasing the temperature always leads to band broadening and loss of spectral structure.

The low-temperature phosphorescence spectra of a variety of PAHs including LPAHs have been measured.[61,63] In most cases the spectra are well structured, and both the vibrational structure and the relative intensity distribution of bands clearly indicate that the phosphorescence transitions are symmetry allowed. The phosphorescence of few hydrocarbons of high symmetry, however, such as coronene (D_{6h}) is both spin- and symmetry forbidden, as follows from the very low intensity of the 0–0 band and the vibrational structure. For these hydrocarbons it is characteristic that the intensity of the 0–0 band strongly increases when heavy-atom perturbers (e.g., methyl iodide) are present in the matrix.[82,83] It has been shown that this effect is purely electronic in nature and due to a second-order mixing of perturber singlet-state character into the triplet state of the phosphorescent molecule.[84,85]

The approximate position of the phosphorescence 0–0 band of a PAH can be predicted by simply subtracting 10,000 cm^{-1} (the value of the S–T splitting of the L_a configuration; see Section 4.3.1) from the energy of the para band observed in the UV/vis absorption spectrum.[63,64]

4.3.2.2. *Quantum Yield and Luminescence Lifetime*

Deactivation of a PAH being in its lowest singlet excited state (S_1) occurs by fluorescence, intersystem crossing into the triplet manifold (T_n), and internal conversion to the electronic ground state (S_0). Under conditions where bimolecular quenching processes as well as photochemical reactions do not take place, the quantum yields of these three deactivation pathways add up to 1.

The quantum yield (Y_f) of fluorescence depends on the rate constants (sec^{-1}) of the different competing deactivation processes. We shall adopt here the rate-parameter notation introduced by Birks,[61] that is, k_{FM} for fluorescence, k_{TM} for S_1–T_n ISC, and k_{GM} for S_1–S_0 IC. Y_f is then given by:

$$Y_f = \frac{k_{FM}}{k_{FM} + k_{TM} + k_{GM}} \tag{4.2}$$

k_{GM} and correspondingly the quantum yield of S_1–S_0 IC is small if $\Delta E(S_1–S_0)$ is larger than approximately 20,000 cm^{-1}, however, in agreement with Siebrand's energy gap rule (see Section 4.3.1) increases with decreasing S_1 energy. For example, the quantum yield of the S_1–S_0 IC of pentacene [$\Delta E(S_1–S_0) = 17,300\,\text{cm}^{-1}$] is about 0.75.[67] Depending on the respective structure and hence state energy diagram, S_1–S_0 IC can become the predominant deactivation pathway of electronically excited LPAHs (an example may be hydrocarbon **39**, Section 4.3.2.1.).

Radiationless deactivation of the S_1 state of PAHs by ISC normally occurs into the nearest lower triplet state T_n. T_n can be the lowest triplet state T_1 or a higher triplet state T_n ($n > 1$). In agreement with Siebrand's energy gap rule, the rate constants k_{TM} decrease exponentially with increasing energy difference of the intercombining states.[62] The size of the S_1–T_n energy gap depends on the state energy diagram of the respective PAH. Three prototypes can be clearly distinguished that are schematically depicted in Figure 4.4.[86]

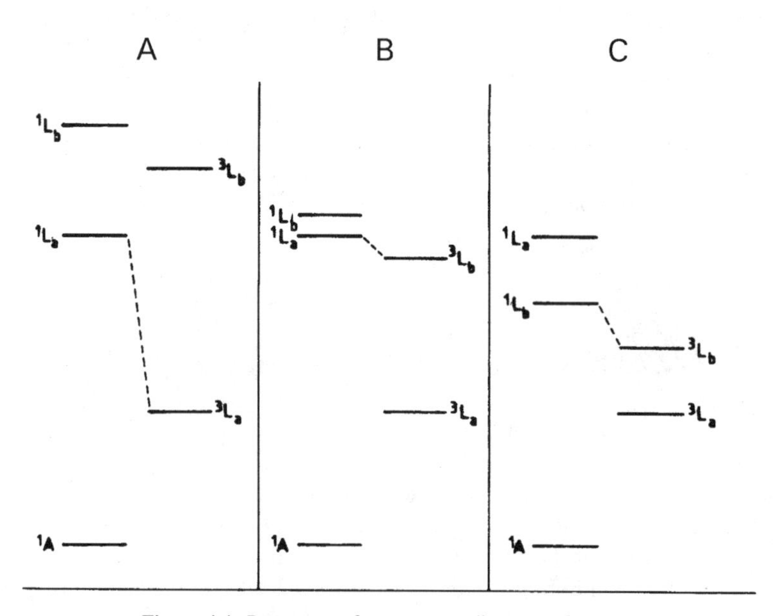

Figure 4.4. Prototypes of state energy diagrams of PAHs.

The three prototypes of state energy diagrams are different with regard to the order of the two lowest singlet and triplet excited states, respectively. In case A both the ^1L_b and ^3L_b state lies above the ^1L_a state. Thus there is a large energy gap between the fluorescent S_1 state and the accepting triplet state T_1 which amounts approximately to 10,000 cm^{-1} corresponding to the S–T split of the L_a configuration (see Section 4.3.1). PAHs to which this type of state energy diagram applies are characterized by very small ISC rates and hence fluorescence yields close to unity. A representative example of class A hydrocarbons is perylene. In some class A hydrocarbons the ^3L_b–^1L_a energy difference is very small so that thermally activated ISC can occur. In those cases the fluorescence yield strongly depends on temperature. It comes close to unity at low temperature (77 K) and decreases significantly with increasing temperature. A well-known example is 9,10-biphenyl-anthracene.[87,88]

Case B and case C hydrocarbons have in common that the energy difference between the fluorescent S_1 state and the accepting triplet state T_n is comparatively small so that efficient ISC can occur. However, case B hydrocarbons emit ^1L_a fluorescence (having a large fluorescence rate constant k_{FM}), while in case C hydrocarbons it is ^1L_b fluorescence (with a much smaller rate constant). Therefore, the fluorescence yield (Eq. 4.2) of case B hydrocarbons is in most cases larger than that of case C hydrocarbons even if the ISC rates are comparable.

Besides the energy difference between the intercombining states, there is another factor that affects significantly the efficiency of ISC in PAHs, viz. deviations from molecular planarity.[89] In most cases the ISC rate constants increase with decreasing molecular planarity and this is very likely due to an increase of spin-orbit coupling.[64]

The quantum yield (Y_p) of phosphorescence of a PAH is given by:

$$Y_p = \frac{k_{TM}}{k_{TM} + k_{FM} + k_{GM}} \cdot \frac{k_{PT}}{k_{PT} + k_{GT}} \tag{4.3}$$

k_{PT} is the rate constant of phosphorescence and k_{GT} the rate constant of the T_1–S_0 ISC. The first term in Eq. 4.3 is the yield of triplet formation, while the second term that regards the competition between the radiative and radiationless deactivation of the triplet state is termed *phosphorescence quantum efficiency.* While k_{PT} is assumed to be relatively constant in (planar, unsubstituted) PAHs (ca. 0.03 sec^{-1}),[90] k_{GT} increases strongly with decreasing $\Delta E(T_1$–$S_0)$. Even in cases where the triplet yield is large, the phosphorescence quantum yield can become vanishingly small due to the strong influence of the T_1–S_0 ISC. An extreme example is provided by the fullerenes C_{60} and C_{70} which according to our present knowledge have to be regarded as large three-dimensional polycyclic aromatic systems;[4,91] although the triplet yields of C_{60} and C_{70} are close to unity, the fullerenes show only very weak phosphorescence due to the extremely small $T_1 - S_0$ energy gaps.[92,93]

Because of the strong effect of the T_1 energy-dependent radiationless T_1–S_0 ISC on the phosphorescence quantum yield, the application of phosphorescence to the analysis of PAHs is restricted to hydrocarbons with T_1 energies larger than approximately 14,000 cm^{-1}. This applies to both low-temperature and room temperature phosphorimetry.[94,95]

The fluorescence and phosphorescence lifetimes,[61] τ_f and τ_p, are related to rate constants according to:

$$\tau_f = (k_{FM} + k_{TM} + k_{GM})^{-1} = Y_f k_{FM}^{-1} \tag{4.4}$$

$$\tau_p = (k_{PT} + k_{GT})^{-1} \tag{4.5}$$

Clearly, the relationships existing between the energy state diagram and the luminescence quantum yields discussed above also apply to the luminescence lifetimes.

Among the LPAHs the all-benzenoid hydrocarbons are best suited for studying phosphorescence properties. This is because the lowest singlet excited state of the all-benzenoid PAHs is always the ^1L_b state (favoring efficient S_1–T_n ISC), and the shift per benzene ring of the phosphorescence transition to longer wavelengths is smaller than for all other PAH classes known. The phosphorescence transitions of the all-benzenoid PAHs so far studied with 7 to 13 rings lie between 20,550 and 17,540 cm^{-1}.[63] The spectra are well structured, the Y_p/Y_f ratios are rather high, and the phosphorescence lifetimes lie between 6 and 8 sec. The largest PAH whose phosphorescence properties have been studied in some detail is the all-benzenoid hexabenzocoronene **21** (Scheme 4.5). **21** has a phosphorescence lifetime (measured in perhydrocoronene at 77 K; see Section 4.4.2.4.) of 6.4 sec. Accordingly the rate constant of the radiationless T_1–S_0 transition very likely is about 0.1 sec^{-1}. This is an unusually low k_{GT} value for a T_1–S_0 transition lying at 17,700 cm^{-1}.[96]

4.3.3. Delayed Fluorescence

P-type delayed fluorescence has been observed in many PAHs under various experimental conditions[61,97] (see Section 4.4.2.2.).

On the other hand, only few PAHs have been studied with regard to the phenomenon of *E*-type delayed fluorescence,[61,97] viz. coronene,[96,98–100] bicoronyl,[101] circum-biphenyl **17** (Scheme 4.4),[102,103] and hexabenzocoronene **21** (Scheme 4.5)[96] (see also Sections 4.4.2.1. and 4.4.2.4.). In all these cases the energy difference between the lowest (phosphorescent) triplet state and the lowest (fluorescent) singlet state is $\leq$4000 cm^{-1}. However, it cannot be excluded that more LPAHs may be found (particularly in the class of the all-benzenoid hydrocarbons) which under suitable conditions show *E*-type delayed fluorescence.

4.4. TECHNIQUES FOR MEASURING LUMINESCENCE SPECTROSCOPIC PROPERTIES OF LPAHs IN RIGID MATRICES

4.4.1. Introductory Remarks

The bandwidths observed in the absorption and emission spectra of organic molecules (e.g., PAHs) embedded in rigid matrices are much larger than the natural linewidths of the electronic transitions of isolated molecules. This is due to interactions between the solute and solvent molecules.[104,105] Two mechanisms of band broadening can be distinguished. In amorphous matrices, characterized by their lack of long-range order, the local conditions (microenvironments) that influence the transition energies of the solute molecules (predominantly through electron–electron interactions) differ from one solute molecule to another. The observed broad bands then are the result of the superposition of many transition lines having slightly different energies. This type of band broadening is called *inhomogeneous broadening*.

There are two strategies for suppressing inhomogeneous broadening: (1) extensive reduction of the number of different microenvironments ("sites") by using crystalline matrices, where only one or few different cage configurations are available for the solute molecules ("Shpol'skii matrices"[104]), or (2) selective excitation of single sites (or a narrow subset of sites) by using a tunable laser as the excitation source. The latter method, called (fluorescence or phosphorescence) "line-narrowing spectroscopy" (LNS) (see Chapter 8), can be applied to both amorphous and crystalline solute–solvent systems, and therefore its application is much less restricted by the type of rigid solvent used than Shpol'skii spectroscopy. Because of the severe solubility problems existing with LPAHs, LNS is expected to be more widely applicable for studying luminescence properties of LPAHs than Shpol'skii spectroscopy, although practical experience is still lacking.

The other type of band broadening, called *homogeneous broadening*, arises from the interaction between the electrons of the solute and the solvent lattice vibrations ("phonons") of the rigid solvent.[104,105] The shape of vibronic bands is determined by the electron–phonon coupling. At low temperature, every vibronic band consists of a narrow zero-phonon line (ZPL) and a relatively broad phonon wing (PW), lying at the low-energy side of the zero-phonon line in emission and its high-energy side in absorption. The ratio of zero-phonon line and phonon wing intensities can be characterized by the Debye–Waller factor

$$\alpha(T) = \frac{I_{ZPL}}{I_{ZPL} + I_{PW}} = \exp[-s(T)] \qquad (4.6)$$

As the Debye–Waller factor decreases exponentially with increasing temperature, narrow bands can only be observed at low temperatures. Therefore, the application of low temperatures is always necessary in order to obtain well-structured spectra consisting of narrow bands.

Even extremely large PAHs, which are almost insoluble in organic solvents (particularly in *n*-alkanes being preferentially used in Shpol'skii spectroscopy) can be vaporized at high temperature and low vacuum and thus may, in principle, be accessible to matrix isolation luminescence spectroscopy. In matrix isolation spectroscopy, the sample is vaporized and the resulting vapor mixed with a large excess of a diluent gas (e.g., nitrogen, argon, or an organic "matrix gas") and the gaseous mixture is then deposited as a solid on a surface at very low temperature (usually 20 K or less).[106,107] Highly resolved emission spectra are obtained; the linewidths can be in the order of 2 cm^{-1} at 4 K. Although matrix isolation luminescence spectroscopy has been widely used in the analysis of low- to medium-molecular-weight PAHs,[108,109] its applicability to LPAHs has obviously not been tested yet.

In the following section the known techniques for measuring luminescence spectroscopic properties of LPAHs are dealt with using the different types of matrix as a guide line.

4.4.2. Matrices

4.4.2.1. Amorphous Matrices

Some liquid branched hydrocarbons, alcohols, ethers, etc., and mixtures of these solvents exhibit large viscosity increases with decreasing temperature, thus preventing crystallization, and finally set to *rigid glasses* at low temperature. Several solvents are available that form clear glasses at 77 K and can be employed in low-temperature luminescence spectroscopy. Extensive lists of suitable nonpolar or polar solvents, respectively, have been published.[110,111] However, for studying luminescence properties of PAHs (with medium to higher molecular weight) in rigid glasses at 77 K, ethanol is (according to the author's experience) the most suitable solvent with regard to solvent power and sample preparation.

In some cases compounds having only modest solubility in organic solvents tend to form microcrystals when the room temperature solution is cooled to 77 K. Formation of microcrystals, often undetectable by visual inspection, can seriously obscure "solution" luminescence properties. Several observations due to microcrystal formation have been reported in the literature.[112,113]

In a special class of solvent mixtures that form rigid glasses at low temperature, one component contains an element having a high atomic number. Due to the external heavy-atom effect (EHAE),[61,63,64] these "heavy-atom solvent systems" affect significantly the luminescence characteristics of the solutes (e.g., PAHs). In the presence of an external heavy-atom perturber the rate constants of the spin-forbidden processes occurring in an electronically excited PAH (i.e., ISC and phosphorescence) increase dramatically in most cases. This is due to a heavy-atom-induced increase of spin–orbit coupling. The EHAE technique was originally developed to improve the sensitivity and selectivity of low-temperature phosphorimetry[114,115] but later proved almost indispensible in room-temperature phosphorimetry.[116] The following characteristics of the EHAE have to be distinguished: (1) With few exceptions the EHAE causes an increase of the triplet yield (first term in Eq. 4.3) and a decrease of the fluorescence quantum yield and lifetime (Eqs. 4.2 and 4.4). This is because the preexponential factor a in Eq. 4.1 becomes larger for the S_1–T_n ISC, while, of course, the size of the S_1–T_n energy gap is not affected. If this energy gap is very large (case A hydrocarbons: see Section 4.3.2.2.), the quenching effect of the external heavy-atom perturber is almost negligible[86]). (2) In the presence of an external heavy-atom perturber both k_{PT} and k_{GT} (Eq. 4.3) become larger, and depending on whether the EHAE affects more strongly the radiative or radiationless deactivation, respectively, of the T_1 state, the phosphorescence quantum efficiency decreases or increases. For most PAHs the latter case applies. (3) On the other hand, however, the EHAE always causes a reduction of the phosphorescence lifetime (Eq. 4.5). (4) In some cases an external heavy-atom-induced change of the vibrational structure of the phosphorescence spectrum is observed (see Section 4.3.2.1.).

Various heavy-atom solvent systems have been examined for possible use in the luminescence spectroscopy of PAHs (and other types of compounds) in rigid glasses at low temperature.[117] The ideal solvent combines a strong EHAE with low absorption in the UV. According to our present knowledge a 1 : 1 mixture (vol/vol) of ethyl bromide and ethanol seems to provide the best compromise.[118]

Solutions of heavy-atom-containing inorganic salts in an organic solvent, preferentially ethanol, have also been studied with regard to the EHAE on the luminescence characteristics of PAHs, (e.g., coronene[83]) at 77 K. Two types of external heavy-atom perturbers can be distinguished: Type-A perturbers (e.g., silver perchlorate) form phosphorescent ground-state complexes with the PAH, whereas associative forces between type-B perturbers and the PAH are negligible.[119] Type-A and type-B perturbers can be best discriminated by their different influence on the emission decay behavior of a phosphorescent compound: In the presence of a type-B perturber the decay curve is multiexponential[120] even at very high perturber concentrations, while at type-A perturber concentrations sufficiently large for complete complexation it is monoexponential with the time constant of the phosphorescent complex.[82]

Also *glassy plastics* are used as matrices in the luminescence spectroscopy of PAHs, particularly for studying photophysical properties at and above room temperature. Various types of plastics such as poly(methyl methacrylate), polystyrene, polyvinyl acetate, and others and different methods to introduce the solute have been described.[121] For example, *E*-type delayed fluorescence (see Section 4.3.3.) of some LPAHs has been examined in plastics [coronene in poly(methyl methacrylate),[99] bicoronyl,[101] and circum-biphenyl **17**[103] (Scheme 4.4) in an epoxide resin]. The activation energy of the thermally activated $T_1 \rightarrow S_1$ ISC can be derived from Arrhenius plots of the intensity ratio of delayed fluorescence and (simultaneously occurring) phosphorescence[96] or from Arrhenius plots of the temperature-dependent rate constant k_{MT} of the $T_1 \rightarrow S_1$ ISC[101]:

$$k_{\mathrm{MT}} = k'_{\mathrm{MT}} \exp[-\Delta E(S_1 - T_1)/kT] = (1/\tau_T^E - 1/\tau_{T(77\,\mathrm{K})}) \qquad (4.7)$$

where k'_{MT} is the rate constant of $T_1 \rightarrow S_1$ ISC between isoenergetic vibronic states, τ_T^E the (observed) lifetime of *E*-type delayed fluorescence (and phosphorescence) at temperature T and $\tau_{T(77\,\mathrm{K})}$ the phosphorescence lifetime at 77 K. The results obtained confirmed the assumption that k_{PT} and k_{GT} (see Eq. 4.5) are to a good approximation temperature independent. The activation energies of *E*-type delayed fluorescence thus obtained agree well with the S_1–T_1 energy intervals derived from the optical spectra.

4.4.2.2. Crystalline Non-Shpol'skii Matrices

Benzophenone is an interesting matrix molecule for studying luminescence properties of PAHs including LPAHs.[102,122] When cooled slowly, liquid benzophenone (M.P. 48°C) sets to a crystalline mass, but rapid refrigeration to 77 K produces a clear, glassy, supercooled melt. The (n,π^*) phosphorescence that benzophenone shows under both conditions is quenched by the presence of guest molecules that have triplet states lower than that of benzophenone, and the sensitized phosphorescence of the guest molecules appears.[123,124] The phosphorescence spectra and lifetimes of PAHs dissolved in benzophenone are (at 77 K) very similar to those measured in rigid glasses (e.g., ethanol). In cases where the energy difference between the lowest triplet state of the benzophenone (host) and the PAH (guest) is relatively small [e.g., naphthalene, $\Delta E(T_1(\text{host})-T_1(\text{guest})) = 3150\,\mathrm{cm}^{-1}$] thermally activated intermolecular triplet–triplet energy transfer into the host takes place, and its phosphorescence is observed at room temperature.[125]

In ternary systems consisting of benzophenone (host), naphthalene (guest I), and a PAH (guest II) whose T_1 and S_1 state lie below the T_1 state of benzophenone, the intense delayed fluorescence of guest II is observed at room temperature. Using time-resolved spectroscopy, the emission has been identified as *P*-type delayed fluorescence and comes about by host–guest II triplet–triplet annihilation, while guest I (naphthalene) acts as trap for the triplet excitation of the host and further increases the delay of the guest II fluorescence emission.[122]

1,2,4,-Trichloro benzene is a strong solvent for PAHs. It is liquid at room temperature and sets to a crystalline mass at 77 K. Trichlorobenzene has been used for measuring fluorescence and phosphorescence spectra of LPAHs, such as hexabenzocoronene **21** (Scheme 4.5) and dicoronylene **12** (Scheme 4.3).[63,126,127] In cases where the phosphorescence transition is not only spin but also symmetry forbidden, and accordingly the intensity of the phosphorescence 0–0 band is very low, a dramatic change in the relative intensity distribution and vibrational structure of the spectra is observed in trichlorobenzene. The 0–0 band becomes the most intense band, and the vibrational structure is significantly simplified. This is due to the external heavy-atom effect of the chlorine substituents (see Section 4.3.2.1.). There is some indication that the "vibronic effect" is the most sensitive probe known for detecting an external heavy-atom effect.[83]

Bi- and polycyclic aromatic hydrocarbons are, not unexpectedly, the most powerful solvents for LPAHs. Luminescence characteristics of LPAHs at room and low temperature have been examined in, for example, 1-methyl-naphthalene (coronene and bicoronyl[101]), phenanthrene (coronene[128]) and *p*-terphenyl [pentacene[129] and terrylene **37** (Fig. 4.2, R = H)[130]].

p-Terphenyl, for example, has been used as a crystalline matrix in single-molecule luminescence spectroscopy of pentacene and terrylene at 1.6 K,[129,130] while phenanthrene served as matrix for studying temperature-activated *P*-type delayed fluorescence of PAHs.[128,131,132]

Methyl derivatives of PAHs, because of their lower melting points, should be even more suitable matrix molecules for examining luminescence characteristics of very large PAHs, and narrow-line highly resolved luminescence spectra should be obtainable using site-selective laser excitation (line-narrowing spectroscopy; see Section 4.4.1.).

4.4.2.3. Shpol'skii Matrices

Lamp- or laser-excited Shpol'skii spectroscopy[104] (including absorption, fluorescence, phosphorescence, and luminescence excitation) in suitable *n*-alkanes or cycloalkanes at low temperature (63 K or less) belongs to the most powerful and widely applied techniques in PAH analysis. In recent years much progress with regard to methodology and instrumentation of Shpol'skii spectroscopy has been achieved, and many examples of the application of the method for both qualitative and quantitative purposes in environmental studies have been reported.[133,134]

As the quantum yield of fluorescence (at a given temperature) is rather solvent independent, the signal heights in Shpol'skii spectra are, because of the line-narrowing effect, much larger (by approximately two orders of magnitude) than that of the broadband spectra observed in amorphous matrices. Hence Shpol'skii spectroscopy is both a highly selective and sensitive method. The minimum solubility of PAHs required for spectra in n-hexane was calculated to be in the range of 0.1–10 $\mu g\,L^{-1}$.[135] This limit is well exceeded by many PAHs, particularly those having a nonplanar structure.

Well-resolved Shpol'skii fluorescence spectra (in n-hexane at 63 K, monochromatic lamp excitation) of a series of katacondensed PAHs with seven rings have been reported.[135] An example is given in Figure 4.5.

It is interesting to note that well-resolved Shpol'skii spectra were obtained even though the dimensions of the solvent molecule were much smaller than those of

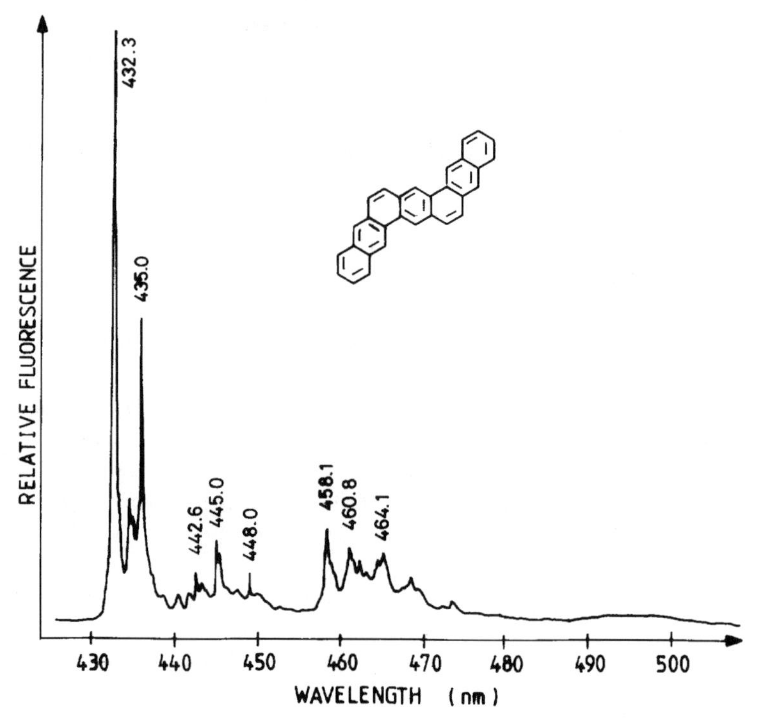

Figure 4.5. Fluorescence spectrum of naphtho[2,3-*c*]pentaphene in n-hexane at 63 K (replotted with permission from Ref. 135).

the solutes. Compounds with moderately resolved spectra, when dissolved in an *n*-alkane with a longer chain (*n*-undecane) gave in most cases the same or lower resolution of the spectrum. This observation, and many others, indicates that the Shpol'skii effect is by no means a mechanistically "simple" but a rather complex phenomenon, the theory of which has not been clarified in every detail.[104]

Also the [4]-, [5]-, [6]-, and [7]helicenes have been studied under Shpol'skii conditions at 4.2 K (fluorescence, phosphorescence, and excitation spectra).[136] However, highly resolved spectra were only observed for [4]-, [5]-, and [6]helicene, whereas for [7]helicene only broad bands were monitored in emission and excitation spectra.

The Shpol'skii spectra of few pericondensed LPAHs with 7–9 rings have been reported.[26,137,138] An example (naphtho[8,1,2-*abc*]coronene in hexane at 63 K)[137] is shown in Figure 4.6.

In the course of a study of the chemical composition of carbon black, Shpol'skii fluorescence spectra of some additional LPAHs with molecular weights

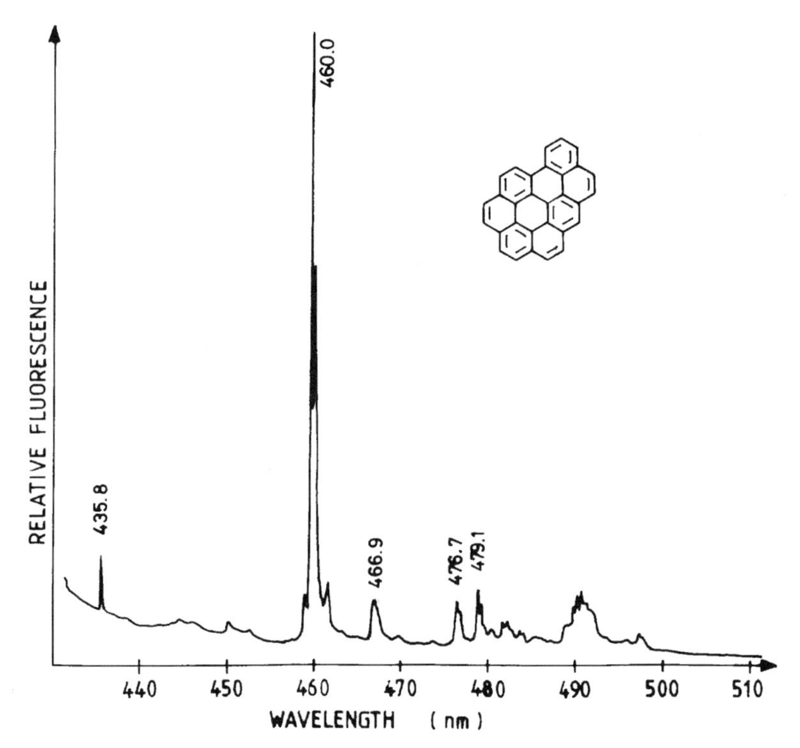

Figure 4.6 Fluorescence spectrum of naphtho[8,1,2-*abc*]coronene in *n*-hexane at 63 K (replotted with permission from Ref. 137).

up to 400 were obtained, but due to the lack of reference compounds these could not be assigned to known structures.[137]

In many cases a characteristic feature of Shpol'skii spectra is the appearance, for a given PAH molecule, of several absorption and fluorescence spectra that are quite identical but shifted slightly in energy from one to another.[139] It is now commonly accepted that this "multiplet" structure is caused by molecules occupying different sites in the host matrix and thus experiencing different host–guest interactions. By using narrow-band laser excitation single host–guest sites can be excited to fluoresce, thus markedly simplifying the spectral structure.[139] However, more recently it has been shown that a multiplet structure can also arise from molecules of the same solute but having different geometries, that is, from different solute confomers.[104] An example is the LPAH **34** (Scheme 4.8), whose Shpol'skii absorption and fluorescence spectra have been studied in *n*-hexane and *n*-octane at 5 K.[138] The fluorescence spectra of the two conformers of this molecule could be measured by using selective laser excitation. The results obtained clearly confirm previous conclusions derived from fluorescence measurements of **34** in fluid solution[68,69] (see Section 4.3.2.1.).

Although the fullerenes C_{60} and C_{70} (which have to be considered aromatic molecules[4]) are not hydrocarbons, and hence in a strict sense do not belong to the class of compounds dealt with in this chapter, we should mention the interesting fact that recently Shpol'skii luminescence spectra of C_{60} and C_{70} have been obtained. Despite the fact that the fluorescence and phosphorescence quantum yields of C_{60} and C_{70} are very low, Russian researchers succeeded in measuring laser-excited Shpol'skii spectra of these molecules both in the fluorescence and phosphorescence mode, showing nicely resolved vibronic transitions.[141,142] A rather exceptional Shpol'skii matrix was used, that is, toluene instead of the familiar *n*-alkanes, and the temperature was as low as 2 K. In another study[143] the influence of the solvent used (*n*-heptane, *n*-hexane, and *n*-pentane) and the rate of cooling on the Shpol'skii fluorescence and phosphorescence spectra of C_{70} has been examined. The luminescence spectrum of a low-concentration solution of C_{70} in *n*-pentane after rapid cooling to 1.5 K (liquid helium temperature) shows superior resolution. Site-selective laser excitation results in linewidths of 3 cm^{-1} and largely simplifies both the fluorescence and phosphorescence spectrum. Undoubtedly, the results obtained add important knowledge to our understanding of the electronic structure and the resulting photophysical properties of the fullerenes.[144]

4.4.2.4. Perhydrocoronene

In order to measure fluorescence, phosphorescence, and luminescence excitation spectra of very large virtually insoluble PAHs, a special technique has been developed that uses perhydrocoronene as a solid spectroscopic matrix.[145]

Perhydrocoronene (PHC), $C_{24}H_{36}$, is a seven-ring saturated hydrocarbon that is essentially flat. The remarkably stable compound forms white needles with a melting point of 370°C.[146] PHC can be prepared from coronene by hydrogenation (temperature 270°C; hydrogen pressure: 300–600 kp/cm^2) using tungsten sulfide as a catalyst.[146] More recently, however, it has been found that PHC is formed as a by-product during the catalytic hydrocracking of petroleum feedstocks and has thus become available at the kilogram scale.[147] For use as a spectroscopic matrix the crude PHC has to be carefully purified (by adsorption chromatography, vacuum sublimation, and crystallization); the purification of PHC has been described in detail.[148] The coronene content of PHC samples used as a spectroscopic matrix was less than 10^{-4}%.[96]

The preparation of solid solutions of LPAHs in PHC is very simple and does not require exclusion of air or humidity. Mixtures of the LPAH and PHC are thoroughly grounded, quickly melted (this can be done in the cylindrical cells used for luminescence measurement) and afterwards cooled down to room temperature.[96] It is assumed that the LPAH/PHC mixtures form highly supersaturated solutions on melting that stay supersaturated on rapid cooling.

It is very likely that the well-resolved fluorescence, phosphorescence, and luminescence excitation spectra of PAHs in PHC stem from molecularly dispersed solid solutions. An example is shown in Figure 4.7.

Curve a is the fluorescence spectrum of pyrene (1%) in PHC at 77 K, while curve b is the fluorescence spectrum of the hydrocarbon in an *n*-hexane/cyclohexane mixture (9 : 1, vol/vol) measured under otherwise identical conditions.[81] Monochromatic lamp excitation and a commercially available luminescence spectrometer were used. It is very likely, however, that still better-resolved spectra can be obtained by increasing the spectral resolution of the emission monochromator and by lowering the sample temperature.

As a further example the phosphorescence spectra of solid solutions of hexabenzocoronene and coronene in PHC at 77 K are given in Figure 4.8.[96] Apart from the fact that the spectrum of hexabenzocoronene (phosphorescence 0–0 band: 17,700 cm^{-1}) lies at longer wavelengths than that of coronene (phosphorescence 0–0 band: 19,210 cm^{-1}; in EPA (ethanol/isopentane/diethyl ether, 2 : 5 : 5, vol/vol): 19,410 cm^{-1}) the spectra are very similar. As follows from the low intensity of the 0–0 band, the radiative transition from the T_1 state to the singlet ground state is symmetry forbidden in both cases (note that both hydrocarbons have D_{6h} symmetry in their electronic ground states).

The PHC technique proved to be suitable for obtaining the spectra of hydrocarbons as large as, for example, circum-para-terphenyl, $C_{52}H_{20}$ (molecular weight: 644).[9] In Figure 4.9 the fluorescence spectrum (on the right) and the fluorescence excitation spectrum (on the left) of circum-para-terphenyl in PHC at room temperature are given.[145] (The α band, not observed in the excitation spectrum, is expected to lie at ≈510 nm).

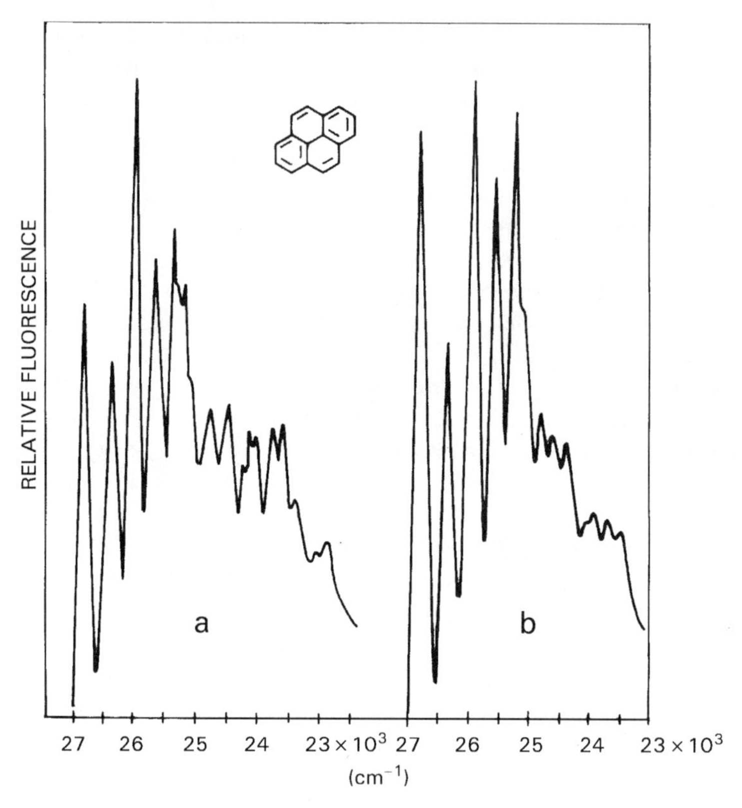

Figure 4.7. Fluorescence spectra of pyrene at 77 K in perhydrocoronene (a) and in *n*-hexane/cyclohexane (b) (replotted with permission from Ref. 81).

Besides its use as a solid matrix for measuring spectra of virtually insoluble LPAHs, PHC can also advantageously be used as a matrix for the study of temperature-dependent photophysical processes of PAHs over a wide temperature range up to approximately 430 K at least.[81,98] Thus, for example, *E*-type delayed fluorescence (Sections 4.3.3. and 4.4.2.1.) of coronene and hexabenzocoronene has been examined in a PHC matrix.[96,98]

The scarce phenomenon of "hot" phosphorescence bands, that is, radiative transitions from higher vibronic levels of the lowest triplet state into the ground state, has been observed in coronene and extensively studied in a PHC matrix.[100] At temperatures significantly above 77 K two additional bands appear at the short-wavelength side of the phosphorescence spectrum of coronene in PHC (and other matrices, e.g., benzophenone or ethylene glycol). These temperature-dependent bands show the characteristic behavior that has to be expected for "hot" phosphorescence bands, that is, the same lifetime as the remaining spectrum

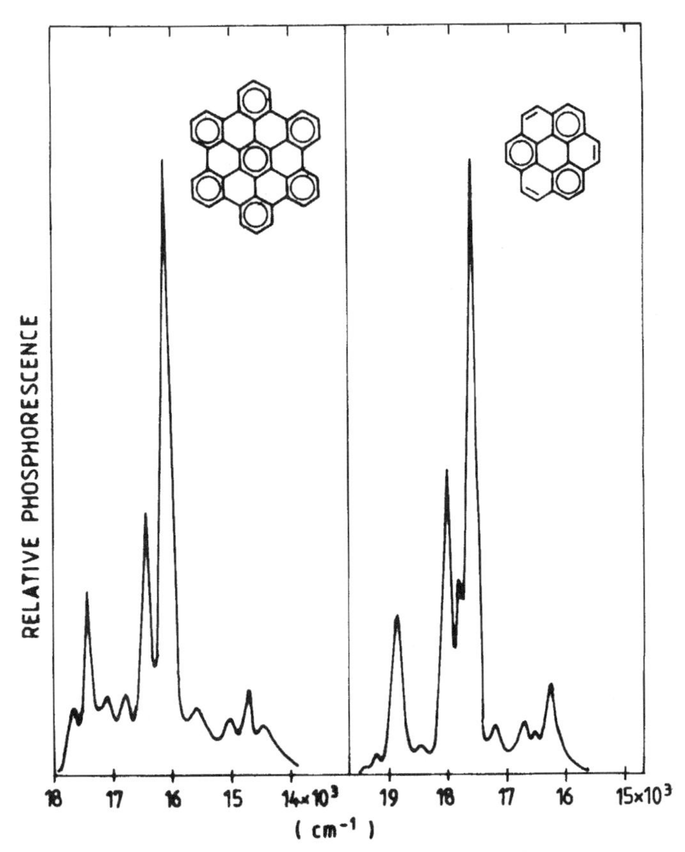

Figure 4.8. Phosphorescence spectra in perhydrocoronene at 77 K of hexabenzocoronene and coronene (replotted with permission from Ref. 96).

and activation energies (obtained from Arrhenius plots) that agree well with the energies of the bands derived from the spectrum.

Besides PHC other perhydrogenated PAHs may also be useful as solid matrices in the luminescence spectroscopy of PAHs. Efficient techniques for the perhydrogenation of PAHs are now available.[149]

4.4.3. Applications

Low-temperature luminescence spectroscopy in rigid matrices plays an important role in PAH analysis and has found many applications in the characterization of environmental samples, including biota.[133] In most cases the spectroscopic

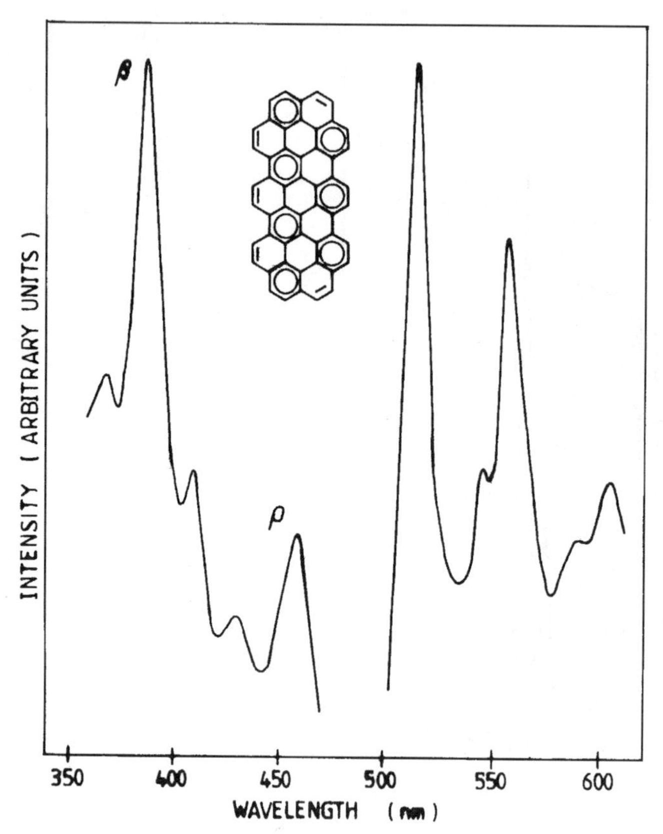

Figure 4.9. Fluorescence emission (on the right) and fluorescence excitation spectrum (on the left) of circum-para-terphenyl in perhydrocorone at room temperature (replotted with permission from Ref. 145).

method is used for identification of PAHs in fractions obtained by chromatography, but some examples where luminescence (e.g., Shpol'skii spectroscopy) has been applied to the entire material (e.g., extracts of coal macerals or coal-tar pitch) have also been reported.[150] For example, coronene and ovalene **14** (Scheme 4.4) were detected in a carbon disulfide extract of a coal maceral by using Shpol'skii luminescence spectroscopy at 77 K.[151]

However, most of the published work is restricted to the analysis of PAHs with fewer than seven rings. The identification and quantitative determination of LPAHs in "real samples" is a challenge because of the following facts: (1) With increasing molecular weight, that is, number of rings, the number of possible PAHs increases dramatically. For example, there are 195 kata- and pericondensed alternant PAHs, respectively, with seven rings (not including unstable free radicals), while the number of possible 8-ring PAHs amounts to 807 and that of the 10-ring PAHs to 16025.[12] (2) In most samples the concentrations of individual PAHs decrease with increasing molecular size as a result of the increasing number of individual isomers. (3) The lack of available standards, partially caused by the small number of synthesized and characterized LPAH isomers, is a major barrier, complicating the analytical problem.[38]

Regarding the "giant" PAHs synthesized more recently (for examples, see Schemes 4.1 and 4.2), luminescence spectroscopic methods will predominantly serve for characterizing the compounds in the course of application-directed research on these molecules. However, few results have been reported yet.

REFERENCES

1. R. G. Harvey, *Polycyclic Aromatic Hydrocarbons/Chemistry and Carcinogenicity*, Cambridge University Press, Cambridge, 1991.
2. E. Clar, *Polycyclic Hydrocarbons*, Vols. 1 and 2, Academic Press, London, and Springer, Berlin, 1964.
3. I. Gutman and S. J. Cyvin, *Introduction to the Theory of Benzenoid Hydrocarbons*, Springer, Berlin, 1989.
4. M. Zander, *Polycyclische Aromaten/Kohlenwasserstoffe und Fullerene*, Teubner, Stuttgart. 1995.
5. R. H. Martin and M. Baes, *Tetrahedron* **31**, 2135 (1975).
6. V. S. Iyer, M. Wehmeier,. J. D. Brand, M. A. Keegstra, and K. Müllen, *Angew. Chem.* **109**, 1675 (1997); *Angew. Chem. Int. Ed. Engl.* **36**, 1603 (1997).
7. B. Schlicke, A.-D. Schlüter, P. Hauser, and J. Heinze, *Angew. Chem.* **109**, 2091 (1997); *Angew. Chem. Int. Ed. Engl.* **36**, 1996 (1997).
8. F. B. Mallory, K. E. Butler, A. C. Evans, E. J. Brondyke, C. W. Mallory, Ch. Yang, and A. Ellenstein, *J. Am. Chem. Soc.* **119**, 2119 (1997).
9. M. Zander and W. Friedrichsen, *Chemiker-Ztg.* **115**, 360 (1991).
10. M. Müller, J. Petersen, R. Strohmaier, Chr. Günther. N. Karl, and K. Müllen, *Angew. Chem.* **108**, 947 (1996); *Angew. Chem. Int. Ed. Engl.* **35**, 886 (1996).
11. M. Zander and W. Friedrichsen, *Z. Naturforsch.* **47b**, 1314 (1992).
12. W. C. Herndon, *J. Am. Chem. Soc.* **112**, 4546 (1990).
13. X. Qiao, D. M. Ho, and R. A. Pascal, Jr. *Angew. Chem.* **109**, 1588 (1997); *Angew. Chem. Int. Ed. Engl.* **36**, 1531 (1997).
14. P. W. Rabideau, A. H. Abdourazak, H. E. Folson. Z. Marcinow, A. Sygula, and R. Sygula, *J. Am. Chem. Soc.* **116**, 7891 (1994).
15. L. T. Scott, M. S. Bratcher, and S. Hagen, *J. Am. Chem. Soc.* **118**, 8743 (1996).
16. M. C. Petty, M. R. Bryce, and D. Bloor, eds., *Introduction to Molecular Electronics*, Arnold, London, 1995.

17. A. Stabel, P. Herwig, K. Müllen, and J. P. Rabe, *Angew. Chem.* **107**, 1768 (1995); *Angew. Chem. Int. Ed. Engl.* **34**, 1609 (1995).
18. A. M. van de Craats, J. M. Warman, K. Müllen, Y. Geerts, and J. D. Brand, *Adv. Mater.* **10**, 36 (1998).
19. K. Jinno, S. Shimura, J. C. Fetzer, and W. R. Biggs, *Polycyclic Aromat. Compds.* **1**, 151 (1990).
20. R. Goddard, M. W. Haenel, W. C. Herndon, C. Krüger, and M. Zander, *J. Am. Chem. Soc.* **117**, 30 (1995).
21. J. C. Fetzer and C. E. Rechsteiner, in M. Cooke, K. Loening, and J. Merritt, eds., *Polynuclear Aromatic Hydrocarbons: Measurements, Means and Metabolism*, Battelle Press, Columbus, 1991, p. 259.
22. J. C. Fetzer and J. R. Kershaw, *Fuel* **74**, 1533 (1995).
23. J. C. Fetzer, W. R. Biggs. and K. Jinno, *Chromatographia* **21**, 439 (1986).
24. J. C. Fetzer and W. R. Biggs, *Polycyclic Aromat. Compnds.* **4**, 19 (1994).
25. J. C. Fetzer, in Absi-Halabi, ed., *Catalysts in Petroleum Refining and Petrochemical Industries*, 1995, Elsevier, Amsterdam 1996, p. 263.
26. J. C. Fetzer, B. R. T. Simoneit, H. Budzinski, and Ph. Garrigues, *Polycyclic Aromat. Compds.* **9**, 109 (1996).
27. B. R. T. Simoneit and J. C. Fetzer, *Org. GeoChem.* **24**, 1065 (1996).
28. D. E. McKinney, D. J. Clifford, L. Hou, M. R. Bogdan, and P. G. Hatcher, *Energy & Fuels* **9**, 90 (1995).
29. A. Léger and J. L Puget, *Astron. AstroPhys.* **137**, L5 (1984).
30. A. Léger, L. d'Hendecourt, and N. Boccara, eds., *Polycyclic Aromatic Hydrocarbons and Astrophysics*, Reidel, Dordrecht, 1987.
31. L. J. Allamandola, in I. Gutman and S. J. Cyvin, eds., *Topics in Current Chemistry* **153**, Springer, Heidelberg, 1990, p. 1.
32. A. Léger, L. d'Hendecourt, and D. Défourneau, *Astron. AstroPhys.* **216**, 148 (1989).
33. M. Zander, *Fuel* **66**, 1459 (1987).
34. J. C. Fetzer, in T. Vo-Dinh, ed., *Chemical Analysis of Polycyclic Aromatic Compounds* Wiley, New York, 1989, p. 59.
35. H. J. Lempka, S. Obenland, and W. Schmidt, *Chem. Phys.* **96**, 349 (1985).
36. C. Reid, *J. Chem. Phys.* **20**, 1214 (1952).
37. E. Clar and M. Zander, *Chem. Ber.* **89**, 749 (1956).
38. W. R. Biggs, and J. C. Fetzer, *Trends Analytic. Chem.* **15**, 196 (1996).
39. A. Bemgard, B. O. Lundmark, and A. Colmsjö, in Ph. Garrigues and M. Lamotte, eds., *Polycyclic Aromatic Compounds: Synthesis, Properties, Analytical Measurements, Occurrence, and Biological Effects* (Suppl. to Volume 3 of *Polycyclic Aromat. Compds.*), Gordon and Breach, Amsterdam, 1993, p. 603.
40. S. M. Shariff, M. M. Robson, and K. D. Bartle, *Polycyclic Aromat. Compds.* **12**, 147 (1997).

41. J. C. Kuei, B. J. Tarbet, W. P. Jackson, J. S. Bradshaw, K. E. Markides, and M. L. Lee, *Chromatographia* **20**, 25 (1985).
42. R. Thoms and M. Zander, *Fresenius' Z. Anal. Chem.* **282**, 443 (1976).
43. G.-P. Blümer and M. Zander, *Fresenius' Z. Anal. Chem.* **288**, 277 (1977).
44. S. A. Wise, in A. Bjørseth and Th. Ramdahl, eds., *Handbook of Polycyclic Aromatic Compounds*, Vol. 2, Marcel Dekker, New York, 1985, p. 113.
45. K. Jinno, J. C. Fetzer, and W. R. Biggs, *Chromatographia* **21**, 274 (1986).
46. S. A. Tucker, A. I. Zvaigzne, W. E. Acree Jr., J. C. Fetzer, and M. Zander, *Appl. Spectrosc.* **45**, 424 (1991).
47. W. E. Acree, Jr., S. A. Tucker. and J. C. Fetzer, *Polycyclic Aromat. Compds.* **2**, 75 (1991).
48. W. Schmidt, *J. Chem. Phys.* **66**, 828 (1977).
49. M. P. Groenewege, *Colloquium Spectroscopicum Internationale VI* (Amsterdam 1956), Pergamon Press, London, 1956, p. 579.
50. M. Zander, *Erdöl Kohle* **15**, 362 (1962).
51. E. Clar, J. M. Robertson, R. Schlögl, and W. Schmidt, *J. Am. Chem. Soc.* **103**, 1320 (1981).
52. M. Zander, U. Breymann, H. Dreeskamp, and E. Koch, *Z. Naturforsch.* **32a**, 1561 (1977).
53. U. Breymann, H. Dreeskamp, E. Koch, and M. Zander, *Chem. Phys. Lett.* **59**, 68 (1978).
54. W. E. Acree, Jr., S. Pandey, S. A. Tucker, and J. C. Fetzer. *Polycyclic Aromat. Compds.* **12**, 71 (1997).
55. G.-P. Blümer and M. Zander, *Fresenius Z. Anal. Chem.* **296**, 409 (1979).
56. P. L. Konash, S. A. Wise, and W. E. May, *J. Liq. Chromatogr.* **4**, 1339 (1981).
57. S.-H. Chen, C. E. Evans, and V. L. McGuffin, *Anal. Chim. Acta* **246**, 65 (1991).
58. J. R. Platt, *J. Chem. Phys.* **17**, 484 (1949).
59. E. Clar, *The Aromatic Sextet*, Wiley, London, 1972.
60. E, Clar and M. Zander, *J. Chem. Soc. (London)* **1958**, 1861.
61. J. B. Birks, *Photophysics of Aromatic Molecules*, Wiley, London, 1970.
62. H. Dreeskamp, E. Koch, and M. Zander, *Ber. Bunsenges. Phys. Chem.* **78**, 1328 (1974).
63. M. Zander, *Phosphorimetry—The Application of Phosphorescence to the Analysis of Organic Compounds*, Academic Press, New York, 1968.
64. S. P. McGlynn, T. Azumi, and M. Kinoshita, *Molecular Spectroscopy of the Triplet State*, Prentice Hall, Englewood Cliffs, NJ, 1969.
65. W. Siebrand, *J. Chem. Phys.* **44**, 4055 (1966).
66. G. W. Robinson and R. P. Frosch, *J. Chem. Phys.* **38**, 1187 (1963).
67. N. J. Turro, *Modern Molecular Photochemistry*, Benjamin/Cummings, Menlo Park, 1978.

68. J. C. Fetzer and W. Schmidt, *Spectrochim. Acta* **45a**, 503 (1989).
69. J. Waluk, J. C. Fetzer, S. J. Hamrock, and J. Michl, *J. Phys. Chem.* **95**, 8660 (1991).
70. J. Waluk and E. W. Thulstrup, *Chem. Phys. Lett.* **123**, 102 (1986).
71. M. Müller, Thesis, University of Mainz, Germany, 1997.
72. K. R. Seddon, *Molten Salt Chemistry: An Introduction and Selected Applications*, NATO ASI Series C: Mathematical and Physical Sciences, D. Reidel, Dordrecht, 1987, p. 365.
73. K.-H. Koch and K. Müllen, *Chem. Ber.* **124**, 2091 (1991).
74. F. Schael and H.-G. Löhmannsröben, *J. PhotoChem. Photobiol. A: Chem.* **69**, 27 (1992).
75. E. Lippert, W. Lüder, and H. Boos, in A. Mangini, ed., *Advances in Molecular Spectroscopy*, Pergamon Press, Oxford, 1962, p. 443.
76. Z. R. Grabowski, R. Rotkiewicz, A. Siemiarczuk, D. J. Cowley, and W. Baumann, *Nouv. J. Chim.* **3**, 443 (1979).
77. W. Rettig, *Angew. Chem.* **98**, 969 (1986); *Angew. Chem. Int. Ed. Engl.* **25**, 971 (1986).
78. M. Zander and W. Rettig, *Chem. Phys. Lett.* **110**, 602 (1984).
79. R. Bunte, K.-D. Gundermann J. Leitich, O. E. Polansky, and M. Zander, *Chem. Ber.* **119**, 1683 (1986).
80. J. Dobkowski, Z. R. Grabowski, B. Paeplow, W. Rettig, K.-H. Koch, K. Müllen, and R. Lapouyade, *New J. Chem.* **18**, 525 (1994).
81. J. C. Fetzer and M. Zander, *Z. Naturforsch.* **45a**, 814 (1990).
82. M. Zander, *Z. Naturforsch.* **39a**, 1145 (1984).
83. M. Zander, *Z. Naturforsch.* **34a**, 1143 (1979).
84. G. G. Giachino and D. R. Kearns, *J. Chem. Phys.* **53**, 3886 (1970).
85. N. Najbar, J. B. Birks, and T. D. S. Hamilton, *Chem. Phys.* **23**, 281 (1977).
86. H. Dreeskamp, E. Koch, and M. Zander, *Chem. Phys. Lett.* **31**, 251 (1975).
87. T. Medinger and F. Wilkinson, *Trans. Faraday Soc.* **61**, 620 (1965).
88. H. Dreeskamp and J. Pabst, *Chem. Phys. Lett.* **61**, 262 (1979).
89. N. I. Nijegorodov and W. S. Downey, *J. Chem. Phys.* **98**, 5639 (1994).
90. W. Siebrand, *J. Chem. Phys.* **47**, 2411 (1967).
91. R. C. Haddon, *Science* **261**, 1545 (1993).
92. J. W. Arbogast, A. O. Darmanyan, C. S. Foote, Y. Rubin, F. N. Diederich, M. M. Alvarez, S. J. Anz, and R. L. Whetten, *J. Phys. Chem.* **95**, 11 (1991).
93. M. R. Wasielewski, M. P. O'Neil, K. R. Lykke, M. J. Pellin, and D. M. Gruen, *J. Am. Chem. Soc.* **113**, 2774 (1991).
94. J. Weijun and L. Changsong, *Anal. Chem.* **65**, 863 (1993).
95. S. H. Ramasamy and R. J. Hurtubise, *Appl. Spectrosc.* **50**, 115 (1996).
96. J. C. Fetzer and M. Zander, *Z. Naturforsch.* **45a**, 727 (1990).
97. C. A. Parker, *Photoluminescence in Solution*, Elsevier, Amsterdam, 1968.
98. M. Zander, *Naturwiss.* **47**, 443 (1960).

99. J. L. Kropp and W. R. Dawson *J. Phys. Chem.* **71**, 4499 (1967).
100. M. Zander, *Z. Naturforsch.* **29a**, 1520 (1974).
101. M. Zander, *Z. Naturforsch.* **30a**, 1097 (1975).
102. M. Zander, *Z. Naturforsch.* **28a**, 1381 (1973).
103. M. Zander, *Z. Naturforsch.* **30a**, 262 (1975).
104. I. Nakhimovsky, M. Lamotte, and J. Joussot-Dubien, *Handbook of Low Temperature Electronic Spectra of Polycyclic Aromatic Hydrocarbons*, Elsevier, Amsterdam, 1989.
105. M. Orrit, J. Bernard, and R. I. Personov, *J. Phys. Chem.* **97**, 10256 (1993).
106. R. C. Stroupe, P. Tokousbalides, R. B. Dickinson, Jr., E. L. Wehry, and G. Mamantov, *Anal. Chem.* **49**, 701 (1977).
107. E. L. Wehry, in A. Bjørseth, ed., *Handbook of Polycyclic Aromatic Hydrocarbons*, Marcel Dekker, New York, 1983, p. 357.
108. J. R. Maple, E. L. Wehry, and G. Mamantov, *Anal. Chem.* **52**, 920 (1980).
109. V. B. Conrad. W. J. Carter, E. L. Wehry, and G. Mamantov, *Anal. Chem.* **55**, 1340 (1983).
110. J. D. Winefordner, S. G. Schulman, and T. C. O'Haver, *Luminescence Spectrometry in Analytical Chemistry*, Wiley, London, 1972
111. S. G. Schulman, *Fluorescence and Phosphorescence Spectroscopy*, Pergamon Press, Oxford, 1977.
112. G. v. Foerster, *J. Chem. Phys.* **40**, 2059 (1964).
113. M. Zander. *Z. Naturforsch.* **38a**, 1146 (1983).
114. L. V. S. Hood and J. D. Winefordner, *Anal. Chem.* **38**, 1922 (1966).
115. M. Zander, *Fresenius Z. Anal. Chem.* **226**, 251 (1967).
116. T. Vo-Dinh, *Room Temperature Phosphorimetry for Chemical Analysis*, Wiley, New York, 1984.
117. W. J. McCarthy and K. L. Dunlap, *Talanta* **17**, 305 (1970).
118. H.-D. Sauerland and M. Zander, *Erdöl Kohle* **25**, 526 (1972).
119. M. Zander, in T. Vo-Dinh, ed., *Chemical Analysis of Polycyclic Aromatic Compounds*, Wiley, New York, 1989, p. 171.
120. S. P. McGlynn, M. J. Reynolds, G. W. Daigre, and N. D. Christodoulas, *J. Phys. Chem.* **66**, 2499 (1962).
121. G. Oster, N. Geacintov, and A. V. Khan, *Nature* **196**, 1089 (1962).
122. H. Dreeskamp and M. Zander, *Z. Naturforsch.* **28a**, 45 (1973).
123. R. M. Hochstrasser, *J. Chem. Phys.* **39**, 3153 (1963).
124. R. M. Hochstrasser and S. K. Lower, *J. Chem. Phys.* **40**, 1041 (1964).
125. E. T. Harrigan and N. Hirota, *J. Chem. Phys.* **49**, 2301 (1968).
126. M. Zander, *Naturwiss.* **52**, 559 (1965).
127. M. Zander, unpublished results (1990).
128. M. Zander, *Ber. Bunsenges. Phys. Chem.* **68**, 301 (1964).
129. W. E. Moerner and L. Kador, *Phys. Rev. Lett.* **62**, 2535 (1989).

130. F. Kulzer, S. Kummer, R. Matzke, C. Bräuchle, and T. Basché, *Nature* **387**, 688 (1997).
131. L. Azarraga. T. N. Misra, and S. P. McGlynn, *J. Chem. Phys.* **42**, 3720 (1965).
132. S. P. McGlynn, B. N. Srinivasan, and H. J. Maria, in J. D. Winefordner, ed., *Spectrochemical Methods of Analysis*, Wiley, New York, 1971, p. 295.
133. F. Ariese, C. Gooijer, and N. H. Velthorst, in D. Barceló ed., *Environmental Analysis: Techniques, Applications and Quality Assurance*, Elsevier, Amsterdam, 1993, p. 449.
134. C. Gooijer, F. Ariese, J. W. Hofstraat, and N. H. Velthorst, *Trends Analytic. Chem.* **13**, 53 (1994).
135. A. L. Colmsjö, *Anal. Chim. Acta* **197**, 71 (1987).
136. K. Palewska and H. Chojnacki, *Mol. Cryst. Liq. Cryst. Sci. Technol., Sect. A* **229**, 31 (1993).
137. A. L. Colmsjö. and C. E. Östman, *Anal. Chim. Acta* **208**, 183 (1988).
138. F. Morgan, Ph. Garrigues, M. Lamotte, and J. C. Fetzer, *Polycyclic Aromat. Compds.* **2**, 141 (1991).
139. T. Vo-Dinh and U. P. Wild, *J. Luminescence* **6**, 296 (1973).
140. K. Palewska, E. C. Meister, and U. P. Wild, *Chem. Phys.* **138**, 115 (1989).
141. B. S. Razbirin, A. N. Starukhin, A. V. Chugreev, Yu. S. Grusko, and S. N. Kolesnik *JETP Lett.* **60**, 451 (1994).
142. B. S. Razbirin, A. N. Starukhin, A. V. Chugreev, D. K. Nelson, Yu. S. Grusko, S. N. Kolesnik, V. N. Zgonnik, L. V. Vinogradova, and L. Federova, *Phys. Solid State* **38**, 522 (1996).
143. J. B. M. Warntjes, I. Holleman, G. Meijer, and E. J. J. Groenen, *Chem. Phys. Lett.* **261**, 495 (1996).
144. C. Gooijer, I. S. Kozin, and N. H. Velthorst, *Microchim. Acta* **127**, 149 (1997).
145. M. Zander, in Ph. Garrigues and M. Lamotte, eds., *Polycyclic Aromatic Compounds: Synthesis, Properties, Analytical Measurements, Occurrence and Biological Effects*, Proc. 13th Int. Symp. Polynuclear Aromatic Hydrocarbons, Bordeaux/France 1991 (Suppl. to the journal *Polycyclic Aromat. Compds.*, Vol. 3), Gordon and Breach, Amsterdam, 1993, p. 143.
146. L. Boente, *Brennstoff-Chemie* **36**, 210 (1955).
147. R. F. Sullivan. M. M. Boduszynski, and J. C. Fetzer, *Energy & Fuels* **3**, 603 (1989).
148. J. C. Fetzer, W. R. Biggs, and M. Zander, *Z. Naturforsch.* **46a**, 291 (1991).
149. *Loc. cit.* **71**, p. 119.
150. J. R. Kershaw, in J. R. Kershaw, ed., *Spectroscopic Analysis of Coal Liquids*, Elsevier, Amsterdam, 1989, p. 155.
151. J. A. G. Drake, D. W. Jones, B. S. Causey, and G. F. Kirkbright, *Fuel* **57**, 663 (1978).

CHAPTER

5

IN-RING AND AT-RING HETEROSUBSTITUTED POLYAROMATIC COMPOUNDS

I. S. KOZIN

Department of Chemistry, Queen's University, Kingston, Ontario, Canada K7L 3N6

S. MATSUZAWA

National Institute for Resources and Environment/NIRE, 16-3 Onogawa, Tsukuba, 305, Japan

5.1. INTRODUCTION

Over the past decades a vast number of analytical applications, based on luminescence detection in low-temperature matrices, have been demonstrated by several research groups (see, for example, a recent review [1]). Using conventional lamp or selective laser excitation combined with fluorescence, time-resolved fluorescence, or, in some instances, phosphorescence detection, extensive spectral atlases have been compiled [2–6] so that reference spectra of a wide range of parent and alkylated polycyclic aromatic hydrocarbons (PAHs) have become available for identification purposes. In addition to the parent compounds, the pyrolysis of fossil fuels and other materials, containing organic matter, may also lead to the formation of various at-ring as well as in-ring heterosubstituted PAH derivatives. Furthermore, when emitted into the environment, parent PAHs undergo specific chemical reactions in the atmosphere, surface waters, soils, sediments, and biota, which result in a variety of heterosubstituted polyaromatic compounds (PACs) that are subsequently further dispersed in the ecosystems [2–4]. A considerable number of PACs have been demonstrated to induce adverse effects when administered to living organisms. In this context, the development of a novel analytical technique capable of isomer-specific detection of PACs at trace levels, even when present in complex mixtures, is highly desirable.

In this chapter, analytical applications of Shpol'skii spectroscopy to the analysis determination of both (at-ring) heterosubstituted PAH derivatives and (in-ring) heterocyclic PACs will be highlighted. A particular emphasis will be placed on those heteropolyaromatic compounds that were found to be the most

Shpol'skii Spectroscopy and Other Site-Selection Methods, Edited by Cees Gooijer, Freek Ariese, and Johannes W. Hofstraat.
ISBN 0-471-24508-9

abundant in the environment, that is, nitrogen-, sulfur-, and oxygen-containing PACs.

5.2. SHPOL'SKII SPECTROSCOPY OF AT-RING HETEROSUBSTITUTED COMPOUNDS

Along with the parent PAHs, nitro- and amino-PAHs, the nitrogen-containing at-ring substituted PAH derivatives, are typical components of aromatic fractions in a wide variety of environmental matrices. These compounds have been detected in atmospheric aerosols, diesel and gasoline exhaust particles, synthetic fuels and shale oil, as well as in coal-derived products [7–13]. The origin of nitro-PAHs is usually related to high-temperature chemical reactions of parent (nonsubstituted) compounds with nitric acid and nitrogen oxides, for instance, during combustion processes. As far as amino-PAHs are concerned, they may be formed from the nitro-substituted analogues through chemical reduction reactions under anaerobic conditions.

The analytical interest in determining both groups of compounds can be clearly understood as many of nitro- and amino-PAHs are known to exhibit strong mutagenic effects [14–16]. Moreover, even greater toxicity is usually observed when living organisms are exposed to a mixture of aromatic compounds, containing both the parent PAHs and the nitro- and amino-substituted derivatives. Since structurally similar compounds exhibit largely dissimilar mutagenic and carcinogenic activities, correct identification of individual isomeric species is of utmost importance in toxicological research. In view of the above considerations, Shpol'skii spectroscopy, as a high-resolution molecular luminescence technique, has a very good perspective.

It should be realized that using fluorescence measurements only amino-PAHs, compounds with relatively high fluorescence quantum yields, can be detected. For nitro-arenes, being poor fluorophores, chemical reduction reactions must be carried out that lead to the formation of the corresponding amino-derivatives. To apply fluorescence detection in liquid chromatography (LC), for instance, different derivatization schemes and reagents have been used. For example, excellent recoveries have been reported for on-line alumina chromatography columns coated with platinum and rhodium catalysts [17] as well as for columns packed with zinc metal particles [18].

In principle, Shpol'skii spectroscopy, as a technique utilizing luminescence detection in cryogenic *n*-alkane matrices, offers significant advantages over conventional analytical methods, since preliminary sample treatment can often be reduced or even completely avoided. In contrast to room temperature luminescence spectra of PAHs, which usually lack detailed spectral structure, Shpol'skii fluorescence and phosphorescence spectra demonstrate a line-narrowed fingerprint character favorable for identification and quantitation of individual compounds, including closely related chemical species. Lamp- and laser-excited cryogenic fluorescence spectra of *n*-alkane solutions of aminopyrene isomers have been known for almost a decade [19, 20]. However, the applicability of Shpol'skii spectroscopy to the analysis of other amino-PAHs and nitro-PAHs has been investigated in more detail only recently.

Using Shpol'skii fluorescence spectra obtained with a scanning photomultiplier (PMT) based detection setup and tunable lamp excitation at 15 K, Matsuzawa et al. [21] have studied a large number of mono- and disubstituted nitroarenes. It has been demonstrated that under Shpol'skii conditions the reduction products of 14 individual mononitrosubstituted PAHs, ranging from 3-nitroacenaphthene to 6-nitrobenzo[*a*]pyrene, exhibit vibrationally resolved fluorescence spectra, which can be used in qualitative and quantitative analysis of extracts of environmental samples. In order to obtain the aminosubstituted derivatives, potassium boronhydride (KBH_4) was used as a reducing agent. The spectra of the reduction products, studied in individual solutions using *n*-hexane and *n*-octane, were practically identical to those of the analogous amino-PAHs, recorded independently under similar experimental conditions.

Quantitative aspects of the measurements were also investigated. The limits of detection, observed for different compounds, were in the range from 174 ng ml^{-1} for 6-aminochrysene down to 0.1 ng ml^{-1} for 1-aminopyrene, which is low enough for the analysis of "real-world" samples. The efficiency of the KBH_4 reduction reaction was estimated by using a standard reference material (SRM 1587), containing a seven-component mixture of nitro-PAHs. Sufficiently high yields (more than 80%) were observed for 1-aminopyrene and 3-aminofluoranthene, while for 2-aminofluorene and 6-aminochrysene only a relatively low conversion (25 and 15%, respectively) was reported. As an interesting detail, formation of unsubstituted PAHs, for example, benzo[*a*]pyrene, as co-products of the reduction reaction was also noticed. This result, however, did not interfere with the measurements on the amino-PAHs, since due to the spectral resolution achieved in *n*-alkane matrices, no significant spectral overlap was observed. Thus, even when applying nonselective broadband lamp excitation, the authors have been able to record Shpol'skii fluorescence spectra of the reduced mixture of nitro-PAHs, and discriminate successfully between the individual components of the mixture (see Fig. 5.1).

In a recent publication focused on the same type of analytes, Kozin et al. [22] have critically examined the experimental parameters, which had been found to be crucial for successful Shpol'skii luminescence detection of nitro- and amino-PAHs. Special emphasis has been given to investigating the photostability of amino-PAHs, their room temperature fluorescence spectra, the linearity of the emission response under Shpol'skii conditions, and the influence of the solidification rate on the spectral features. The authors have been able to extend the spectral library available for identification by publishing vibrationally resolved

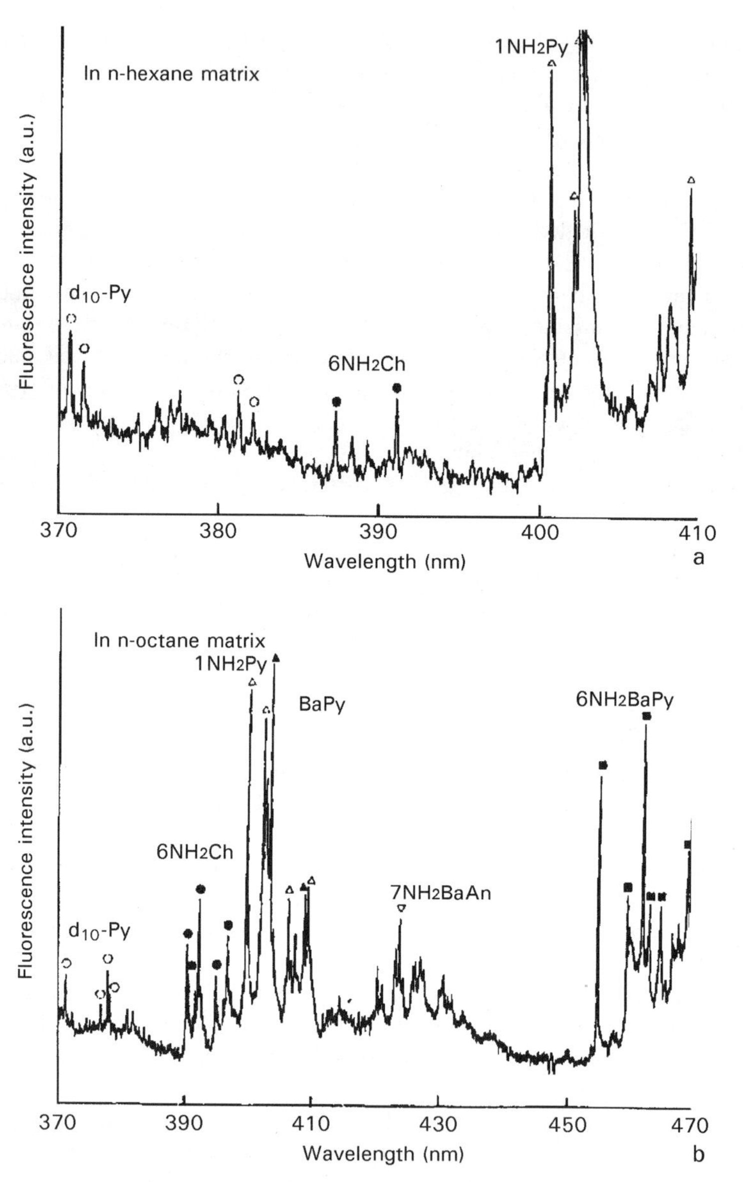

Figure 5.1. Shpol'skii fluorescence spectra of SRM 1587 reduced with potassium boronhydride, using *n*-hexane (a) and *n*-octane (b) as matrices. Abbreviations: Py, pyrene; Ch, chrysene; BaPy, benzo[*a*]pyrene; BaAn, benz[*a*]anthracene (from Ref. 21, with permission).

fluorescence and phosphorescence spectra of a number of individual aminosubstituted naphthalenes, anthracenes, and fluorenes. In order to underline the gain in spectral resolution obtained by carrying out the measurements at cryogenic (helium) temperatures, the Shpol'skii spectra of the examined amino-PAHs were compared with corresponding room temperature fluorescence spectra. As is obvious from Figures 5.2 and 5.3, a tremendous increase in spectral resolution has been obtained.

It has been pointed out, however, that unlike the parent compounds, *n*-alkane solutions of amino-PAHs are prone to efficient intra- as well as intermolecular electron-transfer processes. In line with a study devoted to (laser-based) fluorescence line-narrowing (FLN) experiments with amino-PAHs [23], it was also observed that at ambient temperatures luminescence emission–excitation spectra of amino-PAHs usually possess a significant charge-transfer character, that is, nonstructured S_0–S_1 absorption bands, a large Stokes shift, and broad featureless fluorescence bands. In this respect, the importance of the interaction between the nitrogen lone pair electrons and the π-electron system of the aromatic ring has to be especially emphasized. Furthermore, due to the observed mesomeric effect of the amino group and the increased polarity of aromatic amines, strong analyte–matrix interaction (the so-called electron–phonon coupling, EPC [25]) has to be expected. Moreover, the nonplanarity of amino-PAHs, leading to a decreased compatibility with *n*-alkanes, may cause extensive aggregate formation, especially during the cool-down process.

The above considerations also explain why a number of the examined amino-PAHs undergo rapid phototransformation when exposed to low-level radiation in the UV-vis spectral region, even in the solid state under cryogenic sampling conditions. It has been demonstrated, for example, in experiments with 2-aminofluorene, 9-aminophenanthrene, and 6-aminochrysene [23], that the fluorescence intensity already rapidly decreases within 30–60 sec of sample illumination time. From these observations, it is clear that for the determination of amino-PAHs the use of multichannel detectors, which allow one to register fluorescence emission simultaneously over a wide wavelength region (instead of scanning, i.e., PMT-based, systems), is imperative.

As far as the analyte–matrix interaction is concerned, the experimental data reported in Refs. 22 and 23 demonstrate that the vibronic bands in Shpol'skii spectra are composed of narrow zero-phonon lines superimposed on broad phonon wings (see Fig. 5.3), which indicates intense EPC processes. Furthermore, as is obvious from the spectra of some isomeric compounds, the degree of EPC is strongly influenced by the position of the amino substituent. Hence, for amino-PAHs the choice of the particular solvent to form the cryogenic matrix is much more critical than for parent PAHs. Moreover, in view of the increased analyte–matrix interaction, another critical factor to be considered is the purity of the solvent. It has been reported, for example, that the shape of Shpol'skii

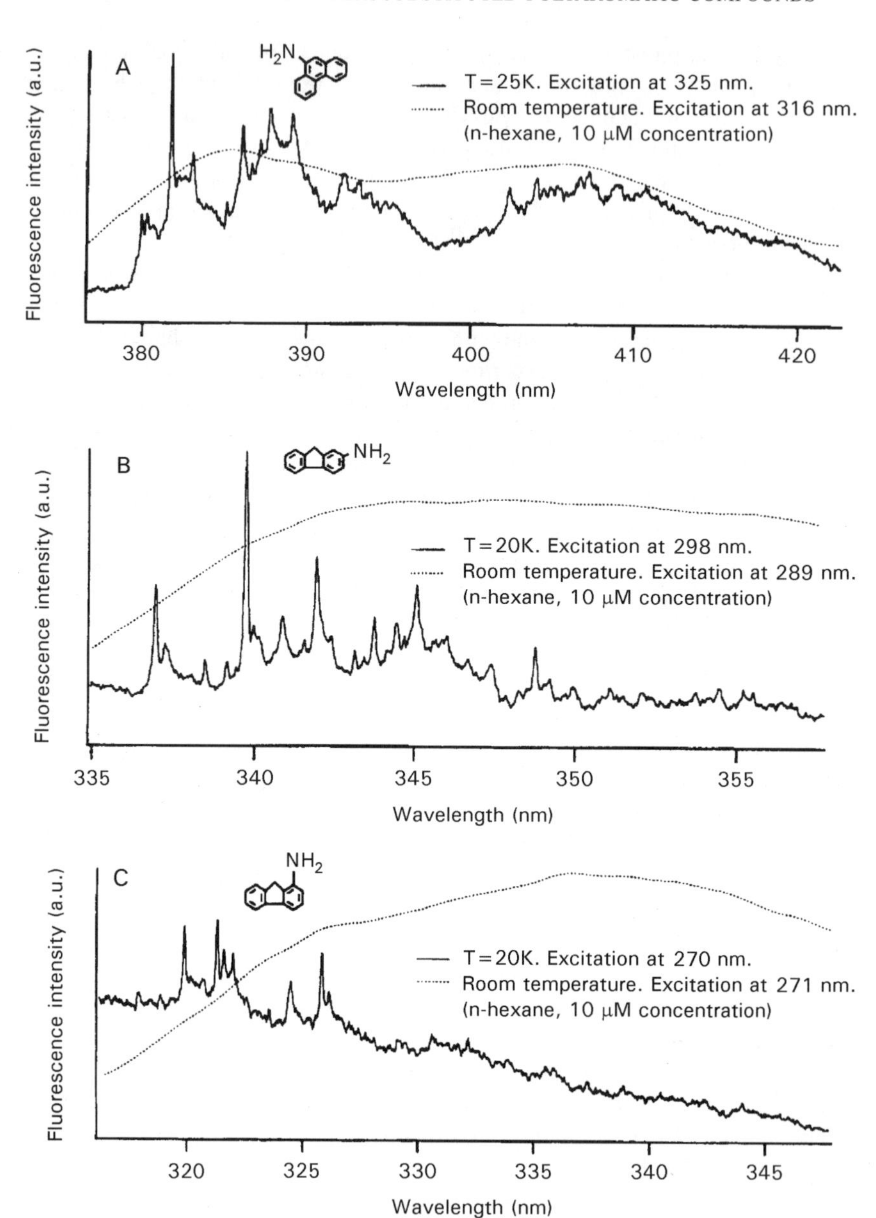

Figure 5.2. Room temperature and Shpol'skii fluorescence spectra of amino-substituted three-ring aromatic PAHs: 9-aminophenanthrene (A), 2-aminofluorene (B) and 1-aminofluorene (C) (from Ref. 22, with permission).

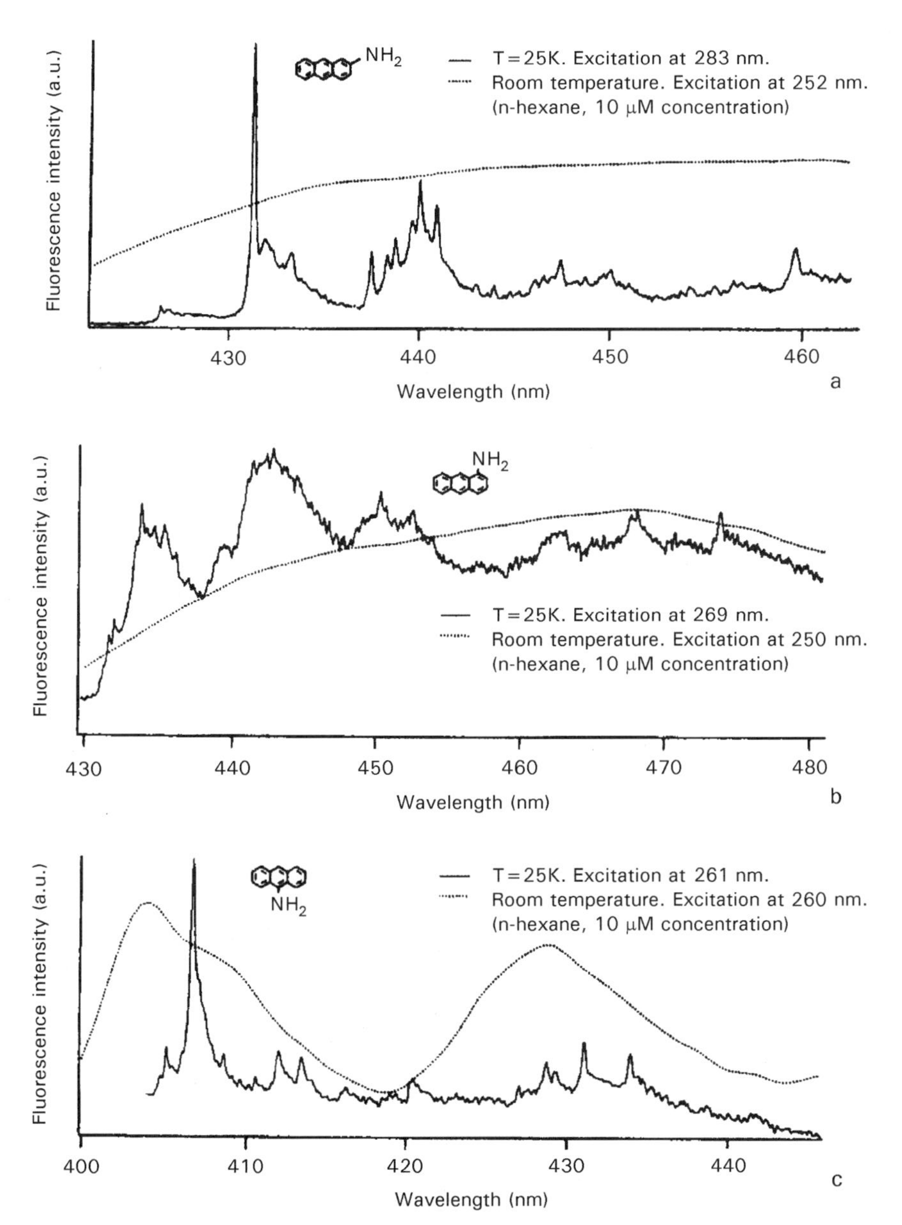

Figure 5.3. Room temperature and Shpol'skii fluorescence spectra of mononitroanthracenes (from Ref. 22, with permission).

fluorescence spectra of a 1 μM *n*-hexane solution of 2-aminofluorene is negatively affected when only a small portion of *n*-octane (0.1% v/v) is added [22].

With regard to possible aggregate formation both at ambient and cryogenic temperatures, a range of amino-PAHs have been examined. At a relatively low concentration level (50 μM and higher) in the liquid state at room temperatures, this negative effect was indeed observed, as significant changes in the shape of fluorescence excitation spectra of the analytes concerned were registered. Consequently, to avoid aggregate formation, the linearity of the fluorescence and the phosphorescence emission responses under Shpol'skii conditions was examined by using several multicomponent solutions with equimolar concentrations of amino-PAHs varied in the range from 10 μM down to 0.1 μM [22]. In this study, straight calibration curves were obtained for most compounds in the concentration range measured. However, 3-aminofluoranthene, 1- and 9-aminoanthracene, exhibiting comparatively bad spectral resolution (i.e., sharp features together with intense phonon wings) and relatively low fluorescence quantum yields, could not be quantified in the mixed solutions. In the separately prepared individual solutions of the examined amino-PAHs, the absolute limits of detection achieved varied from 0.05 ng for 1-aminopyrene to 2 ng for 3-aminofluoranthene. In general, the studies on the applicability of Shpol'skii spectroscopy to the detection of amino-PAHs have clearly demonstrated that, in comparison to the parent PAHs, these compounds are far more difficult to deal with. In fact, using available solvents, the Shpol'skii effect was not observed at all for 1-aminonaphthalene, 9-aminofluorene, and several diaminosubstituted naphthalenes [21, 22].

In spite of these negative results, Kozin and co-workers [22] also attempted Shpol'skii fluorescence spectroscopy with conventional (lamp) excitation for the determination of nitro-PAHs. In order to simulate a "real-world" sample, test solutions containing a mixture of parent and nitro-PAHs (mixed SRM 1647 and SRM 1587) were used. The authors have developed an analytical procedure combining a normal-phase micropreparative chromatographic separation step, an off-line chemical reduction reaction, and Shpol'skii spectroscopic detection. To convert the nitro-PAHs into the luminescent amino derivatives, a well-known reduction reaction with zinc powder was used. In full agreement with the results obtained by Matsuzawa et al. [21], Shpol'skii fluorescence spectra of the reduction products of 2-nitrofluorene, 6-nitrochrysene, 1-nitropyrene, 7-nitrobenzo[*a*]anthracene, and 6-nitrobenzo[*a*]pyrene in the mixed (SRM 1647 and SRM 1587) test solutions were successfully recorded (see Fig. 5.4). For these model samples, however, clearly distinguishable spectra could not be obtained for reduced 9-nitroanthracene and 3-nitrofluoranthene (for reasons, please see the discussion above).

The procedure outlined above was also applied to the analysis of an extremely complex sample of a diesel particulate material SRM 1650, which is

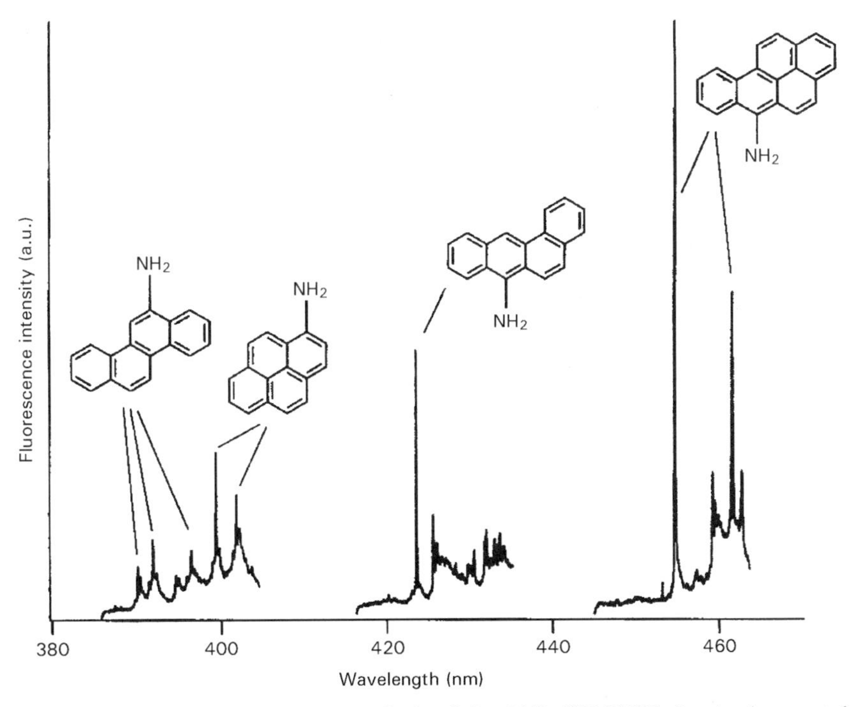

Figure 5.4. Shpol'skii fluorescence spectra of reduced nitro-PAHs (SRM 1587): 6-aminochrysene and 1-aminopyrene, 7-aminobenz[*a*]anthracene, 6-aminobenzo[*a*]pyrene. Lamp excitation at 278, 308, and 313 nm, respectively. *n*-Octane, $T = 25$ K (original authors' data).

representative of heavy-duty diesel engine particulate emissions and known to contain various parent and nitro-PAHs as well as oxidized PAH derivatives [26, 27]. The nitro-PAHs present in the examined sample of SRM 1650 were extracted, chromatographically isolated, and chemically reduced into their respective amino-PAHs. As a confirmation step, two extracted subsamples of SRM 1650 were also subjected to spiking with different amounts of nitro-PAHs contained in a SRM 1587 solution. Positive identification of 1-aminopyrene in the initial and spiked extracts transferred to *n*-octane was carried out by matching the recorded Shpol'skii fluorescence spectra with those obtained from the authentic standard (see Fig. 5.5). For quantification, an internal standard procedure with perylene-d_{12} was used. The quantitative results of the determination of 1-aminopyrene in SRM 1650 by Shpol'skii fluorescence detection were in good agreement with the published certified values obtained by chromatographic methods [22].

Summarizing the results discussed above, it should be stressed that the Shpol'skii method can be used for the analysis of *n*-alkane solutions containing amino- and nitro-PAHs as major constituents. For the successful analysis, it is

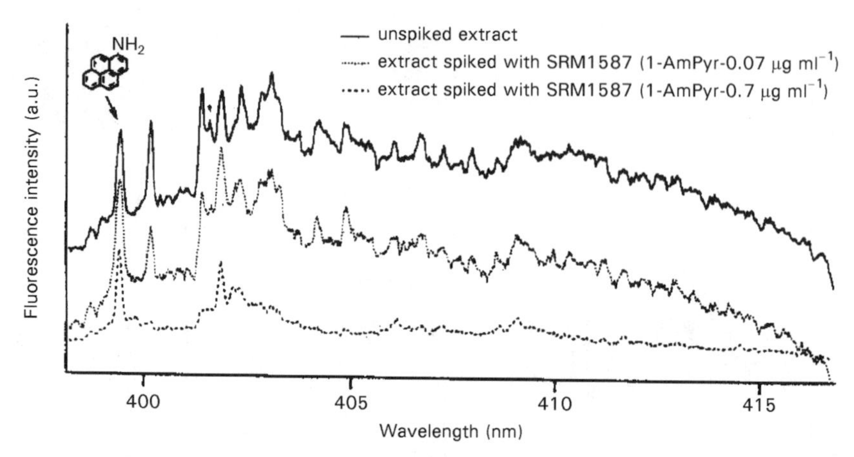

Figure 5.5. Identification of 1-aminopyrene by Shpol'skii fluorescence detection in fractionated methanol extracts of diesel particulate SRM 1650 subjected to chemical reduction. Detector exposure time: 20 sec, unspiked extract; 10 sec, extract spiked with 100-fold-diluted SRM 1587; 2 sec, extract spiked with 10-fold-diluted SRM 1587. Excitation at 364 nm, *n*-octane, $T = 25$ K (from Ref. 22, with permission).

crucial that photostability of amino-PAHs, the choice and the purity of the solvent, as well as possible aggregate formation is taken into account. Although Shpol'skii spectroscopy is not generally applicable to the determination of nitro- and amino-PAHs and the limits of detection achievable are less favorable compared to the parent PAHs, this technique can be used to complement the data obtained by conventional chromatographic methods. With regard to the detection of nitro- and amino-PAHs, the applicability of the Shpol'skii approach may be significantly extended if the problem of the analyte–matrix compatibility is solved, for example, by finding a more appropriate solvent suitable for cryogenic measurements. Another possible way out is to invoke chemical derivatization as it has been developed to improve the detectability of hydroxy-substituted PAHs [28, 29].

5.3. IN-RING SUBSTITUTED HETEROPOLYAROMATIC COMPOUNDS

5.3.1. Nitrogen-Containing Polyaromatic Compounds

Azaarenes and carbazoles are nitrogen-containing aromatic compounds that are wide spread in the environment concomitant with the parent analogues, that is, polycyclic aromatic hydrocarbons. These in-ring heterosubstituted compounds originate either from a variety of geological sources such as petroleum [2, 3 (see the section on occurrence), 30–33], coal products [2, 3, 34–39], shale oil [39, 40], and tar sand bitumen [41] or from anthropogenic emissions related to the discharge of industrial effluents [2, 3, 42–44] and to the incomplete combustion of fuels [45] and organic materials [2, 3, 46]. Unlike relatively nonpolar parent PAHs, azaarenes and carbazoles are usually associated with basic fractions of organic matter. A number of individual azaarenes and carbazoles are known to show mutagenic and carcinogenic activities [2, 35, 47], frequently exceeding that of parent PAHs [47, 48]. As the toxicity of structurally similar compounds may differ greatly [49–51], correct identification and quantification of each isomeric species in complex environmental and biological matrices is of crucial importance.

Conventionally, high-resolution capillary gas chromatography (GC) in combination with flame ionization, nitrogen-selective, or mass spectrometric (MS) detection (e.g., in the multiple-ion-detection, MID, or single-ion-monitoring, SIM, modes) is applied to the analysis of nitrogen-containing compounds in environmental samples [38, 43–45, 52–57]. Furthermore, LC with fluorescence and UV detection has been used [38, 58–61]. However, as a rule, analytical data available from the literature are confined to a range of lower-molecular-weight compounds, whereas often only tentative identification of heavier isomeric compounds is possible. The major difficulties encountered in the analysis of azaarenes and carbazoles are usually referred to co-elution problems and similarities in electron-impact (EI) mass spectra. For example, under GC-MS conditions, despite a large difference in molecular weight and number of fused aromatic rings, similar retention times were observed for acridine and dibenzo [*aj*]acridine [38]. Furthermore, co-elution problems have been reported for benzo[*f*]- and benzo[*c*]quinoline (phenanthridine) [43], dibenzo[*ag*]- and dibenzo[*cg*]-carbazole [31], and dibenzo[*ac*]- and dibenzo[*ah*]acridine [54]. In reversed-phase LC, despite applying various mobile phases, the attempt to distinguish between dibenzo[*aj*]- and dibenzo[*ai*]acridine were not successful either [38]. Although in normal-phase LC-UV some of these problems can be tackled [60], the analysis of the above-mentioned compounds using this technique may be impeded due to insufficient sensitivity.

The studies on the applicability of Shpol'skii spectroscopy to the detection of azaarenes and carbazoles have been initiated by several Russian researchers. The first reports on quasi-linear emission of carbazole and its benzologues appeared in the late 1960s [62]. In the 1970s, vibrational spectral analysis of several azaarenes was carried out in *n*-alkanes at liquid nitrogen (77 K) and liquid helium (4 K) temperatures [63]. These experiments indicated that a range of compounds (e.g., acridine and its derivatives) is amenable to Shpol'skii spectroscopy. However, the first analytical application appeared only in the 1980s.

In 1981, Colmsjö et al. [6] published high resolution luminescence spectra of azafluorene, azaphenanthrenes, azapyrene, and other PACs measured in *n*-alkanes at 63 K under lamp excitation. On this basis, the same group has been able to successfully apply Shpol'skii fluorescence detection to the identification of azapyrene in a contaminated soil extract after high-performance liquid chromatography (HPLC) [64].

As the resolution of Shpol'skii fluorescence spectra is typically improved by lowering the sample temperature below 77 K, Garrigues et al. [30, 65] have carried out a series of measurements at 15 K by using a closed-cycle helium cryogenerator. The authors have recorded vibrationally resolved Shpol'skii fluorescence and phosphorescence spectra of several triaromatic azaarenes in *n*-hexane, and identified isomeric di- and trimethyl-substituted benzo[*h*]quinoline derivatives in a triaromatic fraction of the basic concentrate from a crude oil sample. Figures 5.6 and 5.7, for example, demonstrate the spectra used for identification of di- and trimethylbenzo[*h*]quinolines. By judicious use of the excitation wavelength, each available isomer could be easily identified by comparison of its fluorescence and phosphorescence spectra with those of a synthetic mixture. The results indicated that 2,4-dimethylbenzo[*h*]quinoline and 2,4,6-trimethylbenzo[*h*]quinoline dominate over the other respective isomer in the examined oil samples. Furthermore, by using high-resolution luminescence

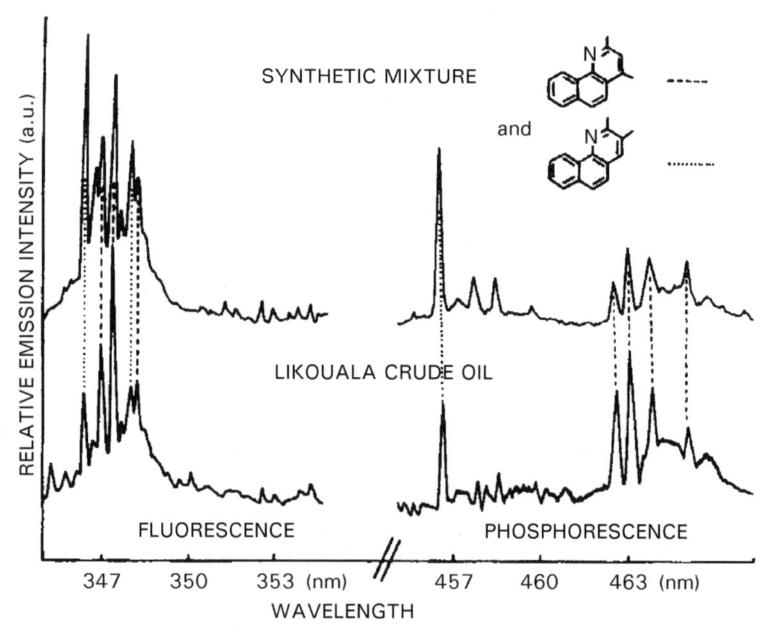

Figure 5.6. Identification of 2,3- and 2,4-dimethylbenzo[*h*]quinolines in a crude oil concentrate ($\lambda_{ex} = 330$ nm, *n*-hexane, $T = 15$ K). From Ref. 30, with permission.

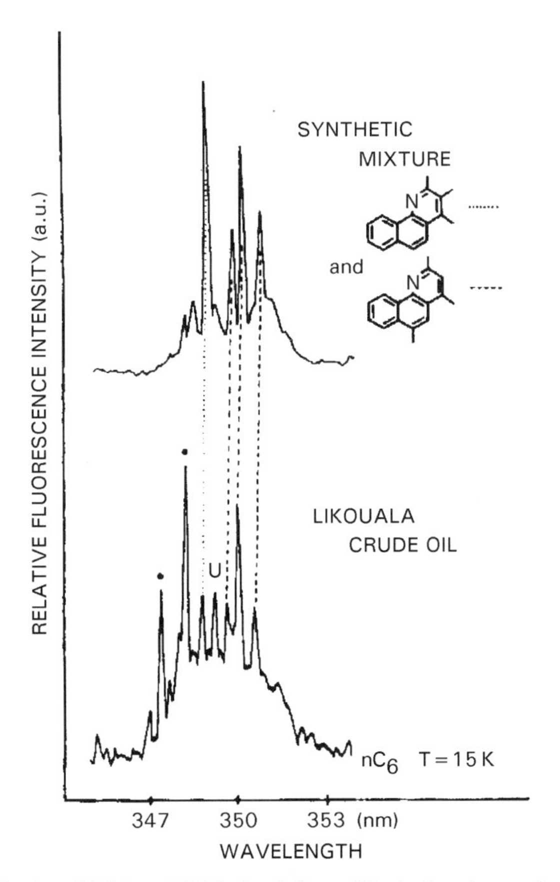

Figure 5.7. Identification of 2,3,4- and 2,4,6-trimethylbenzo[*h*]quinolines in a crude oil concentrate ($\lambda_{ex} = 332.5$ nm, *n*-hexane, $T = 15$ K); peaks (. . .) are attributed to 2,4-dimethylbenzo[*h*]quinoline, peak (U) is attributed to an unknown compound (from Ref. 30, with permission).

detection, it has been revealed that 6-methylbenzo[*c*]phenanthridine is the major compound of azachrysene series in crude oil [32].

Nakhimovsky et al. [5] have significantly contributed to expanding the spectral library available for identification of various PACs by publishing high-resolution Shpol'skii fluorescence and absorption spectra of a wide range of compounds, including nonsubstituted benzoquinolines: benzo[*c*]quinoline (phenanthridine), benzo[*f*]- and benzo[*h*]quinolines, measured at 5 K. Spectra of other azaarenes such as benzoacridines, dibenzoacridines, azachrysene, and azabenzo[*a*]pyrene can be seen elsewhere [2, 3, 5, 67]. Furthermore, a spectrum of a dibenzosubstituted azabenzo[*a*]pyrene was reported in Ref. [66].

Although acridine is considered to be an extremely weak fluorophore even under cryogenic sampling conditions [6], benzacridines and dibenzacridines have been found to exhibit excellent high-resolution Shpol'skii spectra [2, 3, 5, 67]. Obviously, since the planarity of these compounds is practically not affected due to the in-ring nitrogen substituent, many azaarenes are fairly well compatible with cryogenic *n*-alkane solutions. *n*-Octane has been found a suitable solvent for mono- and dibenzologues of acridine, whereas other *n*-alkane matrices, such as *n*-decane and *n*-hexane, have also given high-resolution spectra of some isomeric dibenzacridines [2, 3, 5]. It is interesting to note that *n*-hexane, which has smaller molecular dimensions than those of dibenzacridine, also gives high-resolution spectra [5].

The applicability of Shpol'skii spectroscopy to the analysis of isomeric benzo- and dibenzo-substituted acridine derivatives present in highly complex sediment samples has recently been demonstrated by Kozin et al. [67]. Using conventional lamp excitation, both fluorescence and phosphorescence spectra (32 K) of sample extracts in *n*-octane matrices have been measured, which enabled positive identification of benz[*a*]- and benz[*c*]acridines, dibenz[*ac*]-, dibenz[*aj*]-, dibenz [*ah*]-, and dibenz[*ch*]acridines in river and lake sediments collected from several locations in The Netherlands. Figure 5.8 shows fluorescence spectra of initial extracts, spiked extracts, and synthetic mixtures of acridine derivatives used for identification. As is obvious from Figure 5.8, the spectra of extracts spiked with a standard solution of appropriate azaarenes unambiguously confirm the structural assignments for the compounds detected. Moreover, the above identification carried out by fluorescence detection has been further confirmed by implementing phosphorescence measurements. Using perylene-d_{12} as an internal standard compound, the authors have also been able to determine quantitatively the content of acridine derivatives in the examined sediment samples.

In order to increase the selectivity and to enable direct analysis of target compounds, while avoiding cleanup or intermediate sample treatment, Elsaiid et al. [20] have carried out a series of measurements using a tunable laser as an excitation source. It has clearly been demonstrated that the LESS technique is fully applicable to the analysis of nitrogen-containing heterocyclic PAHs. By using site-selective laser excitation and fluorescence detection, positive identification of dibenzo[*aj*]acridine in a sample of solvent-refined coal (SRC) could be achieved directly, that is, without preliminary separation step.

It has been reported that phenazine and benzo[*c*]cinnoline are present in diesel particulate matter [45]. Although only a few studies have dealt with luminescence detection of these diazapolyaromatic compounds in low-temperature *n*-alkane matrices, the phosphorescence spectrum for phenazine has been reported in relation to theoretical work [68]. Phenazine, for instance, exhibited high-resolution phosphorescence spectra in *n*-pentane and *n*-heptane matrices, but better resolution could be obtained for the latter solvent. For benzo[*c*]cinnoline,

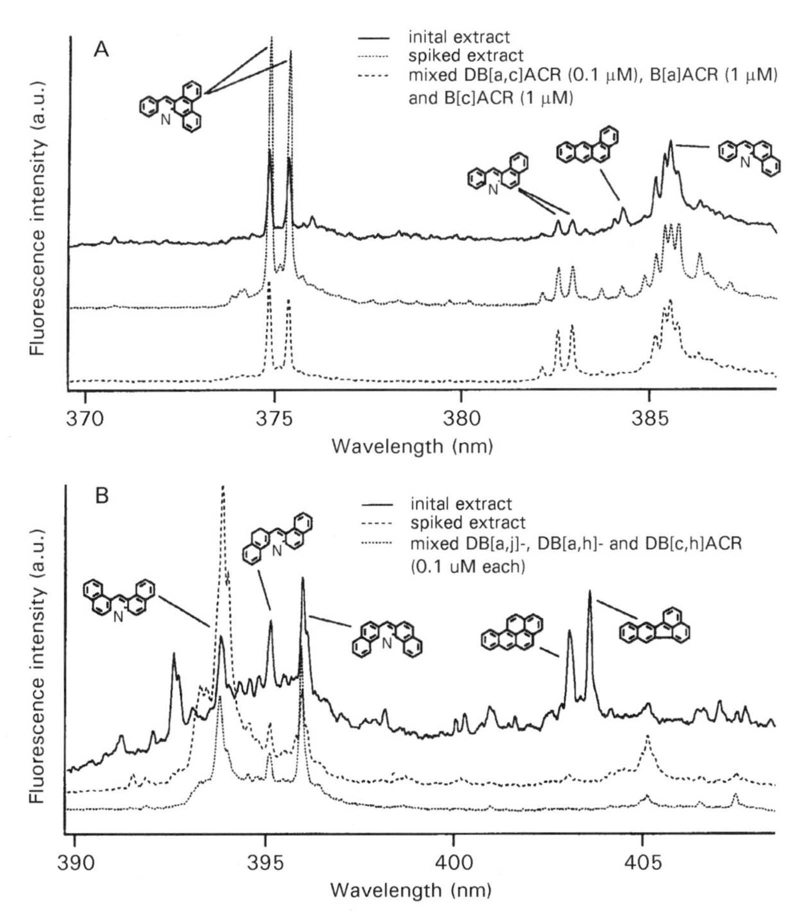

Figure 5.8. Identification of benzo- and dibenzoacridines in a lake sediment extract. Lamp excitation at $\lambda_{ex} = 285$ nm (A), $\lambda_{ex} = 303$ nm (B), *n*-octane, $T = 32$ K (from Ref. 67, with permission).

however, the phosphorescence spectrum has not been reported in the literature. As far as the fluorescence emission of the above-mentioned compounds is concerned, they are both characterized by the extremely low fluorescence quantum yields so that observation of high-resolution fluorescence spectra at low temperature is virtually impossible.

Similarly to the acridine derivatives, the carbazole group of nitrogen-containing heteropolyaromatic compounds also exhibits intense fluorescence and gives rise to high-resolution spectra in Shpol'skii matrices. In 1984, Garrigues and co-workers [69] reported high-resolution spectra of carbazole, three benzocarbazoles, and three dibenzocarbazoles, and showed the applicability of those spectra to the identification of carbazole derivatives in petroleum fractions. *n*-Hexane, *n*-octane, and *n*-decane have been used as solvents for carbazole, benzocarbazoles, and dibenzocarbazoles, respectively. Importantly from the environmental point of view, by applying the Shpol'skii technique, the authors have demonstrated that dibenzo[*cg*]carbazole, a very potent carcinogen, is practically absent in the examined petroleum fractions, whereas benzo[*a*]- and benzo[*c*]-carbazoles, dibenzo[*ai*]- and dibenzo[*ac*]carbazoles have been found to be predominant compounds [31, 69]. The high-resolution fluorescence spectra used for identification of benzocarbazoles and dibenzocarbazoles in extracts of crude oil are depicted in Figures 5.9 and 5.10, respectively. Moreover, the authors have demonstrated that the presence of benzo[*b*]carbazole, which could hardly be

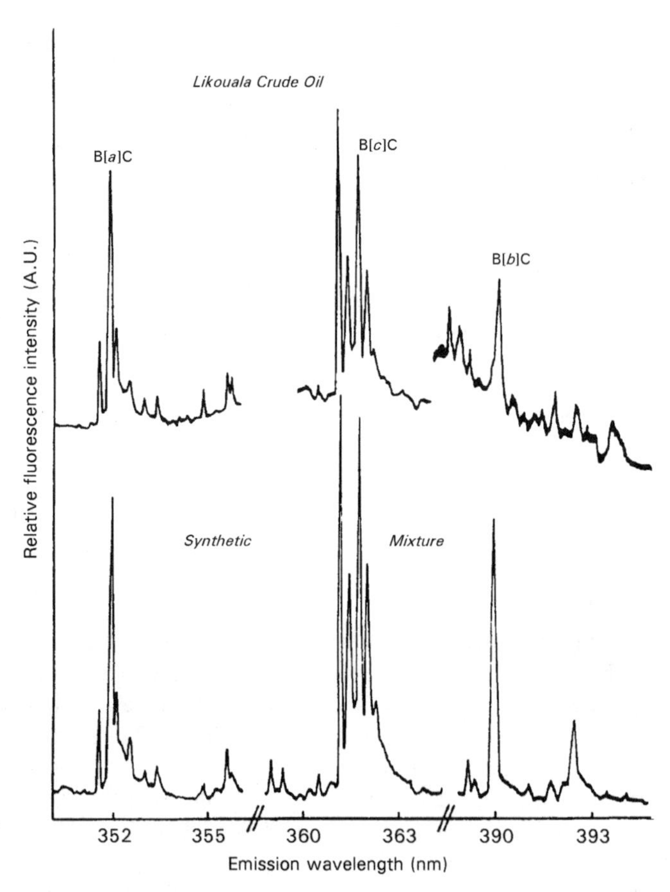

Figure 5.9. Identification of benzo[*a*]-, benzo[*b*]-, and benzo[*c*]carbazole in a crude oil extract ($\lambda_{ex} = 330$ nm, *n*-octane, $T = 15$ K). From Ref. 69, with permission.

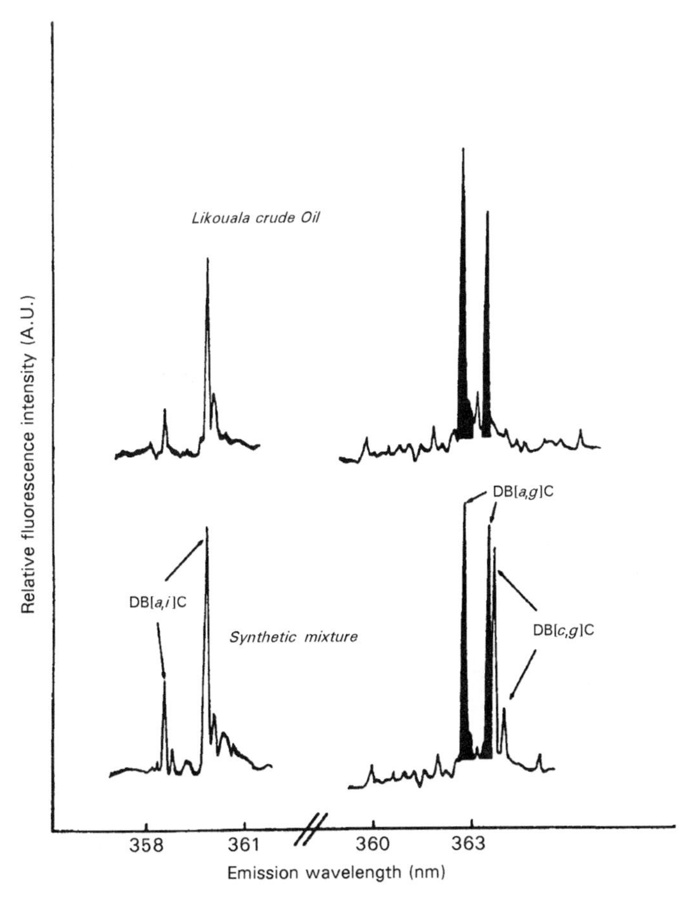

Figure 5.10. Identification of dibenzo[*ai*]- and dibenzo[*ag*]carbazole in a crude oil extract, $\lambda_{ex} = 292$ nm (for dibenzo[*ai*]carbazole), $\lambda_{ex} = 346$ nm (for dibenzo[*ag*]- and dibenzo[*cg*]carbazole), *n*-decane, $T = 15$ K). From Ref. 69, with permission.

detected by gas chromatography, was more easily identified by high-resolution fluorescence spectra (see Fig. 5.9). Reference spectra of the above-mentioned carbazoles can also be found elsewhere [2, 3, 5, 66]. Shpol'skii fluorescence spectra of monomethylcarbazoles have been reported as well [70]. These data clearly demonstrated that the emission spectrum of each compound has a specific quasi-linear pattern, which can be used as a fingerprint to confirm the identification of isomers after routine GC-MS measurements.

As in the case of acridine derivatives, the application of laser excitation to the analysis of carbazoles leads to a significant increase in selectivity and sensitivity.

The LESS approach has been successfully implemented for selective detection of a number of isomeric carbazole derivatives by Elsaiid et al. [20].

Until now, the high analytical potential of Shpol'skii spectroscopy has been demonstrated for azaarenes and carbazoles. If the lack of new standard nitrogen-containing PACs is overcome, the range of compounds amenable to cryogenic luminescence detection in Shpol'skii matrices can be substantially expanded.

5.3.2. Sulfur-Containing Polyaromatic Compounds

It is well known that thiophene, its methylated derivatives, and benzologues, such as, for example, dibenzothiophene, methylated dibenzothiophenes, benzonaphtothiophenes, and phenanthrothiophenes, are present in crude oil and its products and coal products [2, 4 (see section of occurrence), 71, 72]. By utilizing or processing these materials, sulfur-containing compounds disperse into the environment as components of combustion gas, mists, or leakage. Spills from oil tanks and tankers also cause contamination of the river and sea waters, sediments, and coastal sands with sulfur-containing compounds [73]. In fact, sulfur-containing compounds mentioned above have been found in the air, water, sediments, street dust, etc. [2, 4, 74, 75]. Furthermore, it has been reported that some of the sulfur-containing compounds are taken up by marine organisms, including fish [2, 4]. Large sulfur-substituted PAH derivatives are typically found in carbon black, which is manufactured via polycondensation processes [2]. It is known that methylated dibenzothiophenes [76], phenanthrothiophenes [33], and benzophenanthrothiophenes (e.g., benzo[2,3]-phenanthro[4,5-*bcd*]thiophene) [78] induce mutagenic effects. Since sulfur-containing polyaromatic compounds are persistent in the environment (resistant against photochemical [79] and biological [80] degradation), determination of these compounds in environmental compartments and biological systems is highly interesting and important.

Typically, the analysis of thiophene derivatives is performed using GC-MS. However, in the case of complex environmental samples, separation of isomers is far more difficult to handle, even if prior fractionation by LC is applied [71, 72]. For this reason, several research groups have investigated the analytical applicability of Shpol'skii spectroscopy to isomer-specific detection of sulfur-containing compounds at low concentration levels.

A quasi-linear spectrum of dibenzothiophene, a kata-condensed sulfur-containing polyaromatic compound, was reported for the first time in 1980 by Woo et al. [81]. In that work, to obtain phosphorescence spectra of dibenzothiophene, X-ray excited optical luminescence (XEOL) measurements were implemented in frozen *n*-heptane solutions. In general, sulfur-containing polyaromatic compounds exhibit somewhat stronger phosphorescence than fluorescence. For dibenzothiophene, Vial et al. [82] have performed a detailed spectral analysis to study the origin of emission bands in fluorescence and phosphorescence modes.

In the mid-1980s, Garrigues et al. [66, 70] applied high-resolution phosphorescence detection at 15 K in an *n*-hexane matrix to study the distribution of dibenzothiophenes in crude oils. Despite relatively high limits of detection (in the low ppm range), determination of 1- and 2-methyldibenzothiophene in crude oil sample was successful (Fig. 5.11).

Elsaiid and co-workers applied Laser-Excited Shpol'skii spectroscopy (LESS) [20] to the quantitative analysis of dibenzothiophene in complex samples of petroleum crude (SRM 1582) and solvent-refined coal (SRC-II), and obtained

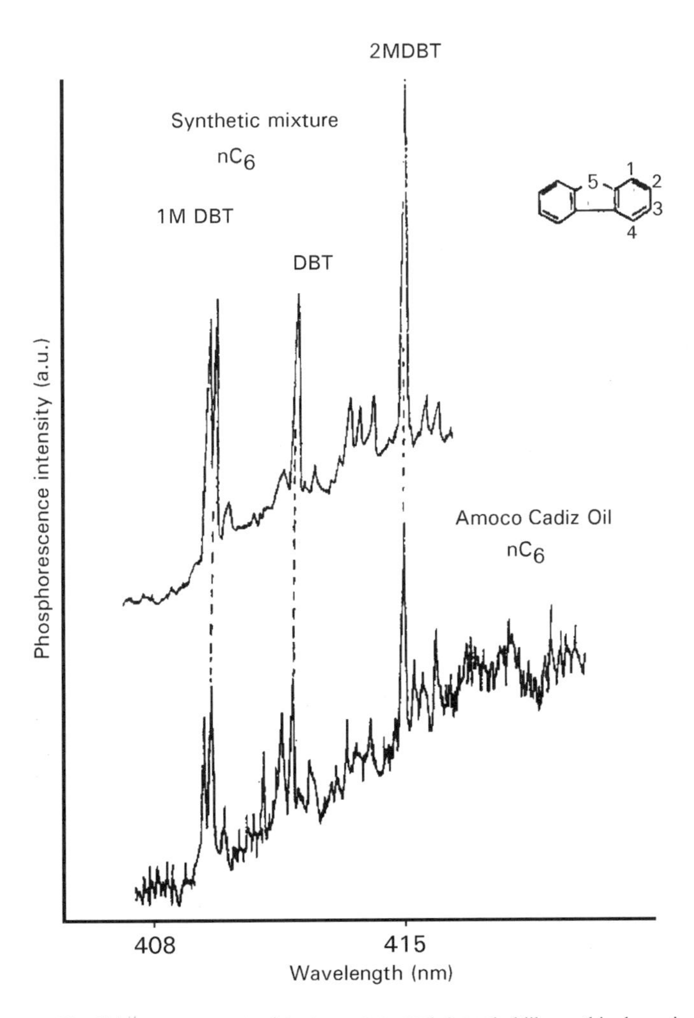

Figure 5.11. Shpol'skii spectroscopic detection of 1- and 2-methyldibenzothiophene in *n*-hexane solutions at $T = 15\,K$ (from Ref. 70, with permission).

Table 5.1. Comparison of Analytical Results for Dibenzothiophene and Dibenzofuran in Petroleum Crude and Solvent Refined Coal (SRC) [20]

	Petroleum crude (SRM 1582)	SRC-II
Dibenzothiophene	45 ± 5 (LESS) 33 ± 2 (NBS data)	1145 ± 25 (LESS) 1020 ± 70 (NBS data) 1182 ± 73 (PETC data)
Dibenzofuran		220 ± 20 (LESS)

Abbreviations: LESS, Laser-excited Shpol'skii spectroscopy; NBS, National Bureau of Standards (currently National Institute of Standards and Technology, NIST); PETC, Pittsburgh, Energy Technology Center.

excellent analytical results (see Table 5.1). In their work, quantification was accomplished by using deuterated dibenzothiophene as an internal standard. In addition to the spectra of the parent compound, Shpol'skii luminescence spectra of 1-, 2-, 3-, and 4-methyldibenzothiophenes have also been published in the spectral atlas [4]. Furthermore, high-resolution spectra of large sulfur-containing compounds: benzo[*b*]-naphthothiophenes, phenanthrothiophenes, and benzophenanthro-thiophene have been measured at low temperature (15 and/or 5 K) [2, 4, 5, 20, 83]. The analysis of the latter two types of compounds is very important since they exhibit strong carcinogenic and/or mutagenic activities [2, 4]. Using LESS reference spectra and site-selective excitation, Elsaiid et al. [20] succeeded in the identification of benzo[*b*]naphtho[1,2-*d*]thiophene, benzo[2,3]-phenanthro[4,5-*bcd*]thiophene, and chryseno[4,5-*bcd*]thiophene in a carbon black sample.

Colmsjö et al. [83] have studied a number of multiring pericondensed thiophenes, ranging from epithiophenanthrene to epithiobenzo[*ghi*]perylene, regarding their luminescence properties upon lamp excitation in Shpol'skii matrices. In their work, it has been demonstrated that most of these bay-region sulfur-substituted polyaromatic compounds exhibit highly resolved cryogenic (63 K) fluorescence spectra. Figure 5.12, for instance, shows fluorescence and phosphorescence spectra of 1,12-epithiobenzo[*e*]pyrene measured at 63 K in an *n*-hexane matrix. As these sulfur containing compounds show similar or better fluorescence efficiency and good analyte–matrix compatibility compared with parent PAH analogues, good spectral resolution has been observed. On the basis of reference spectra obtained, the authors have accomplished positive identification of a number of pericondensed thiophenes in several narrow LC fractions of a complex carbon black soot sublimate sample [83].

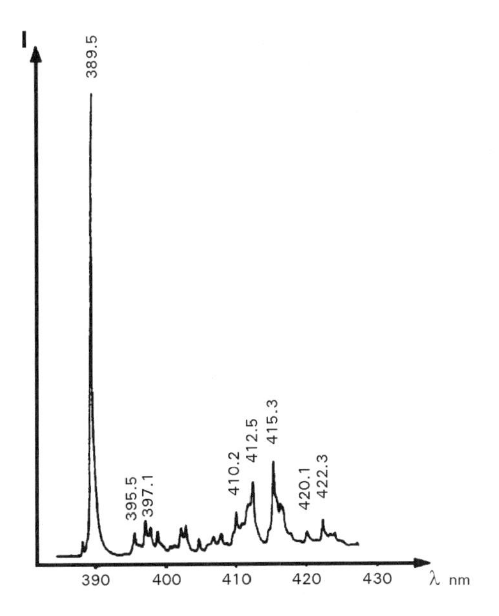

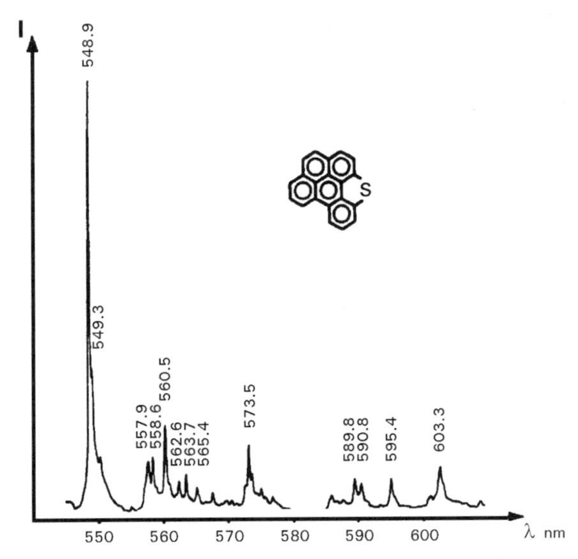

Figure 5.12. Shpol'skii fluorescence (upper graph) and phosphorescence (lower graph) spectra of 1,12-epithiobenzo[*e*]pyrene (*n*-hexane, $T = 63\,K$). From Ref. 33, with permission.

5.3.3. Oxygen-Containing Polyaromatic Compounds

Many researchers have reported that solvent-refined coal (SRC) contains dibenzofuran and its derivatives (see, for example, [4, 72, 84]. For example, as the derivatives, methyldibenzofurans [72, 84] and benzo[*b*]naphthofuran [85] have been detected. Furthermore, Sugimoto et al. [86] reported possible presence of C_2- and C_3-substituted dibenzofurans (GC retention times for these compounds overlap with those of alkylbiphenyls). These dibenzofurans are emitted into the environment by combustion of coal, wood, oils, waste, etc. [4]. Consequently, dibenzofurans have been detected in various environmental samples such as diesel exhaust, air, soil, water, sediment, sewage sludge, and fish [4, 86].

Carbon black and cigarette smoke also contain dibenzofuran and its derivatives [4]. Other oxygen-containing polyaromatic compounds, which have so far been detected in environmental samples, are the derivatives of xanthone [87] and coumarin [88, 89]. These are considered to be oxidation products of PAHs and present at low concentration in environmental samples. Schuetzle et al. [87] have detected 9-xanthone in fractions of diesel particulate extracts. On the other hand, compounds of the coumarin type have been observed in urban air particulate matter [87, 88]. Although the toxicity of dibenzofurans and other oxygen-containing compounds described here has not been investigated sufficiently well, one may expect that some oxygenated PAHs may exhibit even stronger mutagenic or carcinogenic activities than the parent PAHs.

In most cases, GC-MS is applied to the trace analysis of oxygen-containing polyaromatic compounds [72, 84, 86]. Dibenzofuran is separable from methyl-substituted dibenzofurans on a GC capillary column, but the separation of the individual derivatives is more difficult [72, 84]. For dibenzofuran, high-resolution spectra using frozen Shpol'skii matrices have been measured and investigated mainly in the field of physical chemistry [91–93]. To our knowledge, the only analytical application has been reported by Elsaiid and co-workers [20]. Dibenzofuran shows higher fluorescence and phosphorescence quantum yields compared with other oxygen-substituted heterocyclic compounds. With regard to the vibrational analysis, Bree et al. [91] measured high-resolution absorption, fluorescence, and phosphorescence spectra of dibenzofuran at 4.2 K in an *n*-heptane matrix. Shpol'skii luminescence spectra of dibenzofuran can also be found in spectral atlases [4, 5]. Although for this compound highly resolved luminescence spectra can be obtained for *n*-pentane solutions [5], C_6–C_8-alkanes display better analyte–matrix compatibility as the multiplicity of emission bands can be significantly reduced [4, 5, 20].

Tunable laser excitation has been used for low-temperature fluorescence emission and excitation measurements of dibenzofuran and its deuterated analogue (used as an internal standard for quantification) in an *n*-hexane matrix by Elsaiid et al. [20]. In comparison with the data recorded using a conventional lamp excitation source, it was clearly demonstrated that the resolution and selectivity is greatly increased when using the LESS approach. With this technique, direct determination of dibenzofuran in a solvent-refined coal sample (SRC-II) was possible (see Table 5.1 for analytical results). Fluorescence spectra of the dibenzofuran benzologues benzo[*b*]naphtho[1,2-*d*]furan and benzo[*b*]naphtho[2,1-*d*]furan have also been measured. These compounds exhibit better quasi-linear spectra in frozen *n*-octane solutions [4]. Similarly, phosphorescence spectra for xanthone and its 3,6-, 2,6-, and 2,7-dimethyl derivatives [94] have been measured in *n*-hexane matrix at 4.2 K.

It is obvious that mostly due to the increased polarity of the analytes, which may cause poor analyte–matrix compatibility, the applicability of Shpol'skii spectroscopy to the analysis of oxygen-containing heterosubstituted PACs is limited. However, as follows from the above examples, a number of reports have demonstrated that the Shpol'skii approach has some potential in this field. If the planarity of a compound of interest is preserved and the intermolecular energy transfer in *n*-alkane solutions is prevented, the advantages of the cryogenic luminescence techniques, such as excellent selectivity and sensitivity, could be fully exploited in a variety of analytical applications involving an extensive range of heteropolyaromatic compounds.

REFERENCES

1. C. Gooijer, I. Kozin, and N. H. Velthorst. *Mikrochim. Acta* **127**, 142 (1997).
2. *Spectral Atlas of Polycyclic Aromatic Compounds*, *Vol. 1*, W. Karcher, ed., Kluwer Academic Publishers, Dordrecht, The Netherlands, 1985.
3. *Spectral Atlas of Polycyclic Aromatic Compounds*, *Vol. 2*, W. Karcher, ed., Kluwer Academic Publishers, Dordrecht, The Netherlands, 1988.
4. *Spectral Atlas of Polycyclic Aromatic Compounds*, *Vol. 3*, W. Karcher, ed., Kluwer Academic Publishers, Dordrecht, The Netherlands, 1991.
5. L. A. Nakhimovskii, M. Lamotte, and J. Joussot-Dubien, *Handbook of Low Temperature Electronic Spectra of Polycyclic Aromatic Hydrocarbons*, Elsevier, Amsterdam, 1989.
6. A. L. Colmsjö, C. E. Östman, *Atlas of Shpol'skii Spectra and Other Low Temperature Fluorescence Spectra of POM*, University of Stockholm, Stockholm, Sweden, 1981.
7. M. R. Guerin, C.-H. Ho, T. K. Rao, B. R. Clark, and J. L. Epler, *Environ. Res.* **23**, 42 (1980).
8. R. A. Pelroy and D. L. Stewart, *Mutat. Res.* **92**, 297 (1981).
9. G. Lothrop, E. Hefner, 1. Alfheim, and M. Moller, *Science* **209**, 1037 (1980).
10. K. W. Sigvardson, J. M. Kennish, and J. W. Birks, *Anal. Chem.* **56**, 1069 (1984).
11. Z. Jin, Sh. Dong, W. Xu, Y. Li, and X. Xu, *J. Chromatogr.* **386**, 185 (1987).

12. J. Arey, R. Atkinson, B. Zielinska, and P. A. McElroy, *Environ. Sci. Technol.* **23**, 321 (1989).
13. T. Ramdahl, B. Zielinska, J. Arey, and R. Atkinson, *Nature* **321**, 425 (1986).
14. L. M. Ball and L. C. King, *Environ. Int.* **11**, 355 (1985).
15. C.-H. Ho, B. R. Clark, M. R. Guerin, B. D. Barkenbus, T. K. Rao, and J. L. Epler, *Mutat. Res.* **85**, 335 (1981).
16. I. Salmeen, A. M. Durisin, T. J. Prater, T. Riley, and D. Schuetzle, *Mutat. Res.* **104**, 17 (1982).
17. S. B. Tejada, Z. B. Zweidinger, and J. E. Sigsby, *J. Anal. Chem.* **58**, 1827 (1986).
18. W. A. MacCrehan, W. E. May, S. D. Yang, and B. A. Benner, Jr., *Anal. Chem.* **60**, 194 (1988).
19. A. L. Colmsjö, Y. Zebühr, and C. E. Östman, *Chem. Script.* **20**, 123 (1982).
20. A. E. Elsaiid, R. Walker, S. Weeks, A. P. D'Silva, and V. A. Fassel, *Applied Spectroscopy* **42**, 731 (1988).
21. S. Matsuzawa, P. Garrigues, H. Budzinski, J. Bellocq, and Y. Shimizu, *Anal. Chim. Acta.* **312**, 165 (1995).
22. I. S. Kozin, C. Gooijer, and N. H. Velthorst, *Anal. Chim. Acta* **333**, 193 (1996).
23. I. Chiang, J. M. Hayes, and G. J. Small, *Anal. Chem.* **54**, 318 (1982).
24. J. W. Hofstraat, C. Gooijer, and N. H. Velthorst, in *Molecular Luminescence Spectroscopy, Methods and Applications*, *Part 2*, S. G. Schulman, ed., Wiley, New York, 1988, Chapter 4.
25. R. I. Personov, in *Laser Analytical Spectrochemistry*, V. S. Letokhov, ed., Hilger, Bristol, 1985, Chapter 6.
26. Certificate of Analysis. Standard Reference Material 1650, Diesel Particulate Matter, NBS, 1991.
27. W. A. MacCrehan, W. E. May, S. D. Yang, and B. A. Benner, Jr., *Anal. Chem.* **60**, 194 (1988).
28. C. Gooijer, F. Ariese, J. W. Hofstraat, and N. H. Velthorst, *Trends Anal. Chem.* **13**, 53 (1994).
29. S. J. Weeks, S. M. Gilles, R. L. M. Dobson, S. Senne, and A. P. D'Silva, *Anal. Chem.* **62**, 1472 (1990).
30. P. Garrigues, R. De Vazelhes, M. Ewald, J. Joussot-Dubien, J.-M. Schmitter, and G. Guiochon, *Anal. Chem.* **55**, 138 (1983).
31. M. Dorbon, J. M. Schmitter, P. Garrigues, I. Ignatiadis, M. Ewald, P. Arpino, and G. Guiochon, *Org. Geochem.* **7**, 111 (1984).
32. J. M. Schmitter, P. Garrigues, I. Ignatiadis, R. De Vazelhes, F. Perin, M. Ewald, and P. Arpino, *Org. Geochem.* **6**, 579 (1984).
33. J. Mao, C. R. Pacheco, D. D. Traficante, and W. Rosen, *J. Chromatogr. A* **684**, 103 (1994).
34. J.-M. Schmitter, I. Ignatiadis, M. Dorbon, P. Arpino, G. Guiochon, H. Toulhoat, and A. Huc, *Fuel* **63**, 557 (1984).

35. M. Dorbon, I. Ioannis, J.-M. Schmitter, P. Arpino, G. Guiochon, H. Toulhoat, and A. Huc, *Fuel* **63**, 565 (1984).
36. D. W. Later, T. G. Andros, and M. L. Lee, *Anal. Chem.* **55**, 2126 (1983).
37. B. A. Tomkins and C.-H. Ho, *Anal. Chem.* **54**, 91 (1982).
38. M. T. Galceran, M. J. Curto, L. Puignou, and E. Moyano, *Anal. Chim. Acta* **295**, 307 (1994).
39. K. W. Sigvardson, J. M. Kennish, and J. W. Birks, *Anal. Chem.* **56**, 1096 (1984).
40. G. Bett, T. G. Harvey, T. W. Matheson, and K. C. Pratt, *Fuel* **62**, 1445 (1983).
41. S. E. Moschopedis, R. W. Hawkins, and J. G. Speight, *Fuel* **60**, 397 (1981).
42. J. G. Mueller, S. E. Lantz, D. Ross, R. J. Colvin, D. P. Middaugh, and P. H. Pritchard, *Environ. Sci. Technol.* **27**, 691 (1993).
43. S. Sueta, T. Yamashita, N. Shigemori, and K. Haraguchi, *J. Japan Soc. Air Pollut.* **21**, 527 (1986).
44. D. Bodzek, K. Tyrpien, and L. Warzecha, *Intern. J. Environ. Anal. Chem.* **52**, 75 (1993).
45. J. M. Bayona, K. E. Markides, and M. L. Lee, *Environ. Sci. Technol.* **22**, 1440 (1988).
46. R. Atkinson, E. C. Tuazon, J. Arey, and S. M. Aschmann, *Atmosph. Environ.* **29**, 3423 (1995), and references therein.
47. International Agency for Research on Cancer (IARC), Monographs 3 (1973), Lyon.
48. C.-H. Ho, B. R. Clark, M. R. Guerin, B. D. Barkenbus, T. K. Rao, and J. L. Epler, *Mutat. Res.* **85**, 335 (1981).
49. P. L. A. van Vlaardingen, W. J. Steinhoff, P. de Voogt, and W. Admiraal, *Environ. Toxicol. Chem.* **15**, 2035 (1996).
50. M. H. S. Kraak, C. Ainscough, A. Fernández, P. L. A. van Vlaardingen, P. de Voogt, and W. Admiraal, *Aquat. Toxicol.* **37**, 9 (1997).
51. M. H. S. Kraak, P. Wijnands, H. A. J. Govers, W. Admiraal, and P. de Voogt, *Environ. Toxicol. Chem.* **16**, 2158 (1997).
52. M. Blumer, T. Dorsey, and J. Sass, *Science*, 195 (1977).
53. E. T. Furlong and R. Carpenter, *Geochim. Cosmochim. Acta* **46**, 1385 (1982).
54. S. G. Wakeham, *Environ. Sci. Technol.* **13**, 1118 (1979).
55. G. Grimmer and K. W. Naujack, *J. Assoc. Off. Anal. Chem.* **69**, 537 (1986).
56. T. Nielsen, P. Clausen, and F. P. Jensen, *Anal. Chem. Acta* **187**, 223 (1986).
57. W. C. Brumley, C. M. Brownrigg, and G. M. Brilis, *J. Chromatogr.* **558**, 223 (1991).
58. T. Yamauchi and T. Handa, *Environ. Sci. Technol.* **21**, 1177 (1987).
59. F. L. Joe, J. Salemme, and T. Fazio, *J. Assoc. Off. Anal. Chem.* **69**, 218 (1986).
60. H. Carlsson and C. Östman, *J. Chromatogr.* **715** (1995).
61. T. Hanai and J. Hubert, *J. Liq. Chromatogr.* **8**, 2463 (1985).
62. R. N. Nurmukhametov, *Russ. Chem. Rev.* **38**, 180 (1969).
63. L. A. Klimova, G. N. Nersesova, V. A. Prozorovskaya, and G. S. Ter-Sarkisyan, *Russ. J. Phys. Chem.* **50**, 1825 (1976).

64. A. L. Colmsjö and C. E. Östman, *The Sixth Int. Symposium on Polynuclear Aromatic Hydrocarbons. Physical and Biological Chemistry*, M. Cooke and A. J. Dennis, eds., Battelle Press, Columbus, Ohio, p. 201 (1982).
65. P. Garrigues, R. De Vazelhes, J. M. Schmitter, and M. Ewald, *Polynuclear Aromatic Hydrocarbons: Formation, Metabolism and Measurement*, M. Cooke and A. J. Dennis, eds., Battelle, Press, Columbus, Ohio, 1983, p. 545.
66. P. Garrigues and M. Ewald, *Intern. J. Environ. Anal. Chem.* **21**, 185 (1985).
67. I. S. Kozin, O. F. A. Larsen, P. de Voogt, C. Gooijer, and N. H. Velthorst, *Anal. Chim. Acta* **354**, 181–87 (1997).
68. K. P. Dinse and C. J. Winscom, *J. Luminescence* **18/19**, 500 (1979).
69. P. Garrigues, M. Dorbon, J. M. Schmitter, and M. Ewald, *Polynuclear Aromatic Hydrocarbons: Mechanisms, Methods and Metabolism*, M. Cooke and A. J. Dennis, eds., Batttelle Press, Columbus, Ohio, 1984, p. 451.
70. P. Garrigues and H. Budzinski, *Trends Anal. Chem.* **14**, 231 (1995).
71. R. E. Rebbert, S. N. Chesler, F. R. Guenther, and R. M. Parris, *J. Chromatogr.* **284**, 211 (1984).
72. R. V. Schultz, J. W. Jorgenson, M. P. Maskarinec, M. Novotny, and L. J. Todd, *Fuel* **58**, 783 (1979).
73. J. R. Patel, E. B. Overton, and J. L. Laseter, *Chemosphere* **8**, 557 (1979).
74. M. M. Krahn, G. M. Ylitalo, J. Buzitis, S.-L. Chan, U. Varanasi, T. L. Wade, T. J. Jackson, J. M. Brook, D. A. Wolfe, and C.-A. Manen, *Environ. Sci. Technol.* **27**, 699 (1993).
75. H. Takada, T. Onda, and N. Ogura, *Environ. Sci. Technol.* **24**, 1179 (1990).
76. T. McFall, G. M. Booth, M. L. Lee, Y. Tominaga, R. Pratap, M. Tedjamulia, and R. Castle, *Mutat. Res.* **135**, 97 (1984).
77. R. A. Pelroy, D. L. Stewart, Y. Tominaga, M. Iwao, R. N. Castle, and M. L. Lee, *Mutat. Res.* **117**, 31 (1983).
78. W. Karcher, A. Nelen, R. Depaus, J. van Eijk, P. Glaude, and J. Jacob, *Polynuclear Aromatic Hydrocarbons: Chemical Analysis and Biological Fate*, M. Cooke and A. J. Dennis, eds., Battelle Press, Columbus, Ohio, 1981, p. 317.
79. J. R. Payne and C. R. Phillips, *Environ. Sci. Technol.* **19**, 569 (1985).
80. C.-H. Chaîneau, J.-L. Morel, and J. Oudot, *Environ. Sci. Technol.* **29**, 1615 (1995).
81. C. S. Woo, A. P. D'Silva, and V. A. Fassel, *Anal. Chem.* **52**, 159 (1980).
82. M. Vial, J. Jarosz, M. Martin-Bouyer, and L. Paturel, *J. Photochemistry* **33**, 67 (1986).
83. A. L. Colmsjö, Y. U. Zebühr, and C. E. Östman, *Anal. Chem.* **54**, 1673 (1982).
84. Y. Sugimoto, Y. Miki, S. Yamadaya, and M. Oba, *J. Chem. Soc. of Japan* **12**, 1954 (1984).
85. L. E. Paulson, H. H. Schobert, and R. C. Ellman. *Prepr. Pap.-Am. Chem. Soc., Div. Fuel Chem.* **23**, 107 (1978).
86. V. Lopez-Avila, R. Young, and W. F. Beckert, *Anal. Chem.* **66**, 1097 (1994).

87. D. Schuetzle, F. S.-C. Lee, T. J. Prater, and S. B. Tejada, *Intern. J. Environ. Anal. Chem.* **9**, 93 (1981).

88. J. M. Bayona, P. Fernandez, and J. Albaiges, *Polycyclic Aromatic Compounds: Synthesis, Properties, Analytical Measurements, Occurrence and Biological Effects*, P. Garrigues and M. Lamotte, eds., Gordon and Breach, 1993.

89. D. Helmig, J. López-Cancio, J. Arey, W. P. Harger, and R. Atkinson, *Environ. Sci. Technol.* **26**, 2207 (1992).

90. R. Schoeny, *Mutat. Res.* **101**, 45 (1982).

91. A. Bree, V. V. B. Vilkos, and R. Zwarich, *J. Mol. Spectrosc.* **48**, 135 (1973).

92. J. Rima, L. A. Nakhimovsky, M. Lamotte, and J. Joussot-Dubien, *J. Phys. Chem.* **88**, 4302 (1984).

93. C. Amine, L. A. Nakhimovsky, F. Morgan, and M. Lamotte, *J. Phys. Chem.* **94**, 3931 (1990).

94. R. E. Connors and W. R. Christian, *J. Phys. Chem.* **86**, 1524 (1982).

CHAPTER

6

SHPOL'SKII SPECTROSCOPIC ANALYSIS OF PAH METABOLITES

FREEK ARIESE

Department of Analytical Chemistry and Applied Spectroscopy, Vrije Universiteit Amsterdam, De Boelelaan 1083, NL-1081 HV Amsterdam, The Netherlands

6.1. INTRODUCTION: BIOLOGICAL MONITORING OF PAH EXPOSURE

Polycyclic aromatic hydrocarbons (PAHs) can be defined as a group of compounds featuring two or more fused aromatic rings, and containing no other elements than hydrogen and carbon. PAHs are naturally present in crude oil and are consequently found in a number of petrochemical products and waste (e.g., oil spills). Petrochemical PAH mixtures have been formed by successive dehydrogenation of organic material at relatively low temperatures. Alkylated two- or three-ring compounds are typically most abundant. On the other hand, PAHs may also be formed as the result of incomplete combustion (e.g., fires, gasoline and diesel engines, domestic heating, fossil-fuel-fired power plants). Typical combustion-related PAH mixtures contain mainly nonalkylated compounds, and the PAHs of four and more rings are usually more abundant than the smaller ones. Compared to natural background concentrations (for instance, from forest fires), human activity has caused a sharp increase in environmental PAH levels. Attempts are currently being made to reduce emissions by means of cleaner combustion technologies, and by regulating the use of PAH-containing products such as coal tar, creosote, etc.

Many PAHs are proven or suspected carcinogens; well-known examples are benzo[*a*]pyrene (BaP), chrysene, and the most potent PAH known to date, dibenzo[*a*,*l*]pyrene [1]. However, these compounds are relatively inert, and need metabolic activation before they can react with genetic material. The potent carcinogen BaP has been studied extensively and is often used as a model compound to investigate the toxic effects of PAHs [2]. For BaP and many other PAHs there are several metabolic pathways [3]. According to the mono-oxygenation pathway [4], biotransformation starts with the binding to the

Shpol'skii Spectroscopy and Other Site-Selection Methods, Edited by Cees Gooijer, Freek Ariese, and Johannes W. Hofstraat,
ISBN 0-471-24508-9

cytochrome P450 enzyme system, which then catalyzes the addition of an oxygen atom across a double bond of the molecule, forming an epoxide. This epoxide may subsequently be coupled to glutathion, rearrange into a phenol, or be hydrolyzed to yield a saturated dihydrodiol moiety. During phase-II metabolism, phenols and dihydrodiols can be conjugated to polar groups like glucuronic acid or sulfate to further facilitate excretion. The result of this detoxification mechanism is that in mammals and many other classes of higher animals the bulk of PAH molecules is rapidly rendered harmless and excreted. However, some reactive intermediates formed during the detoxification process may form stable adducts with proteins or with DNA (see Chapter 11). The latter could result in cancer initiation if the defective nucleotide is not repaired in time.

Exposure to high PAH levels in the aquatic environment is believed to be responsible for the increased incidence of liver tumors and other hepatic lesions in fish populations at certain locations [5, 6]. Terrestrial animals may be exposed to PAHs via ingestion of polluted plant material or soil. Also humans may be exposed to PAHs via several routes, such as smoking, consumption of smoked or charcoal-broiled food products [7], or the inhalation of soot particles from open fires, car exhausts, or at certain workplaces [8]. From the above it is clear that monitoring of PAH exposure is an important tool in environmental assessment, pollution control, and human health.

Often the determination of PAH levels in environmental matrices is not sufficient, since it provides little information on the actual uptake and subsequent risks. On the other hand, determination of toxicological endpoints such as DNA damage or (pre)cancerous stages has its drawbacks, not only because one would obviously prefer an early warning parameter over a biomarker that indicates that the damage has already been done. Also, it is always difficult to link long-term effects to underlying causes, such as exposures having taken place several years earlier. For these reasons, short-term biomarkers fill an important niche, as they are more easily interpreted in terms of actual exposure levels or body burdens.

As mentioned above, PAHs usually do not show significant accumulation in higher animals. Therefore, the challenge is to find alternative ways to assess the PAH amounts that have actually been absorbed and metabolized. To this end, several methods have been developed to determine the levels of PAH metabolites in specific tissues or excreta [9–12]. Fluorescence-based techniques are frequently used for the analysis of these PAH derivatives. Examples are liquid chromatography with fluorescence detection [9, 10], ion-pair chromatography with fluorescence detection [12], or (synchronous) fluorescence spectrometry [11]. However, in some cases these techniques do not offer the necessary level of selectivity or sensitivity. In this chapter, we will review the development and application of high-resolution fluorescence techniques for the analysis of PAH metabolites. The two spectroscopic methods considered are Shpol'skii spectroscopy [13] and, to a lesser extent, fluorescence line-narrowing spectroscopy

(FLN) [14]. Examples of applications from the fields of metabolism studies, marine monitoring and occupational hygiene will be presented.

6.2. SHPOL'SKII SPECTROSCOPY OF PAH METABOLITES IN *n*-ALKANES

6.2.1. Host–Guest Compatibility

The suitability of the Shpol'skii technique has been demonstrated extensively for parent PAHs and many other types of PAH derivatives (see previous chapters and references therein). Still, publications describing the use of Shpol'skii spectrometry for PAH metabolites are relatively rare. Reviewing the early Russian literature, Nurmukhametov states that "Attempts to obtain an effectively resolved structure for the electronic bands of aromatic alcohols . . . in *n*-hydrocarbon solution did not give positive results" [15]. Nevertheless, a few years later Khesina et al. [16] observed quasi-linear spectra for two phenolic BaP metabolites, 3-OH-BaP and 6-OH-BaP, in frozen *n*-octane. Garrigues and Ewald reported quasilinear spectra for 9-OH-BaP [17]. Among the hydroxy-PAHs included in Karcher's spectral atlas, a Shpol'skii spectrum was included only for 3-OH-BaP [18]. So far, the Shpol'skii effect has only been observed for monohydroxy (phenolic) PAH metabolites. The sensitivity of the Shpol'skii method for phenolic metabolites of BaP is more than an order of magnitude poorer in comparison with the parent compound BaP [19], although at room temperature the fluorescence quantum yields of the two compounds are similar. In the case of 1-hydroxy pyrene the difference in sensitivity is even larger. Experimental evidence suggests that this phenomenon is caused by the limited host–guest compatibility due to the polar hydroxy group, the latter having a stronger influence in the case of the four-ring pyrene derivative than in the case of the five-ring BaP derivative. In principle, two effects can occur, resulting from either the poorer solubility in *n*-alkanes or the poorer compatibility with the *n*-alkane crystal lattice. The first is the formation of aggregates during the cooling procedure, leading to precipitation and reduced quantum yields. The second concerns freezing out of the analyte molecules, at the point of matrix solidification, from the crystalline phase being formed into the amorphous, intergrain phase [20], resulting in broadband emission at the expense of the intensity of the Shpol'skii lines. Both phenomena will also cause the intensity and appearance of the Shpol'skii spectrum to depend critically on cooling rate, sample holder design, traces of other solvents, etc., thus hampering analytical reproducibility and interlaboratory comparability. As an example, Figure 6.1 shows typical Shpol'skii spectra of 3-OH BaP in *n*-octane recorded in different laboratories. Since the experimental conditions such as concentration, temperature and excitation wavelength were similar, the differences in the intensity of the broadband background are, most probably, due to

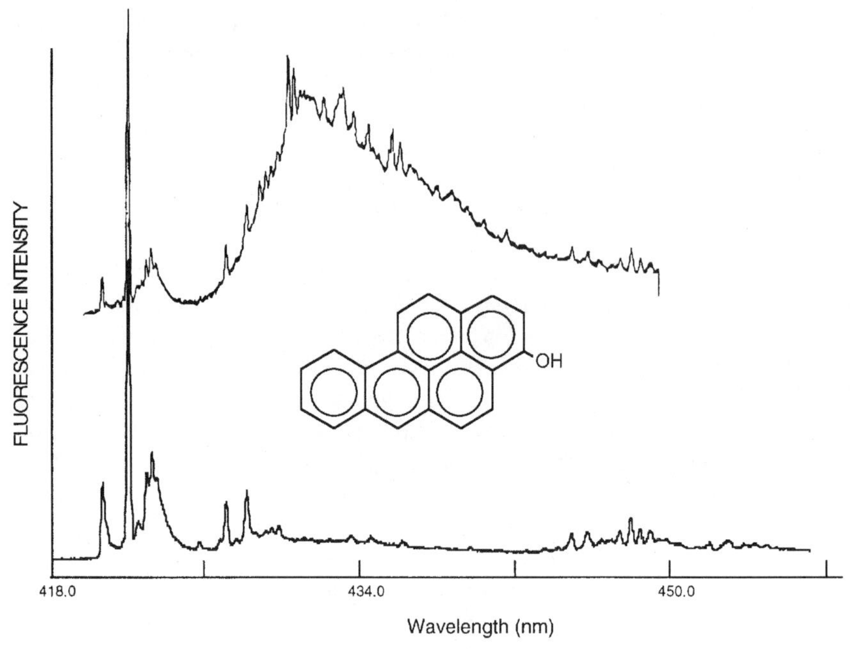

Figure 6.1. Comparison of Shpol'skii spectra, illustrating spectral dependence on experimental conditions for β-type systems. Top: standard reference spectrum of 3-hydroxy BaP in *n*-octane, conc. = 5 m × 10^{-6} M, lamp exc. 307 nm, temp.15 K (from Ref. 18, with permission from Kluwer Academic Publishers). Bottom: spectrum of the same compound, recorded in the author's laboratory; conc. = 1.4 m × 10^{-5} M; lamp exc. 300 nm, temp. = 10 K (from Ref. 21, with permission from Gordon and Breach Publishers).

differences in sample holder design, leading to different cooling rates. The monophenolic BaP metabolites in *n*-octane may be classified as a nonequilibrium, β-type system with intermediate host–guest compatibility. The reader is referred to Chapter 3 for a more detailed discussion on such host–guest combinations.

As a result of the experimental difficulties described above, the direct Shpol'skii analysis of monohydroxy-BaP metabolites can only be applied if trace-level sensitivity is not required. At the onset of a fish-metabolism study, several flounder (*Platichthys flesus*) were given a relatively high dose of BaP (4.1 mg/kg body weight) and bile was collected after 2 days. This material was then used during the initial method optimization phase, and 1-OH-BaP and 3-OH-BaP were directly identified in the hydrolyzed and extracted fish bile sample [21]. Although these results were encouraging, the sensitivity of the method proved insufficient for application to bile from fish exposed to environmentally realistic levels.

In a separate experiment, the freshwater isopod *Asellus aquaticus* was exposed via the water phase to pyrene, and 1-OH-pyrene was extracted and identified based on its Shpol'skii spectrum (P. Cid-Montañés and F. Ariese, unpublished results). Again, as the result of the poor host–guest compatibility and thus poor detection limits, the direct method could not be used for field measurements of 1-OH pyrene.

6.2.2. Chemical Derivatization

In order to improve the compatibility of PAH metabolites with typical *n*-alkane Shpol'skii matrices, Weeks and co-workers [22] first described a procedure to transform monohydroxybenz[*a*]anthracenes into less polar methoxy derivatives, which could subsequently be analyzed by means of laser-excited Shpol'skii spectroscopy (LESS). Later, the methylation and Shpol'skii spectroscopic analysis of several types of BaP metabolites was reported: monohydroxy-BaP derivatives, BaP-dihydrodiols, BaP- dihydrodiolepoxide, and BaP-tetrahydrotetrol [19, 23].

Briefly, the sample extraction and derivatization is carried out as follows: Samples are enzymatically treated to hydrolyze phase-II conjugates, and are subsequently extracted with a nonpolar, volatile solvent such as *n*-hexane. A few mg of sodium hydride is washed with *n*-pentane in a flask under nitrogen atmosphere; dimethyl sulfoxide (DMSO) is added; and the mixture is stirred at 70°C for several minutes until the formation of H_2 bubbles has ceased. After cooling to room temperature, methyl iodide and the sample extract are added; after several minutes of stirring the reaction is quenched with water. The methylated products are quantitatively extracted with *n*-hexane and concentrated, and the solvent is gradually replaced with *n*-octane in a gentle stream of nitrogen.

As an example, Figure 6.2 shows the Shpol'skii spectrum of a methylated fish bile extract from flounder after exposure to moderately polluted harbor sediment in a mesocosm [24]. No further chromatographic separation was applied. Hydroxypyrene is one of the predominant metabolites in fish bile exposed to combustion PAHs, and this compound was positively identified in the methylated extract by comparison of its Shpol'skii spectrum with that of an authentic 1-methoxypyrene calibrant.

The methylation procedure was found to be very practical and quantitative for phenolic metabolites. In the case of 3-methoxy-BaP or 1-methoxy-BaP in *n*-octane the sensitivity of lamp-excited or laser-excited Shpol'skii spectrometry was comparable to that of the parent compound BaP and no longer critically dependent on cooling rate. Apparently, the methylation procedure had turned a β-type system into a fully compatible α-type system (see Chapter 3). Methylated derivatives from all 12 hydroxy-BaP isomers produced good Shpol'skii spectra in *n*-octane [19]. Good results were also obtained for phenolic metabolites when

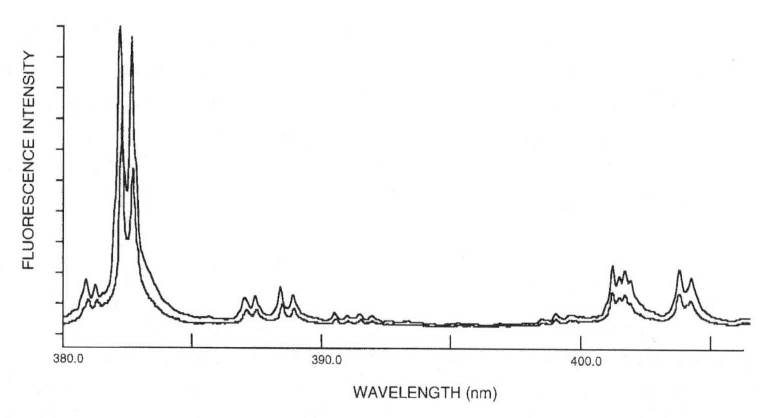

Figure 6.2. Top: Shpol'skii spectrum of hydrolyzed, methylated flounder bile, following exposure to polluted sediment. Bottom: reference spectrum of 1-methoxy-pyrene. Lamp exc. 345; temp. = 28 K; solvent *n*-octane (from Ref. 24, with permission from Gordon and Breach Publishers).

other derivatizing agents were used; suitable Shpol'skii spectra and good detection limits were obtained for acetylated 6-OH-BaP and for octyl derivatives of 3-OH-BaP, 6-OH-BaP, and 9-OH-BaP [23].

Unfortunately, derivatization of polyhydroxylated metabolites proved much more complicated. In the case of methylation of BaP dihydrodiol metabolites a mixture of products was obtained [19]. For instance, BaP 9,10-dihydrodiol yielded 9-methoxy-BaP, 10-methoxy-BaP, as well as the expected dimethylated product 9,10-dihydrodimethoxy-BaP; the relative ratios depended critically on the derivatization conditions. When analyzing methylated fish bile from flounder injected with BaP, small amounts of 7-methoxy- and 8-methoxy-BaP, or of 9-methoxy- and 10-methoxy-BaP, could be observed in some samples, their levels depending on the derivatization conditions [19]. Apparently, at least part of these compounds were formed via elimination/methylation during the derivatization of 7,8-dihydrodiol-BaP and 9,10-dihydrodiol-BaP, respectively.

Other problems are related to the limited sensitivity of Shpol'skii spectra of methylated polyhydroxy metabolites [19, 23]. Table 6.1 lists the detection limits obtained for the various types of (methylated) BaP metabolites, using either lamp or laser excitation. For methylated phenolic BaP metabolites excellent detection limits in the attomole range were reported by Weeks [23] and by Ariese [19], both applying laser excitation and similar detection systems. On the other hand, the detection limits found for permethylated BaP diols, BaP diolepoxide, and BaP tetrol were many orders of magnitude poorer, although the same instrumental setup was used [23]. This observation can only partially be explained by the weaker S_1–S_0 absorption of the pyrene chromophore in comparison with the BaP chromophore. Apparently, in the case of polyhydroxylated metabolites the

Table 6.1. Detection Limits for Selected BaP Metabolites, Illustrating the Effect of Methylation on the Detectability of Mono- and Polyhydroxy Metabolites

Compound	Derivatization	Matrix	Excitation	Detection Limit (femtomole)[a]	Ref.
BaP	—	*n*-octane	Xe Lamp, 300 nm	3	[25]
BaP	—	*n*-octane	Nd:YAG-dye laser	0.05	[26]
3-OH-BaP	—	*n*-octane	Xe lamp, 300 nm	100	[19]
3-OH-BaP	methylated	*n*-octane	Xe lamp, 300 nm	5	[19]
3-OH-BaP	methylated	*n*-octane	Nd:YAG-dye laser, 418.36 nm	0.05	[19]
9-OH-BaP	methylated	*n*-octane	XeCl-dye laser, 392.5 nm	0.05	[23]
BaP 9,10-diol	permethylated	*n*-octane	XeCl-dye laser, 396.0 nm	2000	[23]
BaP 7,8-diol	permethylated	*n*-octane	XeCl-dye laser, 384.0 nm	3000	[23]
BaP diolepoxide	permethylated	*n*-octane	XeCl-dye laser, 356.5 nm	6000	[23]

[a] Concentration detection limits in the final sample can be calculated by dividing the absolute detection limits by the detector cell volume (10 μL for Refs. 19, 25, and 26 and 20 μL for Ref. 23).

permethylated derivatives are still not fully compatible with the *n*-alkane host. This could be related to the nonplanar geometry of these compounds, which precludes incorporation into the planar crystal lattice. In conclusion, Shpol'skii spectrometry, after derivatization, is an excellent method for monohydroxy metabolites, but is less suitable for polyhydroxylated metabolites for reasons of both sensitivity and reproducibility.

6.3. INSTRUMENTAL ASPECTS

For the determination of PAH metabolites the instrumental setup is, in principle, not different from that used for parent PAHs. Schematically, the apparatus consists of the following components: (1) excitation source, which can be a laser or a broadband light source combined with a wavelength selector, (2) cryogenic sample holder, (3) high-resolution emission monochromator, (4) detector and data read out.

1. Excitation sources. Since the Shpol'skii effect is a matrix-induced phenomenon, the use of a highly monochromatic (laser) excitation source is not a prerequisite for observing narrow-band spectra (it is in the case of fluorescence line-narrowing experiments in disordered matrices; see below and other chapters). Mercury and xenon arc lamps fitted to excitation monochromators have been widely used. For reasons of sensitivity, the excitation monochromator is usually operated with relatively broad slit widths (corresponding to bandwidths of 5–20 nm), and the excitation wavelength is tuned to a strong, but relatively broad higher absorption band such as S_2–S_0. More selective excitation into specific, narrow S_1–S_0 absorption lines can be employed to determine particular compounds in complex mixtures, or in order to obtain single-site spectra from systems that otherwise feature multiplet spectra corresponding to different substitutional sites. Narrow-band lamp excitation can be used for high analyte concentrations or for measuring Shpol'skii absorption spectra [27]. However, now tunable laser/dye laser systems are almost exclusively used for selective excitation. Furthermore, the excitation power is typically much higher (several mW) than in the case of a lamp–excitation monochromator combination, and laser beams are easily focused on very small sample volumes (see Chapter 10). Finally, if the laser is of the pulsed type, one can employ time-resolved detection for extra selectivity between different analytes or to remove (Raman) scattering background.

2. Cryostats. The Shpol'skii effect was first observed at a temperature of 77 K, but Shpol'skii spectra show a remarkable improvement when lower temperatures can be used [27]. Closed-cycle helium cryostats have become very popular because of their ease and low cost of operation (no helium consumption). In these cryostats the samples are mounted to a so-called cold station and cooled through thermal conductance; temperatures of 5–20 K can typically be reached. Alternatively, helium bath cryostats can be used for experiments at 4.2 K, or when a high sample throughput is important (instantaneous cooling). As discussed above, the cooling rate and in particular the cooling rate at the point of matrix solidification greatly influences the Shpol'skii spectra of many β-type host–guest combinations. In our laboratory a sample holder was made from a gold-plated copper block, in which four shallow 5-mm holes were drilled. Four 10-μL samples can be added and covered with sapphire windows. The sample thickness is determined by the thickness of teflon spacer rings, typically 0.5 mm. Owing to the small sample volume compared to the mass of the copper plate and the optimal thermal conductivity, the *n*-alkane matrices were observed to solidify practically instantaneously ($\ll$1 sec), thus reducing the probability of analyte molecules freezing out from the crystalline matrix. Contrary to our first expectations, rapid precooling of the sample holder and refrigerator cold station by immersing into liquid nitrogen, followed by cryocooling to 10–20 K, usually had a negative effect on the quality of the Shpol'skii spectra of β-type systems. We presume that under those conditions a strong temperature gradient exists across the sample holder, resulting in a more gradual matrix solidification process.

3. *Emission monochromators*. A good high-resolution monochromator is required to observe the quasi-linear emission spectrum. The spectral resolution of the instrument should be better than 0.2 nm. For optimal sensitivity, the instrument should also have an excellent light throughput and a not too high F/n number. Holographic gratings, optimized for the wavelength area of interest, are preferred. In the case of a single (PMT) detector the resolution is determined by both entrance and emission slits. In the case of multichannel array detectors (see below), the resolution is limited by the detector pixel size and broadening processes in the detector intensifier. If nonselective, short-wavelength excitation is applied, scattered excitation light is easily rejected with an appropriate cutoff filter. If, on the other hand, the excitation wavelength is close to the emission lines of interest, as in the case of selective laser excitation in the S_1–S_0 absorption region, the cutoff functions of such filters are usually not sufficiently steep. In that case, the use of a double or even triple monochromator will be very advantageous. Double monochromators cannot be used with array detectors.

Some attention should be paid to the light collection optics. Usually, the width of the sample holder will be several millimeters, much larger than the typical monochromator entrance slit widths of 50–200 μm. One might therefore be tempted to focus the sample emission onto a small image spot by means of a lens close to the entrance slit, but it should be realized that in that case the strongly diverging light cone inside the monochromator would not match its F/n number, and therefore this approach would still result in serious losses. In practice, we found that different light collecting options with different magnification factors did not make a significant difference in overall sensitivity. Projecting the sample to a spot size larger than the entrance slit is advantageous since it makes the alignment less critical.

4. Detectors. Early high-resolution spectroscopy experiments were carried out with photographic plates mounted to a stationary spectrograph for simultaneous detection of the complete spectral range of interest. Later, single photomultiplier tubes in combination with a scanning monochromator came into use. However, this approach has the obvious disadvantage that the spectrum needs to be recorded point by point. Apart from the time required to carry out the measurements, there are some extra disadvantages when the sample is photochemically unstable, or when the light source is not stable at the time scale of the experiment (flicker noise). More recently, several types of multichannel detectors have become available that can be mounted in the (exit) focal plane of the monochromator after removal of the exit slit. The intensified linear diode array (ILDA) detector typically consists of 512 or 1024 separate photodiodes (pixel size ca. $2\,mm \times 25\,\mu m$). Charge-coupled-device (CCD) detectors typically contain 512×512 pixels of approximately $20 \times 20\,\mu m$ each. ILDA and CCD detectors usually show a somewhat lower quantum efficiency than the best PMTs, but this is more than compensated for by the multiplex advantage. Furthermore, in the case of pulsed laser excitation, time discrimination between a strong excitation scatter pulse and a weak fluorescence signal is more easily accomplished with an intensified multichannel detector than with a PMT. The spectral data are subsequently fed into a computer; a variety of software packages is available for spectral handling.

6.4. APPLICATIONS

6.4.1. Shpol'skii Spectrofluorimetric Analysis of BaP Metabolites in Fish Bile

As mentioned in the Introduction, biological monitoring approaches, such as the determination of a particular compound (or its metabolites) in specific organisms, tissues, or excreta, can provide useful information on its actual uptake. An important environmental issue regards the exposure of fish to PAHs (e.g., through creosoted wood, tar-coated ships, polluted sediments, or oil spills). Fish usually show insignificant accumulation of PAHs [28]; instead, PAHs are rapidly metabolized into more polar derivatives that are stored in the gallbladder prior to excretion [29, 30]. These PAH metabolite levels could provide direct insight into the recent uptake. Krahn and co-workers developed HPLC/fluorescence [9] and GC-MS methods [31] to determine PAH metabolites in fish bile; a rapid synchronous fluorescence method was described by Ariese [11]. These methods were then applied in the field and in mesocosm experiments in order to study the relationships between environmental PAH levels, PAH uptake, and toxicological effects.

BaP is the most widely studied PAH carcinogen, and BaP metabolites are therefore obvious target analytes for exposure monitoring. Bile of fish, exposed to many different PAHs, will contain an even more complex mixture of PAH metabolites. Furthermore, as the result of large solubility differences in water, the uptake of large, carcinogenic PAHs like BaP tends to be orders of magnitude slower than that of smaller, less toxic PAHs [19, 32]. Therefore, the detection of BaP metabolites in field samples requires extremely sensitive and selective methods. With HPLC/fluorescence, the major BaP metabolite, 3-OH-BaP, could only be detected in some bile samples from highly polluted sites [32], but the method was not sufficiently sensitive for moderately polluted areas. As an extra complication, the bile volumes available from many fish species are not sufficient for trace enrichment.

In order to achieve the required detection limits in fish bile samples, a LESS method was developed for the identification and quantification of 3-OH-BaP. The model fish studied was the flatfish species flounder (*Platichthys flesus*). High exposure levels were realized by administering a single dose of BaP via parenteral injection. To simulate semichronic exposure to realistic BaP pollution levels, flounders were kept for four weeks in different mesocosms corresponding to pollution levels commonly found in coastal areas or estuaries in the Netherlands [33]. Field samples were caught by beam trawl at the open North Sea.

Bile samples were enzymatically hydrolyzed and subsequently derivatized with methyl iodide. The product yield of the two major compounds, 1-methoxy-BaP and 3-methoxy-BaP, was independent from the derivatization conditions. A typical Shpol'skii spectrum of a methylated bile extract (from flounder that had received a high dose via injection) is shown in Figure 6.3. Nonselective laser excitation at 348 nm was employed, in order to be able to observe all metabolites simultaneously; the overall dilution factor was 1000. The spectrum is dominated by 3-methoxy-BaP and 1-methoxy-BaP. 3-OH-BaP was chosen as a biomarker for the biomonitoring of BaP uptake in the aquatic environment.

For quantification of 3-methoxy BaP with LESS, an internal standard was added in order to correct for changes in laser power, optical alignment, or sample thickness. Optimal correction can only be achieved if the internal standard and the analyte are simultaneously excited and detected. This means that the internal standard should absorb to some extent at the wavelength chosen for the analyte, and it should have a relatively strong emission line in the emission window covered by the multichannel detector, while not overlapping with the analyte spectrum. Perylene-d_{12} was found to be a suitable internal standard for 3-methoxy-BaP; for correction the peak areas of the 0–0 emission lines of the analyte and the internal standard were ratioed. The sensitivity of the LESS method appeared not to be seriously affected by matrix interferences. When the

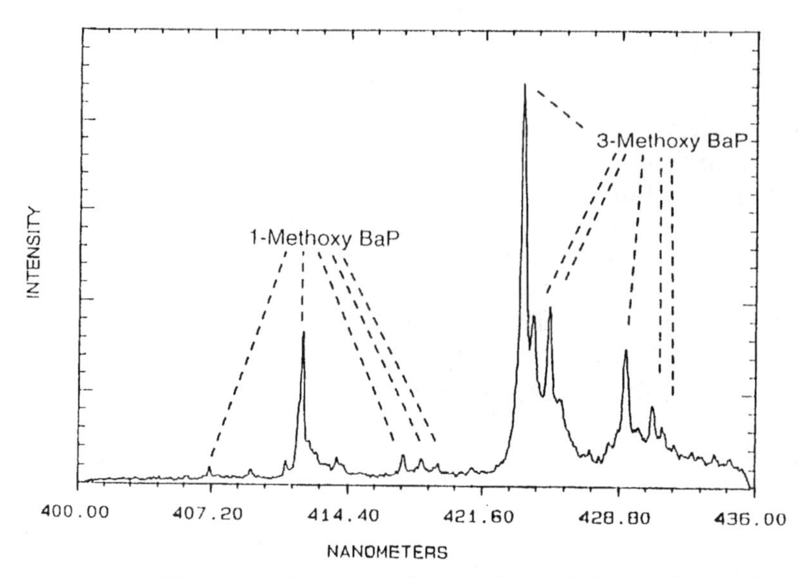

Figure 6.3. Shpol'skii spectrum of methylated flounder bile sample (0.78 mg/kg BaP injected), featuring derivatized 1-hydroxy-BaP and 3-hydroxy-BaP. Nonselective laser excitation at 348 nm; solvent *n*-octane. (Reproduced from Ref. 19, with permission from the American Chemical Society.)

sample treatment was carried out without overall dilution, the detection limit was 2×10^{-11} M (compared to 5×10^{-12} M for standard solutions) or 0.005 ng/ml. The analytical repeatability (four replicates of sample extraction and determination) was 16%.

Figure 6.4 shows the LESS spectrum of 3-methoxy BaP in a flounder bile extract from the Wadden Sea mesocosm. Note that in comparison with Figure 6.3 the compound shows fewer bands. For the spectrum of Figure 6.3 nonselective excitation at 348 nm was used, thus revealing the complete multiplet spectrum (3-methoxy-BaP can occupy several distinct sites in the *n*-octane lattice). For the spectrum of Figure 6.4, however, laser excitation at 418.36 nm was used, selectively tuned to a vibronic transition of the major site. The internal standard perylene d_{12} is also visible in the spectrum of Figure 6.4.

The LESS results from the mesocosm study, summarized in Table 6.2, show that fish exposed to Rotterdam harbor sediment had absorbed and metabolized 40 times more BaP than fish from the Wadden Sea mesocosm. Furthermore, fish from the intermediate mesocosm (same pollution level, but no direct contact with the sediment) showed only a six fold increase, indicating that some uptake of BaP can take place through the water phase or through the diet [34], but that direct contact with the sediment is the major route of exposure for a bottom-dwelling fish like flounder. Direct absorption through skin or gills, or ingestion of PAH-

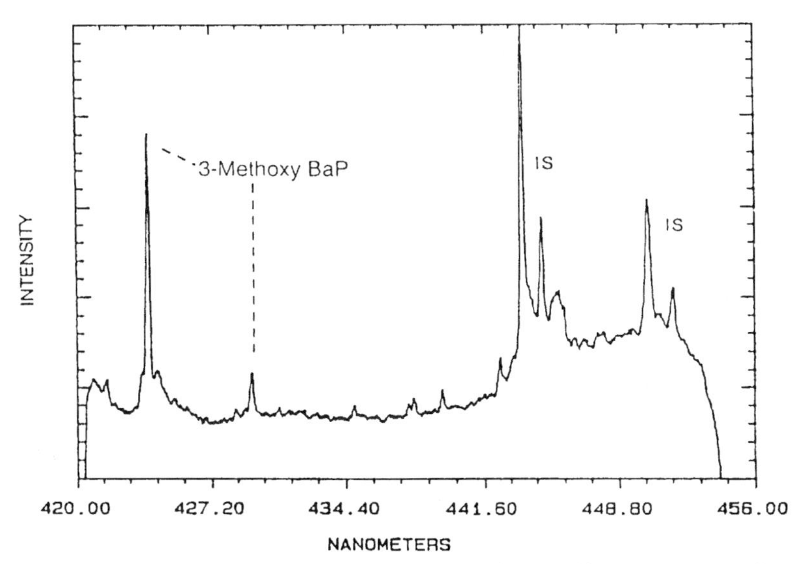

Figure 6.4. Shpol'skii spectrum of methylated bile extract from Wadden Sea mesocosm, featuring derivatized 3-hydroxy-BaP. Selective laser excitation at 418.36 nm; IS = perylene d_{12} (internal standard); solvent *n*-octane. (Reproduced from Ref. 19, with permission from the American Chemical Society.)

Table 6.2. 3-OH-BaP Levels in Fish Bile after Different Exposure Conditions; Data from Refs. 19 and 25

Origin	3-OH-BaP conc. (ng/mL)
Rotterdam harbor sediment (mesocosm)	50 ± 36 ($n = 9$)
Rotterdam harbor sediment; no direct contact	7.7 ± 2.4 ($n = 4$)
Wadden Sea sand (reference mesocosm)	1.2 ± 0.1 ($n = 3$)
Open North Sea (field samples)	0.005 ($n = 2$)

containing particles, may both be important contributors. A few samples from the open North Sea revealed very low BaP exposure levels; 3-OH-BaP was detected at levels close to the detection limit of 0.005 ng/ml.

6.4.2. Shpol'skii Spectrofluorimetric Analysis of BaP Metabolites in Workers' Urine

In certain industrial settings, workers may be exposed to high PAH levels via skin contamination or through inhalation of either volatile PAHs in the gas phase or PAHs bound to soot particles. Epidemiological data have linked the high incidence of lung cancer in workers and ex-workers of steel foundries to specific jobs at coke ovens, carried out without proper protective measures [35]. Jongeneelen developed an HPLC method for the determination of 1-OH-pyrene in urine and found evidence for increased pyrene exposure of workers in certain industries [10]. It would be important to know whether the 1-OH pyrene levels in urine would also be a suitable biomarker for exposure of larger, more carcinogenic PAHs, such as BaP. Unfortunately, this could not be tested by Jongeneelen as the HPLC method was not sufficiently sensitive for urinary 3-OH-BaP (even if pyrene and BaP would be ingested in similar amounts, the fraction excreted via the urine is much smaller for BaP than for pyrene). It was then decided to adapt the LESS method described above to the analysis of methylated urine extracts. Using an overall sample enrichment factor of 20, a detection limit of 0.5 pg/mL urine (0.4 femtomole) was obtained, which was even sufficient for samples from nonexposed control persons [36].

HPLC analysis of urine samples had already shown that for coke-oven workers the 1-OH pyrene levels were elevated by one or two orders of magnitude in comparison to the control group (see Table 6.3). Surprisingly, when applying the LESS method to the same samples to measure 3-OH-BaP, this marker compound was elevated in only two samples (sample codes W69 and W95 in Table 6.3) [36]. Apparently, for the other workers the total BaP uptake had not been significantly higher than the background uptake (e.g., from food, smoking, or traffic) experienced by the control group, even though the 1-OH pyrene data indicated an extra, work-related, source of PAH exposure. A probable explanation for this seeming discrepancy is the fact that the particle filters in the airstream helmets, worn by most workers, very effectively remove particle-bound BaP from the air,

Table 6.3. 1-OH-Pyrene and 3-OH-BaP Levels in Urine Samples from Controls (C-Codes; Afternoon Samples) and Coke-Oven Workers (W-Codes; Post Shift Samples); Data from Ref. 36

Sample Code	Smoking	Airstream Helmet Use	1-OH Pyrene (ng/L)	3-OH-BaP (ng/L)
C39-Fr	−	Not applicable	400	2.3
C40-Fr	+	Not applicable	550	5.7
C41-Fr	−	Not applicable	1240	3.5
W2	+	±	19 000	3.0
W24	−	+	8000	1.7
W39	−	+	10 000	7.6
W61	+	+	16 000	2.0
W69	+	±	4000	39
W83	+	−	56 000	4.0
W89	+	+	20 000	2.0
W95	−	−	17 000	80

while such filters are less effective against smaller, more volatile compounds like pyrene [37]. The level of protection offered by such filters against the heavier, more carcinogenic PAHs is therefore probably better than one would conclude based on the urinary 1-OH pyrene levels.

6.4.3. Fluorescence Line-Narrowing Identification of PAH Metabolites

As discussed above, Shpol'skii spectroscopy after chemical derivatization turned out to be very suitable for the determination of monohydroxy PAHs at extremely low levels. For more polar metabolites, however, such as BaP dihydrodiol, BaP diolepoxide, BaP tetrahydrotetrol, or similar derivatives of other PAHs, even the permethylated derivatization products did not form optimal host–guest combinations with typical Shpol'skii matrices, presumably related to the nonplanarity of the compounds. Obviously, the Shpol'skii approach is not applicable at all to even more polar and less rigid derivatives such as glucuronide, glucoside, or sulfate conjugates or PAH–DNA adducts. For these compounds an alternative approach, fluorescence line-narrowing spectroscopy, may be used to obtain vibrationally resolved emission spectra. The underlying principles will be only briefly discussed here. In disordered matrices, the surroundings of chemically identical analyte molecules are usually very different due to different host–guest interactions and—in the case of flexible compounds—different molecular conformations, affecting the energies of electronic transitions. This results in a Gaussian broadening of absorption and fluorescence bands that remain fairly broad (typically $300\,cm^{-1}$ or more) upon cooling of the sample to cryogenic temperatures. However, when the cooled sample is irradiated with a narrow-band laser, only a small selection of these analyte molecules with matching energy transitions can be excited. When the conformations of the analyte molecules and the orientations of the surrounding (frozen) matrix molecules do not change substantially during the lifetime of the excited state, the same subselection (isochromat) of analyte molecules will fluoresce and a line-narrowed spectrum is observed. FLN spectra can be obtained through excitation into the 0–0 transition or into the vibronic region of the first excited state S_1–S_0. In principle, any disordered matrix can be used for FLN spectroscopy, as long as the electron–phonon coupling is not too strong. This makes the method more widely applicable than the Shpol'skii technique, and sample handling can often be kept to a minimum. A detailed discussion of fluorescence line narrowing (FLN) spectroscopy can be found in Chapter 8 of this volume.

Sanders and co-workers described the use of FLN spectroscopy for the selective determination of polyhydroxy PAH metabolites (1,2,3,4-tetrahydrotetrahydroxy-benz[*a*]anthracene, 8,9,10,11-tetrahydrotetrahydroxy-benz[*a*]anthracene, 1,2,3,4-tetrahydrotetrahydroxy-5-methylchrysene, 1,2,3,4-tetrahydrotetrahydroxy-chrysene, 4,5-dihydrodihydroxy-9-methoxy-benzo[*a*]pyrene, and

7,8,9,10-tetrahydrotetrahydroxy-benzo[*a*]pyrene) in synthetic mixtures in glycerol/water/ethanol glasses at 4.2 K [38]. Also BaP metabolites and BaP–DNA adducts containing the same chromophore could be distinguished. Although the FLN vibrational frequencies were, as expected, very similar, the latter featured a significant red shift compared to the tetrol metabolites [39].

Larsen and co-workers reported how FLN spectroscopy can be applied to study the metabolites formed by the terrestrial isopod *Porcellio scaber* (woodlouse) after exposure to pyrene through its food [40]. The phase-I metabolite 1-hydroxypyrene and a phase-II conjugate (pyrene-1-glucoside) could be identified in the hepatopancreas of the isopod, while in gut samples also unmetabolized pyrene could be detected. Apart from grinding of the organs and adding a glass-forming solvent mixture, glycerol/ethanol/water (50/25/25, v : v : v), no further sample handling was necessary, thus minimizing the risk of degradation processes such as conjugate hydrolysis. By further optimizing excitation conditions (laser power, pulse repetition rate), it was even possible to detect conjugated 1-hydroxy pyrene in the hepatopancreas of isopods collected in the field, downwind from a large steel plant [40].

It should be mentioned that closely related analytes, containing the same chromophoric moiety with different substituents not directly attached to the chromophore, usually show very similar FLN spectra. For instance, pyrene-1-glucoside and pyrene-1-glucuronide could not be distinguished by FLN spectroscopy [40]. Fortunately, because of the presence of an acidic group in the latter, the two compounds have very different chromatographic retention times. In such cases, a combination of HPLC (or another separation technique such as electrophoresis) and FLN spectroscopy can be used for a fully unambiguous analysis (see also Chapters 10 and 11).

6.5. CONCLUSIONS

It is well established that Shpol'skii spectroscopy of parent PAHs is applicable to various types of samples of ecotoxicological interest [42–44]. Because of the excellent identification capacities of the technique, it can be used for qualitative identification or to check peak purity of HPLC fractions [44]. If sufficient care is taken to measure the spectra under reproducible conditions [45] and with the use of proper internal standards [46], the method can also be used for quantitative purposes [25]. For complex analytical problems, laser-excited Shpol'skii spectrometry is particularly powerful.

For metabolites of PAHs the situation is somewhat more complex. Analysis of such metabolites could play an important role in exposure monitoring, both in the environment and at the workplace. Unfortunately, because of their relatively polar character, PAH metabolites are not directly compatible with typical *n*-alkane

Shpol'skii matrices. However, this problem is easily overcome for phenolic hydroxy-metabolites through chemical methylation; the resulting monomethoxy compounds show excellent Shpol'skii spectra in *n*-alkane matrices. The absolute detection limit for methylated 3-OH-BaP in *n*-octane was found to be as low as 50 attomole. Using the LESS approach, 3-OH-BaP could be detected in flounder bile from moderately polluted and relatively clean environments, which had not been possible before with HPLC-based methods. LESS measurements of 3-OH-BaP was also successfully applied to coke oven workers' urine samples and showed that BaP exposure levels at the workplace are not always as directly related to pyrene exposure as previously assumed. For more polar, polyhydroxy metabolites the Shpol'skii approach is less suitable, but in that case FLN spectroscopy, with its much wider range of suitable matrices, can be used. This technique played an important role in elucidating the pyrene metabolite pattern produced by the terrestrial isopod *Porcellio scaber*.

Obviously (laser-excited) Shpol'skii and FLN spectroscopy cannot be regarded as rapid, low-cost, screening techniques. Both require sophisticated, advanced instrumentation, cryogenic temperatures, and are not easily automated. However, as made clear by the above examples, in specific cases these high-resolution molecular luminescence techniques are the methods of choice, in particular if ultra low detection limits or independent spectral identification are required.

ACKNOWLEDGMENTS

Excellent technical assistence from S.J. Kok, M. Verkaik, and G. Ph Hoornweg is gratefully acknowledged.

REFERENCES

1. S. Higginbotham, N. V. S. Ramakrishna, S. L. Johansson, E. G. Rogan, and E. L. Cavalieri, *Carcinogenesis* **12**, 875–78 (1993).
2. D. H. Phillips, *Nature* **303**, 468–72 (1983).
3. E. L. Cavalieri and E. G. Rogan, *Pharmacol. Ther.* **55**, 183–199 (1992).
4. A. H. Conney, *Cancer Res.* **42**, 4875–917 (1982).
5. M. S. Myers, C. M. Stehr, O. P. Olson, L. L. Johnston, B. B. McCain, S.- L. Chan, and U. Varanasi, *Environ. Health Persp.* **102**, 200–15 (1994).
6. P. C. Baumann, W. D. Smith, and W. K. Parland, *Trans. Am. Fish. Soc.* **116**, 79–86 (1987).
7. P. L. Lioy, J. M. Waldman, R. Harkov, C. Pietarinen, and A. Greenberg, *Arch. Environ. Health* **43**, 304–12 (1988).
8. J. C. Chang, S. A. Wise, S. Cao, and J. L. Mumford, *Environ. Sci. Technol.* **26**, 999–1004 (1992).

9. M. M. Krahn, M. S. Myers, D. G. Burrows, and D. C. Malins, *Xenobiotica* **14**, 633–46 (1984).

10. F. J. Jongeneelen, R. B. M. Anzion, P. T. J. Scheepers, R. P. Bos, P. Th. Henderson, E. H. Nijenhuis, S. J. Veenstra, R. M. E. Brouns, and A. Winkes, *Ann. Occup. Hyg.* **32**, 35–43 (1988).

11. F. Ariese, S. J. Kok, M. Verkaik, C. Gooijer, N. H. Velthorst, and J. W. Hofstraat, *Aquat. Toxicol.* **26**, 273–86 (1993).

12. G. J. Stroomberg, C. Reuther, I. Kozin, T. C. van Brummelen, C. A. M. van Gestel, C. Gooijer, and W. P. Cofino, *Chemosphere* **33**, 1905–14 (1996).

13. E. V. Shpol'skii, *Dokl. Akad. Nauk. SSSR* **87**, 935–38 (1952) (*Chem. Abstr.* #47, 4205b).

14. R. I. Personov, E. D. Godyaev, and O. N. Korotaev, *Sov. Phys.-Solid State* **13**, 88 (1971).

15. R. N. Nurmukhametov, *Russ. Chem. Revs.* **38**, 180–93 (1969).

16. A. Ya. Khesina, I. A. Khitrovo, and G. E. Fedoseeva, *J. Appl. Spectr. USSR* **22**, 644–48 (1975).

17. P. Garrigues and M. Ewald, *Intl. J. Environ. Anal. Chem.* **21**, 185–97 (1985).

18. W. Karcher, J. Devillers, P. Garrigues, and J. Jacob, eds., *Spectral Atlas of Polycyclic Aromatic Compounds, Vol. 3*, Reidel/Kluwer, Dordrecht, The Netherlands, 1991, p. 1092.

19. F. Ariese, S. J. Kok, M. Verkaik, G. Ph. Hoornweg, C. Gooijer, N. H. Velthorst, and J. W. Hofstraat, *Anal. Chem.* **65**, 1100–6 (1993).

20. J. W. Hofstraat, I. L. Freriks, M. E. J. de Vreeze, C. Gooijer, and N. H. Velthorst, *J. Phys. Chem.* **93**, 184–90 (1989).

21. F. Ariese, S. J. Kok, C. Gooijer, N. H. Velthorst, and J. W. Hofstraat, in P. Garrigues and M. Lamotte, eds., *PAH: Synthesis, Properties, Analysis, Occurrence, and Biological Effects*. Gordon & Breach, London, 1993, pp. 757–64.

22. S. J. Weeks, S. M. Gilles, R. L. M. Dobson, S. Senne, and A. P. D'Silva, *Anal. Chem.* **62**, 1472–77 (1990).

23. S. J. Weeks, S. M. Gilles, and A. P. D'Silva, *Appl. Spectr.* **45**, 1093–100 (1991).

24. F. Ariese, S. J. Kok, M. Verkaik, C. Gooijer, N. H. Velthorst, and J. W. Hofstraat, in P. Garrigues and M. Lamotte, eds., *PAH: Synthesis, Properties, Analysis, Occurrence, and Biological Effects*. Gordon & Breach, London, 1993, pp. 1039–46.

25. F. Ariese, *Shpol'skii Spectroscopy and Synchronous Fluorescence Spectroscopy: (Bio)monitoring of Polycyclic Aromatic Hydrocarbons and Their Metabolites*, PhD thesis, Vrije Universiteit Amsterdam, 1993.

26. F. Ariese, S. J. Kok, C. Gooijer, N. H. Velthorst, and J. W. Hofstraat, *Fresenius' J. Anal. Chem.* **339**, 722–24 (1991).

27. L. Nakhimovsky, M. Lamotte, and J. Joussot-Dubien, *Handbook of Low Temperature Electronic Spectra of Polycyclic Aromatic Hydrocarbons*, Elsevier, Amsterdam, 1989.

28. U. Varanasi, W. L. Reichert, J. E. Stein, D. W. Brown, and H. R. Sanborn, *Environ. Sci. Technol.* **19**, 836–41 (1985).

29. U. Varanasi, M. Nishimoto, W. L. Reichert, and B.-T. Le Eberhart, *Cancer Res.* **46**, 3817–24 (1986).

30. K. A. Goddard, R. J. Schultz, and J. S. Stegeman, *Drug Metab. Disp.* **15**, 449–55 (1987).
31. M. M. Krahn, D. G. Burrows, G. M. Ylitalo, D. W. Brown, C. A. Wigren, T. K. Collier, S.-L. Chan, and U. Varanasi, *Environ. Sci. Technol.* **26**, 112–26 (1992).
32. M. M. Krahn, D. G. Burrows, W. D. MacLeod, Jr., and D. C. Malins, *Arch. Environ. Contam. Toxicol.* **16**, 511–22 (1987).
33. A. D. Vethaak, J. G. Jol, A. Meijboom, M. L. Eggens, T. ap Rheinallt, P. W. Wester, T. van de Zande, A. Bergman, N. Dankers, F. Ariese, R. A. Baan, J. M. Everts, A. Opperhuizen, and J. M. Marquenie, *Environ. Health Persp.* **104**, 1218–29 (1996).
34. D. C. Malins, M. M. Krahn, D. W. Brown, L. D. Rhodes, M. S. Myers, B. B. McCain, and S.-L. Chan, *J. Nat. Cancer Inst.* **74**, 487–94 (1985).
35. J. W. Lloyd, *J. Occup. Med.* **13**, 53–68 (1971).
36. F. Ariese, M. Verkaik, G. P. Hoornweg, R. J. van de Nesse, S. R. Jukema-Leenstra, J. W. Hofstraat, C. Gooijer, and N. H. Velthorst, *J. Anal. Toxicol.* **18**, 195–204 (1994).
37. P. Leinster and M. J. Evans, *Ann. Occup. Hyg.* **30**, 481–95 (1986).
38. M. J. Sanders, R. Scott Cooper, G. J. Small, V. Heisig, and A. M. Jeffrey, *Anal. Chem.* **57**, 1148–52 (1985).
39. M. J. Sanders, R. Scott Cooper, R. Jankowiak, G. J. Small, V. Heisig, and A. M. Jeffrey, *Anal. Chem.* **58**, 816–20 (1986).
40. O. F. A. Larsen, I. S. Kozin, A. M. Rijs, G. J. Stroomberg, J. A. de Knecht, N. H. Velthorst, and C. Gooijer, *Anal. Chem.* **70**, 1182–85 (1998).
41. I. S. Kozin, C. Gooijer, and N. H. Velthorst, *Anal. Chem.* **67**, 1623–26 (1995).
42. P. Garrigues, G. Bourgeois, A. Veyres, J. Rima, M. Lamotte, and M. Ewald, *Anal. Chem.* **57**, 1068–70 (1985).
43. J. W. Hofstraat, H. J. M. Jansen, G. Ph. Hoornweg, C. Gooijer, N. H. Velthorst, and W. P. Cofino, *Int. J. Environ. Anal. Chem.* **21**, 299–332 (1985).
44. J. W. G. Mastenbroek, F. Ariese, C. Gooijer, N. H. Velthorst, J. W. Hofstraat, and J. W. M. van Zeijl, *Chemosphere* **21**, 377–86 (1990).
45. J. Rima, T. J. Rizk, P. Garrigues, and M. Lamotte, *Polycyclic Aromat. Compnds.* **1**, 161–69 (1990).
46. Y. Yang, A. P. D'Silva, and V. A. Fassel, *Anal. Chem.* **53**, 2107–9 (1981).

CHAPTER

7

AROMATIC BIRADICALS AND CARBENES: HIGH-RESOLUTION ELECTRONIC SPECTRA AND THEIR QUANTUM-CHEMICAL INTERPRETATION

EVA MIGIRDICYAN

Laboratoire de Photophysique Moléculaire du CNRS, Université Paris-Sud, F. 91405 Orsay Cedex, France

OLIVIER PARISEL AND GASTON BERTHIER

Laboratoire d'Etude Théorique des Milieux Extrêmes, Ecole Normale Supérieure, 24, rue Lhomond, F. 75231 Paris CEDEX 05, France

7.1. INTRODUCTION

7.1.1. A Brief History of Biradicals and Carbenes

The first studies on organic compounds containing two unpaired electrons localized on different carbon atoms date back to the beginning of the century with the discovery of Chichibabin [1], Thiele [2], and Schlenk [3] hydrocarbons. The first two compounds, which exhibit diphenylmethyl groups attached in the para positions, can also be represented by quinoid forms in which the two unpaired electrons become paired. In contrast, such mesomeric structures cannot be drawn for the Schlenk hydrocarbon where the diphenylmethyl groups are in meta positions. At the beginning of the century, however, when the requirement for a structure to fulfill Kekule's valence rules was believed to be a prerequisite to its existence, such a lack of closed-shell mesomeric formulae was considered so disturbing that the existence itself of the Schlenk hydrocarbons has been seriously questioned.

According to the classical susceptibility measurements [4], and later to the electron paramagnetic resonance (EPR) experiments carried out over a wide temperature range [5], these three hydrocarbons are characterized by close low-lying singlet and triplet states. Whereas the paramagnetism of the Chichibabin and Thiele compounds decreases as the temperature is lowered, that of the Schlenk species remains significant, even at very low temperatures. Therefore,

Shpol'skii Spectroscopy and Other Site-Selection Methods, Edited by Cees Gooijer, Freek Ariese, and Johannes W. Hofstraat.
ISBN 0-471-24508-9

the first two hydrocarbons are biradicaloids and have a diamagnetic singlet ground state together with a thermally accessible triplet level, while the Schlenk hydrocarbon is a true paramagnetic biradical with a triplet ground state, like dioxygen.

Methylene (CH_2), the prototype of the carbene series, is another species having close low-lying singlet and triplet states. In contrast with biradicals, however, the two unpaired electrons are localized on the *same* carbon atom. In 1959, Herzberg [6–8] was the first to directly produce and observe this carbene in the gas phase, and he characterized it by electronic spectroscopy. Even now there are not so many carbenes known in the gas phase; but interest in studying such species has been recently renewed by their detection in the interstellar medium [9].

A number of photochemical transformations are known to proceed via the intermediary of biradicals and carbenes. Unlike the Chichibabin, Thiele, and Schlenk hydrocarbons, which are stable, these intermediates are often highly reactive species. In fluid systems, they are generally difficult to observe and characterize because of their short lifetimes. Most frequently, these transient species are detected by indirect techniques involving the chemically induced dynamic nuclear polarization (CIDNP) method [10] or by chemical reactions (hydrogen abstraction or electron transfer to paraquat dications, for example the methylviologen dication, namely, 1,1′-dimethyl-4,4′-bipyridinium, the dichloride of which is used as a herbicide) used as monitoring processes [11]. In the particular case of intramolecular hydrogen abstraction reactions, the biradicals have been produced from laser flash photolysis and identified directly by electronic absorption spectroscopy [12, 13].

In the low-temperature matrix isolation technique, the biradicals and carbenes are produced in situ by photolysis of appropriate precursors and then identified by EPR and optical spectroscopies. Relatively high concentrations of fragments can be generated this way because the low temperature and the high viscosity of the surrounding medium prevent diffusion effects and bimolecular reactions.

In glassy matrices, the distribution of molecules isolated in a great variety of sites results in solute spectra consisting of broad bands. Such matrices have been widely used at AT&T Bell Laboratories [14–16] to investigate carbenes by EPR and fluorescence spectroscopy. In contrast, crystalline matrices provide only a small number of distinct lattice sites that can be occupied by guest molecules or trapped species generated from them by photolysis. Substitutional solutions of aromatic carbenes oriented in single crystals have been investigated by optical, EPR, and electron nuclear double resonance (ENDOR) techniques at the University of Chicago [17, 18]. Such experiments, however, are limited to the guests that have similar volume and geometry as the host molecules.

7.1.2. Shpol'skii Matrices and Their Applications

In a fundamental contribution, Shpol'skii and co-workers have shown that sharp electronic spectra can be obtained for planar aromatic molecules dispersed in *n*-alkane polycrystalline matrices [19], when the carbon chain length of the linear paraffin equals the largest dimension of the carbon skeleton of the aromatic guest [20]. A typical Shpol'skii spectrum consists of several sharp *quasi-line* spectra, each of them being associated with guest molecules located in a well-defined lattice site of the matrix [21]. Each quasi-line spectrum can be described in terms of a zero-phonon line, which is predominant, and a phonon band, since the coupling between the aromatic impurity and the *n*-alkane medium is generally very weak [22].

The potentialities of a Shpol'skii environment to accurately study planar aromatic species have been rapidly recognized for the investigation of their triplet states [23], and then extended to aromatic monoradicals [24–27]. Many applications have been reported either in the field of high-resolution spectroscopy of stable or transient species [28, 29] or in the field of analytical chemistry [30] as developed by N. H. Velthorst and collaborators.

o-xylylene 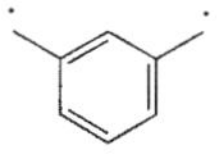 *m*-xylylene

Xylylenes or benzoquinodimethanes are biradicalar species obtained by photochemical dissociation of two hydrogen atoms from xylene (dimethylbenzene), namely, one from each methyl group, as we demonstrated for *o*- and *m*-xylylene and their methylated derivatives in 1968 [31]. Their original characterization by electronic spectroscopy in glassy matrices was later confirmed by the spectral analysis of better-resolved spectra obtained in Shpol'skii matrices at 77 K [32], and then at 5–10 K [33, 34]. The emission and the excitation spectra of *m*-xylylene show roughly the same vibrational structure as those of *m*-xylene, characterized by normal modes of the aromatic ring. Those of *o*-xylylene, however, show a vibrational structure typical of polyenes. The spectroscopic identification of *o*-xylylene has been confirmed by photochemical synthesis [35–37].

In this chapter, we present the fluorescence spectra and decays of the methylated derivatives of *o*- and *m*-xylylenes, as well as those of dibenzocycloheptadienylidene (DBC), of diphenylcarbene (DPC), and of 2-naphthylphenylcarbene (2-NPC) dispersed in *n*-alkane Shpol'skii matrices. The EPR experiments indicate that all these species, except *o*-xylylenes, have a triplet ground state. Therefore, the fluorescence of *m*-xylylene biradicals and of aromatic carbenes

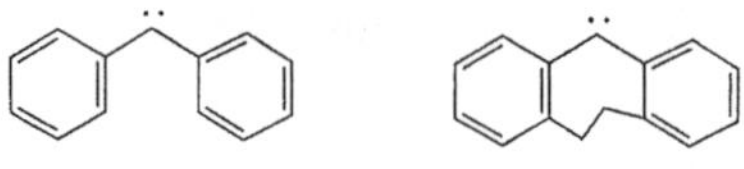

DPC DBC

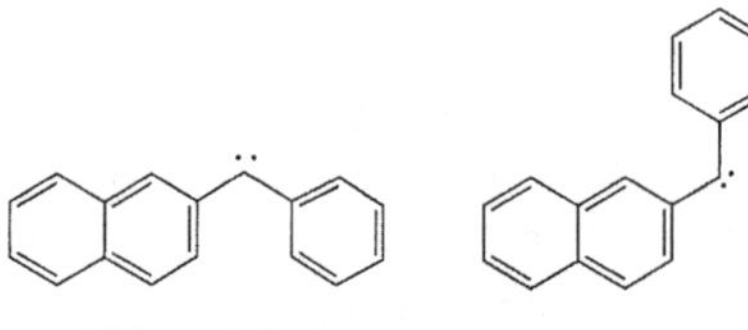

E/trans 2-NPC Z/cis 2-NPC

corresponds to the spin-allowed transition from the first excited triplet state T_1 to the ground triplet state T_0. Whereas *o*- and *m*-xylylenes have a planar structure, DPC and 2-NPC can exist in several planar and nonplanar conformations resulting from the rotation of the aromatic rings around the single bond connecting the aromatic units to the carbenic carbon. Special emphasis will be given to the study of conformational isomerism by fluorescence spectroscopy and quantum chemistry. Then we will describe how the zero-field-splitting (ZFS) parameters, $D(T_1)$, in the first excited triplet state T_1 of the *m*-xylylene biradical and aromatic carbenes, can be determined from the magnetic field effect on the *nonexponential* fluorescence decays. Lastly, recent hole-burning experiments involving triplet–triplet transitions will be discussed.

7.2. SITE-SELECTED FLUORESCENCE AND EXCITATION SPECTRA IN SHPOL'SKII MATRICES

Under conventional broadband excitation, the low temperature (4.2–30 K) fluorescence Shpol'skii spectra of *o*- and *m*-xylylenes, as well as those of carbenes, generally exhibit a multisite structure due to the distribution of guest molecules in several lattice sites. The spectra, with origins shifted by 1–2 nm from each other, can be attributed to geometrically equivalent guests incorporated in different crystallographic sites of the matrix. Such shifts are observed for the planar species like *o*- and *m*-xylylene and DBC. In contrast, the spectra of DPC and 2-NPC, which present origins shifted by 10–12 nm from each other, have been attributed to different conformers of these carbenes. The multisite structure can be considerably simplified by using narrow-band laser excitation where only the species that have an absorption energy coincident with the laser frequency are selectively excited. The sharp electronic spectra presented in this chapter correspond to biradicals and carbenes located in the most prominent sites of Shpol'skii matrices either before or after annealing of the samples.

7.2.1. *o*-Xylylene and Derivatives

In agreement with the qualitative predictions of the Hückel theory, the SCF-CI (configuration interaction) calculations by Baudet [38] establish that *o*-xylylene, a planar C_{2v} species, has a singlet ground state X^1A_1 and two closely spaced excited states, 1^1B_1 and 2^1A_1, as is known in several polyenes [39–41]. We here assume Mulliken's convention to define the molecular symmetry elements. This convention will be used throughout the chapter. The singlet character of *o*-xylylene in its ground state has been experimentally confirmed by Flynn and Michl [35, 36].

The fluorescence and excitation spectra of *o*-durylene (3,4-dimethyl-*o*-xylylene) trapped in two different sites of the *n*-hexane Shpol'skii matrix at 5–10 K are

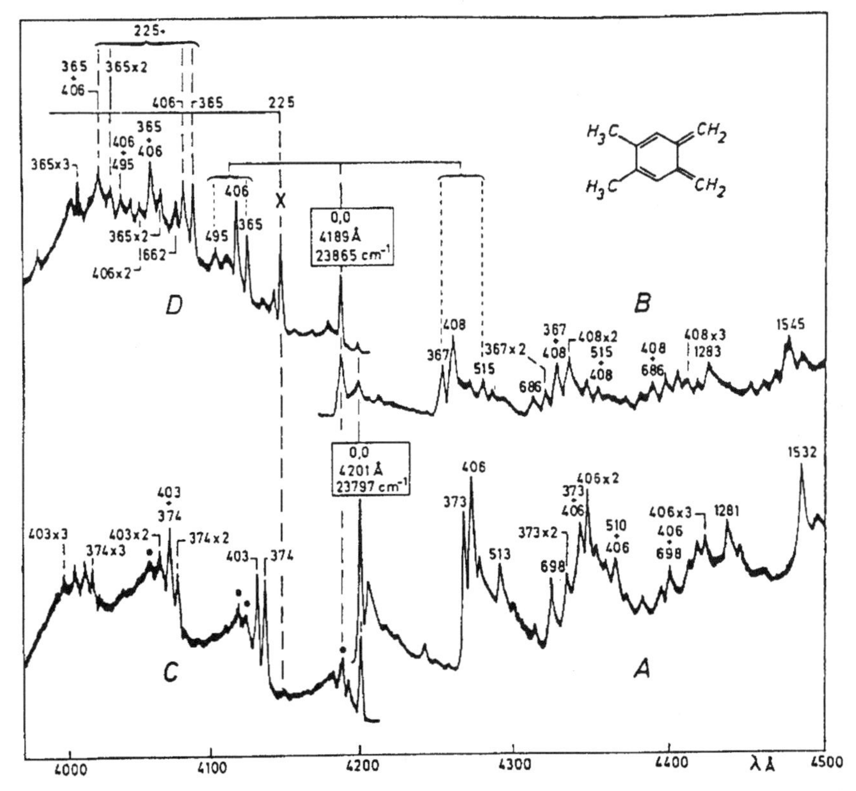

Figure 7.1. Fluorescence (A, B) and excitation (C, D) spectra of *o*-durylene-h_{12} in site I (A, C) and site II (B, D) of *n*-hexane at 5–10 K. The A and B emissions were excited at 4131 Å (0–0 + 403 cm^{-1}) and at 4150 Å (0–0 + 225 cm^{-1}), respectively. The C and D excitation spectra were observed by using the fluorescence lines at 4273 Å (0–0 + 406 cm^{-1}) and at 4254 Å (0–0 + 367 cm^{-1}), respectively. The spectra are not corrected for variations in laser intensity. (Reprinted, with permission, from Ref. 33. Copyright by Elsevier.)

presented in Figure 7.1. The spectra are dominated by progressions of the totally symmetric in-plane C–CH_2 and C–CH_3 bending modes with frequencies at about 400 and 370 cm^{-1}. The totally symmetric C–C and C=C stretching modes with ground-state frequencies at, respectively, 1270 and 1530 cm^{-1} are also present in the fluorescence spectra. Nearly mirror-image symmetry is observed between the emission and the absorption spectra. For a given site, the intense origins of the fluorescence and excitation spectra coincide, which indicates that a *sole* excited level, probably the 1^1B_1 state, is involved in this allowed transition. This

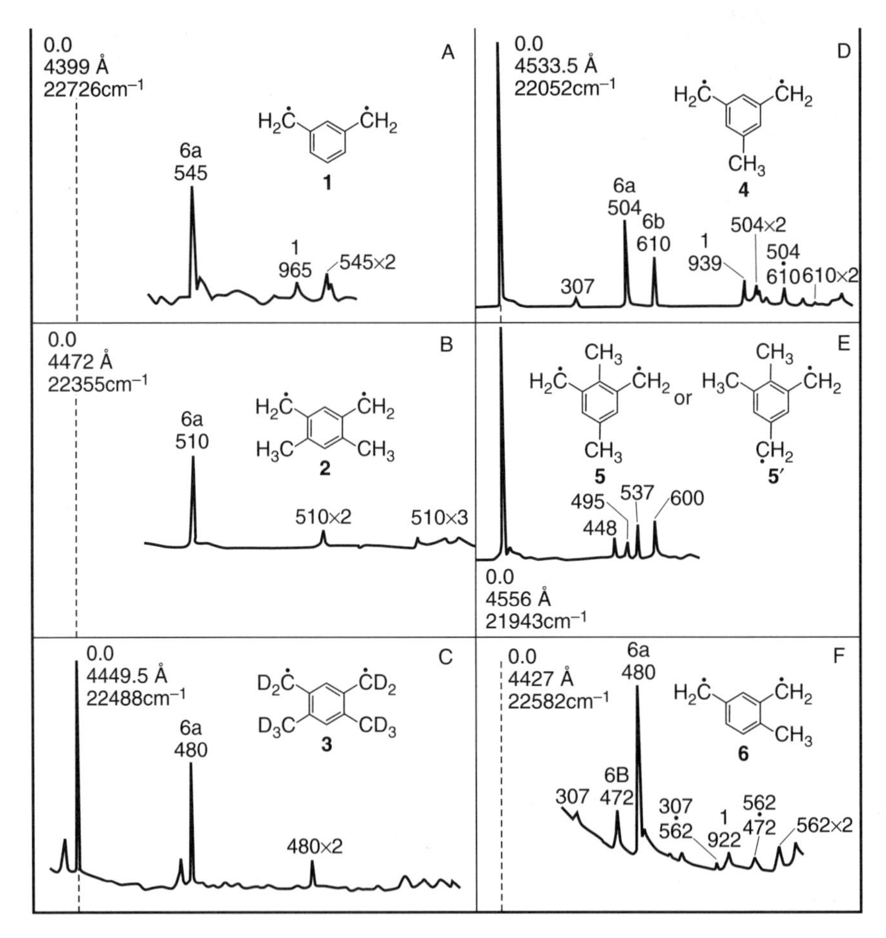

Figure 7.2. Site-selected fluorescence spectra of *m*-xylylene biradicals **1**–**6** isolated in Shpol'skii matrices of *n*-pentane (A and D) and of *n*-hexane (B, C, E, and F) frozen at 5–10 K. The emissions are excited with laser lines corresponding either to the origin of the transition (A, B, and F) or to the most intense excitation band at 4320 Å for C, 4415 Å for D, and 4437 Å for E. (Reprinted, with permission, from Ref. 34. Copyright by the American Chemical Society.)

conclusion is in contradiction with that of a previous report on *o*-durylene, suggesting that *both* the excited 1^1B_1 and 2^1A_1 states were involved in the fluorescence and excitation spectra [42].

7.2.2. *m*-Xylylene and Derivatives

The *m*-xylylene biradical is a planar C_{2v} species. Theory predicts [32, 38] that *m*-xylylene has a triplet ground state of B_2 symmetry and two close-lying triplet

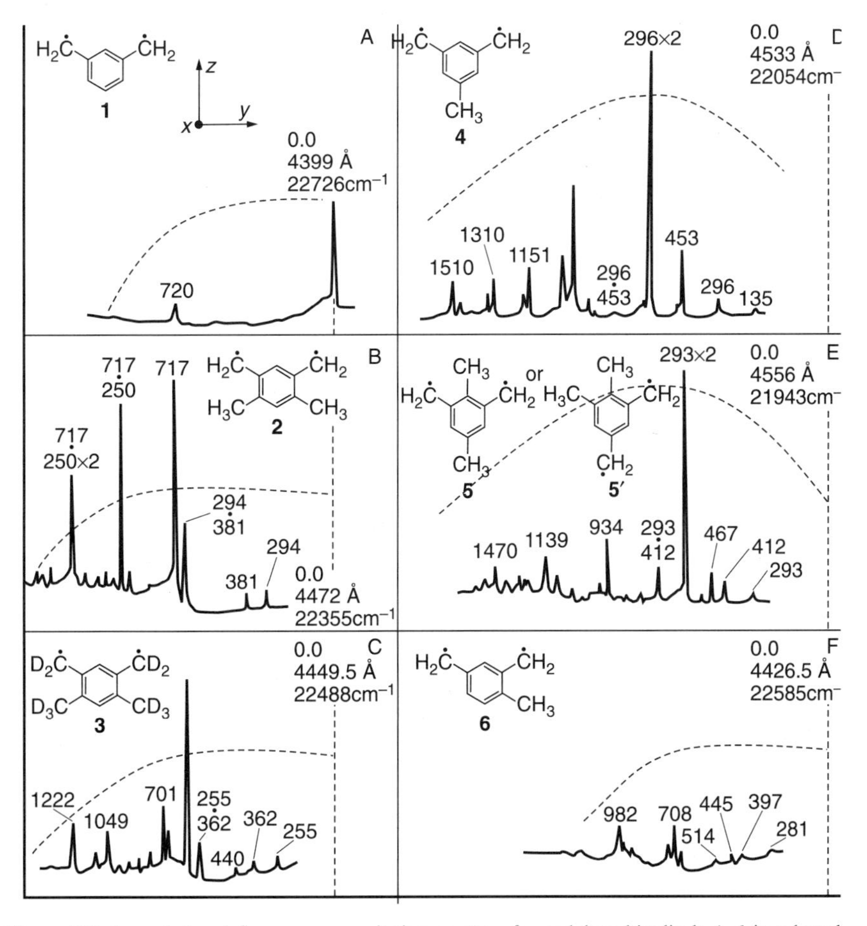

Figure 7.3. Laser-induced fluorescence excitation spectra of *m*-xylylene biradicals **1–6** in selected sites of Shpol'skii matrices at 5–10 K (see caption of Fig. 7.2). Narrow-band observation of the fluorescence origin band (for B–F) or of the first vibronic band at 4507 Å (for A). The dashed curves correspond to the distribution of the laser intensity over this wavelength region. (Reprinted, with permission, from Ref. 34. Copyright by the American Chemical Society.)

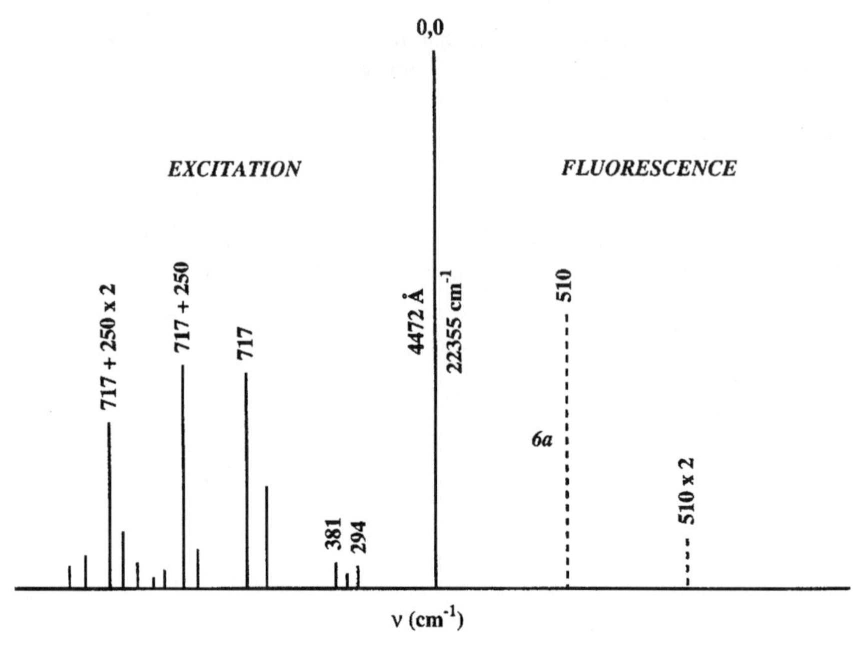

Figure 7.4. Structural asymmetry of the fluorescence and excitation spectra of *m*-durylene-h_{12}. (Reprinted, with permission, from Ref. 34. Copyright by the American Chemical Society.)

excited states of A_1 and B_2 symmetry. The experimental evidence for the triplet character of the ground state has been provided by the EPR experiments of Wright and Platz [43] and then of Goodman and Berson [44]. The fluorescence spectra of biradicals **1**–**6** produced from *m*-xylene (1,3-dimethylbenzene, curve A), durene-h_{14} and -d_{14} (1,2,4,5-tetramethylbenzene, curves B and C), mesitylene (1,3,5-trimethylbenzene, curve D), isodurene (1,2,3,5-tetramethylbenzene, curve E), and pseudo-cumene (1,2,4-trimethylbenzene, curve F) are displayed in Figure 7.2. These emissions can be analyzed by correlations with the vibrational modes and ground-state frequencies of the parent molecules, which indicates that these fragments have an aromatic character. However, the activity of these modes depends on the number and on the position of the methyl groups on the aromatic ring. This is particularly noticeable for modes 6a and 6b deriving from the benzene e_{2g} mode 6 and having frequencies in the 400–600 cm^{-1} range. We use here the Pitzer and Scott notation [45]. The nontotally symmetric mode 6b is clearly absent in spectra A, B, and C, which are dominated by the progression of the totally symmetric mode 6a around 500 cm^{-1}. In addition to these modes, other vibrations are also active in these spectra, such as the ring-breathing mode 1, which is present on spectra A, D, and F, respectively at 965, 939, and

922 cm^{-1}. The excitation spectra of biradicals **1–6** are shown in Figure 7.3. The striking feature is the absence of mirror-image symmetry between the excitation and the fluorescence patterns belonging to a given species [34]. This is illustrated in Figure 7.4 in which the vibrational structures of the spectra corresponding to *m*-durylene-h_{12} (2,4-dimethyl-*m*-xylylene) are schematically reported. While the fluorescence is a single progression in mode 6a at 510 cm^{-1}, the excitation spectrum presents a more complex structure consisting of several weak bands on the low-frequency side and a progression of strong bands starting at 717 cm^{-1}. The absorption–emission asymmetry in vibronic energies, as well as in band intensities, is expected for species presenting two close-lying excited electronic states of different symmetry, as is the case for *m*-xylylene biradicals, according to molecular orbital calculations. The excitation spectra depicted here result probably from the overlap of two electronic manifolds which can interact vibronically [46, 47]. The vibronic levels of one manifold will be mixed with levels of appropriate symmetry of the other manifold, resulting in level shifts and intensity redistribution. At energies close to that of the second excited state, the excitation spectrum will exhibit anomalous spacings and intensities and hence will no longer present mirror-image symmetry with the fluorescence spectrum. The series of strong excitation bands starting around 700 cm^{-1} in spectra A–C and F, and around 930 cm^{-1} in spectra D and E (Fig. 7.3) might correspond to transitions from the ground state to the second excited electronic state.

7.2.3. Aromatic Carbenes

The fluorescence emitted from a triplet carbene was first observed by Trozzolo and Gibbons [48]: the ultraviolet photolysis of diphenyldiazomethane dispersed in a 2-methyltetrahydrofuran (2-MTHF) glass at 77 K gives rise to a broad emission with a maximum at 480 nm, which disappears upon warming the frozen solution. On the basis of the linear correlation between the intensities of the EPR [49] and of the emission spectra, Trozzolo and Gibbons have attributed the broad band at 480 nm to the fluorescence of DPC. This fluorescence corresponds to the spin-allowed transition from the first excited triplet state T_1 to the ground triplet state T_0.

Anderson, Kohler, and Stevenson [50] have investigated the absorption and the fluorescence spectra of DPC in substitutional sites of benzophenone single crystals at 2–30 K. The origin bands of the spectra consist of a sharp zero-phonon line at 470.7 nm and broad phonon side bands extending to high energy in absorption ($\lambda_{max} \approx 465$ nm) and to low energy in emission ($\lambda_{max} \approx 476$ nm). The strong increase of the sharp zero-phonon line relative to the broad phonon band, as the temperature decreases to 2 K, indicates a strong excitation–phonon coupling, attributed to relatively large geometrical changes upon excitation. Particularly noteworthy is the similarity between the broad fluorescence band with $\lambda_{max} = 480$ nm observed in the 2-MTHF glass at 77 K [48] and the broad phonon fluorescence band with $\lambda_{max} = 476$ nm obtained in a benzophenone matrix at 2–30 K [50].

DBC, DPC, and 2-NPC, which will be described now, have all been produced *in situ* by the photolysis of the appropriate diazo precursors dispersed in n-alkane Shpol'skii matrices:

N_2 hv N_2 + DPC

7.2.3.1. *Dibenzocycloheptadienylidene (DBC)*

DBC, which contains a seven-membered ring, cannot be planar. Like 1,4-cycloheptadiene [51], DBC will likely exist in two different nonplanar conformations, C_2 and C_s, the fluorescence spectra of which are expected to lie in different spectral regions and to have different structures. The C_2 conformer, which is close to planarity, is predicted to be more stable than the C_s conformer due to a more favorable π conjugation and a reduced steric hindrance [52].

The fluorescence and excitation spectra of DBC in n-hexane at 20 K are presented in Figure 7.5. The two fluorescence spectra starting at 502.9 nm (I) and 504.7 nm (II), which are sharp and have similar vibrational structures, can both be attributed to the nearly planar C_2 conformer trapped in two different sites of the matrix. The analysis of these spectra is given in Ref. 53 and 54. As expected, the C_s conformer does not show up in the spectra.

7.2.3.2. *Diphenylcarbene (DPC)*

Two distinct fluorescence spectra are emitted from DPC in n-hexane or in n-heptane Shpol'skii matrices for temperatures ranging from 4.2 to 77 K [54, 55]: a broad fluorescence with $\lambda_{max} \approx 482$ nm, very similar to that detected by Trozzolo and Gibbons [48], and a sharp fluorescence starting at 502.8 nm, which was not detected by these authors in their glassy matrix. These two emissions are depicted in Figure 7.6. The relative intensity of the sharp and the broad spectra is concentration dependent [55]. This result may be interpreted by triplet–triplet energy transfer between the two species responsible for these spectra, the efficiency of which increases with concentration. At the highest concentration,

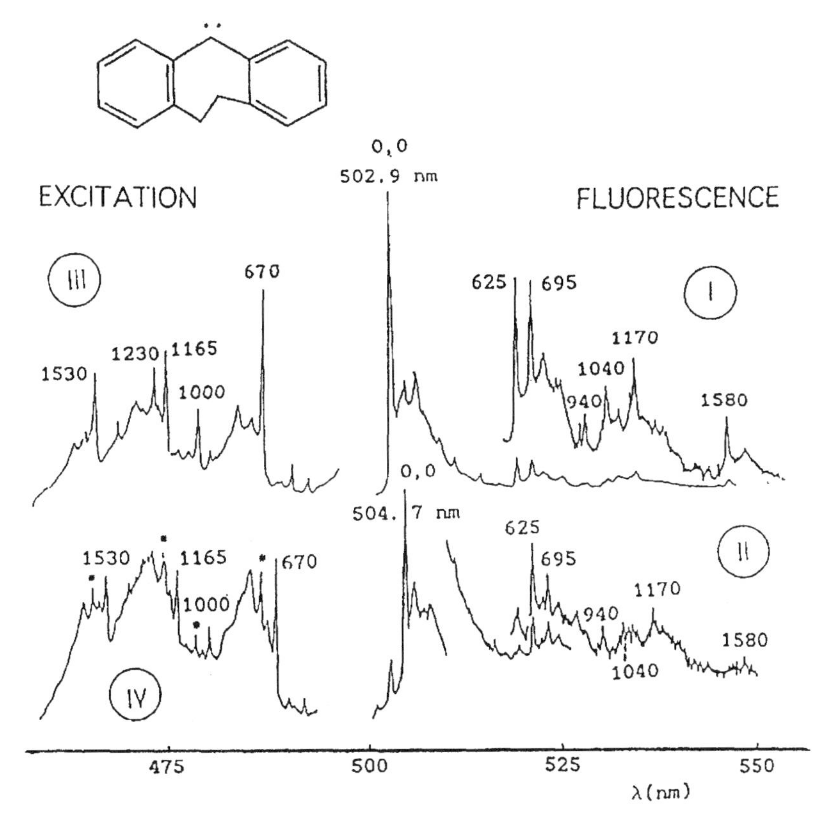

Figure 7.5. Site-selected fluorescence and excitation spectra of DBC in an *n*-hexane Shpol'skii matrix at 20 K. The emissions I and II are, respectively, excited with 486.5- and 488-nm laser lines. The excitation spectra III and IV are obtained by monitoring the fluorescence origins at 502.9 and 504.7 nm, respectively, while scanning the dye laser. (Reprinted, with permission, from Ref. 53. Copyright by the American Chemical Society.)

only the sharp fluorescence is detectable, and its vibronic structure has been analyzed by using the vibrational modes and frequencies of benzenoid compounds [54, 55]. The two fluorescence spectra have been attributed to two different conformations of DPC.

- The *low-energy sharp fluorescence* is attributed to a DPC molecule with a geometry similar to that of DBC, the carbene with the ethano bridge, since DPC and DBC both have their fluorescence origins at 502.8 nm in the *n*-hexane matrix [53–55]. In this geometry, the two aromatic rings of DPC are not too far from being coplanar.

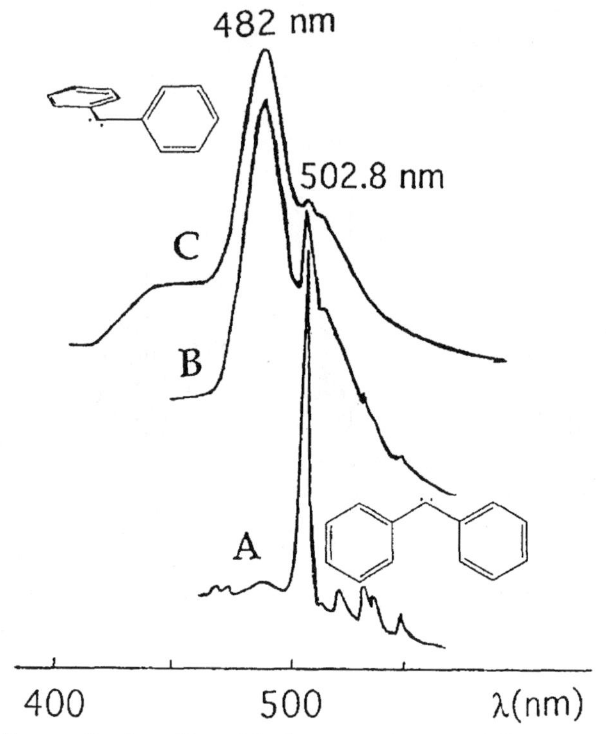

Figure 7.6. Concentration dependence of the fluorescence spectra of DPC in *n*-hexane at 77 K. The spectra are excited with the 315-nm radiation. The concentration of the DPC precursor, the diphenyldiazomethane, is: 10^{-2} M (curve A), 10^{-4} M (curve B), and 10^{-5} (curve C).

- The *broad blue-shifted fluorescence* is attributed to a nonplanar DPC conformer in which the carbenic bond angle (apex angle) amounts to 140°, and the dihedral angles Φ_1 and Φ_2, which are the angles between a benzene ring and the average plane of the molecule, are close to 30°. This geometry corresponds to that determined by Anderson and Kohler [56], who applied the ENDOR technique for DPC in benzophenone ($\Phi_1 = 26.5°$ and $\Phi_2 = 29.5°$). Such an interpretation implies that DPC has nearly the same geometry in frozen *n*-hexane and in benzophenone single crystals. This is consistent with the fact that the broad fluorescence with $\lambda_{max} = 482$ nm observed in *n*-hexane [54, 55] is very similar to the broad structureless phonon band with $\lambda_{max} = 476$ nm observed in benzophenone [50].

These attributions have been confirmed by quantum chemistry calculations [55] using the CS-INDO (conformational and spectroscopic intermediate neglect of differential overlap) level of approximation [57, 58], a semiempirical procedure that has been specially designed for dealing with conformational problems of conjugated molecules in their ground state and in their excited states as well [59], and that refines the well-known INDO scheme. On top of the CS-INDO calculations, CI treatments have been performed within the CIPSI (configuration interaction by perturbative selected iterations [60]) framework.

The energies and the oscillator strengths of the singlet and triplet states have been calculated for DPC in three different geometries: the planar conformer (C_{2v}), the quasi-planar structure (C_2), which was considered as having the same geometry as DBC [53], and the nonplanar conformer (C_1) described by Anderson and Kohler [56]. The corresponding energy diagram is displayed in Figure 7.7 and the data obtained for the triplet states are collected in Table 7.1. The vertical triplet–triplet transitions listed in Table 7.1 compare favorably with the experimental results. In particular, the first T_0–T_1 transition energy computed at 2.62 eV for the C_{2v} structure (i.e., 1^3B_1–1^3A_2) is in good agreement with the energy of 2.49 eV (497 nm) corresponding to the average energy between the fluorescence and the excitation origin bands of the sharp spectrum (see Fig. 4 in Reference 55).

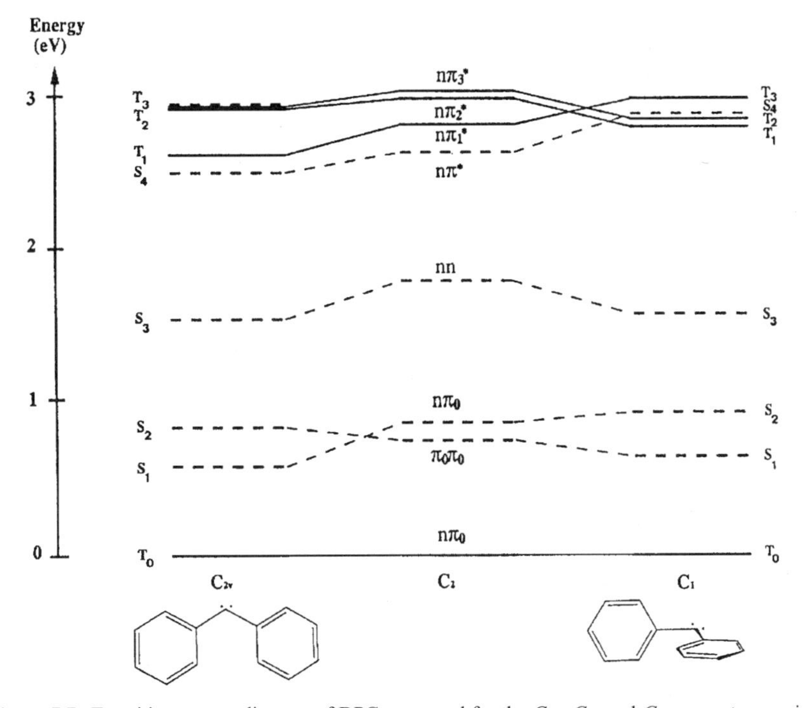

Figure 7.7. Transition energy diagram of DPC computed for the C_{2v}, C_2, and C_1 symmetry species. (Reprinted, with permission, from Ref. 55. Copyright by the American Chemical Society.)

Table 7.1. Calculated Transition Energy (E, eV) and Oscillator Strength (f) of the Lower Triplet States of DPC Conformers [55]

C_{2v}			C_2			C_1		
State	E	f	State	E	f	State	E	f
1^3B_1	0.00		1^3B	0.00		1^3A	0.00	
1^3A_2	2.62	0.122	1^3A	2.82	0.095	4^3A	2.99	0.164
2^3B_1	2.92	0.024	2^3B	2.99	0.005	2^3A	2.80	0.022
2^3A_2	2.94	0.002	2^3A	3.03	0.016	3^3A	2.85	0.012

When the molecular symmetry is lowered, an hypsochromic shift is observed experimentally (2.60 eV) as well as theoretically (2.80 eV, Table 7.1); this is the usual behavior for nonplanar structures as a consequence of a reduced conjugation. In addition, the calculations predict an important change in the nature of the leading configuration of the CI expansion for the first excited triplet state when going from C_{2v} or C_2 species to the C_1 structure (Fig. 7.7), namely, a $n\pi_2^*$ occupancy instead of the $n\pi_1^*$ initial occupancy (Fig. 7.8). As a result of a topological change between π_1^* and π_2^*, the oscillator strength of the first T_0–T_1 transition ($f = 0.022$) in the nonplanar structure C_1 appears to be strongly

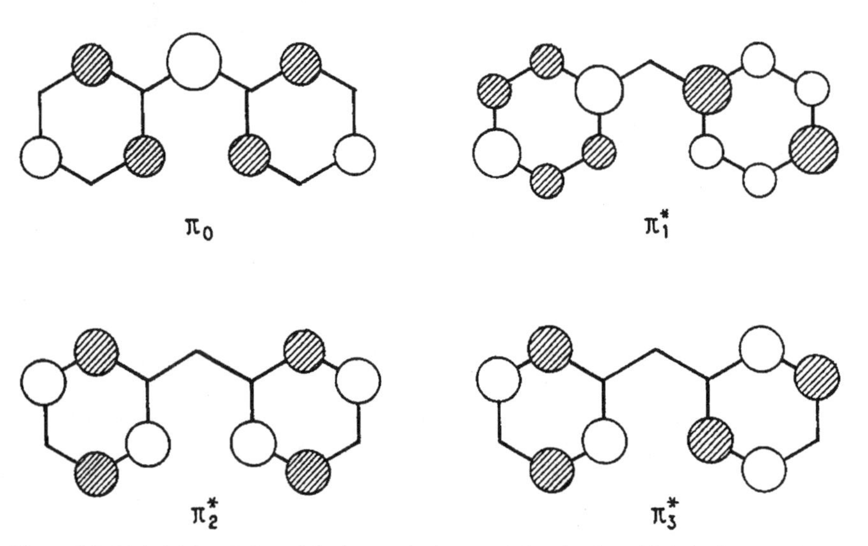

Figure 7.8. Pictorial description of the lowest singly occupied molecular orbitals in the ground state and in the first excited states of DPC. (Reprinted, with permission, from Ref. 55. Copyright by the American Chemical Society.)

reduced when compared to that of the planar conformer ($f = 0.122$); this is in good agreement with the component lifetimes derived from the nonexponential fluorescence decays (*vide infra*). The component lifetimes of the decays measured on the broad fluorescence (nonplanar C_1 species) are 6–12 times longer than those measured on the sharp fluorescence (quasi-planar C_{2v} species) [55].

7.2.3.3. 2-Naphthylphenylcarbene (2-NPC)

The conformational isomerism of carbenes has been the subject of several earlier studies using EPR spectroscopy [61–63]. For example, it was found that 1- and 2-naphthylmethylenes exist in two different conformations, as is the case of the closely related 1- and 2-hydroxynaphthalenes [64], and that these two conformations have different ZFS parameters. The measured D values have been assigned to specific conformers on the basis of calculations that considered the relationship between the ZFS parameters and the π spin densities within the point spin model [62, 63]. Similarly, EPR experiments performed with 2-NPC in a 2-MTHF glass at 77 K indicated that two D values, $D = 0.4044$ and $0.3898\,\text{cm}^{-1}$, can be obtained for the two different conformers of 2-NPC in its triplet ground state [65].

The fluorescence and excitation spectra of 2-NPC in n-hexane at 4.2 K are displayed in Figure 7.9 [66]. The good coincidence between the fluorescence and the excitation origin bands (not shown on Fig. 7.9) indicates that the sharp bands are indeed zero-phonon lines, and consequently 2-NPC is a good candidate for hole-burning experiments ([67], *vide infra*). The fluorescence origins at 600.1 and 588.8 nm are separated by more than 11 nm, which suggests that the two spectra correspond to different conformers [66]:

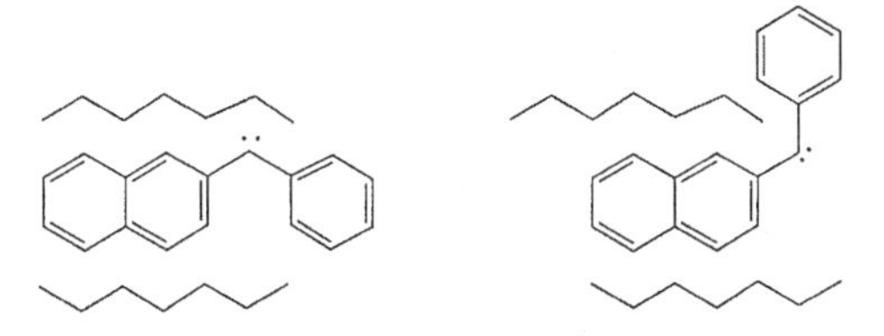

Both spectra, which have slightly different vibrational structures, have been analyzed in Ref. 66.

When 2-NPC is embedded in an n-heptane Shpol'skii matrix, the two sharp fluorescence spectra are again present at 4.2 K, but the origin bands are slightly shifted up to 600.5 and 589.5 nm. However, in this matrix, fluorescence and EPR spectroscopies reveal a light-induced interconversion between the two conformers, which can be reversed on annealing the sample in the dark [66, 68]. The correlation between the light-induced transformation in the EPR and fluorescence spectra established that the conformer responsible for the fluorescence origin at

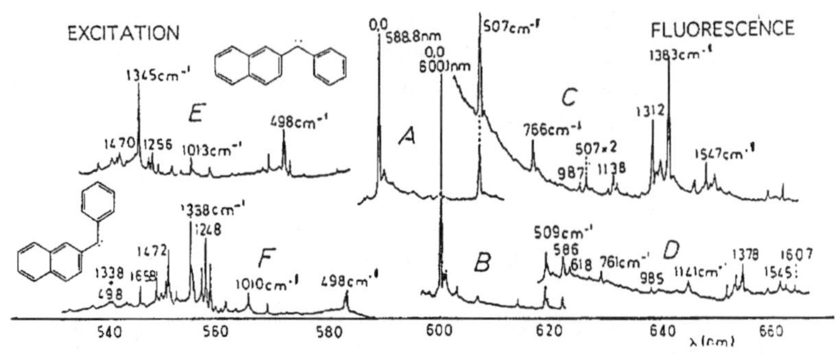

Figure 7.9. Selectively excited fluorescence (A, B, C, and D) and excitation (E and F) spectra of 2-NPC in annealed *n*-hexane Shpol'skii matrix at 4.2 K. The emissions A and B are excited with laser lines at 555.6 and 545.8 nm, respectively. The emissions C and D are excited with laser lines at 588.8 and 600.1 nm, respectively. The excitation spectra E and F are obtained by monitoring the fluorescence origins at 588.8 and 600.1 nm, respectively, while scanning the dye laser. (Reprinted, with permission, from Ref. 66. Copyright by the American Chemical Society.)

600.5 nm has a $D = 0.42 \pm 0.02\ \text{cm}^{-1}$ value in its ground state. The D value of the other conformer amounts to $0.41 \pm 0.01\ \text{cm}^{-1}$.

In order to assign the two fluorescence spectra obtained in Shpol'skii matrices to specific conformers, the energies and the ZFS parameters D of the triplet states of 2-NPC in several planar and nonplanar conformations have been calculated at the CS-INDO/CIPSI level of theory [69], using ROHF (restricted open-shell Hartree–Fock)/3-21G optimized geometries. The variable geometrical parameters are the apex angle α and the two dihedral angles Φ_1 (out-of-plane torsion centered on the carbenic atom for the phenyl group) and Φ_2 (rotation angle of the phenyl around the C_1–C_3 bond). The results collected in Table 7.2 indicate that the

Table 7.2. ROHF/3-21G Optimized Structures of 2-NPC in the Ground State and Corresponding Energies [69]; All Angles are in Degrees (see text for definitions)

$\alpha = (2\ 1\ 3)$

$\Phi_1 = (3\ 1\ 2\ 5)$

$\Phi_2 = (4\ 3\ 1\ 2)$

	Φ_1	Φ_2	α	Energy (au)
E/*trans*	0.0	0.0	149.1	−647.025221
pseudo-*E*/*trans*	30.8	30.6	135.2	−647.033220
	45.0	20.2	134.7	−647.032976
	90.0	1.2	133.6	−647.031422
	135.0	−19.3	135.2	−647.033090
pseudo-*Z*/*cis*	153.2	−33.5	135.6	−647.033507
Z/*cis*	180.0	0.0	148.9	−647.025513

Table 7.3. Calculated and Observed T_0–T_1 Transition Energies (eV) and $D(T_0)$ Values (cm^{-1}) for 2-NPC [69]

	$E(T_0–T_1)$		$D(T_0)$	
	Calculated	Observed	Calculated	Observed
pseudo-*E*/*trans* ($\Phi_1 = 30.8°$)	2.05	2.10	0.3405	0.41 ± 0.01
pseudo-*Z*/*cis* ($\Phi_1 = 153.2°$)	2.01	2.06	0.3461	0.42 ± 0.02

structures corresponding to the minimum energy in the ground state are not the planar species but the nonplanar conformers such that $\Phi_1 = 30.8°$, which we call the pseudo-*E*/*trans* conformer, and $\Phi_1 = 153.2°$, which we call the pseudo-*Z*/*cis* conformer. Among these two conformers, the pseudo-*Z*/*cis* has the lowest ground-state energy. It is important to note that these geometries are very close to the one determined for DPC by the ENDOR technique [56]. The T_0–T_1 transition energies, together with the $D(T_0)$ values calculated for the ground-state T_0, are listed in Table 7.3 for both the pseudo-*E*/*trans* and the pseudo-*Z*/*cis* conformers. This table also contains the energies corresponding to the appropriate fluorescence origins of the two conformers (which are very close in the *n*-hexane or in the *n*-heptane matrices [66]), and the $D(T_0)$ values measured in the *n*-heptane matrix [66]. The correlation between the calculated and the experimental data establishes that the conformer with fluorescence origin at 588.8 nm in *n*-hexane (or 589.5 nm in *n*-heptane) is the pseudo-*E*/*trans*, which has the lowest calculated and observed $D(T_0)$ value. Similarly, the conformer with the fluorescence origin at 600.1 nm in *n*-hexane (or 600.5 nm in *n*-heptane) is the pseudo-*Z*/*cis*, which also has the largest calculated and observed $D(T_0)$ value. This attribution is consistent with the conclusions reported in Ref. 66; the conformer with the larger $D(T_0)$ value fluoresces at the longest wavelength. This interpretation of the experimental results, based on quantum-chemistry calculations, has led to the recent reinterpretation [70] of the EPR data previously obtained for 2-NPC in a glassy 2-MTHF matrix at 77 K [65].

7.3. SITE-SELECTED FLUORESCENCE DECAYS AND ZERO-FIELD SPLITTING OF THE FIRST EXCITED TRIPLET STATE

7.3.1. Fluorescence Decays

The fluorescence decays of *m*-xylylene biradical and its methyl-substituted derivatives [71], as well as those of aromatic carbenes such as DBC [53, 54], DPC [54, 55], and 2-NPC [66, 68] in *n*-hexane at 4.2–30 K, have been recorded

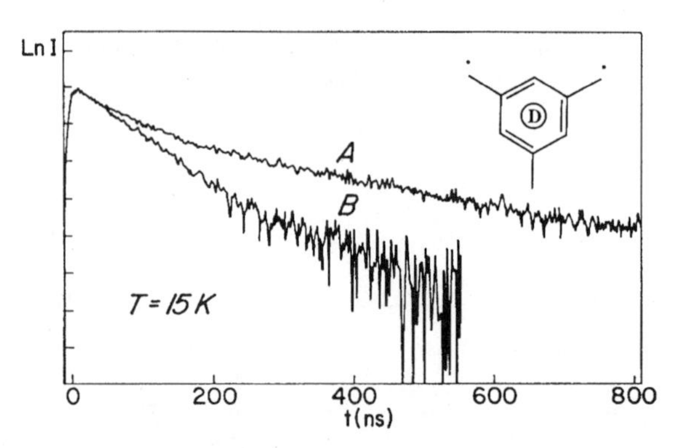

Figure 7.10. Fluorescence decays of mesitylylene-d_{10} biradicals in *n*-hexane at 15 K in the absence (A) and in the presence (B) of a 450 G magnetic field. The decays are excited with a laser line at 4391 Å and measured on the fluorescence origin band at 4502.2 Å. (Reprinted, with permission, from Ref. 71. Copyright by the American Chemical Society.)

in the absence and in the presence of a magnetic field. The species were excited either with radiation selected from a dye laser (*m*-xylylene biradicals and DPC) or with the 337-nm radiation from a N_2 laser (DBC and 2-NPC), and the decays were measured either on the fluorescence origin or on vibronic bands. Examples of such decays recorded for a biradical and for a carbene are presented, on a semilogarithmic scale, in Figures 7.10 and 7.11 respectively.

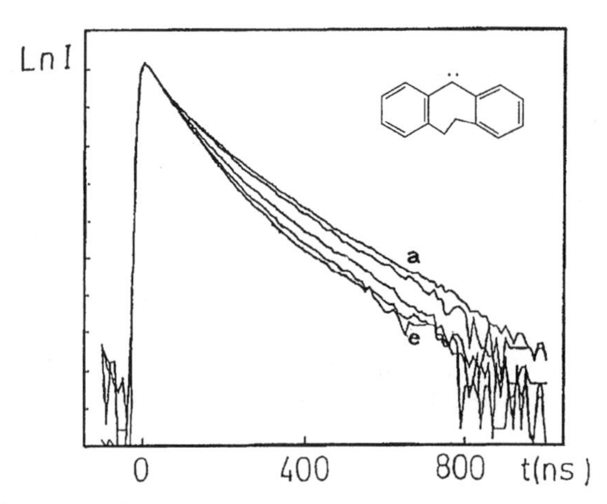

Figure 7.11. Fluorescence decays of DBC in *n*-hexane at 20 K in the absence (a) and in the presence (b–e) of 45, 65, 105, and 145 G magnetic fields. DBC is excited with the 337-nm nitrogen laser line and the decays are measured on the intense fluorescence origin band at 502.9 nm. (Reprinted, with permission, from Ref. 53. Copyright by the American Chemical Society.)

- In the absence of a magnetic field, all the decays are nonexponential. They have been analyzed as a sum of three exponential decays with the same preexponential factor A, since the three sublevels of the emitting triplet are equally populated by the short excitation pulse:

$$I_{\mathrm{calc}} = A \sum_{u=x,y,z} \exp(-t/\tau_u) \tag{7.1}$$

  This treatment leads to the component rate constants k_u, or lifetimes τ_u of the fluorescence decays, which are listed in Table I of Ref. 72 for m-xylylene biradicals, and in Refs. 53, 55, and 66 for aromatic carbenes. These data are consistent with spin-allowed transitions.
- Upon application of a magnetic field, the fluorescence decay curves are significantly modified, as indicated in Figure 7.10 for m-xylylene biradicals and in Figure 7.11 for aromatic carbenes.

The nonexponential fluorescence decays of carbenes are attributed to the emission from the T_1 state spin sublevels, at rates that are faster than the rate of the spin-lattice relaxation between the different sublevels. The modification of the decay curves is due to the mixing of the wave functions of the T_1 sublevels upon application of the field. For randomly oriented species in polycrystalline Shpol'skii matrices, the decay law is expected to be multiexponential. In the presence of a weak magnetic field, however, the decay curves can still be fitted with a convoluted sum of three exponential functions. The results of such an analysis show that the lifetime of the slow component τ_H decreases drastically when the field **H** increases, as indicated in Figures 7.12 and 7.13. This magnetic field effect will be used to estimate the ZFS parameters $D(T_1)$ of biradicals and carbenes in the first excited triplet state T_1.

Interpretation. In the absence of a magnetic field, only the $T_1 \rightarrow T_0$ transitions that involve magnetic sublevels of the same spin function, namely, $T_{1x} \rightarrow T_{0x}$, $T_{1y} \rightarrow T_{0y}$ and $T_{1z} \rightarrow T_{0z}$, are allowed, provided that the ZFS magnetic axes remain the same in both states. Consequently, the radiative decay rate constants k^r from the three T_1 sublevels are equal. This is not true for the corresponding nonradiative decay rate constants k^{nr}, because the three T_1 sublevels, which generally belong to different irreducible representations in the symmetry group of the molecule, are coupled by spin–orbit coupling (SOC) to different singlet states. For example, in planar m-xylylene biradicals of C_{2v} symmetry with y and z axes in the molecular plane as indicated in Figure 7.14, only the T_{1y} and T_{1z} sublevels can interact with the higher-lying $(\sigma\pi)^*$ singlet states. This SOC induces intersystem crossing (ISC), which accelerates the depopulation of the T_{1y} and

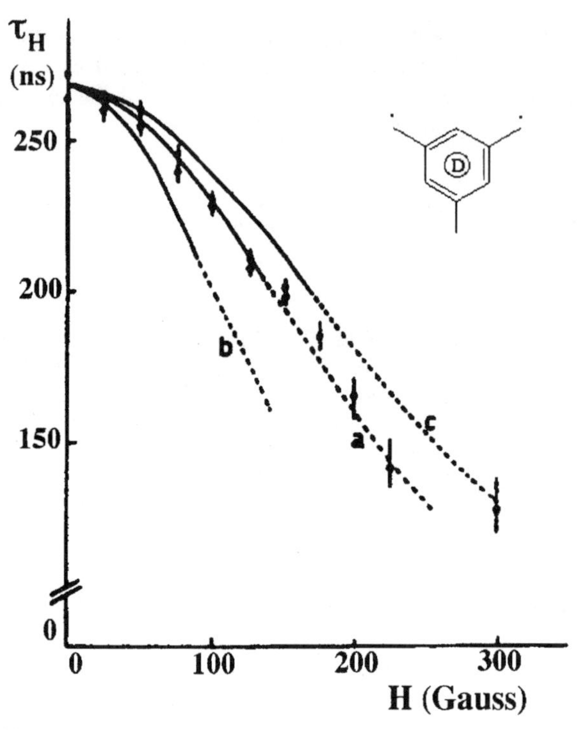

Figure 7.12. Lifetimes of the slow component (τ_H) in the fluorescence decay of perdeuterated mesitylylene in *n*-hexane at 4.2 K as a function of the magnetic field. The points correspond to the observed values. The curves have been calculated for different values of the ZFS parameter: (a) $|D| = 0.042\ \text{cm}^{-1}$; (b) $|D| = 0.030\ \text{cm}^{-1}$; (c) $|D| = 0.050\ \text{cm}^{-1}$. (Reprinted, with permission, from Ref. 73. Copyright by the American Chemical Society.)

T_{1z} sublevels only. The T_{1x} sublevel, which cannot interact with $(\sigma\pi)^*$ singlets, will give rise to the slow component with a total decay rate:

$$k^r + k_x^{nr} < k^r + k_{y\,\text{or}\,z}^{nr} \tag{7.2}$$

As for the *m*-xylylene biradicals, the nonexponential fluorescence decays of carbenes will be interpreted by different ISC rates from the three triplet sublevels, due to different SOC schemes with the singlet states. This interpretation is confirmed by the significant modification of the fluorescence decays observed in the presence of a magnetic field.

7.3.2. The ZFS Parameter $D(T_1)$ in the T_1 State

The ZFS parameter $D(T_0)$ in the triplet ground state T_0 of *m*-xylylene biradicals and of aromatic carbenes, listed in Table 7.4, have been obtained by the

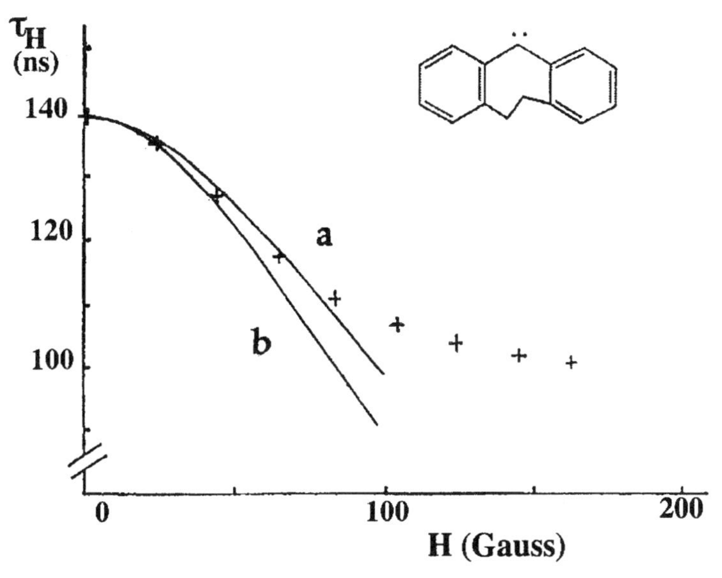

Figure 7.13. Lifetimes of the slow component (τ_H) in the fluorescence decay of DBC in *n*-hexane at 20 K as a function of a weak magnetic field. The crosses correspond to the observed values. The curves have been calculated for two values of the ZFS parameter: (a) $|D| = 0.020\ \mathrm{cm}^{-1}$; (b) $|D| = 0.015\ \mathrm{cm}^{-1}$. (Reprinted, with permission, from Ref. 53. Copyright by the American Chemical Society.)

Table 7.4. Observed and Calculated $D(T_0)$ and $D(T_1)$ Values (cm^{-1}) of *m*-Xylylene Biradicals and Aromatic Carbenes

	$D(T_0)$		$D(T_1)$	
	Experiment	Theory	Experiment	Theory
m-xylylene biradicals	0.011 [43]	0.025 [75]	0.04 [73]	0.049 [75]
DBC	0.3932 [76]		0.02 [53]	
Quasi-planar DPC (C_{2v})		0.2073 [55]	0.007 [55]	0.0155 [55]
Nonplanar DPC (C_1)	0.405 [15, 17]		0.20 [55]	0.0100 [55]
Pseudo-*E*/*trans* 2-NPC	0.3898 [65, 70]	0.3405 [69]	0.038 [67]	0.0619 [69]
	0.41 [66]			
	0.50 [67]			
Pseudo-*Z*/*cis* 2-NPC	0.4044 [65, 70]	0.3461 [69]	0.011 [66]	0.0633 [69]
	0.42 [66]			

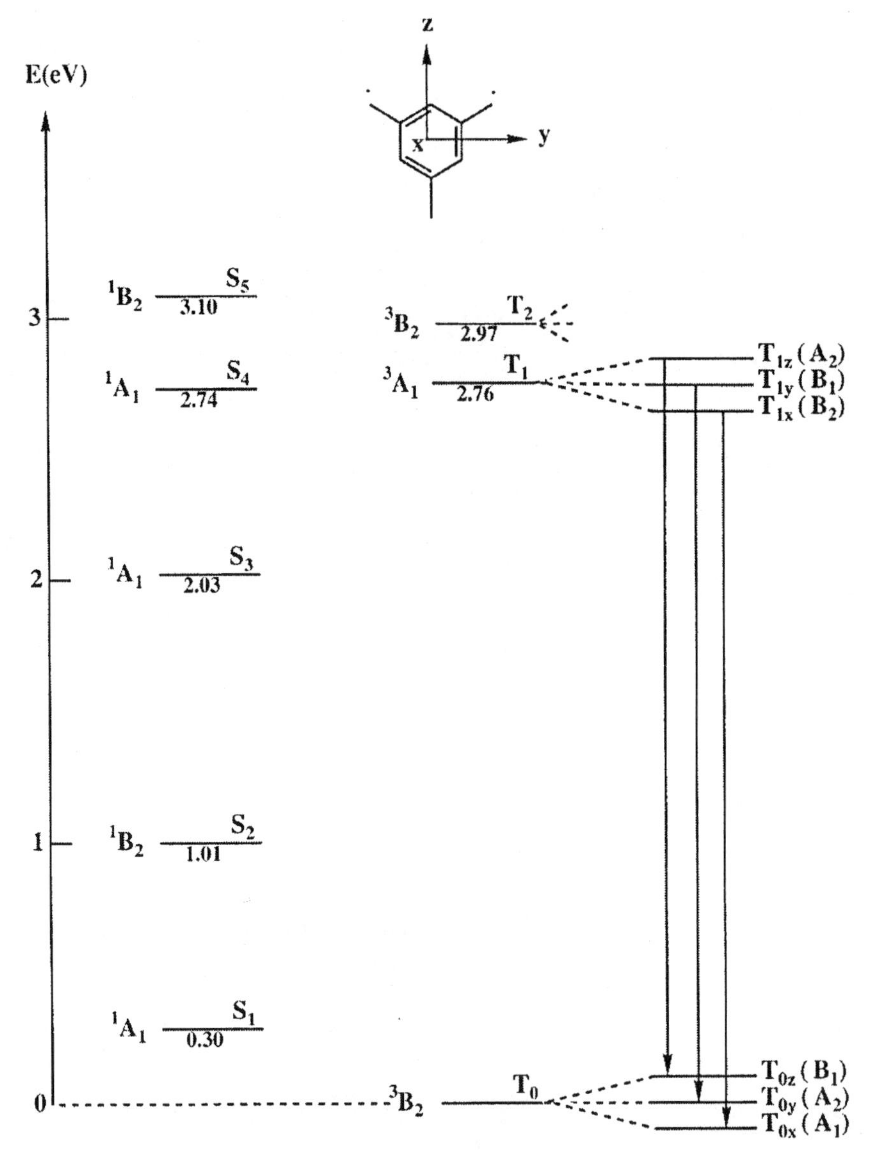

Figure 7.14. Energy-level diagram of the mesitylylene biradical with calculated energies of both triplet and singlet states. The order of the T_x, T_y, and T_z sublevels is arbitrary for both states. (Reprinted, with permission, from Ref. 71. Copyright by the American Chemical Society.)

conventional EPR technique. This method, however, cannot be used to measure the ZFS parameter $D(T_1)$ in the first excited triplet state T_1 of these intermediates, which have much shorter lifetimes and consequently much lower steady-state concentrations. This is why the $D(T_1)$ values of these short-lived species have been determined by a method that uses a magnetic field effect on the emission decays from triplet sublevels [53, 73].

The effect of the Zeeman mixing on the radiative and on the nonradiative decay rates of the $T_1 \rightarrow T_0$ transition has been discussed in detail for *m*-xylylene biradicals [74]. This has led to the following expression for the effective total decay rate $\langle k_{xH} \rangle$ from the T_{1x} sublevel (slow component) in the presence of a weak magnetic field **H** (i.e., $g\beta H \ll D$):

$$\langle k_{xH} \rangle = k_x + \frac{2}{3}\frac{g^2\beta^2H^2}{D^2}\left(\frac{k_y + k_z}{2} - k_x\right) \tag{7.3}$$

where k_x, k_y, and k_z are the total fluorescence decay rates from the T_{1x}, T_{1y}, and T_{1z} sublevels in the absence of the field [73], β is the Bohr magneton and g is the Landé factor.

For planar *m*-xylylene biradicals and for planar (or quasi-planar) aromatic carbenes having the magnetic x axis normal to the molecular plane (or some average plane), the slow component (rate k_x) is assumed to be associated with the T_{1x} sublevel, separated by D from the two other spin sublevels. This assumption is reasonable for all the species studied here, except for nonplanar carbenes such as the DPC conformer emitting the broad fluorescence (*vide supra*). The ZFS parameter $D(T_1)$ in the T_1 state of planar biradicals and carbenes is then estimated from Expression 7.3 where every quantity except D is known from the decay measurements.

The dependence of the slow component lifetime $\tau_H = \langle k_{xH} \rangle^{-1}$ as a function of a weak magnetic field **H** has been calculated for different D values by using Expression 7.3 and the component rate constants determined from the fluorescence decays. The curves thus obtained for several tentative D values are displayed in Figures 7.12 and 7.13. The best fit between the calculated curves and the experimental points gives the $D(T_1)$ values, which have been collected in Table 7.4 together with the corresponding $D(T_0)$ parameters determined by EPR for the ground state T_0. This table also contains the theoretical $D(T_0)$ and $D(T_1)$ values evaluated by quantum-chemistry methods. Despite the deviation between the experimental and theoretical results, examination of the data shows that the D values vary significantly between T_0 and T_1. Moreover, these variations are opposite for biradicals and for carbenes. The dipolar spin–spin interaction is proportional to $\langle 1/r^3 \rangle$ where r is the separation between the two unpaired electrons. The drastic change of the D parameter between the two states suggests that, in the T_1 state, this separation is very different from that in the T_0 state. For all species, this observation is explained by a substantial delocalization of the two unpaired electrons onto the aromatic rings in the T_1 state.

Discussion. In the ground state T_0, *m*-xylylene biradicals contain two unpaired π electrons localized, far from each other, on the two methylene substituents. This picture is compatible with the small $D = 0.011\ \text{cm}^{-1}$ value measured by Wright and Platz [43] for *m*-xylylene and mesitylylene (5-methyl-*m*-xylylene). In the T_1 state, we have found that $D = 0.04 \pm 0.01\ \text{cm}^{-1}$ for mesitylylene-h_{10} and -d_{10}, as well as for 4,6-dimethyl-*m*-xylylene-h_{12} and -d_{12} [73]. This means that, upon $\pi \rightarrow \pi^*$ excitation, the average separation between the two unpaired π electrons of *m*-xylylene biradicals is decreased, indicating their delocalization onto the aromatic ring.

The aromatic carbenes in their ground state T_0 are characterized by a large D value due to the presence of the two unpaired electrons, one in a σ orbital and the other in a π orbital, both localized on the same carbenic carbon atom. The electronic structure of the excited T_1 state involves the promotion of the unpaired electron from the π orbital to a π^* orbital. The calculations performed on DPC [55] and on 2-NPC [69] indicate that this π^* orbital has only a minor contribution from the carbenic $2p\pi$ atomic orbital (see Fig. 7.8). Thus the $\pi \rightarrow \pi^*$ excitation induces a delocalization of the unpaired π electron onto the rings and a decrease of the corresponding spin density on the carbenic center, while the spin density originating from the unpaired σ electron is retained. As a consequence, the average separation between the two unpaired electrons is increased in the excited T_1 state relative to the T_0 ground state, leading to a $D(T_1)$ value much smaller than $D(T_0)$, in agreement with the experimental results listed in Table 7.4.

7.3.3. Concluding Remarks

Magnetic resonance methods used to determine the ZFS parameters D and E require substantial concentrations of molecules in the triplet state. The conventional EPR technique has been successfully used to measure the D and E values of either matrix-isolated carbenes in their triplet ground state [14, 17] or matrix-isolated aromatic hydrocarbons having triplet lifetimes of several seconds [77]. For aromatic carbonyl compounds [78] or N-heterocyclics [79] having triplet lifetimes of the order of milliseconds, the D and E parameters have been determined by ODMR (optically detected magnetic resonance). These two techniques, however, cannot be used for excited species having triplet lifetimes of the order of 10 to 100 ns because their steady-state concentration is too low. This is the reason why the $D(T_1)$ parameter of aromatic biradicals and carbenes in the excited T_1 state has been estimated from the magnetic field effect on the fluorescence decays. This approach, however, has its limitations. In particular, it can only be applied to molecules having triplet sublevels that decay with substantially different rates so that the sublevel mixing induced by the magnetic field changes significantly the fluorescence decays. Furthermore, this method does not give any information on the ordering of the triplet sublevels and therefore on the sign of the D parameter.

In 1977, Al'shitz, Personov, and Kharlamov [80] reported the first measurement of the D parameter of coronene in its lowest triplet state T_1 by using the "phosphorescence line-narrowing" method. In this study, the phosphorescence was selectively excited within the origin band of the $S_0 \rightarrow T_1$ transition of coronene by means of the 514.38-nm narrow line emitted from an argon laser, and was detected with a Fabry–Perot interferometer. This important experiment revealed the possibility of directly investigating the triplet zero-field structure from the optical spectra of molecules.

Hole-burning is another high-resolution technique that can provide the D and E parameters with great precision. In the next section, we will describe a hole-burning study [67, 81], performed by Kozankiewicz et al. at the University of Bordeaux, which led to the determination of the D and E values of a carbene in its first excited triplet state. To our knowledge, this is the first hole-burning experiment carried out on a triplet–triplet transition.

7.4. SELECTIVE LASER EXCITATION: THE HOLE BURNING IN THE T_0–T_1 TRANSITION OF 2-NPC

The absorption and emission spectra of matrix-isolated molecules or radicals at low temperatures are generally composed of inhomogeneously broadened bands having widths of 0.1–10 cm^{-1} in crystalline hosts and 100–500 cm^{-1} in glasses. This is due to the distribution of local environments around the absorbing species. The width of these bands, however, can be considerably reduced by selective laser excitation.

A narrow-band laser light that has a frequency within the inhomogeneous $S_0 \rightarrow S_1$ 0–0 absorption band of the molecule–matrix system can only be absorbed by the molecules that have their 0–0 homogeneous band in resonance with the laser frequency. The selectively excited molecules can then emit a sharp luminescence spectrum detected by a high-resolution monochromator and/or by a Fabry–Perot interferometer. This is the principle of the *fluorescence* or *phosphorescence line-narrowing method*, pioneered by Personov and co-workers [82] and described in more detail in Chapter 8. Alternatively, a narrow-band laser light can create a hole within the inhomogeneous absorption band by removing the absorbing molecules through either photochemical reactions (photochemical hole burning) or a structural rearrangement of the environment (nonphotochemical hole burning). The first *hole-burning* experiments were carried out by the Personov [83, 84] and Rebane [85] groups.

In the case of an $S_0 \rightarrow S_1$ transition, a *single* hole is burned at the same frequency as the laser light, or a *single* fluorescence spectrum is emitted from the selectively excited molecules. The situation is different for $S_0 \rightarrow T_1$ or $T_0 \rightarrow T_1$ transitions, for which *several* holes at different frequencies can simultaneously be burned with a single laser frequency, or a *multiplet* is emitted from the selectively excited molecules, as was observed by Al'shitz, Personov, and Kharlamov in their luminescence line-narrowing experiment [80].

7.4.1. Hole Burning in the T_0–T_1 Transition of 2-NPC

The hole-burning experiments [67, 81] were performed on the fluorescence excitation origin band at $\lambda_{max} = 588.7\,nm$ of the photostable pseudo-$E/trans$ conformer of 2-NPC in n-hexane at 1.8 K. Using a single-mode dye laser with 1 to 3 MHz frequency resolution, a complicated pattern of holes, covering a spectral range of nearly $1\,cm^{-1}$, is burned at 1.8 K, within the inhomogeneous origin band, as shown in Figure 7.15. The central hole, burned at the laser frequency

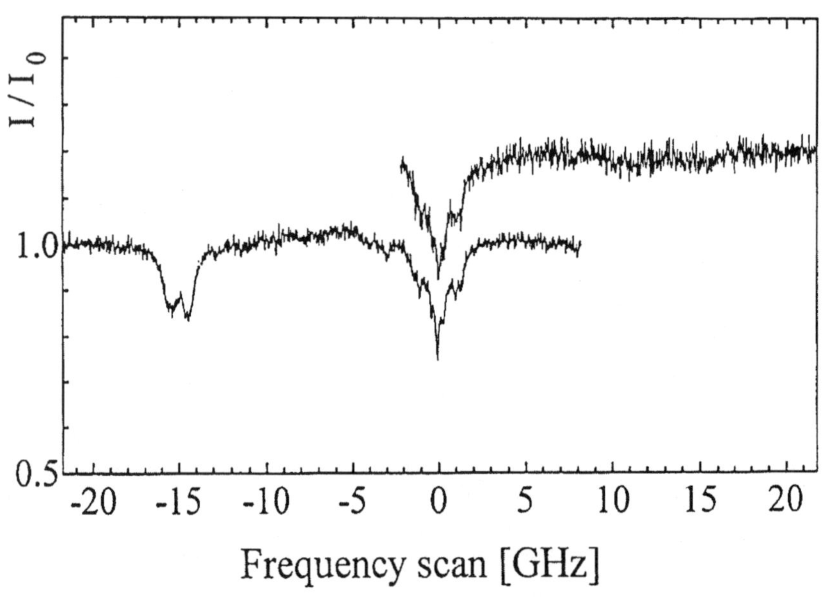

Figure 7.15. Typical holes burned within the inhomogeneous fluorescence excitation origin band of the pseudo-$E/trans$ 2-NPC at 588.7 nm. The spectral position of holes are given with respect to the frequency of the burning light. The spectrum scanning range (30 GHz) requires separate detection of the high- and of the low-energy sides of the hole spectrum. (Reprinted, with permission, from Ref. 67. Copyright by Elsevier.)

(0 GHz in Fig. 7.15) is symmetrically surrounded by two satellite holes (0.3 GHz away) and two satellite pairs of holes (1.0 and 1.3 GHz away). Another satellite pair of holes, deeper and broader than those mentioned before, is located on the low-energy (red) side of the central hole (14.5 and 15.5 GHz away). There are no such satellite holes on the high-energy side of the central hole.

Interpretation. The appearance of satellite holes in Figure 7.15 can be explained by taking into account the three sublevels of the T_0 and T_1 states. Let us first consider the simple case of a D_{2h} molecule with ZFS spin axes having the same direction in both the T_0 and T_1 states. According to the spin selection rules, the burning laser will only excite the three groups (1, 2, 3) of carbenes that

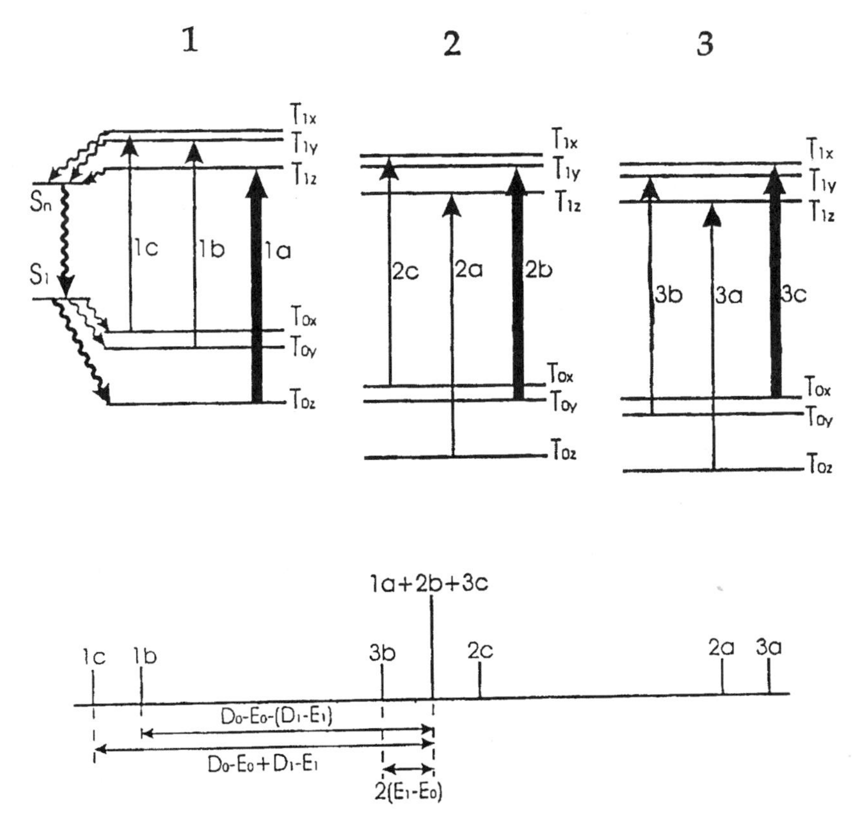

Figure 7.16. Energy-level diagram and electronic transitions of a D_{2h} molecule. The burning transitions, resonant with the laser light, are indicated with thick arrows. The energy separations between sublevels T_z and T_y and between T_y and T_x are, by definition, respectively, $D_0 - E_0$ and $2E_0$ in the T_0 state and $D_1 - E_1$ and $2E_1$ in the T_1 state. The expected spectral pattern of burned holes and their energy separations are shown at the bottom.

have the energy of their $T_{0z} \rightarrow T_{1z}$, $T_{0y} \rightarrow T_{1y}$, and $T_{0x} \rightarrow T_{1x}$ transitions in resonance with the laser frequency, as indicated in Figure 7.16. In this way, these 3 groups are removed from the absorbing set of species, and so are *all* their associated transitions: here, 9 absorption frequencies disappear, leading to 7 experimental holes. In this predicted pattern, the holes are symmetrically distributed with respect to the central hole, whereas in Figure 7.15, the satellite holes around 15 GHz are only detected on the low-energy side of the central hole. The lack of the 2a and 3a holes indicate that groups 2 and 3 are absent, which implies that holes 2c and 3b are equally missing (see Fig. 7.15). Thus the *only* group of carbenes that is effectively burned is group 1, where the species absorb the laser light from the lowest-energy spin sublevel T_{0z}. The absence of holes originating from groups 2 and 3 can be explained as follows.

Since the pseudo-*E*/*trans* conformer of 2-NPC is photochemically stable at 1.8 K, the hole burning should proceed through *nonphotochemical processes* such as the ISC from the T_1 state, $T_1 \rightsquigarrow S_4 \rightsquigarrow S_1 \rightsquigarrow T_0$. The relaxation energy is dissipated into the surrounding lattice and the carbene/matrix system may jump to another energy minimum, shifting the spectral position of the zero-phonon line, and leaving a spectral hole. The most important step of the ISC channel is the $S_1 \rightsquigarrow T_0$ step, which may selectively populate the spin component T_{0z} only. Let us mention that the selective population of the T_{0z} sublevel, where the z axis passes through the carbenic carbon atom and is parallel to the line joining the phenyl centers, has already been demonstrated by the spin-echo experiment on DPC [86]. The carbenes in groups 2 and 3 are selectively relaxed by ISC to the T_{0z} sublevel, where they cannot be burned because they are no longer in resonance with the laser frequency. Therefore, they remain there during the spin-lattice relaxation time which is usually long at 1.8 K. The only group of carbenes that may effectively suppply heat to the surrounding lattice through the ISC process, and thus can be burned, is the group of carbenes (group 1) having their $T_{0z} \rightarrow T_{1z}$ transition in resonance with the laser frequency. We will now focus on this group to explain the existence and the structure of the experimentally observed holes (Figure 7.15).

2-NPC does not belong to the D_{2h} symmetry group. This carbene has C_1 symmetry, and the situation is more complicated. The ZFS principal spin axes x, y, and z in the T_0 state may rotate, leading to new x', y', and z' axes in the T_1 state, since the directions of the spin axes follow the redistribution of the spin density upon electronic excitation. Thus the electronic transitions from every spin sublevel of T_0 to every spin sublevel of T_1 are allowed, as indicated in the energy-level diagram presented in Figure 7.17. The three groups of 2-NPC molecules that can contribute to the hole-burning spectrum are those having their $T_{0z} \rightarrow T_{1z'}$, $T_{0z} \rightarrow T_{1y'}$ and $T_{0z} \rightarrow T_{1x'}$ transitions in resonance with the laser frequency. These three groups derive from group 1 in Figure 7.16 as a result of the loss of spin selection rules when the D_{2h} symmetry is lowered to C_1. The

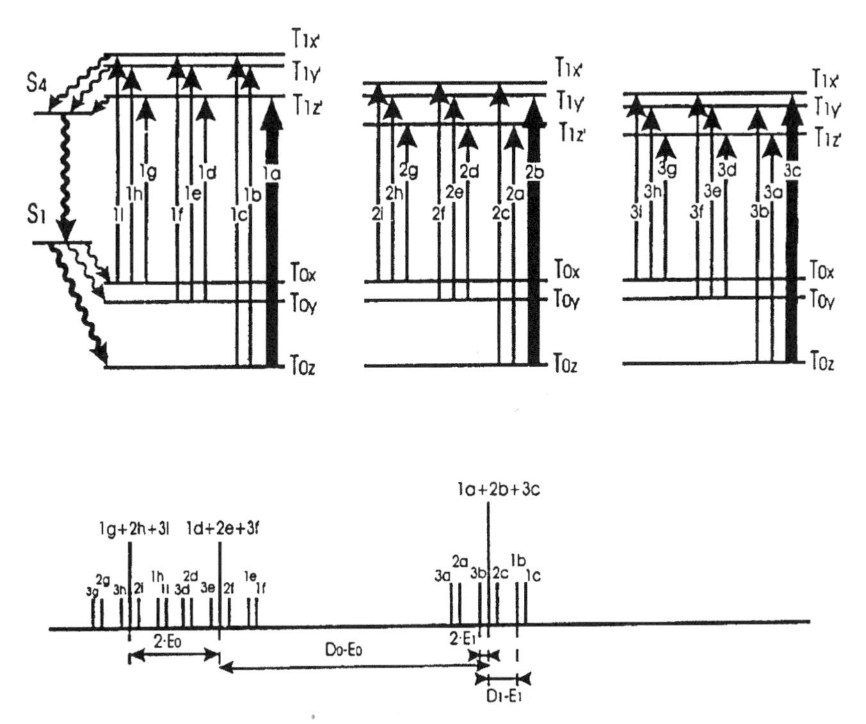

Figure 7.17. Energy-level diagram and electronic transitions of 2-NPC. The burning transitions, resonant with the laser light, are indicated with thick arrows. The energy separations between sublevels T_z and T_y and between T_y and T_x are, by definition, respectively, D_0–E_0 and $2E_0$ in the T_0 state and D_1–E_1 and $2E_1$ in the T_1 state. The expected spectral pattern of burned holes and their energy separations are shown at the bottom. The ordering of the spin sublevels in the ground state is the same as in Ref. 86. (Reprinted, with permission, from Ref. 67. Copyright by Elsevier.)

predicted spectrum of burned holes is shown at the bottom of Figure 7.17. The central hole, at the burning frequency, is symmetrically surrounded by three weaker satellites holes on the low- and three others on the high-energy sides, separated from the center by $2E_1$, $D_1 - E_1$ and $D_1 + E_1$. This substructure around the central hole provides information about the ZFS of the excited T_1 state and closely resembles that observed experimentally (see Figure 7.15). Two other sets of holes, each with the same substructure as that around the central hole, are located on the low-energy side, $D_0 - E_0$ and $D_0 + E_0$ away from the center. They correspond to the satellite holes observed experimentally on the red side of the central hole. The energy positions of these holes provide information about the splitting of the triplet ground state. The comparison between the predicted and experimental spectra of holes allows one to estimate the ZFS parameters of the T_0

and T_1 states of the pseudo-$E/trans$ conformer of 2-NPC:

$$D(T_0) = D_0 = 0.50 \pm 0.01 \text{ cm}^{-1}$$

$$E(T_0) = E_0 = 0.017 \pm 0.003 \text{ cm}^{-1}$$

$$D(T_1) = D_1 = 0.038 \pm 0.003 \text{ cm}^{-1}$$

$$E(T_1) = E_1 = 0.005 \pm 0.001 \text{ cm}^{-1}$$

The above data may be compared with the corresponding parameters shown in Table 7.4.

In summary, the first hole-burning experiment carried out on 2-NPC has led to the precise determination of the ZFS parameters in the T_0 and T_1 states of the photostable conformer. This hole-burning study opens a new field of investigations and hopefully this technique will be applied in the future to many other systems with a triplet ground state, *m*-xylylene biradicals, for example.

7.5. CONCLUSION

In this chapter, we have tried to give a comprehensive survey of studies performed on dimethylenic and carbenic conjugated systems. Since these molecules are elusive compounds, evidence concerning them is generally provided by spectroscopic investigations coupled to quantum-mechanical calculations.

The main achievement is a more exact knowledge of the spin properties of these species, and, above all, of the zero-field splitting parameter D for systems having fluorescent excited triplet states. Whereas the D values can be obtained by EPR or ODMR methods for phosphorescent triplet states, the study of short-lived fluorescent triplets requires the development of a powerful modern laser technique such as hole-burning spectroscopy. Precise quantum-mechanical calculations of the relevant spin-Hamiltonian parameters for large molecules is a challenge for the next years.

Let us finally add that such studies can be considered as part of the current investigations on molecular magnetism, with the quest for the Golden Fleece, namely, molecular compounds from which organic magnets or molecular computing devices could be designed, based on bistability properties [87–89]. They are also expected to be of interest for a better understanding of the reactivity properties of biradicals; this topic has been recently renewed through investigations on enediynes, which are key intermediates in the synthesis of potent antibiotic, antitumoral, and antiviral drugs [90].

REFERENCES

1. A. E. Chichibabin, *Ber. Dtsch. Chem. Ges.* **40**, 1810 (1907).
2. J. Thiele and H. Balhorn, *Ber. Dtsch. Chem. Ges.* **37**, 1463 (1904).

3. W. Schlenk and M. Brauns, *Ber. Dtsch. Chem. Ges.* **48**, 661 (1915).
4. E. Müller and I. Müller-Rodloff, *J. Liebigs Ann. Chemie* **517**, 134 (1935).
5. C. A. Hutchison, Jr., A. Kowalski, R. C. Pastor, and G. W. Wheland, *J. Chem. Phys.* **20**, 1485 (1952).
6. G. Herzberg and J. Shoosmith, *Nature (London)* **183**, 1801 (1959).
7. G. Herzberg, *Proc. Roy. Soc. (London)* **A262**, 291 (1961).
8. G. Herzberg and H. W. C. Johns, *Proc. Roy. Soc. (London)* **A295**, 107 (1966).
9. P. Thaddeus, *J. Chem. Soc. (Faraday Trans.)* **89**, 2125 (1993).
10. F. I. J. de Kanter and R. Kaptein, *J. Am. Chem. Soc.* **104**, 4759 (1982).
11. J. C. Scaiano, *Acc. Chem. Res.* **15**, 252 (1982).
12. G. Porter and M. F. Tchir, *J. Chem. Soc. D* 1372 (1970).
13. G. Porter and M. F. Tchir, *J. Chem. Soc. A* 3772 (1971).
14. R. W. Murray, A. M. Trozzolo, E. Wasserman, and W. A. Yager, *J. Am. Chem. Soc.* **84**, 3213 (1962).
15. A. M. Trozzolo and E. Wasserman, in *Carbenes*, Vol. II, R. A. Moss and M. Jones, Jr., eds., Wiley, New York, 1975, p. 185.
16. A. M. Trozzolo, *Acc. Chem. Res.* **1**, 329 (1968).
17. R. W. Brandon, G. L. Closs, and C. A. Hutchison, Jr., *J. Chem. Phys.* **37**, 1878 (1962).
18. G. L. Closs, in *Carbenes*, Vol. II, R. A. Moss and M. Jones, Jr. eds., Wiley, New York, 1975, p. 159.
19. V. E. Shpol'skii, A. A. Ilina, and L. A. Klimova, *Dokl. Akad. Nauk. SSSR* **87**, 935 (1952).
20. T. N. Bolotnikova, *Opt. Spektrosk.* **7**, 138 (1959).
21. E. L. Wehry and G. Mamantov, in *Modern Fluorescence Spectroscopy*, Vol. 4, p. 193, E. L. Wehry, ed., Plenum, 1980.
22. J. L. Richards and S. A. Rice, *J. Chem. Phys.* **54**, 2014 (1971).
23. A nonexhaustive compilation of earlier references on this topics can be found in *Guide to Fluorescence Litterature*, R. A. Passwater, IFI/Plenum, 1974.
24. K. I. Mamedov and I. K. Nasibov, *Opt. Spektrosk.* **30**, 1052 (1971).
25. C. Branciard-Larcher, E. Migirdicyan, and J. Baudet, *Chem. Phys.* **2**, 95 (1973).
26. W. P. Cofino, J. W. Hofstraat, G. Ph. Hoornweg, C. Gooijer, C. MacLean, and N. H. Velthorst, *Chem. Phys. Lett.* **98**, 242 (1983).
27. W. P. Cofino, S. M. van Dam, D. A. Kamminga, G. Ph. Hoornweg, C. Gooijer, C. MacLean, and N. H. Velthorst, *Mol. Phys.* **51**, 537 (1984).
28. W. P. Cofino, S. M. van Dam, G. Ph. Hoornweg, C. Gooijer, C. MacLean, and N. H. Velthorst, *Spectrochim. Acta* **40A**, 251 (1984).
29. W. P. Cofino, S. M. van Dam, D. A. Kamminga, G. Ph. Hoornweg, C. Gooijer, C. MacLean, and N. H. Velthorst, *Spectrochim. Acta* **40A**, 219 (1984).
30. This topic is the subject of several contributions in this book, to which we refer the interested reader.
31. E. Migirdicyan, *Comp. Rend. Acad. Sci. (Paris)* **266**, 756 (1968).

32. E. Migirdicyan and J. Baudet, *J. Am. Chem. Soc.* **97**, 7400 (1975).
33. A. Després, V. Lejeune, E. Migirdicyan, and W. Siebrand, *Chem. Phys. Lett.* **111**, 201 (1984).
34. V. Lejeune, A. Després, and E. Migirdicyan, *J. Phys. Chem.* **88**, 2719 (1984).
35. C. R. Flynn and J. Michl, *J. Am. Chem. Soc.* **95**, 5802 (1973).
36. C. R. Flynn and J. Michl, *J. Am. Chem. Soc.* **96**, 3280 (1974).
37. K. L. Tseng and J. Michl, *J. Am. Chem. Soc.* **99**, 4840 (1977).
38. J. Baudet, *J. Chim. Phys.* **68**, 191 (1971).
39. B. S. Hudson, B. E. Kohler, and K. Schulten, in *Excited States*, Vol. 6, E. C. Lim, ed., Academic Press, New York, 1982, p. 1.
40. M. Hossain, B. E. Kohler, and P. West, *J. Phys. Chem.* **86**, 4918 (1982).
41. B. E. Kohler and T. A. Spiglanin, *J. Chem. Phys.* **80**, 3091 (1984).
42. W. P. Cofino, G. Ph. Hoornweg, C. Gooijer, C. MacLean, and N. H. Velthorst, *Chem. Phys.* **72**, 73 (1982).
43. B. B. Wright and M. S. Platz, *J. Am. Chem. Soc.* **105**, 628 (1983).
44. J. L. Goodman and J. A. Berson, *J. Am. Chem. Soc.* **107**, 5409 (1985).
45. K. S. Pitzer and D. W. Scott, *J. Am. Chem. Soc.* **65**, 803 (1943).
46. C. Cossart-Magos and S. Leach, *J. Chem. Phys.* **64**, 4006 (1976).
47. A. R. Gregory, W. H. Henneker, W. Siebrand, and M. Z. Zgierski, *J. Chem. Phys.* **65**, 2071 (1976).
48. A. M. Trozzolo and W. A. Gibbons, *J. Am. Chem. Soc.* **89**, 239 (1967).
49. A. M. Trozzolo, E. Wasserman, and J. A. Yager, *J. Chim. Phys.* 1663 (1964).
50. R. J. M. Anderson, B. E. Kohler, and J. M. Stevenson, *J. Chem. Phys.* **71**, 1559 (1979).
51. S. Saebo and J. E. Boggs, *Theochem* **4**, 365 (1982).
52. U. Burkett and N. L. Allinger, *Molecular Mechanics*, ACS Monograph 177, American Chemical Society, Washington, DC, 1982.
53. A. Després, V. Lejeune, E. Migirdicyan, and M. S. Platz, *J. Phys. Chem.* **96**, 2486 (1992).
54. K. W. Haider, M. S. Platz, A. Després, V. Lejeune, and E. Migirdicyan, *J. Phys. Chem.* **94**, 142 (1990).
55. A. Després, V. Lejeune, E. Migirdicyan, A. Admasu, M. S. Platz, G. Berthier, O. Parisel, J.-P. Flament, I. Baraldi, and F. Momicchioli, *J. Phys. Chem.* **97**, 13358 (1993).
56. R. J. M. Anderson and B. E. Kohler, *J. Chem. Phys.* **65**, 2451 (1976).
57. F. Momicchioli, I. Baraldi, and M. C. Bruni, *Chem. Phys.* **70**, 161 (1982).
58. F. Momicchioli, I. Baraldi, and M. C. Bruni, *Chem. Phys.* **82**, 229 (1983).
59. U. Mazzucato and F. Momicchioli, *Chem. Rev.* **91**, 1679 (1991).
60. B. Huron, J.-P. Malrieu, and P. Rancurel, *J. Chem. Phys.* **58**, 5745 (1973).
61. A. M. Trozzolo, E. Wasserman, and J. A. Yager, *J. Am. Chem. Soc.* **87**, 129 (1965).
62. R. S. Hutton and H. D. Roth, *J. Am. Chem. Soc.* **104**, 7395 (1982).
63. H. D. Roth and R. S. Hutton, *Tetrahedron* **41**, 1567 (1985).

64. J. R. Johnson, K. D. Jordan, D. F. Plusquellic, and D. W. Pratt, *J. Chem. Phys.* **93**, 2258 (1990).
65. V. Maloney and M. S. Platz, *J. Phys. Org. Chem.* **3**, 135 (1990).
66. B. Kozankiewicz, A. Després, V. Lejeune, E. Migirdicyan, D. Olson, J. Michalak, and M. S. Platz, *J. Phys. Chem.* **98**, 10419 (1994).
67. B. Kozankiewicz, J. Bernard, E. Migirdicyan, M. Orrit, and M. S. Platz, *Chem. Phys. Lett.* **245**, 549 (1995).
68. A. Després, E. Migirdicyan, K. Haider, V. Maloney, and M. S. Platz, *J. Phys. Chem.* **94**, 6632 (1990).
69. O. Parisel, G. Berthier, and E. Migirdicyan, *Can. J. Chem.* **73**, 1869 (1995).
70. H. D. Roth and M. S. Platz, *J. Phys. Org. Chem.* **9**, 252 (1996).
71. V. Lejeune, A. Després, B. Fourmann, O. Benoist d'Azy, and E. Migirdicyan, *J. Phys. Chem.* **91**, 6620 (1987).
72. V. Lejeune, A. Després, E. Migirdicyan, and W. Siebrand, *J. Phys. Chem.* **95**, 7585 (1991).
73. V. Lejeune, A. Després, and E. Migirdicyan, *J. Phys. Chem.* **94**, 8861 (1990).
74. V. Lejeune, PhD Thesis, Orsay (France), 1990.
75. V. Lejeune, G. Berthier, A. Després, and E. Migirdicyan, *J. Phys. Chem.* **95**, 3895 (1991).
76. I. Moritani, S.-I. Murahashi, M. Nishino, Y. Yamamoto, K. Itoh, and N. Mataga, *J. Am. Chem. Soc.* **89**, 1259 (1966).
77. C. A. Hutchison, Jr. and B. W. Mangum, *J. Chem. Phys.* **34**, 908 (1961).
78. T. H. Cheng and N. Hirota, *J. Chem. Phys.* **56**, 5019 (1972).
79. J. Schmidt, W. S. Veeman, and J. H. van der Waals, *Chem. Phys. Lett.* **4**, 341 (1969).
80. E. I. Al'shitz, R. I. Personov, and B. M. Kharlamov, *JETP Lett.* **26**, 586 (1977).
81. B. Kozankiewicz, J. Bernard, E. Migirdicyan, M. Orrit, and M. S. Platz, *Mol. Cryst. Liq. Cryst.* **283**, 191 (1996).
82. R. I. Personov, in *Spectroscopy and Excitation Dynamics of Condensed Molecular Systems*, V. M. Agranovitch and R. M. Hochstrasser, eds., North- Holland, Amsterdam, 1983, Chap. 10, p. 555.
83. R. I. Personov, E. I. Al'shitz, and L. A. Bykovskaya, *Opt. Comm.* **6**, 169 (1972).
84. B. M. Kharlamov, R. I. Personov, and L. A. Bykovskaya, *Opt. Comm.* **12**, 191 (1974).
85. A. A. Gorokhovski, R. K. Kaarli, and L. A. Rebane, *Zh. Eksp. Teor. Fiz. Pisma Red.* **20**, 474 (1974).
86. D. C. Doetschman, B. J. Botter, J. Schmidt, and J. H. van der Waals, *Chem. Phys. Lett.* **38**, 18 (1976).
87. H. Iwamura, *Adv. Phys. Org. Chem.* **26**, 179 (1990).
88. O. Kahn, *Molecular Magnetism*, VCH Publishers, 1993.
89. R. Chiarelli, A. Rassat, and P. Rey, *J. Chem. Soc., Chem. Comm.* 1081 (1992).
90. K. K. Wang, *Chem. Rev.* **96**, 207 (1996).

CHAPTER

8

FUNDAMENTAL ASPECTS OF FLUORESCENCE LINE-NARROWING SPECTROSCOPY

RYSZARD JANKOWIAK

Ames Laboratory-USDOE, Iowa State University, Ames, IA 50011, USA

In this chapter the basic aspects of a low temperature, laser based fluorescence technique, i.e., fluorescence line-narrowing spectroscopy (FLNS) are discussed. A brief summary of various site selection methods in frequency and time domain and a short discussion of homogeneous and inhomogeneous broadening (sections 8.1 and 8.2.) are presented. Section 8.3 describes the zero-phonon lines and phonon side bands, and illustrates the relation between the quantities that determine the Franck-Condon, Debye-Waller, and Huang-Rhys factors. Model calculations of the fluorescence spectral lineshape are also provided. This is followed by a description of the principles of FLNS and instrumentation. A sampling of applications in a variety of different condensed phase environments and future prospects conclude the chapter.

SYMBOLS AND ABBREVIATIONS

α, Debye–Waller factor

α-OH-TAM, α-hydroxytamoxifen

ω_L, laser frequency

APT, aluminum phthalocyanine tetrasulfonate

B[*a*]P, benzo[*a*]pyrene

CE, capillary electrophoresis

CT, charge transfer

dAMP, deoxyadenosine monophosphate

DB[*a*,*l*]P, dibenzo[*a*,*l*]pyrene

EM, ethanol/methanol

FC, Franck–Condon factor

Shpol'skii Spectroscopy and Other Site-Selection Methods, Edited by Cees Gooijer, Freek Ariese, and Johannes W. Hofstraat.
ISBN 0-471-24508-9

FLN, fluorescence line narrowing
FWHM, full width at half-maximum
HB, hole-burning
HPLC, high-performance liquid chromatography
LLN, luminescence line narrowing
MCF7, human breast adenocarcinoma cells
PAH, polycyclic aromatic hydrocarbon(s)
PE, photon echo
PLN, phosphorescence line narrowing
PSB, phonon sideband
R, R-phonon transition ($R = 1, 2, \ldots$)
S, Huang–Rhys factor
S_0, ground state
S_1, first excited singlet state
TAM, tamoxifen
TLS, two-level systems
ZPL, zero-phonon line
ZSD, zero-site distribution

8.1. INTRODUCTION

Several excellent descriptions of fluorescence line-narrowing spectroscopy (FLNS), sometimes called site-selection spectroscopy, can be found in recently published books [1–7] and reviews [8–12]. Most of them, however, are addressed to the specialists in the field of solid-state spectroscopy. In this chapter, only the basic aspects of FLNS, with an emphasis on the information it provides, are described. By emphasizing diverse applications of FLNS, we strive to address a wide circle of readers in environmental and bioanalytical sciences. Personov summarizes the historical development of FLNS and other site-selective spectroscopies in the general introduction of this book (Chapter 1). Here we mention only that the method emerged with the work of Denisov and Kizel [13], Szabo [14], and Personov et al. [15]. FLNS is a well-established technique [1, 4, 7–12] that has been successfully used for detection and characterization of many different "systems." Examples include a large number of organic molecules [1, 11, 16], inorganic systems [17] (see also Chapter 17), and a wide variety of biomolecular systems, including photosynthetic pigments [18–23] and antenna protein–pigment complexes [24]. Other applications, described in this book, refer to studies of DNA adducts (see Chapter 11), proteins (Chapter 12), chlorophylls (Chapter 13), and porphyrins [25] (Chapters 3 and 12).

Laser-based, solid-state spectroscopy, which is basically energy selective, encompasses experiments in both time [26, 27] and frequency domains [1, 2, 15, 17]. There are three main line-narrowing techniques: FLNS (see below), hole-burning (HB) spectroscopy described in Chapter 9, and various types of photon echoes (PE). The first studies suppressing inhomogeneity effects were echo experiments on ruby crystals by Hartman et al. [28]. Photon echo techniques have been discussed in detail in Refs. 26, 27, 29–31. The ultimate goal of these methods is to measure the homogeneous linewidth of optical transitions. Both frequency- (FLNS and HB) and time-domain methods (PE) can be utilized to obtain information regarding molecular dynamics and electron-phonon coupling. FLNS and HB provide a frequency selection, by exciting the homogeneous ensemble of molecules, while the PE experiments measure a temporal dependence of the dipolar moment of the molecular ensemble [26, 27, 29, 32]. Thus these methodologies provide a means to overcome the inhomogeneous broadening, which is typically about 100–500 cm^{-1} in amorphous solids. In PE, in contrast to FLNS and HB, the elimination of the inhomogeneous broadening has a different origin; it is the result of two carefully designed periods of evolution where the optical inhomogeneous dephasing in the first period is exactly canceled by rephasing the process in the second period [32].

This chapter focuses on FLNS; it will be demonstrated how one can obtain vibrationally resolved spectra of the ground-state molecule, vibrational frequencies in the excited state, and insight into the electron–phonon coupling phenomenon. It will be shown that vibronic excitation within the $S_0 \rightarrow S_1$ absorption spectrum provides the highest selectivity in FLNS, since typically there is no correlation between different excited electronic states. As a result, excitation of isolated molecules in glasses at very high energies (e.g., into the S_2 state) leads to population of all sites (within the inhomogeneously broadened band) and a disappointingly broad fluorescence spectrum [1, 2, 11]. Therefore, in this chapter we will discuss only the vibronic excitation within the envelope of the first excited state, which, as argued below, can be used as a powerful method for chromophore identification and/or structural characterization [9, 11, 33]; see Section 8.3.3.

At present, it is widely appreciated that FLNS is the most powerful optical frequency-domain technique available for fluorescent analytes. It will be shown below that use of the FLN approach not only leads to at least two orders of magnitude improvement in spectral resolution over fluorescence excited under non-line-narrowing conditions, but also provides a sensitive "microscopic" probe of the interaction of a molecule with the local solvent environment. To address the fundamental aspects of this powerful methodology, we begin by examining the qualitative picture of inhomogeneous and homogeneous broadening.

8.2. INHOMOGENEOUS AND HOMOGENEOUS BROADENING

In solid molecular systems all broadening mechanisms for absorption bands can be categorized as either homogeneous or inhomogeneous. Homogeneous broadening is that which is the same for each and every chemically identical molecule in the ensemble (see below). Inhomogeneous broadening is the result of the fact that an analyte in a disordered host (glasses, polymers, proteins, DNA) can generally adopt a very large number of energetically inequivalent sites (different individual microenvironments). This is why optical bands of chromophores (ions and/or molecules) embedded in amorphous solids have, as a rule, large inhomogeneous broadening. Mechanisms leading to broadening are strains, random electric fields, field gradients of charged defects, interactions with collective excitations (such as phonons), and interactions between centers/chromophores [34]. A simple pictorial way to illustrate such broadening is shown in Figure 8.1. Consider an ensemble of chromophores in a rigid (low-temperature) matrix, possessing statistically different environments, as shown schematically in the left frame of Figure 8.1. The immediate environments of the identical guest molecules (labeled in Figure 8.1 as site 1, 2, and 3) in the disordered matrix are different, and as a result, the molecules absorb at somewhat different energies, ω_1, ω_2, and ω_3, respectively. This is illustrated by the three narrow spectra (solid lines) shown to the right in Figure 8.1. In an absolutely perfect crystal all three transitions would have the same frequency, $\omega_1 = \omega_2 = \omega_3$, and the resulting narrow absorption band would be only homogeneously broadened [35]. These narrow spectra have a Lorentzian line shape and reflect the homogeneous chromophore spectra with a line width, Γ_{hom}. The latter width is determined by excited-state dephasing interactions and lifetime effects, which are the same for all chromophores; the value of Γ_{hom} depends on the specific transition under consideration (see Chapter 2 and Section 8.3 for details). For inhomogeneous

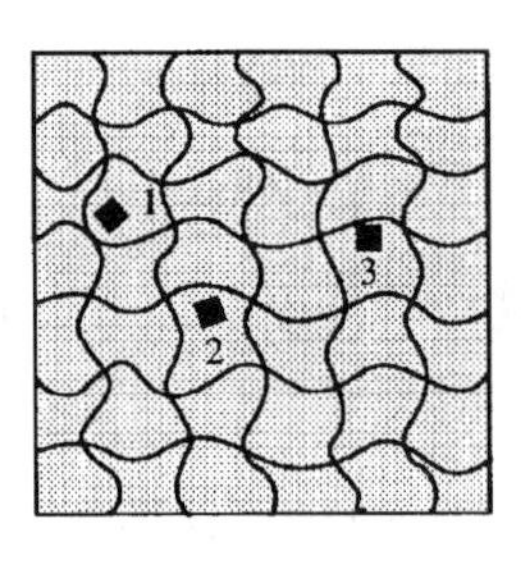

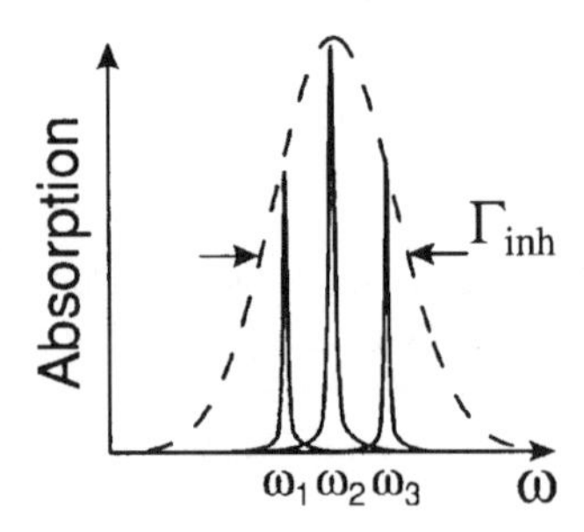

Figure 8.1. Schematic of absorbers dispersed in an amorphous solid matrix and the resulting, low-temperature, inhomogeneously broadened absorption profile (dashed line; see text for details). (Based on Ref. 35.)

broadening, Γ_{inh}, the statistical distribution of the inequivalent sites leads to a broad absorption profile, often being Gaussian [36, 37] (dashed line in Fig. 8.1), with a typical width of several hundred wavenumbers. Assuming very weak coupling with phonons (see below), the full width at half-maximum of the Gaussian distribution (in Fig. 8.1) refers to the inhomogeneous broadening, Γ_{inh}. These broad bands usually reflect the configurational statistics of the rigid solvent cage around the chromophore rather than the dynamics of the cage reorganization after excitation [35]. The strength of the broadening is measured by the ratio of $\Gamma_{\text{inh}}/\Gamma_{\text{hom}}$, which can change from approximately 1 to about 10^5, depending upon the host material [1, 35, 38–41]. Thus the essential feature of FLNS is the use of a narrow laser source for excitation of chromophores, either directly within the inhomogeneously broadened origin absorption band (referred to below as 0–0 excitation) or into a vibronic region of the first electronically excited state S_1 (i.e., vibronic excitation; see below).

8.3. FLUORESCENCE LINE-NARROWING SPECTROSCOPY

8.3.1. Zero-Phonon Lines and Phonon Side Bands

The main properties of narrow ZPLs in the spectra of crystal impurities were first discussed in detail by Kane [41]. Shortly thereafter, the principal analogy between the ZPL in the optical spectrum of an impurity center and the Mössbauer line in the spectrum of γ quanta, absorbed or emitted by a radioactive nucleus in a crystal lattice, was demonstrated [42–44]. A comprehensive theory of impurity spectra in solids can be found, for example, in an excellent book written by Rebane [41].

Briefly, an inhomogeneously broadened fluorescence origin band (0–0 band) has no structure at low temperature (as shown by the solid line in Fig. 8.2), but consists of many sharp bands (dashed lines), which correspond to the fluorescence zero-phonon lines (ZPL) of the "guest" molecules occupying inequivalent sites. A zero-phonon transition is one for which no net change in the number of phonons accompanies the electronic transition. Building to lower energy on each ZPL is a broad phonon (lattice vibrational) wing, referred to below as the phonon side band (PSB). As mentioned above, each single-site ZPL carries a homogeneous linewidth, Γ_{hom}, which is determined by the total dephasing time of the optical transition (vide infra). Since the PSBs contribute to the absorption and/or fluorescence origin bands, the width of the Gaussian shown in Figure 8.2A (solid lines) is approximately given by $\Gamma_{\text{inh}} + S\omega_{\text{m}}$ [40, 45, 46], where ω_{m} is the mean phonon frequency (for organic molecules typically about 20–30 cm^{-1}) and S is the Huang–Rhys factor defined in Figure 8.3 (see below). If excitation occurs at ω_{L} using a spectrally narrow laser, as schematically shown in Figure 8.2, only

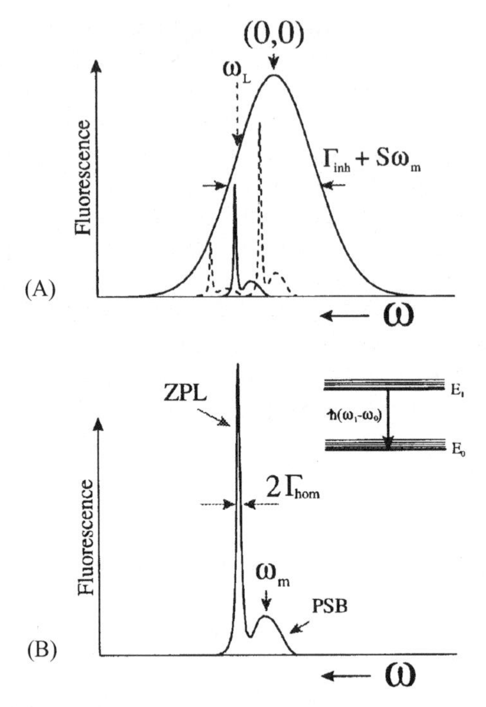

Figure 8.2. (A) Schematic of the inhomogeneously broadened fluorescence origin band; FWHM equals ($\Gamma_{inh} + S\omega_m$). ω_B is the laser excitation frequency. (B) Schematic view of the resulting fluorescence zero-phonon line (ZPL) and its phonon sideband (or phonon wing, PW). The PW has the intensity $(1 - \alpha)$ (see Eq. 8.9) and is displaced to lower energies by ω_m, which corresponds to the mean phonon frequency. Γ_{hom} and Γ_{instr} are the homogeneous and instrument linewidths, respectively.

those chromophores can be excited whose transition energies coincide with the frequency of the laser. If the concentration of chromophores is sufficiently low, no energy transfer to other chromophores occurs. Subsequently, resonant fluorescence occurs, providing a fluorescence line-narrowed spectrum, as shown in Figure 8.2B. Often one speaks of selecting an "isochromat." It will be shown below that for $\Gamma_{instr} \ll \Gamma_{hom}$, the width of the ZPL is $2\Gamma_{hom}$. A comprehensive treatment of the nature of a spectral band can be found in Refs. 1, 2, 44, and 47. Briefly the transition from state $|0\rangle$ with energy E_0 to state $|1\rangle$ with energy E_1 (see Fig. 8.2B) is accompanied by the absorption (or emission) of a light quantum of frequency $\omega = (E_1 - E_0)/\hbar$ [1]. The probability of this transition is proportional to the square of the matrix element $|\langle 1|\hat{V}|0\rangle|^2$, where $\hat{V}$ is the operator of the interaction between the electrons of the chromophores and the light. Expressions for the intensity of the spectral band, $I(\omega)$, as shown in Figure 8.2B, have been derived [1, 47]. Here we note only that $(E_1 - E_0)$ defines the purely electronic transition, that is, without electron–phonon coupling (see next section) or coupling to the tunneling degrees of freedom that exist in glasses, polymers, and proteins [12, 35, 40]. The resulting ZPL would have the form of a Lorentzian, with a width defined by lifetime broadening. However, due to the presence of electron–phonon coupling and interactions with the low-energy excitations (i.e., two-level systems, TLS), the width of the ZPL increases and is temperature dependent [1, 2, 46, 48, 49]. The TLS in amorphous solids have a high spectral density at very low energies, and are responsible for the observed anomalous temperature dependence of the ZPLs. Typically, the homogeneous linewidth varies with temperature as $\sim T^{1+\mu}$ [31, 39, 40, 50]. The value of $\mu \sim 0.3$ and its origin is still a matter of debate [50]. The latter power-law and time dependence of the homogeneous linewidth, due to so-called spectral diffusion, are discussed in more detail in Chapter 9 (see also Refs. 29–32, 50–54). Thus the band shape of the single-site fluorescence profile, as shown in Figure 8.2B, consists of two portions:

$$F(\omega, T) = \phi_{\mathrm{ZPL}}(\omega, T) + \phi_{\mathrm{PSB}}(\omega, T) \tag{8.1}$$

where the first term describes the ZPL and the second describes the phonon side band (phonon wing). For a mathematical description of this topic see Refs. 1 and 2.

8.3.2. Electron–Phonon Coupling

Before discussing the spectral line shape in more detail, the coupling of the electronic excitation with external nuclear modes (phonons) has to be addressed. In general, the coupling is strong for molecules which upon electronic excitation undergo large geometry changes and/or large changes of the electronic charge distribution. This is illustrated in Figure 8.3, where the configurational coordinate diagram of the ground and excited states of a hypothetical molecule is shown. The quantities that describe the Franck–Condon (FC) and Huang-Rhys (S) factors are defined. The two parabolas represent the potential energy; the electronic ground state and the first excited state are labeled as E_0 and E_1, respectively. The energy levels of a guest molecule are depicted in the interaction with the local phonon i (energy $\hbar\omega_i$). A harmonic oscillator model is used to describe a number of phonon quanta. The potential may be written in the following form [55]:

$$V = \sum_i A_i q_i + \sum_{ij} B_{ij} q_i q_j \tag{8.2}$$

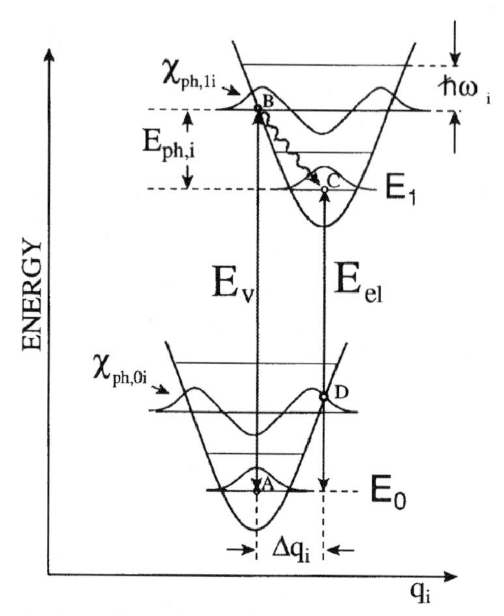

Figure 8.3. Configurational coordinate diagram for the analysis of the optical transition between the ground (E_0) and the excited (E_1) electronic states. At low temperature the peak of the absorption and fluorescence band corresponds to the transition (*A*–*B*) and (*C*–*D*), respectively. The energy separation is the Stokes shift. $S = (E_v - E_{el})/\hbar\omega_i$; $\hbar\omega_i$ describes the phonon energy levels. q_i represents a normal coordinate of the lattice. $\chi_{ph,1i}$ and $\chi_{ph,0i}$ represent the phonon wavefunctions for mode *i* in the E_1 and E_0 states, respectively.

The linear coupling coefficients A_i (in Eq. 8.2) describe the change in energy of the excited state, due to shifts in the equilibrium positions of the normal oscillator that occur during the optical transition. The quadratic coupling coefficients (B_{ij}), for $i = j$, describe the change in the energy of the excited state resulting from the change in the normal oscillator frequencies during an excitation. For $i \neq j$ they describe the change in the energy due to the mixing of normal coordinates. The quadratic electron–phonon coupling is discussed in detail elsewhere [2, 32, 56–58].

Thus, within the linear electron–phonon coupling approximation only the equilibrium position, but not the frequency ω_i, of the phonon modes is changed between the two electronic states. In this case q_i represents the lattice normal coordinate belonging to phonon mode *i*, and Δq_i corresponds to the change of the equilibrium position. This difference in sensitivity to changes in the environment is characterized by a dimensionless parameter, *S* (see Eq. 8.5); E_{el} and E_v in Figure 8.3 correspond to pure electronic and vertical transition energies, respectively. Under these conditions, $E_{ph,i} = E_v - E_{el}$, and the strength of the electron–phonon coupling (*S*) is given by $S = E_{ph,i}/\hbar\omega_i$. The FC factor for the ZPL is defined as: FC = exp($-S$). Thus, depending on the electron–phonon coupling strength (as shown in Fig. 8.4), the selectively excited fluorescence will consist of a sharp feature, which is the zero transition for all lattice phonons, followed by a broad fluorescence PSB (to the "red"), which is a superposition of all lattice phonon transitions. The sharp feature (see Figs. 8.2 and 8.4) is referred to as a ZPL, since, as mentioned above, no lattice phonons are created at this frequency. ZPLs identify the adiabatic (nonvertical) transition energy, which simultaneously takes the molecule to its excited state and the surrounding solvent to its new equilibrium position. The transition occurring at the maximum of the phonon wing corresponds to the vertical solvent transition [59]. In order to assess the value of S, one has to estimate the integrated intensity of the ZPL and the PSB. The fractional intensity of the sharp ZPL and the broad PSB is expressed as the Debye–Waller factor, α [1, 2, 44, 60], where

$$\alpha = I_{\mathrm{ZPL}}/(I_{\mathrm{ZPL}} + I_{\mathrm{PSB}}) = \sum_i |\langle \chi_{\mathrm{ph},1i}(0)|\chi_{\mathrm{ph},0i}(0)\rangle|^2 \tag{8.3}$$

$\langle \chi_{\mathrm{ph},1i}|\chi_{\mathrm{ph},0i}\rangle$ is the overlap integral, and $\chi_{\mathrm{ph},1i}$ and $\chi_{\mathrm{ph},0i}$ represent the phonon wave functions for mode i in the excited and ground electronic state [60], respectively. It has been shown that α is a decreasing function of temperature [1, 2]. As shown in Figure 8.4, the relative intensities of these two contributions, ZPL and PSB, depend upon the strength of the electron–phonon coupling. At the low-temperature limit ($T \sim 0\,\mathrm{K}$) α is defined as:

$$\alpha = \exp(-S) \tag{8.4}$$

where $S(T = 0\,\mathrm{K})$ is given [44] by

$$S(T = 0) = \frac{M_i \omega_i}{2\hbar} \sum_i (\Delta q_i)^2 \tag{8.5}$$

M_i and ω_i are the reduced mass and frequency of the mode i, respectively. $S \sim (\Delta q_i)^2$ is small for the weak electron–phonon coupling ($S \ll 1$; see upper frame of Fig. 8.4), while $S \gg 1$ is observed in the strong-coupling limit. In the latter case only weak ZPLs are observed, even at very low temperature (see Fig. 8.5). For *very* strong coupling ($S \gtrsim 10$) no ZPL are observed, since they are FC forbidden. Thus α, expressed in terms of S (see below for the definition of the apparent Huang–Rhys factor) measures the strength of the electron–phonon coupling [1, 2]. Figure 8.4 shows that an excited-state displacement, Δq_i, suppresses the intensity of the ZPL, while Eq. 8.4 indicates that the decrease is exponential. The PSB is usually structureless; for discrete phonons the intensity

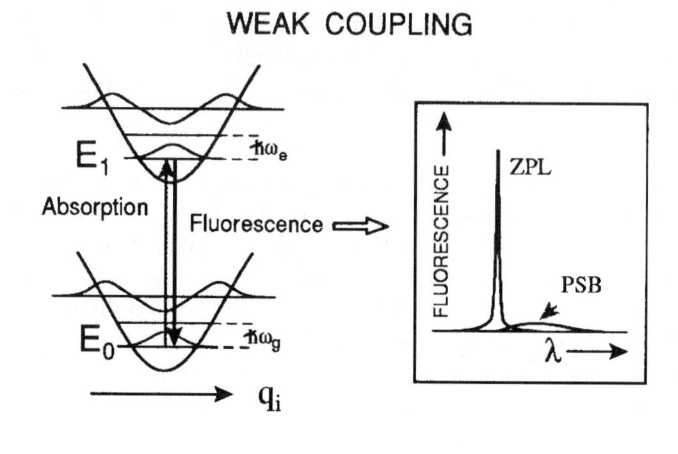

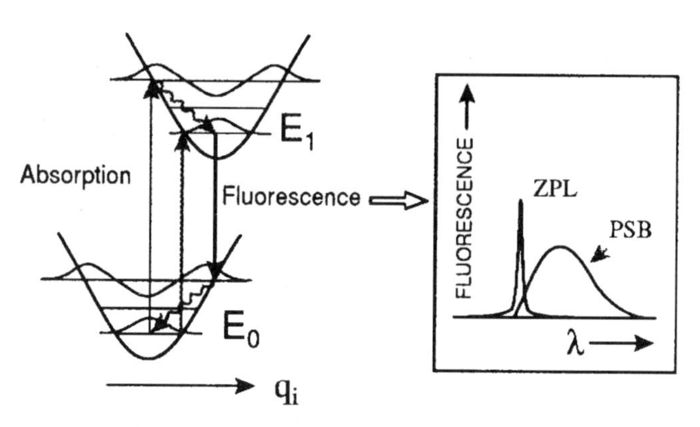

Figure 8.4. Diagram of potential curves in the case of linear electron–phonon coupling ($\hbar\omega_g = \hbar\omega_e$), and molecular lineshapes for a transition between two vibronic states of a guest molecule embedded in an amorphous host. Optical transitions are shown as vertical arrows in accordance with the Franck–Condon principle. Both weak- and strong-coupling limits are illustrated. If $\hbar\omega_g \neq \hbar\omega_e$, then the transition energies are different and the 0–0 transition is accompanied by a change in the vibrational energy of the zero-point level; see text.

distribution in a phonon band follows the Poisson distribution [1, 2, 44–46]; that is, $f_R = S^R \exp(-S)/R!$, see Eq. 8.12, where R is the number of phonons.

Unfortunately, the strength of the electron–phonon coupling cannot be directly estimated from Eq. 8.4. This is because the spectrum is always a convolution of homogeneous absorption and fluorescence spectra. This is because the width of the ZPL of the line-narrowed fluorescence is determined by the linewidth of both the initial excited-state levels and the final ground-state levels [61]. For illustration, we assume first that $\phi_{ZPL}(\omega - \omega_0)$ and $\varepsilon_{ZPL}(\omega' - \omega_0)$ are the spectral distribution functions for fluorescence and absorption, respectively, with the maximum of the ZPL at frequency ω_0. At the moment, the PSB contribution is neglected; this means that spectra of various centers (chromophores) are represented by "pure" ZPLs and differ only in the value of ω_0. If $G(\omega_0)$ corresponds to a Gaussian-distributed ensemble of chromophores, and $L(\omega' - \omega_{ex})$ describes a laser profile at frequency ω_{ex}, the fluorescence spectrum generated at an excitation frequency (ω_{ex}) can be expressed as the following convolution [1]:

$$F(\omega, \omega_{ex}) = \int_0^{\infty}\int_0^{\infty} \phi_{ZPL}(\omega - \omega_0)\varepsilon_{ZPL}(\omega' - \omega_0)G(\omega_0)L(\omega' - \omega_{ex})\, d\omega'\, d\omega_0 \tag{8.6}$$

In the first approximation, for illustrative purposes, one can assume that $G(\omega_0)$ is constant (infinitely large inhomogeneous broadening) and that the laser linewidth is much narrower than Γ_{hom} and $\Gamma_{hom} \ll \Gamma_{inh}$ [1]. Then, instead of Eq. 8.6, the site-selected fluorescence spectrum can be simply written as:

$$F(\omega, \omega_{ex}) \approx \int_0^{\infty} \phi_{ZPL}(\omega - \omega_0)\varepsilon_{ZPL}(\omega_{ex} - \omega_0)\, d\omega_0 \tag{8.7}$$

Since the homogeneous band consists solely of ZPL (i.e., $S \ll 1$), and both $\phi_{ZPL}(\omega - \omega_0)$ and $\varepsilon_{ZPL}(\omega' - \omega_0)$ have a Lorentzian line shape with a width of Γ_{hom}, it is easy to show that after selective excitation, the shape of the FLN spectrum can be described as:

$$F(\omega, \omega_{ex}) \approx 2\Gamma_{hom}/[(\omega - \omega_{ex})^2 + (2\Gamma_{hom})^2] \tag{8.8}$$

Equation 8.8 indicates that despite large inhomogeneity, the FLN spectrum consists of a line that has the shape of the ZPL (a Lorentzian shape) and the width $\Delta\omega \sim 2\Gamma_{hom}$ [2].

However, as illustrated in Figures 8.2–8.4, the absorption and emission spectra of chromophores in glasses consist of ZPLs and broad phonon wings. To include the phonon transitions, Eq. 8.7 has to be modified. The terms $\phi_{ZPL}(\omega - \omega_0)$ and $\varepsilon_{ZPL}(\omega_{ex} - \omega_0)$ of Eq. 8.7 have to be rewritten in the form of two terms to account for the ZPL plus PSB spectral shape. Taking into account the DW factor for the

ZPL (i.e., α) and for the phonons ($1-\alpha$), we have [62] the following expressions for the single-site absorption (Eq. 8.9) and single-site fluorescence (Eq. 8.10):

$$\varepsilon_{\mathrm{ABS}}(\omega_{\mathrm{ex}}-\omega_0)=\alpha\varepsilon_{\mathrm{ZPL}}(\omega_{\mathrm{ex}}-\omega_0)+(1-\alpha)\varepsilon_{\mathrm{PSB}}(\omega_{\mathrm{ex}}-\omega_0) \tag{8.9}$$

$$\phi_{\mathrm{FL}}(\omega-\omega_0)=\alpha\phi_{\mathrm{ZPL}}(\omega-\omega_0)+(1-\alpha)\phi_{\mathrm{PSB}}(\omega-\omega_0) \tag{8.10}$$

where ϕ_{ZPL} and $\varepsilon_{\mathrm{ZPL}}$ describe the ZPL, and ϕ_{PSB} and $\varepsilon_{\mathrm{PSB}}$ describe the corresponding vibronic parts of the spectra (i.e., the phonon sideband). ω_0 is the center frequency of the ZPL of a particular site. With $G(\omega_0)$, defined above, also called the zero-site-distribution (ZSD) function, the resulting fluorescence spectrum, excited at ω_{ex}, can be written as:

$$F(\omega,\omega_{\mathrm{ex}})=\int_0^{\infty}\phi_{\mathrm{FL}}(\omega-\omega_0)G(\omega_0)\varepsilon_{\mathrm{ABS}}(\omega_{\mathrm{ex}}-\omega_0)\,d\omega_0 \tag{8.11}$$

8.3.3. Model Calculations

For the linear Franck–Condon approximation, and the case where phonons are sufficiently delocalized, more physically transparent formulas for the single-site absorption (emission), in the $T\rightarrow 0\,\mathrm{K}$ limit, have been derived [45, 46]. The single-site fluorescence profile is described by the following equation:

$$F(\omega)=\underbrace{e^{-S}l_0(\omega-\Omega_0)}_{\mathrm{ZPL}}+\underbrace{\sum_{R=1}^{\infty}S^R\frac{e^{-S}}{R!}l_R(\omega-\Omega_0+R\omega_{\mathrm{m}})}_{\mathrm{PSB}} \tag{8.12}$$

where ω_{m} is the phonon mean frequency, $l_0(\omega-\Omega_0)$ is the Lorentzian line shape of the ZPL with peak position Ω_0 and homogeneous width Γ_{hom}, and $l_R(\omega-\Omega_0+R\omega_{\mathrm{m}})$ is the normalized line-shape function of the R-phonon transition ($R=1,2,\ldots$) peaking at $\Omega_0-R\omega_{\mathrm{m}}$. As mentioned above, a Poisson weighting factor for every number R determines the intensity distribution of the PSB. Within the framework of the mean frequency approximation, the one-phonon profile $l_1(\omega-\Omega_0+R\omega_{\mathrm{m}})$ represents the product $g(\omega)D(\omega)$, with $g(\omega)$ and $D(\omega)$ being the phonon density of states and an electron–phonon coupling term, respectively. The profile l_R is obtained by folding l_1, R times, with itself. Various line shapes for l_1 can be used [45, 46]; for example, if l_1 is Gaussian with a full width at half-maximum (FWHM), σ, the profile l_R is also Gaussian with $\mathrm{FWHM}=\sqrt{R}\sigma$ [45]. Finally, the spectrum for the whole ensemble of chromophores can be obtained by convolution with the ZSD function; the shape of the latter is typically taken as Gaussian with a FWHM corresponding to Γ_{inh}.

Of course, at higher temperature creation (and annihilation) of phonons takes place. Therefore, for $T > 0\,K$ a thermal population of phonon levels, according to Bose–Einstein statistics, has to be taken into account [44]. The temperature-dependent expression for the single-site absorption spectrum has been obtained by Hayes et al. [46], which in the case of fluorescence can be written as:

$$F(\omega) = e^{-S(2\bar{n}+1)} \sum_{R=0}^{\infty} \sum_{P=0}^{R} \frac{[S(\bar{m}+1)]^{R-P}[S\bar{n}]^{P}}{(R-P)!P!} l_{R,P}[\omega - \Omega_0 + (R-2P)\omega_m] \quad (8.13)$$

$\bar{n}$ is the phonon occupation number, ω_m is the mean phonon frequency, while $S(\bar{n}+1)$ and $S\bar{n}$ represent phonon creation and annihilation factors, respectively. As above, S is defined so that $S\omega_m$ (optical reorganization energy) $= \Sigma_i S_i\omega_i$. By analogy to Eq. 8.12, the value of R corresponds to the zero-, one-, ..., R-phonon transition, while the second sum over P leads to a redistribution of intensity within the PSB according to the number of created or annihilated phonons. An increase of temperature not only leads to more intense PSBs at the expense of the ZPL, but also gives rise to the anti-Stokes part of the PSB.

Model calculations, performed at $T = 4.2\,K$ with the above-described theory, are shown in Figure 8.5. For the phonon line shape function a Gaussian of width Γ is used. Below, only the effect of S on the calculated single-site (solid lines) and convoluted fluorescence spectra (dashed lines) is illustrated. Calculations were performed for $S = 0.5$, 1.5, 5.0, and 8.0, as shown in frames A, B, C, and D, respectively. In agreement with the schematic representation shown in Figure 8.4, the shape of the fluorescence spectra strongly depends on the strength of electron–phonon coupling. This example demonstrates that an increase of the coupling strength (S) leads to an increase in PSB intensity compared to the ZPL. Increasing the value of S also leads to a red shift of the total fluorescence peak position and, as a result, to larger Stokes shifts given by approximately $2S\omega_m$ [1, 2, 45].

The homogeneous linewidth (Γ_{hom}) is related to the T_2 relaxation time by the following relation:

$$\frac{1}{T_2} = \pi c \Gamma_{hom} = \frac{1}{2T_1} + \frac{1}{T_2'} \quad (8.14)$$

where T_1 is the excited-state lifetime and T_2' is the pure dephasing time. For the purpose of this chapter it suffices to say that T_2' is due to the modulation of the single-site transition frequency, which results from the interaction of the excited state with bath phonons (and other low-energy excitations in glasses [40, 63, 64]). This interaction does not lead to electronic relaxation of the excited state, but

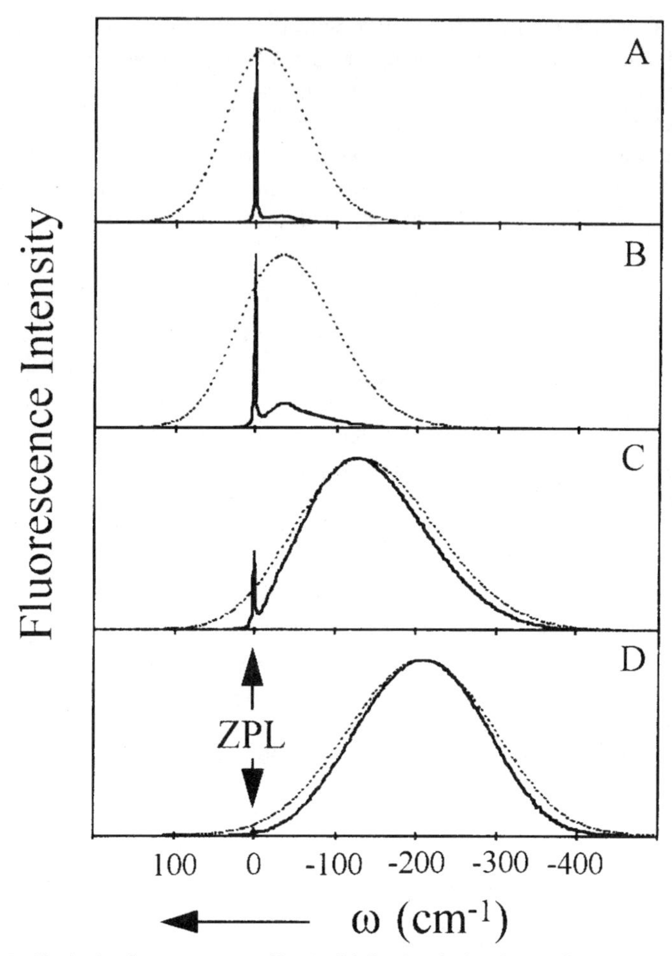

Figure 8.5. Single-site fluorescence profiles (solid lines) calculated according to Eq. 8.13 (with $\omega_m = 30\ \text{cm}^{-1}$, $\Gamma = 40\ \text{cm}^{-1}$, $\Gamma_{\text{hom}} = 2\ \text{cm}^{-1}$, and $\Gamma_{\text{inh}} = 100\ \text{cm}^{-1}$) for different values of S. For frames A, B, C, and D the values of S are 0.5, 1.5, 5.0, and 8.0. The dashed lines represent the corresponding total (convoluted) fluorescence spectra.

rather to a decay of the phase coherence of the superposition state initially created by the photon. In units of cm^{-1}, Γ_{hom} can be expressed as:

$$\Gamma_{\text{hom}} = (\pi T_2 c)^{-1} \tag{8.15}$$

where c is the speed of light in cm s^{-1} and T_2 is the total dephasing time [39, 40]. The typical value of the homogeneous width is $< 1\ \text{cm}^{-1}$ (at 4.2 K), while the phonon side band is broad with a width ranging between 15 to hundreds of cm^{-1}. This is why the largest selection and fluorescence line narrowing is observed for an excitation within the low-energy band of the ZSD. Recall that the latter corresponds to the envelope of the ZPLs without the contribution from the PSBs. Fidy and Vanderkooi [7, 65], based on the work of Funfschilling et al. [22], developed a methodology of generating the ZSD function, and gave its explicit definition [65]. In HB spectroscopy, the ZSD can be obtained via hole-burning action spectroscopy, as demonstrated by Small et al. [66]. It has also been shown that the line intensity of a $(1, 0) \rightarrow (0, m)$ transition can be described [65] as:

$$I_m = I_{\text{exc}}(\text{PDF})A_k V F_m K \tag{8.16}$$

where I_{exc} is the excitation intensity, PDF is the population distribution function, A_k is the $(0, 0) \rightarrow (1, k)$ absorption transition probability, V is related to the vibrational transition probability, F_m is the $(1, 0) \rightarrow (0, m)$ emission probability, and K is an arbitrary constant.

Since FLN and spectral HB (see Chapter 9) are complementary, a broad ZSD may influence the quality of high-resolution spectra, see Subsection 8.3.5. In concluding this section, two points should be made. First, one has to remember that the ZPLs also contain a contribution from the molecules excited via the phonon side band, and, second, the Debye–Waller factor (see Eq. 8.3) in a high-resolution fluorescence spectrum is involved twice: in the excitation and emission process. As a result, the "apparent" DW factor (α') is approximately the square of the true DW factor [1]:

$$\alpha^1 \approx \alpha^2 \tag{8.17}$$

The latter indicates that the "apparent" S (S_{app}) approximately equals $2S$; for details see Refs. 1 and 2. The physical significance of S is, perhaps, most easily appreciated by recalling that $2S\omega_{\text{m}}$ corresponds roughly to the Stokes shift for relaxed fluorescence, where ω_{m} is the mean frequency of the coupled phonons. Therefore, in the limit of strong electron–phonon coupling, the ZPL is FC-factor forbidden, meaning that much of the selectivity of FLN is lost.

8.3.4. Ground- and Excited-State Vibrational FLN Spectra

0–0 Excitation. Figure 8.6 (upper frame) depicts the situation where a narrow frequency laser at ω_{L} (width $\Gamma_{\text{L}} \ll \Gamma_{\text{inh}}$) provides an excitation into the inhomogeneously broadened 0–0, or origin band, of the chromophore. As in Figure 8.2, the laser selects (excites) only those chromophores, or sites, whose 0–0 transitions, or ZPLs, overlap the laser profile. The key question is how to measure the ground-state vibrational frequencies. The principle is illustrated in Figure 8.6.

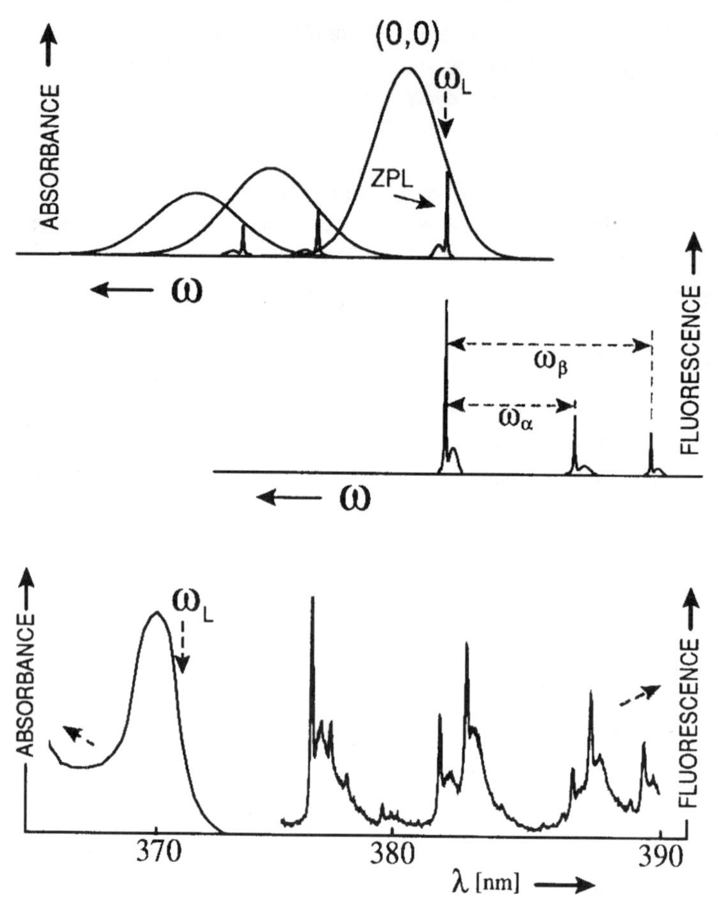

Figure 8.6. Origin band excitation in FLN spectroscopy. Top: Laser frequency, ω_L, excites only those chromophores whose 0–0 transition overlap the laser profile (see text for details). ω_α and ω_β correspond to the ground-state vibrational frequencies, in cm^{-1}. Bottom frame: 0–0 absorption band (left-hand spectrum) and FLN spectrum obtained for pyrene in ethanol glass at $T = 4.2\,K$, $\lambda_{ex} = 371.5\,nm$. The sharp peaks correspond to the pyrene ground-state vibrational frequencies.

The top frame (lower spectrum) clearly reveals that the origin excited FLN spectrum is composed of the highest-energy ZPL, corresponding to the transition between the zero-point vibrational levels of the fluorescent (S_1) and ground electronic (S_0) states, and a series of vibronic ZPLs corresponding to transitions terminating at vibrational levels of S_0. To summarize the top frame of Figure 8.6, it may be stated that the origin band excited FLN spectrum may provide ground-state (S_0) vibrational frequencies (ω_α and ω_β) and relative vibronic intensities, which, in combination, may serve as a fingerprint of studied chromophores.

An illustration of the inhomogeneously broadened absorption band and the ground-state–excited FLN spectrum of pyrene embedded in ethanol glass are shown in the bottom frame of Figure 8.6. For clarity, only part of the FLN spectrum is shown; all narrow peaks are ZPLs, and their frequencies correspond to the ground-state vibrations. Shifting of the laser excitation to lower energy would result in an equal shift of the complete fluorescence spectrum, showing the same vibrational frequencies. Due to superposition of all lattice phonon transitions, the PSBs appear broad, as discussed above.

Luminescence Line Narrowing (LLN). It is well known that the distinction between fluorescence and phosphorescence only holds true for species possessing "light" atoms, that is, atoms derived from the first or second row of periodic table. For "heavy" atoms, or molecules containing such atoms, one may often speak of emission in a general sense as "luminescence." Thus one may also talk about LLN experiments; an example is shown in Figure 8.7, which shows the 4.2 K luminescence spectra of Ru(II)tris(3,3′-biisoquinoline) in an ethanol/methanol (EM; 4 : 1, v/v) glass [67]. The nonresonantly excited spectrum is broad as expected, while the LLN spectrum shows a rich line structure. In the narrowed spectrum, the electronic origin (as well as the vibronic bands) consists of a ZPL and long-wavelength PSBs ($\omega_m \sim 23$ cm^{-1}). From the narrow spectrum

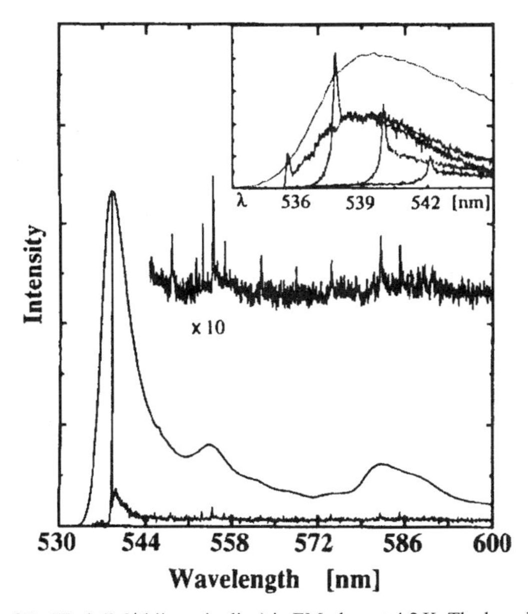

Figure 8.7. LLN of Ru(II)tris(3,3′-biisoquinoline) in EM glass at 4.2 K. The broad and narrow spectra have been excited at 440.0 and 538.45 nm, respectively. The insert shows resonant LLN spectra obtained at 15 K; see text for details. (From Ref. 67, with permission.)

of Figure 8.7, Riesen and Krausz [67] estimated the value of the DW factor as $\sim$0.6. The insert shows some resonant LLN spectra (at $T = 15\,\mathrm{K}$) with the inhomogeneously broadened band (instrumental resolution: 0.06 nm). The insert indicates that the DW factor is a rough estimate, since the number of nonresonantly excited centers is considerably increased by exciting at the high-energy region of the inhomogeneously broadened absorption band. This is because the approximation discussed in Section 8.3.2 (see Eq. 8.8) is only valid if the inhomogeneous broadening is much larger than the width of the PSB. Figure 8.8 shows a comparison of the resonant ZPLs in the LLN and excitation line-narrowed experiment. Again, distinct features of ZPLs and PSBs are observed. In the non-line-narrowed spectra (dashed curves) the presence of the PSB manifests itself by a Stokes shift (vide infra) of the broad origins in the excitation and luminescence spectra.

Phosphorescence Line Narrowing (PLN). PLN experiments have received relatively little attention, since the low oscillator strength of the $S_0 \rightarrow T_1$ transition (the transition is forbidden with respect to spin) makes such experiments more difficult. Nevertheless, PLN spectra have been obtained for a number of molecules [1, 68–72]. The mechanism for occurrence of PLN spectra is the

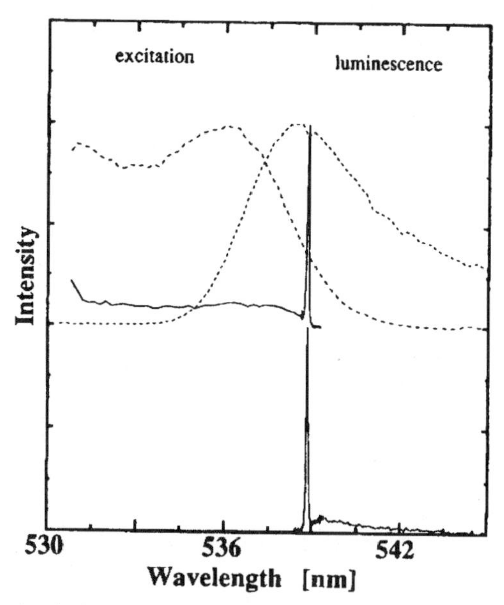

Figure 8.8. LLN and excitation line narrowing in the region of the electronic origin of Ru(II)tris(3,3′-biisoquinoline) in EM glass at 4.2 K. Solid and dashed lines represent the luminescence and excitation spectra obtained under line-narrowing and non line-narrowing conditions, respectively. $T = 4.2\,\mathrm{K}$. (From Ref. 67, with permission.)

same as the one discussed above for the FLN spectra. The only difference is that instead of the singlet state S_1, the triplet state T_1 is involved in the PLN experiments. Thus the vibrational frequencies of molecular vibrations in the triplet state can also be measured. The side bands in the PLN spectra are separated from the resonant feature by $\pm D$ for molecules with $D \neq 0$ and $E = 0$, and by $\pm(D+E)$, $\pm(D-E)$ and $\pm E$ for a molecule with $D \neq 0$ and $E \neq 0$ [68]. D and E correspond to the zero-field-splitting parameters. In PLN experiments, the side bands will only be observed if the spin-lattice relaxation rate is reasonably fast compared to the phosphorescence lifetime of the spin levels involved. For more details see Refs. 67–72.

Vibronic excitation. We return now to the FLN spectra generated by excitation into the S_1 vibronic region. It has been established that the S_1 state vibrational dynamics, frequencies, and intensities are by far the most sensitive to subtle changes in the structure of the chromophore and its environment [9, 11]. This sensitivity has a firm theoretical understanding based on the Duschinsky effect (a mixing of the normal coordinates in the excited state [73] and vibronically induced anharmonicity). Thus this mode of excitation has great potential in analytical applications. To probe the S_1-state vibrational structure one employs vibronically excited FLN, as illustrated in Figure 8.9. In this case an energy-level diagram is used to illustrate the principles of the vibronic excitation. For simplicity, ω_L excites only two overlapping one-quantum vibronic transitions ($1_{\alpha,0}$) and ($1_{\beta,0}$), which will not be resolved in the absorption spectrum. The laser excites two different isochromats, one for α and one for β, and because of the correlation, the two isochromats undergo vibrational relaxation (wiggly arrows) to two different points (A and B) in the zero-point distribution of the S_1 state. Following population of the zero-point level, fluorescence occurs from these two energetically distinct isochromats and produces a "doubling" of every line in the FLN spectrum. As a result, two strong ZPLs [$(0–0)_B$ and $(0–0)_A$] and two weaker vibronic lines [$(0–1)_B$ and $(0–1)_A$] are observed. The displacements between ω_L and the doublet components of the origin transition, [$(0–0)_B$ and $(0–0)_A$], yield the excited-state vibrational frequencies (i.e., ω'_α and ω'_β, as shown in frame B of Figure 8.9). The structure of the ZPLs [$(0–0)_B$ and $(0–0)_A$ lines] is referred to as *multiplet origin structure*. Vibronically excited FLN is now used exclusively for identification and conformational analysis of DNA adducts (see Chapter 11) because of its vastly superior selectivity. A key point is that by tuning ω_L across the $S_1 \leftarrow S_0$ absorption spectrum, one can determine the frequencies and FC factors of all active excited-state vibrations [1, 2, 4]. Furthermore, each ω_L yields a distinct ZPL "fingerprint."

An example of the vibronically excited FLN spectra obtained for the *trans*- and *cis-anti*-dibenzo[*a,l*]pyrene (DB[*a,l*]P)-tetrols derived from the most potent carcinogen, DB[*a,l*]P diolepoxide [74], is shown in Figure 8.10. The results for the *cis-anti*- isomers are considered first since they are of the type expected when

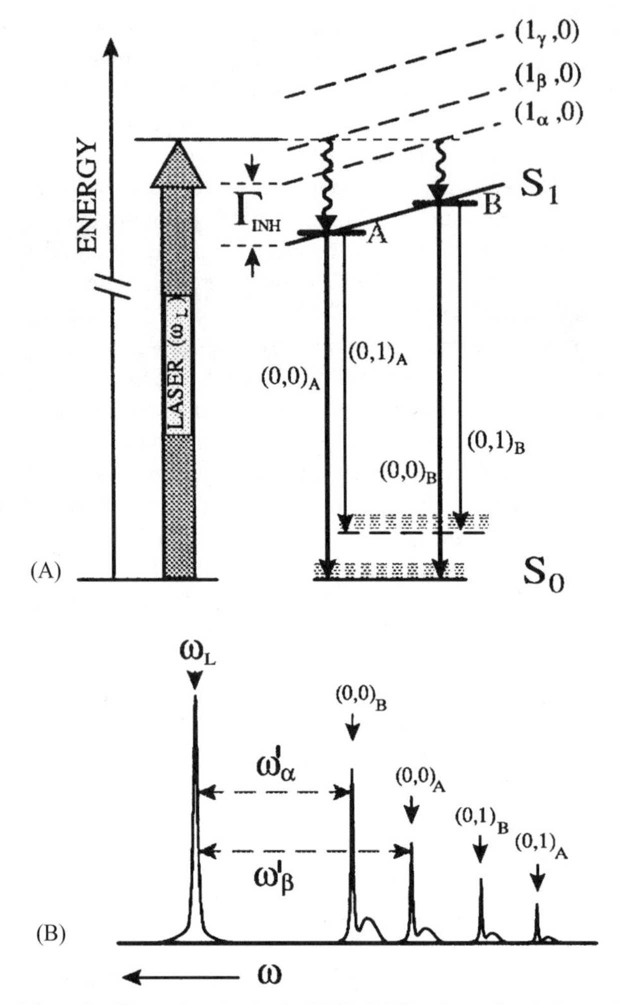

Figure 8.9. Schematic of laser site selection in FLNS. (A) The slope of excited-state levels represents the variation of their energies as a function of the site. Γ_{inh} denotes the inhomogeneous broadening of the 0–0 transition. Using the laser excitation, ω_L, two subsets of molecules within Γ_{inh} are selectively excited (see text for details). (B) Schematic of the resulting fluorescence spectrum; two 0–0 transitions as well as the corresponding 0–1 transitions are well resolved. ω'_α and ω'_β are the vibrational frequencies in the excited state.

only one conformation is present and it is the same in the two glass hosts. The FLN multiplet origin structures for the *cis-anti* isomer in ethanol (solid line, a) and glycerol/water (dashed line, b) are shown in Figure 8.10A. The same excitation wavelength, 373.0 nm, was used for both glasses since the 0–0

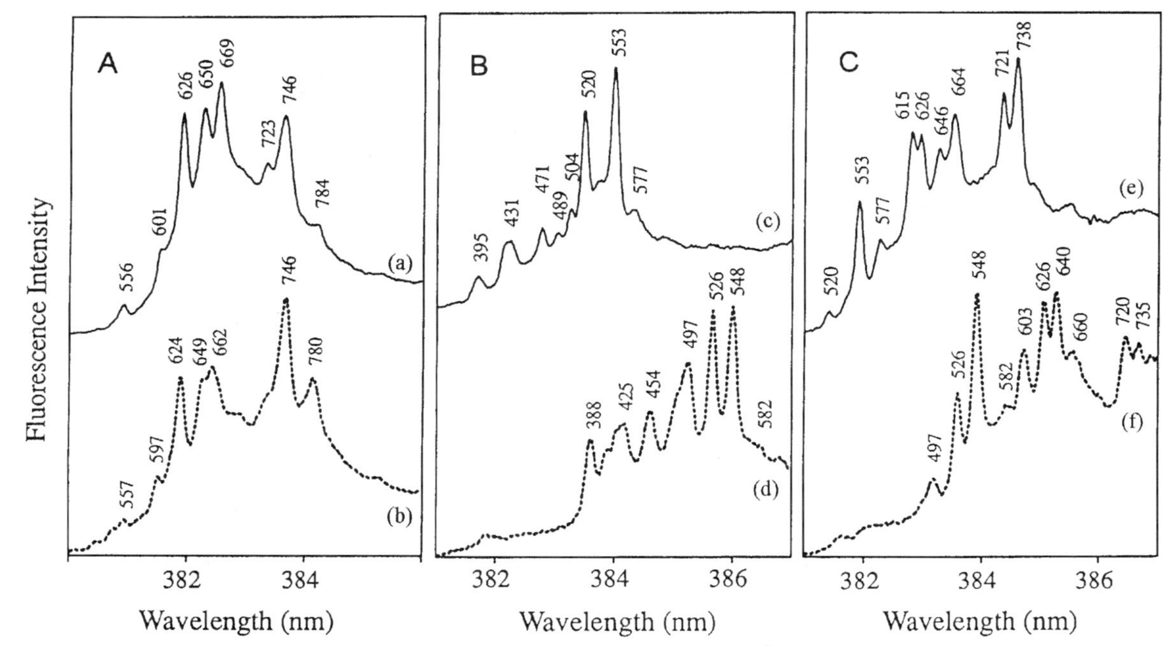

Figure 8.10. Frame A: FLN spectra for *cis-anti*-DB[*a,l*]P tetrol in ethanol (a) and in 50/50 glycerol/water (b); laser excitation at 373.0 nm. Frames B and C: FLN spectra for *trans-anti*-DB[*a,l*]P tetrol in ethanol (c and e) and in 50/50 glycerol/water (d and f). Laser excitation wavelengths are 376.0, 378.0, 374.0, and 376.0 nm, respectively, for spectra (c) through (f). Due to the red shift of the origin band of *trans-anti*-DB[*a,l*]P tetrol in glycerol/water compared to ethanol glass [74], different laser wavelengths were used in order to excite similar vibrational modes with approximately the same efficiency. $T = 4.2$ K. The FLN peaks are labeled with their excited-state vibrational frequencies, in cm^{-1}. (From Ref. 74 with permission.)

bands of these tetrols (see Ref. 74) are displaced from each other by only 0.4 nm. This wavelength excites vibrations of the S_1 state with frequencies in the 550–800 cm^{-1} region. The FLN bands are labeled with their S_1 vibrational frequencies in cm^{-1}. The vibrational frequencies in the two glasses are very similar (± 3 cm^{-1} uncertainty). The intensity distributions are also similar, with the higher relative intensities of the 746 and 780 cm^{-1} modes in glycerol/water explicable in terms of the 0.4 nm red shift of the 0–0 band relative to ethanol. If two molecular conformations existed for the *cis-anti*-DB[*a*,*l*]P-tetrols, one would not expect the striking similarity between spectra a and b in Figure 8.10A. This point was emphasized in Refs. 33 and 75 and will become clearer when the results for the *trans-anti* isomer are considered. Thus there is only one dominant conformation for *cis-anti*-DB[*a*,*l*]tetrol, and its structure is not significantly different in the two solvents [74]. The same was observed for the *cis-syn*-isomer [74] (data not shown).

The FLN spectra for the *trans-anti*-DB[*a*,*l*]P tetrols are shown in Figure 8.10 (frames B and C). Spectra (c) and (e) were obtained in ethanol, and spectra (d) and (f) were measured in glycerol/water glass, respectively. Frames B and C correspond to different excitation wavelengths (λ_{ex}) that expose different regions of the vibronic spectrum. In ethanol glass, spectra for $\lambda_{ex} = 376.0$ and 374.0 nm are shown in Frames B and C, respectively; for glycerol/water, FLN spectra obtained for $\lambda_{ex} = 378.0$ and 376.0 nm are shown. The λ_{ex} values for glycerol/water glass are 2 nm higher because the 0–0 bands in this solvent are red shifted by this amount relative to the ethanol matrix (for details see Ref. 74). That is, it is important that the excitation wavelengths for the two glasses excite the same excited-state vibrational region. Comparison of spectra c and d in frame B reveals that there are significant differences in vibrational frequencies and intensities between ethanol and glycerol/water glasses. This is also the case for spectra e and f (frame C). Spectra obtained with other λ_{ex} values (not shown) were consistent with this. Thus the FLN results for the *trans-anti* isomer strongly indicated that its molecular conformations in the two glasses are different, in agreement with ^{1}H NMR spectroscopy and molecular-dynamics simulations [74]. A similar conclusion was reached in Ref. 75 for the *trans-syn* isomer. The distinct FLN spectra in Figure 8.10 (frames B and C) correspond to two conformations of *trans-anti*-DB[*a*,*l*]P-tetrol, both with a half-chair structure for the cyclohexenyl ring, though with different orientations of the hydroxyl groups. This example demonstrates that vibronically excited FLN spectroscopy is a powerful method for vibrational structure analysis of excited electronic state and/or possible conformational changes in analytes of interest.

To conclude this section, it should be noted that FLNS can easily distinguish between *trans-anti*- and *cis-anti*-DB[*a*,*l*]P-tetrols, as shown in Figure 8.10; the latter are also distinguishable from the corresponding FLN spectra of *trans-syn*- and *cis-syn*-DB[*a*,*l*]P-tetrols (see Refs. 74 and 75). Finally, it is important to note that an understanding of the conformational behavior of the stereoisomeric tetrols at the 11,12,13,14-positions of DB[*a*,*l*]P is essential for the spectroscopic identification of DNA adducts derived from the biologically highly active fjord region of *syn*- and *anti*- DB[*a*,*l*]P-11,12-diol 13,14 epoxides. For FLN spectra of DB[*a*,*l*]P-derived DNA adducts, see Chapter 11.

8.3.5. Selectivity and Detection Limits

One complication with FLN spectroscopy is that prolonged irradiation can lead to a severe degradation in the quality (resolution) of FLN spectra. This is because FLNS is inherently connected with HB, and in glasses Γ_{inh} is much larger than the average phonon frequency, ω_m (see Fig. 8.2). In systems where HB is efficient, the ZPL at ω_L responsible for FLN (and/or ZPLs within the 0–0 band) when excited vibronically [11]) will be removed prior to recording of the FLN spectrum, resulting in a broad emission from PSB excitation. Although at first sight HB seems detrimental to FLNS, it can actually be used to improve the resolution and/or selectivity of FLN spectra. This has been demonstrated by Bogner and Schwartz [76], Funfschilling et al. [77], Hofstraat et al. [78], Jankowiak et al. [11], and recently by Myers' group [79, 80]. Such a method was also used to obtain pure single-molecule fluorescence spectra [81]. The method was designed to eliminate the broad fluorescence contribution that arises from excitation of sites via PSBs. This elimination results in a spectrum originating mostly from sites excited by zero-phonon absorption, and as a consequence leads to FLN spectra with improved selectivity. Since the methodology employs both FLN and HB, it has also been called a double spectral selection technique [11]. The difference spectra methodology can also be applied to short-lived fluorescing species for which gated fluorescence detection is not possible, and scattered laser light poses a problem [11, 82]. Subjecting the sample to white light restoration pulses [83] can also decrease zero-phonon hole-burning. Very recently Myers et al. [79, 80] employed a complementary technique of "saturation subtracted FLN," which is useful for systems that have reasonably long excited-state lifetime, but burn holes efficiently. In this approach, laser pulses shorter than the excited-state lifetimes (but very narrow to accomplish selective excitation) are used to partially deplete the ZPL-resonant molecules by bottlenecking them in the excited electronic state. As shown in Ref. 79, subtraction of a properly scaled high-intensity emission spectrum from an undepleted low-intensity spectrum leaves behind a nearly pure ZPL-excited emission that agrees well with that obtained via HB-subtraction methodology (vide infra). An example of the phonon subtraction method, [0–0 excitation], for squaraine in polyethylene film [79] is shown in Figure 8.11. This figure compares the original, undepleted spectra with the HB-subtracted (top) and saturation-subtracted (bottom) spectra obtained at 7 K. Both subtraction methods yield

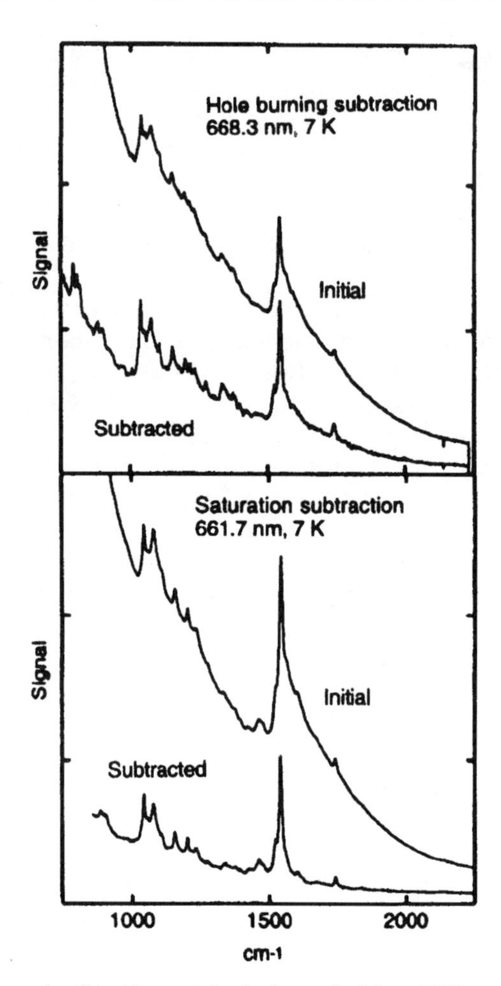

Figure 8.11. An example of double spectral selection methodology. FLN spectra were obtained for squarine in polyethylene film ($d \sim 100\,\mu m$) at $T = 7\,K$. Top: Hole-burning subtraction at $\lambda_B = 668.3\,nm$; bottom: saturation subtraction at a wavelength of 661.7 nm. In both cases the upper curve corresponds to the "undepleted" FLN spectrum, and the lower curve is obtained by subtracting from the upper one an appropriately scaled "ZPL depleted" spectrum; see text for details. (From Ref. 79 with permission.)

nearly the same FLN spectra with enhanced vibrational resolution. Another example of hole-burning subtracted FLN spectra is shown in Figure 8.12 for BO-IMI labeled dAMP nucleotide (R. Jankowiak et al., unpublished results). In this case vibronic excitation was utilized. For details on phosphate-specific fluorescence labeling under aqueous conditions, see Ref. 84. Briefly, BO-IMI was

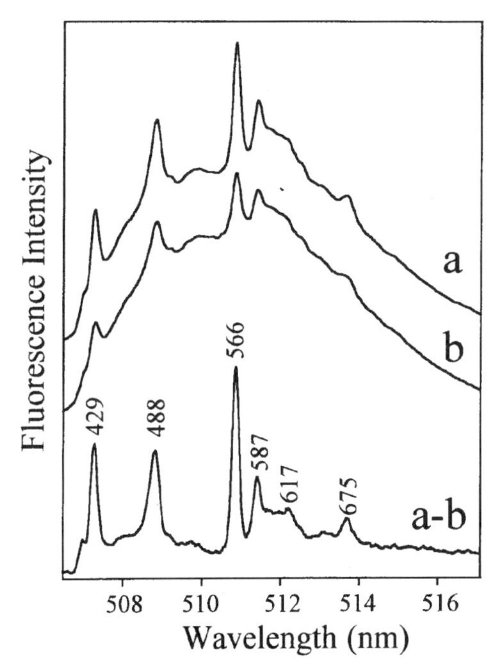

Figure 8.12. FLN spectra of BO-IMI-dAMP, $\lambda_{ex} = 496.5\,\text{nm}$, $T = 4.2\,\text{K}$. Spectrum (a) is obtained with a laser excitation power density, $I = 30\,\text{mW/cm}^2$ and an exposure time of ~60 s. Spectrum (b) is obtained under experimental conditions as in (a), followed a 300 s exposure to $50\,\text{mW/cm}^2$ excitation power density. The lowest spectrum is the difference spectrum (a, b) obtained, as discussed in text (Jankowiak and Zamzow, unpublished data).

prepared by coupling N-acetyl-L-histidine to the nucleophilic fluorescent dye BODIPY FL C_3 hydrazide followed by labeling a terminal phosphate group with BO-IMI. The spectra shown in Figure 8.12 were obtained by acquiring an initial FLN spectrum (curve a), exposing the sample to higher laser excitation energies for a few minutes, acquiring a second FLN spectrum (curve b), and subtracting the second from the first. Spectrum c is the difference spectrum, which is considerably sharper than spectra a and b. The numbers correspond to the excited-state vibrational frequencies, in cm^{-1}. Thus, as discussed above, the process has the effect of removing the underlying broad band fluorescence (from the PSBs), highlighting the ZPL transitions. Continuing with the topic of selectivity, it should be emphasized again that FLNS can provide all the ground- (origin excitation) and the excited-state (S_1) vibrational frequencies (and intensities) of studied molecules. Finally, turning to the question of detection limits, with the FLN instrumentation described in this chapter (employing low laser power densities to minimize HB effects), a detection limit for moderately fluorescent chromophores is in the subfemtomole range [11]. Such a detection level is sufficient for many practical toxicological and biological applications [8, 11, 33]; see also Chapters 10 and 11. To improve the limits of detection, methods for circumventing HB are often required [11]. For example, it was found that the simultaneous and synchronous scanning of the laser excitation and monochromator wavelengths provided better-resolved ZPLs [85]; the scan interval, however, has to be much smaller than the width of Γ_{inh}. Furthermore, white light restoration pulses [83], and/or performing the FLN experiments at elevated temperatures ($\sim$10–15 K), may also reduce the quantum efficiency of HB, leading to improved detection limits. Unfortunately, by increasing the sample temperature, the decreased site depletion is at the cost of less intense and broader fluorescence signals [86]. Finally, it should be noted that in addition to being highly selective and sensitive, FLNS is a very practical technique. For a laser-based method, FLNS is simple (all apparatus components are available commercially), and the procedure for rapid sample cooldown (in relatively nonexpensive Dewars) ensures a high sample turnover rate. Interesting analytical aspects of FLNS are discussed in Chapters 10 and 11.

8.3.6. Instrumentation

The experimental apparatus for FLNS can be relatively simple [5, 9, 11, 33, 62]. The equipment needed to perform FLN experiments can be divided into four major components: (1) spectrally narrow excitation source (preferably a tunable laser); (2) optically accessible low-temperature sample chamber (e.g., cryostat and/or a simple glass Dewar); (3) dispersion device (grating monochromator or high-resolution pressure-scanned Fabry–Perot interferometer), and (4) detection system. A large number of pulsed and CW tunable laser sources are available. The cost of optical multianalyzer systems, with diode arrays and/or charge-coupled devices (CCD), has been drastically reduced in recent years, although they are still expensive. Cooled photomultiplier tubes in conjunction with various signal processing units (e.g., photon counting systems, gated integrators, and/or lockin amplifiers) can also be utilized. Both types of cryostats, immersion or conduction, can be used for cooling to helium temperature; in the latter case, however, it is important to monitor the sample temperature to avoid local heating by the laser. A schematic of the instrumentation used in this laboratory is shown in Figure 8.13. The excitation source is a Lambda Physik (Acton, MA) Lextra XeCl excimer laser–(FL-2002) dye laser system. Depending on the position of the 0–0 absorption band of the chromophores under study, the analytes can be probed with either the excimer and/or argon ion laser, under non-line-narrowing (NLN) conditions, or with the dye laser system under FLN (or LNN/PLN) conditions (4.2 K). A McPherson model 2061 1-m monochromator disperses the fluorescence. For FLN measurements, a Princeton Instruments (PI) IRY-

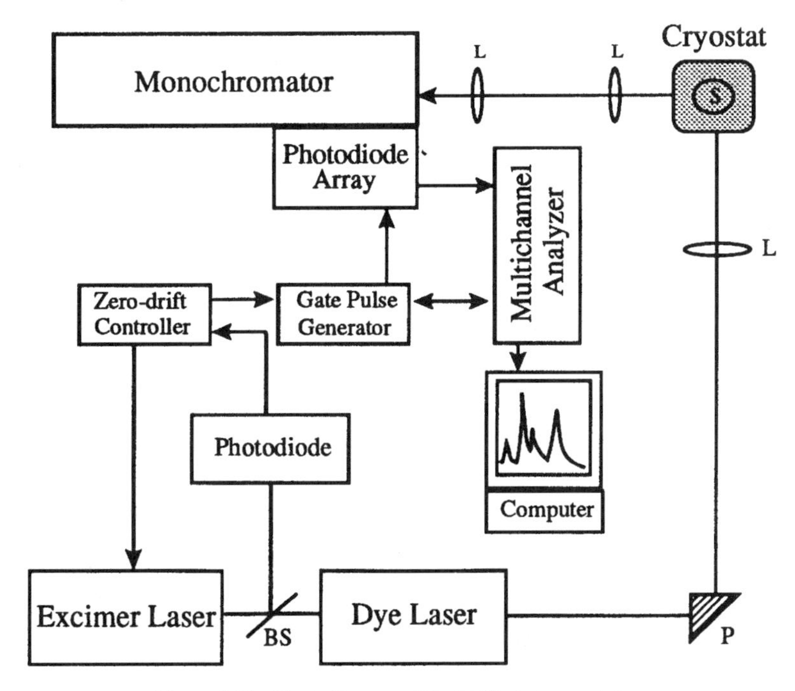

Figure 8.13. Block diagram of the FLNS instrumentation.

1024/GRB intensified diode-array detector or CCD is utilized. For vibronically excited FLN spectra, a resolution of 0.06 nm (2400 G/mm grating and 200 μm slit), providing a ~10 nm window is sufficient, while for FLN/LLN/PLN spectra excited into the origin band, the highest available resolution has to be utilized. The diode array may be operated either in CW or gated detection mode. In the latter mode the output of a reference photodiode is used to trigger a high-voltage pulse generator. Fluorescence spectra may be acquired using different delays and widths of the fluorescence observation window (gate). CW detection, used with various continuous-wave lasers, can significantly improve detection levels; attomole detection limits are achievable.

8.3.7. A Sampling of FLNS Applications

Before showing examples of FLNS applications to chromophores in a variety of different condensed-phase environments, we summarize the information that can be provided by FLN spectroscopy:

- Inhomogeneous broadening (a distribution of 0–0 transition energies) and homogeneous linewidth, Γ_{hom}, which contains information regarding molecular dynamics.
- Electron–phonon coupling, which provides a sensitive probe of the interaction of the molecule with the local matrix environment.
- Temperature dependence of Γ_{hom} provides insight on interaction with tunneling modes associated with two-level systems (TLS) (see Chapter 9 and Refs. 29–31, 39, 40, 50).
- Vibrational frequencies in the ground and excited states for both singlet and triplet states can be measured. The latter can be used as an analytical tool to provide a "fingerprint" of a molecule and/or to identify compounds in complex mixture [9, 11, 87].
- Various aspects of electron–vibrational interactions can be addressed.
- For sufficiently high concentration of impurities/chromophores embedded in amorphous matrices, incoherent energy transfer in the inhomogeneously broadened systems can be studied; this topic is not discussed in this chapter; for details see Refs. 5, 6, and 88.
- Using pulsed excitation and time-resolved spectroscopy, site-to-site variation in radiative and nonradiative transition probabilities can also be measured [88].

Thus, considering the versatility and richness of information provided by FLN spectroscopy, the prospects for continuous applications in the field of environmental sciences are promising, with new applications still emerging (see Chapters 10 and 11). Below, for illustrative purposes, only four examples are briefly discussed.

FLNS Analysis of Perprotio- and Perdeuterio-Benzo[a]pyrene. Often fluorescence analysis of complex mixtures meet with serious difficulties. Even after preliminary extraction and chromatographic separation of a complex mixture, the obtained fractions contain many components. As a result the individual components can not be properly characterized. Recently, capillary electrophoresis (CE) has been interfaced with FLNS for on-line separation and spectral characterization of analytes [87, 89]; for a description of the apparatus and other applications of the CE-FLNS system see Chapter 10. Here, in Figure 8.14, only one example of separation and spectral characterization of a mixture of deuterated and protonated benzo[*a*]pyrene (B[*a*]P) by CE-FLNS is shown. Frame A of Figure 8.14 is a portion of the room-temperature fluorescence electropherogram for the B[*a*]P-d_{12}/B[*a*]P mixture ($c \sim 10^{-5}$ M), generated from a three-dimensional plot of fluorescence emission (intensity and wavelength) versus time [89]. The excimer laser was used for excitation (308 nm), and the total fluorescence emission signal in the 350–500 nm region was integrated. The wavelengths of

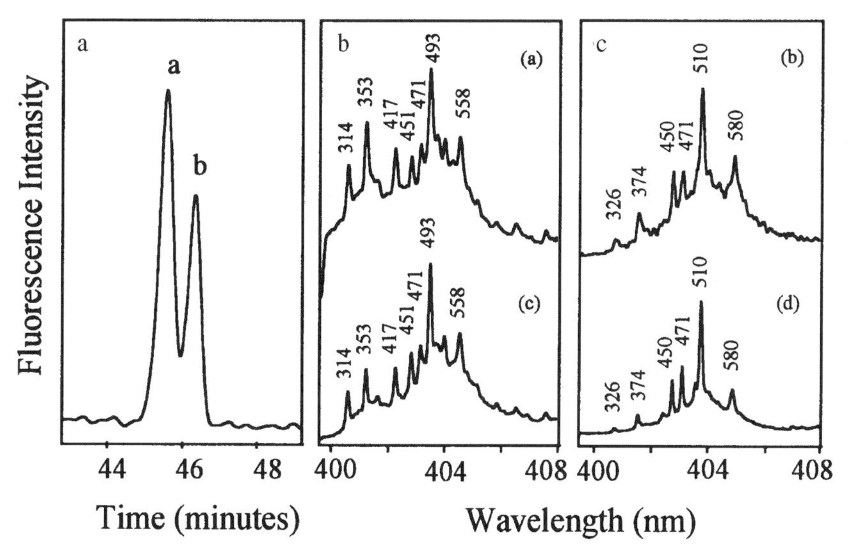

Figure 8.14. A: Room-temperature fluorescence electropherogram for a mixture of (a) B[a]P-d_{12} and (b) B[a]P using a CE buffer consisting of 40 mM sodium bis(2-ethylhexyl) sulfosuccinate and 8 mM sodium borate in acetonitrile–water (30 : 70, v/v); pH = 9; capillary 85 cm × 75 μm I.D.; applied voltage; 25 kV; current, 50 μA. Frames B and C: FLN spectra in the CE-buffer matrix at $T = 4.2$ K, $\lambda_{ex} = 395.7$ nm, obtained for the CE-separated analytes (a) and (b). Spectra (c) and (d) are from the library of FLN spectra of PAHs for B[a]P-d_{12} and B[a]P, respectively. The FLN peaks are labeled with their S_1 vibrational frequencies, in cm^{-1}. (From Ref. 89 with permission.)

the fluorescence origin bands obtained for peaks (a) and (b) were ~403 nm (data not shown).

FLN spectra for CE-separated peaks (a) and (b) using selective laser excitation at 395.7 nm at 4.2 K are shown in frames B (spectrum a) and C (spectrum b), respectively. The peaks in the FLN spectra correspond to excited-state vibrational modes and are labeled with their S_1 vibrational frequencies in cm^{-1}. While some modes are similar in the two spectra, there are some obvious differences. For B[a]P-d_{12}, there are strong modes at 353, 493, and 558 cm^{-1}; for B[a]P, the strong modes are at 510 and 580 cm^{-1}. Thus spectra (a) and (b) are clearly distinguishable with a different pattern of vibrational frequencies and intensities, revealing differences in analyte composition. These data (and other FLN spectra obtained using different laser excitation wavelengths, not shown here) can be used as "fingerprints" for spectral identification of a compound. The identity of CE-separated analytes is obtained by matching acquired FLN spectra with the available library of FLN spectra for PAHs. Comparison of the CE-FLN spectra in Figure 8.14 shows that spectrum (a) is virtually indistinguishable from spectrum (c), the B[a]P-d_{12} standard, and spectrum (b) is identical to spectrum (d), the

B[*a*]P standard. Therefore, peaks (a) and (b) of the electropherogram can be unambiguously assigned as deuterated B[*a*]P and protonated B[*a*]P, respectively. In conclusion, FLNS coupled to liquid separation techniques such as HPLC and CE (see Chapter 10 for details) has the potential to become a powerful tool for molecular analyte characterization.

FLN Spectra of TAM Metabolites. The "antiestrogen" tamoxifen (TAM) is widely used for the treatment of advanced breast cancer and as a prophylactic agent in healthy women with a family history of breast cancer. A controversy surrounds its prophylactic use since TAM has been shown to cause liver cancer in rats [90, 91]. Preliminary data from this laboratory indicate that FLNS can distinguish between TAM metabolites and TAM-derived adducts, and therefore could be used to establish the full spectrum of metabolites/adducts formed by TAM. An example of FLN spectra of TAM and its two major metabolites is shown in Figure 8.15.

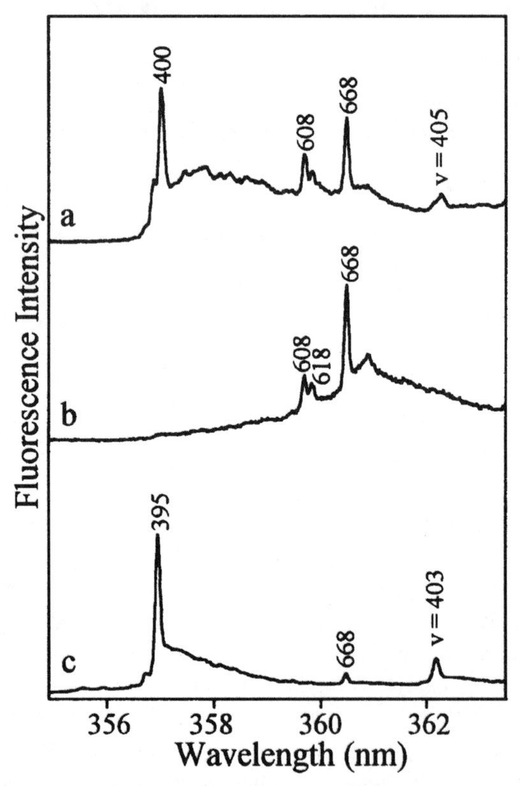

Figure 8.15. FLN spectra of photocyclized TAM (a), dihydro-TAM (b), and α-OH-TAM (c) in ethanol glass. $T = 4.2\,\mathrm{K}$, $\lambda_{ex} = 353.0\,\mathrm{nm}$. $v = 405$ and $403\,\mathrm{cm}^{-1}$ correspond to the ground-state vibrational frequencies, which build on the strong ZPL at 400 and $395\,\mathrm{cm}^{-1}$, respectively. Other numbers correspond to the vibrational frequencies in the excited state, in cm^{-1}.

FLN spectra of these compounds can be obtained only after photocyclization, that is, irradiation with a 308-nm laser for some minutes to induce rapid conversion to phenanthrene derivatives. Figure 8.15 shows the FLN spectra of the photocyclized TAM (spectrum a), dihydro-TAM (spectrum b), and α-OH-TAM (spectrum c), obtained at 4.2 K and 352 nm excitation. The sharp peaks correspond to specific vibrations of the S_1 state of the analytes. Various excitation wavelengths (data not shown) reveal that these analytes can be distinguished via FLN spectroscopy. In addition, these results indicate that also a substitution at the α-position in various TAM–DNA-derived adducts could be detected due to differences in the FLN vibrational frequencies. FLN spectra of TAM are also sufficiently selective to distinguish TAM–DNA adducts from traces of unreacted TAM. It is anticipated that more detailed studies, utilizing high-resolution FLN spectroscopy coupled with either HPLC (off-line) and/or CE (on-line), should allow detection of trace amounts of TAM, TAM metabolites, and/or TAM–DNA adducts in biological samples. These data demonstrate that substituted triphenylethylenes (e.g., TAM) and substituted di-phenylethylenes (e.g., the synthetic estrogen diethylstilbestrol, data not shown), upon irradiation can be converted into analytes which can be studied via FLN spectroscopy.

FLN Spectra of Analytes in Acidified Glasses. Recent data obtained in this laboratory show that it is also possible to "tag" catechol estrogen derived adducts with a fluorescent marker. For example, a nonfluorescent estrogen adduct (7[4-hydroxyestron-1(α,β)-yl]guanine) and its corresponding nonfluorescent metabolite were labeled with dansyl chloride in acetone (basic solution) to form didansyl or monodansyl products (R. Jankowiak et al.; unpublished results). At low temperatures the products showed very broad CT-type fluorescence with a maximum at ~500 nm. The didansylated and monodansylated products revealed strong and moderate electron–phonon coupling, respectively (data not shown). The 0–0 band of the monodansylated and didansylated products was around 319 and 323 nm, respectively. However, protonation eliminated the CT character of the S_1-fluorescent state, allowing for FLN studies. This is not new; Chiang et al. [92] first showed that FLN could be observed for the amino derivatives of polycyclic aromatic compounds in acidified glass. Other data, recently obtained in this laboratory, show that dansyl chloride labeled amino acids give sharp FLN spectra, albeit with very small differences in vibrational frequencies. Thus the key requirement is to find fluorescent markers, which upon labeling of structurally similar analytes will provide sufficiently different FLN spectra due to specific interactions between the fluorescent marker and the analyte, thereby allowing for structural identification.

FLN Spectra of Dye Molecules in Cells. The last example deals with dye molecules that can be detected within the cell. Milanovich et al. [93] reported recently that HB of aluminum phthalocyanine tetrasulfonate (APT) inside MCF-10F human mammary epithelial cells [94] is possible. The HB mechanism was shown to be nonphotochemical [93]. The observed HB of APT in cells is strong evidence that the APT is in a glassy environment. By comparison of the fluorescence excitation spectra and hole-burning characteristics in various matrices, Milanovich et al. [93] suggested, based on the peak position and band shape for APT-treated (stained) cells, that the APT in the MCF-10F cells is localized in an acidic aqueous environment, most probably in lysosomes. An example of vibronically excited FLN spectra of APT in ethanol glass and APT-treated human breast adenocarcinoma (MCF-7) cells at 4.2 K (curves a and b, respectively) is shown in Figure 8.16. The staining procedure and sample preparation is described elsewhere [93]. A comparison of the spectrum of APT in ethanol glass (curve a) with that in MCF-7 cell (curve b) indicates that both matrices show similar inhomogeneity and excited-state vibrational frequencies. The demonstration of HB [93] and FLN of dye molecules within cells opens new possibilities to study cellular and subcellular structures by low-temperature high-resolution spectroscopies.

Finally, it should be noted that FLN spectra can be generated for many ions/molecules in standard organic (or inorganic) glasses, polymers, and proteins

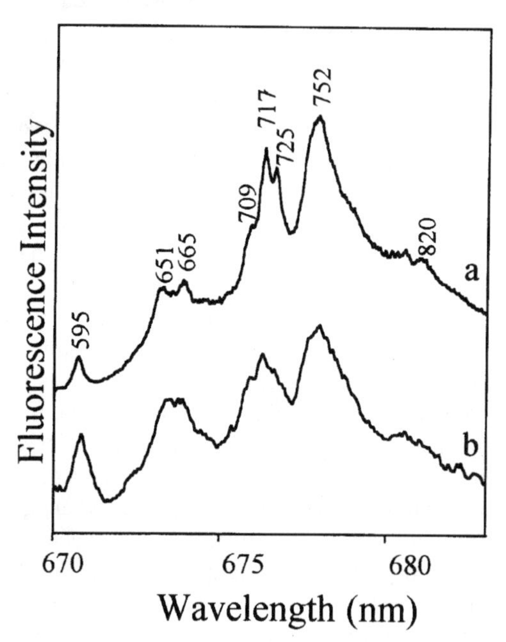

Figure 8.16. Comparison of the vibrationally excited FLN spectra for APT in two different condensed-phase environments. Curve a: APT in ethanol glass; curve b: APT in MCF-7 cells in 1 : 1 freezing medium and glycerol. $T = 4.2\,\text{K}$, $\lambda_{ex} = 645.0\,\text{nm}$ (Jankowiak and Milanovich, unpublished results).

[1–12]. Other examples of matrices that reveal fluorescence line narrowing include: solid pellet DNA [11], etiolated leaves [24], analytes adsorbed on quartz [Jankowiak, unpublished results], analytes spotted or sorbed on TLC silica plates [85, 95], vapor-deposited amorphous organic matrices composed of aromatic molecules [96], urine [11], and dye-stained breast tissue (unpublished results). This indicates that FLNS might have a variety of applications in chemical separation, catalysis, and/or biomedical applications where spectral characterization could be done on a solid surface, in various cell lines, and/or directly on a tissue sample.

8.4. CONCLUDING REMARKS AND FUTURE PROSPECTS

FLNS is the most widely used method for probing the inhomogeneous profile. This phenomenon can only be observed when the laser frequency, ω_L, excites the origin or vibronic bands of the fluorescent state (S_1). Excitation of higher energy states ($S_n, n \geq 2$) results in S_1 fluorescence spectra, which exhibit, at best, only a slight degree of line narrowing. This is a consequence of the site excitation energies of different electronic states being largely uncorrelated. Due to the selectivity and sensitivity of time-resolved FLNS, a wealth of information regarding molecular dynamics and interaction with the local environment can be obtained. With such attributes, and rapid sample throughput, FLNS is ideally suited for the analysis of many fluorescent molecules. The appearance of FLN spectra depends on chromophore–solvent coupling. As shown above, at one extreme, FLN spectra display sharp ZPL associated with very weak PSBs, at the other extreme spectra are composed of broad but distinct PSBs. These limiting cases depend on the strength of electron–phonon coupling, that is, the extent to which the electronic structure of the chromophores interacts with the local low-frequency degrees of freedom of the environment [97]. Thus the strength of electron–phonon coupling dictates how strongly the electrons of the fluorescent molecule interact with the surrounding solvent.

FLNS provides unprecedented selectivity for fluorescent molecular and biomolecular species. For analytes that are nonfluorescent, tagging with fluorescent chromophores is always an option. Based on preliminary results obtained in this laboratory, it is anticipated that FLNS studies of fluorescently labeled analytes should lead to sensitive identification of many originally nonfluorescent compounds. Recent interfacing of FLNS with CE provides another valuable tool for the analysis of closely related compounds and/or complicated mixtures. With the subfemtomole detection levels and on-line characterization by FLNS, many applications in chemical, biological, and toxicological sciences can be anticipated. Other (off-line) applications of FLNS to selectively stained biological condensed-phase environments (e.g., cells, tissues, etc.), present a new challenge for interdisciplinary research, and promise new insights into local structure and dynamics of complex biological systems.

ACKNOWLEDGMENT

Iowa State University operates Ames Laboratory for the U.S. Department of Energy under contract no. W-7405-Eng-82. The work was supported by the Office of Health and Environmental Research, Office of Energy Research, and in part by the National Cancer Institute (grant PO1 CA49210).

REFERENCES

1. R. I. Personov, in V. M. Agranovich and R. M. Hochstrasser, eds., *Spectroscopy and Excitation Dynamics of Condensed Molecular Systems*, North-Holland, Amsterdam, Vol. 4, 1983, pp. 555–619.
2. I. S. Osadko, in K. Dusek, ed., *Advances in Polymer Science*, Springer-Verlag, Berlin, Heidelberg, Vol. 114, 1994, pp. 123–86.
3. B. E. Kohler, in C. B. Moore, ed., *Chemical and Biochemical Applications of Lasers*, Academic Press, New York, 1979, pp. 31–51.
4. J. W. Hofstrat, C. Gooijer, and N. H. Velthorst, in S. G. Schulman, ed., *Molecular Luminescence Spectroscopy: Methods and Applications, Part 2*, John Wiley & Sons, New York, 1988, pp. 383–459.
5. P. M. Selzer, in W. M. Yen and P. M. Selzer, eds., *Topics in Applied Physics*, Springer Verlag, New York, Heidelberg, Berlin, Vol. 49, 1986, p. 113.
6. M. J. Weber, in W. M. Yen and P. M. Selzer, eds., *Topics in Applied Physics*, Springer Verlag, New York, Heidelberg, Berlin, Vol. 49, 1986, pp. 189–240.
7. J. Fidy and J. M. Vanderkooi, in R. H. Douglas, J. Moan, and Gy. Rontó, eds., *Light in Biology and Medicine*, Plenum Press, New York, Vol. 2, 1991, pp. 367–74.
8. J. M. Weber, *J. Lum.* **36**, 179 (1987).
9. R. Jankowiak and G. J. Small, *Anal. Chem.* **61**, 1023A (1989).
10. B. P. Price and J. C. Wright, *Anal. Chem.* **62**, 1989 (1990).
11. R. Jankowiak and G. J. Small, *Chem. Res. Toxicol.* **4**, 256 (1991).
12. H. Riesen, *Comments Inorg. Chem.* **14**, 323 (1993).
13. Yu. V. Denisov and V. A. Kizel, *Opt. Spectrosc.* **23**, 251 (1967).
14. A. Szabo, *Phys. Rev. Lett.* **25**, 924 (1970).
15. R. I. Personov, E. I. Al'Shitz, and L. A. Bykovskaya, *Opt. Commun.* **6**, 169 (1972).
16. L. Nakhimovski, M. Lamotte, and J. Joussot-Dubien, *Handbook of Low Temperature Electronic Spectra of Polycyclic Aromatic Hydrocarbons*, Elsevier, Amsterdam—New York—Oxford—Tokyo (1989).
17. L. A. Riseberg, *Phys. Rev. A* **7**, 67 (1973).
18. I. Renge, K. Mauring, P. Sarv, and J. Avarmaa, *Chem. Phys.* **90**, 6611 (1986).

19. J. Fünfschilling and D. F. Williams, *Photochem. Photobiol.* **26**, 109 (1977).
20. R. A. Avarmaa and K. K. Rebane, *Spectrochim. Acta* **41A**, 1365 (1985).
21. J. Hala, I. Pelant, M. Ambroz, P. Pancoska, and K. Vacek, *Photochem. Photobiol.* **41**, 643 (1985).
22. K. K. Rebane and R. A. Avarmaa, *Chem. Phys.* **68**, 191 (1982).
23. J. Fünfschilling and I. Zschokke-Gränacher, *Chem. Phys. Lett.* **91**, 122 (1982).
24. R. A. Avarmaa, I. Renge, and K. Mauring, *FEBS Lett.* **167**, 186 (1984).
25. P. J. Angiolillo, J. S. Leigh, and J. M. Vanderkooi, *Photochem. Photobiol.* **63**, 133 (1982).
26. W. H. Hesselink and D. A. Wiersma, in V. M. Agranovich and A. A. Maradudin, eds., *Modern Problems in Condensed Matter*, Vol. 4, North-Holland, Amsterdam, 1983, pp. 249–63.
27. C. A. Walsh, M. Berg, L. R. Narasimhan, and M. D. Fayer, *Acc. Chem. Res.* **20**, 120 (1987).
28. N. A. Kurnit, I. D. Abella, and S. R. Hartmann, *Phys. Rev. Lett.* **13**, 567 (1964).
29. L. R. Narasimhan, K. A. Littau, D. W. Pack, Y. S. Bai, A. Elschner, and M. D. Fayer, *Chem. Rev.* **90**, 439 (1990).
30. H. C. Meijers and D. A. Wiersma, *J. Chem. Phys.* **101**, 6927 (1994).
31. J. M. A. Koedijk, R. Wannemacher, R. J. Silbey, and S. Völker, *J. Phys. Chem.* **100**, 19945 (1996).
32. S. Mukamel, *Principles of Nonlinear Optical Spectroscopy*, New York: Oxford University Press, 1995, Chapters 9–11, pp. 261–344.
33. R. Jankowiak and G. J. Small, in A. H. Neilson, ed., *Handbook of Environmental Chemistry* (3J), Springer, 1998, pp. 119–45.
34. A. M. Stoneham, *Rev. Mod. Phys.* **41**, 82 (1969).
35. J. Friedrich and D. Haarer, *Angewandte Chemie, Int. Ed.* **23**, 113 (1984).
36. A. Elschner and H. Bässler, *Chem. Phys.* **112**, 285 (1987).
37. R. Jankowiak, K. D. Rockwitz, and H. Bässler, *J. Phys. Chem.* **87**, 552 (1983).
38. D. Haarer, in W. E. Moerner, ed., *Topics in Current Physics*, *Persistent Spectral Hole Burning: Science and Applications*, Chapter 3, Springer-Verlag, New York, 1987, pp. 79–123.
39. S. Völker, in J. Fünfschilling, ed., *Relaxation Processes in Molecular Excited States*, Kluwer Academic Publishers, Dordrecht, 1st ed., 1989, pp. 113–242.
40. R. Jankowiak, J. M. Hayes, and G. J. Small, *Chem. Rev.* **93**, 1471 (1993).
41. A. O. Kane, *Phys. Rev.* **119**, 40 (1960).
42. I. P. Dzyub and A. F. Lubchenko, *Fiz. Tverd. Tela* **3**, 3602 (1961).
43. R. H. Silsbee, *Phys. Rev.* **128**, 1726 (1962).
44. K. K. Rebane, *Impurity Spectra of Solids*, Plenum, New York, 1970.
45. J. M. Hayes, J. K. Gillie, D. Tang, and G. J. Small, *Biochim. Biophys. Acta* **932**, 287 (1988).

46. J. M. Hayes, P. A. Lyle, and G. J. Small, *J. Phys. Chem.* **98**, 7337 (1994).
47. M. Lax, *J. Chem. Phys.* **20**, 1742 (1952).
48. G. Flecher and J. Friedrich, *Chem. Phys. Lett.* **50**, 32 (1977).
49. J. Kikas, *Chem. Phys. Lett.* **57**, 511 (1978).
50. R. Jankowiak and G. J. Small, in A. Blumen and R. Richert, eds., *Disorder Effects on Relaxation Processes*, Chapter 15, Springer-Verlag, Heidelberg, Berlin, 1994, pp. 425–448.
51. K. A. Littau and M. D. Fayer, *Chem. Phys. Lett.* **176**, 551 (1991).
52. D. T. Leeson, D. A. Wiersma, K. Fritsch, and J. Friedrich, *Phys. Chem.* **101**, 6331 (1997).
53. R. Jankowiak and G. J. Small, *Phys. Rev.* **B97**, 14805 (1993).
54. H. C. Meijers and D. A. Wiersma, *Phys. Rev. Lett.* **68**, 381 (1992).
55. M. N. Sapozhnikov, *Phys. Stat. Sol. (b)* **75**, 11 (1976).
56. J. L. Skinner, *Annu. Rev. Phys. Chem.* **39**, 463 (1988).
57. I. S. Osadko, in Dusek, ed., *Advances in Polymer Science*, Springer-Verlag, Berlin, Vol. 114, 1994, pp. 123–86.
58. B. D. Di Bartolo, *Optical Interactions in Solids*, John Wiley and Sons, New York 1968, Chap. 15, pp. 222–48.
59. L. E. Brus and V. E. Bondybey in Sheng Hsein Lin, ed., *Radiationless Transitions*, Chap. 6, 1980, pp. 259–88.
60. T. Keil, *Phys. Rev.* **140**, A601 (1965).
61. R. M. Hochstrasser and C. A. Nyi, *J. Chem. Phys.* **70**, 1112 (1979).
62. I. I. Abram, R. A. Auerbach, R. R. Birge, B. E. Kohler, and J. M. Stevenson, *J. Chem. Phys.* **63**, 2473 (1974).
63. R. Jankowiak and G. J. Small, *Science* **237**, 618 (1987).
64. J. M. Hayes, R. Jankowiak, and G. J. Small, in W. E. Moerner, ed., *Topics in Current Physics*, *Persistent Spectral Hole Burning: Science and Applications*, Springer-Verlag, New York, Vol. 44, 1987, pp. 153–99.
65. A. D. Kaposi, V. Logovinsky, and J. M. Vanderkooi, *The International Society for Optical Engineering* **1640**, 792 (1992).
66. N. R. S. Reddy, H. van Amerongen, S. L. S. Kwa, R. van Grondelle, and G. J. Small, *J. Phys. Chem.* **98**, 4729 (1994).
67. H. Riesen and E. Krausz, *Chem. Phys. Lett.* **172**, 5 (1990).
68. H. Riesen and E. Krausz, *Chem. Phys. Lett.* **182**, 266 (1991).
69. R. M. Kharlamov, E. I. Al'shitz, and R. I. Personov, *Soviet Phys. JETP* **60**, 428 (1984).
70. K. K. Rebane and A. A. Gorokhovskii, *J. Lum.* **36**, 237 (1987).
71. A. Z. Genack, R. M. Macfarlane, and R. G. Brewer, *Phys. Rev. Lett.* **37**, 1078 (1976).
72. E. I. Al'shitz, R. I. Personov, and B. M. Kharlamov, *Opt. Spectrosc. (USSR)* **41**, 474 (1977).

73. F. Dushinsky, *Acta Physiochem.* **7**, 551 (1937).
74. R. Jankowiak, F. Ariese, D. Zamzow, A. Luch, M. Kroth, A. Seidel, and G. J. Small, *Chem. Res. Toxicol.* **10**, 677 (1997).
75. F. Ariese, G. J. Small, and R. Jankowiak, *Carcinogenesis* **17**, 829 (1996).
76. U. Bogner and R. Schwarz, *Phys. Rev.* **B24**, 2846 (1981).
77. J. Fünfschilling, D. Glatz, and I. Zschokke-Gränacher, *J. Lum.* **36**, 85 (1986).
78. J. W. Hofstraat, M. Engelsma, C. Gooijer, and N. H. Velthorst, *Spectrochim. Acta* **45A**, 491 (1989).
79. J. Wolf and A. B. Myers, *Mol. Cryst. Liq. Cryst.* **291**, 135 (1996).
80. J. Wolf, K.-Y. Law, and A. B. Myers, *J. Phys. Chem.* **100**, 11870 (1996).
81. P. Tchnio, A. B. Myers, and W. E. Moerner, *Chem. Phys. Lett.* **213**, 325 (1993).
82. R. Jankowiak, R. S. Cooper, D. Zamzow, G. J. Small, G. Doskocil, and A. M. Jeffrey, *Chem. Res. Toxicol.* **1**, 60 (1988).
83. L. A. Bykowskaya, A. T. Gradynshko, R. I. Personov, Y. U. Ramanovskii, K. N. Solov'ev, A. J. Starukhin, A. M. Shulga, *Zhur. Prikl. Spectrosk.* **27**, 1088 (1978).
84. P. Wang and R. W. Giese, *Anal. Chem.* **65**, 3518 (1993).
85. R. S. Cooper, R. Jankowiak, J. M. Hayes, P. Lu, and G. J. Small, *Anal. Chem.* **60**, 2692 (1988).
86. J. Friedrich and D. Haarer, *J. Chem. Phys.* **76**, 61 (1981).
87. R. Jankowiak, D. Zamzow, W. Ding, and G. J. Small, *Anal. Chem.* **68**, 2549 (1996).
88. D. L. Huber, in W. M. Yen and P. M. Selzer, eds., *Topics in Applied Physics; Laser Spectroscopy of Solids*, Springer-Verlag, Vol. 49, Chapter 3, 1986, pp. 83–112.
89. D. Zamzow, C.-H. Lin, G. J. Small, and R. Jankowiak, *J. Chromatogr. A* **781**, 73 (1997).
90. D. H. Phillips, A. Hewer, I. N. M. White, and P. B. Farmer, *Carcinogenesis* **15**, 793 (1994).
91. M. R. Osborne, A. Hewer, I. R. Hardcastle, P. L. Carmichael, and D. H. Phillips, *Cancer Res.* **56**, 66 (1996).
92. I. Chiang, J. M. Hayes, and G. J. Small, *Anal. Chem.* **54**, 315 (1982).
93. N. Milanovich, T. Reinot, J. M. Hayes, and G. J. Small, *Biophys. J.* **74**, 2680 (1998).
94. H. D. Soule, T. M. Maloney, S. R. Wolman, W. D. Peterson, Jr., R. Brenz, C. M. McGrath, J. Russo, R. J. Pauley, R. F. Jones, and S. C. Brooks, *Cancer Res.* **50**, 6075 (1990).
95. J. W. Hofstraat, M. Englesma, P. W. Cefino, G. P. Hornweg, C. Gooijer, and N. H. Velthorst, *Anal. Chim. Acta* **159**, 359 (1984).
96. R. Jankowiak and H. Bässler, *J. Mol. Electr.* **1**, 73 (1985).
97. R. M. Hochstrasser and P. N. Prasad, in E. C. Lim, ed., *Excited States*, Plenum Press, New York, Vol. 1, 1974.

CHAPTER

9

DYNAMICS OF GLASSES AND PROTEINS PROBED BY TIME-RESOLVED HOLE BURNING

TIJSBERT M. H. CREEMERS

Center for the Study of Excited States of Molecules, Huygens and Gorlaeus Laboratories, University of Leiden, P.O. Box 9504, 2300 RA Leiden, The Netherlands

SILVIA VÖLKER

Center for the Study of Excited States of Molecules, Huygens and Gorlaeus Laboratories, University of Leiden, P.O. Box 9504, 2300 RA Leiden, The Netherlands and Department of Biophysics, Faculty of Exact Sciences, Free University, De Boelelaan 1081, 1081 HV Amsterdam, The Netherlands

9.1. OUTLINE AND INTRODUCTION

This chapter is focused on the dynamics of doped organic glasses and photosynthetic pigment–protein complexes at low temperatures, studied in our group by time-resolved hole burning. While glasses are "true" disordered materials, proteins are only partly disordered since they are supposed to have a "hierarchical organization" [1, 2]. Hence it is important to learn about their potential energy surfaces, which govern the dynamics. Low-temperature laser experiments carried out over more than ten orders of magnitude in time yield a distribution of relaxation rates. These, in turn, provide insight into the energy landscape of these complex systems when measured over a wide range of temperatures.

In Section 9.1.1, we give a general introduction to two-level systems (TLSs), the low-frequency modes assumed to be responsible for the dynamics of glasses, and discuss their influence on the optical linewidth. We then consider the dynamics of proteins in terms of distributions of relaxation rates and compare it to that of glasses (Section 9.1.2).

Next we introduce the concept of hole-burning (HB) and show how the shape of a hole is affected by the time evolution of the glassy system (Section 9.2.1). The mechanisms leading to hole-burning are mentioned in Section 9.2.2. We further describe the pulse sequence and time-resolved HB experimental setup and the procedure used to determine the "effective" homogeneous linewidth Γ'_{hom} from the holewidth (Sections 9.2.3 and 9.2.4). Results from our group on optical

Shpol'skii Spectroscopy and Other Site-Selection Methods, Edited by Cees Gooijer, Freek Ariese, and Johannes W. Hofstraat.
ISBN 0-471-24508-9

dephasing, spectral diffusion (SD), and energy-transfer-induced spectral diffusion (ET $\rightarrow$ SD) in doped organic glasses are presented in Sections 9.3.1 and 9.3.2.

In the last part of the chapter we turn to the dynamics of proteins at low temperature and examine what can be learned about them from experiments on photosynthetic complexes, such as subunits of the light-harvesting complex LH1 of purple bacteria (Section 9.4.1) and the subcore complexes of the photosystem II reaction center of green plants (Section 9.4.2). Finally, we compare our time-resolved hole-burning results with three-pulse photon-echo and temperature-cycling hole-burning experiments reported on other protein complexes and indicate some future trends (Section 9.4.3).

9.1.1. Doped Organic Glasses

Glasses are solids in nonequilibrium, obtained by fast cooling of a liquid to an amorphous state. Although structurally disordered, their mechanical properties in many respects are similar to those of crystals. Glasses lack long-range order, and, in contrast to crystals, their potential energy surface is aperiodic with varying minima and barrier heights. A one-dimensional cross-section through such a multidimensional energy surface is plotted in Figure 9.1 (top) as a function of a conformational coordinate.

At low temperature, glasses show anomalous thermal and optical properties that one conventionally describes with the two-level-system model [3, 4]. A TLS can be pictured as a double-well potential (see Fig. 9.1, bottom), in which each well represents a distinct (local) configuration of the glass; the TLS is characterized by a barrier height V and asymmetry Δ, yielding an energy splitting E, and a relaxation rate R. A change in the glass structure is modeled by transitions or "flips" from one potential well to the other. If these transitions occur at very low temperature, they are caused by phonon-assisted tunneling. TLSs are assumed to have a broad distribution of tunneling parameters and energy splittings E, leading to a broad distribution of relaxation rates R varying over many orders of magnitude [5–7]. Although the TLS model describes the anomalous low-temperature properties of glasses rather well, the microscopic nature of TLSs is not really understood.

The dynamic properties of a glass at low temperature can be studied by doping it with a probe molecule or chromophore (see Fig. 9.1, bottom). Because the TLSs are coupled to the chromophore, the "effective" homogeneous linewidth Γ'_{hom} (for a definition of Γ'_{hom}, see Section 9.2.4) of the electronic transition will depend on parameters like temperature T and the time scale of the experiment t_d. The glass dynamics comprises very fast and very slow TLS fluctuations. The fast ones, which occur around an equilibrium structure during the excited-state lifetime, lead to optical dephasing. As a consequence, Γ'_{hom} will increase with T. The slow fluctuations are responsible for the time evolution of the glass

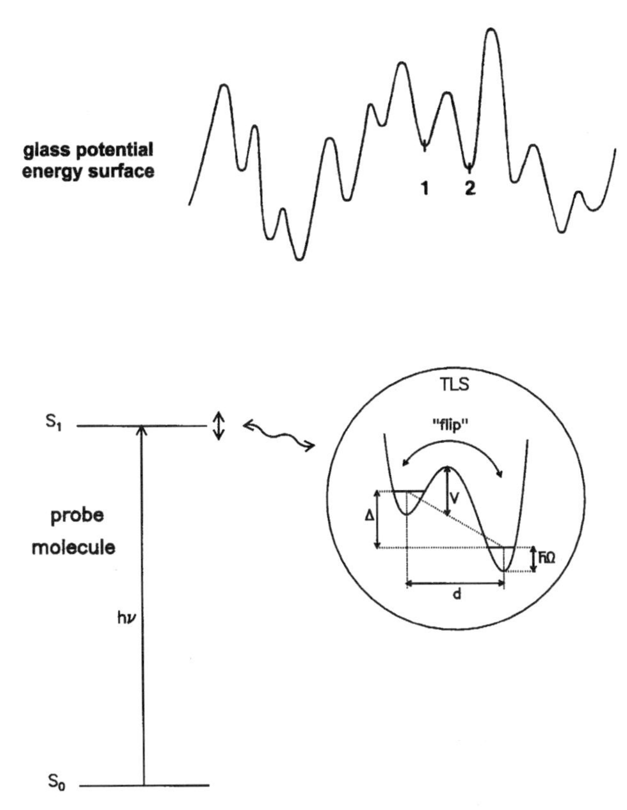

Figure 9.1. Top: Schematic potential energy surface of a glass as a function of a conformational coordinate. At low temperature, a glass can be described by a broad distribution of two-level systems (TLSs). Bottom, right: A TLS is characterized by an asymmetry Δ, a barrier height *V*, and a barrier thickness *d*. Bottom, left: The electronic transition of a probe molecule interacts with a "flipping" two-level system. This causes fluctuations of the transition frequency and broadening of the optical transition.

structure as a whole. They cause the optical transition to shift in time, which, in turn, leads to an increase of Γ'_{hom} with t_{d} [8–10]. This process, called *spectral diffusion* (SD) [11], is intrinsic to the nature of the TLSs [4, 5].

The absorption bands of doped glasses are generally very broad, a few $100\,\mathrm{cm}^{-1}$, even at liquid He temperatures. They are inhomogeneously broadened with a width Γ_{inh} that is orders of magnitude larger than the homogeneous linewidth Γ_{hom} (see Fig. 9.2a, top). The large value of Γ_{inh} arises from the slightly different environments of the individual chromophores within the amorphous host. This leads to a statistical distribution of the electronic transition energies and to a Gaussian profile of the spectral band.

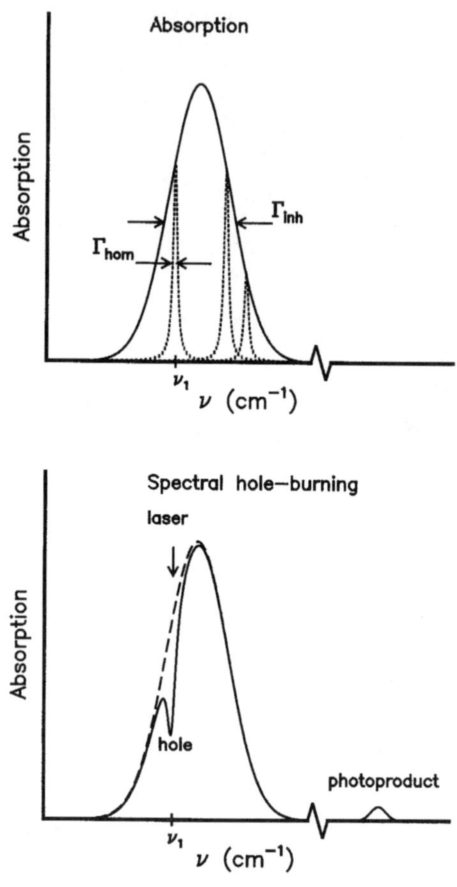

Figure 9.2. (a) Top: Diagram of an inhomogeneously broadened absorption band of width Γ_{inh}. The homogeneous bands of width Γ_{hom} of the individual electronic transitions are hidden under the broad inhomogeneous absorption band. Bottom: Laser-induced spectral hole burned at excitation frequency ν_1. The photoproduct absorbs at a different frequency, here outside the inhomogeneous band.

Information on the dynamics of the system is contained in the homogeneous linewidth, hidden under the large inhomogeneously broadened band. The value of Γ_{hom} can be obtained with coherent laser techniques, either in the time domain by means of two- and three-pulse photon echoes, or in the frequency domain by means of fluorescence line narrowing, single-molecule spectroscopy, and spectral hole-burning [12, 13].

The lineshape of a homogeneously broadened transition is usually Lorentzian; it represents the Fourier transform of an exponential decay curve. In a *crystalline*

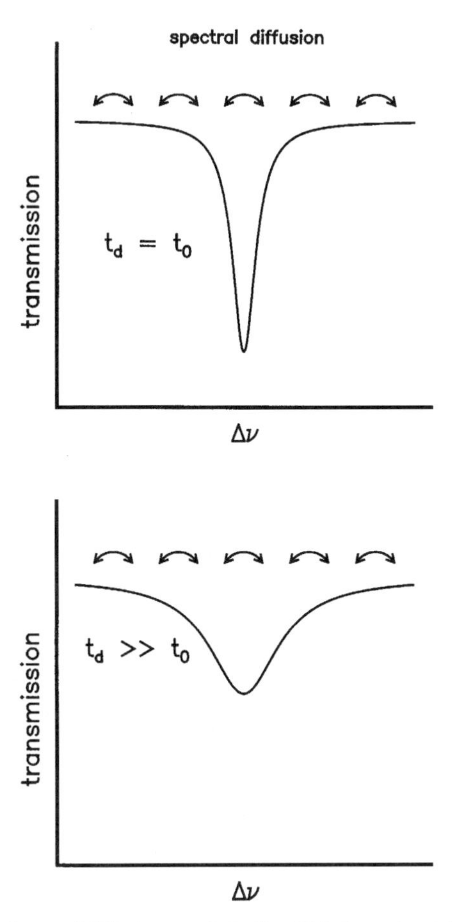

Figure 9.2. (b) Effect of spectral diffusion on a hole. Top: Hole burned and detected at time $t_d = t_0$. Jumps in transition frequency, indicated by arrows, cause SD. Bottom: After a waiting time $t_d \gg t_0$, the width of the hole increases and the depth decreases due to SD. The area of the hole stays constant if the number of molecules involved remains constant.

host, the homogeneous linewidth Γ_{hom} of the electronic transition of a probe molecule is represented by an expression of the following form:

$$\Gamma_{\text{hom}}(T) = \frac{1}{2\pi T_1} + \frac{1}{\pi T_2^*(T)} \tag{9.1}$$

where T_1 is the lifetime of the excited state (or, more precisely, the time the excitation resides on a given molecule), and T_2^* is the "pure" dephasing time.

$T_2^*(T)$ represents the time interval during which the coherence of the electronic transition is lost. For $T \to 0$, Γ_{hom} is given by the first term in Eq. (9.1), Heisenberg's uncertainty relation. For $T > 0$, $T_2^*(T)$ is determined by guest–host interactions induced by phonon scattering. They cause frequency fluctuations of the optical transition.

In a doped *glass*, Γ_{hom} will additionally increase due to the presence of "flipping" TLSs. One then obtains the "effective" homogeneous linewidth Γ'_{hom} (see Section 9.2.4), which will depend on temperature and waiting (or delay) time between excitation and probing. For $T \leq 10\,\text{K}$, the temperature dependence of Γ'_{hom} has been found to obey a $T^{1.3}$-power law [12, 14, 15], independent of the glass and the probe (see Sections 9.3.1 and 9.3.2):

$$\Gamma'_{\text{hom}}(T) = \Gamma_0 + a(t_d)T^{1.3\pm0.1} \tag{9.2}$$

The residual linewidth Γ_0 for $T \to 0$ is given by $\Gamma_0 = (2\pi T_1)^{-1}$, with T_1 the excited-state lifetime. At *low* concentrations, $\Gamma_0 = (2\pi\tau_{\text{fl}})^{-1}$, with τ_{fl} the fluorescence lifetime of the probe molecule (see Section 9.3.1). The second term in Eq. (9.2) not only accounts for "pure" dephasing, as in crystals, but also for spectral diffusion [16–22].

At *higher* concentrations, energy transfer between probe molecules may occur. The time a given molecule is in the excited state will then decrease because energy transfer opens extra channels for relaxation (see Section 9.3.2). By studying energy transfer, we have discovered a remarkable effect: spectral diffusion induced by energy transfer. This process, which we call ET → SD, gives additional contributions to the residual linewidth and to the temperature-dependent term in Γ'_{hom} [17, 23–25].

A phenomenon that intrigued us is the effect of high pressure on the dynamics of TLSs. Assuming that the TLSs are low-frequency modes associated with voids or excess free volume in the glass [26], one would expect to be able to change the number of TLSs and, therefore, the amount of SD by applying pressure. We have performed this type of experiments up to $\Delta p = 30\,\text{kbar}$ and found that high pressure indeed reduces, or even eliminates in some cases, the number of TLSs. The effect observed is a decrease of SD and a change in the functional dependence of Γ'_{hom} on temperature. The results are not discussed in this chapter but can be found in Refs. 17 and 61.

9.1.2. Protein Dynamics

Proteins are materials that display both crystalline and glassy properties. In their native state they have well-defined tertiary structures that are reflected in their crystalline properties. In contrast to crystals, they are not static but undergo conformational changes between a large number of slightly different intermediate structures. These are called conformational substates (CSs) [1]. The CSs are separated by a wide distribution of energy barriers with multiple minima on the potential energy surface [27], reminiscent of the TLSs in glasses. The main difference is that glasses are random, whereas CSs of proteins are assumed to be hierarchically organized. Within each CS, a number of CSs exist that include the next tier or rank of CSs [2]. Whether conformational changes in proteins may be explained by "flipping" TLSs with a continuous distribution of relaxation rates as glasses [17, 18, 21, 22], or whether they are characterized by discrete and sharp rates [28, 29] because of their hierarchical organization, is still a controversial issue (see Section 9.4.3). Since some regions of the energy landscape of proteins are supposed to be rough and others smooth, it has been suggested that the randomness in their energy landscape is organized [30, 31]. One of the aims of studying the dynamics of proteins is to relate the structure to the biological function by following the time evolution of their conformational rearrangements. We will show that spectral diffusion in photosynthetic proteins at low temperature (Sections 9.4.1 and 9.4.2) leads to a distribution of relaxation rates that appears to be different from that reported in the literature for Zn- and H_2-substituted myoglobin [28, 29] (Section 9.4.3). A qualitative interpretation of the results will be given.

9.2. HOLE BURNING

9.2.1. Basic Concepts and Spectral Diffusion of a Hole

In a spectral hole-burning experiment, the inhomogeneously broadened absorption band is irradiated at a given wavelength with a narrow-band laser. Whenever the molecules resonant with the laser wavelength undergo a phototransformation (photochemical or photophysical), a hole is created in the original absorption band (see Fig. 9.2a, bottom) [12, 32–36]. The width of the hole, under given conditions, is then proportional to the homogeneous linewidth (see Section 9.2.4). The photoproduct will absorb at a different wavelength, either within the absorption band, or outside. Because the laser selects molecules absorbing at a given frequency ν_1, and not molecules in a specific environment, the correlation between transition energy and environmental parameters is, in general, different for the photoproduct and the original molecule. As a consequence, the width of the photoproduct band or antihole is larger than that of the hole. The optical resolution that can be reached with hole-burning is 10^3–10^5 times higher than reached by conventional techniques, which makes it a powerful tool for spectroscopy in the MHz regime [12].

The effect of spectral diffusion (SD) on a hole is illustrated in Figure 9.2b. Assume that a hole is burned and probed at time t_0 (see Fig. 9.2b, top). After a delay time $t_d \gg t_0$, structural relaxation will have taken place in the glass, causing TLSs to "flip" (see Section 9.1.1). Due to the coupling of the TLSs to the optical transition, this "flipping" will induce spectral diffusion, which will be manifested as a simultaneous increase of the hole width and a decrease of the hole depth with delay time t_d. If the number of molecules involved in the process remains constant, the area of the hole will stay constant (see Fig. 9.2b, bottom). To study the effect of SD, hole shapes have to be measured as a function of delay time at various temperatures, excitation wavelengths, and concentrations of probe molecules.

9.2.2. Hole-Burning Mechanisms

Hole-burning mechanisms can be divided in two general categories: persistent HB and transient HB (THB). Within the first category, one may distinguish between photochemical hole-burning (PHB) and nonphotochemical hole-burning (NPHB). The time scales involved in the PHB and NPHB processes are usually seconds to minutes, whereas THB often lasts micro- or milliseconds [12].

In the PHB mechanism, the irradiated molecules undergo a photochemical reaction (either intra- or intermolecular, reversible or irreversible) and transform into a photoproduct absorbing at a different wavelength from that of the original molecule. If the temperature is kept sufficiently low, the hole remains for hours or days [14–17, 32, 33, 36–38]. For example, in free-base porphin [32, 37] and chlorin (H_2Ch) [23–25, 38], persistent holes arise from phototautomerization of the two inner protons. Systems with other PHB mechanisms are mentioned in Ref. 12.

NPHB is a photophysical reaction by which the relative orientation of the excited molecule changes with respect to its surroundings. This creates a photoproduct that absorbs at a wavelength very close (a few cm^{-1}) to that of the original molecule [39, 40]. Since the potential barriers between the different orientations are small, NPHB holes are usually persistent only at very low temperatures. Most stable probe molecules, when embedded in a glass, undergo NPHB [39]. A model has been proposed for NPHB in glasses [39–41] that assumes that the probe molecules are coupled to different TLSs in the ground and excited states. NPHB occurs then through tunneling in the excited state and subsequent relaxation to a different ground-state configuration. Another type of NPHB reaction is that observed by us in glasses doped with bacteriochlorophyll-*a* (BChl-*a*) [16, 20, 42]. In such a system the holes, which are persistent at low temperature, are probably due to a reorientation, after excitation, of the central magnesium atom of BChl-*a* with respect to the glassy host. The photoproduct absorbs at $\sim 100\,cm^{-1}$ from the originally excited BChl-*a* [42]. We think that a similar mechanism is valid for the B820 and B777 subunits of the light-harvesting complex LH1 of purple bacteria [43].

In THB, population is transferred from the ground state through the excited state of interest to a metastable state. The latter may be the same excited state [44], a long-living triplet state [16, 20], or a hyperfine level of the ground state [45]. The lifetime of the hole (usually microseconds to seconds) is determined by the decay of the metastable state [16, 20, 25, 35, 46, 47].

Dynamic processes such as optical dephasing, energy transfer or spectral diffusion, which determine the "effective" homogeneous linewidth and, therefore, the hole width, depend on the interaction of the optical transition of the probe with its surroundings and are independent of the hole-burning mechanism [16, 17, 20].

9.2.3. Pulse Sequence and Experimental Setup for Time-Resolved HB

9.2.3.1. Pulse Sequence

Figure 9.3a shows the pulse sequence used to burn and probe holes after a variable delay time t_d. It consists of three steps. First, a base line is measured at low laser intensity. During this time the frequency of the laser is scanned over the part of the inhomogeneous absorption band of interest. Then a hole is burned during a time t_b at a fixed frequency, at much higher intensity than used for the base line (factor 10–10^3). Finally, the hole is probed after a delay time t_d, varied for each experiment. During probing at time t_p the frequency of the laser is scanned again at low intensity over the region of the hole. To obtain the hole profile, the difference between the signals before and after burning is recorded [16, 17, 20].

9.2.3.2. Experimental Setup for Time-Resolved HB

To perform time-resolved hole-burning experiments, we use various types of cw single-frequency lasers, in combination with acousto-optic modulators (AOMs; see Fig. 9.3b) to create the pulse sequence described in Figure 9.3a. The choice of the laser depends on the sample and the time scale of the experiment.

For delay times shorter than a few seconds and down to microseconds, we use current- and temperature-controlled single-mode diode lasers [48]. The type of diode laser depends on the wavelength needed. The main advantage of these semiconductor lasers is that their frequency can be scanned very fast, up to ~10 GHz/μs, by sweeping the current through the diode. A disadvantage is their restricted tunable wavelength region (5–10 nm, by changing the temperature of the laser).

For delay times longer than ~100 ms, either a cw single-frequency titanium : sapphire ring laser (bandwidth ~0.5 MHz, tunable from ~700 to 1000 nm) or a dye laser (bandwidth ~2 MHz, tunable between ~570 and

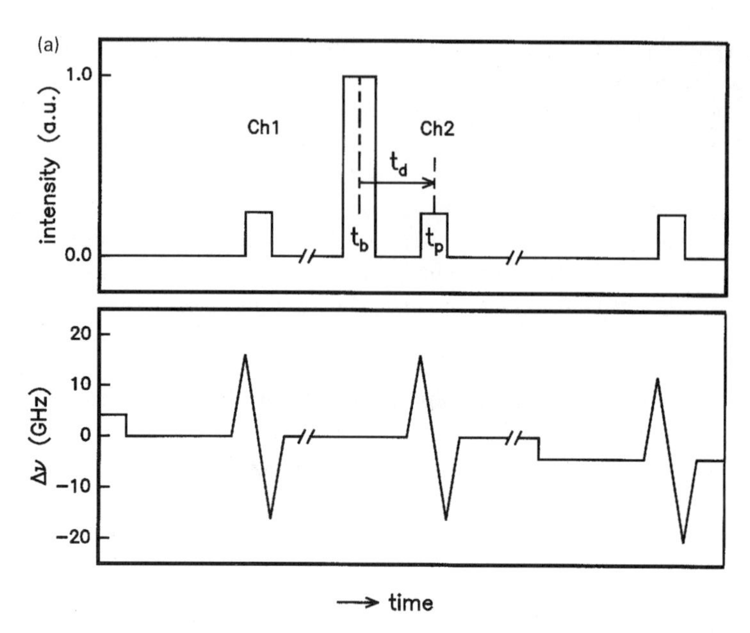

Figure 9.3. (a) Pulse sequence used in time-resolved hole burning. Top: Timing of the laser pulses. Bottom: frequency ramp and steps; t_b: burn time, t_d: delay time, t_p: probe time, $\Delta\nu$: change in laser frequency.

700 nm), both pumped by an Ar^+ laser (2–15 W), are used. The frequency of these lasers can be scanned with a maximum speed of ~100 MHz/ms. This speed is about 100,000 times slower than that of the diode lasers. The wavelength of the lasers is calibrated with a Michelson interferometer (resolution ~50 MHz), and its mode structure is monitored with a confocal Fabry–Perot (FP) etalon (FSR = 300 MHz, 1.5 and/or 8 GHz).

Burning-power densities between 1 μW/cm^2 and 20 mW/cm^2, with burning times from 1 μs to ~100 s, are generally used. The holes are either probed in fluorescence excitation or in transmission through the sample with the same laser, but with its power attenuated by a factor of 10 to 10^3. The delay time t_d between burning and probing the holes usually varies from 1 μs to 24 h. For delay times shorter than ~100 s, the burn and probe pulses are produced with two AOMs in series (two instead of one to reduce the laser light leaking through when switched off, suppression better than 10^6). For delay times longer than ~100 s, the intensity of the probe pulse is reduced with an optical density filter.

The fluorescence or transmission signal of the hole is detected with a photomultiplier and subsequently amplified. The signals are averaged in different ways, depending on the delay time. For $t_d < 100$ ms, a sequence of probe–burn–probe cycles is applied with a repetition rate of about 10 Hz using home-built

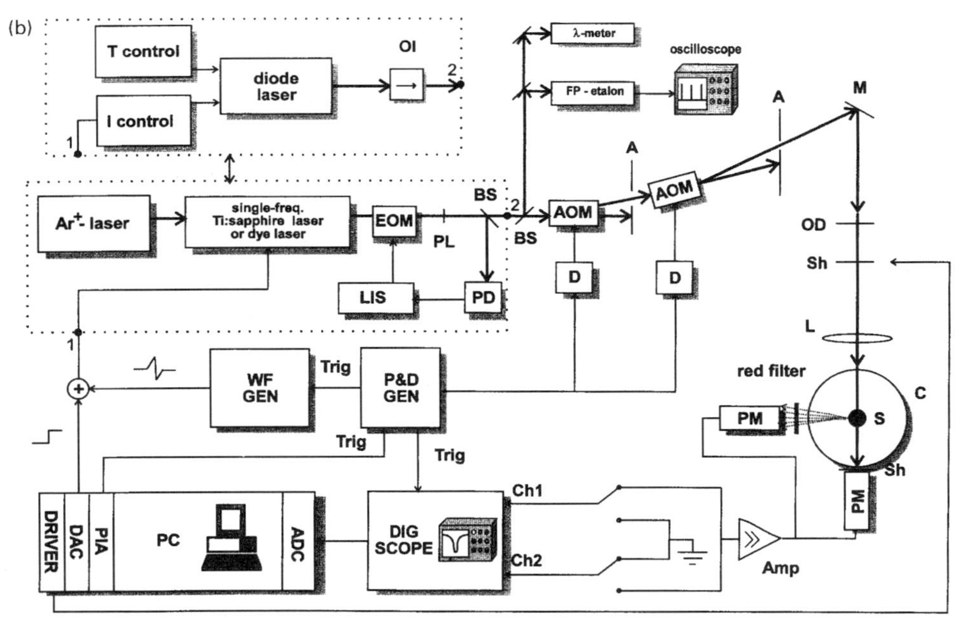

Figure 9.3. (b) Time-resolved hole-burning setup. Three types of cw single-frequency lasers were used: either a temperature- and current-controlled (*T*- and *I*-control) diode laser, a dye laser, or a titanium : sapphire laser. OI: optical isolator, λ meter: Michelson interferometer, FP: Fabry–Perot etalon, EOM: electro-optic modulator, LIS: electronic control circuit for light intensity stabilization, BS: beam splitter, PD: photodiode, AOM/D acousto-optic modulator and driver, P&D GEN: pulse- and delay generator, WF GEN: waveform synthesizer, + : summing amplifier, A: diaphragm, M: mirror, OD: neutral density filters, Sh: shutter, L: lens, S: sample, C: ^{4}He cryostat, PM: photomultiplier, Amp: amplifier, DIG SCOPE: digital oscilloscope, PIA: peripheral interface adapter, ADC: analog-to-digital converter, DAC: digital-to-analog converter [47–49].

electronics (see Fig. 9.3b). For longer delay times and up to $t_d = 100$ s, the repetition rate is lower. The signals before and after burning are stored in two channels of a digital oscilloscope. After each probe–burn–probe cycle, the frequency of the laser is slightly shifted (by a few times the holewidth) to obtain a fresh base line for each burnt hole. Transient holes, which decay with a triplet lifetime of $\tau_T \sim 100\,\mu$s to a few milliseconds in the systems studied, are averaged 10^3–10^4 times, whereas persistent holes with delay times $\tau_T \leq t_d \leq 100$ s are averaged 50–100 times because of their better signal-to-noise ratio. For delay times $t_d > 100$ s, the signals are averaged point by point about 1000 times with the PC, with a total number of 200–1000 points, depending on t_d. In this case, the pulse scheme shown in Figure 9.3a is used only once and not cycled through. The experiments are controlled with a personal computer (PC) [16, 17, 47, 49].

9.2.4. Determination of the "Effective" Homogeneous Linewidth Γ'_{hom}

To determine the shape of the hole and the value of the "effective" homogeneous linewidth Γ'_{hom}, holes are burned for a given set of parameters (temperature, delay time, wavelength, concentration and pressure) as a function of burning-fluence density Pt_b/A (where P is the burning power of the laser, t_b the burning time, and A the area of the laser spot on the sample), and the holewidths are extrapolated to $Pt_b/A \rightarrow 0$ to take into account the effect of power broadening. The holes are generally well fitted with Lorentzian curves. The extrapolated value of Γ_{hole}, which we call $\Gamma_{hole,0}$, consists of three terms [16,17]:

$$\Gamma_{hole,0}(t_b, t_d) = \Gamma'_{hom}(t_b) + \Gamma'_{hom}(t_d) + 2\Gamma_{laser} \tag{9.3}$$

where $\Gamma'_{hom}(t_i)$ represents the "effective" homogeneous linewidth at time t_i. Γ_{laser} is equal to the bandwidth of the laser, which may vary from 0.5 MHz (for the titanium : sapphire laser) to ~50 MHz (for some diode lasers [48]). The sum in Eq. (9.3), a convolution in general, is only valid if the line shapes are Lorentzian. To determine $\Gamma'_{hom}(t_d)$, the quantity of interest, a hole-burning experiment is first performed at a delay time $t_d = t_b$,

$$\Gamma_{hole,0}(t_b, t_b) = 2\Gamma'_{hom}(t_b) + 2\Gamma_{laser} \tag{9.4}$$

Inserting $\Gamma'_{hom}(t_b)$ from Eq. (9.4) into Eq. (9.3) yields

$$\Gamma'_{hom}(t_d) = \Gamma_{hole,0}(t_b, t_d) - \tfrac{1}{2}\Gamma_{hole,0}(t_b, t_b) - \Gamma_{laser} \tag{9.5}$$

For $t_d \leq 100$ s, usually $t_b = t_d$ and Eq. (9.5) reduces to

$$\Gamma'_{\text{hom}}(t_d) = \tfrac{1}{2}\Gamma_{\text{hole},0}(t_b, t_b) - \Gamma_{\text{laser}} \tag{9.6}$$

From Eq. (9.5) we see that for $t_d \gg t_b$, $\Gamma'_{\text{hom}}(t_d)$ is significantly larger than the value derived from Eq. (9.6) conventionally used.

9.3. DOPED ORGANIC GLASSES

9.3.1. Dephasing and Spectral Diffusion at Low Concentrations

In order to examine the validity of the theoretical models for optical dephasing in glassy systems [7–10], we have carried out time-resolved hole-burning experiments on the $S_1 \leftarrow S_0$ 0–0 transition of bacteriochlorophyll-*a* (BChl-*a*) in a series of organic glasses: 2-methyltetrahydrofuran (MTHF), (protonated and deuterated) ethanol, diethyl ether, and triethylamine (TEA). The delay time t_d was varied over 10 orders of magnitude in time, from 10^{-6} to 10^4 s, and the temperature from 1.2 to 4.2 K [16, 20].

Values of the "effective" homogeneous linewidth Γ'_{hom} as a function of temperature are shown in Figure 9.4 for two delay times, $t_d = 300$ s (NPHB) and 15 μs (THB), for four glassy systems. We see that Γ'_{hom} follows a $T^{1.3\pm0.1}$ dependence for both THB and NPHB data. Furthermore, all curves extrapolate to $\Gamma_0 = (2\pi\tau_{\text{fl}})^{-1}$ for $T \to 0$, with $\tau_{\text{fl}} \approx 4$ ns the fluorescence lifetime of BChl-*a* [50]. This is to be expected because spectral diffusion is not active at $T \to 0$. Whereas no spectral diffusion is observed for TEA [16,17,20], the increase of t_d from 15 μs to 300 s leads to an increase of Γ'_{hom} by a factor of 2 for MTHF and ethanol! Further, it is striking that Γ'_{hom}, even for $t_d = 15$ μs, is more than one order of magnitude larger for the latter than for TEA. The data suggest that the more disordered the host the stronger the SD. Thus it looks as if TEA, which shows no SD, is highly ordered at liquid-He temperatures [16, 17, 20]. The results prove that SD strongly depends on the nature of the glass. The $T^{1.3}$-power law, characteristic for optical dephasing in organic glassy systems [12–14, 20, 51–55], however, is independent of t_d [16, 20], which according to current theoretical models indicates that the same TLSs are responsible for optical dephasing and spectral diffusion [9, 10, 16, 18, 21, 22].

TEA doped with BChl-*a* is an intriguing system because it undergoes "pure" dephasing without observable SD, like in crystals. From the $T^{1.3}$ dependence, on the other hand, we conclude that TEA is glasslike. The small value of Γ'_{hom} for TEA as compared to other glasses, and the results obtained at high pressure [17], suggest that only a few TLSs contribute to the dephasing in TEA [16]. Under a pressure of 30 kbar, the $T^{1.3}$-dependence transforms into an $\exp(-E/kT)$

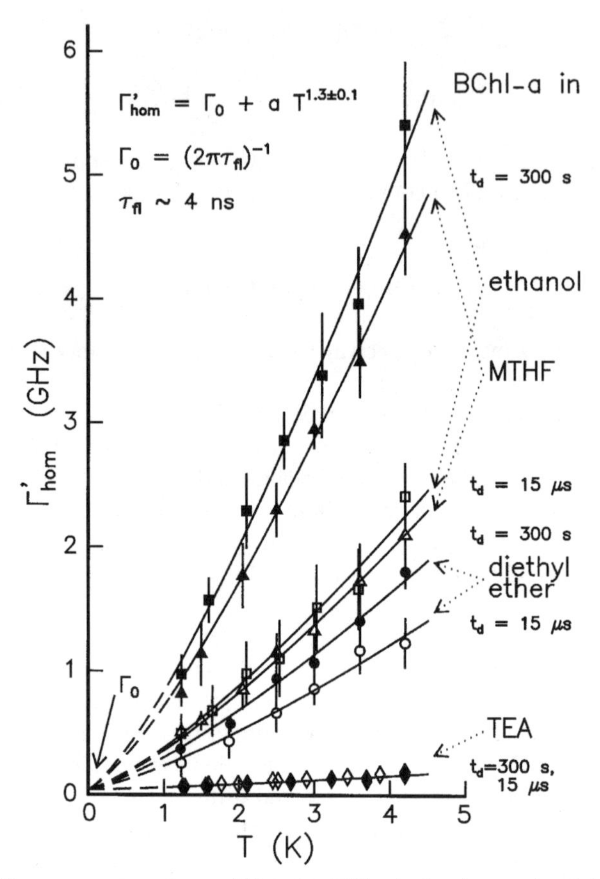

Figure 9.4. Temperature dependence of Γ'_{hom} for BChl-*a* in the glasses ethanol (squares), MTHF (triangles), diethyl ether (circles), and TEA (diamonds). $t_{\text{d}} = 15\,\mu\text{s}$ (open symbols) and $t_{\text{d}} = 300\,\text{s}$ (closed symbols). All curves follow a $T^{1.3}$ power law and extrapolate to the fluorescence lifetime-limited value $\Gamma_0 = (2\pi\tau_{\text{fl}})^{-1}$ for $T \to 0$, with $\tau_{\text{fl}} \sim 4\,\text{ns}$ [20].

dependence, characteristic for crystalline systems, proving the disappearance of TLSs [17].

The delay time dependence of Γ'_{hom} for BChl-*a* in MTHF at 1.2 and 4.2 K is shown in Figure 9.5 between 10^{-6} and 10^{4} s. The dependence is linear in the logarithm of t_{d}. Furthermore, SD increases with temperature, the slope being much steeper at 4.2 K. By investigating this dependence for various glasses, we have found that not only "pure" dephasing, but also the amount of SD, increases with temperature according to the $T^{1.3}$-power law [16].

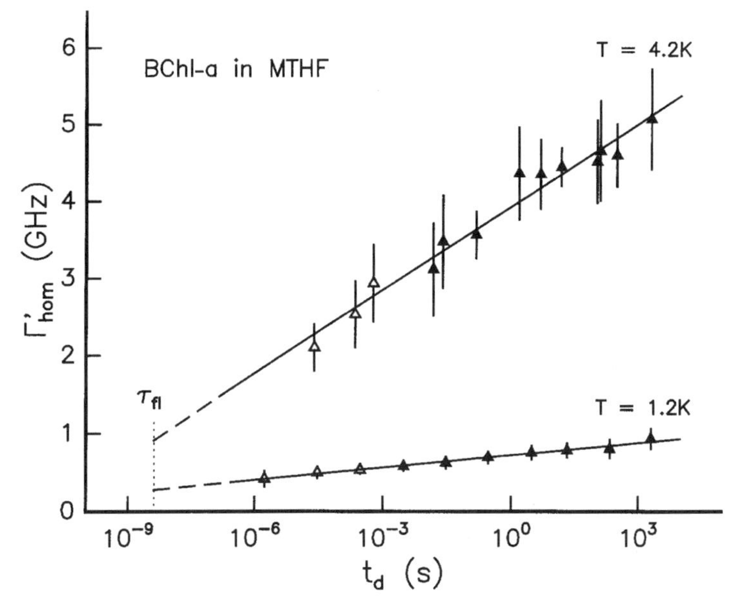

Figure 9.5. Γ'_{hom} as a function of $\log(t_d)$ for BChl-*a* in MTHF at 1.2 and 4.2 K. The data follow $\Gamma'_{\text{hom}} \propto \log(t_d)$ from 10^{-6} to 10^4 s. The amount of spectral diffusion, given by the slope $d\Gamma'_{\text{hom}}/d\log(t_d)$, increases with temperature [16].

Figure 9.6 shows plots of Γ'_{hom} as a function of $\log(t_d)$ for BChl-*a* in ethanol, MTHF, diethyl ether, and TEA at 1.2 K. The results confirm that SD depends markedly on the glass chosen. By comparing results in ethanol doped with various probe molecules (not shown), it was concluded that the number of active TLSs in the glass and the amount of SD are not influenced by the nature of the probe [16].

All our experiments for doped organic glasses at low concentrations can be cast in the following form:

$$\Gamma'_{\text{hom}} = \Gamma_0 + a(t_d)T^{1.3} = (2\pi\tau_{\text{fl}})^{-1} + [a_{\text{PD}} + a_{\text{SD}}(t_d)]T^{1.3} \tag{9.7}$$

where the first term is determined by the fluorescence lifetime τ_{fl} and the second term consists of two contributions, a "pure" dephasing one, $a_{\text{PD}}T^{1.3}$, and a delay-time-dependent contribution determined by "normal" spectral diffusion caused by "flipping" TLSs, $a_{\text{SD}}(t_d)T^{1.3}$ [16, 17].

The logarithmic dependence of Γ'_{hom} on t_d fitted through the data in Figures 9.5 and 9.6 implies that the distribution of TLS relaxation rates $P(R)$ is proportional to $1/R$ [18]. Within the standard TLS model [3–7, 9, 10] this means that the distribution of TLS-tunneling parameters must be flat and,

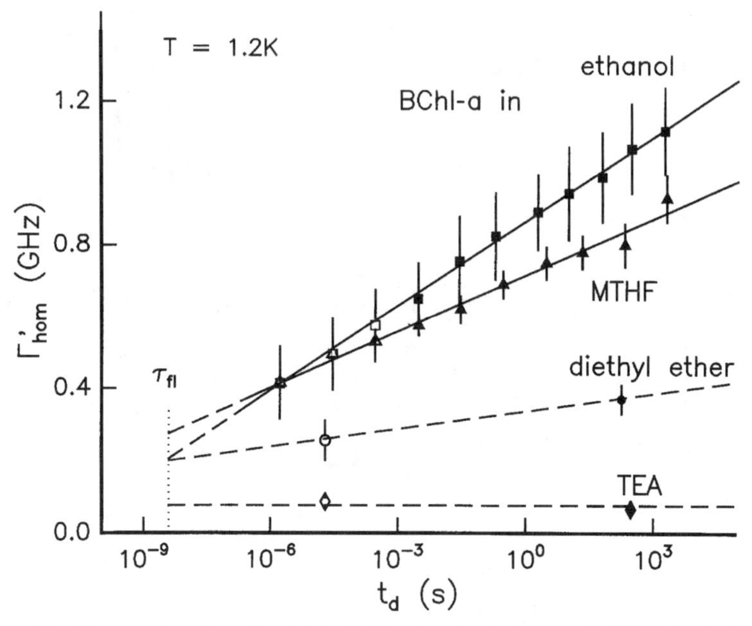

Figure 9.6. Γ'_{hom} as a function of $\log(t_d)$ for BChl-*a* in ethanol (squares), MTHF (triangles), diethyl ether (circles), and TEA (diamonds). The amount of spectral diffusion strongly depends on the glass [16].

therefore, the TLS density of states $\rho(E) = \text{constant}$, which leads to $\Gamma'_{hom} \propto \Gamma_0 + T\log(t_d)$ [5–7, 18]. Thus, according to this theory, there is an intertwined linear dependence on T and $\log(t_d)$. Experimentally, however, one invariably observes a proportionality with $T^{1.3\pm0.1}$, which, so far, had been explained by assuming $\rho(E) \propto E^{0.3}$, where E is the energy splitting of the eigenstates of the TLS [8–10, 18, 54, 56].

In an attempt to solve this contradiction and to explain the gap in the distribution of relaxation rates of the TLSs reported for zinc porphin in ethanol by three-pulse photon echoes [21], the standard model of TLSs has been modified [18]. By taking a different distribution of the TLS asymmetries and tunneling matrix elements [57], an expression was obtained in [18] for the time and temperature dependence of Γ'_{hom} that combines the $T^{1.3}$ dependence on temperature with a slightly nonlinear dependence on the logarithm of the delay time. The model of Ref. 18 not only fits our dephasing data for BChl-*a* in MTHF and ethanol as a function of both temperature and delay time [16], but also fits time-dependent hole-burning data from the literature for cresyl violet in deuterated ethanol [22], and stimulated three-pulse photon echo data for zinc and magnesium porphin in ethanol [21]. A discrepancy, however, remains in the region between 10^{-5} and 10^{-1} s between the hole-burning [16] and the stimulated photon-echo data [21] which has not been solved yet [16]. It has recently been proposed that laser heating artifacts in three-pulse photon-echo experiments may account for the contradictions reported [58, 59].

9.3.2. Spectral Diffusion Induced by Energy Transfer at High Concentrations

At concentrations above 10^{-4} M, energy transfer between probe molecules may take place. This leads to a shortening of the time during which the excitation resides on individual molecules and, therefore, to an increase of Γ'_{hom} through the T_1 term, which is then equal to $\Gamma'_0 = \Gamma_0 + \Gamma_0^{ET}$, where $\Gamma_0^{ET} = (2\pi\tau_{ET})^{-1}$ is the energy transfer rate. So far, there had been no hint in the literature that energy transfer (ET) in glassy systems may trigger spectral diffusion far in excess of that expected for a given transfer time τ_{ET}. To our surprise, we have recently discovered such a process, which we call ET $\rightarrow$ SD [17, 23–25, 60]. This phenomenon has by now been demonstrated in a number of glasses, like polystyrene (PS) doped with free-base chlorin (H_2Ch) [23–25, 60], and TEA and MTHF doped with BChl-*a* [17, 61].

Let us illustrate this effect for the system H_2Ch in PS, which has been investigated as a function of concentration c, excitation wavelength λ_{exc}, temperature T, and delay time t_d. Figure 9.7 shows the dependence of Γ'_{hom} on λ_{exc} for four concentrations, at $T = 1.2$ K, superimposed on the $S_1 \leftarrow S_0$ 0–0 absorption band. At the lowest concentration, Γ'_{hom} has a small value and is independent of λ_{exc}. At higher concentrations, the data follow S-shaped curves, which become more pronounced the higher the concentration. This behavior indicates that we observe an effect caused by "downhill" energy transfer from molecules absorbing in the blue side of the absorption band to molecules in the red. The S-shaped curves can be fitted with an error function found to coincide with the normalized area of the 0–0 band at wavelengths longer than λ_{exc}. The qualitative trend of the data suggests that the effect measured involves *all* molecules absorbing to the red of λ_{exc} [23, 24, 60].

We have further found that Γ'_{hom} increases linearly with concentration for a given λ_{exc}, and that the hole shapes are Lorentzian (not shown) [23]. If Γ'_{hom} were given by direct energy transfer of the Förster type [62], it would not depend linearly on c, but on c^2, and the holes would not be Lorentzian. Furthermore, to be consistent with a reasonable Förster's radius of 60–80 Å [62], we would expect values of Γ'_{hom} significantly smaller than those measured. Thus the major part of the effect observed in Figure 9.7 is not directly caused by energy transfer, but only *indirectly*. We have proposed that Γ'_{hom} is predominantly determined by spectral diffusion triggered by energy transfer, in addition to "normal" SD [23,

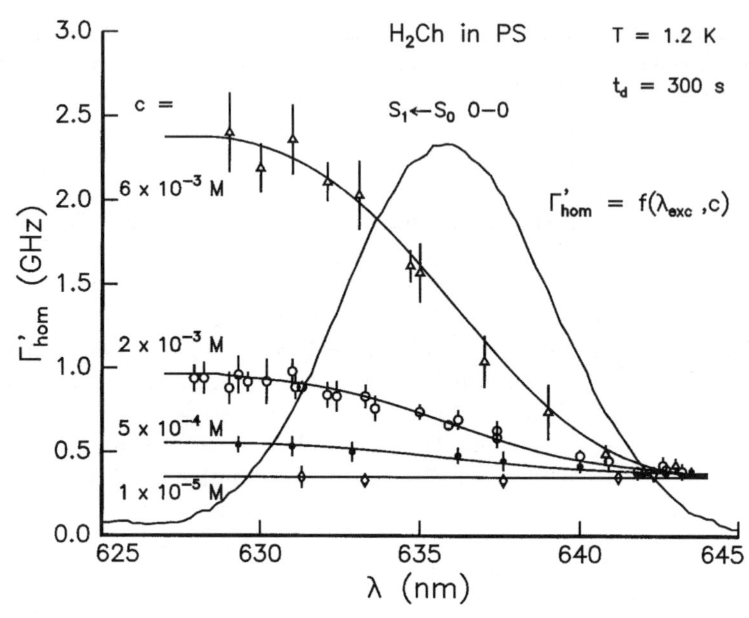

Figure 9.7. Dependence of Γ'_{hom} on excitation wavelength for H_2Ch in PS at 1.2 K, for four concentrations, together with the $S_1 \leftarrow S_0$ 0–0 absorption band. Γ'_{hom} increases towards the blue following an error function. The latter represents the area of the 0–0 absorption band to the red side of λ_{exc} [23].

24]. The essence of our idea is that during the excited-state lifetime excitation energy is transferred from one probe molecule to another absorbing further to the red. At each step, the mismatch of the excitation energies of donor and acceptor molecules ΔE will be released as "heat" into the glass. The number of transfer steps and the available energy will increase with the concentration of probe molecules absorbing at energies lower than the initial excitation wavelength (i.e., to the red of it). In this way, a large part of the glass is explored. If during this exploration the excitation finds itself close to one or more TLSs, energy can be stored in them. These activated TLSs may subsequently relax and give rise to "extra" spectral diffusion, in addition to "normal" SD, during many orders of magnitude in time. In the absence of energy transfer, that is, in the absence of exploration of the glass, there is only a small probability for the excitation to be close to a TLS. This explains why radiationless decay, which is localized on a given molecule, is ineffective in producing SD [17].

As a consequence of ET → SD, one would expect that Γ'_{hom} increases with parameters such as delay time (even at $T \to 0$), concentration, and shifts toward the blue within the 0–0 band. We have indeed confirmed these expectations.

Figures 9.8a,b show plots of $\Gamma'_{\rm hom}$ versus temperature, for various delay times. For low concentrations, the value of $\Gamma'_{\rm hom}$ increases with delay time at a given temperature, and extrapolates to $\Gamma_0 = (2\pi\tau_{\rm fl})^{-1}$, as expected for "normal" SD (compare Figs. 9.4 and 9.8a). For higher concentrations and in the blue wing of the band (see Fig. 9.8b), however, $\Gamma'_{\rm hom}$ extrapolates to $\Gamma'_0 > \Gamma_0$, with Γ'_0 increasing with delay time, as expected from our ET → SD model [23, 24].

Also Figure 9.9, which is a plot of $\Gamma'_{\rm hom}$ versus $\log(t_{\rm d})$ for three concentrations, measured in the blue wing of the band, supports the ET → SD model. The data follow $\Gamma'_{\rm hom} \propto \log(t_{\rm d})$, as for "normal" SD [16] (see Section 9.3.1), but the slope $d\Gamma'_{\rm hom}/d\log(t_{\rm d})$ increases with concentration. The results prove that the amount of "extra" SD is indeed enhanced by concentration and towards the blue part of the band. At the red onset of the band there is no "extra" SD, not even at higher concentrations (not shown). We conclude that this "extra" SD is indeed triggered by "downhill" ET within the 0–0 band.

Our results for doped glasses at high concentrations can be summarized as follows:

$$\Gamma'_{\rm hom} = \Gamma'_0(c, \lambda_{\rm exc}, t_{\rm d}) + a(c, \lambda_{\rm ex}, t_{\rm d})T^{1.3\pm0.1} \tag{9.8}$$

with

$$\Gamma'_0 = \Gamma_0 + \Gamma_0^{\rm ET}(c, \lambda_{\rm exc}) + \Gamma_0^{\rm ET\to SD}(c, \lambda_{\rm exc}, t_{\rm d}) \tag{9.9}$$

$$a = a_{\rm PD} + a_{\rm SD}(t_{\rm d}) + a_{\rm ET\to SD}(c, \lambda_{\rm exc}, t_{\rm d}) \tag{9.10}$$

In Eq. (9.9), the residual linewidth Γ'_0 is given by the sum of three terms: the fluorescence lifetime-limited value $\Gamma_0 = (2\pi\tau_{\rm fl})^{-1}$, the direct energy transfer rate given by Förster's mechanism $\Gamma_0^{\rm ET} \propto c^2$, and the residual linewidth arising from "extra" SD triggered by energy transfer $\Gamma_0^{\rm ET\to SD} \propto c\log(t_{\rm d})$ [25]. In Eq. (9.10), the coupling constant a is also given by the sum of three terms: the "pure" dephasing contribution $a_{\rm PD}$, which depends only on the guest–host coupling strength, the contribution of "normal" spectral diffusion $a_{\rm SD} \propto \log(t_{\rm d})$, and the contribution of "extra" SD induced by energy transfer $a_{\rm ET\to SD} \propto c\log(t_{\rm d})$. For the glassy system H_2Ch in PS at high concentrations, all the terms contributing to Eqs. (9.9) and (9.10) have been determined [25]. The effect of ET → SD is particularly noticeable in the glass TEA, because it does not undergo SD at low concentrations [16]. In contrast to normal glasses, an energy barrier of $\sim 100\,{\rm cm}^{-1}$ has to be crossed first in TEA in order to create TLSs that will be active in SD. To explain these results, we have proposed a qualitative molecular model [17], which still needs to be tested experimentally.

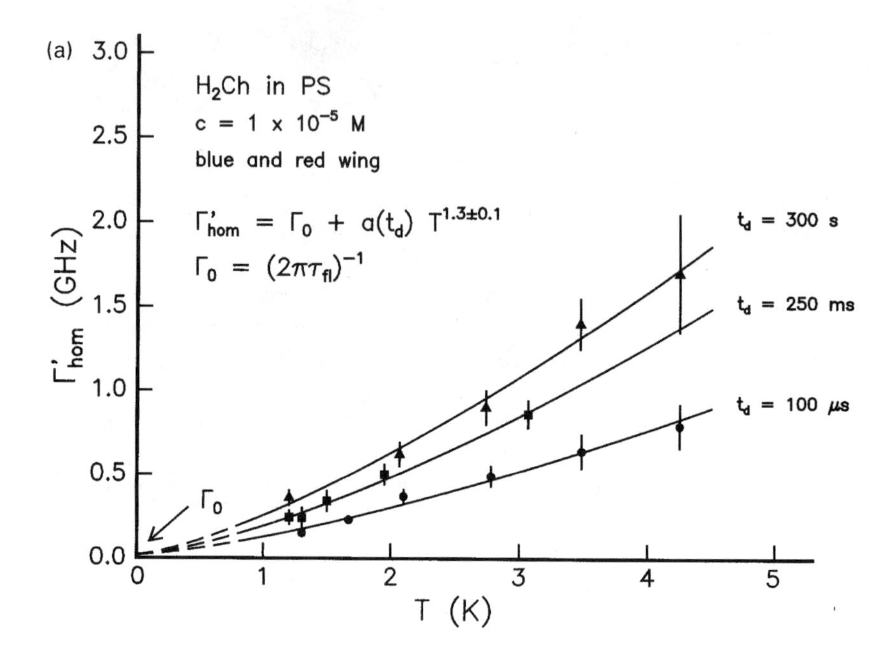
(a)
H_2Ch in PS
c = 1 x 10^{-5} M
blue and red wing
$\Gamma'_{hom} = \Gamma_0 + a(t_d)\ T^{1.3\pm0.1}$
$\Gamma_0 = (2\pi\tau_{fl})^{-1}$
t_d = 300 s
t_d = 250 ms
t_d = 100 μs
Γ_0
Γ'_{hom} (GHz)
T (K)

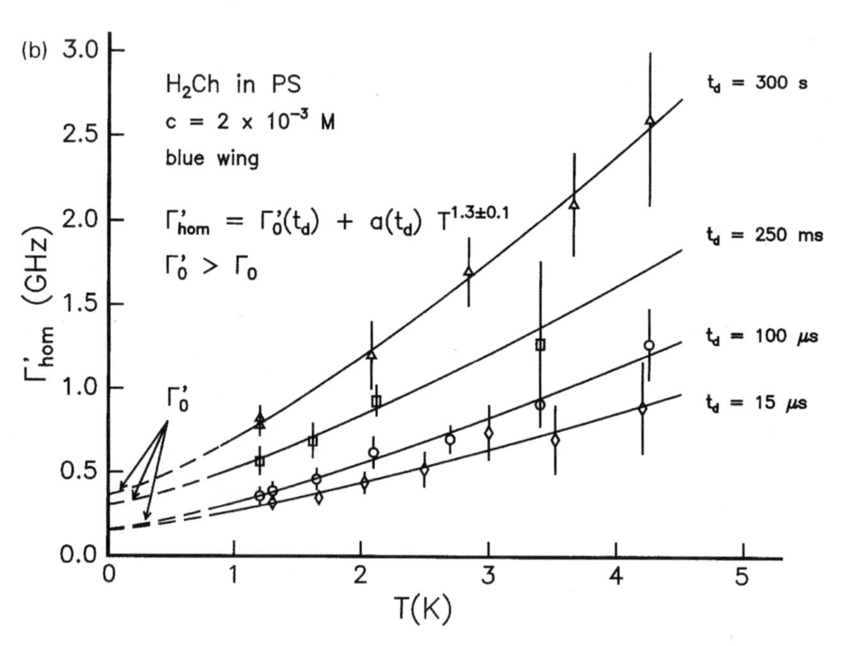
(b)
H_2Ch in PS
c = 2 x 10^{-3} M
blue wing
$\Gamma'_{hom} = \Gamma'_0(t_d) + a(t_d)\ T^{1.3\pm0.1}$
$\Gamma'_0 > \Gamma_0$
t_d = 300 s
t_d = 250 ms
t_d = 100 μs
t_d = 15 μs
Γ'_0
Γ'_{hom} (GHz)
T(K)

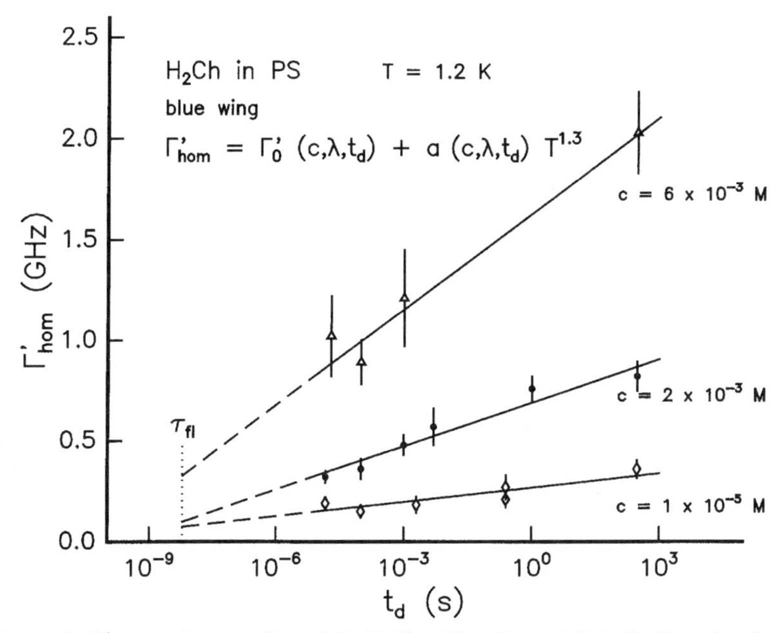

Figure 9.9. Γ'_{hom} as a function of $\log(t_d)$ for H_2Ch in PS at $T = 1.2$ K, in the blue wing, for three concentrations. The data show a linear dependence of Γ'_{hom} on $\log(t_d)$. The slope $d\Gamma'_{\text{hom}}/d\log(t_d)$, representing the amount of spectral diffusion, becomes steeper with concentration. The results prove that spectral diffusion is induced by energy transfer [25].

9.4. PROTEIN DYNAMICS

Optical dephasing and spectral diffusion in pigment–protein complexes at low temperatures is the subject of this section. As an example, we will discuss results from our laboratory dealing with the temperature- and delay-time dependence of the "effective" homogeneous linewidth Γ'_{hom} for two types of photosynthetic protein complexes: the B820-dimer and B777-monomer subunits of the light-harvesting complex LH1 of purple bacteria (Section 9.4.1) [49], and subcore complexes of the photosystem II reaction center (PSII RC) of green plants (see

---

Figure 9.8. (a) Temperature dependence of Γ'_{hom} for H_2Ch in PS at low concentration, for various delay times, in the red and blue wing of the 0–0 band. The data extrapolate to the fluorescence lifetime-limited value Γ_0 for $T \to 0$. The increase of Γ'_{hom} with delay time is caused by "normal" SD. (b) The same as for (a), for $c = 2 \times 10^{-3}$ M, in the blue wing of the 0–0 band, for four delay times. Γ'_0 increases with delay time, proving that spectral diffusion is correlated with energy transfer (ET → SD) [24].

Section 9.4.2) [47, 63]. We will show that spectral diffusion in these proteins is different from that in doped organic glasses [47, 49, 64]. The relation between optical linewidths and protein dynamics represents a challenging problem also because our results differ from those reported for Zn-substituted myoglobin and cytochrome c obtained by stimulated three-pulse photon echoes [28, 29, 65], and H_2-substituted myoglobin obtained by temperature-cycling hole burning [29–31]. Finally, we will interpret the results qualitatively in terms of distributions of relaxation rates (Section 9.4.3).

9.4.1. The Subunits B820 and B777 of the Light-Harvesting Complex LH1 of Purple Bacteria

Light-harvesting (LH) complexes in photosynthetic purple bacteria are responsible for the efficient collection of sunlight and for the transfer of excitation energy to the reaction center (RC). The primary charge separation occurs in the RC and leads to the subsequent conversion of excitation energy into a chemically useful form. Most purple bacteria contain two types of LH complexes: the LH1 core complex surrounding each RC, and peripheral LH2 complexes, which absorb slightly further to the blue and transfer the energy to LH1 [66].

Both the LH1 and LH2 complexes have ringlike structures [67,68] containing a basic repeating subunit called B820. This subunit consists of a dimer of bacteriochlorophyll-*a* (BChl-*a*) molecules bound to two helical polypeptides (α and β), each comprising ~50 amino acids [69] with a protein mass of 6 kDa. The circular arrangement of LH1 in *Rhodospirillum rubrum G9* is built up from 16 B820 subunits. It has an outer diameter of 116 Å and an inner one of 68 Å enclosing the RC [67]. The LH1 complex can be reversibly dissociated into its constituent subunits by detergent titration [70]. The detergent, *n*-octyl-β-glucopiranoside (OG), forms micelles that enclose either the B820-dimer subunit (if ~1.2% OG is used) or the B777-monomer subunit (when ~5% OG is used). Reassociation of the subunits yields an aggregate with a spectrum nearly identical to that of LH1 [71, 72]. Although the structure of the LH1 and LH2 complexes is known, there is currently much debate regarding the mechanism by which the excitation energy is transferred within the LH complexes and whether the excitations are localized or delocalized [73, 74].

In order to get an insight into the conformational changes of these proteins, and to compare their dynamics to that of glasses, we have performed time-resolved hole-burning experiments on the B820 and B777 subunits and on BChl-*a* embedded in the same OG detergent without the protein [43, 49]. The results have further cleared up a long-standing problem, namely, whether the BChl-*a* molecule in B777 is bound or not to the protein [75, 76].

The absorption spectra of the B777 monomer, the B820 dimer, and the LH1 aggregate (B873) at 1.2 K are shown in Figure 9.10. The absorption band of B820

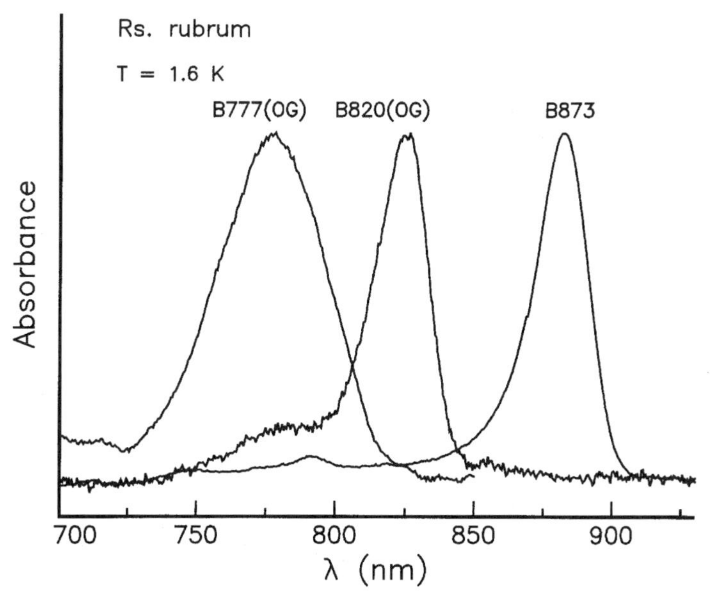

Figure 9.10. Absorption spectra of B873 (LH1) and the B820 and B777 subunits of LH1 of *Rs. rubrum* at 1.6 K. The maxima of the bands lie at ~882, 825, and 778 nm, respectively. The width of the B777 monomer band is significantly larger than that of the B820 dimer and the B873 aggregate [43].

has its maximum at 825 nm and a width of $\sim 340\,\text{cm}^{-1}$. B777 (absorption maximum at 778 nm) has a much larger width of $\sim 800\,\text{cm}^{-1}$. Narrow holes of a few 100 MHz to a few GHz, depending on temperature, were burned in B820 and B777, which proves that their absorption bands are inhomogeneously broadened [43,49]. Since the hole widths at a given T are constant over the whole band in both subunits (not shown), we concluded that there is no energy transfer within the band [77]. If "downhill" energy transfer were present, as in LH2 [78], we would expect an increase of Γ'_{hom} toward the blue side of the absorption band.

The dependence of Γ'_{hom} on temperature for B820 is shown in Figure 9.11 for various delay times. As in doped organic glasses [12,16,20], Γ'_{hom} follows a $T^{1.3\pm0.1}$ power law, indicating that two-level systems are also responsible for the dephasing in this protein [18]. The extrapolation value $\Gamma_0 = (2\pi\tau_{\text{fl}})^{-1}$ for $T \rightarrow 0$ is consistent with $\tau_{\text{fl}} = (2.3 \pm 0.1)$ ns, the fluorescence lifetime of B820 [79]. We have found similar results for B777, with a value of Γ_0 corresponding to $\tau_{\text{fl}} \approx 3$–$4$ ns (not shown) [49]. The increase of Γ'_{hom} with delay time, at a given T, is an indication that conformational changes occur in B820.

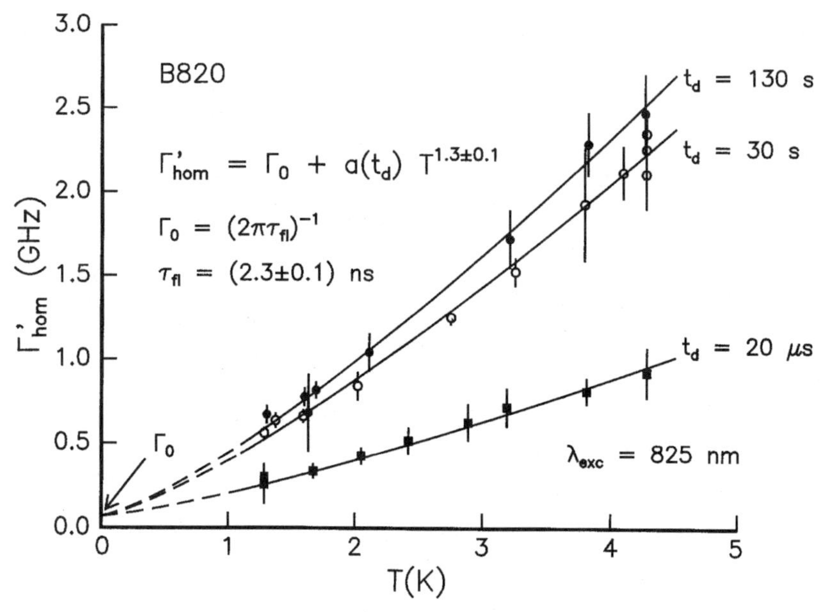

Figure 9.11. Temperature dependence of Γ'_{hom} for B820 at three delay times t_d. The curves follow a $T^{1.3}$ power law and extrapolate to the fluorescence lifetime-limited value Γ_0 for $T \to 0$. The value of Γ'_{hom} increases with t_d, indicating the presence of spectral diffusion in B820 [49].

The spectral diffusion behavior of B820 between 10^{-6} and 10^5 s is shown in Figure 9.12 in which Γ'_{hom} is plotted versus the logarithm of t_d, for two temperatures. For delay times $t_d < 100$ ms, Γ'_{hom} is constant at a given T; that is, there is no spectral diffusion, as in crystalline systems. We think, therefore, that BChl-*a* is rigidly built into the B820 protein. For $t_d \geq 100$ ms up to at least 10^5 s, $\Gamma'_{hom}/(t_d)$ increases logarithmically with t_d, as in doped glasses, with a slope $d\Gamma'_{hom}/d\log(t_d)$ proportional to the amount of spectral diffusion. The onset of SD, which occurs at $t_{d,0} \sim 100$ ms, appears to be independent of T at least for $T \leq 4.2$ K. For glasses, the onset occurs at $t_d \sim \tau_{fl}$, immediately after the excited molecules have decayed to the ground state. Since there is more SD at 4.2 K than at 1.2 K, we conclude that a larger number of TLSs is activated at higher temperatures. The T dependence of Γ'_{hom} is always $T^{1.3\pm0.1}$, at short and long delay times.

To check whether the peculiar behavior of Γ'_{hom} versus $\log(t_d)$ observed for B820 is characteristic for the protein or caused by the surrounding OG detergent, buffer, and/or glycerol, we have performed similar experiments on B777 and compared the results to those obtained for BChl-*a* embedded in a mixture of OG, buffer, and glycerol (OG-glass) without protein [49]. The results are plotted in

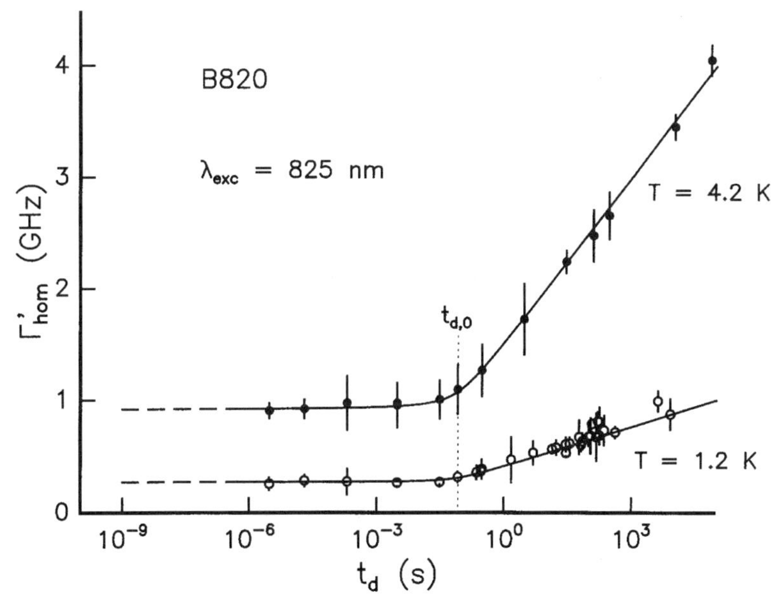

Figure 9.12. Dependence of Γ'_{hom} on delay time for B820 at 1.2 and 4.2 K. For delay times $t_d < 100\,ms$, Γ'_{hom} is constant at a given *T*, like in crystalline systems. For $t_d \geq 100\,ms$, Γ'_{hom} increases linearly with $\log(t_d)$, as for glassy systems. The amount of SD increases with temperature [49].

Figure 9.13 in a normalized form as $a(t_d) = (\Gamma'_{hom} - \Gamma_0)/T^{1.3}$ versus $\log(t_d)$. There is a fundamentally different behavior for the two samples. BChl-*a* in the OG mixture has a typical glasslike behavior, with Γ'_{hom} increasing linearly with $\log(t_d)$ over at least 10 orders of magnitude in time. The distribution of relaxation rates *P*(*R*) for BChl-*a* in OG glass is, thus, continuous and proportional to $1/R$ [18]. B777 shows qualitatively the same type of curve as B820, only the onset of SD occurs at $t_{d,0} \sim 30\,ms$. We conclude that the delay time at which SD sets in is characteristic for the protein in these subunits. The results of Figure 9.13 prove that BChl-*a* in B777 is bound to the protein and solve the long-standing controversy in the literature about this issue [75, 76].

We further wanted to know whether the spectral diffusion behavior observed in B820 and B777 is characteristic for these subunits or is more general for photosynthetic complexes. To this purpose we have performed time-resolved HB experiments on larger protein complexes of green plants. They are described in the next section [47].

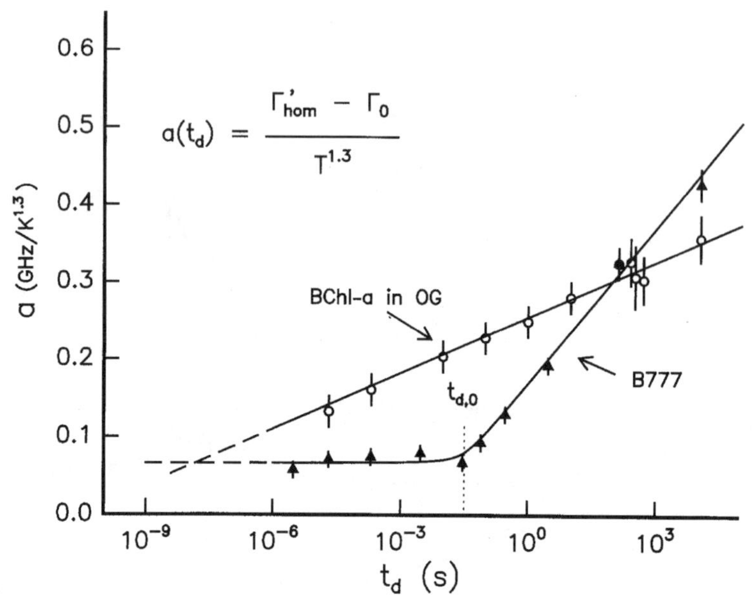

Figure 9.13. Coupling constant $a(t_d)$ as a function of $\log(t_d)$ for BChl-*a* in OG glass and B777. The data for B777 are qualitatively similar to those for B820, but the onset of SD occurs here at $t_{d,0} \approx 30\,\text{ms}$. The results demonstrate that the BChl-*a* molecules in B777 are bound to the protein [49].

9.4.2. Subcore Complexes of the Photosystem II Reaction Center of Green Plants

As mentioned above, we have also studied protein motions in photosynthetic complexes larger than the B820 and B777 subunits. They comprise three subcore complexes of the photosystem II reaction center (PSII RC) of green plants: the isolated reaction center RC, the CP47 core antenna, and the CP47 RC complexes [47, 63]

The PSII core includes the reaction center RC, the smallest unit that shows photochemical activity, and several core antenna complexes. The RC binds six chlorophyll-*a* (Chl-*a*), two pheophytins-*a* (Pheo-*a*), and one or two β-carotenes [80]. It has a protein mass of ~110 kDa. The core antenna CP47, with a mass of ~70 kDa, is a protein attached to the RC that binds 13–15 Chl-*a* molecules [81, 82] and two β-carotenes. The CP47 RC complex [83] contains 21–23 Chl-*a* molecules and has a mass of ~180 kDa.

The absorption spectra of CP47 RC, CP47, and the RC are shown in Figure 9.14, together with the spectral distributions of their respective "trap" pigments [63]. The "traps" are not involved in energy transfer, their distributions are

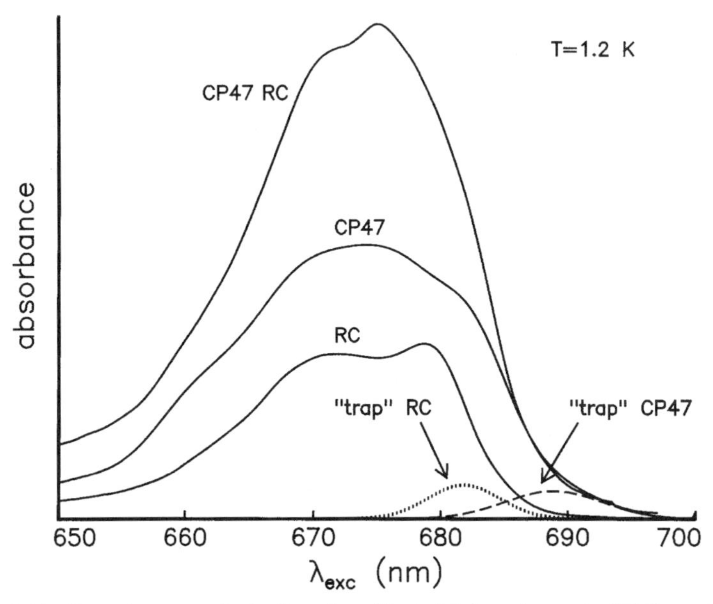

Figure 9.14. Absorption spectra in the Q_y-region of the isolated reaction center RC, the core antenna complex CP47, and the CP47 RC complex of PSII of green plants at 1.2 K. The bands are heterogeneously broadened. The spectral distributions of the RC and CP47 "trap" pigments, as determined by hole burning, are also indicated [63].

characterized by their fluorescence decay time [84]. We have performed hole-burning experiments in the spectral region of the "traps" and determined the "effective" homogeneous linewidth Γ'_{hom} as a function of temperature for various delay times, and as a function of delay time between 10^{-5} and 10^5 s (a day) for various temperatures [47]. As for the LH1 subunits and the glasses, Γ'_{hom} follows again a $T^{1.3\pm0.1}$ power law (not shown), suggesting that the low-temperature dynamics is determined by TLSs. It has recently been reported that even picosecond infrared vibrational-echo experiments on myoglobin between 60 and 180 K obey the same power law [85].

The variation of Γ'_{hom} with delay time is qualitatively similar to that of the LH1 subunits. For $t_d < 300$ ms, there is no spectral diffusion, $\Gamma'_{hom} = \text{constant}$. For $t_d \geq 300$ ms, SD sets in differently for each protein complex [47]. This is illustrated in Figure 9.15, where $a_{SD}(t_d)$ is plotted versus $\log(t_d)$:

$$a_{SD}(t_d) = \frac{\Gamma'_{hom} - \Gamma_0}{T^{1.3}} \quad a_{PD} \tag{9.11}$$

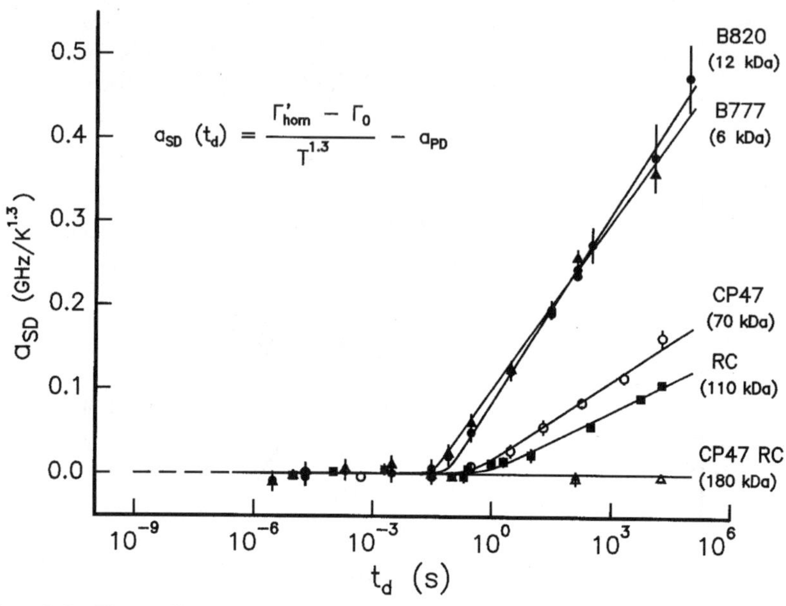

Figure 9.15. The coupling constant $a_{SD}(t_d)$ as a function of $\log(t_d)$ for the B777 and B820 subunits of LH1 of purple bacteria [49] and the RC, CP47 and CP47 RC complexes of PSII of green plants [47]. The protein masses are indicated. All complexes show a qualitatively similar behavior with respect to spectral diffusion. The onset and the amount of SD appear to be correlated with the mass or size of the protein [47].

Not only the results for the three subcore complexes of the PSII RC are shown [47], but for comparison, also those for B820 and B777 [49]. CP47 RC, the largest protein, does not undergo SD at any delay time measured. It only shows "pure" dephasing, that is, fast fluctuations during the excited-state lifetime. The RC and CP47 complexes, with smaller protein masses, do show spectral diffusion for delay times $t_d \geq 1$ s and ≥ 300 ms, respectively. The lighter B820 and B777 subunits (6 and 12 kDa) show much more SD than the heavier PSII complexes. From the onset and amount of SD, we conclude that only very slow motions, involving the whole protein or substantial parts of it, occur in these photosynthetic complexes at low temperature. All these systems appear to be rigid (crystallinelike) for $t_d \leq 10$ ms–1 s. For longer delay times, the SD behavior is similar to that in glasses. The distribution of relaxation rates $P(R)$ is very broad and continuous with $P(R) \propto 1/R$ but, contrary to glasses, it has a cutoff frequency in the region of 1–100 Hz. We further conclude that SD appears to be correlated with the mass of the protein. The lighter the protein, the greater the amount of SD [47].

The nature of the protein motions involved are not known yet. Time-resolved HB experiments on CP47 RC and larger complexes at higher temperatures are planned to verify at what temperature and delay time spectral diffusion sets in.

9.4.3. Comparison of Protein Motions in Photosynthetic Complexes and Other Systems at Low Temperature

In addition to our time-resolved hole-burning experiments [47, 49, 64], only a few studies on protein dynamics by optical techniques have been reported [28–31, 65, 85, 86]. These studies have been performed by three-pulse photon echoes [28, 29, 65], infrared photon echoes [85], and temperature-cycling hole burning [29–31, 86]. The systems investigated were Zn-mesoporphyrin IX-substituted myoglobin from 1.7 to 23 K [28, 29] and cytochrome-c at 1.8 K [65], free-base-protoporphyrin IX-substituted myglobin in the 100 mK temperature range [29–31] and between 4 and 70 K [86], and wild-type myoglobin and its mutants from 60 to 300 K [85].

In particular, the three-pulse photon-echo results on Zn-substituted myoglobin between 10^{-10} and 10^{-1} s showed a stepwise behavior of Γ'_{hom}, which was interpreted in terms of a sharp distribution of relaxation rates that shifted with temperature according to an Arrhenius law [28], and a potential energy surface was subsequently constructed [29]. The discrete rates reported are very different from the continuous distribution of rates with a cutoff at 1–100 Hz that we obtained for photosynthetic complexes. On the other hand, temperature-cycling hole-burning experiments in the 100 mK region carried out on free-base-substituted myoglobin between minutes and 240 h indicated a continuous broadening of the holes with time, a curve that was fitted with an exponential function on a logarithmic time scale [30]. It is not clear yet whether the different results arise from the nature of the proteins studied or from the differences in techniques and time scales used. Obviously more experiments are needed before these questions can be answered.

To illustrate the functional relation between a given distribution of relaxation rates $P(R)$ and the variation of Γ'_{hom} with delay time [21, 22, 28, 29, 65], we have plotted the curves shown in Figure 9.16. On the left we see $P(R)R$ versus $1/R$, whereas on the right, the corresponding Γ'_{hom} versus $\log(t_{\text{d}})$ is plotted. In the top row, we see the distribution of rates for a glass, $P(R) \propto 1/R$, leading to a linear dependence of Γ'_{hom} on $\log(t_{\text{d}})$ (see Sections 9.1.1 and 9.3.1). In the middle, a delta function represents a sharp, discrete rate, with $P(R) \propto (1/R)\,\delta(R - R_0)$. This yields a sharp step function at $t_{\text{d}} = 1/R_0$ in a Γ'_{hom} versus $\log(t_{\text{d}})$ plot. The results of Refs. 28 and 29 were interpreted in terms of several of such step functions that shifted to shorter times t_{d} with increasing temperature. In the bottom row, a log-normal function with a given distribution of rates is depicted. It leads to a smoothened step function centered around $t_{\text{d}} = 1/R_0$ in a Γ'_{hom} versus $\log(t_{\text{d}})$

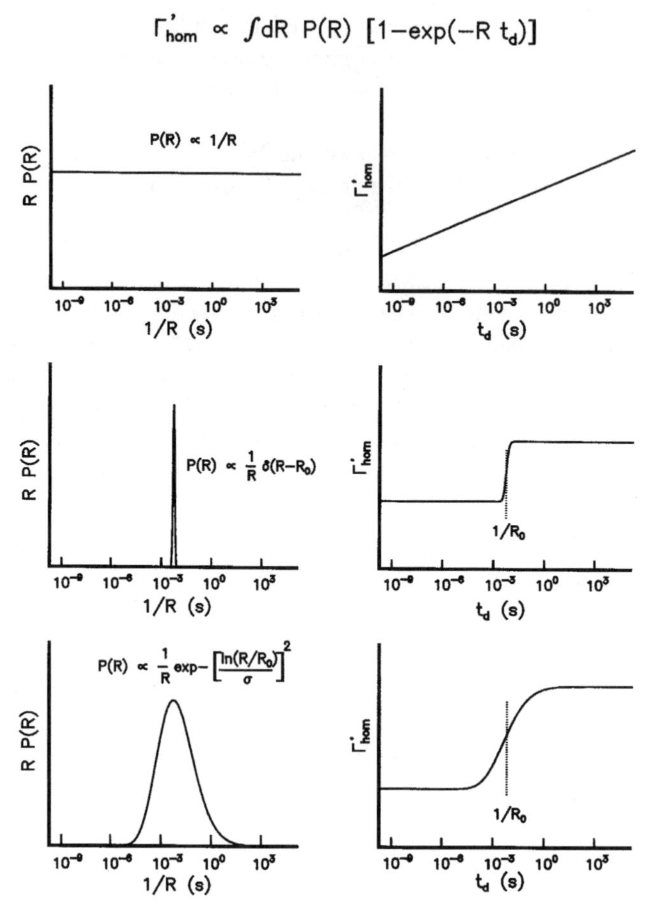

Figure 9.16. Relation between the distribution of relaxation rates $P(R)$ and the dependence of Γ'_{hom} on $\log(t_\text{d})$. Top: Rate distribution $P(R) \propto 1/R$, as found for glasses. Middle: A delta-function distribution with one rate R_0. Bottom: A log-normal distribution of rates with a maximum at R_0.

plot. Log-normal functions have previously been used to describe the relaxation rates in some glassy systems at low temperatures [22] and in Zn-substituted cytochrome-c at 2 K [65].

Our results on photosynthetic complexes cannot be explained by any of the three distributions of Figure 9.16. We have, therefore, assumed that two sequential processes play a role. The first one, at short delay times, is a crystallinelike process determined by "pure" dephasing. In this time span, there is no spectral diffusion and only fast fluctuations localized around the excited chromophore occur. After $t_\text{d} \approx 10\,\text{ms} - 1\,\text{s}$, depending on the protein complex, the second process sets in (onset of SD). For $t_d > t_{d,0}$ up to at least 24 h, this spectral diffusion process appears to be glasslike. It is in this time span that conformational rearrangements involving the whole protein take place. At present, we neither know what causes the onset of SD nor what the mechanism is that correlates the amount of SD with the protein mass.

To find out whether the cutoff in the relaxation rates is a general phenomenon or specific to the photosynthetic proteins studied here, it is important to investigate a wider class of proteins. In this way it will be possible to get an insight into the relation between the time evolution of conformational changes and their biological function. Also to understand the folding processes of proteins it is necessary to have information on the structural barriers and the organization of the energy landscape [1, 2, 27, 28, 87].

Acknowledgments

We thank J. H. van der Waals for reading of the manuscript and useful comments. The investigations were supported by the Netherlands Foundation for Physical Research (FOM) and Chemical Research (SON) with financial aid from the Netherlands Organization for Scientific Research (NWO).

REFERENCES

1. H. Frauenfelder, S. G. Sligar, and P. G. Wolynes, *Science* **254**, 1598 (1991); H. Frauenfelder and P. G. Wolynes, *Physics Today* **47**, 58 (1994).
2. A. Ansari et al., *Proc. Natl. Acad. Sci. USA* **82**, 5000 (1985).
3. P. W. Anderson, B. I. Halperin, and C. M. Varma, *Phil. Mag.* **25**, 1 (1972); W. A. Phillips, *J. Low Temp. Phys.* **7**, 351 (1972).
4. W. A. Phillips, ed., *Amorphous Solids: Low Temperature Properties*, Springer, Berlin, 1981; W. A. Phillips, *Rep. Prog. Phys.* **50**, 1657 (1987).
5. J. L. Black and B. I. Halperin, *Phys. Rev. B* **16**, 2879 (1977).
6. P. Hu and L. R. Walker, *Solid State Commun.* **24**, 813 (1977); *Phys. Rev. B* **18**, 1300 (1978).
7. R. Maynard, R. Rammal, and R. Suchail, *J. Phys. Letters* **41**, L-291 (1980).
8. D. L. Huber, *J. Lumin.* **36**, 307 (1987); W. O. Putikka and D. L. Huber, *Phys. Rev. B* **36**, 3436 (1987).
9. Y. S. Bai and M. D. Fayer, *Chem. Phys.* **128**, 135 (1988); *Phys. Rev. B* **39**, 11066 (1989).
10. A. Suarez and R. J. Silbey, *Chem. Phys. Lett.* **218**, 445 (1994).
11. J. Klauder and P. Anderson, *Phys. Rev.* **125**, 912 (1962).
12. S. Völker, in J. Fünfschilling, ed., *Relaxation Processes in Molecular Excited States*, Kluwer, Dordrecht, 1985, pp. 113–242, and references therein; *Annu. Rev. Phys. Chem.* **40**, 499 (1989), and references therein.

13. J. L. Skinner and W. E. Moerner, *J. Phys. Chem.* **100**, 13251 (1996), and references therein.
14. H. P. H. Thijssen, A. I. M. Dicker, and S. Völker, *Chem. Phys. Lett.* **92**, 7 (1982); H. P. H. Thijssen, R. van den Berg, and S. Völker, *Chem. Phys. Lett.* **97**, 295 (1983).
15. R. van den Berg, A. Visser, and S. Völker, *Chem. Phys. Lett.* **144**, 105 (1988).
16. J. M. A. Koedijk, R. Wannemacher, R. J. Silbey, and S. Völker, *J. Phys. Chem.* **100**, 19945 (1996).
17. T. M. H. Creemers, J. M. A. Koedijk, I. Y. Chan, R. J. Silbey, and S. Völker, *J. Chem. Phys.* **107**, 4797 (1997).
18. R. J. Silbey, J. M. A. Koedijk, and S. Völker, *J. Chem. Phys.* **105**, 901 (1996), and references therein.
19. W. Breinl, J. Friedrich, and D. Haarer, *Chem. Phys. Lett.* **106**, 487 (1984); J. Friedrich and D. Haarer, in I. Zschokke, ed., *Optical Spectroscopy of Glasses*, Reidel, Dordrecht, 1986, p. 149, and references therein.
20. R. Wannemacher, J. M. A. Koedijk, and S. Völker, *Chem. Phys. Lett.* **206**, 1 (1993).
21. H. C. Meijers and D. A. Wiersma, *J. Chem. Phys.* **101**, 6927 (1994).
22. K. A. Littau, M. A. Dugan, S. Chen, and M. D. Fayer, *J. Chem. Phys.* **96**, 3484 (1992), and references therein.
23. F. T. H. den Hartog, M. P. Bakker, J. M. A. Koedijk, T. M. H. Creemers, and S. Völker, *J. Lumin.* **66&67**, 1 (1996).
24. F. T. H. den Hartog, M. P. Bakker, R. J. Silbey, and S. Völker, *Chem. Phys. Lett.* **297**, 314 (1998).
25. F. T. H. den Hartog, C. van Papendracht, R. J. Silbey, and S. Völker, *J. Chem. Phys.* **110**, 1010 (1999).
26. M. H. Cohen and G. S. Grest, *Phys. Rev. Lett.* **45**, 1271 (1980); *Solid State Commun.* **39**, 143 (1981).
27. R. Elber and M. Karplus, *Science* **235**, 318 (1987).
28. D. Thorn Leeson and D. A. Wiersma, *Phys. Rev. Lett.* **74**, 2138 (1995); D. Thorn Leeson and D. A. Wiersma, *Nature Struct. Biol.* **2**, 848 (1995).
29. D. Thorn Leeson, D. A. Wiersma, K. Fritsch, and J. Friedrich, *J. Phys. Chem. B* **101**, 6331 (1997).
30. J. Gafert, H. Pschierer, and J. Friedrich, *Phys. Rev. Lett.* **74**, 3704 (1995).
31. K. Fritsch, J. Friedrich, F. Parak, and J. L. Skinner, *Proc. Natl. Acad. Sci. USA* **93**, 15141 (1996).
32. S. Völker and J. H. van der Waals, *Mol. Phys.* **32**, 1703 (1976).
33. A. A. Gorokhovskii, R. K. Kaarli, and L. A. Rebane, *J. Exp. Theor. Phys. Lett.* **20**, 216 (1974); *Opt. Commun.* **16**, 282 (1976).
34. B. M. Kharlamov, R. I. Personov, and L. A. Bykovskaya, *Opt. Commun.* **12**, 191 (1974).
35. A. Szabo, *Phys. Rev. B* **11**, 4512 (1975).
36. H. De Vries and D. A. Wiersma, *Phys. Rev. Lett.* **36**, 91 (1976).

37. S. Völker, R. M. Macfarlane, A. Z. Genack, H. P. Trommsdorff, and J. H. van der Waals, *J. Chem. Phys.* **67**, 1759 (1977).
38. S. Völker and R. M. Macfarlane, *J. Chem. Phys.* **73**, 4476 (1980).
39. J. M. Hayes and G. J. Small, *Chem. Phys.* **27**, 151 (1978).
40. R. Jankowiak and G. J. Small, *Science* **237**, 618 (1987), and references therein.
41. J. M. Hayes, R. P. Stout, and G. J. Small, *J. Chem. Phys.* **74**, 4266 (1981).
42. H. van der Laan, H. E. Smorenburg, Th. Schmidt, and S. Völker, *J. Opt. Soc. Am. B* **9**, 931 (1992).
43. T. M. H. Creemers, C. A. DeCaro, R. W. Visschers, R. van Grondelle, and S. Völker, *J. Phys. Chem B* **103**, 9770 (1999).
44. R. Wannemacher, J. M. A. Koedijk, and S. Völker, *J. Lumin.* **60&61**, 437 (1994).
45. Th. Schmidt, R. M. Macfarlane, and S. Völker, *Phys. Rev. B* **50**, 15707 (1994).
46. R. M. Shelby and R. M. Macfarlane, *Chem. Phys. Lett.* **64**, 545 (1979).
47. F. T. H. den Hartog, C. van Papendrecht, U. Störkel, and S. Völker, *J. Phys. Chem. B* **103**, 1375 (1999).
48. U. Störkel et al., to be published.
49. T. M. H. Creemers, U. Störkel, S. Musa, R. W. Visschers, R. J. Silbey, and S. Völker, to be published.
50. J. S. Connolly, E. B. Samuel, and A. F. Janzen, *Photochem. Photobiol.* **36**, 565 (1982).
51. H. Fidder, S. de Boer, and D. A. Wiersma, *Chem. Phys.* **139**, 317 (1989).
52. D. W. Pack, L. R. Narasimhan, and M. D. Fayer, *J. Chem. Phys.* **92**, 4125 (1990), and references therein.
53. R. van den Berg and S. Völker, *Chem. Phys. Lett.* **127**, 525 (1986); *Chem. Phys. Lett.* **137**, 201 (1987).
54. *Optical Linewidths in Glasses*, special issue of *J. Lumin.* **36**, 179–329 (1987), and references therein.
55. R. Jankowiak and G. J. Small, *Chem. Phys. Lett.* **207**, 436 (1993), and references therein; *Phys. Rev. B* **47**, 14805 (1993).
56. Th. Schmidt, J. Baak, D. A. van der Straat, H. B. Brom, and S. Völker, *Phys. Rev. Lett.* **71**, 3031 (1993).
57. A. Heuer and R. J. Silbey, *Phys. Rev. Lett.* **70**, 3911 (1993); *Phys. Rev. B* **49**, 1441 (1994); *Phys. Rev. B.* **53**, 609 (1996).
58. P. Neu, R. J. Silbey, S. J. Zilker, and D. Haarer, *Phys. Rev. B* **56**, 11571 (1997).
59. S. J. Zilker and D. Haarer, *Chem. Phys.* **220**, 167 (1997).
60. J. M. A. Koedijk, T. M. H. Creemers, F. T. H. den Hartog, M. P. Bakker, and S. Völker, *J. Lumin.* **64**, 55 (1995).
61. A. J. Lock, T. M. H. Creemers, and S. Völker, *J. Chem. Phys.* **110**, 7467 (1999).
62. Th. Förster, *Ann. Phys. (Leipzig)* **2**, 55 (1948); Th. Förster, in O. Sinanoglu, ed., *Modern Quantum Chemistry*, Part III, Academic Press, New York, 1965, p. 93.

63. F. T. H. den Hartog, J. P. Dekker, R. van Grondelle, and S. Völker, *J. Phys. Chem. B* **102**, 11007 (1998).
64. U. Störkel, T. M. H. Creemers, F. T. H. den Hartog, and S. Völker, *J. Lumin.* **76&77**, 327 (1998).
65. D. Thorn Leeson and D. A. Wiersma, *J. Phys. Chem.* **98**, 3913 (1994).
66. R. van Grondelle, J. P. Dekker, T. Gillbro, and V. Sundstrom, *Biochim. Biophys. Acta* **1187**, 1 (1994); G. R. Fleming and R. van Grondelle, *Phys. Today* **47**, 48 (1994).
67. S. Karrasch, P. A. Bullough, and R. Ghosh, *EMBO J.* **14**, 631 (1995).
68. G. McDermott, S. M. Prince, A. A. Freer, A. M. Hawthornwaite–Lawless, M. Z. Papiz, R. J. Cogdell, and N. W. Isaacs, *Nature* **374**, 517 (1995).
69. J. F. Miller et al., *Biochem.* **26**, 3033 (1987).
70. R. W. Visschers et al., *Biochem.* **30**, 5734 (1991).
71. P. S. Parkes-Loach, J. R. Sprinkle, and P. A. Loach, *Biochem.* **27**, 2718 (1988).
72. R. Ghosh, H. Hauser, and R. Bachofen, *Biochem.* **27**, 1004 (1988).
73. R. Kumble, S. Palese, R. W. Visschers, P. L. Dutton, and R. M. Hochstrasser, *Chem. Phys. Lett.* **261**, 396 (1996).
74. J.-Y. Yu, Y. Nagasawa, R. van Grondelle, and G. R. Fleming, *Chem. Phys. Lett.* **280**, 404 (1997).
75. P. S. Parkes-Loach, T. J. Michalski, W. J. Dacs, U. Smith, and P. A. Loach, *Biochem.* **29**, 2951 (1990).
76. J. N. Sturgis and B. Robert, *J. Mol. Biol.* **238**, 445 (1994).
77. C. A. De Caro, R. W. Visschers, R. van Grondelle, and S. Völker, *J. Lumin.* **58**, 149 (1994).
78. C. De Caro, R. W. Visschers, R. van Grondelle, and S. Völker, *J. Phys. Chem.* **98**, 10584 (1994), and references therein.
79. V. Helenius, R. Monshouwer, and R. van Grondelle, *J. Phys. Chem. B* **101**, 10554 (1997).
80. C. Eijckelhoff and J. P. Dekker, *Biochim. Biophys. Acta* **1231**, 21 (1995).
81. S. L. S. Kwa et al., in N. Murata, ed., *Research in Photosynthesis*, Vol. I, Kluwer, Dordrecht, 1992, p. 263.
82. H.-C. Chang et al., *J. Phys. Chem.* **98**, 7717 (1994).
83. C. Eijckelhoff, J. P. Dekker, and E. J. Boekema, *Biochim. Biophys. Acta* **1321**, 10 (1997).
84. M. L. Groot, J. P. Dekker, R. van Grondelle, F. T. H. den Hartog, and S. Völker, *J. Phys. Chem.* **100**, 11488 (1996).
85. C. W. Rella et al., *Phys. Rev. Lett.* **77**, 1648 (1996); K. D. Rector et al., *J. Phys. Chem. A* **103**, 2381 (1999).
86. Y. Shibata, A. Kurita, and T. Kushida, *J. Chem. Phys.* **104**, 4396 (1996); *Biochemistry* **38**, 1789 (1999).
87. H. Frauenfelder, *Nature Struct. Biol.* **2**, 821 (1995); H. Frauenfelder and B. McMahon, *Proc. Natl. Acad. Sci. USA* **95**, 4795 (1998).

CHAPTER

10

COUPLING OF FLUORESCENCE LINE-NARROWING SPECTROSCOPY AND LIQUID SEPARATION TECHNIQUES

CEES GOOIJER AND STEVEN J. KOK*

Analytical Chemistry and Applied Spectroscopy, Vrije Universiteit, De Boelelaan 1083, 1081 HV Amsterdam, The Netherlands

10.1. INTRODUCTION

As detailed elsewhere in this volume, high-resolution luminescence techniques such as laser- and lamp-excited Shpol'skii fluorescence (and phosphorescence) and fluorescence line-narrowing (FLN) spectroscopy have a distinct potential for analyte identification and/or characterization at low concentration levels in complex samples. Especially laser-excited Shpol'skii spectroscopy (LESS) combines an extremely high sensitivity with an extremely high selectivity since the (highly monochromatic) laser radiation can be tuned to the sharp vibronic transitions in the S_1-S_0 absorption region and, furthermore, used for site-selective excitation. LESS has been successfully involved as a standalone technique, for example, for the analysis of a target compound like the highly carcinogenic dibenzo[*a*,*l*]pyrene in complex environmental samples [1]. Its main disadvantage is a limited applicability range since the analytes concerned should be compatible with the *n*-alkane solvent, though significant improvements have been achieved by enhancing the solidification rate of the solution [2].

In FLN spectroscopy the line-narrowing effect is based on selective excitation of analyte molecules out of a broad S_1-S_0 absorption band. In contrast to Shpol'skii spectroscopy, where the sharp lines are only observed for certain solute–solvent combinations, the solute–solvent compatibility is less critical, though—ideally—the phonon coupling should be weak and the solution should form a clear glass upon freezing to cryogenic temperatures. This freedom in solvent choice prompted some research groups to explore the possibility of 'hyphenation' between FLN spectroscopy and liquid separation techniques

**Current Address: Organon, Div. Pharmaceutics, P.O. Box 20, 5340 BH Oss, The Netherlands*

Shpol'skii Spectroscopy and Other Site-Selection Methods, Edited by Cees Gooijer, Freek Ariese, and Johannes W. Hofstraat.
ISBN 0-471-24508-9

wherein the solvent composition is determined by the operative separation conditions. The term *hyphenation* implies that the analytical characteristics of the coupled system are improved in comparison to the off-line combination of the techniques involved.

Interesting developments concerning column liquid chromatography (LC), thin-layer chromatography (TLC), and capillary electrophoresis (CE) will be discussed below. Of course, the hyphenation to FLN spectroscopy should not be treated independently from other spectrometric techniques. In the analytical chemistry world, wherein FLN spectroscopy is still generally considered as rather exotic, the FLN-coupled techniques will only be involved if, in relevant problems, other techniques like LC–mass spectrometry (MS) and CE–MS do not provide adequate information.

For this reason, the coupling to FLN spectroscopy will be discussed within the general context of hyphenated techniques. Emphasis will be on LC since, for this separation method, hyphenation to various spectrometric techniques is in a mature state.

10.2. HYPHENATION OF LC AND SPECTROMETRIC DETECTION

Column liquid chromatography is a powerful tool for the separation of constituents of complex mixtures and can handle a wide range of sample types and analyte classes. Because of its versatility, LC finds extensive application in divergent fields of interest such as environmental, biomedical, pharmaceutical, and polymer chemistry [3]. LC analysis is usually combined with on-line UV absorption detection. Since most organic compounds absorb radiation in the (deep) UV region of the spectrum, this detection principle has nearly universal applicability but rather limited selectivity. Other well-known, and more selective, methods are on-line fluorescence and electrochemical detection. These three detection methods are very useful for the reliable quantification of LC-separated compounds; however, they are limited as far as the identification of individual components is concerned.

Spectrometric techniques like MS or Fourier-transform infrared spectrometry (FT–IR) are a rich source of qualitative information from which analyte identity may be inferred. The usefulness of these techniques including Raman and FLN spectroscopy as a standalone method in the direct analysis of mixtures is, however, often rather limited because the spectra of (even simple) mixtures frequently show severe overlap. In other words, mixture constituents have to be separated prior to spectrometric detection in order to allow their identification with a reasonable degree of certainty. This is the main impetus for the development of hyphenated systems that combine efficient analyte separation with specific detection [4–6].

Traditionally, spectrometric analysis of an LC eluate was performed off-line via collection of fractions [7]. This approach allows that the chromatographic and spectroscopic parts of the system be optimized independently. However, off-line methods generally are tedious and time consuming, especially when the entire chromatogram and, thus, a large number of fractions, have to be analyzed. Besides, maintaining chromatographic resolution may be problematic and off-line procedures are susceptible to sample contamination and analyte loss. Therefore, spectrometric detection in LC is preferably carried out in an on-line fashion using an appropriate interface that directly couples the chromatograph to the spectrometer. In this way, the column effluent is continuously monitored, and spectral data are acquired on the fly on a, typically, 0.1–1-sec time scale.

On-line coupling of LC and a spectrometric technique is not always straightforward. The conditions required for proper detection are often not compatible with the physical state of the analytes as they are eluted from the column. The volumetric flow rate and/or the composition of the eluent may well complicate the coupling. Moreover, as the spectral acquisition time available under flow conditions is limited, spectrometric detection has to be fast, which frequently is not possible without serious loss of spectral resolution and/or sensitivity. In order to bridge the gap between the techniques involved, on-line coupling almost invariably requires a sophisticated interface.

For LC–MS this problem has been solved. Its present stage is that of a mature, routinely used on-line technique based on dedicated interfaces [8]. Unfortunately, for FT–IR spectroscopy, FLN spectroscopy, and Raman spectroscopy, the design of an on-line combination is problematic, although not impossible:

1. The on-line coupling of LC and FT–IR using a flow cell to continuously record the IR transmission of the effluent is seriously hampered by the fact that most common LC solvents have intense absorption bands in the mid-IR region. This necessitates the use of flow cells with a very small optical path length, which strongly limits the sensitivity of on-line FT–IR detection. Improving the signal-to-noise ratio by signal averaging is not possible because of the short residence time of the analytes in the flow cell. Furthermore, since the absorption bands of the eluent take up wide regions of the spectrum, the spectral information that can be obtained is restricted.
2. Fluorescence line-narrowing spectroscopy requires cryogenic temperatures (< 20 K) to provide vibrationally resolved emission spectra. These spectra are highly analyte specific and therefore well suited for identification purposes. Obviously, on-line FLN detection is not a realistic option, but has been accomplished in a delayed mode (see below).
3. Considering Raman spectroscopy, because of the inherently low efficiency of the Raman scatter process, its usefulness as a detection technique in LC

is limited, unless long-pathway detector cells can be used. One has to make use of resonance and/or surface-enhancement effects. Until now, only few studies on the on-line LC–resonance Raman combination have been reported [9, 10]. Surface-enhanced Raman (SER) spectroscopy seems to open the widest perspectives: It can be observed upon laser excitation of analytes adsorbed on (rough) metallic surfaces, usually a colloid silver solution (silver sol). On-line LC–SER studies have been published [11, 12]. Problems to deal with are influences of LC eluent conditions (presence of organic modifiers, buffers, and other additives) on silver sol activity, as well as clogging and memory effects caused by deposition of silver on tubing and flow cell walls (although the latter problem can be overcome by using a windowless flow cell).

10.3. SEMI-ON-LINE COUPLING OF LC AND SPECTROSCOPIC TECHNIQUES

The above considerations forced the research group of N. H. Velthorst, in close cooperation with U. A. Th. Brinkman, to develop a semi-on-line coupling of LC and spectroscopic detection techniques. A schematic drawing of the instrumental setup (which will be described in more detail below) is presented in Figure 10.1 [13]. In this indirect approach, the eluent is evaporated and the LC-separated compounds are immobilized on the substrate prior to the collection of spectral data. The immobilization of the chromatogram is accomplished by using an

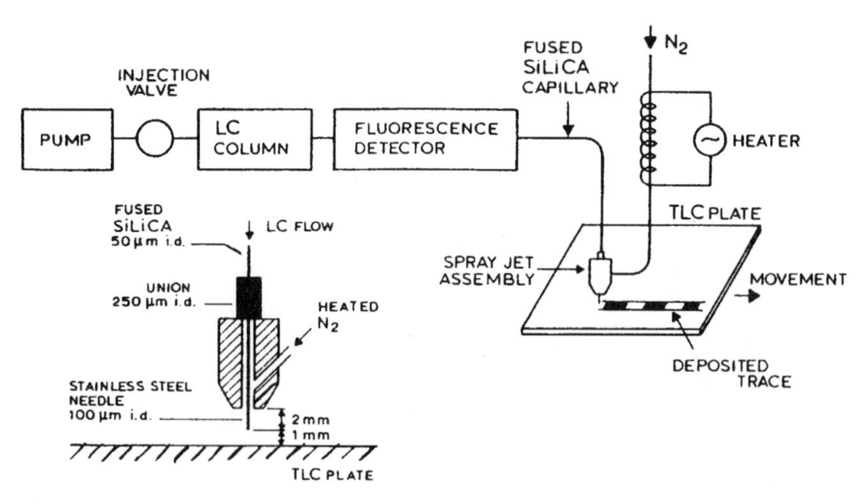

Figure 10.1. Schematic of the LC–TLC system and the coupling interface. (From Ref. 13 with permission.)

interface that evaporates the eluent and continuously deposits the LC column effluent on the moving substrate. Furthermore, since the chromatogram is stored as a continuous trace, subsequent analysis can be performed without any time constraints. This means that, if detection is nondestructive, analyte spots can be examined repeatedly. For instance, after rapid screening of the complete deposited trace under low-performance conditions, optimal spectra can then be recorded for a few interesting parts of the chromatogram. The deposited analytes may also be subjected to several spectrometric detection procedures in order to enhance the identification potential. Spectrometric analysis of the stored chromatogram can, in principle, be carried out at any convenient time or place, which may be helpful when suitable facilities for spectrometric measurements are limited and spectrometers have to be shared.

Principal parts of a semi-on-line coupled system are the solvent-elimination interface and the substrate. The interface should adequately effect the evaporation of the eluent and maintain the chromatographic resolution obtained during the deposition process. In this respect, as with on-line coupling, the column effluent flow rate, the composition, and the nature of the analytes are important factors. Small volumes of a volatile solvent may be readily evaporated by a stream of nitrogen over the substrate. Rapid elimination of aqueous eluents, on the other hand, requires a sophisticated interface with enhanced solvent elimination power (e.g., a pneumatic nebulizer). Elimination of the eluent may also be hampered by the presence of nonvolatile additives such as buffer salts and ion-pairing reagents. The analytes, of course, must be considerably less volatile than the eluent in order to achieve their deposition. Since LC is particularly used for nonvolatiles, this condition is generally met. The choice of the substrate is important as well. It should not be affected by either the eluent or the deposited compounds, and should be compatible with the spectrometric technique selected, without introducing additional interferences. For example, thin-layer chromatographic plates are excellent substrates for fluorescence experiments, but are less suited for IR studies because of the background adsorption of the TLC sorbent. Furthermore, the physico-chemical characteristics of the substrate, such as wettability and packing density, may influence the immobilization efficiency; residual eluent easily spreads over a substrate with a hard and smooth surface (e.g., a KBr window), while it may be effectively sorbed by a TLC plate.

To date, the semi-on-line LC–FTIR coupling has been fully evaluated. A spray-jet assembly, as depicted in Figure 10.1, has been developed to eliminate the solvent (based on a heated nitrogen flow) and to deposit the effluent on the substrate zinc selenide (instead of a TLC plate) [14]. Thus FT–IR transmission spectra can be recorded, which can be directly used in a library search [15]. Limitations are the flow rate which (in the case of aqueous methanol) should not exceed 30 μl/min while the water fraction should be lower than 30 vol %. These eluent conditions imply that narrow-bore LC columns (internal diameter typically

1.0 mm) have to be used. Another constraint is that buffer salts in the eluent cannot be applied since they interfere both with the deposition and the detection. A significant extension of the applicability range of the semi-on-line LC–FTIR system has been achieved by using an on-line post-column extraction module in conjunction with the solvent elimination interface [16, 17]. It consisted of a liquid-phase segmentor, an extraction coil, and a phase separator while dichloromethane (appropriate because of its favorable volatility) was used as extraction solvent. Thus both high water contents as well as nonvolatile buffer salts can be applied for the LC separation, while flow rates of 0.2 ml/min and columns of 2.1 mm internal diameter can be handled.

Compared to the LC–FTIR combination, until now, little attention has been paid to the semi-on-line coupling with SER and FLN spectroscopy [18, 19]. Nonetheless, it has been shown that the same (spray-jet) interface as in LC–FTIR can be applied. The substrate of choice in both SER and FLN is a TLC plate (instead of ZnSe, the preferred substrate in LC–FTIR). In the following, attention will be focused on recent achievements obtained for semi-on-line LC–FLN spectroscopy, wherein, of course, the choice of the TLC material is of main importance; the technique will be denoted as LC–TLC–FLN. Besides it is noted that exploratory experiments of FLN for on-column detection in LC have been reported as well [20]. Coupling of capillary electrophoresis to FLN is discussed in Section 10.5.

10.4. LC–TLC-FLUORESCENCE

10.4.1. General

The feasibility of the on-line coupling of LC and TLC has been demonstrated by several research groups [21–24] whose primary goal was to enhance the chromatographic separation efficiency. LC–TLC is a two-dimensional technique in which the deposited LC trace serves as the starting point for a second separation by means of TLC. When combining, for example, a C-18 modified silica LC column and a bare silica TLC plate, the chromatographic selectivity can be quite high. Van de Nesse et al. [13] followed this approach to identify polycyclic aromatic compounds in a marine sediment sample. Isocratic LC was not sufficient to separate all PAHs present in the sample. Therefore, the LC chromatogram was deposited on a 30% acetylated cellulose plate, which was subsequently developed in a perpendicular direction; this enabled complete resolution of the overlapping LC peaks (Fig. 10.2). All PAHs were identified with a conventional fluorescence spectrometer on the basis of fluorescence excitation/emission (FEE) spectra recorded from the spots. The identification limits were in the low-picogram range.

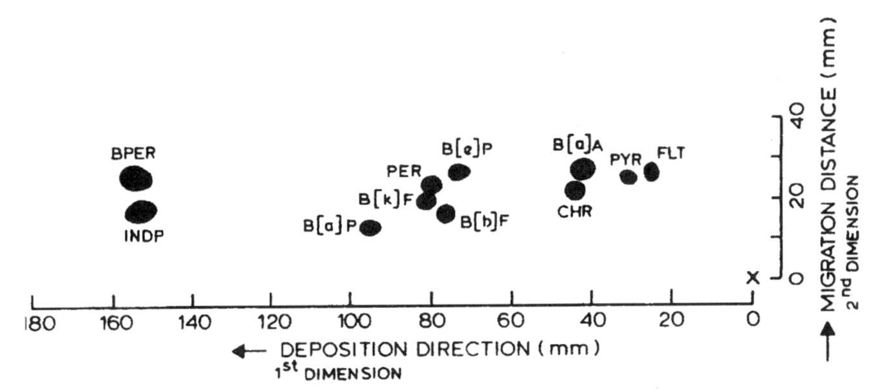

Figure 10.2. Two-dimensional LC–TLC of a marine sediment sample. First dimension: isocratic reversed-phase LC separation (methanol–water, 90 : 10, v/v; injected amount, 10 μl); LC effluent deposited on a 30% acetylated cellulose TLC plate; translation table speed 3.20 mm min^{-1}. Second dimension: TLC perpendicular to the deposition trace with methanol–diethyl ether–water (6 : 4 : 1, v/v/v). The cross marks the start of the translation table 8 min after injection. Abbreviations: PYR, pyrene; FLT, fluoranthene; B[*a*]A, benz[*a*]anthracene; B[*e*]P, benzo[*e*]pyrene; CHR, chrysene; PER, perylene; B[*b*]F, benzo[*b*]fluoranthene; B[*k*]F, benzo[*k*]fluoranthene; INDP, indeno[1,2,3-*cd*]pyrene; and BPER, benzo[*ghi*]perylene. (From Ref. 13 with permission.)

Of course the spectral selectivity of FEE will be insufficient for structurally highly similar fluorescent analytes. For these types of analytes, the FLN technique has to be involved: In principle it provides detailed information about vibrational frequencies (not only in the electronic ground state but also in the S_1 state) FLN spectroscopy requires only tiny amounts of analytes, levels that cannot be easily handled by FTIR and Raman spectroscopy. Earlier, it had been shown that FLN spectra can be obtained from analytes on TLC plates [25, 26], though PEI-cellulose plates, as used in the ^{32}P autoradiography post-labeling technique for DNA adducts, are less appropriate [27].

10.4.2. LC–TLC–FLN of Chlorinated Pyrenes

To explore the potential of LC–TLC–FLN, a sample of 1-chloropyrene containing several isomeric impurities was analyzed [28]. The same sample was also studied by LC–ZnSe–FTIR [29].

The separations were performed on a 150 × 1.1 mm I.D. column packed with 5-μm C-18 silica with methanol–water (95 : 5, v/v) as the eluent at a flow rate of 30 μl/min. A conventional fluorescence detector was inserted to enable a comparison of the chromatogram recorded on the fly and the immobilized chromatogram obtained after deposition (see Fig. 10.1). Such a comparison directly reveals whether the chromatographic integrity is conserved during the deposition process or that significant broadening of the chromatographic peaks reduces the separation efficiency. To prevent extra band broadening by the volume of the interconnecting parts of the system, the LC effluent was led to the spray jet assembly through a 40 cm × 50 μm I.D. fused-silica capillary. The capillary was connected by a 250 μm I.D. (1/16 in. O.D.) union to a 100 μm I.D. (475 μm O.D.) stainless-steel syringe needle with a conically shaped tip. One side of the union was glued to the needle; the other end was attached to the capillary by a fingertight connection. A nitrogen flow around the tip of the needle heated to about 130°C was applied to enhance evaporation of the eluent. During deposition, the TLC plate was moved by a Camag (Muttenz, Switzerland) Linomatt III translation table at a speed of 1.62 mm/min. The immobilized chromatogram was scanned with a densitometer in the fluorescence mode.

For the measurement of FLN spectra, parts of the TLC plate with typical dimensions of 20 × 10 mm, containing the deposited trace, were cut out and placed in the cryogenic sample holder, which consisted of two round sapphire glass windows fitted in a 2.5-cm-I.D. gilded copper ring. Indium wire was used as a spacer gasket between the windows. For thermal conduction, the TLC plate was immersed in glycerol. The sample holder was mounted on a closed-cycle helium refrigerator and cooled to 10 K. Excitation was achieved with a dye laser that was pumped by a XeCl excimer laser operating at 308 nm; the butyl-PBD dye covered a spectral range of 359–386 nm. The laser power output was strongly attenuated by neutral-density filters to achieve pulse energies of 1 – 10 μJ. Most spectra were measured at a repetition rate of 50 Hz. The TLC plate was illuminated at a 30° angle, and fluorescence was dispersed by a 1-m monochromator with a linear dispersion of 8 Å/mm. For detection, a blue-enhanced intensified linear photodiode array was used. Signals were processed by an optical multichannel analyzer. The spectral range viewed by the array (25 mm wide; 1024 diodes) was 20 nm. Time resolution was achieved with a pulse generator at a fixed gate of 40 ns; the pulser was triggered by a photodiode that detected a small portion of the excitation light. To compensate for the 30-ns intrinsic delay of the pulser, an optical delay line was created by retarding the excitation light pulse through a 30-m optical fiber before reaching the sample. The illuminated spot size was about 25 mm^2.

Figure 10.3 shows the LC chromatograms of 20 mg/l impure 1-chloropyrene (injection volume 5 μl) recorded on the fly (A) and after deposition (B) [28]. Apart from peak 1, which is substantially broadened, it is obvious that the chromatographic integrity is reasonably well preserved in the stored chromatogram (differences in peak heights are caused by different excitation/emission settings). The broadening of peak 1 is attributed to the fact that, in the present setup, the TLC plate material and the LC column packing material are the same (i.e., both are C-18-modified silica). As a result the analyte spot (peak 1) migrates after deposition on the still-wet TLC plate, causing additional band broadening.

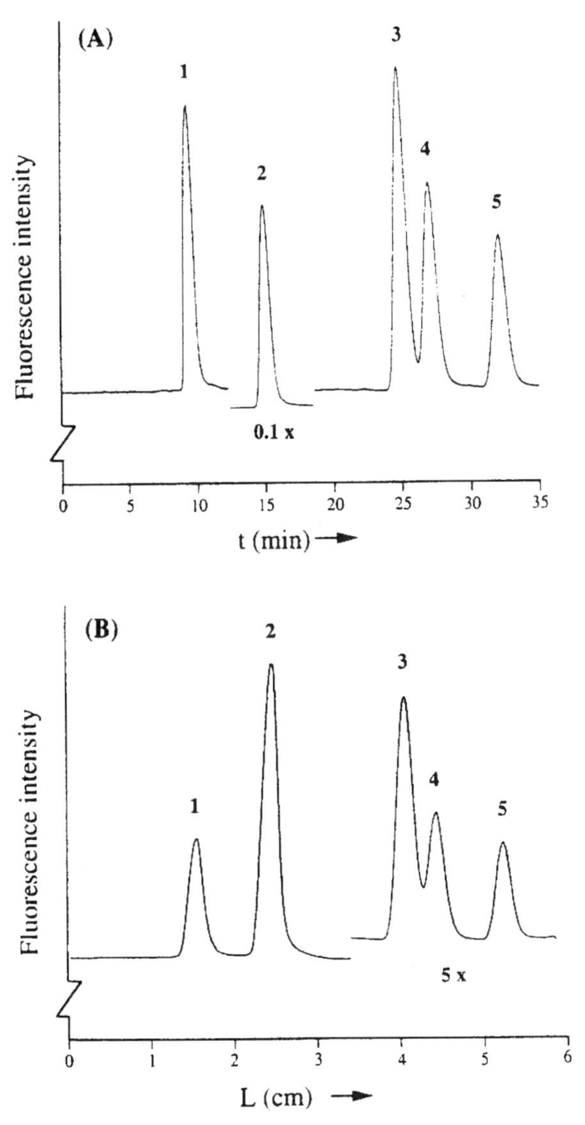

Figure 10.3. LC chromatograms of 20 mg/l of impure 1-chloropyrene. (A) On-line fluorescence detection. Excitation envelope: 255–370 nm with a maximum at 320 nm. Emission wavelength: >380 nm. (B) Off-line fluorescence detection with densitometry after deposition on a C-18-modified silica HPTLC plate. Translation table speed: 1.62 mm/min. Excitation wavelength: 313 nm. Emission wavelength: >390 nm. Spatial resolution: 0.14×2.0 mm. (From Ref. 28 with permission.)

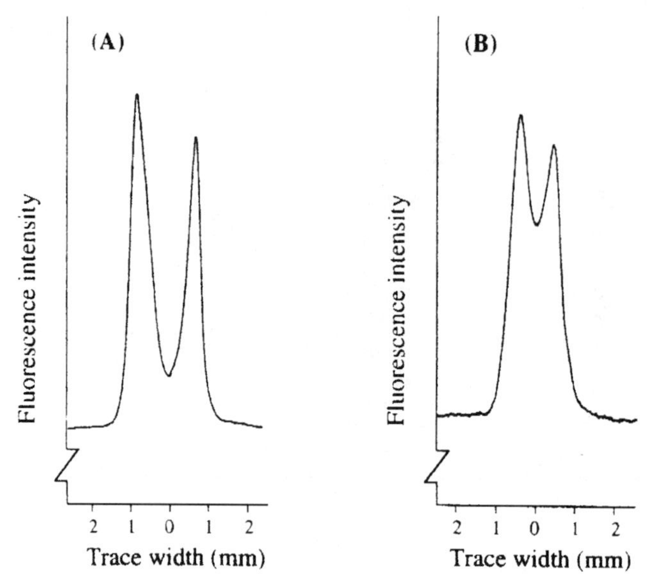

Figure 10.4. Densitometric scans perpendicular to the deposited trace of (A) TLC spot 1 and (B) TLC spot 5 of the LC chromatogram of Figure 10.3B. Excitation wavelength: 313 nm. Emission wavelength: >390 nm. Spatial resolution: 0.05×2.0 mm. (From Ref. 28 with permission.)

This is also obvious from Figure 10.4, showing densitometric scans perpendicular to the deposited trace for spots 1 and 5. The bifurcated peak shape in this figure is the result of migration of the effluent from the center of the trace toward both sides. The first eluting component has a higher R_f value on the TLC plate than the last eluting one.

In regard to identification of the five observed spots, spots 3, 4, and 5 are of particular relevance since they underline the identification potential of FLN. For these compounds the FEE technique is not appropriate since the room-temperature excitation and emission spectra are identical indeed. To exploit fully the identification power of the FLN technique, it is of main importance to establish the proper excitation conditions. This is illustrated in Figure 10.5 showing spectra measured with excitation at 361.00 nm (A) and 367.50 nm (B), respectively [28]. The FLN spectra measured for spots 3 and 5 with 361.00 nm excitation are rather similar. The 0–0 lines occur at the same wavelength (382.4 nm), and although there are slight differences in the vibronic part of the spectrum, it is relatively difficult to distinguish between the two compounds since the vibronic bands are much less intense than the 0–0 line. However, when 367.50 nm excitation is used, assignment of the spots is more facile. The FLN spectra of the 380–385 nm region of spots 3, 4, and 5 show that

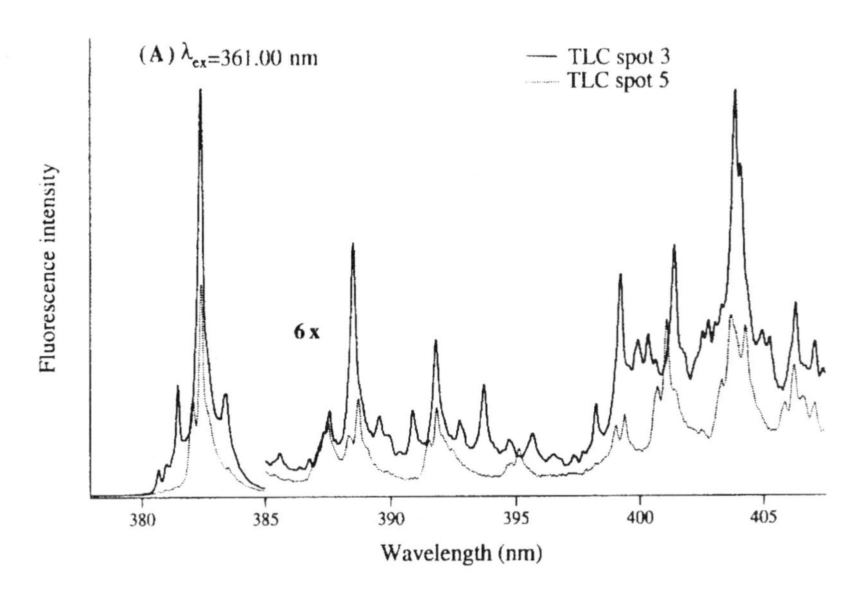

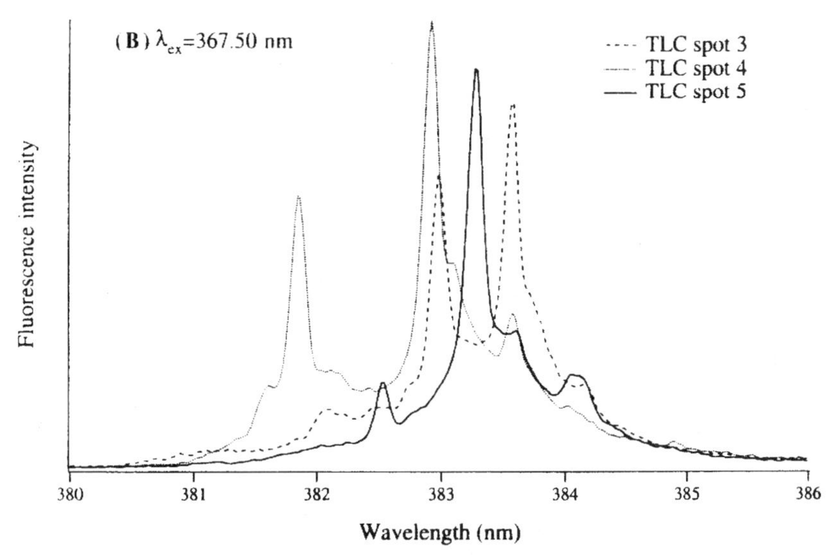

Figure 10.5. FLN spectra of the TLC spots of the LC chromatogram of Figure 10.3B (A) Spots 3 and 5 measured with excitation at 361.00 nm. (B) Spots 3, 4, and 5 using 367.50-nm light for excitation. (From Ref. 28 with permission.)

the positions of the lines, which are all due to vibronically excited 0–0 transitions, are clearly different. Comparison with the FLN spectra of the pure compounds reveals that spots 3 and 4 are due to 1,6-dichloropyrene and 1,8-dichloropyrene, respectively. Similarly, spectral assignments could be carried out for pyrene (spot 1), 1-chloropyrene (spot 2), and 1,3-dichloropyrene (spot 5).

As for the analyte detectability in the LC–TLC–FLN setup, the amount of sample that is deposited on the TLC plate is primarily determined by the injection volume of the LC system. Since relatively low flow rates are required for successful LC–TLC coupling, narrow-bore LC columns should be used, and injection volumes should not exceed ca. 5 μl (unless use could be made of techniques as post-column liquid–liquid extraction as developed in the context of LC–ZnSe–FTIR; see above). When the 0–0 transition is considered, the concentration detection limit for pyrene ($S/N = 3$) on a C-18-modified silica high-performance TLC plate was 1×10^{-8} M (5 μl injection, $\lambda_{ex} = 363.00$ nm), which corresponds to a deposition of 10 pg. Similarly, for 1,6- and 1,8-dichloropyrene ($\lambda_{ex} = 361.00$ nm) and 1-chloropyrene ($\lambda_{ex} = 369.00$ nm) detection limits were found to be 0.4, 0.4, and 3 pg, respectively. Typical acquisition times for these measurements were 5–7 min. The present detection limits are about 100-fold better than those in earlier work in which the 363.80-nm line of an argon-ion laser was used for excitation of the 1-chloropyrene sample [26].

The improved detection limits for the higher substituted chloropyrenes are caused by two factors. First, the molar absorptivities of the S_0–S_1 transitions increase with the number of substituents. Second, the fluorescence lifetimes decrease, from 315 to 185 to 85 ns on going from pyrene to 1-chloropyrene and the dichloropyrenes, respectively. The shorter lifetimes are more favorable in light of the fixed 40-ns gate width.

To conclude this section, it can be stated that the hyphenation of LC and FLN through analyte storage on a TLC plate combines the potential of an efficient separation method and a sensitive and selective detection technique. LC–TLC–FLN can be used successfully for the investigation of real samples that contain spectrally interfering species. It should be recalled that the FLN spectra provide spectral information about the ground-state vibrations as well as the vibrations of the first excited electronic state. If only one excited-state vibration is probed, the resulting spectrum contains a single 0–0 transition with corresponding ground-state vibrations and can be used as a fingerprint for comparison with spectra of pure compounds. Simultaneous probing of more vibrations in the excited state provides a multiplet structure; the energy distance(s) between the excitation wavelength and the 0–0 transitions are directly related to the excited-state vibrations and thus yield a second fingerprint (see also Chapter 8). This spectral information is especially important for analytes with relatively strong 0–0 lines and weak vibronic transitions like polycyclic aromatic hydrocarbons. The position of the 0–0 transition itself does not provide more information than do conventional broadbanded emission spectra, since this position is not fixed—as it is in Shpol'skii spectroscopy—but shifts with the laser wavelength selected. As will be detailed below, the vibrations in the S_1 state are especially important in distinguishing between structurally similar compounds such as stereoisomeric benzo[*a*]pyrene tetrols [30].

10.4.3. Identification of Stereoisomeric Benzo[*a*]pyrene Tetrols by LC–TLC–FLN

Identification of stereoisomeric benzo[*a*]pyrene tetrols is quite important in environmental science, especially when these compounds are the hydrolysis products of adducts formed between benzo[*a*]pyrene, B[*a*]P, and DNA. B[*a*]P is one of the most studied polycyclic aromatic hydrocarbons (PAHs), a class of well-known environmental pollutants [31]. Their uptake by organisms can occur directly via air or water, or with particles on which PAHs are adsorbed. To enable excretion of the water-insoluble PAHs from living cells, they are metabolized to oxygenated products. Some of these products can react with DNA under formation of stable adducts, which seem to be the initiators of tumorigenesis and mutagenesis.

The metabolism of B[*a*]P to the DNA-reactive B[*a*]P-7,8-diol-9,10-epoxide (BPDE) and the binding of B[*a*]P and BPDE to DNA have been studied in some detail [31]. Two mechanisms for the formation of this adduct have been proposed in the literature: one involves formation of BPDE, which reacts with DNA, forming adducts with pyrenelike chromophoric structures [32, 33]. The other involves one-electron oxidation, forming adducts with B[*a*]P chromophoric structures, which generally cause depurination (see also Chapter 11) [34]. Both reaction products are referred to as B[*a*]P–DNA adducts. The concentration of stable B[*a*]P–DNA adducts in organisms is generally determined by radiometry using ^{32}P postlabeling [35]. In this technique, which is extremely sensitive, the various adducts are separated by means of multidimensional TLC using polyethylene imine (PEI) cellulose plates. The main disadvantage is that the chromatographic resolution is low, migration distances are difficult to control, and the radioactivity detection provides no further information on analyte identity. As an alternative, identification of B[*a*]P–DNA adducts in both hydrolyzed and intact DNA at low adduct levels (1 adduct per 10^8 bases) can be performed using FLN spectroscopy [34,36–38]. Unfortunately, the combination of TLC and FLN using PEI plates is quite problematic due to the spectroscopic characteristics of the PEI-cellulose material [27].

Another approach to determine adduct levels is the hydrolysis of B[*a*]P–DNA adducts utilizing either enzymatic or acidic conditions [39] and subsequent analysis of the products. In the case of BPDE binding to DNA, this approach results in the formation of four pairs of stereoisomeric free B[*a*]P tetrols [40];

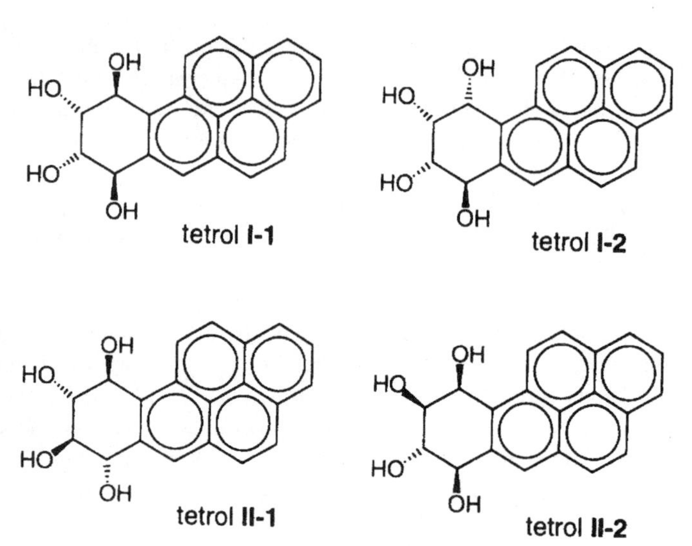

Figure 10.6. Structures of the four stereoisomeric benzo[*a*]pyrene tetrols and their abbreviated descriptions. Only one structure is drawn for each enantiomeric pair.

their structures are depicted in Figure 10.6. These isomers can be easily separated by reversed-phase column liquid chromatography (RPLC) [41]. The tetrols have been determined in B[*a*]P-treated tissue using various techniques, such as RPLC with conventional [42] or laser-induced [43] fluorescence detection, microcolumn RPLC, micellar electrokinetic capillary chromatography (MEKC) with conventional fluorescence detection [44], and RPLC combined with synchronous scanning fluorometry [39]. Shugart and Harvey [42, 44] notified the existence of stereoisomerism; they have identified the four stereoisomeric forms on the basis of retention times using standard compounds.

Hurtubise and co-workers performed spectroscopic studies in batch experiments using cyclodextrin (CD)/NaCl matrices to enhance fluorescence and phosphorescence efficiencies, both at low and at room temperature [40, 45]. They determined the fluorescence as well as phosphorescence lifetimes and quantum yields of the tetrols with the objective to utilize these data for identification purposes. When time-resolved fluorescence spectra are recorded using a proper detection gate width, the tetrols can be identified on the basis of differences in peak height ratios [45]. To enhance such differences, derivative spectra were presented. Unfortunately, although the differences can be observed, they are relatively small. It has to be realized that in derivative spectra signal-to-noise ratios decrease. Unambiguous identification of the individual isomers in a mixture undoubtedly requires the involvement of a separation technique. Within this context it should be noted that the compounds concerned cannot be studied by Shpol'skii spectroscopy as has, for instance, successfully been done for monohydroxy B[*a*]P [46]; because of their polarity, they are not compatible with *n*-alkane matrices (see Chapter 6).

The instrumental setup used by Kok et al. to perform the LC–TLC–FLN experiments [30] was essentially the same as described above [28]. Instead of the excimer laser a Nd : YAG laser system was used for excitation, and instead of the 1-m monochromator a triple monochromator was used for collecting the fluorescence. As discussed in the previous section, the mobile-phase composition used to perform the separation is rather critical: It should be a compromise between LC separation efficiency and volatility to achieve adequate deposition on the TLC plate. A methanol–water mixture of 60 : 40 (v/v) appeared to be a good compromise, though 20% better LC resolution is obtained applying a 50 : 50 (v/v) mixture. An on-the-fly chromatogram of the four B[*a*]P tetrols separated by microcolumn LC is given in Figure 10.7. The separation efficiency is relatively poor, which is due to both the rather large injection volume and the column quality. No attempts were made to further improve the LC results.

As regards the effluent deposition, the high water content of the mobile phase poses some problems. When methanol–water (60 : 40, v/v) is sprayed on a C-18-modified TLC plate, discrete eluent drops are formed because the sorption by this matrix is relatively low. The bare silica HPTLC plates displayed much better

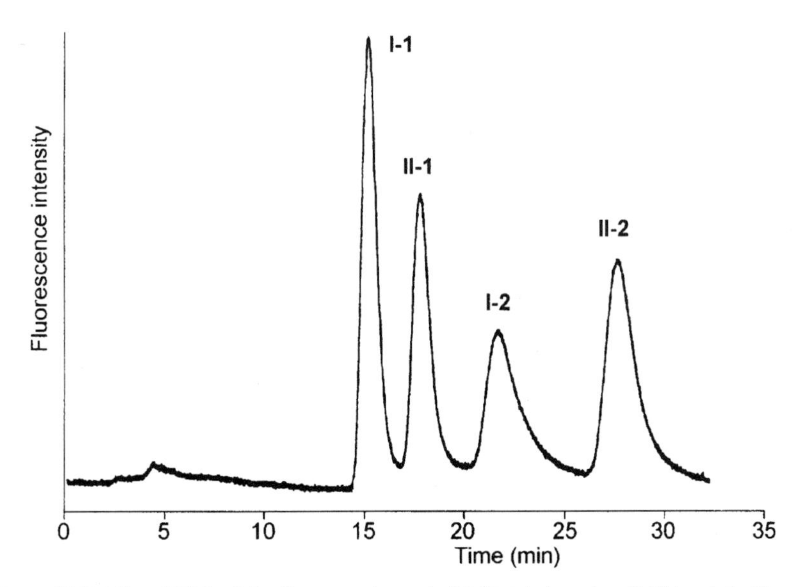

Figure 10.7. Micro-RPLC of the four stereoisomeric B[*a*]P tetrols, using C-18-bonded silica and methanol–water (60 : 40, v/v) with conventional fluorescence detection. (From Ref. 30 with permission.)

wetting characteristics, which adequately solved this part of the problem. However, when the eluent is not completely evaporated, the deposited analytes will start to migrate on the TLC plate. This will cause the spots to spread in all directions and, thus, create loss of resolution as well as analyte detectability. In the present instance, complete evaporation was obtained by raising the temperature of the nitrogen gas surrounding the spray needle to 175°C, and decreasing the distance of the needle to the TLC plate from the conventional 2.0 to 0.5 mm. No analyte degradation was observed, probably owing to evaporation processes reducing the actual temperature in the sprayed liquid.

The areas of the analyte spots on the TLC plate must be kept small, because FLN spectroscopy is typically performed using a 3-mm^2 laser beam. Evidently, spot sizes will decrease when the translation speed of the TLC plate is lowered. At too low a speed, however, chromatographic resolution will be lost and spectroscopic identification will be hampered. The minimum translation speed, which resulted in a resolution larger than unity between all the peaks, was 4.0 mm min^{-1}. At this speed, a 10-cm-long TLC plate can contain the relevant part of the LC chromatogram (i.e., between 10 and 35 min; see Fig. 10.7).

The areas containing the separate B[*a*]P tetrols were cut out of the TLC plate, cooled to 12 K, and illuminated with either the 355-nm light of the Nd : YAG laser or the 360–368 nm output of the dye laser. Examples of FLN emission spectra of the four tetrols using different laser excitation wavelengths are shown in Figure 10.8A–C. In the spectra the energy differences (in cm^{-1}) between the excitation line and a number of emission bands are indicated. These differences represent the vibrations of the S_1 state for the four isomeric tetrols. As mentioned, the spectral details are significantly influenced by the laser excitation wavelength used: The vibrations of the S_1 state differ most for the four isomers in the 650–950 cm^{-1} range. Whereas at 355-nm excitation (Fig. 10.8A) the emission spectra are practically identical, they are obviously different at 364-nm excitation (Fig. 10.8B). At 368-nm excitation, differences are less pronounced (Fig. 10.8C). Hence the 364-nm excitation wavelength is the most appropriate one for identification purposes. The results, which were determined using both the 355-nm output of the Nd : YAG laser and the 360–368 nm dye laser output, are schematically depicted in Figure 10.9. In this scheme both the identical and the characteristic vibrations are marked. Tetrols I-1, I-2, and II-2 all show vibrations close to 825 cm^{-1}, but tetrol I-2 shows two bands in the 825–950 cm^{-1} range (884 and 922 cm^{-1}), compared to one for tetrol II-2 (902 cm^{-1}), and none for tetrol I-1. Only tetrol II-1 shows the 860 cm^{-1} band. The 950 cm^{-1} band can be used to ensure proper wavelength calibration. The 1106 cm^{-1} band, which can be seen for all four tetrols, was determined using dye laser excitation in the 360–363.5 nm range. The resolution of the technique is sufficient to distinguish between vibrations that differ by 10 cm^{-1}.

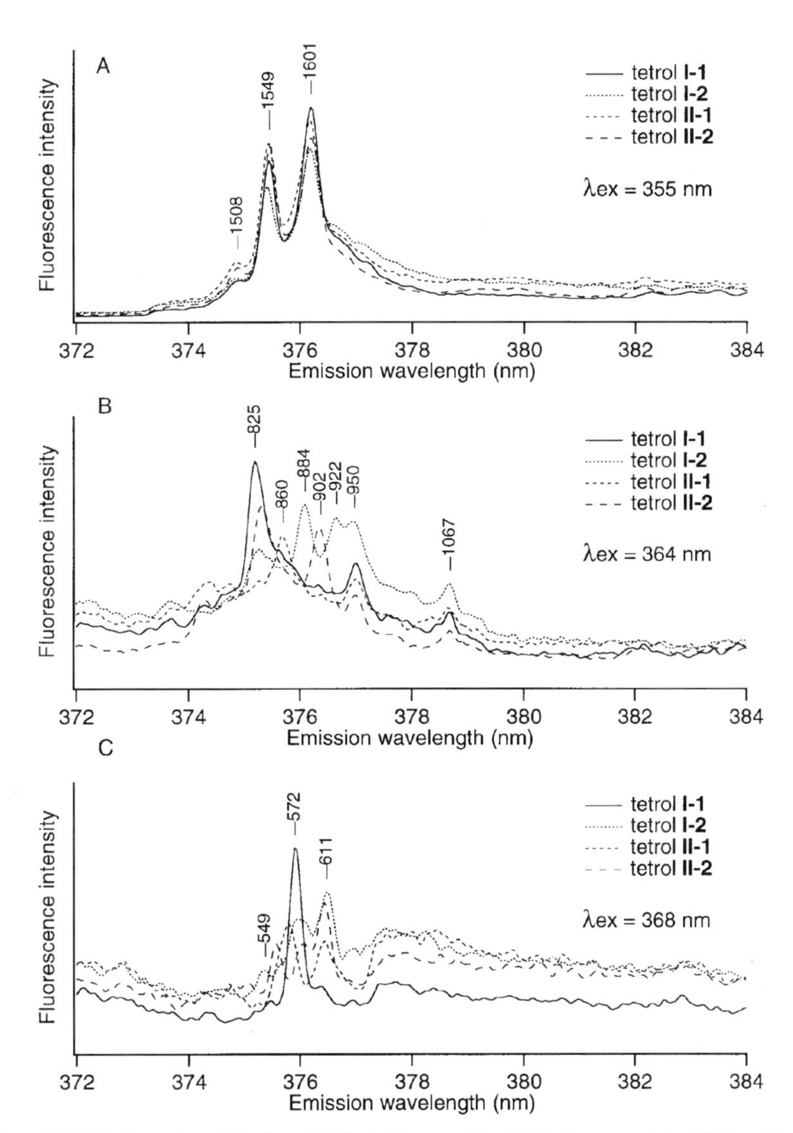

Figure 10.8. FLN spectra of the four B[*a*]P tetrols on a silica TLC plate recorded at different laser wavelengths 355 nm (A), 364 nm (B), and 368 nm (C). The energy differences between the excitation and the 0–0 emission wavelengths, which represent the vibrations of the S_1 state are given in cm^{-1}. (From Ref. 30 with permission.)

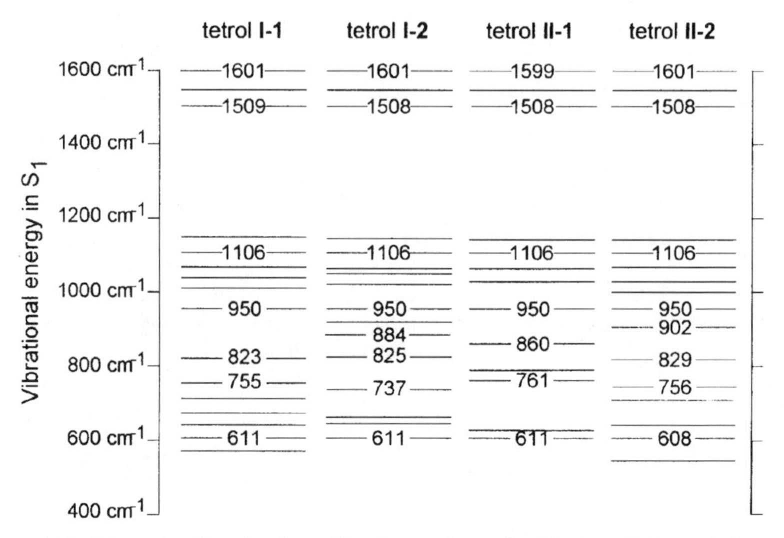

Figure 10.9. Schematic of the vibrations of the S_1 state (in cm^{-1}) of the four B[*a*]P tetrols. Data were determined using both the 355-nm output of the Nd : YAG and the 360–368 nm output of the dye laser. (From Ref. 30 with permission.)

10.5. CE–FLN OF PAH–DNA ADDUCTS

10.5.1. Characteristics of CE

As extensively outlined in Chapters 8 and 11, the usefulness of FLN spectroscopy for studying PAH–DNA adducts has been shown by the group of Small and Jankowiak at Iowa State University [36, 47, 48]. Unfortunately, coupling of FLNS to the PEI-cellulose TLC plates used in the ^{32}P radiometric postlabeling technique is not straightforward [27], while—at present—the appropriateness of LC–TLC–FLN to characterize DNA–PAH adducts still remains to be demonstrated. Interestingly, an alternative hyphenated technique has been successfully developed (i.e., the CE–FLN system), in which FLN is coupled to capillary electrophoresis [49–51].

CE is a relatively new electrodriven separation technique that has gained popularity in analytical chemistry because of its extremely high separation efficiency. Its conventional mode, denoted as capillary-zone electrophoresis (CZE), is well suited for the determination of charged as well as ionogenic compounds. Other CE modes involve other separation principles as well, such as micellar electrokinetic chromatography (MEKC), capillary gel electrophoresis (CGE) and electrochromatography (CEC), using particle-filled capillaries. As a result, CE techniques are not limited to ionogenic compounds but have been used for analysis of amino acids, peptides, proteins, nucleic acid bases, DNA oligonucleotides, and numerous organic molecules. Both small ions and large biomolecules can be separated [52]. Electrophoresis is a powerful approach for gene mapping and DNA sequencing; the chemical analysis of individual cells by CE has attracted much attention [53]. CE is a microseparation technique using separation capillaries with internal diameters of typically 50–75 μm. Injection volumes that can be used without affecting the separation performance are usually a few nanoliters only; the same holds for the detection volume. These characteristics imply that detection limits that can be achieved in CE terms of absolute amounts can be quite impressive (for example, 600 fg using commercially available absorption detectors [54]) but nevertheless in terms of concentration rather unfavorable. This explains why most applications of CE reported to date deal with biological/bioanalytical and pharmaceutical samples [55]; its potential in environmental analysis is limited since the detection of pollutants generally has to be performed at much lower concentration levels. It also explains why in the recent literature improvement of concentration detection limits based on laser-induced fluorescence received special attention [56, 57].

In favorable cases, CE provides extremely high separation efficiencies. Unfortunately, all CE modes indicated above have in common that migration times and separation efficiencies depend on parameters such as pH, temperature, ionic strength, and the condition of the capillary walls, which are not easily controlled [58], certainly not when real samples containing unknown interferences have to be analyzed [59]. This means that, even more than in LC, there is an urgent need for hyphenation of CE and spectrometric detection techniques. Only such hyphenation enables unambiguous identification of analytes by combining structural, spectral information, with the observed migration. A recent paper reviews the state-of-the-art of analyte identification in CE [60]. Compared to LC, hyphenation in CE is still in its infancy. The perspectives of CE–MS are undoubtedly good and wide ranging. Nevertheless, despite of the overwhelming number of papers devoted to coupling of CE and mass spectrometry, CE–MS is still not yet suited for routine application and only in exceptional cases applied to the analysis of real-life samples.

10.5.2. CE–FLN Coupling

The major problems envisaged in combining FLNS with CE were the design of a compact and reliable liquid helium capillary cryostat capable of rapid cooling and the ability to form disordered matrices for FLN to be operative in typical CE buffers. These problems could be overcome [49–51]. The experimental setup is schematically depicted in Figure 10.10. The CE capillary can be cooled to either 77 or 4.2 K by a flow of liquid nitrogen or helium through the capillary cryostat. To allow the sequential characterization of the separated analyte zones in the

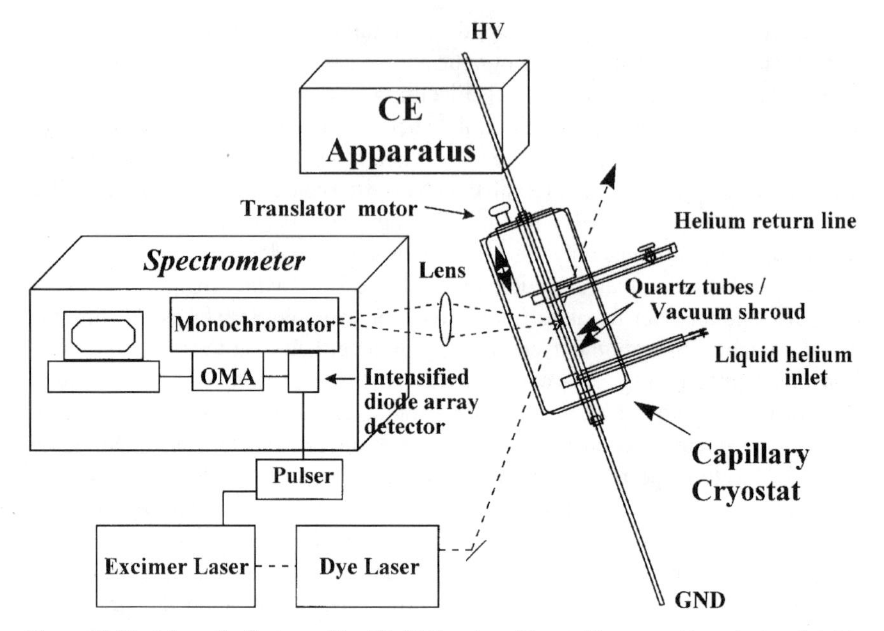

Figure 10.10. Schematic diagram of the CE–FLN system. The capillary cryostat can be translated to enable detection of separated analyte zones. HV = high voltage, GND = ground. (From Ref. 49 with permission.)

capillary, the cryostat is attached to a translation stage, which moves the capillary through the detection region in the direction of the capillary axis.

The capillary cryostat (CC) consists of a double-walled quartz cell with inlet and return lines for introducing liquid nitrogen or liquid helium. The outer portion of the CC is evacuated. The capillary is positioned in the central region of the CC. The low thermal capacity of the capillary section to be cooled and the small dimensions of the CC (inner portion, 4 mm I.D. × 22 cm length) ensure rapid cooling to 4.2 K. Warming can be achieved by closing the coolant flow regulator or by replacing the cryogenic liquid with nitrogen or helium gas. The length of the CC (22 cm) and its quartz cell (7 cm optical window) is greater than the travel distance of the translation stage (5 cm); if the 5 cm automated translation is insufficient, the frozen capillary can be manually positioned to a new location. Fluorescence is collected at a right angle with respect to the excitation laser beam. To discriminate against scattered and reflected laser light, the CC is tilted by 20°. Further discrimination against scattered laser light and background fluorescence from the capillary walls is obtained by spatial filtering.

Fluorescence of CE-separated analytes is detected first at 300 or 77 K using high-energy laser excitation at 308 or 351.1 nm (from an argon ion laser). The fluorescence spectra acquired during translation of the CC provide the basis for the integrated fluorescence electropherograms as a function of time (or capillary position). It should be noted that under these conditions the spectra exhibit no line narrowing but show broad bands, since the temperature is not low enough for FLN spectroscopy. Nevertheless, they provide useful spectral information for each of the separated analytes. After the electropherogram is generated, the temperature of the capillary is lowered to 4.2 K for high-resolution FLN characterization. Selective excitation using the excimer-pumped dye laser provides FLN spectra for the analytes at various vibronic ($S_1 - S_0$) excitation wavelengths. The use of low temperatures alleviates the problems with photodegradation of analytes.

As noted, one of the concerns in developing CE–FLN was that the CE buffer solutions would not form a glassy matrix or would lead to sudden and violent cracking, which would destroy the capillary. Such problems are not encountered. Consistently, a glassy matrix is formed, presumably due to the presence of salts, which assist in glass formation and to the small internal diameter of the CE capillary, which, in combination with its very low thermal capacity, leads to rapid cooling. It has been observed that the capillary and CC can be cooled to 4.2 K in less than 1 min [51].

10.5.3. Analysis of PAH–DNA Adducts

The analytical potential of the CE–FLN technique was demonstrated for the analysis of mixtures of benzo[*a*]pyrene- and dibenzo[*a,l*]pyrene (DB[*a,l*]P) adducts [50, 51]. The latter PAH, one of the most potent carcinogens [1], is known to yield a variety of adducts formed via both one-electron oxidation and diolepoxide pathways. The attention was confined to the three one-electron oxidation standard adducts denoted as DN[*a,l*]P-10-N7Ade, DB[*a,l*]P-10-N1Ade, and DB[*a,l*]P-10-N3Ade, which have identical structures except for the binding position of DB[*a,l*]P to adenine. These three, structurally very similar, individual DB[*a,l*l]P adducts can be distinguished by FLNS, by monitoring the S_1 excited-state vibronic frequencies. In a mixture, however, this is not possible (a similar problem as described above for the four isomeric benzo[*a*]pyrene tetrols, which was tackled by LC–TLC–FLN [30]).

Figure 10.11A shows the electropherogram obtained for such a mixture of adducts recorded using LIF detection. In fact, MEKC was applied: The buffer was composed of acetonitrile : water (30 : 70, v/v) containing 40 mM sodium bis(2-ethyl)sulfosuccinate and 8 mM sodium borate adjusted to pH 9. The three major peaks in the electropherogram represent the three adducts concerned; the smaller one was attributed to an impurity.

The FLN spectra obtained for the three major peaks (using the CE–FLN setup) are depicted in Figure 10.11B. The peaks in the spectra are labeled with their

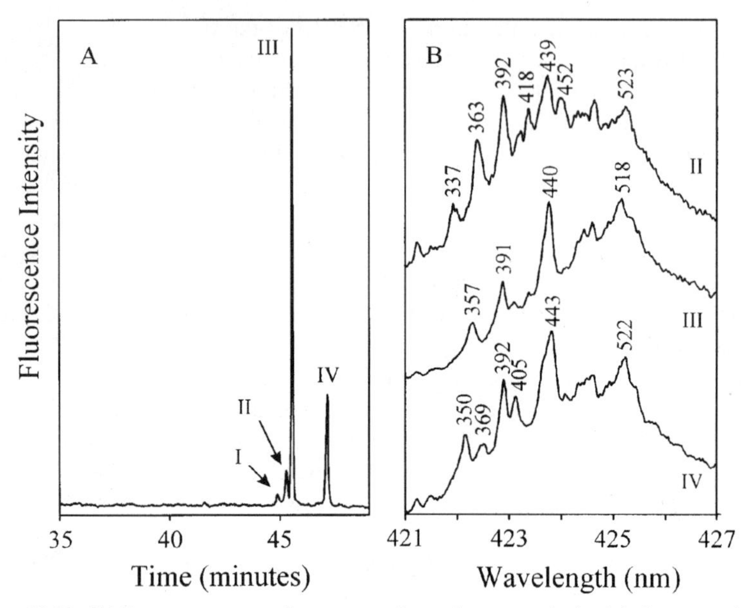

Figure 10.11. (A) Room-temperature fluorescence electropherogram obtained during separation of a mixture of (II) DB[*a*,*l*]P-10-N7Ade, (III) DB[*a*,*l*]P-10-N1Ade, and (IV) DB[*a*,*l*]P-10-N3Ade. An unidentified impurity is labeled as peak (I). (B) FLN spectra for the three CE-separated adducts, obtained at 4.2 K using selective laser excitation at 416.0 nm. The FLN peaks are labeled with their S_1 vibrational frequencies, in cm^{-1}. (From Ref. 51 with permission.)

S_1-state vibrational frequencies. There are some significant differences for the three adducts. It should be noted that identification of the three adducts on the basis of these data required the availability of library spectra of the individual (pure) compounds recorded for exactly the same solvent composition. Such spectra were recorded using a regular helium immersion dewar. Thus it was unambiguously shown that peaks II, III, and IV correspond to the adducts briefly denoted as DN[*a*,*l*]P-10-N7Ade, DB[*a*,*l*]P-10-N1Ade, and DB[*a*,*l*]P-10-N3Ade, respectively.

10.6. CONCLUDING REMARKS

The achievements discussed above illustrate the analytical potential of combining liquid-state separation techniques and FLN spectroscopy. In fact, the freedom of solvent (matrix) composition is larger than originally assumed: Good-quality spectra were obtained for analytes deposited on various types of TLC plates (although PEI-cellulose, unfortunately, appeared to be less appropriate), as well as for analytes in frozen CE-buffer solutions.

Nonetheless, at present the research on hyphenation of TLC, LC, and CE with FLN spectroscopy is still in an exploratory stage. The field of applicability—though interesting—is still very small; attention seems to be paid exclusively to PAHs and PAH derivatives, thus far ignoring other analytes exhibiting native fluorescence. Also, the number of researchers active in this field is small, and several practical problems still need to be overcome: (1) FLN spectroscopy setups are not commercially available but need to be assembled in the laboratory, requiring laser spectroscopy expertise, (2) the interfaces as used in LC–TLC–FLN and in CE–FLN need to be laboratory-made as well, and (3) CE and LC experience is a prerequisite to successfully operate and exploit the hyphenated techniques considered.

The challenge is to show the appropriateness of CE–FLN and/or LC–TLC–FLN in solving real-life analytical problems that cannot be handled by alternative methods. Robustness of instrumentation needs to be (further) improved and special attention should be paid to quantitative aspects, focusing on reproducibility of results [61].

REFERENCES

1. I. S. Kozin, C. Gooijer, and N. H. Velthorst, *Anal. Chem.* **67**, 1623 (1995).
2. I. S. Kozin, C. Gooijer, and N. H. Velthorst, *Anal. Chim. Acta* **333**, 193 (1996).
3. E. Heftman, *Chromatography, Part B: Applications* 5th ed., Elsevier, Amsterdam, 1992.
4. T. Hirschfeld, *Anal. Chem.* **52**, 297A (1980).
5. R. Smits, *LC–GC Int.* **7**, 505 (1994).
6. U. A. Th. Brinkman, *Analusis* **24**(4), M12 (1996).
7. J. F. K. Huber, A. M. van Urk-Schoen, and G. B. Sieswerda, *Z. Anal. Chem.* **264**, 257 (1973).
8. W. M. A. Niessen and A. P. Tinke, *J. Chromatogr. A* **703**, 37 (1995).
9. M. D'Orazio and U. Schimpf, *Anal. Chem.* **53**, 809 (1981).
10. C. Kong Chong, C. K. Mann, and T. J. Vickers, *Appl. Spectrosc.* **46**, 249 (1992).
11. L. M. Cabalin, A. Rupérez, and J. J. Laserna, *Anal. Chim. Acta* **318**, 203 (1996).
12. R. Sheng, F. Ni, and T. M. Cotton, *Anal. Chem.* **63**, 437 (1991).
13. R. J. van de Nesse, G. J. M. Hoogland, J. J. M. de Moel, C. Gooijer, U. A. Th. Brinkman, and N. H. Velthorst, *J. Chromatogr.* **552**, 613 (1991).
14. G. W. Somsen, *Solvent-Elimination-Based Coupling of Column Liquid Chromatography and Vibrational-Spectroscopic Detection Techniques*, Ph.D. Thesis, Vrije Universiteit, Amsterdam (1997).

15. G. W. Somsen, R. J. van de Nesse, C. Gooijer, U. A. Th. Brinkman, N. H. Velthorst, T. Visser, P. R. Kootstra, and A. P. J. M. de Jong, *J. Chromatogr.* **552**, 635 (1991).
16. G. W. Somsen, E. W. J. Hooijschuur, C. Gooijer, U. A. Th. Brinkman, N. H. Velthorst, and T. Visser, *Anal. Chem.* **68**, 746 (1996).
17. G. W. Somsen, I. Jagt, C. Gooijer, N. H. Velthorst, U. A. Th. Brinkman, and T. Visser, *J. Chromatogr. A* **756**, 145 (1996).
18. G. W. Somsen, P. G. J. H. ter Riet, C. Gooijer, N. H. Velthorst, and U. A. Th. Brinkman, *J. Planar Chromatogr.* **10**, 10 (1997).
19. G. W. Somsen, S. K. Coulter, C. Gooijer, N. H. Velthorst, and U. A. Th. Brinkman, *Anal. Chim. Acta* **349**, 189 (1997).
20. J. W. Hofstraat, C. Gooijer, and N. H. Velthorst, *Appl. Spectrosc.* **42**, 614 (1988).
21. J. W. Hofstraat, M. Engelsma, R. J. van de Nesse, C. Gooijer, N. H. Velthorst, and U. A. Th. Brinkman, *Anal. Chim. Acta* **186**, 247 (1986).
22. D. E. Jaenchen and H. J. Issaq, *J. Liq. Chromatogr.* **11**, 1941 (1988).
23. C. T. Banks, *J. Pharm. Biomed. Anal.* **11**, 705 (1993).
24. E. Müller and H. Jork, *J. Planar Chromatogr.* **6**, 21 (1993).
25. J. W. Hofstraat, M. Engelsma, W. P. Cofino, G. Ph. Hoornweg, C. Gooijer, and N. H. Velthorst, *Anal. Chim. Acta* **159**, 359 (1984).
26. J. W. Hofstraat, H. J. M. Jansen, G. Ph. Hoornweg, C. Gooijer, and N. H. Velthorst, *Anal. Chim. Acta* **170**, 61 (1985).
27. S. J. Kok, R. Evertsen, U. A. Th. Brinkman, N. H. Velthorst, and C. Gooijer, *Anal. Chim. Acta* **4–5**, 1 (2000).
28. R. J. van de Nesse, I. H. Vinkenburg, R. H. J. Jonker, G. Ph. Hoornweg, C. Gooijer, U. A. Th. Brinkman and N. H. Velthorst, *Appl. Spectrosc.* **48**, 788 (1994).
29. G. W. Somsen, L. L. P. van Stee, C. Gooijer, U. A. Th. Brinkman, N. H. Velthorst, and T. Visser, *Anal. Chim. Acta* **290**, 269 (1994).
30. S. J. Kok, R. Posthumus, I. Bakker, C. Gooijer, U. A. Th. Brinkman, and N. H. Velthorst, *Anal. Chim. Acta* **303**, 3 (1995).
31. N. E. Geacintov, in S. K. Yang and B. D. Silverman, eds., *Polycyclic Aromatic Hydrocarbon Carcinogenesis Structure–Activity Relationships*, CRC Press, Boca Raton, FL, 1988.
32. P. Sims, P. L. Grover, A. Swaisland, K. Pal, and A. Hewer, *Nature* **252**, 326 (1974).
33. M. Hall and P. L. Grover, in C. S. Cooper, P. L. Groover, eds., *Chemical Carcinogenesis and Mutagenesis*, Springer-Verlag, Berlin, 1990, p. 327.
34. E. G. Rogan, P. D. Devanesan, N. V. S RamaKrishna, S. Higginbotham, N. S. Padmavathi, K. Chapman, E. L. Cavalieri, H. Jeong, R. Jankowiak, and G. J. Small, *Chem. Res. Toxicol.* **6**, 356 (1993).
35. A. C. Beach and R. C. Gupta, *Carcinogenesis* **13**, 1053 (1992).
36. R. Jankowiak, P. Lu, G. J. Small, and N. E. Geacintov, *Chem. Res. Toxicol.* **3**, 39 (1990).
37. P. Lu, H. Jeong, R. Jankowiak, G. J. Small, S. K. Kim, M. Cosman, and N. E. Geacintov, *Chem. Res. Toxicol.* **4**, 58 (1991).

38. P. D. Devanesan, N. V. S. RamaKrishna, R. Todorovic, E. G. Rogan, E. L. Cavalieri, H. Jeong, R. Jankowiak, and G. J. Small, *Chem. Res. Toxicol.* **5**, 302 (1992).
39. A. Weston, M. L. Rowe, D. K. Manchester, P. B. Farmer, D. L. Mann, and C. C. Harris, *Carcinogenesis* **10**, 251 (1989).
40. J. Corley and R. J. Hurtubise, *Anal. Chem.* **65**, 2601 (1993).
41. E. H. J. M. Jansen, R. H. van de Berg, and E. D. Kroese, *Anal. Chim. Acta* **290**, 86 (1994).
42. L. Shugart, J. McCarthy, B. Jimenez, and J. Daniels, *Aquat. Toxicol.* **9**, 391 (1987).
43. R. Wang and J. W. O'Laughlin, *Environ. Sci. Technol.* **26**, 2294 (1992).
44. S. D. Harvey, R. M. Bean, and H. R. Udseth, *J. Microcol. Sep.* **4**, 191 (1992).
45. L. Shu and R. J. Hurtubise, *Appl. Spectrosc.* **47**, 1892 (1993).
46. F. Ariese, S. J. Kok, M. Verkaik, G. Ph. Hoornweg, C. Gooijer, and N. H. Velthorst, *Anal. Chem.* **65**, 1100 (1993).
47. M. Suh, F. Ariese, G. J. Small, R. Jankowiak, and N. E. Geacintov, *Biophys. Chem.* **56**, 281 (1995).
48. K.-M. Li, R. Todorovic, E. G. Rogan, E. L. Cavalieri, F. Ariese, M. Suh, R. Jankowiak, and G. J. Small, *J. Biochem.* **34**, 8043 (1995).
49. R. Jankowiak, D. Zamzow, W. Ding, and G. J. Small, *Anal. Chem.* **68**, 2549 (1996).
50. D. Zamzow, G. J. Small, and R. Jankowiak, *Mol. Cryst. Liq. Cryst.* **291**, 155 (1996).
51. D. Zamzow, C. H. Lin, G. J. Small, and R. Jankowiak, *J. Chromatogr. A* **781**, 73 (1997).
52. J. Tehrani, R. Macomber, and L. Day, *J. High Resolut. Chromatogr.* **14**, 10 (1991).
53. B. L. Hogan and E. S. Yeung, *Anal. Chem.* **64**, 3841 (1992).
54. M. W. F. Nielen, *Trends Anal. Chem.* **12**, 345 (1993).
55. R. L. St. Claire III, *Anal. Chem.* **68**, 569R (1996).
56. S. J. Kok, E. M. Kristenson, C. Gooijer, N. H. Velthorst, and U. A. Th. Brinkman, *J. Chromatogr. A* **771**, 331 (1997).
57. S. J. Kok, G. Ph. Hoornweg, T. de Ridder, U. A. Th. Brinkman, N. H. Velthorst, and C. Gooijer, *J. Chromatogr. A* **806**, 355 (1998).
58. H. T. Chang and E. S. Yeung, *Electrophoresis* **16**, 2069 (1995).
59. J. Y. Chai and J. Henion, *J. Chromatogr. A* **703**, 667 (1995).
60. S. J. Kok, N. H. Velthorst, C. Gooijer, and U. A. Th. Brinkman, *Electrophoresis* **19**, 2753 (1998).
61. S. J. Kok, I. Bakker, U. A. Th. Brinkman, N. H. Velthorst, and C. Gooijer, *Anal. Chim. Acta* **389**, 77 (1999).

CHAPTER

11

HIGH-RESOLUTION FLUORESCENCE ANALYSIS OF POLYCYCLIC AROMATIC HYDROCARBON DERIVED ADDUCTS TO DNA AND PROTEINS

FREEK ARIESE

Dept. of Analytical Chemistry and Applied Spectroscopy, Vrije Universiteit Amsterdam, De Boelelaan 1083, NL-1081 HV Amsterdam, the Netherlands

RYSZARD JANKOWIAK

Ames Laboratory USDOE, Iowa State University, Ames, IA 50011, USA

11.1. INTRODUCTION

Fluorescence techniques are being used extensively for the analysis of polycyclic aromatic hydrocarbons (PAHs) and their derivatives. However, as is the case for most organic compounds, conventional fluorescence spectra are of limited use for identification or analysis of mixtures as a result of severe spectral broadening. In the first section of this volume it has been shown for parent PAHs and a number of planar, relatively nonpolar derivatives that inhomogeneous broadening can be strongly reduced by freezing the sample to low temperatures in crystalline (usually *n*-alkane) matrices. However, this approach is not suitable for more polar derivatives, such as polyhydroxylated metabolites (see Chapter 6) or PAHs bound to biological macromolecules. Instead, polar, amorphous matrices and site-selection methods can be used. In this chapter, it is shown how fluorescence line-narrowing (FLN) spectroscopy can be applied to the study of PAHs adducted to DNA or proteins. FLN spectroscopy can be used for chemical identification, conformational analysis, and/or to probe the microenvironment of the adduct under study.

PAHs continue to be widely studied, one of the main reasons being the carcinogenic potency of many polycyclic aromatic compounds [1–3]. Some well-known examples are benzo[*a*]pyrene (B[*a*]P), 7,12-dimethylbenz[*a*]anthracene (DMBA), and the most potent carcinogenic PAH known to date, dibenzo[*a*,*l*]pyrene (DB[*a*,*l*]P) [4]. PAHs are present in many petrochemical products and are formed during incomplete combustion processes. In the environment, animals

Shpol'skii Spectroscopy and Other Site-Selection Methods, Edited by Cees Gooijer, Freek Ariese, and Johannes W. Hofstraat.
ISBN 0-471-24508-9

may be exposed to PAHs as the result of oil spills or exposure to soils or sediments polluted by combustion-related PAHs. There is convincing evidence that the occurrence of liver tumors in various fish populations is related to high PAH exposure levels [5]. Also humans may be exposed to PAHs through smoking, traffic exhausts, grilled or smoked foods, or at certain work places [6, 7]. For instance, the very high incidence of lung cancer among women in the Chinese province of Xuan Wei has been linked to the high PAH content of the particular type of coal used in that region for cooking [8].

Most PAHs are chemically fairly inert compounds. Upon uptake, mammals and also many other animal species metabolize PAHs in order to facilitate excretion. However, certain metabolic pathways lead to reactive intermediates that can bind to biologically important molecules. Two major mechanisms have been identified that lead to DNA adduct formation, and these will briefly be discussed here. According to the monooxygenation pathway [9], metabolism starts with the enzymatic addition of an oxygen atom across a double bound, leading to an epoxide. This epoxide may then be isomerized into a phenolic group, or hydrolyzed into a dihydrodiol derivative. In the case of B[*a*]P, it has been shown that one of most potent carcinogenic intermediates results from the subsequent oxidation of 7,8-dihydrodiol B[*a*]P into B[*a*]P-7,8-dihydrodiol-9,10 epoxide (BPDE) [10]. BPDE may bind to proteins, or form stable adducts to DNA bases, in particular the purine bases adenine (Ade) or guanine (Gua). As will be shown below, BPDE and other reactive metabolites may also bind to these DNA bases in a fashion that leads to loss of the adducted base from the DNA helix (depurination). A second metabolic pathway involves the formation of a radical cation through one-electron oxidation. Through this mechanism, several PAHs have been shown to bind to DNA bases, typically resulting in the depurination of Ade or Gua bases [3]. Stable adducts formed via the radical cation pathway have not yet been identified. Chemical structures of the above mentioned compounds are shown in Figure 11.1.

There can be several reasons to carry out PAH adduct measurements, ranging from elucidation of the chemical mechanisms of DNA binding to biological monitoring of PAH exposure. Since in many cases the reactive metabolites exist as a mixture of different isomers, which can bind with different stereochemistries to different nucleophilic sites of different DNA bases, the analytical techniques involved should be extremely selective. FLN spectroscopy, owing to the highly resolved spectra that can be obtained and the specificity of the vibrational patterns, has been shown to offer the required level of selectivity [11–13]. Furthermore, in the case of strong fluorophores (as many PAHs are), detection limits in the low femtomole range or even lower can be obtained.

Briefly, the FLN technique is based on site or energy selection by means of a laser. For a chemically pure compound in a disordered matrix, a broad distribution of absorption spectra exists, since each individual molecule will experience a

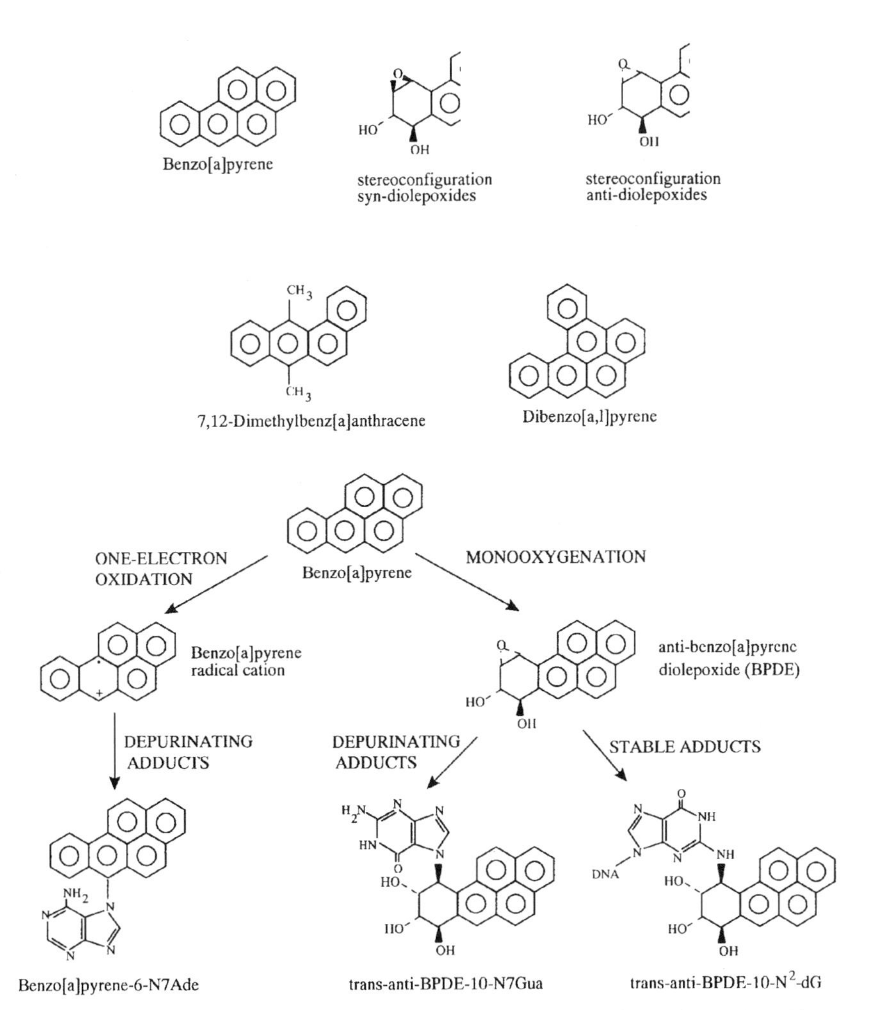

Figure 11.1. Chemical structures of selected PAHs, nomenclature of diolepoxide stereoconfigurations, and examples of adduct formation through the one-electron oxidation and monooxygenation mechanisms.

different environment. However, when the sample is irradiated with a narrow-banded laser, only a narrow subset of molecules, with absorption transitions matching the photon energy of the laser, will be excited. Provided no molecular movements occur at cryogenic temperatures, only this energetically equivalent subset or "isochromat" will fluoresce, resulting in a line-narrowed spectrum. FLN spectra reveal highly specific vibrational patterns which can be used for fingerprint identification. As an illustration, Figure 11.2 shows fluorescence

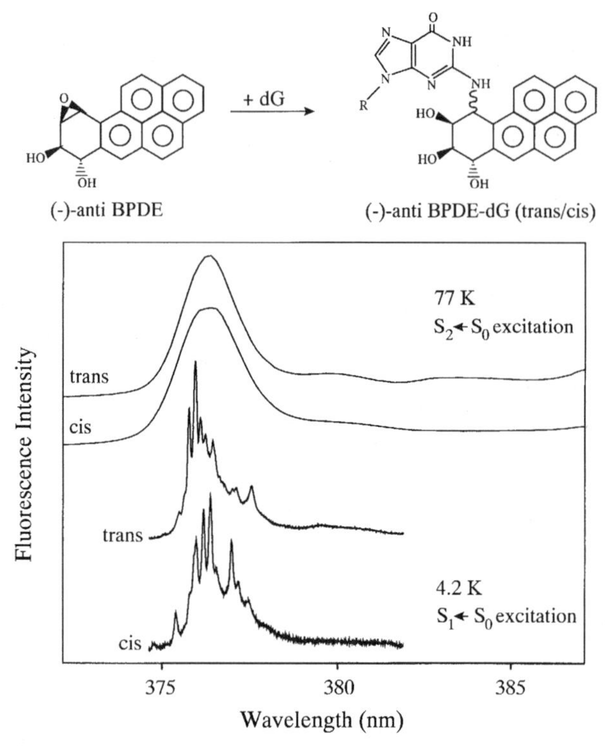

Figure 11.2. Illustration of FLN selectivity for two closely related isomers: *trans* vs. *cis* adducts of $(-)$-*anti*-BPDE to dGMP. The top spectra were recorded under non-line-narrowing conditions ($T = 77\,\text{K}$, $S_2 \leftarrow S_0$ excitation). The bottom spectra were recorded under FLN conditions (4.2 K, $S_1 \leftarrow S_0$ vibronic excitation) and show the different vibrational patterns in the origin multiplet. The peaks correspond to excited-state vibrational frequencies.

spectra of two closely related stereoisomers. The reaction of $(-)$-*anti*-BPDE with deoxyguanosine monophosphate (dGMP) results in a mixture of *trans* and *cis* adducts. Isomer-specific identification is important, since these adducts will be embedded in different ways in the DNA helix, and hence their mutagenic potential and biological repair rates will be different (see Section 11.4). Under conventional conditions these two compounds show identical fluorescence spectra, and even at 77 K the two mononucleotide adducts cannot be distinguished (upper two spectra). However, under FLN conditions, sharp zero-phonon lines (ZPLs) appear, corresponding to specific vibronic transitions. The vibrational fingerprint of the *trans*-adduct is completely different from that of the *cis* adduct, as shown in the two lower spectra. The fundamental principles of FLN spectroscopy are explained in more detail in Chapter 8. The instrumentation and methodologies used for FLN studies of adducts are described in Section 11.2. The application of FLN spectroscopy to the isomer-specific analysis of stable and depurinating DNA adducts, originating from both in vitro and in vivo experiments, will be the topic of Section 11.3. We will also illustrate how FLN spectroscopy can be used to probe the interactions of a PAH adduct with the DNA helix (Section 11.4) or to study complex conformational equilibria (Section 11.5). FLN characterization of PAH–protein adducts is discussed in Section 11.6. Earlier reviews covering several of these subjects have been published by Jankowiak and Small [11–13].

11.2. METHODOLOGY

In this section, we will briefly discuss the analytical methodology and instrumentation used for FLN spectroscopic analysis of PAH-DNA adducts.

11.2.1. Sample Handling

PAH adducts to DNA or protein may be obtained from a variety of experiments (in vitro, in vivo, or field measurements). Obviously, the optimal sample handling method will depend on the adduct levels and degree of complexity of the samples. In the case of low exposures, some form of analyte preconcentration will be necessary. Stable adducts bound to macromolecular DNA or protein can, in principle, be analyzed without any further digestion procedures, provided that the number of different adducts in the sample is limited [14–16]. Whole DNA samples, obtained via conventional DNA isolation techniques, are typically analyzed in aqueous buffer solution; the addition of other solvents such as glycerol can provide insight into the helix stability and adduct conformation [17]. FLN spectra of whole DNA samples can also be recorded by freezing dry centrifuge pellets, contained in a quartz sample tube without any solvent, to liquid helium temperatures [14].

When analyzing more complex mixtures, it may be necessary to carry out some form of chromatographic separation prior to FLN analysis. Gel techniques such as polyacrylamide gel electrophoresis (PAGE) are widely used to separate DNA fragments, while thin-layer chromatography (TLC) separation of radioactively labelled adducts (^{32}P postlabeling) is used for single nucleotides from digested DNA. In both cases the DNA bands or necleotide spots need to be extracted from the matrix and redissolved in a suitable solvent for optimal sensitivity and background reduction. With some other types of TLC material, such as C-18 coated plates, FLN spectroscopy can be carried out directly on the plate (see Chapter 10). A new and promising approach is the on-line analysis of PAH–DNA adducts after separation by capillary electrophoresis and cooling of the separated analytes inside the capillary with liquid helium (Chapter 10).

Many of the samples described in this chapter were obtained in an off-line fashion after semipreparative high-performance liquid chromatography (HPLC) separation. Fractions were collected before, during, and after the most relevant chromatographic peaks (as detected with a conventional fluorescence detector), evaporated to dryness, and stored at −20°C. For FLN analysis the samples were thawed, dissolved in a glass-forming mixture of water/glycerol/ethanol (in most cases 2:2:1), transferred to a 20-μL quartz tube, and sealed with a rubber septum. A quick screening of the samples under low-resolution conditions (77 K, short-wavelength excitation at 308 nm and a broad, low-resolution detection window) is helpful to select the most promising fractions and to check for potential interferences.

11.2.2. Instrumentation

In FLN spectroscopy, energetically equivalent sites are excited by means of a narrow-banded laser source. Both continuous-wave or pulsed lasers can be used; the latter type offers the possibility of time-resolved detection for improved selectivity or background reduction. Lasers offering nanosecond pulses are suitable, but the photon energy of subpicosecond lasers is not sufficiently well defined, as follows from the Heisenberg uncertainty principle. Only part of the analyte's absorption spectrum is suitable for FLN spectroscopy: The chromophore can only be excited via its inhomogeneously broadened origin absorption profile (origin band excitation) or via vibronic $S_1 \leftarrow S_0$ absorption bands (vibronic excitation), as explained in detail in Chapter 8. Excitation into the S_2 or higher electronic state does not lead to line-narrowed spectra, since higher electronic states are not correlated to the emitting state S_1. This means that different excitation ranges are required for different chromophores. For versatility, solid-state lasers with broad tuning ranges or dye lasers with easily exchangeable dye circuits are preferred. In order to reduce hole-burning phenomena, alignment and test runs are carried out with an excitation wavelength slightly different from that of the actual measurement.

Samples can be cooled using either a closed-cycle refrigerator (minimum temperature typically 10 K) or a helium bath cryostat (4.2 K). The first type does not require an extensive cryogenic infrastructure, while sample throughput will be higher with the latter. Usually, several samples will be cooled at the same time, and samples and sample holder may be precooled in liquid nitrogen to reduce the cooldown time (closed-cycle refrigerators) or to save helium (bath cryostats). In our experience, the cooling procedure does not affect the shape, resolution, or intensity of the FLN spectra of most PAH adduct samples, but it may influence the degree of unwinding in short double-stranded DNA oligomers [18].

Fluorescence from the sample is collected by means of a positive lens and projected on the entrance slit of a high-resolution monochromator. Care should be taken to optimize the collection efficiency of the lens system, taking into account the direction of specular reflections and the F/n number of the monochromator. Scanning monochromators and photomultiplier detectors (PMTs) can be used, but the analysis time is greatly reduced if an intensified linear diode array or CCD camera is available and a complete spectrum can be recorded simultaneously. Such multiplex detectors have the additional advantage that temporal changes due to analyte degradation, hole burning, or laser instability affect all points of the spectrum to the same extent. The resolution required for observing vibrationally resolved FLN spectra is about 0.1 nm. The wavelength calibration of both the laser and the monochromator/detector should be carefully checked. When comparing PAH–DNA adduct spectra of closely related compounds (for instance, see below the case of DMBA depurinating adducts), differences in vibrational frequencies of only 3–6 cm^{-1} need to be distinguished. In the case of pulsed excitation, pulser units that switch the intensifier chip of the detector on and off at specific times following the excitation pulse can be applied for time-resolved detection. In our experience, PMTs with boxcar integrators do not offer the same level of discrimination between instant (scatter or Raman) emission and longer-living luminescence.

With origin-band excitation, no vibrational relaxation takes place, and the subset of excited molecules (or isochromat) will show line-narrowed fluorescence when returning to different vibrational levels of the S_0 ground state. This way, only vibrational information of the ground state is obtained. On the other hand, when exciting into the vibronic region of the first excited state S_1, the excited analyte molecules will first decay to the lowest vibrational level of the first excited state before undergoing radiant decay. Thus the observed FLN spectra contain information on the vibrational levels of both the excited state and the ground state, which, when combined with the relative vibronic intensities, serve as a fingerprint. In our studies, we practically always use vibronic excitation, and record the emission in the 0–0 band region. The energy difference between the laser excitation source and the observed 0–0 multiplet is equal to vibrational energy levels of the S_1 excited state. By probing a series of excitation wavelengths, all excited-state vibrational frequencies are obtained, and these can be used for unambiguous analyte identification. Often the low-energy vibrations (300–800 cm^{-1}) are most useful when similar compounds need to be distinguished.

11.2.3. Standard Adduct Samples

Unfortunately, it is currently not possible to calculate and predict the S_1 vibrational levels with the necessary degree of accuracy. Therefore, FLN spectra of unknown compounds are difficult to interpret. In practice, standard spectra need to be available for comparison, and one of the most challenging aspects in this field of research is the preparation of (isomerically) pure, well-characterized standard adducts. Dozens of different PAH adduct standards have been obtained in recent years from various sources; without these reference standards this research would not have been possible. All standards have been purified extensively and were characterized using several different techniques, such as NMR, mass spectrometry, and circular dichroism. For proper comparison, it is important to record standard spectra and spectra from real samples under identical conditions (excitation wavelength, solvent composition). Especially when closely related compounds need to be distinguished (for example, B[*a*]P tetrols and B[*a*]P–DNA adducts with the same stereochemistry), minor differences in vibrational patterns or small shifts in the position of the 0–0 envelope should not be obscured by experimental inconsistencies.

11.3. FLN IDENTIFICATION OF PAH–DNA ADDUCTS

The first application of FLN spectroscopy to the characterization of PAH–DNA adducts was published in 1984 [19]. Since then, FLN spectroscopy has been employed extensively for identification of both stable and depurinating adducts from a variety of PAHs, including B[*a*]P, DMBA, and DB[*a*,*l*]P (see Fig. 11.1 for chemical structures). Over the years, the FLN instrumentation and methodology has been improved significantly in terms of sensitivity, selectivity, and reproducibility. The sensitivity of the technique is particularly useful for in vitro and in vivo studies, where often only picomoles of a particular adduct (or even less) are available. In the case of BPDE adducts (which contain a moderately strong pyrene-type fluorophore), we can currently detect 1 adduct per 10^8 base pairs in 100 μg of DNA which corresponds to ca. 1 femtomole of adduct. B[*a*]P derivatives with a full B[*a*]P-type chromophore and a stronger S_1-S_0 absorption can be detected at even lower levels. The spectral resolution of $5\,cm^{-1}$ is often sufficient to distinguish between closely related species, such as adducts bound to different nucleophilic centers of a given base. Another unique feature of the FLN technique is that it can also be used directly to compare adducts bound to single nucleotides with adducts bound to oligonucleotides or whole, intact DNA. Here, we will provide an overview of the types of adducts that were positively identified by comparing their high-resolution FLN spectra with those of appropriate standard adducts. We will also briefly describe some of the new insights in the field of cancer initiation that this research has helped to develop.

11.3.1. Benzo[*a*]pyrene-DNA Adducts

The first FLN study of PAH–DNA adducts concerned (+/−)-*anti*-BPDE, covalently bound to the exocyclic NH_2 group of deoxyguanosine [19]. As illustrated in Figure 11.2, *cis*- and *trans*-adducted BPDE species (the terminology refers to the stereochemistry of the nucleophilic addition to the 9,10-epoxide) yield very different FLN spectra. Also, adducts from *anti*-BPDE (7-hydroxy group and 9,10-epoxide on different sides of the saturated ring; see Fig. 11.1) are easily distinguished from *syn*-BPDE adducts (7-hydroxy group and 9,10-epoxide on the same side of the saturated ring) [12]. On the other hand, diastereoisomers like (+)-*trans-anti*-BPDE-N^2-dG and (−)-*trans-anti*-BPDE-N^2-dG, in which the chromophoric moieties are enantiomers that are not directly attached to further asymmetric centers, yield very similar spectra that can only be distinguished if they adopt significantly different conformations in a DNA or oligonucleotide helix [17]. Pure enantiomers cannot be distinguished by FLNS alone.

FLN spectra of intact DNA samples showed that the *trans-anti*-BPDE adduct was the major stable DNA adduct from B[*a*]P in a number of *in vitro* and *in vivo* studies with rodent systems [14, 15]. Unexpectedly, the major adduct observed in liver DNA from English sole exposed to B[*a*]P was of *syn*-BPDE configuration [11].

Adducts bound to the exocyclic aminogroups of Ade (N^6) or Gua (N^2) are relatively stable in the DNA helix. On the other hand, adducts may also be formed via binding to other nucleophilic centers of the DNA bases, such as N3 or N7 of Ade or N7 or C8 of Gua. This will result in destabilization of the glycosylic bond and subsequent loss of the purine base from the DNA helix (depurination; see Fig. 11.1 for some key structures). Chen et al. and Devanesan et al. quantified the amounts of stable adducts and depurinating adducts formed in vivo in mouse skin and in vitro in rat liver microsomes after exposure to B[*a*]P, 7,8-dihydrodiol B[*a*]P, or *anti*-BPDE [20,21]. The depurinating adducts BPDE-10-N7Gua and BPDE-10-N7Ade were found in all samples. A first identification was based on HPLC retention times. Subsequently, FLN spectra of collected HPLC fractions were recorded and compared with authentic standards for structure confirmation. An example illustrating how FLN spectroscopy allows for fingerprint identification of BPDE-10-N7Gua is shown in Figure 11.3. The spectra show origin multiplet emission resulting from laser excitation into an S_1 vibronic region (see also Chapter 8). In the spectra of the HPLC fractions from in vitro (middle curve) and in vivo (bottom curve) experiments, both the vibrational frequencies and their relative intensities are identical to those of the synthetic adduct standard (top curve).

A second class of depurinating adducts concerns reaction products of the B[*a*]P radical cation, formed via the one-electron oxidation pathway, and subsequently bound at its C6 position to Ade or Gua. For our studies, five synthetic depurinating adducts were available: B[*a*]P-6-N1Ade, B[*a*]P-6-N3Ade, B[*a*]P-6-N7Ade, B[*a*]P-6-N7Gua and B[*a*]P-6-C8Gua. While the first adduct was not detected and the B[*a*]P-6-N3Ade adduct was only observed in a horseradish peroxidase-induced in vitro system, the other three adducts were found in large amounts in all systems studied [20]. In fact, both in the rat liver microsome

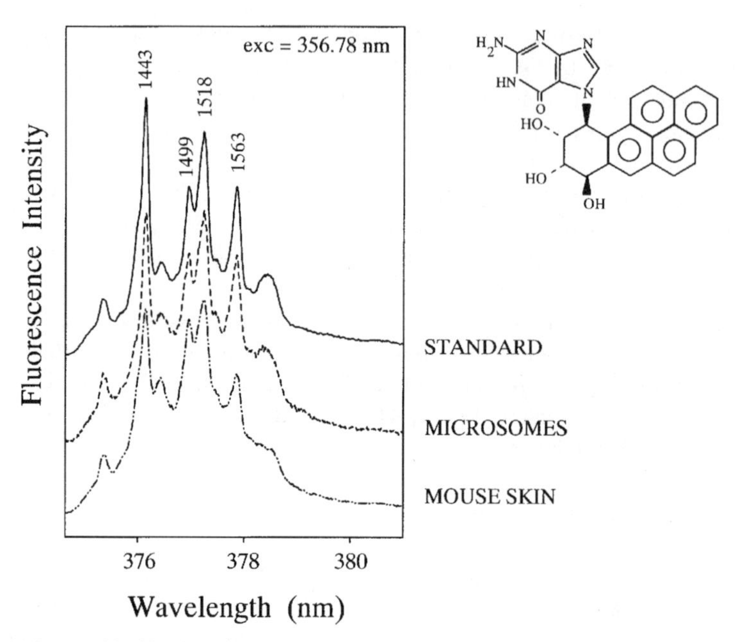

Figure 11.3. FLN identification of the depurinating adduct *trans-anti*-BPDE-6-N7Gua. Top spectrum: synthetic adduct standard; middle spectrum: product isolated from microsome-induced in vitro reaction; bottom spectrum: product isolated from a mouse skin in vivo experiment [20]. Peaks are labeled with their excited-state vibrational frequencies in cm^{-1}; $\lambda_{ex} = 356.78$ nm.

experiment and in the mouse skin experiment, the total amount of depurinating, one-electron oxidation adducts was larger than that of stable adducts. Two of the above-mentioned depurinating adducts, B[*a*]P-6-N7Gua and B[*a*]P-6-C8Gua, were detected in another in vivo system, after application of B[*a*]P to rat mammary glands [22]. In this case, FLN spectroscopy proved especially important, since some HPLC fractions happened to co-elute with synthetic standards, but actually contained other unknown B[*a*]P derivatives that could only be distinguished based on their FLN spectra.

The presence of the B[*a*]P-6-N7Gua adducts in urine of rats, treated with B[*a*]P, has been demonstrated [23]. The same target analyte was recently used to study human exposure to PAHs. B[*a*]P-6-N7Gua adducts were identified in urine of individuals exposed to coal smoke. The analysis was carried out by a combination of solid-phase extraction/HPLC and CE-FLNS [24]. It was found that individuals exposed to coal smoke excreted about 150–250 pmol of B[*a*]P-6-N7Gua daily.

While for many years the monooxygenation pathway and formation of stable adducts was believed to be the only relevant mechanism in PAH carcinogenesis, these recent results established the importance of depurination and the one-electron oxidation pathway for PAH-induced DNA damage. As shown in the next sections, several other PAHs were found to react in a similar fashion to DNA.

11.3.2. 7,12-Dimethylbenz[*a*]anthracene–DNA Adducts

For the potent carcinogen DMBA, the relative abundance of depurinating adducts was even more striking. In mouse skin DNA, virtually all (99%) adducts were depurinating adducts formed via the one-electron oxidation mechanism [3]. The most accessible electrophilic site in the DMBA radical cation appears to be the C-12 methyl group: The adducts identified were 7-methyl-B[*a*]A-12-CH_2-N7Ade (79%) and 7-methyl-B[*a*]A-12-CH_2-N7Gua (20%) [3, 22]. Both purine moieties are covalently bound at the N7 nitrogen, and the bases are separated from the chromophore by a methylene link (see Fig. 11.4 for structures). Nevertheless, the two adducts can be distinguished based on small but significant differences of the order of 3–6 cm^{-1} in their low-energy S_1 vibrational frequencies [22]. These minor spectral differences are shown in Figure 11.4, featuring also the FLN spectra of the same two adducts isolated from rat mammary gland following in

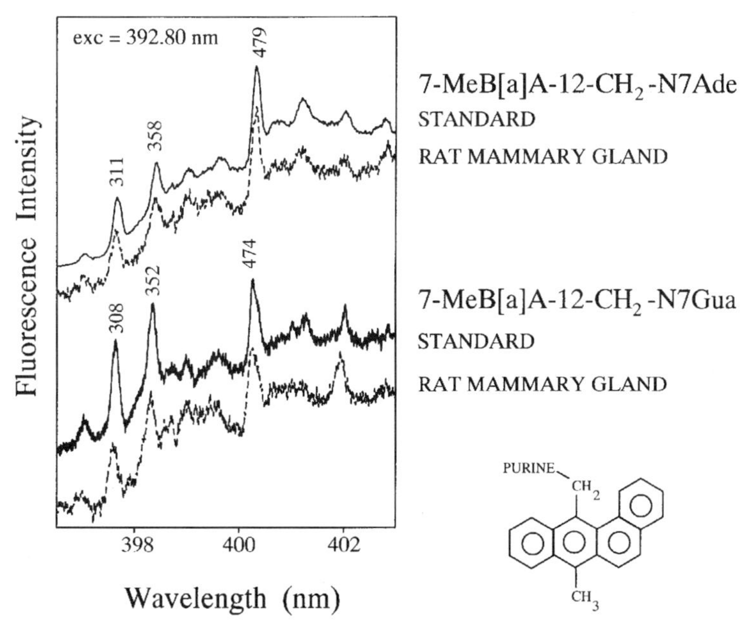

Figure 11.4. FLN spectra of the depurinating adducts 7-MeB[*a*]A-12-CH_2-N7Ade and 7-MeB[*a*]A-12-CH_2-N7Gua isolated from rat mammary gland [21] and comparison with synthetic adducts. Peaks are labeled with their excited-state vibrational frequencies in cm^{-1}; $\lambda_{ex} = 392.80$ nm.

vivo exposure to DMBA. The major adduct detected was the one formed with adenine, and this agrees very well with the fact that an A → T transversion is the major mutation observed in the H-*ras* oncogene in mouse skin papillomas [25].

11.3.3. Dibenzo[*a,l*]pyrene–DNA adducts

Toxicological studies have shown that DB[*a,l*]P is the most potent carcinogenic PAH known to date [4]. Both stable and depurinating adducts can be formed. FLN spectroscopy played a major role in unraveling the extremely complex mixture of stable and depurinating adducts produced from DB[*a,l*]P [26, 27]. Similar to B[*a*]P, stable adducts from DB[*a,l*]P are formed via a diolepoxide intermediate, of which several stereoisomers exist. Standard adducts of *syn*- and *anti*-DB[*a,l*]PDE bound at the C-14 position to N^6-dA, N^6-dAMP, N^2-dG, and N^2-dGMP have been synthesized [26]. These adducts can be formed via either *trans*- or *cis* addition to the epoxide. Furthermore, because of the chirality of the deoxyribose moiety, the (+)- and (−)-DB[*a,l*]PDE enantiomers yield diastereomeric adducts, rendering the analysis of these adducts even more complex. *syn*- and *anti*-DB[*a,l*]PDE-derived standard adducts to –dG (or –Gua) and –dA (or –Ade), as well as DB[*a,l*]P-derived one-electron oxidation adduct standards, were fully characterized. The library of FLN and low-resolution spectra of four pairs of isomers of stable –dG and –dA adducts, and depurinating adduct standards formed by electrochemical oxidation of DB[*a,l*]P in the presence of adenine and guanine, was completed. The purity of standards was checked with CE–FLNS. FLN spectra of DB[*a,l*]P standard adducts have been recorded and serve as a database for the identification of biologically produced stable adducts. As was the case for BPDE, adducts with different stereochemistries (*cis*- vs. *trans* addition, *syn* vs. *anti* configurations), show sufficiently different spectra.

For the analysis of stable adducts, macromolecular DNA is enzymatically digested. The mononucleotides are subsequently fractionated by multidimensional HPLC, and identified by means of FLN spectroscopy. A major complicating factor is the fact that most DB[*a,l*]PDE adducts exhibit dual conformations, which are often solvent dependent. Spectroscopic studies and molecular-dynamics calculations of model depurinating DB[*a,l*]PDE adducts and DB[*a,l*]P tetrols were carried out to help us understand the underlying conformational changes of the saturated ring and characterize the stereochemistry of more complex adducts (see further Section 11.5). DB[*a,l*]P was also found to yield a variety of depurinating adducts via both the one-electron oxidation and the diolepoxide pathway. Two-dimensional HPLC and FLN spectroscopy was used for analysis of depurinating adducts, formed by exposing DNA to DB[*a,l*]P in a rat liver microsome in vitro system. The following adducts were produced: DB[*a,l*]PDE-10-N3Ade, DB[*a,l*]PDE-10-N7Ade, DB[*a,l*]PDE-10-N7Gua, and DB[*a,l*]PDE-10-C8Gua (one-electron oxidation products), as well as *trans-syn*-

DB[*a*,*l*]P diolepoxide-14-N7Ade and *anti*-DB[*a*,*l*]P diolepoxide-14-N7Gua (monooxygenation products) [27]. In total, the depurinating adducts constituted 84% of total adducts. More recently, we succeeded in identifying most of these adducts in rat mammary gland tissue and in mouse skin, following in vivo exposure [13]. The major depurinating adduct formed in rat mammary gland was the one-electron oxidation product DB[*a*,*l*]P-10-N7Ade. An example of FLN spectra that confirmed the formation of DB[*a*,*l*]P-10-N7Ade adducts in HPLC fractions from these two in vivo experiments is shown in Figure 11.5.

In order to illustrate the toxicological implications of the results shown in this section, Table 11.1 summarizes experimental data obtained in recent years on the in vivo formation in mouse skin of stable and depurinating adducts from the PAHs discussed in this chapter. The results can be directly compared to the carcinogenicity of the PAH or PAH metabolite concerned and the clonal H-*ras* mutations found in the PAH-induced papillomas, [3, 25, 28]. The data demonstrate that the strongest carcinogens included in Table 11.1, DB[*a*,*l*]P, DB[*a*,*l*]P-11,12-dihydrodiol, 7,12-DMBA, B[*a*]P, and B[*a*]P-7,8-dihydrodiol, induced a substantial fraction (37–99%) of depurinating adducts. A moderate carcinogen, *syn*-DB[*a*,*l*]PDE induced 32% of depurinating adducts. The weaker PAH carcinogens (*anti*-DB[*a*,*l*]PDE and *anti*-BPDE) formed comparatively more (>90%) stable adducts.

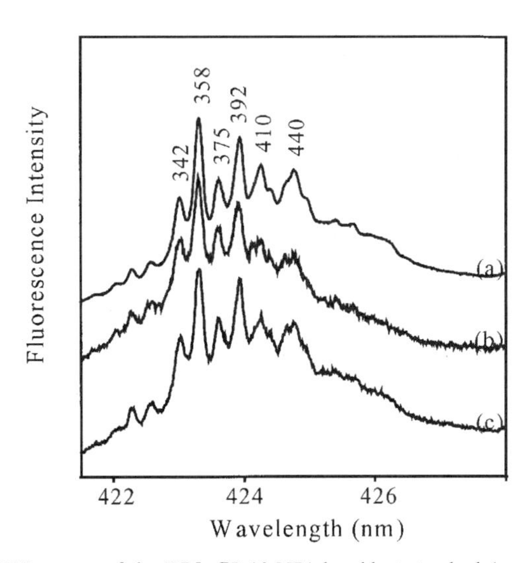

Figure 11.5. FLN spectra of the DB[*a*,*l*]P-10-N7Ade adduct standard (spectrum a), the biological fraction from rat mammary gland tissue (in vivo, spectrum b) and mouse skin experiment (spectrum c). $\lambda_{ex} = 416.00$ nm; $T = 4.2$ K. The FLN peaks are labeled with their excited-state vibrational frequencies in cm^{-1}.

Table 11.1. Relative Abundance of Stable and Depurinating Adducts, Major H-*ras* Mutations and Carcinogenic Potency of Several PAHs and PAH Metabolites in Mouse Skin

	Adducts Formed in Mouse Skin[a]			Relative
PAH–carcinogen	Stable adducts	Depurinating adducts	H-*ras* Mutations[b]	Carcinogenicity[c]
B[*a*]P	N^2dG (23%)	N7Gua+C8Gua (46%) N7Ade (25%)	GGC → GTC (condon 13) CAA → CTA (condon 61)	+++
B[*a*]P-7,8-dihydrodiol	N^2dG (61%)	N7Gua (12%) N7Ade (25%)	GGC → TTC (codon 13) CAA → CTA (codon 61)	+++
anti-BPDE	N^2dG (98%) N^6dA (0.7%)	N7Gua (0.1%) N7Ade (1.2%)	GGC → TTC (codon 13)	+/−
7,12-DMBA	Unidentified (~1%)	N7Ade (79%) N7Gua (20%)	CAA → CTA (codon 61)	++++
DB[*a*,*l*]P	N^2dG (0.52%) N^6dA (0.44%)	N7Gua+C8Gua (18%) N7Ade+N3Ade (81%)	CAA → CTA (codon 61)	+++++
DB[*a*,*l*]P-11,12-dihydrodiol	N^6dA (8%) N^2dG (10.8%)	N7Ade (75%) N7Gua (5%)	CAA → CTA (codon 61)	++++
syn-dB[*a*,*l*]PDE	N^6dA (15.3%) N^2dG (49.6%)	N7Ade (25%) N7Gua (7%)	−	++
anti-DB[*a*,*l*]PDE	N^6dA (32%) N^2dG (59%)	N7Ade (2.5%) N7Gua (0.5%)	CAA → CTA (codon 61)	+

[a] Results from Cavalieri and Rogan [3].
[b] From Chakravarti et al. [25, 28].
[c] Evaluation of carcinogenicity is based on initiation–promotion experiments, which measure the tumor-initiating potential in mouse skin. Carcinogenic activity: +++++ (extremely active); ++++ (very active); +++ (active); ++ (moderately active); + (weakly active); +/− (very weakly active); − (inactive).

Interestingly, Chakravarti et al found a strong correlation between the major depurinating base (81% Ade from DB[*a*,*l*]P, 75% Ade from DB[*a*,*l*]P-11,12-dihydrodiol, and 79% Ade from 7,12-DMBA) with the major clonal H-*ras* mutation in the PAH-induced papillomas, being an A $\rightarrow$ T transversion of the central adenine base in codon 61 [25]. In those cases where a relatively large proportion of depurinating adducts was formed involving Gua (i.e., B[*a*]P, B[*a*]P-7,8-dihydrodiol), a second major mutation (G $\rightarrow$ T) was found in codon 13 of the induced papillomas. The relative proportion of Ade/Gua depurination correlated with the relative incidence of these two key mutations. These results are consistent with the idea that the formation of depurinating adducts constitutes to a major pathway of tumor initiation [3, 24, 28, 29].

11.4. PROBING THE ENVIRONMENT OF PAH–DNA ADDUCTS

It has been known for some time that different stereoisomers of a given PAH metabolite often have strongly different mutagenic or carcinogenic properties, even when the adduct formation rates are similar. Mutagenicity is also strongly base sequence dependent. In order to explain these phenomena, it has been suggested that adduct conformations play an important role in key genetic processes like DNA repair and DNA duplication [30]. In turn, the energetically most favorable conformation within a double helix will be strongly dependent on the stereochemistry of the adduct and the particular base sequence of the adducted strand and the partner strand. In this section, we will provide some examples of how FLN spectroscopy can be used to establish these conformations by probing the interactions between the PAH chromophore and its microenvironment.

Low-temperature fluorescence studies of DNA adducts, formed by reaction of (+)- and (−)-*anti*-BPDE with sequence-defined polynucleotides, revealed that the *trans* and *cis* adducts from both diolepoxides can adopt three DNA conformations: external, base stacked, and intercalated [31, 32]. These conformations can be distinguished on the basis of the fluorescence origin band, shifting to the red in going from external to internal conformations. This redshift is accompanied by an enhanced electron–phonon coupling strength, related to the fact that the S_1-state of the pyrene-type chromophore acquires increasing charge-transfer character, due to mixing of its $\pi\pi^*$ state with pyrene-base charge-transfer states. Independent confirmation of the external versus intercalated nature of these adducts had been obtained from quenching studies [33]. The approach of using FLN spectroscopy to assess the degree of chromophore–DNA base interaction was then used to study the conformations as a function of adduct stereochemistry (Section 11.4.1), as a function of base sequence (Section 11.4.2), and the influence of adduct conformation on DNA repair rates in vivo (Section 11.4.3).

11.4.1. Conformational Studies of B[*a*]P-DNA Adducts as a Function of Adduct Stereochemistry

A combination of polyacrylamide gel electrophoresis with low-temperature spectroscopy was used to investigate the conformations of adducts formed by the reaction of (+)- or (−)-*anti*-BPDE by either *trans* or *cis* addition at N^2-dG to the duplex oligonucleotide 5′-d(CCATC<u>G</u>CTACC)·(GGTAGCGATGG) [17]. The underlined <u>G</u> denotes the base carrying the adduct. For these adducted oligomers the solution NMR structures of the (+)-*trans*-, (−)-*trans*- and (+)-cis adducts had been reported [34–36].

These three adducts were then used to test our approach to determine adduct conformations in DNA by FLN spectroscopy. In addition, the results might lead to a conformational assignment for the (−)-*cis* adduct and the detection of minor conformations that cannot be observed by solution NMR because of dynamic equilibrium effects. Only the main findings will be described here. Figure 11.6 shows FLN spectra obtained for the four adduct stereoisomers in aqueous buffer matrix. The (+)-*trans-anti* adduct yields a well-resolved FLN spectrum with strong ZPLs in the 378 nm region (curve a). This indicates that there are no strong stacking interactions with the bases, which is in agreement with the solution conformation established for this adduct by Cosman et al. [34]. However, the broad emission band near 380 nm indicates the presence of a second, minor conformation with stronger stacking interactions. The FLN spectrum of the (−)-*trans-anti* adduct (curve b) does not show any broadband emission. This adduct exists in a purely external conformation, in agreement with the solution conformation reported in the literature [35]. Spectrum c shows that the FLN spectrum of the (+)-*cis-anti* adduct is distinctly different. The intensities of the ZPLs are relatively weak and superimposed on a strong broadband emission at the low-energy side of the spectrum. This is due to strong electron–phonon coupling, as expected for an intercalated conformation, and is in agreement with the solution conformation established for this adduct by Cosman et al. [36]. Finally, curve d shows that the FLN spectrum of the (−)-*cis* adduct is very similar to that of its (+)-*cis* counterpart; apparently also this adduct exists predominantly in an intercalated conformation.

11.4.2. Flanking Base Effects on the Conformation of B[*a*]P–DNA Adducts

As a second example, the conformations of the *trans* adduct of (+)-*anti*-BPDE to N^2-dG, the major stable DNA adduct formed from B[*a*]P, have been studied as a function of flanking bases in short oligonucleotides [18]. Three 11-mer oligonucleotides d(5′-CTATG$_1$G$_2$G$_3$TATC-3′) were synthesized containing the (+)-*trans-anti* BPDE adduct at one specific guanine base of the GGG sequence,

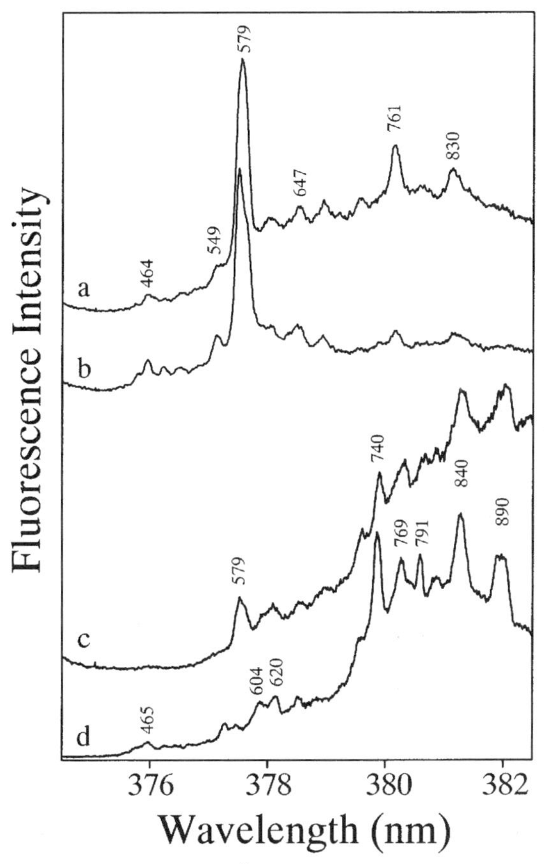

Figure 11.6. FLN spectra of *anti*-BPDE-N^2-dG adducts of different stereochemistries in duplex oligonucleotides. Curve a: (+)-*trans* adduct; curve b: (−)-*trans* adduct; curve c: (+)-*cis* adduct; curve d: (−)-*cis* adduct. $\lambda_{ex} = 369.48$ nm. The FLN peaks are labeled with their excited-state vibrational frequencies in cm^{-1}. (Reproduced from Ref. 17.)

a known mutational hot spot [30]. Spectra were recorded of the single-stranded (ss) oligomers or after duplex formation with the complementary strand d(GATACCCATAG).

Figure 11.7 shows the FLN spectra for the BPDE adduct bound to the above-mentioned 11-mer ds-oligonucleotide at G_3 (a), G_2 (b) and G_1 (c), obtained with $\lambda_{ex} = 365.58$ nm. The FLN spectra show the same vibrational frequencies, which is to be expected since all three oligomers contain the same adduct of the same *trans-anti* stereochemistry. However, there are significant differences between the spectra when one compares the relative intensities of the narrow ZPLs and the

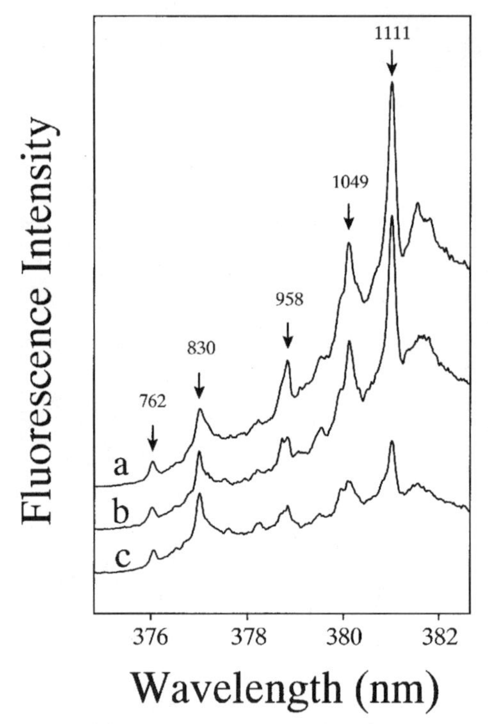

Figure 11.7. FLN spectra of (+)-*trans-anti*-BPDE-N2-dG adducts bound to different guanines in duplex oligonucleotides. Curve a: adduct bound to (d(... $G_2\mathbf{G}_3$T...) · (...ACC...); curve b: adduct bound to d(... $G_1\mathbf{G}_2G_3$ ...) · (...CCC...); curve c: adduct bound to d(...T$\mathbf{G}_1G_2$...) · (...CCA...). Aqueous buffer matrix, $T = 4.2$ K; $\lambda_{ex} = 365.58$ nm. The FLN peaks are labeled with their excited-state vibrational frequencies in cm^{-1}. (Reproduced from Ref. 18.)

broadbanded, red-shifted emission at 380–383 nm. As argued above, the latter is due to electron–phonon coupling and is a clear indication of strong stacking interactions with the DNA bases. Curves a and b are very similar, but distinct from spectrum c. The similarity of spectra a and b suggests that the nature of the 3′-flanking base (T or G, respectively) has little effect on the adduct conformation. In both cases the pyrene-type chromophore of the BPDE adduct experiences strong electron–phonon coupling, presumably due to intercalation. For the adduct bound to G_1 (spectrum c) no intense, broadbanded emission is observed, indicating that this adduct exists in a helix-external conformation. The differences between spectra b and c indicate that the nature of the 5′-flanking base (G or T, respectively) strongly influences the preferred conformation of this particular adduct, in agreement with the directionality of this adduct found by Cosman [34].

Interestingly, chromophore–base interactions were also observed in the single-stranded samples [18].

PAGE separation followed by FLNS was also used to study the binding and conformations of (−)-*anti*- and (+)-*anti*-BPDE to several sequence-defined double-stranded oligomers [37]. Two of the oligomers contained central 5′-RAGGAR-3′ sequences (R = purine), which appear to be frequently mutated by racemic (±)-*anti*-BPDE. Two other oligomers contained a central 5′-CCGG-3′ or 5′-TGGT-3′ sequence, which are strongly preferred for covalent binding but are less frequently mutated. Importantly, the (+)-*anti*-BPDE bound to the more mutagenically inclined 5′-RAGGAR-3′ sequences yielded few external-type adducts and an unusually high proportion of base-stacked adducts in comparison to the other two sequences. These results suggest a possible role of (partially) base-stacked adduct conformations in mutagenesis.

11.4.3. Persistence of PAH–DNA Adducts in Mouse Skin in Vivo

As mentioned above, the three-dimensional structure of the PAH adduct within the DNA helix may have an important impact on the efficiency with which the cell recognizes and repairs such lesions. The time evolution of BPDE–N^2-dG adducts in mouse skin DNA, following a single dose of 1 μmol B[*a*]P/mouse, was studied using HPLC and FLN spectroscopy [14]. Total B[*a*]P–DNA binding reached a maximum at 24 hours after treatment, then declined rapidly until 4 days after treatment and more slowly thereafter. Using selective laser excitation of macromolecular DNA, it was found that the formation rates were similar for helix-external and for base-stacked/intercalated adducts, but the rates of repair were very different [14]. As discussed above, the relative ratio of sharp line-narrowed emission and broad, structureless emission at lower energy provides a measure of the relative amounts of external versus base-stacked adducts. The DNA sample obtained 48 h after in vivo application of B[*a*]P contains much larger amounts of base-stacked/intercalated adducts than the 8-h sample, indicating a relatively slow repair rate for these adducts. The possible biological significance of these observations of conformation-dependent rates of DNA-adduct repair and their possible dependence on DNA sequence is discussed in Ref. 14.

Following a similar approach, the conformations and repair rates of stable DB[*a*,*l*]P-derived DNA adducts produced in vitro and in vivo (mouse skin) were studied using 32P-postlabeling and laser-based fluorescence techniques [38]. These studies were performed by reacting polynucleotides and CT-DNA with DB[*a*,*l*]PDE metabolites or by exposing native DNA from mouse epidermis to DB[*a*,*l*]P. The results showed that the adducts are heterogeneous, possess different structures, and adopt different conformations. External and intercalated adduct conformations were observed both in CT-DNA and in mouse skin DNA samples. Differences in adduct repair rates were observed; namely, the analysis of mouse skin DNA samples obtained at 24 and 48 h after exposure to DB[*a*,*l*]P clearly showed that external adducts are repaired more efficiently than intercalated adducts. These results, taken together with those mentioned above for B[*a*]P–DNA adducts, indicate that the repair of DNA damage resulting from PAH diolepoxides is conformation dependent.

11.5. CONFORMATIONAL ANALYSIS BY MEANS OF FLN SPECTROSCOPY

The application of FLN spectroscopy to another type of conformational studies will be illustrated in this chapter. In FLN spectroscopy, vibrationally resolved fluorescence spectra are obtained by selecting energetically equivalent sites from an inhomogeneously broadened ensemble of analyte molecules. For a solution of a single, pure compound, the inhomogeneous broadening is due to different solute–solvent interactions in the disordered matrix, as well as to the existence of energetically inequivalent conformations in the case of flexible molecules. In most cases, these flexible analyte molecules can adopt a continuum of conformations, which—in combination with a continuum of host–guest interactions—leads to a Gaussian distribution of $S_1 - S_0$ transitions. Different isochromats selected by the laser usually show identical vibrational patterns under FLN conditions. However, some compounds can adopt two or more different conformations, separated by energy barriers, and each featuring distinct FLN spectra. In this section we will describe how FLN spectroscopy, in combination with computational chemistry, was used to study the conformational equilibria of several depurinating adducts of the fjord-region carcinogen DB[*a*,*l*]P.

Low-resolution spectra recorded at 77 K of *trans-syn*-DB[*a*,*l*]PDE bound to N7Ade, N7Gua, or N3Ade showed an unexpected diversity of fluorescence spectra [39]. In our experience, diolepoxide adducts bound to different DNA bases usually yield fluorescence spectra that are indistinguishable under low-resolution conditions and show only minor differences under FLN conditions. In contrast, the 0–0 origin bands of these DB[*a*,*l*]PDE adducts showed a completely different pattern: the N7Ade adduct showed a strong 0–0 transition at 382 nm, while the N3Ade adduct showed a much broader spectrum with a relatively weak origin band at 389 nm. These origin bands will be referred to as origin I and origin II, respectively. The N7Gua adduct showed both bands, the relative intensities depending on the solvent matrix. More polar solvents, such as water/glycerol 50/50, were found to increase the intensity of origin II. This solvent dependence was a strong confirmation that the different spectra did not originate from a mixture of two structurally different compounds, but from a mixture of species originating from the same compound.

FLN spectroscopy of the three adducts in different matrices yielded two distinct sets of vibrational patterns [39]. Under FLN conditions, the N7Ade adduct did not only show a strong origin I multiplet, but also displayed a weak origin II band at 387 nm, which had not been observed in the low-resolution spectra. The FLN spectra of the two origin multiplets are shown in Figure 11.8, curves a and d, respectively. In both cases the excitation wavelengths were selected to excite the 700–800 cm^{-1} S_1-vibronic absorption region. Interestingly, origin multiplets I and II both show clear ZPLs, but the frequencies, corresponding to S_1 vibrational levels, are very different. Three major vibrational modes, 730, 751, and 798 cm^{-1}, are observed in the first origin region (spectrum a), while origin band II shows only two strong modes at 748 and 772 cm^{-1} (spectrum d). This indicates that the red shift of origin band II could not be solely the result of a stacking interaction between the adenine base and the aromatic moiety. The

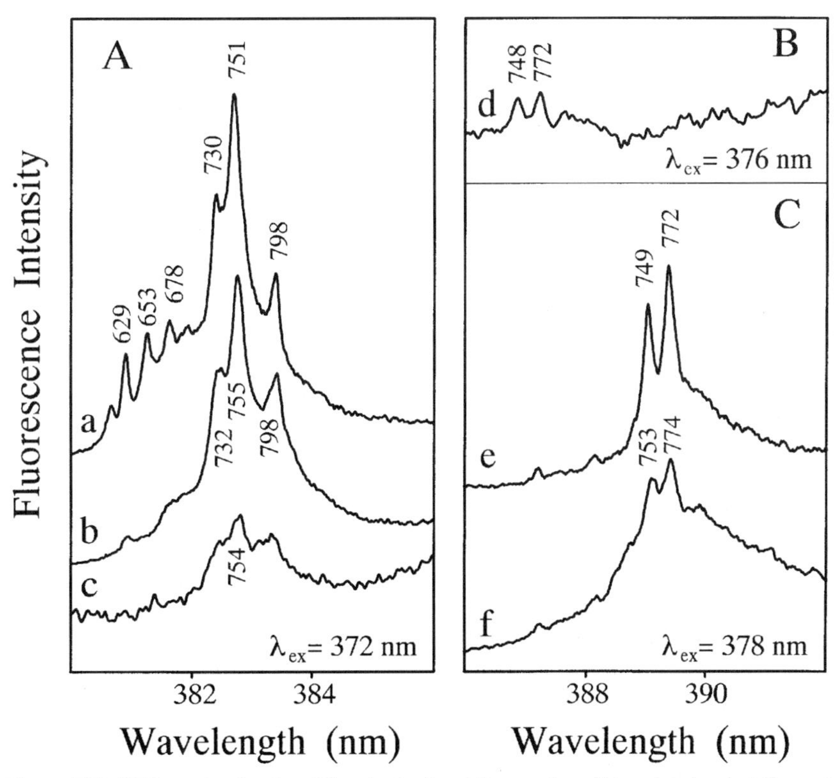

Figure 11.8. FLN spectra showing different vibrational frequencies within origin bands I (frame A, ethanol matrix) and II (Frames B and C, water/glycerol 50 : 50 matrix) of *trans-syn*-DB[*a*,*l*]PDE depurinating adducts. Spectrum a and d: DB[*a*,*l*]PDE-14-N7Ade; spectrum b and e: DB[*a*,*l*]PDE-14-N7Gua; spectrum c and f: DB[*a*,*l*]PDE-14-N3Ade. (Reproduced from Ref. 39.)

latter is not expected to result in significant changes in vibrational frequencies [17]. Instead, the FLN results indicate that the *trans-syn*-DB[*a*,*l*]PDE-N7Ade adduct must possess two distinct molecular conformations. The FLN spectra recorded for the N7Gua (curves b and e) and N3Ade (curves c and f) adducts showed origin I and origin II multiplets, similar to those of *trans-syn*-DB[*a*,*l*]PDE-N7Ade (curves a and d). This indicates that these adducts do not only have the same absolute stereochemistry, but also exist as similar conformational species, albeit with different origin I/origin II intensity ratios [39].

Using molecular mechanics, dynamical simulations, and semiempirical quantum-mechanical calculations, the hypothesis was tested that the occurrence of two distinct 0–0 origin bands for DB[*a*,*l*]PDE-14-N7Ade is due to a conformational equilibrium [39]. The conformational space indeed yielded two potential energy minima; in both structures the five-ring aromatic system is severely distorted. In conformation I, the proximity of the distal ring forces the adenine base into a pseudoaxial position and the cyclohexenyl ring adopts a half-boat structure. In conformation II, the distal ring is bent in the opposite direction, allowing the cyclohexenyl ring to adopt a half-chair structure with the base in a pseudo-equatorial position, partially stacked over the distal ring. In these two conformations the orientations of the saturated ring, especially the bonds directly attached to the chromophore, are completely different, explaining the observation of totally different S_1 vibrational frequencies. The difference in $S_1 - S_0$ transition energies calculated for the two conformers agreed very well with the spectroscopic data, and the relative orientations of the hydrogens bound to the cyclohexenyl ring in the major (half-boat) conformation I were in agreement with the measured proton NMR coupling constants [40]. The optimized ground-state structures are shown in Figure 11.9. The insets show the conformations of the cyclohexenyl ring from a different angle. Of course, the above phenomena can only be observed in special cases, that is, when two different conformations are separated by an energy barrier and are sufficiently close in energy to have both conformations populated.

Later, dual conformations were observed for many other DB[*a*,*l*]PDE derivatives, and it became clear that a full understanding of the conformational equilibria of all DB[*a*,*l*]PDE stereoisomers would be required before attempting to analyze complex adduct mixtures from biological experiments.

Low-temperature fluorescence studies and molecular-dynamics simulations were then carried out for all four stereoisomeric DB[*a*,*l*]P tetrols [41]. These tetrols could be synthesized in relatively large amounts, and their solution NMR spectra were available. The molecular calculations indicated the potential existence of dual conformations for all four stereoisomers, but under FLN conditions this behavior was only observed for the *trans-syn* and the *trans-anti* tetrols. However, the low-temperature fluorescence spectra of *trans-syn*-, *cis-syn*-, *trans-anti*-, and *cis-anti*-DB[*a*,*l*]PDE adducts to deoxyadenosine showed a more

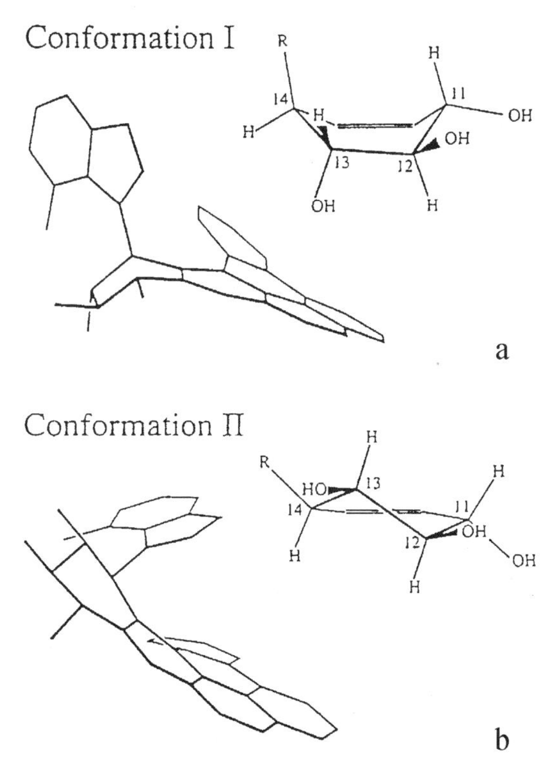

Figure 11.9. Optimized 0 K ground-state conformations of *trans-syn*-DB[*a*,*l*]PDE-14-N7Ade. Conformer I shows the cyclohexenyl ring in a half-boat conformation and the Ade moiety in a pseudoaxial position. Conformer II shows the cyclohexenyl ring in a half-chair conformation and the Ade moiety in a pseudo-equatorial position. The insets show the conformation of the cyclohexenyl ring from a different angle. (Reproduced from Ref. 39.)

complex conformational behavior [42]. All stereoisomeric dA adducts adopt two different conformations with either half-chair or half-boat structures for the cyclohexenyl ring, and an "open"- or "folded"-type configuration between the adenine base and the aromatic system. The stacking interaction in the folded conformation adds stability to the structure, and this is presumably the main reason why in dA adducts the folded conformations are more populated than in the case of the corresponding tetrols. The fluorescence line-narrowed spectra reveal that the *trans-syn-*, *cis-syn-*, *trans-anti-*, and *cis-anti*-DB[*a*,*l*]PDE–14-N^6dA adducts can be distinguished, and their spectra have proven to be very useful for the identification of DB[*a*,*l*]P–DNA adducts formed in biological samples [43, 44]. However, one should be aware of these conformational changes and solvent dependencies when identifying DB[*a*,*l*]P diolepoxide adducts, since adducts showing completely different FLN spectra could still be of the same stereochemistry.

11.6. FLN STUDIES OF PAH–PROTEIN ADDUCTS

Electrophilic intermediates, such as BPDE and other activated PAH metabolites, do not only form adducts with DNA bases; they can also bind to other nucleophilic cellular targets. Binding to DNA has been studied most intensively, as it is generally assumed that the resulting DNA damage is the initial step of PAH-induced chemical carcinogenesis. However, binding to other biologically relevant molecules is of scientific relevance as well. Fluoranthene was found to bind through its 2,3-diol-1,10-epoxide intermediate to rat hemoglobin (Hb) [45]. Protein adducts of B[*a*]P could not be identified directly, but B[*a*]P tetrahydrotetrol hydrolysis products were observed in digested Hb [46]. Upon in vivo administration of thirteen different tritiated PAHs to mice, Singh and Weyand [47] detected binding to blood protein in all cases. Relative adduct patterns were similar for serum protein and globin, but for serum levels were a factor of 10 higher, probably due to the direct accessibility of serum proteins in circulating blood [47]. For a given PAH, the extent of binding to proteins should be directly related to the binding to DNA, and the presence of PAH–protein adducts could serve as a monitoring tool to assess PAH body burdens [47, 48]. An attractive feature of PAH–protein adduct analysis is the ease with which sufficient amounts of sample can be obtained in a relatively noninvasive manner. For example, for an assessment of PAH uptake by humans, one could take blood samples from test subjects. One ml of blood would yield approximately 40 mg of serum albumin and 150 mg of hemoglobin. In comparison, the same amount of blood would yield only 10 μg of lymphocyte DNA [48].

Several methods have been developed for the analysis of PAH–protein adducts. Usually, the protein is isolated and purified, after which the adducted carcinogen is released by acidic or enzymatic hydrolysis, extracted, and analyzed by chromatographic means [46]. However, through this procedure the nature of the carcinogen–protein bond can no longer be determined. Myers and Pinorini [49] reacted human Hb in vitro with tritiated BPDE and separated the nondigested proteins by means of cellulose ion-exchange liquid chromatography. Several product peaks were obtained, and the retention behavior provided some qualitative information on the basic/acidic nature of the adducted proteins. Below, we will describe how FLN spectroscopy can be used to identify further the protein adducts of B[*a*]P in terms of binding sites.

Proteins possess several nucleophilic groups that could be potential target sites for reactive PAH metabolites. BPDE, for instance, could bind to the amino or carboxylic functionality at the protein chain ends, or to certain side chains. Model compounds were synthesized in which BPDE was bound to a carboxylic group (alanyl ester), to sulfur (S-BPDE-glutathione thioether and S-BPDE thioethylamine) or to an amino group (ethylene diamine). All derivatives were *trans* addition products of racemic (±) *anti*-BPDE. Their spectra were recorded under high-resolution conditions and showed that FLN spectroscopy can distinguish between the oxygen, nitrogen, and sulfur adducts [16]. On the other hand, the two sulfur (thioether) adducts yielded very similar spectra. The same approach was then used to analyze human hemoglobin samples from nonsmoking volunteers, reflecting in vivo exposure to ambient B[*a*]P levels. Figure 11.10 (top) shows the high-resolution FLN spectrum of intact (nonhydrolyzed) human hemoglobin obtained using vibronic excitation at 356.9 nm. The spectrum is identical to that

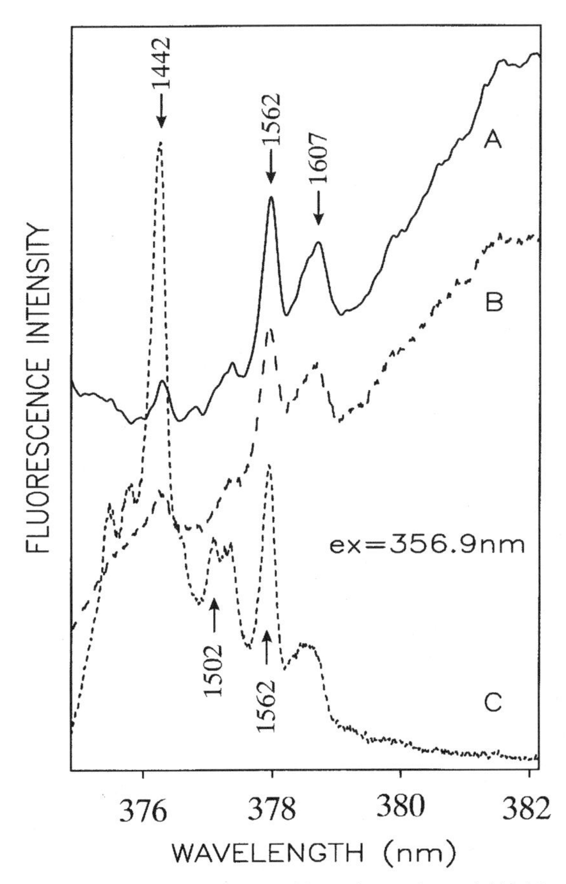

Figure 11.10. Comparison of the FLN spectra of intact human hemoglobin (A) and a synthetic C-10 ester (*trans anti* BPDE 10 alanate) (B). The spectrum is easily distinguished from that of the B[*a*]P tetrol-globin physical complex (curve C). $\lambda_{ex} = 356.9$ nm. (Reproduced from Ref. 16.)

of the ester model compound (middle spectrum), but different from those of the thioethers or amino derivatives (not shown), and also easily distinguished from the BPDE hydrolysis product, *trans-anti*-7,8,9,10-tetrahydrotetrol B[*a*]P (bottom spectrum). These observations show in a very straightforward manner that the main adduction sites for BPDE in hemoglobin are carboxylic groups, either from acidic side chains (aspartic acid, glutamic acid) or the carboxylic endgroup of the protein backbone. The protein adducts bound to O, S, or N are distinguished by FLN spectroscopy on the basis of small, but under high-resolution conditions significant, differences in $S_1 - S_0$ energies, rather than differences in vibrational frequencies. This means that great care should be taken to guarantee that samples and standards are measured under identical conditions, in particular regarding solvent composition. The amount of adduct in the sample of Figure 11.10 (top) was about 30 fmol, which illustrates the sensitivity of the method, even for derivatives with a relatively poor S_1 absorption, such as the pyrene chromophore of BPDE adducts. These results show the potential of FLN spectroscopy for PAH–protein adduct identification.

11.7. CONCLUDING REMARKS

In the previous sections we have illustrated the important role that FLN spectroscopy can play in the identification of PAH adducts and their conformational analysis. For most PAH–DNA adducts studied to date, the method provides excellent selectivity combined with sensitivities in the low femtomole range; for the strongest fluorophores even subfemtomole detection limits can be obtained. The method is applicable to PAH–protein adducts as well.

Over the years we have collaborated with several research groups, working with different types of PAH carcinogens. Especially in cases where new compounds were detected that had not been observed before, the independent confirmation of its identity based on the FLN fingerprint spectra was indispensable. The unambiguous identification of depurinating adducts, often amidst much higher concentrations of metabolites, was only possible with the combined selectivities of HPLC and FLN spectroscopy. Now that the abundance of depurinating adducts has been shown for several PAHs, the biological implications of this type of DNA damage deserves closer study. The data compiled in Table 11.1 strongly suggest that there is a correlation between the carcinogenic potency, the type of mutations, and the relative abundance of depurinating adducts to guanine and adenine bases.

This chapter also provides some recent examples, illustrating how FLN spectroscopy can be applied to study the nature of PAH–DNA interactions, and how adducts in a helix-external conformation can be distinguished from intercalated adducts, based on the extent of electron–phonon coupling. It was shown, for instance, that the conformation of *anti*-BPDE–dGMP adducts depend on the stereochemistry and the base sequence around the adducted base, and which type of adduct is preferentially repaired in in vivo systems. For some DB[*a,l*]P adducts, solvent-dependent changes in conformations were observed; the distinct conformers could be identified based on the FLN vibrational patterns, NMR coupling constants, and molecular-dynamics calculations. This information is currently being used when interpreting FLN spectra of DB[*a,l*]P adducts from biological systems.

At present, the combination of FLN spectroscopy with (multidimensional) HPLC is the most powerful analytical approach to study PAH–DNA adduct formation. There will certainly be perspectives for the coupling of LC and FLN via deposition on a TLC plate, as outlined in Chapter 10. In addition, the recent marriage of CE and FLN, also discussed in Chapter 10, provides another promising approach for the *on*-line detection and spectroscopic identification of such analytes.

Finally, it should be noted that FLN spectroscopy can also be applied to nonfluorescent adducts after derivatization with a fluorescent label (e.g., labeled catechol estrogen-derived DNA adducts; [13, 50]). We are encouraged to think that the FLN and/or CE-FLN spectroscopic studies of labeled DNA adducts (fluorescent upon derivatization) will lead to further insights into the formation and identification of DNA adducts and will expand the applicability of FLN spectroscopy.

Acknowledgment

Ames Laboratory is operated for the U.S. Department of Energy by Iowa State University under contract no. W-7405-Eng-82. This work was supported by the Office of Health and Environmental Research, Office of Energy Research, and partly by Grant POI CA49210 from the National Cancer Institute. The excellent technical assistence from M. Suh is gratefully acknowledged. Thanks are also due to the research groups of Professors E. L. Cavalieri (University of Nebraska Medical Center) and N. E. Geacintov (New York University) for providing us with numerous samples and adduct standards.

REFERENCES

1. R. G. Harvey, *Acc. Chem. Res.* **14**, 218–26 (1981).
2. D. H. Phillips, *Nature* **303**, 468–72 (1983).
3. E. L. Cavalieri and E. G. Rogan, in A. H. Neilson, ed., *The Handbook of Environmental Chemistry, Vol. 3. PAHs and Related Compounds*, Springer-Verlag, Heidelberg, 1998, pp. 81–117.

4. S. Higginbotham, N. V. S. Ramakrishna, S. L. Johansson, E. G. Rogan, and E. L. Cavalieri, *Carcinogenesis* **12**, 875–78 (1993).
5. M. S. Myers, C. M. Stehr, O. P. Olson, L. L. Johnston, B. B. McCain, S.-L. Chan, and U. Varanasi, *Environ. Health Persp.* **102**, 200–15 (1994).
6. P. L. Lioy, J. M. Waldman, R. Harkov, C. Pietarinen, and A. Greenberg, *Arch. Environ. Health* **43**, 304–12 (1988).
7. F. J. Jongeneelen, R. B. M. Anzion, P. T. J. Scheepers, R. P. Bos, P. Th. Henderson, E. H. Nijenhuis, S. J. Veenstra, R. M. E. Brouns, and A. Winkes, *Ann. Occup. Hyg.* **32**, 35–43 (1988).
8. J. C. Chuang, S. A. Wise, S. Cao, and J. L. Mumford, *Environ. Sci. Technol.* **26**, 999–1004 (1992).
9. A. H. Conney, *Cancer Res.* **42**, 4875–4917 (1982).
10. E. Huberman, L. Sachs, S. K. Yang, and H. V. Gelboin, *Proc. Natl. Acad. Sci. USA* **73**, 607–11 (1976).
11. R. Jankowiak and G. J. Small, *Anal. Chem.* **61**, 1023A–1029A (1989).
12. R. Jankowiak and G. J. Small, *Chem. Res. Toxicol.* **4**, 256–69 (1991).
13. R. Jankowiak and G. J. Small, in, A. H. Neilson, ed., *The Handbook of Environmental Chemistry, Vol. 3, PAHs and Related Compounds*, Springer-Verlag, Heidelberg, 1998, pp. 119–145.
14. M. Suh, F. Ariese, G. J. Small, R. Jankowiak, A. Hewer, and D. H. Phillips, *Carcinogenesis* **16**, 2561–2569 (1995).
15. P. D. Devanesan, S. Higginbotham, F. Ariese, R. Jankowiak, M. Suh, G. J. Small, E. L. Cavalieri, and E. G. Rogan, *Chem. Res. Toxicol.* **9**, 1113 (1996).
16. R. Jankowiak, B. W. Day, P.-Q. Lu, M. M. Doxtader, P. L. Skipper, S. R. Tannenbaum, and G. J. Small, *J. Am. Chem. Soc.* **112**, 5866–5869 (1990).
17. M. Suh, F. Ariese, G. J. Small, R. Jankowiak, T. Liu, and N. E. Geacintov, *Biophys. Chem.* **56** 281–296 (1995).
18. M. Suh, R. Jankowiak, F. Ariese, B. Mao, N. E. Geacintov, and G. J. Small, *Carcinogenesis* **15**, 2891–98 (1994).
19. V. Heisig, A. M. Jeffrey, M. J. McGlade, and G. J. Small, *Science* **223**, 288–91 (1984).
20. L. Chen, P. D. Devanesan, S. Higginbotham, F. Ariese, R. Jankowiak, G. J. Small, E. G. Rogan, and E.L. Cavalieri, *Chem. Res. Toxicol.* **9**, 897–903 (1996).
21. P. D. Devanesan, N. V. S. RamaKrishna, R. Todorovic, E. G. Rogan, E. L. Cavalieri, H. Jeong, R. Jankowiak, and G. J. Small, *Chem. Res. Toxicol.* **5**, 302 (1992).
22. R. Todorovic, F. Ariese, P. D. Devanesan, R. Jankowiak, G. J. Small, E. G. Rogan, and E. L. Cavalieri, *Chem. Res. Toxicol.* **10**, 941–47 (1997).
23. E. G. Rogan, N. V. S., Ramakrishna, S. Higginbotham, E. L. Cavalieri, H. Jeong, R. Jankowiak, and G. J. Small, *Chem. Res. Toxicol.* **3**, 441–44 (1990).
24. K. P. Roberts, C.-H. Lin, M. Singhal, G. P. Casale, G. J. Small, and R. Jankowiak, On-line Identification of Depurinating DNA Adducts in Human Urine by CE-FLNS Spectroscopy, *Electrophoresis* 2000, **21**, 779–806 (2000).

25. D. Chakravarti, J. C. Pelling, E. L. Cavalieri, and E. G. Rogan, *Proc. Natl. Acad. Sci. USA* **92**, 10422–426 (1995).
26. K.-M. Li, N. V. S. RamaKrishna, N. S. Padmavathi, E. G. Rogan, and E. L. Cavalieri, *Polycyclic Aromat. Compd.* **6**, 207–13 (1994).
27. K.-M. Li, R. Todorovic, E. G. Rogan, E. L. Cavalieri, F. Ariese, M. Suh, R. Jankowiak, and G. J. Small, *Biochemistry* **34**, 8043–49 (1995).
28. D. Chakravarti, P. Mailander, S. Higginbotham, E. L. Cavalieri, and E. G. Rogan, *Proc. Am. Assoc. Cancer Res.* **40**, 507 (1999).
29. E. L. Cavalieri and E.G. Rogan, *Pharmacol. Ther.* **55**, 183–99 (1992).
30. H. Rodriguez and E. L. Loechler, *Biochemistry* **32**, 1759–69 (1993).
31. R. Jankowiak, P. Lu, G. J. Small, and N. E. Geacintov, *Chem. Res. Toxicol.* **3**, 39–46 (1990).
32. P. Lu, H. Jeong, R. Jankowiak, G. J. Small, S. K. Kim, M. Cosman, and N. E. Geacintov, *Chem. Res. Toxicol.* **4**, 58–69 (1991).
33. N. E. Geacintov, in R. G. Harvey, ed., *Polycyclic Aromatic Hydrocarbon Carcinogenesis*, American Chemical Society, Washington, D.C., 1985, pp. 107–124.
34. M. Cosman, C. de los Santos, R. Fiala, B. E. Hingerty, S. B. Singh, V. Ibanez, L. A. Margulis, D. Live, N. E. Geacintov, S. Broyde, and D. J. Patel, *Proc. Natl. Acad. Sci. USA* **89**, 1914–18 (1992).
35. C. de los Santos, M. Cosman, B. E. Hingerty, V. Ibanez, L. A. Margulis, N. E. Geacintov, S. Broyde, and D. J. Patel, *Biochemistry* **31**, 5245–5252 (1992).
36. M. Cosman, C. de los Santos, R. Fiala, B. E. Hingerty, V. Ibanez, E. Luna, R. G. Harvey, N. E. Geacintov, S. Broyde, and D. J. Patel, *Biochemistry* **32**, 4145–55 (1993).
37. G. A. Marsh, R. Jankowiak, M. Suh, and G. J. Small, *Chem. Res. Toxicol.* **7**, 98–107 (1994).
38. R. Jankowiak, F. Ariese, D. Zamzow, A. Luch, A. Seidel, D. H. Phillips, F. Oesch, and G. J. Small, *Chem. Res. Toxicol.* **11**, 674–85 (1998).
39. F. Ariese, G. J. Small, and R. Jankowiak, *Carcinogenesis* **17**, 829–37 (1996).
40. A. Luch, *Ph.D. Thesis*, University of Mainz, 1995, p. 127.
41. R. Jankowiak, F. Ariese, D. Zamzow, A. Luch, H. Kroth, A. Seidel, and G. J Small, *Chem. Res. Toxicol.* **10**, 677–86 (1997).
42. K. Roberts, C.-H. Lin, R. Jankowiak, and G. J. Small, *J. Chrom. A* **853**, 159–70 (1999).
43. P. D. Devanesan, F. Ariese, R. Jankowiak, G. J. Small, E. G. Rogan, and E. L. Cavalieri, *Chem. Res. Toxicol.* **12**, 796–801 (1999).
44. P. D. Devanesan, F. Ariese, D. Zamzow, R. Jankowiak, G. J Small, E. G. Rogan, and E. L. Cavalieri, *Chem. Res. Toxicol.* **12**, 789–95 (1999).
45. D. A. Hutchins, P. L. Skipper, S. Naylor, and S. R. Tannenbaum, *Chem. Res. Toxicol.* **1**, 22–24 (1988).
46. S. Naylor, L.-S. Gan, B. W. Day, R. Pastorelli, P. L. Skipper, and S. R. Tannenbaum, *Chem. Res. Toxicol.* **3**, 111–117 (1990).
47. R. Singh and E. H. Weyand, *Polycyclic Arom. Compnds.* **6**, 135–142 (1994).

48. M. M. Doxtader, B. W. Day, R. Tannenbaum, and R. R. Dasari, in S. RadhaKrishna and B. C. Tan, eds., *Laser Spectroscopy and Non-Linear Optics*. Springer Verlag, Heidelberg, 1990.
49. S. R. Myers and M. T. Pinorini, *Polycyclic Arom. Compnds.* **6**, 143–150 (1994).
50. R. Jankowiak, D. Zamzow, D. E. Stack, E. L. Cavalieri, and G. J. Small, *Chem. Res. Toxicol.* **11**, 1339–45 (1998).

CHAPTER

12

BEYOND STRUCTURE: FLUORESCENCE LINE-NARROWING SPECTROSCOPY USED TO STUDY PROTEINS

JANE M. VANDERKOOI

Johnson Research Foundation, Department of Biochemistry and Biophysics, School of Medicine, University of Pennsylvania, Philadelphia, PA 19104, USA

12.1. INTRODUCTION

Since the first elucidation of the three-dimensional structure of proteins, now about half a century ago, the words *structure* and *function* have been closely intertwined. The implicit assumption is that structure to a large extent determines function and hence by elucidating structure we should gain insight into function. Although this premise has certainly held for most cases, even after structure is known, the scientist is in the position where scientists always are after solving one problem: The answers provided by one set of experiments (in this case structure determination) opens up a whole new set of questions.

Some questions go *beyond* the paradigm that protein structure equates to function. There are basic questions about the dynamics of the protein and nature of the interactions between its atoms that are not revealed from solely structure determination. Various types of spectroscopy can give information in this area. In this chapter, what new questions can be addressed about the nature of proteins are explored using the high-resolution optical technique fluorescence line-narrowing (FLN).

First, what about proteins make them especially interesting for study by this physical technique is considered. Then some examples from the literature using FLN to study proteins are described.

12.2. RATIONALE OF THE EXPERIMENT

12.2.1. The Nature of Proteins

An experiment using fluorescence to study proteins follows this procedure: A fluorescent moiety (either intrinsic or introduced into the protein) is identified,

Shpol'skii Spectroscopy and Other Site-Selection Methods, Edited by Cees Gooijer, Freek Ariese, and Johannes W. Hofstraat
ISBN 0-471-24508-9

and then from the details of the emission behavior, inferences about the environment is made. The polypeptide chain forms the "solvent" or "matrix" for the fluorescent chromophore, and elucidation of the nature of the polypeptide interaction with the fluorescent molecule is the experimental quest.

What, in general, do we want to know about the protein using this type of experiment? Broadly, we would like to know about the protein's "state of matter." This conceptual problem is related to the question of how proteins fold into a definite structure and maintain specificity in function. Many proteins are small and irregular in shape. As such they have a large surface area. The role of surface charges and their interactions with water is long recognized [1]. Yet at the same time their interiors are highly structured, and amino acid placement around an active site is often rigid and very precise.

Two models are used to describe solid condensed matter: crystalline or glass. The crystal structures of about literally thousands of proteins are now known. One can get the impression that the protein is rigid, with the atoms in defined positions, as for a crystal. But in terms of motion, proteins exhibit behavior consistent with glasses. At room temperature, proteins undergo fluctuations on the time scale ranging for those of vibrations (ps) to large-scale local folding and unfolding of the secondary and tertiary structure (μsec to minutes, even hours). The range of time scales for relaxation spans some 10 orders of magnitude and fluctuations ensure that, at a given instant, there is a distribution of molecular subconformations.

These apparent glasslike properties of proteins can be successfully explained using the two-level-system (TLS) model [2, 3]. In the model of glasses, a TLS comprises a group of atoms or molecules that can be in either of two quantum-mechanical potential wells along a conformational coordinate separated by an energy barrier. The subconformations or conformational substates show strong similarity with the TLS, but a difference is that for "pure" glassy materials the energy barriers are assumed to have a smooth distribution, whereas in proteins the conformational substates have a punctuated barrier distribution and thus a tiered hierarchy [4]. Each subconformation of the protein is more or less closely related to the time-averaged structure revealed by techniques such as X-ray or NMR. This model of proteins is able to explain many data concerning notably the diffusion of small ligands in myoglobin [5, 6] and is also consistent with low-temperature specific heat measurements (linear in T below 1 K), anomalous thermal conductivity and dielectric response [7].

Just as the size and shape of proteins vary dramatically, the "crystalline" or "glass" properties also vary from protein to protein [8]. Each individual protein type possesses some properties suggestive of crystalline solids, as shown by their highly regular X-ray diffraction patterns, but also exhibit disorder characteristic of amorphous or glassy structures.

12.2.2. Chromophore/Protein Interactions

Here are some questions about proteins that are "beyond" protein structure but are related to the question of the nature of the protein: What are the forces that bind the fluorescent molecule to the neighboring amino acids, what are the dynamics, what are the energy barriers between dynamic forms, how do these affect the reactivity of the prosthetic group, how do these affect the specificity of the reaction?

In order for the chromophore to report on the environment, neighboring groups must be coupled to the chromophoric molecule. Absorption of light causes a dipole change in the environment, and the ground and excited state interact differently with their surroundings. The neighboring atoms exert field effects on the molecule, and shift the absorption and fluorescence spectra by raising or lowering the energy level of the ground- and excited-state molecule.

The forces within the molecule that affect spectral transitions are summarized by a recent review [9]. Both fixed and induced charges of the surrounding groups in the amino acids produce the milieu for the chromophore. Amino acids in the protein include those with fixed charges (Lys, Asp, Glu), polarizable groups (Trp, Phe), dipolar groups (peptide linkage). They also include those that can form H bonds (Try, Thr, Ser, Cys, peptide linkage).

From the spectroscopist's point of view the study of chromophores in proteins is interesting because so much is known about the chromophore's environment. In a protein with known structure all the time-averaged positions of neighboring atoms are defined. There has been a recent explosion in computer power, and programs to compute local electric fields in proteins are commercially available. This means that proteins provide a surrounding for chromophores that is better characterized than even a pure solvent.

From the protein chemist's point of view the spectroscopy of protein-bound molecules is important because spectroscopy gives information on the forces that influence the chromophore, and this knowledge can greatly lead to understanding of reactivity. It can also lead to clues on the structure and disorder in proteins. How proteins fold so specifically remains an outstanding problem in molecular biology, and the contribution of spectroscopy to understanding of the nature of protein interiors is valuable.

12.2.3. FLN as Applied to Chromophore/Protein Studies

The principles and technical development of FLN spectroscopy are described in other contributions of this volume, as well as in comprehensive reviews [10]. Specific application of FLN measurements to study proteins [11] and a recent review describing protein studies [12] are available.

To briefly describe the FLN technique, a laser is used to excite one subpopulation within the inhomogeneously broadened absorption spectrum, and emission from this subpopulation is monitored. The sample is maintained at cryogenic temperature to ensure that it is in the lowest vibrational ground state. A population distribution diagram (Fig. 12.1) can be used to describe the situation of photoselection into an inhomogeneously broadened sample [13]. The energy levels of given subpopulation of molecules are given by the column

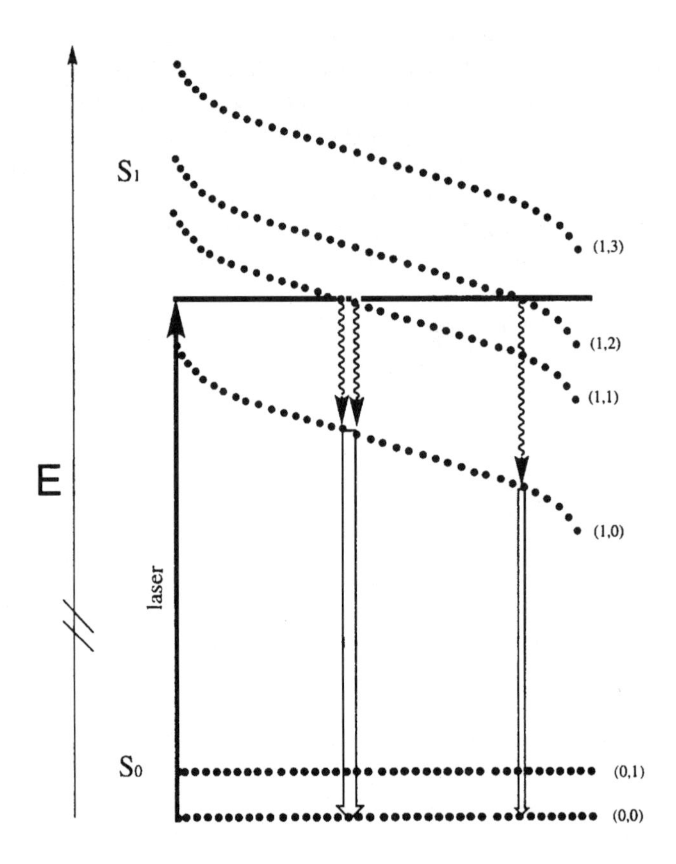

Figure 12.1. Energy map. The molecules are arranged according to the difference in energy from the lowest vibrational level of the S_0 and S_1 state, with the ground-state molecules being placed at zero energy, by convention. Curved arrows indicate vibrational relaxation; solid arrow shows electronic excitation and open arrows show electronic emission. In the situation shown, two subpopulations of molecules are excited, yielding two 0–0 emission lines. The picture illustrates that more molecules are near the mean value of the electronic energy gap than at the extremes and that one excitation frequency can excite subsets of molecules to different vibronic levels. At the excitation frequency diagrammed, more molecules in the 1–1 level are excited than in the 1–2 level. The energy map has been previously described [13].

of dots. In this diagram, we consider that the electronic transitions are inhomogeneously distributed, but not the vibrational levels. Consequently, the vibronic levels remain parallel to each other.

Information that can be obtained from FLN include the following:

1. One can get vibrationally resolved spectra of the ground-state molecule. This information can be obtained by resonance Raman techniques for nonfluorescent molecules. Fluorescence and resonance Raman spectroscopy can be considered as essentially variants of the same process [14]. Like resonance Raman spectroscopy, FLN can give vibrationally resolved spectra for molecules in very low concentrations, and therefore information on atomic positions and bonds can be determined for a chromophore that is part of a larger complex.
2. The vibrational levels of the excited-state molecule can also be acquired. The vibrational levels depend upon the structure of the excited-state molecule, important data for understanding photoreactions.
3. One can obtain the difference in energy between the ground and excited electronic states, independent of other broadening mechanisms including phonon wing contributions. For chromophores in proteins, there is some disorder around chromophores, and this generates a distribution of energy gaps. The distribution of 0–0 energies is related to the uniqueness of the environment.
4. Finally, in ideal cases one can get some insight to the mechanisms that produce homogeneous broadening. One such mechanism is electron–phonon coupling. This value tells how strongly the surrounding atoms are influenced by the electrons of the fluorescent molecule.

We are discussing here only conventional fluorescence-narrowing techniques, but variations have been carried out. A promising technique is single-molecule fluorescence spectroscopy [15–17]. If the laser width is narrower than the homogeneous linewidth, excitation at any frequency within the inhomogeneous absorption band will excite a very small fraction of the total molecules in the inhomogeneous envelope. As the laser is moved farther away from the absorption maximum, the density of absorbing centers becomes sufficiently low so that at any excitation frequency, there should be only one molecule in resonance.

FLN must be carried out at low temperature, but this needs not be a disadvantage. Temperature itself can be a variable in the experiment, and low temperature also allows for the examination of low barriers in protein reactions.

The resolution that one obtains for chromophores can vary from highly resolved to nonresolved, depending upon the electron–phonon coupling and other parameters. Unlike conventional fluorescence spectroscopy, the resolution of FLN spectra depend upon the wavelength of excitation. An explanation for this

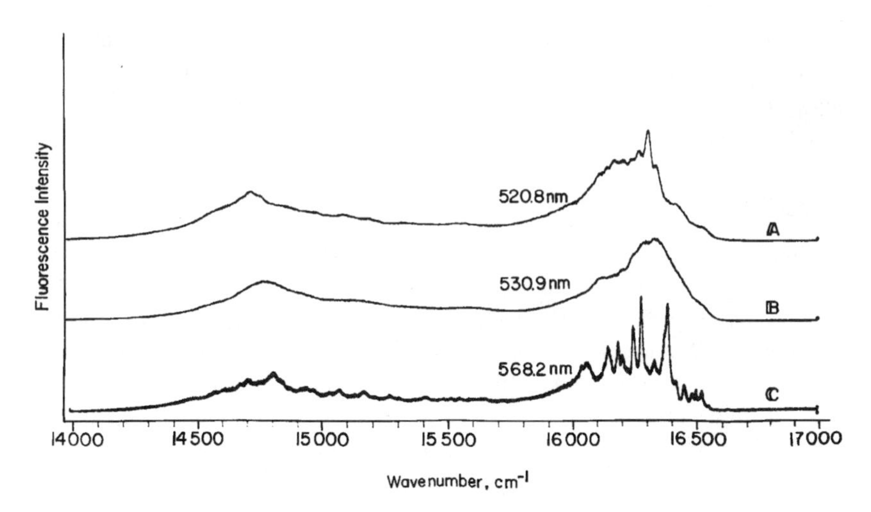

Figure 12.2. FLN spectra: dependence upon excitation. The heme of horse myoglobin was replaced by metal-free mesoporphyrin. The excitation wavelength is indicated in the figure. Data are unpublished results, but details of the experiment we described previously [30].

is as follows: Coupling to the matrix (which in this case is the polypeptide chain) produces a broad absorption, called the phonon wing, to the high-energy side of the zero-phonon line [18]. When excitation occurs to the high-energy side of the inhomogeneously distributed transition, proportionately more molecules are excited via this phonon wing, leading to loss of resolution. In Figure 12.2 spectra are presented for myoglobin, in which Fe was substituted by metal-free porphyrin. In the spectral region of 16000–16500 cm^{-1} multiple zero-phonon lines are observed [19] in the lower spectrum. These indicate excited-state vibrational frequencies. (Unlike in Fig. 12.1, where excitation occurred into two vibronic bands, which would lead to two observed 0–0 lines, in this case >6 vibronic bands were excited). In the upper two spectra, which used higher-energy excitation, the spectra showed broad bands, comparable in resolution to conventional spectra.

12.3. PROTEINS STUDIED BY FLN

Vibrationally resolved spectra can be used analytically to identify molecules, and since very weak fluorescence signals can be detected, FLN can detect molecules even in the presence of many other absorbing molecules. This feature has been exploited for studying particular compounds in tissues. For example, fluorescence with vibrational fine details has been obtained that led to the unequivocal identification of porphyrins in muscle slices [20] and in parasites [21]. High resolution also allowed for the identification of the metabolic products of carcinogenic aromatic hydrocarbons [22, 23]. In another application, FLN was used to characterize the binding site of an antibody to a pyrene derivative [24].

In plants, chlorophyll is protein-bound and line-narrowed fluorescence spectra were early on shown to be obtained from leaves [25–28]. High-resolution spectroscopy of chlorophyll is described in this issue.

Heme proteins remain a favorite subject of FLN following the report of such spectra from a heme protein derivative [29]. A basic problem facing the study of heme proteins is this: The active group of heme proteins is the same—yet this large protein class exhibits a variety of functions, including electron transfer, proton transfer, degradative and biosynthetic functions, cellular regulation, and molecule transportation. This suggests that the polypeptide chain has a large role in modifying the reactivity of the heme group—and it is worthwhile to try to figure out how this is accomplished. The heme itself is a chromophore, and therefore one is directly examining the active center of the protein. Heme is not fluorescent and so the native Fe-porphyrin is substituted with a fluorescent metal-porphyrin or with free-base porphyrin (i.e., no metal). Controls are carried out to ascertain that the structures of the derivatives and the native proteins are the same.

Early work has established that FLN spectra of metal derivatives of heme proteins depend in many ways upon the polypeptide chain [30]. Spectral changes occur when the protein conformation changes due to binding of substrate [31] and due to changes in ionization of the protein [32]. A photochemical tautomerization reaction occurring in free-base porphyrin can be monitored by FLN [33], and this has proved useful to monitor protein conformational barriers [34, 35]. These and other work are reviewed in detail elsewhere [12].

12.4. SPECIFIC QUESTIONS ADDRESSED BY FLN

12.4.1. Structure of the Excited State: Comparison of Ground- and Excited-State Vibrational Frequencies

In the remainder of the chapter we look at particular applications of FLN to the study of heme proteins. One of FLN's uses for protein systems is the extraction of vibrational information from a chromophore. Infrared absorption and Raman scattering are usually used in vibrational spectroscopy. The first is limited in aqueous samples by the strong absorption of water in much of the spectral region. In addition, in proteins there are many IR absorbing modes; so it is often difficult to specifically look at prosthetic groups with IR absorption. As for Raman, fluorescent materials are not good candidates for this work. Therefore, FLN has an experimental niche—it is able to provide vibrational information for fluorescent compounds in a mixture.

As noted, FLN spectra can give the vibrational frequencies of both the ground and excited state, and a comparison tells how the chromophore changes in the excited state, information that is important to photochemistry. Comparison of ground- and excited-state vibrational levels for a compound in a proteins are given in Table 12.1. There is a general downshift in the frequency of the excited state relative to the ground-state molecule, which is consistent with a relative increase in size upon excitation. Similar shifts in frequency were seen for the excited state of Zn cytochrome *c* relative to its ground state [36, 37] and for lumiflavin in *n*-decane matrix [38]. When a means to scan the excitation spectrum at high resolution is available, these data can also be obtained from comparing the excitation and emission spectra as recently shown by Herenyi et al. [39].

A continuing interest of the authors is to correlate spectral distributions with parameters within the protein. As mentioned above, the physical characteristics of porphyrins are greatly influenced by the polypeptide environment of heme proteins, and this is the basis of the specificity of the many particular heme proteins, in spite of the fact that the central player, the heme, is the same. Since excited states are always more nonbonding than the ground state, they can be relatively more sensitive to environment.

Heme proteins with different function can show quite different vibrational levels. We were interested to see whether proteins with the same apparent function would provide the same environment. We used Zn-substituted hemoglobin samples for this study [40]. Hemoglobin has two types of subunits, called the alpha and beta, and although both transport O_2, their binding affinities and ligand-induced conformational response are different. In our experiment the fluorescent derivative of each subunit was compared. In Figure 12.3 we show

Table 12.1. Vibrational Frequencies of Ground and Singlet Excited-State Mg-Myoglobin[a]

Mg-myoglobin Ground (cm^{-1})	Mg-myoglobin Excited (cm^{-1})
670	667
752	734
990	988
1072	1077
1134	1141
1206	1207
1524	1530

[a] Data from Kaposi et al. [55]. The sample measured was Mg-protoporphyrin IX, which had replaced heme in horse myoglobin.

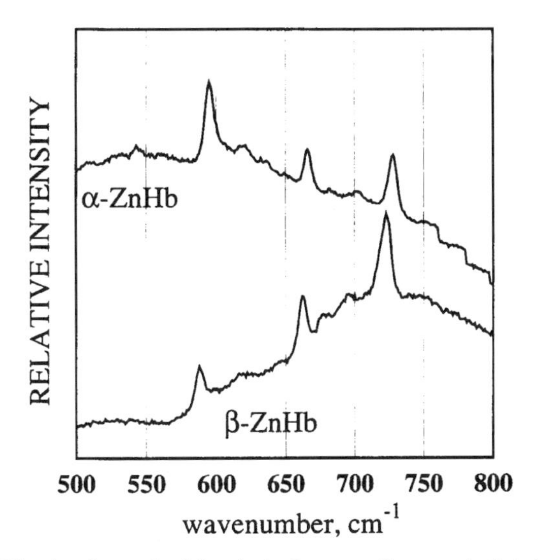

Figure 12.3. Vibrationally resolved "excitation" spectra. Fe was substituted for Zn in the α or β subunit of human hemoglobin, as indicated, and the excitation spectrum of each subunit was determined. The sample was at pH 8.0. Data are unpublished results of K. Sudhakar and J. Vanderkooi. To obtain these spectra, emission spectra were summed to approximate the excitation spectrum. Experimental detail has been described previously [40].

that there are subtle differences in the vibrations of the excited-state porphyrin in two polypeptide environments.

12.4.2. Distribution of 0–0 Energy and Relationship to Protein Structure

The influence of local electric fields in proteins has received attention with regard to protein stability and enzymatic specificity [41–43]. When the positions of the neighboring atoms of the protein change, the chromophore experiences different electric fields. In addition, torsional strains imposed by the protein matrix will also change the transition energies. It follows that the observed spectral dispersion in the transition frequency should be sensitive to the protein electrostatic properties and protein disorder. Cooling of the sample allows the subconformations to be "frozen in," either by being energetically trapped into local minima [44, 45] or kinetically trapped by the increasing viscosity of the solvent [46].

In the FLN technique zero-phonon line and phonon-wing contributions to the spectrum are separated. Again, using the system described in Figure 12.3, we wanted to see whether the environments provided by the two heme pockets were different such that it affected the electronic energy gap. The population distribution function (PDF), which is obtained from the distribution of zero-phonon lines, reflects the distribution of the energies between ground and excited states. The procedure to obtain this function has been described [47]. The low-temperature absorption and emission spectra and the PDF for the fluorescent porphyrin in the beta subunit are shown in Figure 12.4. The figure illustrates that the absorption and emission spectra are shifted to higher and lower energy, respectively, relative to the PDF. There were distinct differences in the PDF for the two subunits, not shown, and for these proteins we also observed that there were splittings in the absorption spectra [40], again indicating that there are differences in the environment for the porphyrin in the two protein subunits.

Experimentally determined values of the PDFs for porphyrins bound to the heme pocket in several types of heme proteins are summarized in Table 12.2. The examples illustrate that the protein polypeptide chain influences the distribution of the electronic transitions in the sample. For instance, the same polypeptide

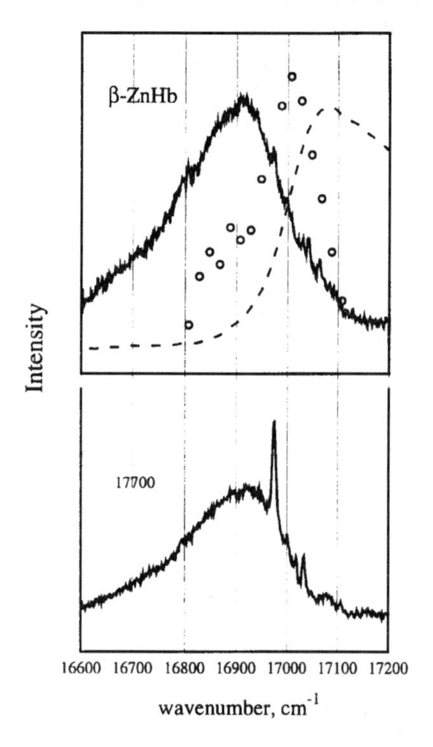

Figure 12.4. PDF for hemoglobin at pH 6.0. Zn was substituted for Fe in the β-subunit of human hemoglobin. Since Zn protoporphyrin in the β-subunit is fluorescent and Fe-protoporphyrin IX in the α-subunit is not, the fluorescent probe is specific to the β-subunit only. Top: Dotted and solid lines are the conventional absorption and fluorescence spectra, respectively. Points are the energy distribution of the zero phonon line showing the true distribution in the electronic energy gap. Bottom: representative FLN spectrum, excitation at 17660 cm^{-1}. Details of this experiment were previously described [40].

Table 12.2. Inhomogeneous broadening: Widths of the 0–0 Transition

Sample	2σ, (μ) $(cm^{-1})^a$	Ref.
Chromophores		
Single crystal	≤ 1	[56]
Polycrystalline host	1–10	[56]
Disordered solids (glasses, polymers)	100–1000	[56]
Covalently bound porphrin/protein		
Zn(II) horse cytochrome *c* (folded)	65	[48]
Sn(IV) horse cytochrome *c* (folded)	65	[57]
Zn(II) horse cytochrome *c* (unfolded)	361	[48]
Metal-free horse cytochrome *c*	48, 76	[36]
Noncovalently bound pophyrin/protein		
Mg horse myoglobin	50, 76	[55]
Metal-free horse myoglobin	25, 38	[55]
α-ZnHb	75, 80	[40]
β-ZnHb	50, 65	[40]

[a] Data obtained from fits to the Gaussian distribution function $I(\bar{\nu}) = I_{max}(\bar{\nu})e^{-[(\bar{\nu}-\mu)^2/2\sigma^2]}$. Distributions were obtained from samples at ~10 K or lower.

chain (cytochrome *c*) but different chromophore (Zn porphyrin or Sn porphyrin) had the same width, suggesting about the same disorder due to the protein polypeptide chain. The width broadens dramatically when the protein is made to unfold by addition of a denaturant [48]. When the charge on a residue near the chromophore was altered by changing pH, the mean position of the distribution of energies shifted [32]. The inhomogeneous broadening can also change when substrate is bound to the protein [31, 49].

The distribution functions are continuous, that is to say, as the laser excitation is shifted the emission is observed to shift in parallel. This result can be expected when one considers that there are many neighboring atoms, and hence many atomic contacts. This supports the glasslike model of proteins, indicating that there is a continuous distribution of atomic positions and hence electronic energies. However, the PDFs are nonglasslike in other ways. Although the distributions of state electronic levels are wider than that found in crystalline matrices, they are narrower than for completely amorphous materials (Table 12.2). In addition, the distributions found in many proteins can rarely be fit using a mono-Gaussian and in most cases require two or more Gaussians—unlike what would be expected for a glassy material. The structure of the components that give rise to the multiple distributions remains obscure. The PDF data suggest that, although glasslike by some experimental criteria, proteins can not be completely described by models of glasses.

What then in the protein determines the shift in the spectrum? Analysis of the local electric fields in the heme pockets of several proteins are now being carried out. We have used available software to calculate the local field at the heme in cytochrome *c* [50], and some conclusions can be drawn from this protein. These values for the contribution of the electric field by the protein were calculated at the central pyrrole nitrogen. In Figure 12.5 values for the fields exerted by the protein are given for two conditions—when the heme propionates are ionized (i.e., negatively charged) or protonated (i.e., neutral). (In this calculation only formal charges were considered. A later calculation also considered partial charges [51].) Several conclusions can be made regarding the data of local fields in proteins. First, it can be seen that fields experienced by the porphyrin are directional—different field values are seen for the *x* and *y* direction. Second, the electric field within the protein and experienced by the central porphyrin is remarkably nonuniform: When the carboxylates are ionized, the A and D rings experience a much larger negative field than B and C rings. This means that there is a potential drop across the porphyrin ring. Third, changing the fixed charge near the porphyrin by changing the ionization of a nearby group alters the field experienced by the heme.

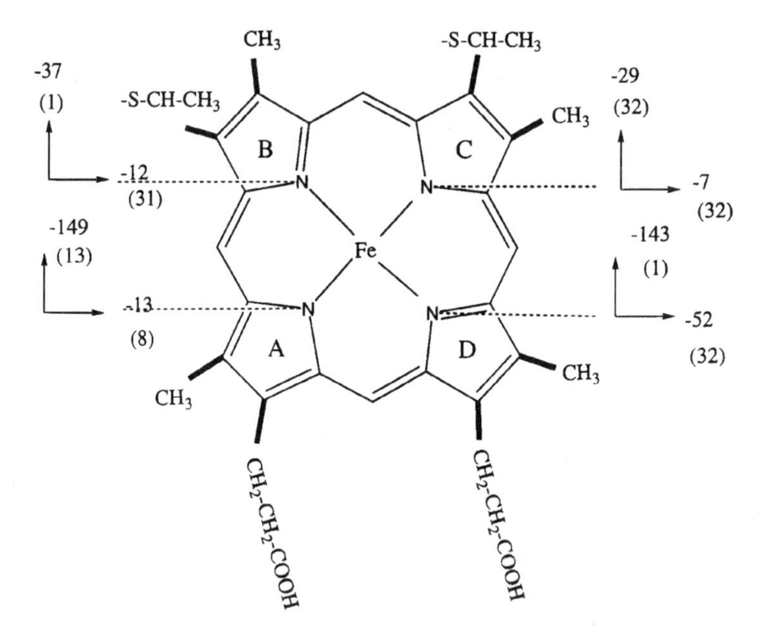

Figure 12.5. Fields at porphyrin sites in ferrous horse cytochrome *c*. In the coordinate used the heme lies in the *x*, *y* plane and the *x* and *y* directions are indicated. The fields (mV/Å) at the nitrogen positions in the *x* and *y* direction are indicated. The unlabeled numbers are the fields when the propionates are ionized and the numbers in parentheses are the values when this group is the propionic acid. Details in Ref. 50.

12.4.3. Relaxation Processes in Proteins and FLN Temperature Dependence

The third type of experimental data discussed here relates to relaxation processes. As we mentioned, proteins exhibit relaxation processes over a large time-scale range. Fluorescence measurements give access to the study of wide time ranges. At the shortest times, in the femtosecond range, dephasing occurs. These times are influenced by the protein matrix in metal-substituted heme proteins [52, 53]. At longer times, there is relaxation from the S_1 to S_0 states. For excited-state porphyrins these times are in the nanosecond regime. Fluorescence lifetimes have long been used to characterize excited states, and the electronic relaxation is affected by various energy-transfer mechanisms.

Less is known about vibrational relaxation times, which should be intermediate between these two times. (In Fig. 12.1, vibrational relaxation is indicated by the wavy arrow.) Recently, Herenyi and co-workers, using a combination of FLN and vibronic hole-burning spectroscopy, measured the relaxation time from vibronic states of mesoporphyrin in myoglobin [39]. In this elegant study, the researchers examined distinct tautomeric forms of mesoporphyrin within the protein, which they had previously identified and had shown to be photochemically interconvertible [33, 54]. In this way they were able to interrogate different interactions with the protein. The relaxation times from excited vibronic levels, which ranged from 1 to 10 psec, are dependent both upon the mode and upon the given tautomer. The latter information shows that there is coupling between the vibronically excited-state molecule and the protein matrix.

Relaxation processes are evident in measuring the temperature dependence of FLN spectra. Again, using Zn-substituted hemoglobin as an example, the temperature dependence is shown in Figure 12.6. The spectra show increased background and widening of the vibronic bands as temperature is raised, until at above ~60 K, line-narrowed spectra were no longer seen. The origins of the line broadening is of interest. As the temperature increases, the contributions of the phonon wing increase, and this could change the resolution. The electronic excited-state lifetime is ~10 nsec, and motion of the surrounding atoms on this time scale can obliterate the energy selection of excitation. The mechanism of temperature dependence of the line narrowing for chromophores in proteins requires more investigation.

12.5. SUMMARY REMARKS

An important contribution of FLN is that it provides an experimental approach to relate physical changes in the protein with effects on the chromophore. FLN showed that the spectra of chromophores in proteins are inhomogeneously

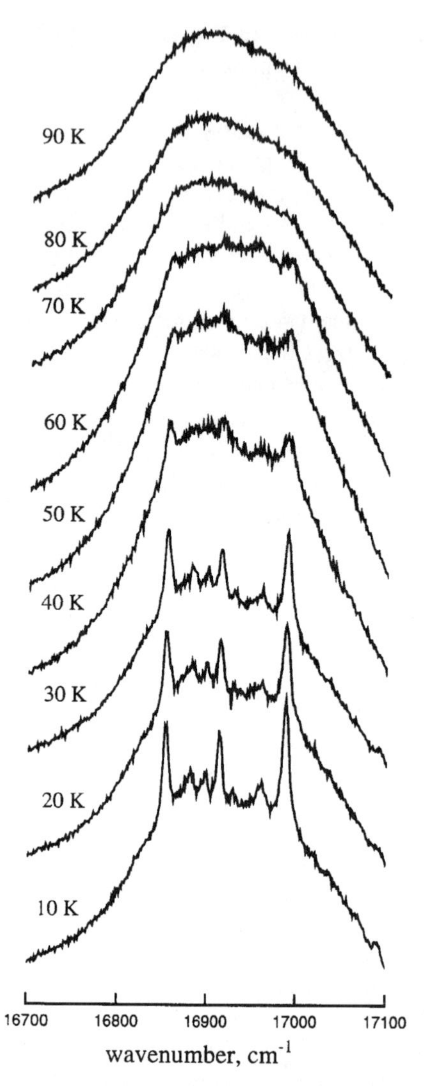

Figure 12.6. Temperature dependence of the FLN spectra of Zn-substituted hemoglobin. The sample was the same as described in Fig. 12.4 and details of the experiment were described previously [40]. Excitation was at 17580 cm^{-1}. Temperatures are indicated in the figure.

broadened in a continuous manner, consistent with the damped motion of proteins. The inhomogeneity for the chromophore in the protein is dependent upon the protein conformation, and is intermediate between that of a crystal and a glass. At the same time configurational substates can be selected, suggesting that there is indeed a hierarchy of protein motion and structure. It is clear that the electric field exerted by neighboring atoms shifts the electronic transition, and the inhomogeneity is greater when the surrounding disorder is greater. The phonon coupling also depends upon the chromophore and the protein. FLN additionally provides ground- and excited-state vibrational frequencies, thereby allowing for structural differences between the excited state and the ground-state molecule to be detected using a single form of spectroscopy.

ACKNOWLEDGMENTS

This work was supported by NIH grants PO1 GM48130 and RO1 GM 55004. Dr. Monique Laberge is thanked for helpful discussion.

REFERENCES

1. C. Tanford, *The Hydrophobic Effect*, John Wiley & Sons, New York, 1980.
2. P. W. Anderson, B. I. Halperin, and C. M. Varma, *Philos. Mag.* **25**, 1 (1972).
3. W. A. Phillips, *J. Low Temp. Phys.* **7**, 351 (1972).
4. A. Ansari, J. Berendzen, S. F. Bowne, H. Frauenfelder, I. E. T. Iben, T. B. Sauke, E. Shyamsunder, and R. D. Young, *Proc. Natl. Acad. Sci. USA* **82**, 5000 (1985).
5. R. H. Austin, K. Beeson, L. Eisenstein, H. Frauenfelder, I. C. Gunsalus, and V. P. Marshall, *Science* **181**, 541 (1973).
6. H. Frauenfelder and G. Petsko, *Biophys. J.* **10**, 465 (1980).
7. L. R. Narashimhan, K. A. Littau, D. W. Pack, Y. S. Bai, A. Elscher, and M. D. Fayer, *Chem. Rev.* **90**, 439 (1990).
8. J. M. Vanderkooi, *Biochim. Biophys. Acta* 1 (1998).
9. M. Laberge, *Biochim. Biophys. Acta* **1386**, 305 (1998).
10. R. I. Personov,in V.M. Agranovich and R.M. Hochstrasser, eds., *Site Selection Spectroscopy of Complex Molecules in Solutions and Its Applications. Spectroscopy and Excitation Dynamics of Condensed Molecular Systems*, North-Holland, Amsterdam, 1983, p. 555.
11. J. M. Vanderkooi, P. J. Angiolillo, and M. Laberge in L. Brand and M.L. Johnson, eds., *Meth. Enzymol. Fluorescence Spectroscopy*, Vol. 278, John Wiley & Sons, New York, 1997, p. 71.
12. J. Fidy, M. Laberge, A. Kaposi, and J. M. Vanderkooi. *Biochem. Biophys. Acta* **1386**, 331 (1998).
13. A. D. Kaposi and J. M. Vanderkooi, *Proc. Natl. Acad. Sci.* **89**, 11371 (1992).
14. R. M. Hochstrasser and C. A. Nyi, *J. Chem. Phys.* **70**, 1112 (1979).
15. W. E. Moerner snd T. Basche, *Angew. Chemie-Int. Ed.* **32**, 457 (1993).

16. A. B. Myers, P. Tchenio, M. Z. Zgierski, and W. E. Moerner, *J. Phys. Chem.* **98**, 10377 (1994).
17. E. Betzig and R. J. Chichester, *Science* **262**, 1422 (1993).
18. I. S. Osad'ko, E. I. Al'Shits, and R. I. Personov, *Sov. Phys. Solid State* **16**, 1286 (1975).
19. R. I. Personov, *Spectrochimica Acta* **38B**, 1533 (1983).
20. A. V. Novikov, *Photochem. Photobiol.* **59**, 12 (1994).
21. C. Larralde, S. Sassa, J. M. Vanderkooi, H. Koloczek, J. P. Laclette, F. Goodsaid, E. Sciutto, and C. S. Owen, *Molecular Biochem. Parasitology* **22**, 203 (1987).
22. R. Jankowiak, P. Lu, and G. J. Small, *J. Pharm. Biomed. Anal.* **8**, 113 (1990).
23. R. Jankowiak and G. J. Small, *Anal. Chem.* **61**, 1023 (1989).
24. K. Singh, P. L. Skipper, S. R. Tannenbaum, and R. R. Dasari, *Photochem. Photobiol.* **58**, 637 (1993).
25. R. Avarmaa, I. Renge, and K. Mauring, *FEBS Lett.* **167**, 186 (1984).
26. R. Avarmaa, R. Jaaniso, K. Mauring, I. Renge, and R. Tamkivi, *Mol. Phys.* **57**, 605 (1986).
27. I. Renge, K. Mauring, and R. Avarmaa, *Biochem. Biophys. Acta* **766**, 501 (1984).
28. I. Renge, K. Mauring, and R. Vladkova, *Biochim. Biophys. Acta* **935**, 333 (1988).
29. P. J. Angiolillo, J. S. Leigh, Jr., and J. M. Vanderkooi, *Photochem. Photobiol.* **36**, 133–137 (1982).
30. J. M. Vanderkooi, V. T. Moy, G. Maniara, H. Koloczek, and K. G. Paul, *Biochemistry* **24**, 7931 (1985).
31. J. Fidy, K.-G. Paul, and J. M. Vanderkooi, *Biochemistry* **28**, 7531 (1989).
32. H. Anni, J. M. Vanderkooi, K. A. Sharp, T. Yonetani, S. C. Hopkins, L. Herenyi, and J. Fidy, *Biochemistry* **33**, 3475 (1994).
33. J. Fidy, J. M. Vanderkooi, J. Zollfrank, and J. Friedrich, *Biophys. J.* **61**, 381 (1992).
34. L. Herenyi, J. Fidy, J. Gafert, and J. Friedrich, *Biophys. J.* **69**, 577 (1995).
35. J. Fidy, J. M. Vanderkooi, J. Zollfrank, and J. Friedrich, *Biophys. J.* **63**, 1605 (1992).
36. V. Logovinsky, A. D. Kaposi, and J. M. Vanderkooi, *J. Fluorescence* **1**, 79 (1991).
37. V. Logovinsky, A. D. Kaposi, and J. M. Vanderkooi, *Photochem. Photobiol.* **57**, 235 (1992).
38. R. J. Platenkamp, A. J. W. G. Visser, and J. Koziol, in V. Massey and C. H. Williams, eds., *Flavins and Flavoproteins*, Elsevier/North Holland, Amsterdam, 1982.
39. L. Herenyi, A. Suisalu, K. Mauring, K. Kis-Petik, J. Fidy, and J. Kikas, *J. Phys. Chem,* B, **102**, 5932 (1998).
40. K. Sudhakar, M. Laberge, A. Tsundeshige, W. W. Wright, and J. M. Vanderkooi, *Biochemistry* **37**, 7177 (1998).
41. J. Wendoloski and J. B. Matthew, *Proteins* **5**, 313 (1989).
42. S. Northrup, T. G. Wensel, C. F. Meares, J. J. Wendoloski, and J. B. Matthew, *Proc. Natl. Acad. Sci. USA* **87**, 9503 (1990).

43. K. Langsetmo, J. A. Fuch, C. Woodward, and K. A. Sharp, *Biochemistry* **30**, 7609 (1991).
44. H. Frauenfelder, S. G. Sligar, and P. Wolynes, *Science* **254**, 1598 (1991).
45. R. Elber and M. Karplus, *Science* **235**, 318 (1987).
46. A. Ansari, C. M. Jones, E. R. Henry, J. Hofrichter, and W. Eaton, *Science* **256**, 1196 (1992).
47. J. Fuenfschilling and D. F. Williams, *Photochem. Photobiol.* **26**, 109 (1977).
48. V. Logovinsky, A. D. Kaposi, and J. M. Vanderkooi, *Biochim. Biophys. Acta* **1161**, 149 (1993).
49. J. Fidy, G. R. Holtom, K.-G. Paul, and J. M. Vanderkooi, *J. Phys. Chem.* **95**, 4364 (1991).
50. M. Köhler, J. Gafert, J. Friedrich, J. M. Vanderkooi, and M. Laberge, *Biophys. J.* **71**, 77 (1996).
51. M. Laberge, J. M. Vanderkooi, and K. A. Sharp, *J. Phys. Chem.* **100**, 10793 (1996).
52. D. T. Leeson, O. Berg, and D. A. Wiersma, *J. Phys. Chem.* **98**, 3913 (1994).
53. D. T. Leeson and D. A. Wiersma, *Phys. Rev. Lett.* **74**, 2138 (1995).
54. J. Fidy, K. G. Paul, and J. M. Vanderkooi, *J. Phys. Chem.* **93**, 2253 (1989).
55. A. D. Kaposi, J. Fidy, S. S. Stavrov, and J. M. Vanderkooi, *J. Phys. Chem.* **97**, 6317 (1993).
56. K. K. Rebane, *Chem. Phys.* **189**, 139 (1994).
57. M. Laberge and J. M. Vanderkooi, *Biospectroscopy* **6**, 413 (1996).

CHAPTER

13

HIGH-PRESSURE AND STARK HOLE-BURNING STUDIES OF PHOTOSYNTHETIC COMPLEXES

MARGUS RÄTSEP AND GERALD J. SMALL

Ames Laboratory–USDOE and Department of Chemistry, Iowa State University, Ames, IA 50011, USA

In this chapter we review how the combinations of high-resolution spectral hole-burning spectroscopy with high pressure and external electric (Stark) fields leads to new insights on the excited-state electronic structures and excitation energy-transfer processes of light-harvesting complexes. Since the above combinations are a recent development from our laboratory, considerable discussion is given to experimental details including the high-pressure system (~800 MPa at liquid helium temperatures) and the Stark cell (~100 kV/cm). Recently obtained results for the bacteriochlorophyll *a* antenna complexes of *Rhodobacter sphaeroides*, *Rhodospirillum molischianum*, *Rhodopseudomonas acidophila*, and *Chlorobium tepidum* are reviewed.

13.1. INTRODUCTION

Understanding in detail the mechanisms for electronic excitation and electron transfer in photosynthetic units represents a formidable problem whose solution will depend on X-ray structural data, ultrafast time-domain kinetic measurements, accurate electronic structure calculations and dynamical simulations, and high-resolution optical frequency-domain spectra. Spectral hole-burning spectroscopy (SHB) is a versatile and general methodology for the generation of highly structured optical spectra. SHB is capable of eliminating or significantly reducing the contribution from site inhomogeneity to vibronic linewidths in amorphous systems. It provides an improvement in spectral resolution of 2–3 orders of magnitude. The attributes of SHB important to the study of the excited-state electronic structure and dynamics of photosynthetic complexes have been the subject of several review articles [1–3]. These reviews provide a broad coverage of the complexes that have been studied and discussion of the importance of the

Shpol'skii Spectroscopy and Other Site-Selection Methods, Edited by Cees Gooijer, Freek Ariese, and Johannes W. Hofstraat.
ISBN 0-471-24508-9

results obtained. However, little attention in the above reviews is given to the experimental aspects of SHB. In this chapter the focus will be on the application of spectral hole burning to the study of excited electronic state structure and dynamics of light-harvesting photosynthetic complexes using high-pressure and electric-field techniques. The experimental setups for the study of the effects of high pressure ($\sim$100 MPa) and electric field ($\sim$100 kV/cm) are described, and recent results are presented. We assume that the reader is familiar with the basic principles of SHB (see Chapter 9). For coverage of "non-photosynthetic" applications of SHB, both basic and technological, the reader is referred to the book edited by Moerner [4] and a review article by Jankowiak et al. [3]. The latter also includes a section on photosynthetic complexes.

Briefly, SHB involves the excitation of a narrow isochromat of chromophores within an inhomogeneously broadened absorption band. SHB occurs when the decay of the excited isochromat to the ground state is either blocked (persistent hole-burning) or delayed (transient hole-burning). Blocking of the decay to the ground state can be due to a photochemical reaction initiated in the excited state such as the inner proton tautomerization of the porphyrins or due to a rearrangement of the host environment (typically a glass or protein). These two mechanisms are commonly referred to as photochemical and nonphotochemical hole-burning. In transient hole-burning, an intermediate state or a short-lived product (that reverts to the initial state) in a photochemical reaction is used to store (temporarily) the depleted ground-state population. In addition to the zero-phonon holes (ZPH) at the burn laser frequency, all these mechanisms give rise, due to coupling of the chromophore to the matrix phonons, to real- and pseudo-phonon sideband holes at energies higher and lower than the ZPH. In a similar manner, real- and pseudo-vibronic holes appear due to loss of absorption at the excited-state vibrational frequencies of the chromophore.

Next we briefly discuss the types of information spectral hole-burning can provide on photosynthetic protein–chlorophyll (Chl) complexes. First, it can provide a detailed analysis of the active intramolecular Chl modes (frequencies, Franck–Condon factors) for the $S_1(Q_y) \leftarrow S_0$ absorption system. This linear electron–vibration coupling analysis for the protein-bound Chl can be compared with that for Chl monomer in "unnatural" hosts, providing information on spectral perturbations from the protein and/or coupling between Chl molecules in the protein complex. Second, SHB can be used to characterize the linear electron–phonon coupling of the optical transition where by phonon is meant the low-frequency ($\leq 150\ \mathrm{cm}^{-1}$) intermolecular modes of the protein–pigment complex. It is to be appreciated that knowledge of the linear electron–phonon and –vibration coupling strengths is essential for an understanding of excitation transport in light harvesting complexes. In this regard, knowledge of the contribution of site inhomogeneity to the absorption linewidths can also be important. Determination of this type of broadening (diagonal energy disorder) can also be accomplished with hole-burning, the third type of information. Fourth, at sufficiently low temperature the width of the ZPH coincident with the laser burn frequency (ω_B) can be used to determine the lifetime of the excited state due to rapid transport (e.g., energy transfer or primary charge separation). Fifth, spectral hole-burning provides a novel window on excitonic interactions of strongly coupled Chl aggregates. The superior resolution afforded by SHB can often lead to resolution of closely spaced Q_y states.

The antennae of photosynthetic units most often comprise two or more structurally and electronically distinct light-harvesting protein–chlorophyll complexes that act in concert to channel solar excitation energy to the reaction center [5, 6]. These complexes usually contain two or more strongly coupled Chl molecules, which means that the $Q_y(S_1)$ states are extended with the nature of the delocalization, or occupation numbers of the contributing Chl molecules, determined by the structure of the complex. An early example of where delocalized or nanoexcitonic states are important for understanding the absorption, circular dichroism, and hole-burned spectra is the Fenna–Matthews–Olson (FMO) bacteriochlorophyll (BChl) *a* complex of *Prosthecochloris aestuarii* [7, 8]. This complex is a C_3 trimer of subunits, with each subunit containing seven symmetry-inequivalent BChl *a* molecules [9, 10]. The BChl *a* antenna network of purple bacteria such as *Rhodobacter sphaeroides*, *Rhodospirillum molischianum*, and *Rhodopseudomonas acidophila* comprise a light harvesting 2 (LH2) and LH1 protein complex. The LH2 and LH1 complexes of the first two species and strain 10050 of *Rps. acidophila* are often referred to as B800–850 and B875, respectively, because of the approximate locations, in nm, of their BChl *a* $Q_y(S_1)$-absorption bands at room temperature.

Recently, the X-ray structure of the LH2 complex of *Rps. acidophila* (strain 10050) was reported at a resolution of 2.5 Å [11]. The structure revealed that this complex is a cyclic 9-mer of α,β-polypeptide pairs. The arrangement of the BChl *a* molecules is shown in Figure 13.1, where the arrows, which lie nearly in the membrane plane, indicate the direction of Q_y-transition dipoles. Several $\mathrm{Mg} \cdots \mathrm{Mg}$ separation distances are indicated. The relatively large separation of 21 Å between adjacent B800 molecules results in weak coupling, $V \sim -20\ \mathrm{cm}^{-1}$ [12], consistent with hole-burning data that indicated that excitonic effects are unimportant for the B800 ring [13–15]. However, the B850 nearest-neighbor distances of 8.9 and 9.6 Å lead to large coupling energies, $V \sim +300\ \mathrm{cm}^{-1}$ [12, 16]. Figure 13.1 shows that the B850 ring of 18 BChl *a* molecules should be viewed as a 9-mer of dimers. More recently, the X-ray structure of the isolated LH2 or B800–850 complex of *Rs. molischianum* was reported [17]. Interestingly, LH2 was shown to be an 8-mer of α,β-polypeptide pairs. A comparison of the two structures is given by Koepke et al. [17]. Suffice it to say that the picture that has the B800 molecules weakly coupled and the B850 molecules strongly coupled remains intact for LH2 of *Rs. molischianum* with the distances indicated

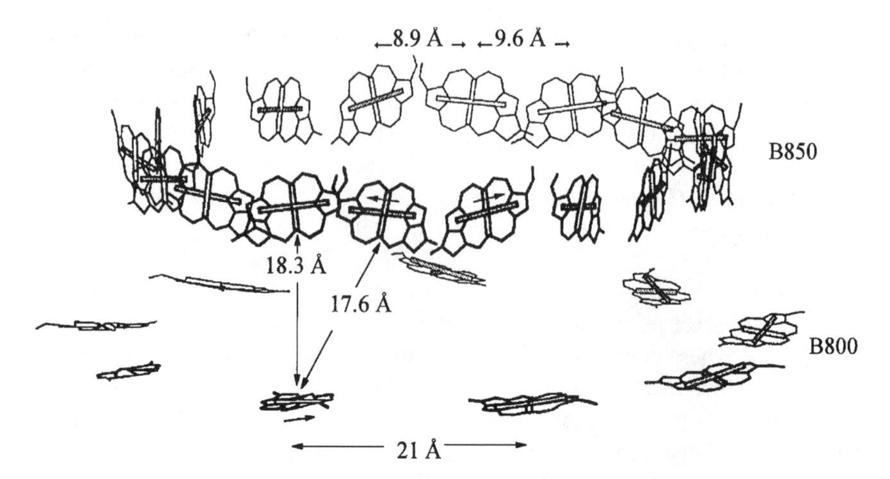

Figure 13.1. A schematic (based on Fig. 1 in Ref. 12) showing the arrangement of the 18 B850 (upper ring) and 9 B800 (lower ring) BChl *a* molecules in the LH2 antenna complex from *Rps. acidophila*. Within the circular array, nearest-neighbor distances (Mg···Mg) between B850 molecules are either 8.9 or 9.6 Å. The 9.6 Å distance is that of the two BChl *a* molecules associated with the $\alpha\beta$-polypeptide pair. The two B850 BChl *a* molecules nearest to a B800 molecule are separated by distances of 17.6 and 18.3 Å. The horizontal arrows indicate the directions of the Q_y-transition dipoles.

in Figure 13.1 for *Rps. acidophila* and orientations of the B850 molecules relative to each other very similar to those of *Rs. molischianum*. An X-ray structure of the LH2 complex from *Rb. sphaeroides* has yet to be determined. Concerning the LH1 complex of *Rhodospirillum rubrum*, it is a cyclic 16-mer of BChl *a* dimers with an inner core large enough to house the RC complex [18]. Unfortunately, the resolution of the electron micrographs is too low to permit electronic-structure calculations of the type reported for the LH2 complex of *Rps. acidophila* [12, 16]. However, the energy-minimization and molecular-dynamics simulations of Hu et al. [19] indicate that the structural arrangement of LH1's BChl *a* molecules is similar to that shown in Figure 13.1, the main difference being the larger ring size of LH1 [20].

It has been known for some time that pressure is an important tool for studying the structure and dynamics of proteins [21]. Windsor and co-workers [22, 23] had already demonstrated that pressures of ~350 MPa have a significant effect on the kinetics of secondary charge separation in the reaction center of *Rb. sphaeroides* and *Rhodopseudomonas viridis*. These works and those on the Q_y-absorption spectra of the LH2 complex of *Rb. sphaeroides* [15, 24] and *Rps. acidophila* [24, 25] as well as of the FMO antenna complex of *Chlorobium tepidum* [26] have clearly demonstrated that the linear pressure shifts (to the red) of Q_y-states can be large (-0.5 to -0.7 cm^{-1}/MPa) relative to those of isolated monomers in polymers and proteins (-0.05 to -0.15 cm^{-1}/MPa) [27–30]. In addition, the pressure shifts for different Q_y states of a given complex are generally not the same [15, 31]. This is important, since one can alter the energy gap(s) associated with energy and electron transfer. In recent years, a method of spectral hole burning at low temperatures has been introduced in these studies, which has improved spectral sensitivity by several orders of magnitude and has allowed experiments to be performed at very moderate pressure changes of only several tens of MPa [28–30]. In addition to energy gap, pressure may affect other parameters important to transfer such as the reorganization energy (electron–phonon coupling), the electronic coupling, and the structural heterogeneity, which leads, for example, to a distribution of values for donor–acceptor energy gaps. Fortunately, spectral hole-burning can often be used to examine the pressure dependencies of these quantities. A simple theory, which predicted a pressure dependence of the frequency shift of the spectral holes and their broadening data, was described by Haarer and co-workers [27]. The pressure shift of the optical transitions depends in a linear fashion on the hydrostatic compressibility, which is characteristic of the host material and the solvent shift describing the dye–matrix interaction. A fully statistical microscopic theory of pressure effects on spectral holes and of the inhomogeneous line shape itself, was developed by Laird and Skinner [32]. This theory assumes that (1) the transition frequency of a solute molecule can be described by a sum of pairwise interactions with each solvent and (2) each solvent molecule occupies a position that is independent of other solvent molecules. Their theory predicted that the hole shift with pressure is burn-frequency dependent (color effect); that is, it depends on where in the inhomogeneous line the hole is burned. This dependence is utilized to obtain the compressibility values of the host matrix. It should be noted that the Laird–Skinner theory is not applicable to photosynthetic complexes characterized by strong excitonic coupling such as the special BChl pair of the reaction center and the B850 and B875 rings of LH2 and LH1. However, when the chlorophyll pigments are well separated, as in the case for the B800 molecules (Fig. 13.1), the electrostatic interpigment couplings are weak and charge-transfer state effects are negligibly small. Thus Laird–Skinner theory is still applicable to the B800 band of the LH2 complex and to the FMO antenna complex.

Reddy et al. [15] have studied the effect of pressure on the B800→B850 energy transfer rate in the LH2 complex of *Rb. sphaeroides*. They burned zero-phonon holes into the B800 band of *Rb. sphaeroides* from ambient pressure to 680 MPa. Hole profiles and absorption spectra are shown in Figure 13.2. The ZPH widths of 5.8 and 5.3 cm^{-1} yielded B800→B850 transfer times of 1.8 and 2.0 ps, respectively. The B800→B850 transfer at ambient pressure has been reported to occur in 0.7 ps at room temperature [33] and 2.4 ps at liquid helium temperatures [13, 34]. This transfer rate in the low-temperature limit cannot be understood in terms of the Förster energy-transfer mechanism based on spectral overlap between the B800 fluorescence *origin* band and B850 absorption *origin* band. The Stokes shift for the B800 band is only $\sim 12\,cm^{-1}$ in the low-temperature limit [34]. Based on hole-burned spectra and the Franck–Condon factors for BChl *a*, Reddy et al. concluded that at ambient pressure the B800→B850 transfer is significantly contributed to by an acceptor mode near $750\,cm^{-1}$ with a Franck–Condon factor of 0.05. (They were aware that conventional Förster theory with its spectral overlap criterion is invalid when the relevant fluorescence and absorption bands of the donor and acceptor are inhomogeneously broadened.) Figure 13.2 reveals that the increase of pressure from ambient to 680 MPa at 4.2 K is accompanied by an increase in the B800–B850 energy gap from ~ 750 to $\sim 900\,cm^{-1}$. Therefore, at 680 MPa the $750\,cm^{-1}$ mode is no longer optimally poised to act as the primary acceptor mode in the transfer. However, there are three closely spaced modes near $920\,cm^{-1}$ whose combined Franck–Condon factor equals that of the $750\,cm^{-1}$ mode (0.05) [34, 35]. At high

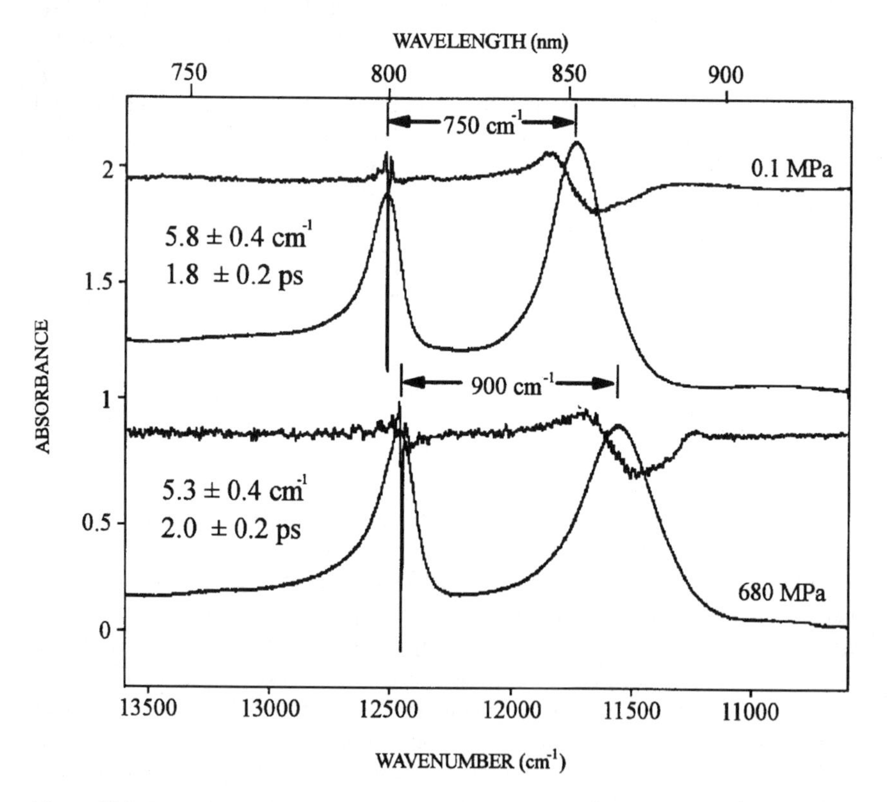

Figure 13.2. Low-temperature (4.2 K) hole-burned spectra obtained by burning in the B800 band of *Rb. sphaeroides*. The depths of the ZPH hole at 0.1 and 680 MPa are 12% and 8%, respectively. A burn intensity of $\sim 250\,mW/cm^2$ and burn time of 2 min were used.

pressure (680 MPa) the 920 cm^{-1} modes, instead of the 750 cm^{-1} mode, assume primary importance in energy transfer. When both the 750 and 920 cm^{-1} modes are taken into account, Reddy et al. [15] calculate that the energy-transfer rate in the low temperature limit at ambient pressure and 680 MPa is $(1.6\ ps)^{-1}$ and $(2.0\ ps)^{-1}$, respectively. If there were no Franck–Condon activity near 900 cm^{-1}, then it follows that there should be a reduction in the rate by at least a factor of 2.

In a similar way as strain fields, an externally applied electric (Stark) field using the SHB technique has proven to be a powerful approach for studies of photosynthetic complexes. Classical Stark modulation (CSM) spectroscopy, in which the response of an absorption band to the applied field is determined, has been extensively applied to the special pair band of the bacterial reaction center (see, for example, Refs. 36–40). The resolution of Stark spectroscopy is significantly improved by utilizing spectral hole-burning. This combination has proven to be powerful for analysis of dipole moment changes, $\Delta\mu$, associated with the $S_1 \leftarrow S_0$ transitions of isolated molecular chromophores in amorphous polymer films at liquid helium temperatures [41–52]. Stark hole-burning has also been reported for mesoporphyrin IX in horseradish peroxidase [53]. Because the zero-phonon holes of the chromophores in these systems are very narrow ($\lesssim$1 GHz) at liquid helium temperatures, modest field strengths (few tens of kV/cm) could be used to determine the matrix-induced ($\Delta\mu_{ind}$) and molecular ($\Delta\mu_0$) contributions to the change in dipole moment between the ground and excited electronic states. In some of these studies the Stark cell configuration allowed for propagation of the laser burn and read beams perpendicular to the external field ($\mathbf{E}_S$) so that the Stark effect on the ZPH could be determined with burn laser polarization parallel and perpendicular to $\mathbf{E}_S$. Use of both polarizations is important for determination of $\Delta\mu_{ind}$ and $\Delta\mu_0$. We note that a dependence of $\Delta\mu_{ind}$ on the location of the burn wavelength within the inhomogeneously broadened absorption band yields additional information on the polarizability of the chromophore and the internal electric field ($\mathbf{E}_{int}$) it experiences [51].

13.2. EXPERIMENTAL METHODS

The basic experimental setup for a hole-burning is relatively simple. It consists of a laser (burn) for exciting a narrow isochromat of an inhomogeneously broadened absorption band and a means for probing the absorption band prior to and after the burning. Modifications to this basic setup are dictated by the types of information sought and whether the type of hole-burning is persistent or transient (vide infra). Absorption changes are monitored via either the transmission of a probe beam or by fluorescence excitation (provided the sample is fluorescent) in any one of several ways. When one is concerned only with the ZPH, the same laser as used for burning (after suitable attenuation) can be tuned to measure the ZPH in the transmission mode. However, in the study of photosynthetic systems, the probe laser is often replaced by light from a monochromator. In our laboratory, a Fourier transform spectrometer operating in the visible and near-infrared region has been used. There are several reasons for this, including simplicity and that for the maximum benefits from hole burning one needs to acquire the entire hole spectrum, not just the ZPH of antenna complexes. In the studies to be briefly reviewed here, a Coherent CR899-21 Ti : sapphire laser (linewidth of 0.07 cm^{-1}) pumped by a 15-W Coherent Innova 200 Ar-ion laser was used for hole burning. The preburn and postburn absorption spectra were obtained by using a Bruker IFS120 HR Fourier transform spectrometer, the difference of such spectra being the persistent nonphotochemical hole-burned spectrum. High-resolution spectra were recorded at a resolution of 0.2 cm^{-1}.

High pressures were generated by a three-stage compressor (Model U11, Unipress-equipment Division, Polish Academy of Sciences, Warsaw, Poland). Helium gas is used as the pressure-transmitting medium. Gas flow between compressor stages is regulated by needle valves with the first stage connected to a helium gas cylinder. At the heart of the compressor is a pump capable of generating pressures of up to 80 MPa with manually controlled output. Compression ratios in the first and second stages are close to one and five, respectively. The multiplication ratio for the third state is 79, so that the maximum pressure is 1.5 GPa. A flexible thick-walled beryllium–copper capillary (o.d./i.d. = 3.0 mm/0.3 mm) connects the third stage of the compressor with a specially designed high-pressure optical cell (Unipress-equipment) for variable temperatures, which has a maximum pressure rating of 800 MPa. A manganin resistance gauge (Model MPG10, Unipress-equipment) was placed inside the third stage of the compressor to measure the gaseous helium pressure.

Figure 13.3 shows a section of the cylindrical high-pressure cell in the plane of the windows. The main body of the pressure cell was made of Be–Cu alloy (Brush alloy 25) with a diameter o.d. and height of 80 and 76 mm, respectively. Samples contained in gelatin capsule (sample volume of 0.12 ml) are placed in the cell cavity, which has a diameter o.d. and height of 7 and 8 mm, respectively. Optical access to the sample is provided by four sapphire windows (thickness of 4 mm). Windows are supported by window plugs (Ni–Cr alloy) having a conical light port with 2.5-mm-I.D. aperture and a 14° opening angle. Each conical plug is held in place by an annular brass plug. Tightening of the brass plugs (using specially designed O-rings covered with a thin layer of indium) isolates the high-pressure sample region from the outside. For low-temperature measurements the pressure cell is placed in a custom-made liquid helium cryostat with a 90-mm-I.D. tail section (Model 11 DT supervaritemp, Janis Research Co., Wilmington, MA). Sample temperature was monitored using a calibrated (1.4–360 K) silicon diode thermometer (Lake Shore Cryogenic Model DT-470) mounted on top of the pressure cell. Performing high-pressure experiments at low temperatures presents some difficulties since helium solidifies at a pressure of 18 MPa at 4.2 K and, for example, at 800 MPa at 55 K. Thus, at 4.2 K the sample needs to be warmed to effect a pressure change to higher pressures. Furthermore, once the sample is cooled below the freezing point of helium, there is a slight pressure drop from the third compressor stage to the pressure cell due to blockage of the capillary (prior to freezing of He in the cell). This pressure drop was corrected for using a calibration curve, which was provided by Unipress. Calibration was obtained by comparison of pressures measured at the cell cavity with a silicon diode with whose measured in the third stage of the compressor, vide supra. Note: Utilization of high pressures poses a potential danger to the operator. To minimize danger we employed safety shields (6-mm-thick polycarbonate, Lexan^{TM}) of suitable heights around the cryostat and compressor so as to block/redirect flying debris in the case of accidental decompression of either the pressure cell or the compressor. The Be–Cu high-pressure capillary (length of ~2 m) was encased in a touch flexible metal shroud and secured to the laboratory ceiling.

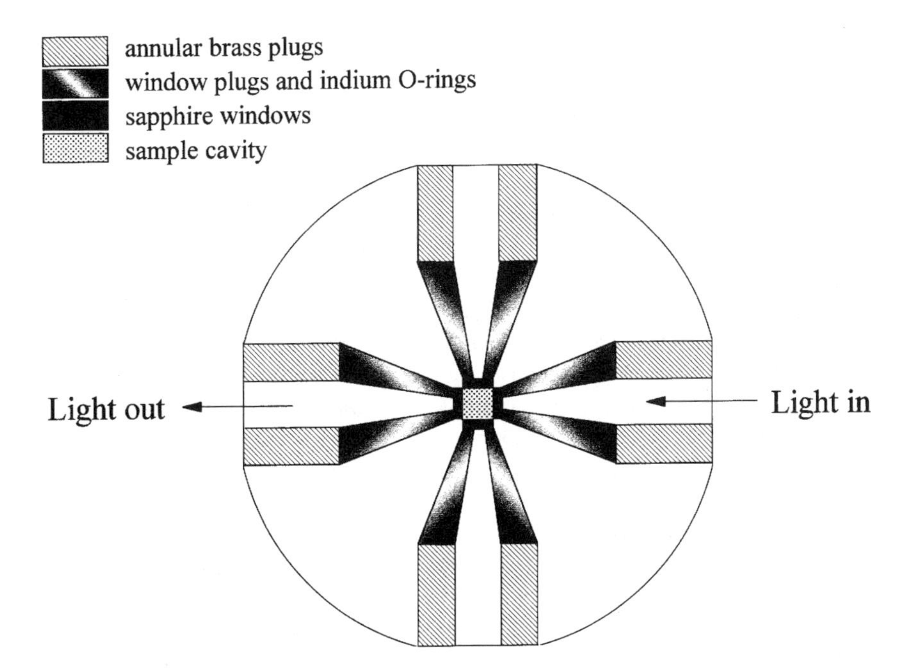

Figure 13.3. Top view of the high-pressure cell.

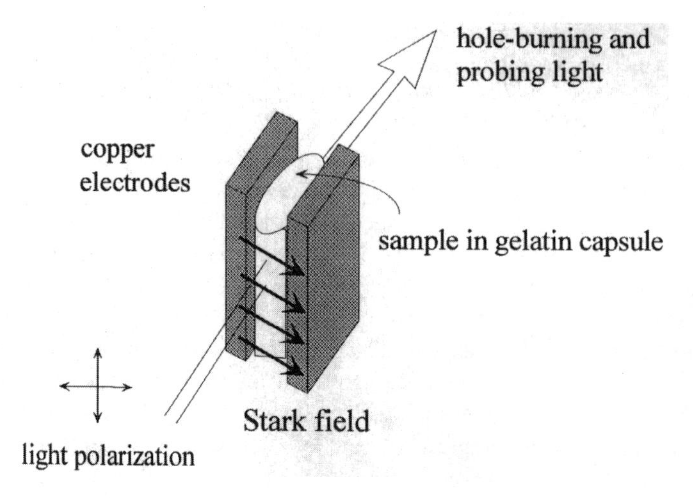

Figure 13.4. Geometry of the Stark cell. The light can be polarized parallel or perpendicular to the electric field.

For electric field measurements a specially designed Stark cell was used (Fig. 13.4). It allows for measurement of the dependence of the Stark signal on the angle between the electric field vector of the light and the applied electric field by turning the polarization direction of the incident light. The electric field is applied by means of two copper electrodes. Teflon spacers were used to set the distance between the electrodes (± 0.05 mm). Samples were contained in gelatin capsules (0.10 ml/O.D. $= 4.5$ mm) purchased from Torpac, Inc. Prior to insertion into the Stark cell, the gelatin capsule filled with sample was allowed to soften for about 5 min at room temperature so that it could be mechanically squeezed by the two copper electrodes of the Stark cell. This procedure yielded an optical path length perpendicular to the applied field of $\sim$6 mm with a distance of $\sim$2 mm between the electrodes. The electric field could be applied parallel or perpendicular to the burn laser polarization by positioning a polarizer placed in front of the Stark cell. The probing light (unpolarized) was collinear with the burning beam. A Trek, Inc., Model 610C DC high-voltage power supply (0 to ± 10 kV) was used. By changing the polarity of the power supply a maximum Stark field of $\sim$100 kV/cm was achievable. All measurements were made at 1.8 K in a Janis 10 DT liquid helium cryostat. With the sample immersed in pumped liquid helium, arcing never occurred at our maximum voltage. With an optical path length of $\sim$6 mm, the concentration of the chromophores and complexes was sufficiently low to ensure formation of high-quality glasses. In this regard, use of gelatin capsules was important. Optical densities of the samples were adjusted to $\sim$0.5 in the spectral regions where hole-burning was performed.

13.3. SOME RECENT RESULTS ON PHOTOSYNTHETIC COMPLEXES

13.3.1. High-Pressure Studies

In previous studies of a number of photosynthetic complexes it was shown that the pressure dependencies of the Q_y-absorption spectrum and the ZPH of selected Q_y states provide considerable insight into the strengths of nearest-neighbor BChl–BChl couplings [15, 26, 54]. Earlier results on pressure-dependent hole-burning studies of isolated chromophores in glasses, polymers, and proteins [27, 30] were important to the interpretations of the results for photosynthetic complexes. In these latter references the dependencies of the position and width of Q_y-absorption bands and ZPH were shown to be linear in pressure as observed for the isolated chromophores studied in Refs. 27 and 30. However, the linear pressure shifts of Q_y bands for complexes with strong BChl–BChl coupling, such as the special pair band of bacterial reaction centers and LH1 of purple bacteria, were observed to be about a factor of 5 larger than for isolated chromophores, about -0.5 versus about -0.1 cm^{-1}/MPa.

In the study [24] one of our objectives was to determine the dependence of the B870's pressure shift of the light harvesting 2 (LH2) antenna complexes of *Rps. acidophila* and *Rb. sphaeroides* on position within its inhomogeneously broadened profile. Specifically, it is the lowest-energy exciton level (B870) associated with the B850 ring, which is reported on since sharp ZPH cannot be burned into the adjacent level, which are the main contributors to the intensity of the B850 band. This is a consequence of their ultrafast (~100 fs) downward interexciton level relaxation dynamics [14, 55, 56]. To this end, ZPH action spectroscopy has been used to characterize the weak B870 absorption band. In ZPH action spectroscopy one burns a series of persistent ZPH holes across the inhomogeneously broadened absorption profile at constant burn fluence, see Ref. 55 for details. Here we also present very recent data from the pressure studies of LH1 antenna complex of *Rb. sphaeroides* [57].

The lower absorption spectrum of Figure 13.5 is that of the isolated LH2 complex of *Rps. acidophila* at 4.2 K. The bands near 11500 and 12500 cm^{-1} are B800 and B850, respectively, with widths (fwhm) of 200 and 125 cm^{-1}. The absorption profile shapes for *Rb. sphaeroides* are very similar, as are their widths [24]. However, the B800–B850 energy gap for *Rb. sphaeroides* is significantly smaller than that of *Rps. acidophila* (also *Rs. molischianum*) at all temperatures, by ~200 cm^{-1} at 4.2 K. The diminution in this energy gap is mainly due to the B850 band of *Rps. acidophila* being red shifted relative to that of *Rb. sphaeroides* [24]. At the bottom of Figure 13.5 is the exciton energy-level diagram for the B850 ring for *Rps. acidophila* from Ref. 58 calculated in the nearest dimer–dimer coupling approximation. In the absence of energy disorder, the E_2, E_3, and E_1 levels are symmetry forbidden in absorption from the ground state. However, the structure of the complex, which leads, in part, to the Q_y transition dipoles lying nearly in the membrane plane (perpendicular to the C_9-axis), results in the lowest-energy E_1 level carrying almost all the absorption intensity. For example, the lowest-energy A level (B870) is predicted to carry less than 1% of the B850 band intensity [12, 16]. Thus in Figure 13.5 the lowest E_1 level is placed at the B850 band maximum. Set off to the upper right is the B870 ZPH action spectrum whose overall profile faithfully represents the B870 absorption profile with $\Gamma_{\text{inh}} = 120 \pm 10\ \text{cm}^{-1}$ and band maximum indicated at the position of the upward solid arrow. The displacement of the A level (B870) below the B850 maximum (or E_1 level) is defined as ΔE and equals $200\ \text{cm}^{-1}$, The weak shoulder on the low-energy side of the B850 band is due to B870, which carries 3–5% of the intensity of the entire B850 band [24, 58]. The 4.2 K absorption profile of B875 for the LH1 only mutant is shown at the top of Figure 13.5 along with ZPH action spectrum of B896 for which $\Gamma_{\text{inh}} = 150 \pm 10\ \text{cm}^{-1}$ and $\Delta E = 140 \pm 10\ \text{cm}^{-1}$. B896 of LH1 is analogous to B870 of the B850 ring.

The pressure shifting ZPH holes burned into the B870 band of *Rps. acidophila* at a burn temperature of 12 K are shown in Figure 13.6. The linear pressure shifts in $\text{cm}^{-1}/\text{MPa}$ range from -0.45 to -0.56 for *Rps. acidophila* in going from the high- to low-energy side of B870. The corresponding range for *Rb. sphaeroides* is from -0.47 to -0.60, essentially identical to that for *Rps. acidophila*. The increase in the pressure shift rate in going from the high- to low-energy sides of B870 is substantial, ~25% for the highest and lowest burn frequencies. Pressure broadening of the ZPH was observed to be strong, $\sim 0.1\ \text{cm}^{-1}/\text{MPa}$ for both species. This broadening prevented measurements to higher pressures than those indicated in the Figure 13.6. Since the hole spectra were recorded with a Fourier transform spectrometer, we are able to obtain the linear pressure shifts and broadenings of the B800 and B850 absorption bands for both species (Table 13.1). The -0.39 and $-0.38\ \text{cm}^{-1}/\text{MPa}$ linear pressure shifts for the B850 band of *Rps. acidophila* and *Rb. sphaeroides* are identical within experimental uncertainty. Additional experiments were performed using much higher pressures than those indicated in Figure 13.6. Results for the pressure shifting and broadening for *Rps. acidophila* (85 K) are given in Figure 13.7. The linear shift and broadening rates for both *Rps. acidophila* and *Rb. sphaeroides* are given in Table 13.1 along with those for the B800 band. The B850 linear shift rates at these higher temperatures are only slightly higher in magnitude than at 12 K, as one might have anticipated on the basis of earlier works [15, 26], which point out that the temperature dependence of compressibility of polymers and proteins is weak. High-pressure data for the B875 absorption band of an LH1 only mutant of *Rb. sphaeroides* yielded a very large linear pressure shifting of -0.60 and a broadening rate of $0.12\ \text{cm}^{-1}/\text{MPa}$. High-pressure hole-burning data for the lowest exciton level (B896) of the B875 band lead to a linear pressure shifting of $-0.67\ \text{cm}^{-1}/\text{MPa}$ and to an average broadening rate of

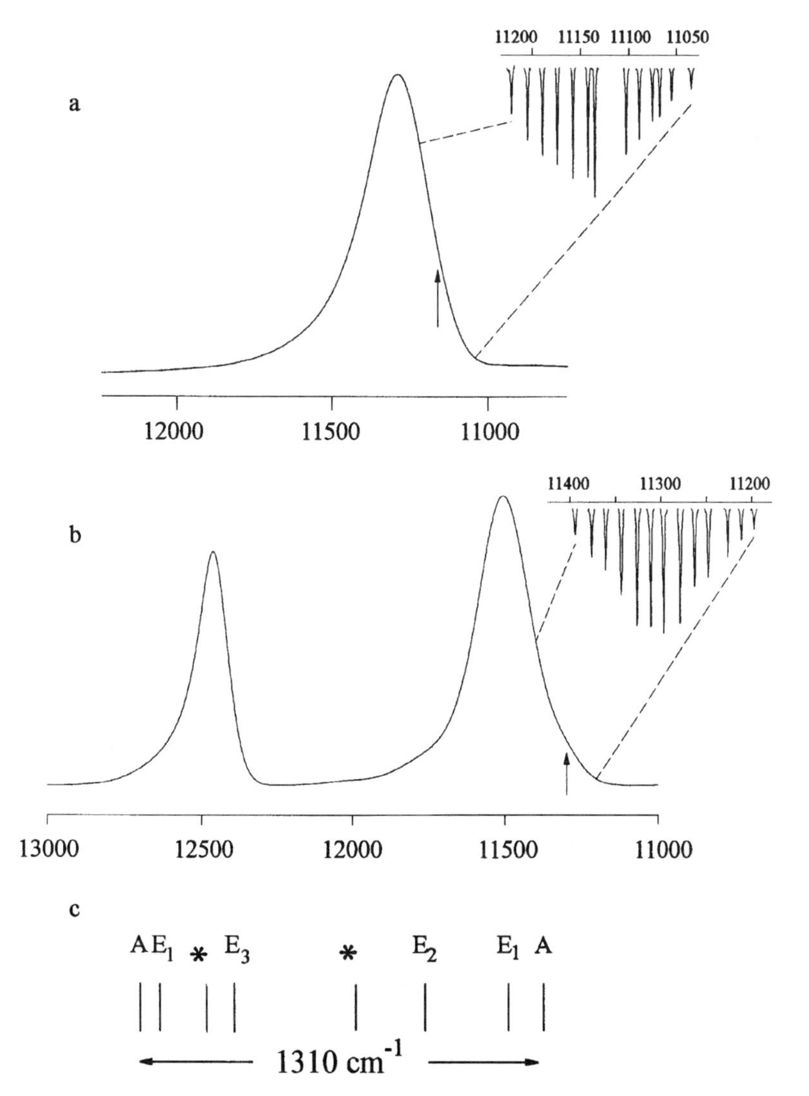

Figure 13.5. (a) 4.2 K B875 absorption and B896's zero-phonon hole (ZPH) action spectrum (read resolution = 0.8 cm^{-1}) for LH1-only mutant of *Rb. sphaeroides*. (b) 4.2 K B850 absorption and B870's zero-phonon hole action spectrum (read resolution = 0.5 cm^{-1}) for LH2 of *Rps. acidophila*. The ZPH action spectra in (a) and (b) were generated with a constant burn fluence of 50 and 100 J/cm^2, respectively. The arrows locate the center of the action spectra (140 and 200 cm^{-1} below the B875 and B850 band maximum, respectively). The fractional OD changes for ZPH at the arrows are 0.11 and 0.15 for B896 and B870, respectively. The action spectrum carries an inhomogeneous width of $150 \pm 10\,cm^{-1}$ in (a) and $120 \pm 10\,cm^{-1}$ in (b). Also shown together is the calculated exciton manifold of the B850 ring (c). The strongly absorbing E_1 level in the lower manifold is placed at B850 absorption maximum. The asterisks indicate two closely spaced doubly degenerate levels. For easy comparison, the absorption spectra and the exciton levels are shown in the same wavenumber scale, while the x axis of the action spectra has been expanded three-fold.

Table 13.1 Pressure-Dependent Data for Isolated LH2 Complex from *Rps. acidophila* and *Rb. sphaeroides*[a]

		Pressure Shift		Pressure Broadening	
Temperature		12 K	85 K	12 K	85 K
Rps. acidophila	B800	−0.09	−0.11	~0	~0
	B850	−0.39	−0.40	0.13	0.14
Temperature		12 K	100 K	12 K	100 K
Rb. sphaeroides	B800	−0.15	−0.14	~0	~0
	B850	−0.38	−0.41	0.15	0.23

[a] The uncertainties for the shift rate are $\pm 0.003\ \text{cm}^{-1}/\text{MPa}$, and for the broadening ± 0.001 and $\pm 0.004\ \text{cm}^{-1}/\text{MPa}$ for *Rps. Acidophila* and *Rb. sphaeroides*, respectively.

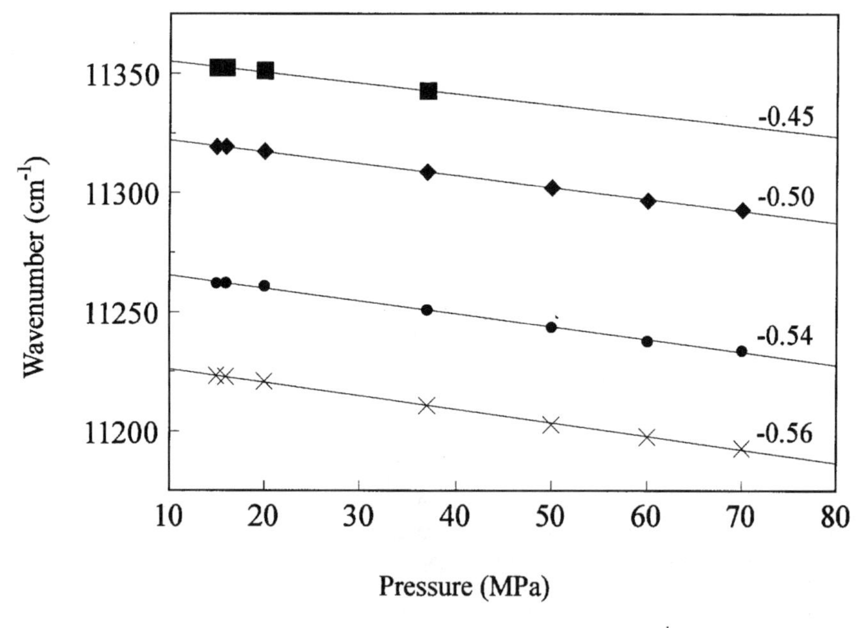

Figure 13.6. Linear pressure shifting of B870 ZPH (read resolution = $1\ \text{cm}^{-1}$) for *Rps. acidophila*. Four holes were burned at 11352, 11319, 11262, and $11223\ \text{cm}^{-1}$, respectively, at 15 MPa and 12 K. With increasing pressure, all holes shifted linearly (see the solid lines) to the red with rates (in $\text{cm}^{-1}/\text{MPa}$, uncertainty = $\pm 0.01\ \text{cm}^{-1}/\text{MPa}$ for the bottom three lines and $\pm 0.03\ \text{cm}^{-1}/\text{MPa}$ for the uppermost one) indicated in the figure. For the pressure range used, no irreversible pressure shifting of B870 ZPH was observed (elastic behavior).

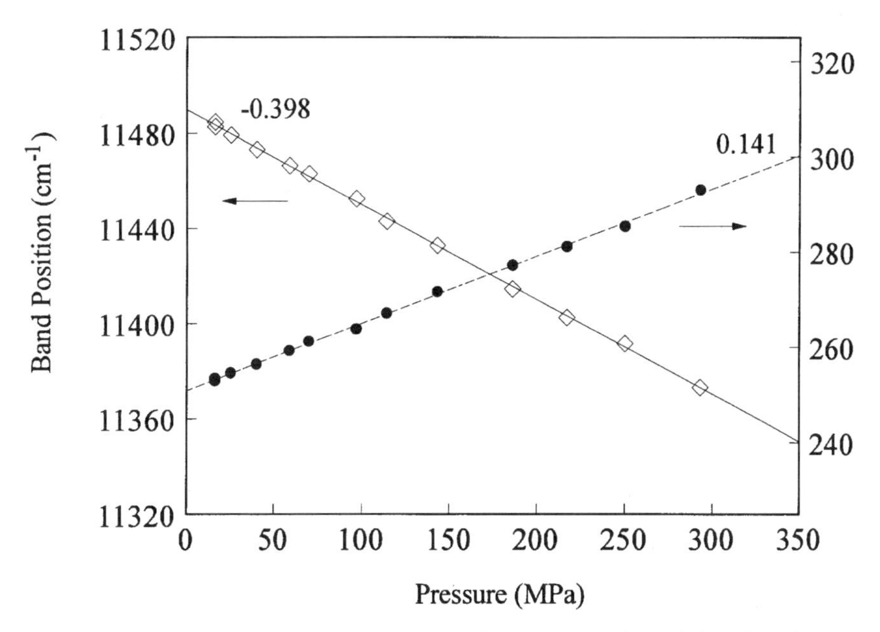

Figure 13.7. B850 band pressure shifting (diamonds) and broadening (circles) data for *Rps. acidophila* at 85 K. Linear shift (solid line) and linear broadening (dashed line) rates are -0.398 ± 0.003 and $0.141 \pm 0.001\ \mathrm{cm}^{-1}/\mathrm{MPa}$, respectively.

$0.07\ \mathrm{cm}^{-1}/\mathrm{MPa}$. To conclude this subsection, we note that the much smaller pressure shifting and broadening rates of the B800 band relative to the B850 and B870 bands is consistent with the B800 molecules being weakly coupled, as pointed out in Ref. 15.

We turn next to the question of why the linear pressure shifts of the ZPH associated with the lowest exciton level (B870) of the B850 ring are, on average, about 25% greater than that of the B850 absorption band for which the shift is $\sim -0.40\ \mathrm{cm}^{-1}/\mathrm{MPa}$ (Table 13.1). Following that, we consider the question of why the pressure shift of the B870 ZPH increases by ~25% in going from the high- to low-energy side of the inhomogeneously broadened B870 absorption profile.

It is proposed that the answer to the first question has its origin in energy disorder (diagonal and/or off-diagonal) stemming from the glasslike structural heterogeneity of proteins. The most detailed study of how energy disorder leads to splitting of degeneracies in a C_n-ring system and redistribution of oscillator strength between exciton levels is that of Wu and Small [59]. They employ symmetry-adapted energy defect patterns for systematic analysis of the problem. The focus of their work was the mixing of the allowed E_1 and forbidden A (B870) levels due to energy disorder and the relationship between the energy gap of these two levels and the absorption intensity of B870; see also Ref. 60. What emerges, in part, from their calculations is that disorder-induced splitting of the allowed E_1 level, which is the dominant contributor to the B850 band, results in the lower- and higher-energy E_1 components shifting to the red and blue, respectively. Given that both components carry comparable absorption intensity [59], it follows that the shifting of the B850 absorption band should be smaller than that of the B870 exciton level (which is selectively interrogated by hole-burning) if the extent of energy disorder increases with increasing pressure. We note that this reasoning provides an explanation for why the pressure broadening of the B850 band, $\sim 0.15\ \mathrm{cm}^{-1}/\mathrm{MPa}$, is a factor of seven times larger than that of the special pair band (P960) of the bacterial reaction center [15] since only a single exciton (dimer) level contributes to P960.

The second question, which is why the linear pressure shift of the B870 complex increases as its absorption frequency shifts to the red, is a difficult one. Although a theory of linear pressure shifts exists for isolated chromophores in isotropic, homogeneous amorphous solids [32], a theory does not exist for anisotropic excitonically coupled arrays. The theory of Ref. 32 yields

$$\frac{\Delta\tilde{\nu}}{\Delta P} = \frac{n\kappa}{3}(\tilde{\nu}_{\mathrm{max}} - \tilde{\nu}_{\mathrm{vac}}) \tag{13.1}$$

for the shift rate with κ the compressibility, $\tilde{\nu}_{\mathrm{max}}$ the frequency of the absorption maximum or hole at ambient pressure, $\tilde{\nu}_{\mathrm{vac}}(>\tilde{\nu}_{\mathrm{max}})$and is the gas-phase absorption frequency. The factor n is the power of the attractive chromophore–solvent interaction ($\propto R^{-n}$). When applied to the linear shifts of ZPH of isolated chromophores in polymers with $n = 6$, this equation yielded values for κ in reasonable agreement with bulk values [61]. For polymers and proteins κ is in the 0.05–0.15 GPa^{-1} range. In earlier works we applied Eq. 13.1 to relatively weakly coupled BChl complexes such as the FMO complex [26] and for this complex determined a value of $\kappa = 0.1\ \mathrm{GPa}^{-1}$. Since the B800 molecules are weakly coupled, we used Eq. 13.1 to arrive at estimates for κ. From Table 13.1, the 12 K linear pressure shifts of the B800 band for the isolated complexes of *Rps. acidophila* and *Rb. sphaeroides* are -0.09 and $-0.15\ \mathrm{cm}^{-1}/\mathrm{MPa}$. Their $\tilde{\nu}_{\mathrm{max}}$ (B800) values are 12445 and 12505 cm^{-1}. With $\tilde{\nu}_{vac} = 13340\ \mathrm{cm}^{-1}$ [62] and $n = 6$ for the BChl *a*–protein interactions, Eq. 13.1 yields $\kappa = 0.05$ and 0.09 GPa^{-1} for *Rps. acidophila* and *Rb. sphaeroides*, respectively.

Returning to the problem of the dependence of B870's pressure shift on excitation frequency, one might suggest that the dependence is due to an increase in κ as the excitation frequency is tuned from high to low energy. Here, one would have to argue that the tightness of packing of the α,β-polypeptide pairs of the ring [11, 17] is negatively correlated with decreasing B870 excitation frequency. The calculations of Wu and Small [59] provide support for such correlation. Their results indicate that the location of the inhomogeneously broadened B870 absorption profile (as well as its intensity) is determined, in part, by energy disorder (diagonal and/or off-diagonal) of the B850 ring and that the red-most absorbing B870 levels are associated with greater energy disorder and, as a consequence, stronger coupling with the allowed E_1 level. If increasing energy disorder is associated with greater structural disorder, which seems physically reasonable, it follows that the compressibility κ should increase with decreasing B870 excitation frequency. It should be pointed out that increasing energy disorder leads to greater electronic localization effects for the Q_y states of the B850 ring. The same is true for the B875 ring of LH1.

In summary, the large pressure shifting of the B850 absorption band and its lowest energy exciton level, B870, cannot be understood in terms of electrostatic interactions between BChl *a* molecules. Charge-transfer states or, equivalently, electron exchange interactions between neighboring BChl *a* molecules, are mainly responsible for the largeness of the pressure shift. We believe that the independence of the B800 bandwidth of pressure is most likely linked to the fact the B800 molecules are weakly coupled. The results of calculations on the effects of static, random diagonal energy disorder indicate that the states of the B800 ring are highly localized on individual molecules while the exciton levels of the B850 ring still exhibit considerable delocalization [63]. The B850 absorption band is contributed to mainly by the two components of the strongly allowed E_1 level, which are split apart by energy disorder. The results of Ref. 60 for the B850 ring show that the red shifting of the lower-energy E_1 component ($E_{1,1}$) with increasing disorder is larger than that of the higher E_1 component ($E_{1,h}$). Thus, if an increase in pressure increases the energy disorder, one has an explanation for the pressure broadening of the B850 band.

13.3.2. Stark Hole-Burning on LH2 Complexes

Stark hole-burning spectroscopy at 1.8 K was used to determine the dipole moment changes $f\Delta\mu$ (f, the local field correction factor) for the B800 and B850 absorption bands of the LH2 complexes of *Rb. sphaeroides*, *Rsp. acidophila* (strain 10050), and *Rs. molischianum* [64, 65]. Specifically, it is the lowest-energy exciton level (B870) of the B850 ring that was reported on. Hole-burning values of $f\Delta\mu$ are summarized in Table 13.2. The results of the 825-nm band of the BChl *a* (FMO) antenna complex of *Cb. tepidum* are also included. For each band, $f\Delta\mu$ was determined for burn-laser polarization parallel and perpendicular to the Stark field $\mathbf{E}_S$. All values of $\Delta\mu_\perp$ and $\Delta\mu_\parallel$ in Table 13.2 were obtained by averaging the values obtained with different burn frequencies. Results from classical Stark modulation (CSM) studies at 77 K of the B800, B850, and B875 bands of *Rb. sphaeroides*, *Rps. acidophila*, and *Rs. molischianum* have recently been reported [66, 67]. For comparison, values of $f\Delta\mu_{CSM}$ and

Table 13.2. Values of Diple Moment Change ($\Delta\mu$) and Polarizability Change ($\Delta\alpha$) as Determined by Stark Hole-burning and Classical Stark Modulation (CSM) Spectroscopy

Species	Band	$f\Delta\mu_{\parallel}$(D)	$f\Delta\mu_{\perp}$(D)	Band	$f\Delta\mu_{CSM}$(D)[a]	f^2 Tr($\Delta\alpha$)(A^3)[a]
Rb. sph.	B870	1.10±0.1	1.44±0.1	B850	4.2 (3.3)[b]	619
Rps. acid.	B870	1.0 ±0.05		B850	3.2	1250
Rs. moli.	B870	1.2±0.1		B850	3.2	1420
Rb. sph.	B800	0.9 ±0.1	0.7 ±0.1	B800	1.1 (0.9)[b]	5
Rps. acid	B800	0.62±0.06	0.55±0.06	B800	1.5	—
Rs. moli	B800	1.2±0.2		B800	2.8	290
Cb. tepid.	B825	0.51±0.06	0.72±0.04	B825	1.4[b]	—

[a] Results from [67] except for *b*, which are from Ref. 38. Typical stated uncertainties for $\Delta\mu_{CSM}$ are ±0.1 D.

f^2 Tr($\Delta\alpha$) for B850 and B800 of the LH2 complexes are given on the right side of Table 13.2.

The effect of electric field on ZPH of B870 of *Rb. sphaeroides* is shown in Figure 13.8 for both parallel and perpendicular orientations between the burn-laser polarization and Stark field $\mathbf{E}_S$ ($E_S = 0, 44, 88$ kV/cm). For both orientations, only symmetrical broadening of the ZPH was observed. It was determined that the Stark broadening was fully reversible; that is, field turnoff reduced the hole width to its original value. This is not the case for the hole depth or integrated hole area, but it was determined that this is the result of spontaneous hole filling associated with nonphotochemical hole-burning [68] rather than an electric field effect (results not shown). At the resolution of the experiment there was no evidence for Stark splitting or shifting of the holes for either polarization. The hole broadening rate was higher in the perpendicular direction. The broadening of spectral holes versus applied electric field for both perpendicular and parallel orientation is plotted in Figure 13.9. The absence of Stark splitting for both parallel and perpendicular laser polarizations is informative [42, 53], as we now briefly discuss.

Consider first the case of isolated molecules in glassy matrices and let θ be the angle between the molecular dipole moment difference vector $\Delta\boldsymbol{\mu}_0$ and the transition dipole vector **D**. Linearly polarized light preferentially burns out those molecules with **D** parallel to the polarization vector **e** of the light. The experimentally observed dipole moment change can be written as $\Delta\boldsymbol{\mu} = \Delta\boldsymbol{\mu}_0 + \Delta\boldsymbol{\mu}_{ind}$, where $\Delta\boldsymbol{\mu}_{ind}$ is the matrix-induced contribution. Then $\Delta\boldsymbol{\mu}_0$ is dominant, and the photoselection phenomenon enables one to probe molecules for which the angle between $\Delta\boldsymbol{\mu}_0$ and **e** is well defined (shallow-hole limit). As discussed in detail in Ref. 42, Stark splitting of the hole can be observed for **e**

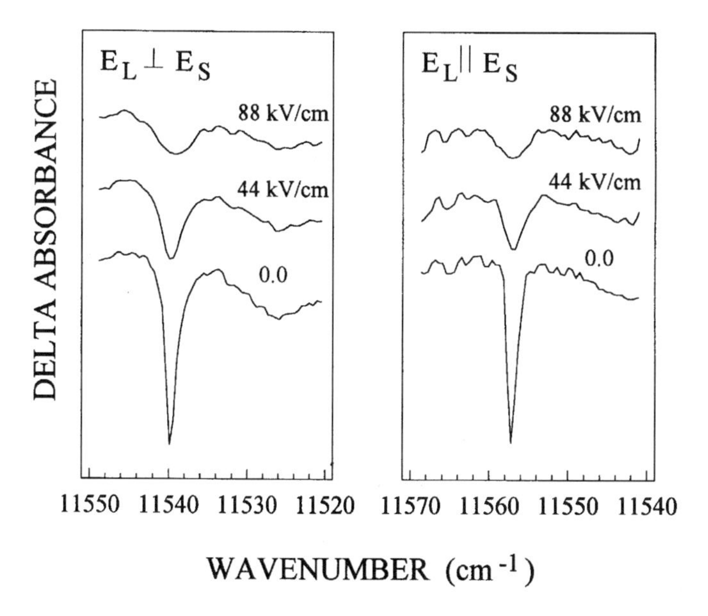

Figure 13.8. Stark effect on zero-phonon holes burned into B870 of *Rb. sphaeroides*. Left and right frames for burn-laser polarization perpendicular and parallel to the Stark field, $\mathbf{E}_S$, respectively.

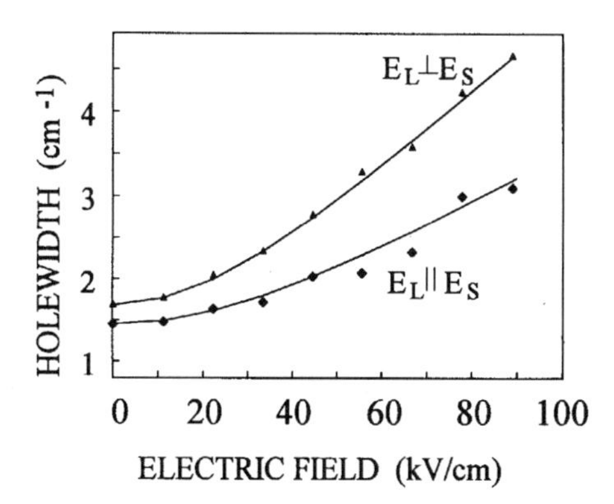

Figure 13.9. Dependence of the B870 holewidth of *Rb. sphaeroides* on electric field for perpendicular (triangles) and parallel (diamonds) burn-laser polarizations. The burn frequencies were 11539 and 11557 cm^{-1}, respectively. The solid curves are theoretical fits calculated using Eq. 13.3, which yielded $f\,\Delta\mu_{\perp} = 1.48$ D and $f\,p\Delta\mu_{\parallel} = 1.06$ D.

parallel or perpendicular to $\mathbf{E}_S$, depending on the value of θ. For example, for $\theta = 0$ or π Stark splitting occurs for parallel polarization, while broadening occurs for perpendicular polarization. The situation is reversed for $\theta = \pm\pi/2$. However, when $\Delta\boldsymbol{\mu}_{\text{ind}}$ is dominant, only Stark broadening is expected for both polarizations. This is because the orientation of $\Delta\boldsymbol{\mu}_{\text{ind}}$ relative to $\mathbf{D}$ or $\mathbf{e}$ is random for a glassy matrix; that is, the matrix field varies significantly from site to site. The assumption of random orientations of $\Delta\boldsymbol{\mu}_{\text{ind}}$ for Chl molecules in photosynthetic complexes is questionable, since the structure of the protein around these chromophores is well defined even though the Q_y-absorption bands do suffer from significant inhomogeneous broadening. In other words, one might have expected to observe a Stark splitting of B870 holes for one of the two polarizations. Indeed, Gafert et al. [53] observed Stark hole splitting for two of the three sites of mesoporphyrin substituted in horseradish peroxidase. For the same molecule in a glass, only Stark broadening was observed.

The Stark effect was investigated at different positions in the inhomogeneously broadened B870 band. A 10–20% increase of dipole moment difference from the blue to the red edge of the B870 band was observed. Near the center of the band ($\lambda = 866.5$ nm) we obtained $f\Delta_{\perp} = 1.44 \pm 0.1$ D and $f\Delta\mu_{\parallel} = 1.10 \pm 0.1$ D for perpendicular and parallel orientations, respectively. The solid curves through the Stark broadening data in Figure 13.9 are the theoretical fits obtained using the theory of Kador et al. [43], which is valid when the variance of $\Delta\mu$ is small. They define the parameter

$$F = \frac{2f\Delta\mu\, E_S}{\hbar(\gamma' + \gamma'_d)} \tag{13.2}$$

where γ' is the homogeneous width of the zero-phonon line and γ'_d the additional width associated with artifacts such as saturation broadening. The sum $(\gamma' + \gamma'_d)$ is the width at zero field. When $F \lesssim 3.5$, their general expression for Stark broadening simplifies considerably to

$$\Gamma(F) = 2(\gamma' + \gamma'_d)(1 + F^2)^{1/2} \tag{13.3}$$

The above inequality is satisfied in our experiments. For sufficiently large F, $\Gamma(F) \propto F$, which depends linearly on E_S. The good agreement between the experimental points and the theoretical curve indicates that the observed broadening is due to the linear Stark effect up to 88 kV/cm. Moreover, in all measurements the broadening was symmetric and the center frequencies of the holes were not affected by the applied electric field, indicating that the quadratic Stark effect can be neglected. We emphasize that it is only the frequency shift of an individual absorber that is linear in field. As can be seen in Figure 13.9, the hole broadening as a function of field strength becomes linear in the high-field region.

Stark hole-burning studies [65] of the B896 level of the B875 antenna complex of the wild-type chromophore and an LH1-only mutant of *Rb. sphaeroides* yield dipole moment changes $f\Delta\mu = 0.84 \pm 0.08$ and 0.78 ± 0.06 D, respectively; that is, the results are identical within experimental uncertainty. No dependence on wavelength or laser polarization was observed.

13.3.3. BChl *a* Antenna Complex of *Cb. tepidum*

In this subsection, Stark hole-burning data are presented for the 825-nm band of the FMO complex of *Cb. tepidum* [65]. The 4.2 K Q_y-absorption spectrum of BChl *a* or FMO complex from *P. aestuarii* and *Cb. tepidum* exhibit prominent bands at 825, 814, and 805 nm [7, 26]. Stark hole-burning is not possible for exciton levels higher in energy than the 825-nm band due to hole broadening from subpicosecond interexciton-level relaxation processes.

Stark hole-burning was performed at several burn frequencies (ω_B) between 12161 and 12076 cm^{-1}, from the high- to low-energy sides of the 825-nm band. No dependence of the Stark results on ω_B was observed. Some hole profiles obtained with $\omega_B \approx 12115$ cm^{-1} are shown in Figure 13.10 for laser polarization

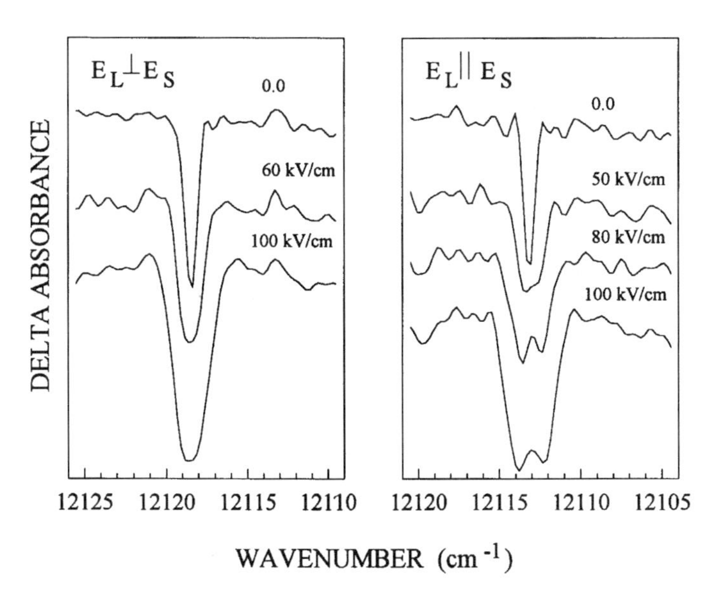

Figure 13.10. Stark effect on zero-phonon holes burned into the 825-nm band of the FMO BChl *a* antenna complex from *Cb. tepidum*. Left and right frames correspond to burn laser polarization perpendicular and parallel to the Stark field, $\mathbf{E}_S$, respectively.

perpendicular and parallel to the Stark field $\mathbf{E}_S$. These polarizations lead to hole broadening and splitting, respectively. The broadening and splitting of spectral holes versus applied electric field is plotted in Figure 13.11. The observation of Stark splitting for $\mathbf{e}\|\mathbf{E}_S$ but not for $\mathbf{e} \perp \mathbf{E}_S$ means that the angle γ between $\Delta\boldsymbol{\mu}$ and the transition dipole $\mathbf{D}$ of B825 is smaller than 45° [53]. To arrive at an upper limit for γ, the Stark effect on the hole profile for angles (χ) of 0, 30, 60, and 90° between $\mathbf{e}$ and $\mathbf{E}_S$ were determined with $\mathbf{E}_S = 100$ kV/cm (Figure 13.12). Stark splitting is not observed for $\chi = 60°$ and the splitting for $\chi = 30°$ is considerably smaller than the splitting for $\chi = 0°$ (1.1 and 1.4 cm^{-1}, respectively). Comparison of the hole profiles in Figure 13.12 with the results of theoretical simulations in Ref. 69 led to $\gamma \lesssim 15°$. The smallness of γ and the observation that the Stark splitting ($\Delta\omega_{SS}$) is maximum for $\chi \approx 0$ means that in the well-known expression (see, for example, Ref. 70)

$$\Delta\omega_{SS} = (29.8)^{-1} f \Delta\mu\, E_S \cos\theta \tag{13.4}$$

the angle θ between $\Delta\boldsymbol{\mu}$ and $\mathbf{E}_S$ is $\lesssim 15°$, so that $\cos\theta \gtrsim 0.97$. In Eq. 13.4 the units of $\Delta\omega_{SS}$, $\Delta\mu$, and E_S are cm^{-1}, Debye, and kV/cm. This equation with the data in Figure 13.11 yields $f\Delta\mu = 0.42 \pm 0.05$ D, where $\Delta\mu$ is the total dipole moment change. That the Stark effects do not depend on ω_B means that statistical fluctuations in structure from complex to complex are too small to be reflected in $\Delta\mu$. The Stark splittings plotted in Figure 13.11 were measured directly from the hole profiles. Since the Stark split components overlap and broaden with increasing field strength, it is necessary to simulate the hole profiles so as to

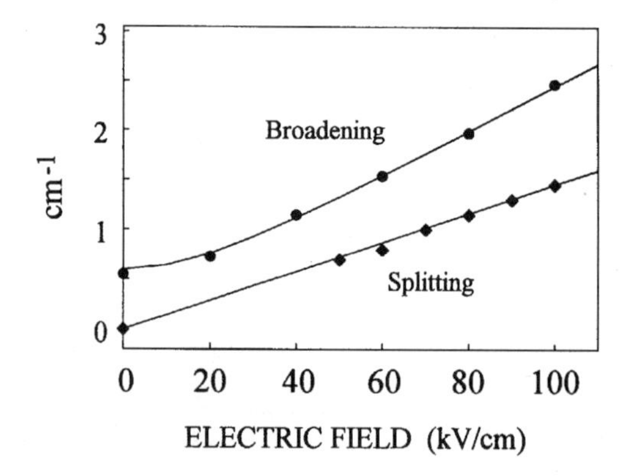

Figure 13.11. Dependence of the hole width and Stark splitting for the zero-phonon holes of the 825-nm band of the FMO complex of *Cb. tepidum* on electric field for perpendicular (circles) and parallel (diamonds) burn-laser polarization. The burn frequencies were 12119 and 12113 cm^{-1}, respectively. The solid curves are theoretical fits calculated using Eq. 13.3, which yielded $f\,\Delta\mu_{\perp} = 0.72$ D (broadening) and Eq. 13.4 with $f\,\Delta\mu_{\|} = 0.42$ D (splitting).

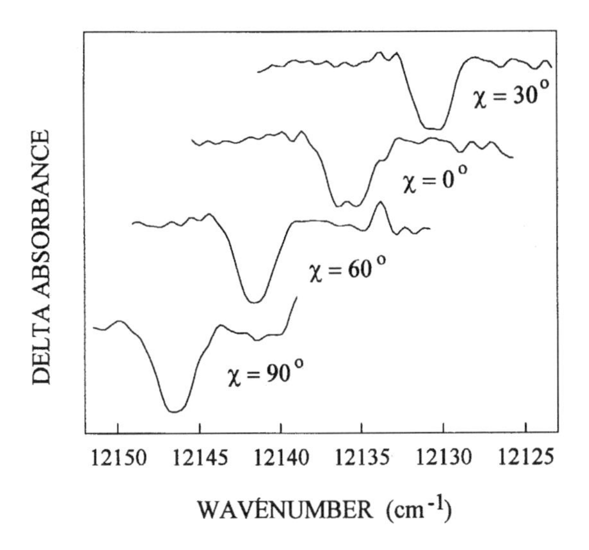

Figure 13.12. Stark effect on the hole profile of the 825-nm band of the FMO complex of *Cb. tepidum* for angles (χ) of 0, 30, 60, and 90° between the burn laser polarization and electric field of $E_S = 100\ \text{kV/cm}$.

arrive at a more accurate value for $f\Delta\mu_{\|}$. The simulations with identical Gaussian line shapes for the split components led to $f\Delta\mu_{\|} = 0.51 \pm 0.06$ D (the 80 and 100 kV/cm profiles were used). The Stark broadening data on Figure 13.11 for laser polarization perpendicular to $\mathbf{E}_S$ led to $f\Delta\mu_{\perp} = 0.72$ D.

Observation of Stark splitting means that the angle γ between $\Delta\boldsymbol{\mu}$ and the transition dipole, **D**, is quite well defined. That $f\Delta\mu_{\perp}$ is somewhat larger than $f\Delta\mu_{\|}$ indicates that there may be a random contribution to $\Delta\mu$. Consistent with this is that the two components of the Stark-split hole profile broaden with increasing field strength (vide infra). The broadening suggests that there is a random contribution to $\Delta\boldsymbol{\mu}_{\text{ind}}$ in addition to a contribution to the induced dipole moment change vector that is relatively well defined with respect to the molecular frame. (We consider it unphysical to assume that the 0.51 D dipole moment change is entirely molecular, i.e., equal to $f\Delta\mu_0$). The well-defined contribution might be associated with the "inner shell" of the protein matrix with the random contribution coming from the "outer shell" and, perhaps, even the glass-forming solvent. Fitting of the Stark broadening data with Eq. 13.3 led to $f\Delta\mu_{\text{ind}}$ (random) ≈ 0.5 D. Whether or not the model in which $\Delta\boldsymbol{\mu}_{\text{ind}} = \Delta\boldsymbol{\mu}_{\text{ind}}$ (nonrandom) + $\Delta\boldsymbol{\mu}_{\text{ind}}$(random) can explain why $f\Delta\mu_{\perp}$ is slightly larger than $f\Delta\mu_{\|}$ is unclear. We note that one is dealing with vectorial quantities; that is, $f\Delta\boldsymbol{\mu}_{\|} = f\Delta\boldsymbol{\mu}_0 + f\Delta\boldsymbol{\mu}_{\text{ind}}$ (nonrandom). Although it is reasonable to assume that

$f\Delta\boldsymbol{\mu}_{\text{ind}}$ (nonrandom) lies mainly in the plane of the BChl *a* molecule, it is not obvious that it is parallel to $f\Delta\mu_0$.

There are two findings from the Stark hole-burning experiments that were unanticipated. First, the values of $f\Delta\mu$ for the B800 band and the B870 band of the B850 BChl *a* ring are not very different (Table 13.2). The values fall in the range of $\sim$0.6–1.2 D. Such values are not much larger than those of the 825-nm band of the FMO complex. Second, the CSM values for $f\Delta\mu$ are significantly larger than the hole-burning values. This is also the case for the 825-nm band of the FMO complex (vide supra). In attempting to understand the apparent discrepancies, it is important to know that with CSM spectroscopy the theory of Liptay [71] is used, in approximate form, to analyze the response of an absorption band to the external field. This involves fitting the Δ-absorbance (ΔA) spectrum to the sum of the first and second derivatives of the absorption spectrum and, when necessary, also the zeroth derivative. These derivatives relate, respectively, to changes in the polarizability, dipole moment, and oscillator strength. This requires an absorption spectrum with very high S/N ratio. In Refs. 66 and 67 the low S/N ratios for the B800 and B850 bands necessitated fitting each band with a sum of skewed Gaussians. By necessity, the Gaussians of each band are assumed to have identical $\Delta\boldsymbol{\alpha}$ and $\Delta\boldsymbol{\mu}$. We have argued that this assumption for the B850 band is doubtful because of its underlying exciton level structure and energy disorder. With Stark hole-burning, as reported on here and in the references cited, one is concerned only with the linear Stark effect, and the high resolution of the technique allows one to probe single excited states. Furthermore, the theory used to analyze the data was designed for the problem at hand, that is, inhomogeneously broadened bands across which a variation in $\Delta\boldsymbol{\mu}$ is taken into account. The ability to study this variation is an important attribute of Stark hole burning. A disadvantage of the technique is that, with typical attainable field strengths, it cannot be applied to states with ultrashort lifetimes (e.g., the E_1 and E_2 levels of the B850 and B875 bands) because of excessive hole-broadening.

13.4. SUMMARY

We have described a quite unique facility at the Ames Laboratory that combines high pressure and external electric fields with spectral hole-burning for the study of excitation energy transfer and charge separation processes in light-harvesting and reaction-center complexes. Results recently obtained for several antenna protein–bacteriochlorophyll complexes were reviewed. The advantages gained by using pressure are as follows: First, linear pressure shifts for the photocatalytic Q_y-states greater than about $-0.2\ \text{cm}^{-1}/\text{MPa}$ indicate that excitonic coupling between Chls of a complex are important; second, pressure shifts greater than about $-0.4\ \mathrm{cm}^{-1}/\mathrm{MPa}$ are indicative of strong excitonic coupling; and third, the linear pressure shifts of weakly coupled Chls in complexes, $\leq -0.1\ \mathrm{cm}^{-1}/\mathrm{MPa}$ can be used to determine the "isotropic" compressibility, κ, of the complex. With a value of κ at hand, the contributions from electrostatic and electron-exchange interactions between strongly coupled Chls of complexes can be determined. Knowledge of these contributions provides important new benchmarks for electronic-structure calculations. An accurate description of the Q_y-excited electronic states is essential for understanding the excitation energy and electron-transfer dynamics. Stark hole-burning spectroscopy yields the dipole moment changes associated with the $S_1(Q_y) \leftarrow S_0$ transitions. These dipole moment changes also serve as new benchmarks for electronic-structure calculations. In the case of strongly coupled B850 and B875 rings of the LH1 and LH2 complexes of purple bacteria, respectively, it seems clear that the dipole moment changes associated with the B850 and B875 absorption bands cannot be understood without taking into account energy disorder from structural heterogeneity, which serves to destroy the cyclic C_n symmetries of the B850 and B875 rings. The Stark effect, as well as pressure-induced effects, provides new insights on the effects of structural heterogeneity on excited-state electronic structure and transport dynamics. Currently, the effects of energy disorder on the excitonic structures and energy-transfer dynamics of antenna complexes are of considerable interest.

ACKNOWLEDGMENTS

Hsing-Mei Wu, Ryszard Jankowiak, John Hayes, and Raja Reddy of the Ames Laboratory contributed very significantly to the experiments on the antenna complexes discussed in this chapter. Research at the Ames Laboratory was supported by the Division of Chemical Sciences, Office of Basic Energy Sciences, U.S. Department of Energy. Ames Laboratory is operated for USDOE by Iowa State University under Contract W-7405-Eng-82.

REFERENCES

1. S. G. Johnson, I.-J. Lee, and G. J. Small, in H. Scheer, ed., *Chlorophylls*, CRC Press, Boca Raton, Florida, 1991, p. 739.
2. N. R. S. Reddy, P. A. Lyle, and G. J. Small, *Photosynthesis Research* **31**, 167 (1992).
3. R. Jankowiak and G. J. Small, in J. Deisenhofer and J. Norris, eds., *Photosynthetic Reaction Centers*, Academic Press, Boston, 1993, p. 133.
4. W. E. Moerner, ed., *Topics in Current Physics, Persistent Spectral Hole Burning: Science and Applications*, Springer Verlag, New York, Vol. 44, 1987.

5. H. Zuber and R. A. Brunisholz, in H. Scheer, ed., *Chlorophylls*, CRC Press, Boca Raton, Florida, 1991, p. 627.
6. H. Zuber and R. Cogdell, in R. E. Blankenship, M. T. Madigan, and C. E. Bauer, eds., *Anoxygenic Photosynthetic Bacteria*, Kluwer Academic Publishers, Dordrecht, 1995, p. 315.
7. S. G. Johnson and G. J. Small, *J. Phys. Chem.* **95**, 471 (1991).
8. R. M. Pearlstein, *Photosyn. Res.* **31**, 213 (1992).
9. B. W. Matthews and R. E. Fenna, *Acc. Chem. Res.* **13**, 309 (1980).
10. D. E. Tronrud, M. F. Schmid, and B. W. Matthews, *J. Mol. Biol.* **188**, 443 (1986).
11. A. Freer, S. Prince, K. Sauer, M. Papiz, A. Hawthornthwaite-Lawless, G. McDermott, R. Cogdell, and N. W. Isaacs, *Structure* **4**, 449 (1996).
12. K. Sauer, R. J. Cogdell, S. M. Prince, A. Freer, N. W. Isaacs, and H. Scheer, *Photochem. Photobiol.* **64**, 564 (1996).
13. H. van der Laan, Th. Schmidt, R. W. Visschers, K. J. Visscher, R. van Grondelle, and S. Völker, *Chem. Phys. Lett.* **170**, 231 (1990).
14. N. R. S. Reddy, R. J. Cogdell, L. Zhao, and G. J. Small, *Photochem. Photobiol.* **57**, 35 (1993).
15. N. R. S. Reddy, H.-M. Wu, R. Jankowiak, R. Picorel, R. J. Cogdell, and G. J. Small, *Photosyn. Res.* **48**, 277 (1996).
16. R. G. Alden, E. Johnson, V. Nagarajan, W. W. Parson, C. J. Law, and R. J. Cogdell, *J. Phys. Chem. B* **101**, 4667 (1997).
17. J. Koepke, X. Hu, C. Muenke, K. Schulten, and H. Michel, *Structure* **4**, 581 (1996).
18. S. Karrasch, P. A. Bullough, and R. Ghosh, *EMBO J.* **14**, 631 (1995).
19. X. Hu, T. Ritz, A. Damjanovic, and K. Schulten, *J. Phys. Chem. B* **101**, 3854 (1997).
20. M. Z. Papiz, S. M. Prince, A. M. Hawthornthwaite-Lawless, G. McDermott, A. A. Freer, N. W. Isaacs, and R. J. Cogdell, *Trends in Plant Science* **1**, 198 (1996).
21. H. Frauenfelder, N. A. Aldering, A. Ansari, D. Braunstein, B. R. Cowen, M. K. Hong, I. E. T. Iben, J. B. Johnson, S. Luck, M. C. Marden, J. R. Mourant, P. Ormos, L. Reinisch, R. Scholl, A. Schulte, E. Shyamsunder, L. B. Sorensen, P. J. Steinbach, A. Xie, R. Young, and K. T. Yue, *J. Phys. Chem.* **94**, 1024 (1990).
22. N. Redline, M. W. Windsor, and R. Menzel, *Chem. Phys. Lett.* **186**, 204 (1991).
23. N. Redline and M. W. Windsor, *Chem. Phys. Lett.* **198**, 334 (1992).
24. H.-M. Wu, M. Rätsep, R. Jankowiak, R. J. Cogdell, and G. J. Small, *J. Phys. Chem. B* **101**, 7641 (1997).
25. H.-M. Wu, N. R. S. Reddy, R. J. Cogdell, C. Muenke, H. Michel, and G. J. Small, *Mol. Cryst. Liq. Cryst.* **291**, 163 (1996).
26. N. R. S. Reddy, R. Jankowiak, and G. J. Small, *J. Phys. Chem.* **99**, 16168 (1995).
27. Th. Sesselmann, W. Richter, D. Haarer, and H. Morawitz, *Phys. Rev. B* **36**, 7601 (1987).
28. J. Zollfrank, J. Friedrich, J. Fidy, and J. M. Vanderkooi, *J. Chem. Phys.* **94**, 8600 (1991).
29. J. Zollfrank, J. Friedrich, and F. Parak, *Biophys. J.* **61**, 716 (1992).
30. J. Zollfrank and J. Friedrich, *J. Phys. Chem.* **96**, 7889 (1992).

31. A. Freiberg, A. Ellervee, P. Kukk, A. Laisaar, M. Tars, and K. Timpmann, *Chem. Phys. Lett.* **214**, 10 (1993).
32. B. B. Laird and J. L. Skinner, *J. Chem. Phys.* **90**, 3274 (1989).
33. A. P. Shreve, J. K. Trauman, H. A. Frank, T. G. Owens, and A. C. Albrecht, *Biochim. Biophys. Acta* **973**, 93 (1991).
34. N. R. S. Reddy, G. J. Small, M. Seibert, and R. Picorel, *Chem. Phys. Lett.* **181**, 391 (1991).
35. I. Renge, K. Mauring, and R. Avarmaa, *J. Lumin.* **37**, 207 (1987).
36. D. J. Lockhart and S. G. Boxer, *Proc. Natl. Acad. Sci. USA* **85**, 107 (1988).
37. T. J. DiMagno, E. J. Bylina, A. Angerhofer, C. D. Youvan, and J. R. Norris, *Biochemistry* **29**, 899 (1990).
38. D. S. Gottfried, J. W. Stocker, and S. G. Boxer, *Biochim. Biophys. Acta* **1059**, 63 (1991).
39. S. Krawczyk, *Biochim. Biophys. Acta* **1056**, 6444 (1991).
40. T. R. Middendorf, L. T. Mazzola, K. Lao, M. A. Steffen, and S. G. Boxer, *Biochim. Biophys. Acta* **1143**, 223 (1993).
41. U. Bogner, P. Schätz, R. Seel, and M. Maier, *Chem. Phys. Lett.* **102**, 267 (1983).
42. A. J. Meixner, A. Renn, S. E. Bucher, and U. P. Wild, *J. Chem. Phys.* **90**, 6777 (1986).
43. L. Kador, D. Haarer, and R. Personov, *J. Chem. Phys.* **86**, 5300 (1987).
44. L. W. Johnson, M. D. Murphy, C. Pope, M. Foresti, and J. R. Lombardi, *J. Chem. Phys.* **86**, 4335 (1987).
45. A. Renn, S. E. Bucher, A. J. Meixner, E. C. Meister and U. P. Wild, *J. Lumin.* **39**, 181 (1988).
46. R. B. Altmann, I. Renge, L. Kador and D. Haarer, *J. Chem. Phys.* **97**, 5316 (1992).
47. A. J. Meixner, A. Renn, and U. P. Wild, *Chem. Phys. Lett.* **190**, 75 (1992).
48. E. Vauthey, K. Holliday, C. Wei, A. Renn, and U. P. Wild, *Chem. Phys.* **171**, 253 (1993).
49. E. Vauthey, J. Voss, C. de Caro, A. Renn, and U. P. Wild, *Chem. Phys.* **184**, 347 (1994).
50. R. B. Altmann, L. Kador, and D. Haarer, *Chem. Phys.* **202**, 167 (1996).
51. L. Kador, S. Jahn, D. Haarer, and R. Silbey, *Phys. Rev. B* **41**, 12215 (1990).
52. R. B. Altmann, D. Haarer, and I. Renge, *Chem. Phys. Lett.* **216**, 281 (1993).
53. J. Gafert, J. Friedrich, J. M. Vanderkooi, and J. Fidy, *J. Phys. Chem.* **99**, 5223 (1995).
54. G. J. Small, *Chem. Phys.* **197**, 239 (1995).
55. N. R. S. Reddy, R. Picorel, and G. J. Small, *J. Phys. Chem.* **96**, 6458 (1992).
56. S. Savikhin and W. S. Struve, *Chem. Phys.* **210**, 91 (1996).
57. H.-M. Wu, M. Rätsep, R. Jankowiak, R. J. Cogdell, and G. J. Small, *J. Phys. Chem. B*, **102**, 4023 (1998).
58. H.-M. Wu, N. R. S. Reddy, and G. J. Small, *J. Phys. Chem. B* **101**, 651 (1997).
59. H.-M. Wu and G. J. Small, *Chem. Phys.* **218**, 225 (1997).
60. H.-M. Wu, M. Rätsep, I.-J. Lee, R. J. Cogdell, and G. J. Small, *J. Phys. Chem. B* **101**, 7634 (1997).

61. I. Perepechko, ed., *Low Temperature Properties of Polymers*, Pergamon, Oxford, 1980.
62. I. Renge, *Chem. Phys.* **167**, 173 (1992).
63. H.-M. Wu and G. J. Small, *J. Phys. Chem. B* **102**, 888 (1998).
64. M. Rätsep, H.-M. Wu, J. M. Hayes, and G. J. Small, *Spectrochim. Acta* **54**, 1279 (1998).
65. M. Rätsep, H.-M. Wu, J. M. Hayes, R. E. Blankenship, R. J. Cogdell, and G. J. Small, *J. Phys. Chem. B*, **102**, 4035 (1998).
66. L. M. P. Beekman, M. Steffen, I. H. M. van Stokkum, J. D. Olsen, C. N. Hunter, S. G. Boxer, and R. van Grondelle, *J. Phys. Chem. B* **101**, 7284 (1997).
67. L. M. P. Beekman, R. N. Frese, G. J. S. Fowler, R. Picorel, R. J. Gogdell, I. H. M. van Stokkum, C. N. Hunter, and R. van Grondelle, *J. Phys. Chem. B* **101**, 7293 (1997).
68. L. Shu and G. J. Small, *J. Opt. Soc. Am. B* **9**, 738 (1992).
69. P. Schätz and M. Maier, *J. Chem. Phys.* **87**, 809 (1987).
70. R. M. Hochstrasser, *Accts. Chem. Res.* **6**, 263 (1973).
71. W. Liptay, in E. C. Lim, ed., *Excited States*, Academic Press, New York, 1974, p. 129.

CHAPTER

14

SITE-SELECTIVE SPECTROSCOPY OF AMORPHOUS POLYMERS

S. J. ZILKER AND D. HAARER

Physics Department and Bayreuther Institut für Makromolekülforschung, University of Bayreuth, D-95440 Bayreuth, Germany

14.1. INTRODUCTION

Among the first systematic studies of polymer glasses at low temperatures were thermal measurements performed by Reese [1]. He pointed out that at low temperatures ($T < 1\,\mathrm{K}$) their specific heat contains—apart from the Debye T^3 dependence—an additional term linear in T. Further anomalies exist in the heat conductivity [2], the ultrasound absorption [3], and the sound velocity [4, 5]. It is very intriguing that these phenomena are present in many different organic and inorganic amorphous solids, irrespective of their chemical composition. This universality has been puzzling researchers for a long time. Following a seminal paper by Zeller and Pohl [6], two independent publications by Anderson, Halperin, and Varma [7] and by Phillips [8] were able to explain these anomalies by introducing the concept of two-level systems.

To a good approximation, the disordered energy landscape of an amorphous solid can be split into many individual double-well potentials, each constituting a two-level system (TLS). At low temperatures, phonon-assisted tunneling through the barrier is then responsible for these additional degrees of freedom present in glasses but not in crystals. Figure 14.1 shows a TLS that is essentially characterized by two parameters, the asymmetry Δ and the tunneling matrix element Δ_0 given by

$$\Delta_0 \approx \hbar\Omega \exp(-\lambda) \tag{14.1}$$

Here, $\hbar\Omega$ is on the order of a vibrational energy in a given single potential, whereas λ denotes the so-called tunneling parameter, which is a measure of the overlap between the wave functions of the two localized states:

$$\lambda = \frac{d}{\hbar}\sqrt{2mV_0} \tag{14.2}$$

Shpol'skii Spectroscopy and Other Site-Selection Methods, Edited by Cees Gooijer, Freek Ariese, and Johannes W. Hofstraat
ISBN 0-471-24508-9

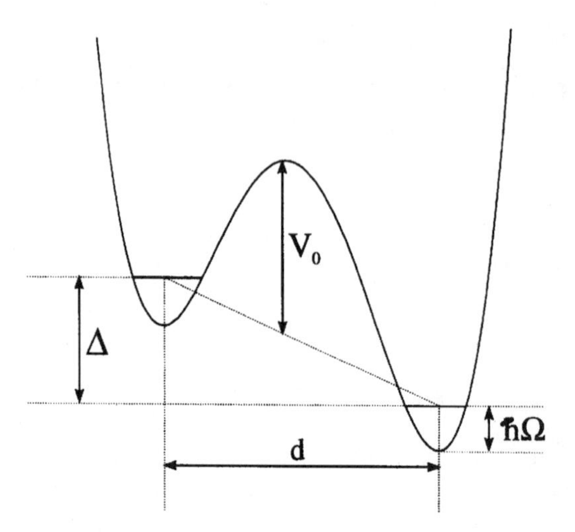

Figure 14.1. Two-level system in an amorphous solid. V_0 is the height of the energy barrier, Δ the asymmetry, d the distance between the minima, and $\hbar\Omega$ the ground-state energy.

d is the distance between the potential minima, and m the mass of the tunneling particle. In order to describe amorphous solids on a macroscopic scale, the whole TLS ensemble has to be considered. Thus the distribution of the TLS parameters has to be either known or has to be derived from reasonable assumptions. Since the microscopic nature of the TLS was not known, the latter approach was taken in the original papers [7, 8] and a flat distribution of both λ and Δ was postulated:

$$P(\Delta, \lambda)\, d\Delta\, d\lambda = \bar{P}\, d\Delta\, d\lambda \tag{14.3}$$

This assumption can be justified for the Δ parameter by a comparison of energy scales. The tunneling systems are frozen in at the glass transition temperature T_g, which is of the order of 10^2 K. Hence, TLS will have asymmetries Δ ranging from zero up to $k_B T_g$: At low temperatures ($T \approx 1$ K), however, only a small part of this range is thermally accessible. In zeroth-order approximation the distribution of Δ is constant. In contrast, a simple physical argument for a flat distribution of λ does not exist. Only the lack of reasons for a distinct functional dependence and the simplicity of the proposed model support this assumption. The broad distribution of λ (on which Δ_0 depends exponentially; see Eq.14.1) gives rise to a wide dispersion of TLS relaxation rates that extends, for example, in polymethylmethacrylate (PMMA), over at least 15 orders of magnitude [9–11]. Transformed into a distribution for the energy splitting E between both wells of the TLS and the tunneling matrix element Δ_0, the distribution function is often written in the literature as:

$$P(E) = \text{const}$$
$$P(\Delta_0) = \frac{1}{\Delta_0} \qquad (14.4)$$

The TLS model can be successfully applied to many amorphous solids with very different structure and composition such as amorphous alcohols [12, 13], water [14, 15], polymers [16], solidified gases [17], etc. Experimentally, however, the microscopic nature of TLS was only identified in a few special cases. A rotation of SiO_4 groups is suspected to form TLS in silicate glasses [18]; furthermore, water molecules embedded in a polymer host were found to switch between different states under infrared irradiation [19]. Monte Carlo simulations [20] in a Ni–P glass have shown that the free volume in the matrix enables configurational changes of single atoms or, even more likely, of groups of atoms that can be described by double–minimum potentials. Since the existence of free volume is characteristic for glasses, these simulations provide a possible explanation for the universality of the TLS model, but an experimental proof of its validity does not yet exist.

14.2. EXPERIMENTAL TECHNIQUES

14.2.1. Site-Selective Spectroscopy

Polymers are optically transparent in the visible wavelength regime. Thus, one can incorporate an optical probe, a chromophore, into the matrix in order to perform spectroscopic studies. Due to the strongly disordered local environments of the individual dye molecules, strong inhomogeneous line broadening leads to broad absorption bands (around $200\,\text{cm}^{-1}$) even at liquid helium temperatures.

The three most common experimental techniques, spectral hole burning (HB) [21, 22], photon-echo spectroscopy (PE) [23], and single-molecule spectroscopy (SMS) [24], rely on their ability to overcome the inhomogeneous broadening: HB by marking a subset of chromophores that absorb at the same frequency, PE by rephasing the electronic dye oscillators and, thus, eliminating their different frequencies, and SMS by studying individual dye molecules at very low concentrations and, hence, avoiding any ensemble average. The different approaches already hint to an important point: The linewidths measured by these three techniques in one given system will not be identical.The first reason is the difference of the involved experimental time scales: HB and SMS are rather slow techniques (on the order of seconds), since a frequency scan of the laser is required for recording the spectrum, and, in the case of HB, a spectral hole has to be burned. Transient hole-burning techniques, however, have been able to resolve TLS dynamics down to a few microseconds [25, 26], whereas studies in continuous dilution refrigerators have observed TLS dynamics for time spans of several weeks [9]. Photon echo spectroscopy, working with picosecond lasers, covers relaxation times from the nanosecond up to millisecond regime [10, 11, 27, 28].

The second reason for different linewidths is based on the observed specimen: HB and PE are ensemble-averaging techniques, whereas SMS can resolve single molecules. SMS experiments showed the existence of a dispersion of the individual linewidths. Thus one can not expect to obtain the same results even on identical time scales. The linewidth $\Gamma(t)$ determined by site-selective spectroscopy as a function of the experimental time scale t is given by [29]:

$$\Gamma(t) = \frac{1}{\pi T_2(t)} = \frac{1}{2\pi T_1} + \frac{1}{\pi T_2^*} + \Gamma_{SD}(t) \tag{14.5}$$

where T_2 is the dephasing time, which includes population relaxation via T_1 (the fluorescence lifetime) and dephasing due to TLS or local vibrations via the so-called *pure* dephasing time T_2^*. The latter contains all processes occurring on time scales faster than T_1 and depends on the temperature. $\Gamma_{SD}(t)$, in contrast, represents slower line-broadening processes, so-called spectral diffusion (SD), and is a function of the experimental temperature and time scale. SD is a random walk of the absorption line of the chromophore in the frequency domain due to the coupling to TLS relaxations. The latter is mediated by elastic (or electric) dipolar strain fields. The theoretical framework of spectral diffusion was originally developed for spin–resonance experiments [30]. Black and Halperin [31] were the first to use the analogy between these spin–$\frac{1}{2}$ systems and the TLS in amorphous solids in order to explain ultrasound experiments on glasses. Reinecke [32] finally extended the model to optical linewidths. For small TLS densities and a dipolar interaction between TLS and chromophores a Lorentzian diffusion kernel is obtained. Thus spectral holes have Lorentzian shapes. Photon echo decays, being their Fourier analog, are consequently single exponential.

The time dependence of the spectral diffusion contribution to the holewidth is given by [32, 33]:

$$\Gamma_{SD}(t) \sim \log(R_{eff}t) \tag{14.6}$$

if one assumes a distribution of the TLS parameters as in the standard model (see Eq. 14.3). Any different choice of the distribution function, however, is likely to yield an algebraic behavior of the line broadening. In the above formula, R_{eff} is the so-called effective maximum rate of spectral diffusion, being the energy average over the maximum flip rate R_{max} of the TLS in a given matrix. For polymer glasses, R_{eff} is typically on the order of $10^{10}\,s^{-1}$ [10, 34] at 1 K and it scales as T^3.

14.2.2. Specifics of Polymers

Amorphous polymers are on the one hand highly suitable for optical experiments at low temperatures. They are optically transparent in the visible region and very homogeneous on the scale of optical wavelengths so that high-quality samples can be prepared. This is especially advantageous for pulsed measurements. Furthermore, their glass transition point is above room temperature, which facilitates the cooling process: No special care has to be taken to avoid crystalline phases such as, for example, in ethanol [35]. Furthermore, some monomers, such as, for example, methylmethacrylate, are good solvents for a wide variety of dye molecules.

One the other hand, however, polymers have a very small specific heat c. In the case of PMMA, c is given by [2]

$$c(T) = 4.6T\ \mu J/gK^2 + 29.2T^3\ \mu J/gK^4 \tag{14.7}$$

corresponding to $100\,\mu J/gK$ at 1.5 K. Thus, especially photon echo experiments, which work with nJ pulses focused to spot sizes of less than $100\,\mu m$, are sensitive with respect to heating artifacts. This was first observed by Wiersma and co-workers in accumulated photon echo experiments using pulse cycles at kHz repetition rates [37]. The heat generated by earlier cycles created a thermal nonequilibrium that is probed by subsequent pulses. In three-pulse (stimulated) photon echo spectroscopy, heating was first observed by Zilker et al. [10] in PMMA. Here, the heat dumped by the first two excitation pulses (on the order of $20\,\mu J/g$) created a cloud of phonons through radiationless transitions of the excited chromophores. This causes additional dephasing probed by the third pulse of the same sequence. Figure 14.2 shows the maximum temperature that can be reached in the focus spot of a laser in a PMMA sample for four given initial temperatures. At 0.75 K, an external heat load of $20\,\mu J/g$ can raise the sample temperature by 0.5 K and, at 1.5 K, ΔT still amounts to 0.15 K! Neu et al. were subsequently able to model the laser heating process theoretically [11]. Their summary states that three-pulse photon echo spectroscopy below 2 K is almost certainly affected by heating artifacts (see Fig. 14.2), since most glasses have a specific heat on the same order of magnitude as PMMA. Also hole-burning experiments can be susceptible to heating, especially if the burning process is performed by a pulsed laser in order to increase the time resolution. This was first shown by Kharlamov et al. [25], again for samples of PMMA.

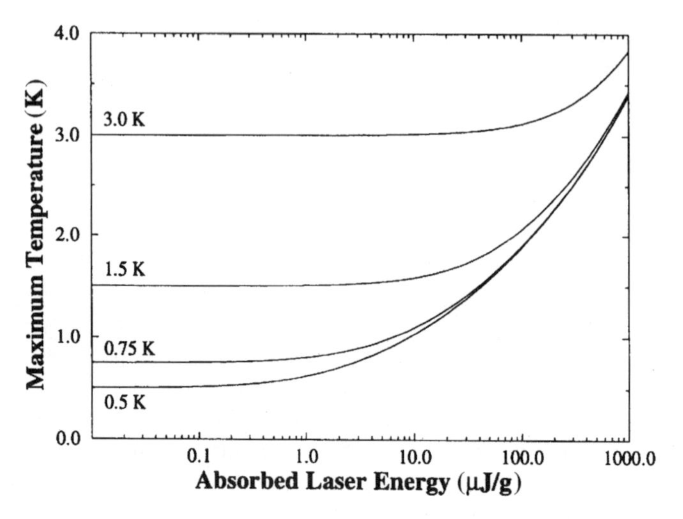

Figure 14.2. Estimate of the maximum temperature of a PMMA sample reached in the focal spot as a function of the deposited laser energy (for four given initial temperatures) [36].

14.3. CHEMICAL PHYSICS OF AMORPHOUS POLYMERS AT LOW TEMPERATURES

This section highlights some recent results in the field of chemical physics of polymer glasses at low temperatures. In its first part, time- and temperature-resolved measurements of the relaxation processes at low temperatures will be presented, whereas the second part is devoted to local fields in polymers, such as electric fields and hydrostatic pressure.

14.3.1. Relaxation Dynamics

Glasses show relaxation dynamics over more than 15 orders of magnitude in time. Thus different experimental techniques are necessary to cover the whole range. Over the past years, fast relaxation processes were mainly studied by stimulated photon echo spectroscopy, despite its inherent problems with sample heating. Hence, there are few reliable data for polymer systems available at this time. Zilker and Haarer [10] performed measurements on PMMA, doped with the chromophore zinc tetraphenylporphin (ZnTPP). Their data are plotted in Figure 14.3 for two different temperatures, 0.75 and 3.0 K. At the lower temperature, the effect of sample heating can be clearly seen: For small waiting times t (being the time between the second and the third—the probing—pulse) there is an enhanced line broadening due to the light-induced creation of phonons. After about 300 μs, the phonon cloud has already spread over a relatively large sample volume (of about 1 mm radius) so that the focal spot (radius 50 μm) has essentially returned to thermal equilibrium [11]. This leads to a narrowing of the line observed in the data.

The 3.0 K curve—not affected by heating due to the higher specific heat—also shows a surprising result: In contrast to the logarithmic line broadening predicted by the standard TLS model (see Eq. 14.6), an algebraic time dependence is observed. The latter was theoretically proposed by Silbey et al. in 1996 [34]. Based on the above-mentioned Monte Carlo simulation of a Ni–P glass, Silbey introduced a new distribution function for the tunneling matrix element Δ_0 (see Eq. 14.4 for comparison):

$$P(\Delta_0) = \frac{1}{\Delta_0^{1-v}} \tag{14.8}$$

Inserted into the formalism for spectral diffusion, the latter expression yields an algebraic time dependence of the line broadening:

$$\Gamma_{SD}(t) = aT^{1+v}\left[1 - (R_{\text{eff}}\, t)^{-v/2}\right] \tag{14.9}$$

The logarithmic time dependence of the standard model is now replaced by an algebraic law. The solid line in Figure 14.3 is a fit to the experimental data with the above formula, yielding $v = 0.18$ and $R_{\text{eff}} = 2.6 \times 10^{11}\ \text{s}^{-1}$. The value of v,

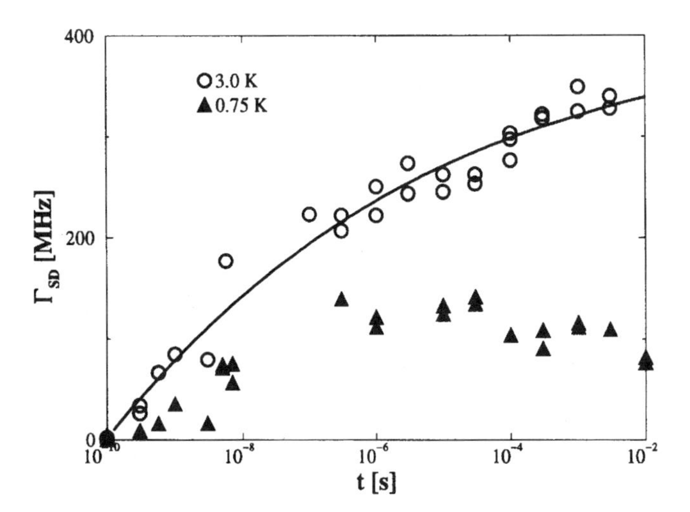

Figure 14.3. Contribution of spectral diffusion Γ_{SD} (see Eq. 14.6) to the linewidth of ZnTPP in PMMA as determined by three-pulse photon echo spectroscopy at two different temperatures [10]. The solid line is a fit with a theoretical model [34], as given in Eq. 14.9.

however, strongly depends on the investigated system; transient hole-burning studies yielded numbers ranging between 0 and 0.25 [34, 38]. Furthermore, it was shown that at waiting times of a few milliseconds a transition from $\nu = 0.15$ to 0 takes place in ethanol and deuterated ethanol [38], corresponding to a crossover to the standard form of the TLS distribution function (see Eq. 14.4). This agrees with earlier hole-burning studies of alcohol [39] and polymer glasses [16] that showed logarithmic line broadening ranging from seconds to several hours.

Recently, however, Maier et al. demonstrated a second deviation from the logarithmic behavior [9, 40]: For waiting times larger than approximately 10^3 s, they also observed an algebraic line broadening. But the time dependence has now a positive curvature contrary to the negative value of the three-pulse echo data. The results of the hole-burning study of Maier et al. are shown in Figure 14.4. Whereas the hole broadening is logarithmic up to about 10^3 s, a strong increase of the slope can be observed at longer times. Maier et al. [9] described this behavior with a modified distribution function for the TLS parameters, which is assumed to be valid on long time scales:

$$P(\Delta, \Delta_0) = \bar{P}\left[\frac{1}{\Delta_0} + \frac{A}{\Delta_0^2}\right] \tag{14.10}$$

The first term in the parentheses is the familiar distribution of the standard model. The second term is assumed to arise from strong interactions between the two-level systems, as previously proposed by Burin and Kagan [41, 42]. Based on a

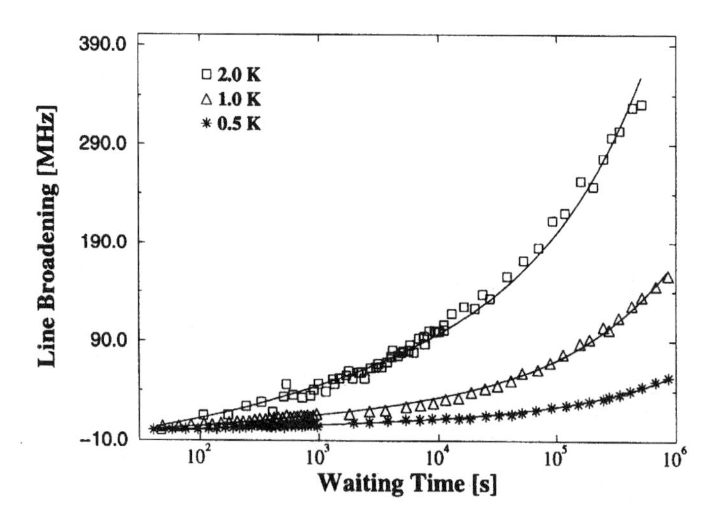

Figure 14.4. Broadening of spectral holes in PMMA with time. The data at 0.5 and 1.0 K were obtained using the organic dye phthalocyanine, the 2.0 K data using tetraphenylporphin [40].

similar concept, Coppersmith introduced a scenario where tunneling is suppressed by dipolar TLS–TLS interactions [43]. The TLSs that do tunnel are only those whose interactions are to some degree frustrated. Thus the tunneling rates of a large fraction of TLS are substantially reduced, leading to an overall shift of the relaxation processes of coupled TLS to longer times. This is reflected by an apparent increase of the density of states. Therefore, the line broadening observed at $t \leq 10^3$ s is mainly due to frustrated TLS, whereas for later times, tunneling of coupled TLS takes place. The broadening is not logarithmic with time, but follows an algebraic $\sqrt{t}$ law (see Fig. 14.4). The parameter A determines the crossover between both regimes.

However, this interpretation is not the only possible one and is not yet confirmed by other types of experiments. Heuer and Neu [44] successfully described the same data with the assumption that for long waiting times the glass explores configurations that are not characterized by the uniform distribution function in λ given by the standard tunneling model. Thus the energy landscape of a glass is not structureless, but contains high barriers in addition to uniformly distributed lower barriers within each deep basin. This is similar to the Frauenfelder picture, which was suggested for the energy landscape of proteins [45]. Heuer and Neu then ascribed relaxation phenomena with high barriers to degrees of freedom (rotations or switching processes) being caused by the side chains of the polymer glass or of the protein. In proteins, however, the distribution of small (TLS-like) barriers is much more restricted to lower values than in a glass. Hence, proteins show little or almost no spectral diffusion for waiting times between 10^0 and 10^2 s at 100 mK. This is demonstrated in Figure 14.5, which shows a fit of the Heuer model to two systems, a glycerol/water glass (upper curve), and the protein myoglobin (lower curve), both measured by Friedrich and co-workers [46] at a temperature of 100 mK. The protein data can be fit without a logarithmic TLS term reflecting the absence of TLS relaxations. However, the ad hoc introduction of high-energy barriers in polymers still remains a speculative assumption.

Thus the true reason for the increased density of states at long waiting times is not yet known. Even if it were known, there would be one problem left. Up to now, there has been no distribution function that is able to describe the whole range of relaxation rates present in glasses. The Silbey model with its algebraic time dependence works well from nanoseconds up to milliseconds. Then the standard TLS model, predicting a logarithmic behavior, is valid in an intermediate range, but it is replaced by an increased density of states (and a $\sqrt{t}$ dependence) beyond 10^3 s.

Finally, the main question is the microscopic nature of the two-level systems. Their identification has not made much progress in the past 25 years. However, promising approaches have been published recently: From the experimental point of view, there are two remarkable new results: Maier et al. [47] showed in a hole-

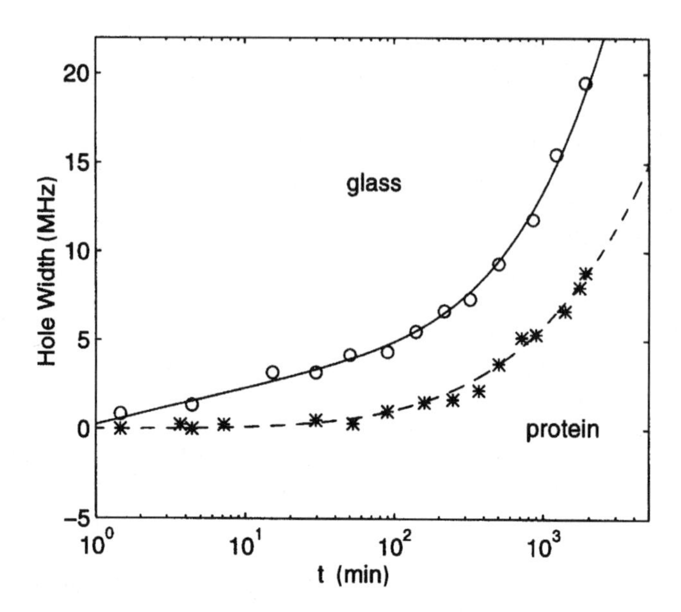

Figure 14.5. Broadening of a spectral hole in a glycerol/water glass (upper curve) and in the protein myoglobin (lower curve) at a temperature of 100 mK. The solid lines are fits with a model proposed by Heuer and Neu [44]; the experimental data are taken from Friedrich and co-workers [46]. (Reprinted from Ref. 44 with permission from the American Institute of Physics.)

burning investigation that there are TLSs in PMMA that have electric dipole moments on the order of 0.3 D. Hence, electric and elastic TLS exist. Pohl et al. [48] have investigated hydrogenated amorphous silica films. At a concentration of 1% H, the TLS density was reduced by a factor of 200 as compared to pure silica. Thus for the first time a system is available in which the amount of low-energy excitations is under experimental control. Furthermore, the hydrogenated films suggest that motions of atoms are responsible for the formation of TLS in these systems. Still, a lot of work remains to be done to explore the true microscopic nature of the two-level systems taking into account the universal applicability of the TLS model.

14.3.2. Temperature-Dependent Dephasing: Local Modes

Time-resolved measurements can be used to probe the distribution of the tunneling matrix element Δ_0. Temperature-dependent investigations, on the other hand, are sensitive to the distribution of the TLS energy splitting E, which is supposed to be flat in the standard TLS model. This leads to a nearly linear temperature dependence of the homogeneous linewidth [33]. However, there were several photon echo studies that found superlinear temperature laws of the form

$$\Gamma_{\rm h}(T) = aT^{1+\mu} \tag{14.11}$$

with $\mu = 0.2$–0.7 [37, 49, 50]. This was phenomenologically modeled by modifying the flat distribution of E and introducing

$$P(E) = E^{\mu} \tag{14.12}$$

Alternatively, the same result can be obtained by assuming a weak energy dependence for the average value of the TLS asymmetry Δ [51]. However, most of these studies were performed at temperatures above 1 K, where additional dephasing processes have to be taken into account, namely, dephasing due to local modes (often called boson peak), which is a characteristic process in disordered amorphous solids. Depending on the dye/matrix combination, local modes can be either due to optical phonons of the host or to vibrations and librations of the chromophore. In the case of PMMA, echo experiments identified a local mode as being connected with an optical phonon of the host by using different chromophores [50]. For tetra-*tert*-butylterrylene (TBT) in PMMA, echo measurements by Zilker et al. [52] found a local mode with an energy of 11 cm^{-1}. Figure 14.6 shows the inverse pure dephasing time (equivalent to the homo-

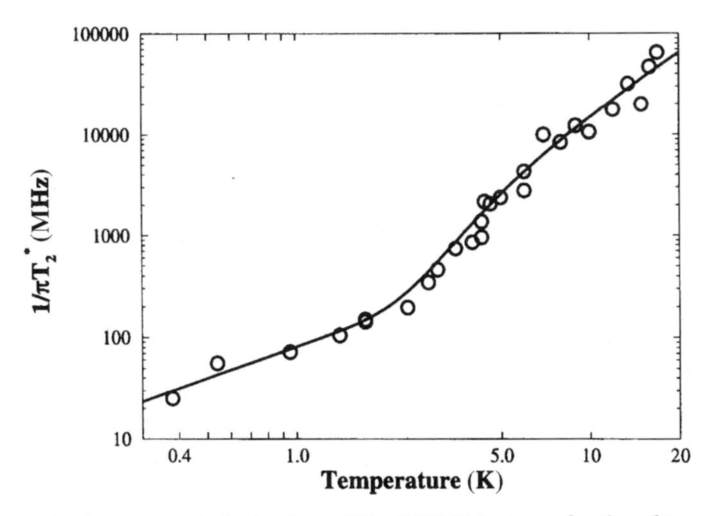

Figure 14.6. Inverse pure dephasing time $1/T_2^*$ of TBT/PMMA as a function of temperature. The solid line is a fit with Eq. 14.13. (Reprinted from Ref. [76] with permission from Elsevier Science.)

geneous linewidth minus the lifetime contribution) of TBT/PMMA. The solid line is a fit with an equation first derived by Silbey et al. [29, 53]:

$$\frac{1}{\pi T_2^*}(T) = aT^\alpha + b\frac{\exp(-\Delta E/kT)}{[1-\exp(-\Delta E/kT)]^2} \tag{14.13}$$

The fit result for α is 1.0; thus the nearly linear temperature dependence predicted by the standard TLS model is confirmed by the experimental data. However, a superlinear law can be easily obtained if the data are only fit over a limited temperature interval above 1 K, since the contribution of the local mode is then difficult to separate from TLS dephasing. Already at about 4 K, the linewidth is dominated by the local mode.

In spite of their importance, local modes have not yet been investigated in great detail. Hopefully, this will change in the near future. Local modes are located between the low-energetic TLS dynamics and the higher-lying α-relaxation process. Thus they may help to provide a unified picture of the amorphous state including both the TLS physics at low temperatures and the physics of the glass transition.

14.3.3. External Fields

The application of external fields to polymer matrices yields a wealth of information about the dye–matrix interaction and about matrix parameters. Since hole-burning can detect very small frequency shifts, small external perturbations can be registered. In this section a historic perspective will be used, starting with Stark and pressure experiments. Toward the end of the chapter, single-molecule results will be presented that show that most important material parameters are subject to broad distributions.

14.3.3.1. Electric Field Experiments

The application of a static electric field on a dye-doped polymer leads to a shift of the absorption lines (Stark effect) of those chromophores that have different static dipole moments in the ground and excited state [54]. For polar and, surprisingly, also for nonpolar (centrosymmetric) chromophores embedded in polymers a linear field dependence of this effect is observed. Nonpolar chromophores show matrix-induced dipole moment differences $\Delta\vec{\mu}_{\text{mat}}$ [55]. Since the random electric fields in the polymer and the isotropic orientation of the chromophores cause a random orientation of the dipole moment differences $\Delta\vec{\mu}$, a spectral hole will broaden symmetrically when an electric field is applied. Moreover, the matrix-induced component of the $\Delta\vec{\mu}$ vector also shows a distribution of its absolute value. The resulting bell shape of a hole spectrum is plotted in Figure 14.7.

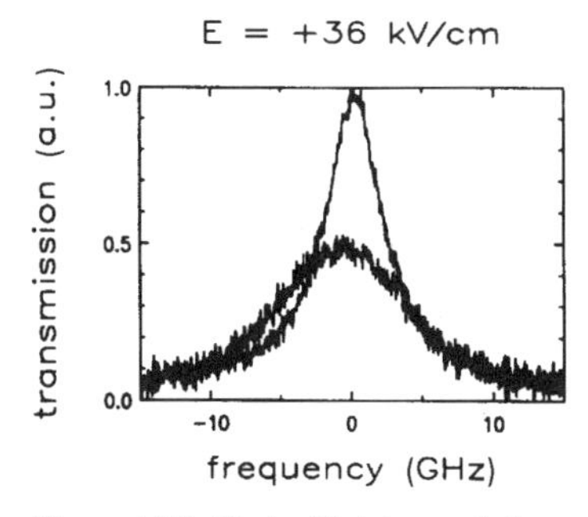

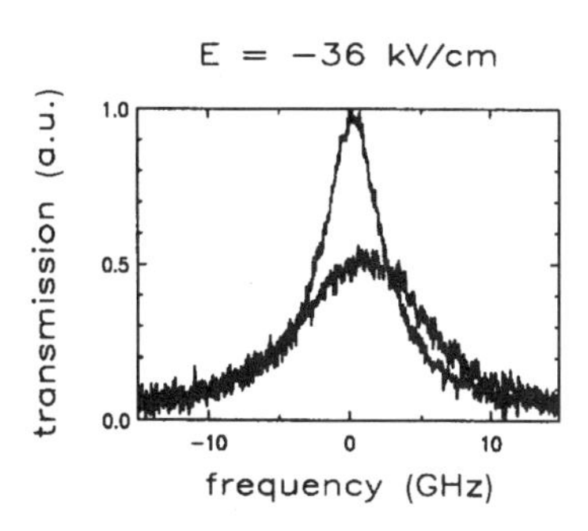

Figure 14.7. Stark effect in a poled sample of DANS/PMMA doped with tetra-*tert*-butyl-phthalocyanine (TBPc) [58]. The narrow profiles are the original hole spectra before applying the external field.

Apart from determining chromophore properties such as the intrinsic dipole moment difference $\Delta\mu_{\text{int}}$ [56, 57], Stark experiments can be used to evaluate the internal fields of polymers by measuring matrix-induced dipole moment differences of nonpolar chromophores such as tetra-*tert*-butyl-phthalocyanine (TBPc) in Ref. 58. In the latter study (see Fig. 14.7;), the sample was a 3-μm-thick film of PMMA with a polar stilbene derivative (DANS, 4-dimethylamino-4′-nitrostilbene) as side group. These systems are commonly used for second-harmonic measurements [59]. In order to obtain second-harmonic generation, an anisotropic orientation of the nonlinear DANS chromophores has to be induced. This is achieved by electrical poling at elevated temperatures [59]. Thus the matrix-induced dipole moment difference of the hole-burning molecule TBPc can be split into two parts

$$\Delta\vec{\mu} = \Delta\vec{\mu}_{\text{mat}} + \Delta\vec{\mu}_{\text{pol}} \tag{14.14}$$

since TBPc is nonpolar. The additional contribution $\Delta\vec{\mu}_{\text{pol}}$ arises from the poling field, which leads to a preferential direction of the internal field. Hence, $\Delta\vec{\mu}_{\text{pol}}$ is not randomly oriented, but is oriented parallel to the electric field applied during poling. Choosing the latter parallel to the laser polarization, one obtains an additional, *unidirectional* frequency shift $\Delta\nu$ of all chromophores. This results in a shift of the center frequency of the hole (see Fig. 14.7), which can be expressed as

$$\Delta\nu = -\frac{\kappa}{h}\Delta\mu_{\text{pol}} E \tag{14.15}$$

where $\kappa = (\varepsilon + 2)/3$ (ε being the low-temperature dielectric constant of the matrix) and E is the external electric field. Since $\Delta\mu_{\text{pol}}$ is given by

$$\Delta\mu_{\text{pol}} = \Delta\alpha\, E_{\text{pol}} \tag{14.16}$$

(with $\Delta\alpha$ being the difference in molecular polarizability between excited and ground state), one can determine the strength of the field E_{pol} induced by the poling process as 3×10^5 V/cm from the frequency shift. From the observed line broadening, on the other hand, the average strength of the randomly oriented matrix field can be calculated as 2×10^6 V/cm. Thus the randomly oriented internal fields in nonlinear-optical polymers are about one order of magnitude larger than the macroscopic internal field strengths achievable by electric poling, as was also pointed out by Kador and co-workers in 1987 [60].

Therefore, spectral hole-burning yields interesting results about polymer properties and about the dipolar interactions between the matrix components which cannot be directly obtained from other optical techniques such as second-harmonic generation.

14.3.3.2. Hydrostatic Pressure

Spectral hole-burning is sensitive to pressure changes of less than 0.1 MPa. At this level, one can assume that only intermolecular distances will experience small variations by the applied pressure, but that covalent intramolecular bonds will remain unchanged. Pressure effects will consequently lead to a minor change of the dipolar interactions between the matrix components which determine the solvent shift $\Delta\nu_{\mathrm{s}}$ of the chromophores. Hence, the solvent shift—being the change from the vacuum absorption line of the chromophore to the frequency when it is embedded in the solid—will now depend on the pressure, $\Delta\nu(p)$. This leads to a frequency shift of the spectral hole, as was first reported by Richter et al. [61;] in 1984. An example is depicted in Figure 14.8, which shows a spectral hole in PMMA before and after application of a hydrostatic pressure of 15 bar.

The pressure-dependent solvent shift $\Delta\nu(p)$ can be expressed in a simple and very intuitive model introduced by Sesselmann et al. [62] as:

$$\Delta\nu(p) = \Delta\nu_{\mathrm{s}} + \left(\frac{\partial \nu_{\mathrm{s}}}{\partial p}\right)_{p=0} p \tag{14.17}$$

It is known from early solvent-shift theories [63] that the dispersive part of the solvent shift $\Delta\nu_{\mathrm{s}}$ is given by:

$$\Delta\nu_{\mathrm{s}} = -K\rho_{\mathrm{w}}\frac{1}{V_{\mathrm{c}}} \tag{14.18}$$

where K contains a collection of constants, V_{c} is the volume of the solvent cage, and ρ_{w} the number density of the matrix molecules. Then the derivative in Eq. 14.17 yields:

$$\left(\frac{\partial \nu_{\mathrm{s}}}{\partial p}\right)_{p=0} = -K\left(\frac{1}{V_{\mathrm{c}}}\frac{\partial \rho_{\mathrm{w}}}{\partial p} + \rho\mathrm{w}\frac{\partial}{\partial p}\frac{1}{V_{\mathrm{c}}}\right) \tag{14.19}$$

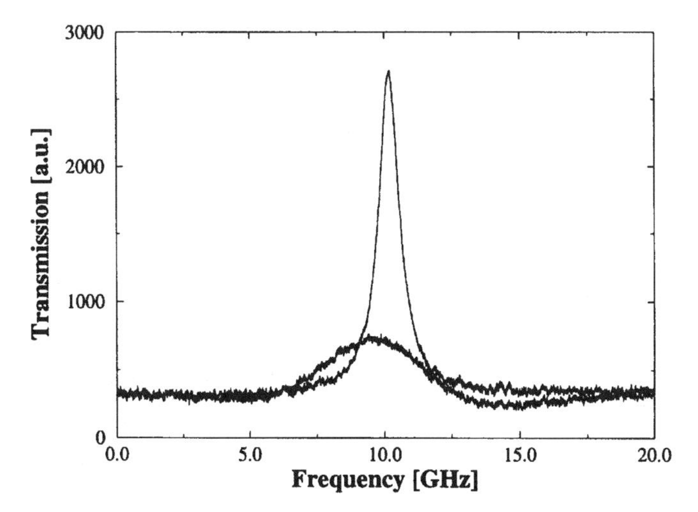

Figure 14.8. Spectral hole in tetraphenylporphin/PMMA before and after application of a hydrostatic pressure of 15 bar. The sample temperature is 2.2 K; the data were obtained courtesy of Reinhard Wunderlich, University of Bayreuth.

By defining a local compressibility κ_{loc} for the change of the volume of the solvent cage

$$\kappa_{\text{loc}} = -\frac{1}{V_{\text{c}}}\left(\frac{\partial V_{\text{c}}}{\partial p}\right) = V_{\text{c}}\frac{\partial}{\partial p}\frac{1}{V_{\text{c}}} \tag{14.20}$$

and a bulk compressibility κ_{bulk}

$$\kappa_{\text{bulk}} = -\frac{1}{V}\left(\frac{\partial V}{\partial p}\right) = \frac{1}{\rho_{\text{w}}}\frac{\partial \rho_{\text{w}}}{\partial p} \tag{14.21}$$

one obtains the pressure-dependent solvent shift as

$$\Delta\nu(p) = \Delta\nu_{\text{s}}\left[1 + \left(\kappa_{\text{bulk}} + \kappa_{\text{loc}}\right)p\right] \tag{14.22}$$

This equation therefore describes the shift of a chromophore line in a matrix with respect to its vacuum absorption frequency. Laird and Skinner derived this formula from a model that considers the changes of the intermolecular interactions in the glass [64]. Then it is not necessary to assume a special form of the solvent cage in order to obtain the proportionality factor between κ_{bulk} and κ_{loc}. Their result is:

$$\Delta\nu(p) = \Delta\nu_s(1 + 2\kappa_{\text{bulk}}\, p) \tag{14.23}$$

Hence it is possible to measure compressibilities in a purely optical experiment at low temperatures. The results of these investigations agree with mechanical studies within about 15% [62]. Furthermore, one can obtain information about the intermolecular interactions in the amorphous solid within the framework of the model by Laird and Skinner.

Pressure-dependent hole-burning studies may also be used to determine compressibilities of matrices that do not allow large mechanical stress due to their structural composition, such as sol–gel glasses and xerogels. Furthermore, local stress fields can be imaged due to the spatial resolution achievable by optical means. This may be especially interesting in certain types of morphologies such as copolymers and in phase boundaries of block copolymers.

14.3.4. Single-Molecule Spectroscopy

In the two previous subsections, only the average shift or broadening, respectively, of absorption lines was considered. However, it is obvious that the strong disorder present in amorphous solids will lead to broad distributions of the important physical parameters. With the discovery of single-molecule spectroscopy [65–67] it became feasible to resolve the behavior of single chromophores, providing insight into the underlying distributions for the characteristic parameters such as linewidth, Stark and pressure shift, electron–phonon coupling,etc.

Stark-effect experiments were performed in polyethylene by Orrit et al. [68]. Polyethylene is a special case, since it consists of amorphous and crystalline domains. Figure 14.9 shows a histogram of $\Delta\mu$ (more exactly of its component parallel to the applied field) of 60 terrylene molecules in a polyethylene matrix at 1.75 K. All the line shifts were found to follow a linear field dependence, thus the chromophores are assumed to be mainly located in amorphous domains. In organic crystals, such as *p*-terphenyl doped with pentacene [69], one often observes a quadratic Stark effect for unpolar chromophores. Furthermore, the line shape and the linewidth did not change under variation of the electric field. This is easily understood: Since many chromophores with different orientations of $\Delta\vec{\mu}$ contribute to the Stark effect observed in spectral hole-burning, the line broadening is only an inhomogeneous effect.

The average value of $|\Delta\mu|$ in polyethylene is relatively large, namely, 0.44 Debye. As expected, the histogram is symmetrical around $\Delta\mu = 0$, with the exception of four molecules that have extraordinarily large values. The authors of Ref. 68 suggest that these chromophores are strongly distorted.

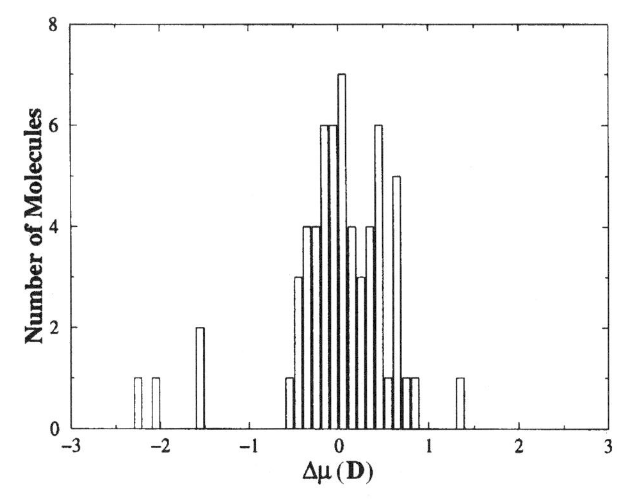

Figure 14.9. Matrix-induced dipole moment differences Δμ of 60 single terrylene molecules in a polyethylene matrix at a temperature of 1.75 K. (Adapted from Ref. 68.)

Unfortunately, there have not been any single-molecule experiments on polymers with hydrostatic pressure. Studies in crystals, however, suggest a similar behavior as in the case of Stark effect experiments [70–72]: There is a dispersion of the pressure shifts of individual molecules due to their different local environments in the matrix.

14.3.5. Comparison of Different Spectroscopic Methods

As the preceding subsection has shown, single-molecule spectroscopy can provide a lot of important information that is buried under the ensemble average carried out by techniques such as hole-burning and photon echo spectroscopy. Thus it is very interesting to compare the results of the three types of experiments in the same samples. The first study of this type was performed by Orrit and co-workers [73] with SMS and HB on three polymer systems, namely, polyvinylbutyral, polyethylene, and polymethylmethacrylate. With the exception of polyethylene—where the holes showed strong saturation broadening—the linewidth determined by hole-burning agreed with the mean value of the linewidths of the single molecules that have been studied.

Vainer et al. reported the first comparison of photon-echo and single-molecule spectroscopy [75] on the semiamorphous polymer polyethylene. Zilker et al. [76] performed a photon-echo study of the amorphous polymer polyisobutylene (PIB) doped with tetra-*tert*-butylterrylene (TBT). The latter was previously investigated by Basché et al. with SMS [74, 77]. Figure 14.10 depicts a histogram of the

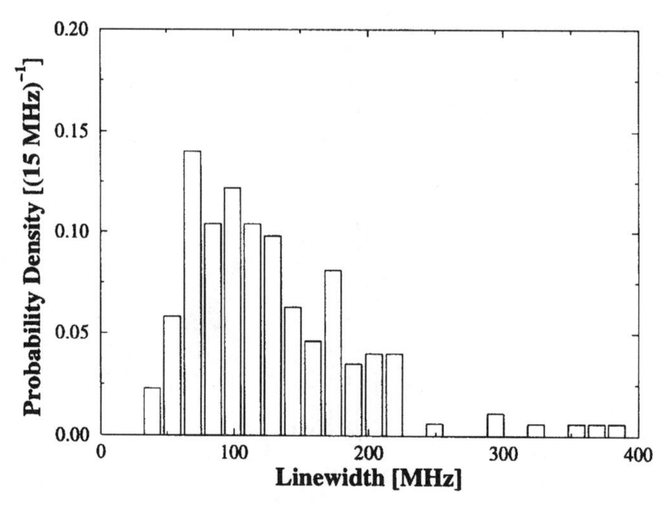

Figure 14.10. Histogram of the linewidths of 173 TBT molecules in PIB at 1.4 K. (Reprinted from Ref. 52 wih permission from Gordon and Breach Publishers.)

linewidths of 173 TBT molecules in PIB at a temperature of 1.4 K. The histogram is clearly asymmetric, with a sharp cutoff at about 50 MHz, which coincides with the inverse of the fluorescence lifetime of TBT [76]. The mean of the histogram is located at about 125 MHz. It is obvious that the linewidth dispersion arises from the different environments of the individual chromophores in the matrix. However, two interesting questions can be asked: First, why are ensemble-averaging techniques not able to detect the presence of these dispersions, and second, what causes the dispersion, spectral diffusion or pure dephasing?

Pure dephasing is caused by relaxation processes (such as TLS or local vibrations) which occur on time scales faster than T_1. Thus pure dephasing determines the homogeneous linewidth Γ_h. Spectral diffusion, in contrast, is a random walk of the absorption line in the frequency domain due to TLS relaxations. The contribution of this random-walk effect to the total linewidth is $\Gamma_{SD}(t)$, which is a function of the observation time t. The only way to distinguish between both processes is to perform two-pulse photon-echo spectroscopy, which measures the true homogeneous linewidth and to compare the result with the SMS histogram. Figure 14.11 shows the result of such a study on TBT/PIB by Zilker et al. [52, 76, 78]. The authors determined the homogeneous linewidth Γ_h as a function of temperature from 0.4 to 22 K. The solid line in Figure 14.11 is a fit to the data with the expression

$$\Gamma_h(T) = \frac{1}{2\pi T_1} + aT^{\alpha} + b\frac{\exp(-\Delta E/kT)}{[1 - \exp(-\Delta E/kT)]^2} \tag{14.24}$$

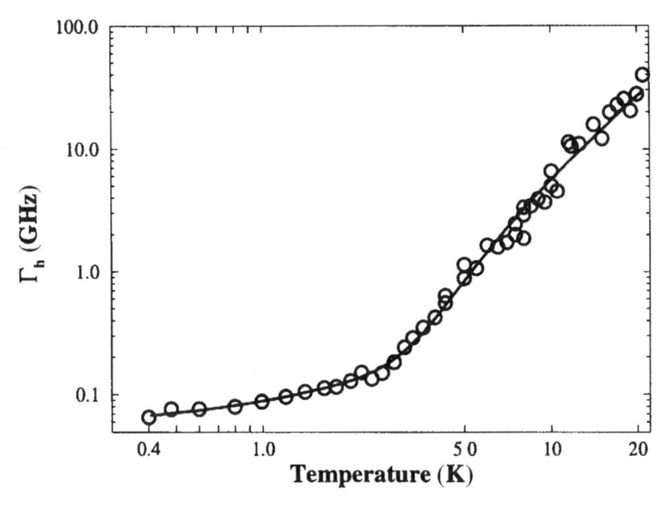

Figure 14.11. Homogeneous linewidth of TBT in PIB as a function of temperature. The solid line is a fit to the data with Eq. 14.24.

The first term amounts to 55 MHz and is due to the fluorescence lifetime T_1. The second term contains the contribution of TLS dephasing with $\alpha = 1.1$ and $a = 34\,\text{MHz/K}^{1.1}$. The third term arises from an exponentially activated local vibration of the matrix with an energy of $\Delta E = 13.4\,\text{cm}^{-1}$ [52]. At 1.4 K, the temperature of the SMS experiments on TBT/PIB, the homogeneous linewidth is 106 MHz. Hence, two conclusions can be drawn: First, SMS measures molecules that have smaller linewidths than the photon-echo result. Therefore, a distribution of the homogeneous linewidth Γ_h must exist. Second, the homogeneous linewidth of 106 MHz is very close to the average of the SMS distribution (125 MHz). Thus the SMS histogram in Figure 14.10 is mainly caused by a distribution of Γ_h and only to a lesser extent by a dispersion of $\Gamma_{SD}(t)$ arising from spectral diffusion.

It is interesting to investigate why ensemble-averaging techniques are not capable of resolving linewidth distributions if they are as broad as the one depicted in Figure 14.10. A distribution of T_2^* leads to a decay of the photon-echo signal, which is not single exponential. Vainer et al. [75] devised a simple way of calculating this effect: A subensemble of chromophores with identical linewidth Γ_i causes an echo signal with an amplitude of the electric light field of

$$P_i(\tau) = \rho_i \exp(-2\pi\Gamma_i\tau) \tag{14.25}$$

where ρ_i is the number of molecules in the subensemble divided by the total number of chromophores and τ is the delay between the two excitation pulses. In order to obtain the echo intensity $I(\tau)$, one has to sum over the amplitudes of all subensembles:

$$I(\tau) = \left(\sum_i P_i(\tau)\right)^2 = \left(\sum_i \rho_i \exp(-2\pi\Gamma_i\tau)\right)^2 \tag{14.26}$$

Figure 14.12 shows the two-pulse echo decay calculated for the case of a distribution of Γ. Hereby it was assumed that the SMS distribution obtained for TBT/PIB represents only a distribution of homogeneous linewidths and that no spectral diffusion contributes to the histogram. The dashed line is an exponential fit to the curve. The deviations are rather small as compared to typical signal-to-noise ratios of experimental data plotted as a reference in Figure 14.12. (The experimental data were shifted downward for better visibility.) Thus in a typical experiment the curve would still be considered as single exponential. Only a higher dynamical range of the echo experiment would make a discrimination possible. The same conclusion is valid for hole-burning experiments, as was shown by Rebane et al. [79], since the small deviations will occur predominantly in the wings of the Lorentzian spectrum, where they are difficult to detect.

Thus only a combination of different spectroscopic techniques is able to provide complete information about the dynamics in an amorphous solid. Dispersions of characteristic parameters can only be detected by SMS. On the other hand, however, SMS lacks the capability to determine ensemble averages in an efficient way and to resolve the glass dynamics over many decades in

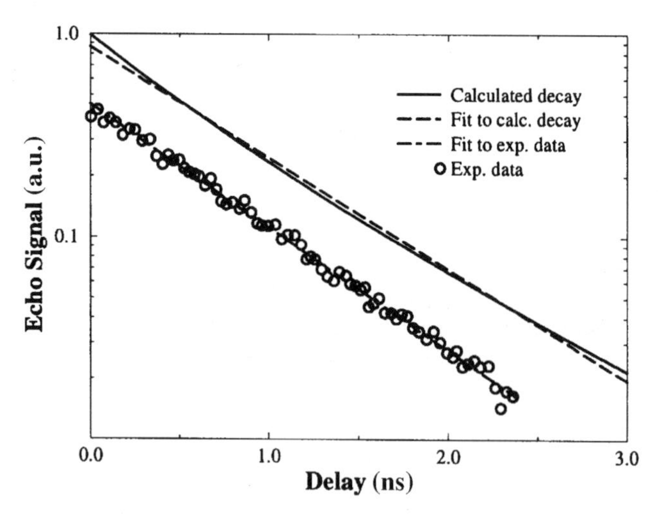

Figure 14.12. Comparison of the experimental photon echo decay in TBT/PIB at 1.4 K with a calculation (Eq. 14.26) assuming a distribution of homogeneous linewidths as in Figure 14.10.

observation time. In this respect, a combination of photon echo and hole-burning spectroscopy, covering 15 orders of magnitude in time, probably remains the most powerful tool to study the low-temperature properties of amorphous polymers.

14.4. TECHNICAL APPLICATIONS

The field of site-selective spectroscopy of polymers was boosted in the late 1970s by the interest in optical data storage. (See e.g., U.S. Patent No. 4,101,976, July 18, 1978, G. Castro, D. Haarer, R. M. MacFarlane, and H.-P. Trommsdorff, *Frequency Selective Optical Data Storage System.*) A hole-burning memory makes use of the resolution enhancement obtained by line-narrowing spectroscopy: The width of a spectral hole at low temperature ($T \approx 2\,\mathrm{K}$) is about 10^4 times smaller than the width of the inhomogeneous band. Hence, one can increase the storage density of an optical device (such as a compact disc) by a corresponding factor F via the extension of the storage capacity into the frequency domain. Figure 14.13 illustrates the principle [80]: In the focus spot of the laser beam, spectral holes represent a bit sequence. However, there are major technical obstacles that prevented the breakthrough of the hole-burning memory:

1. The multiplexing factor F is strongly temperature dependent, since the width of spectral holes increases with T. At 100 K, F is already reduced to less than 100 [81].

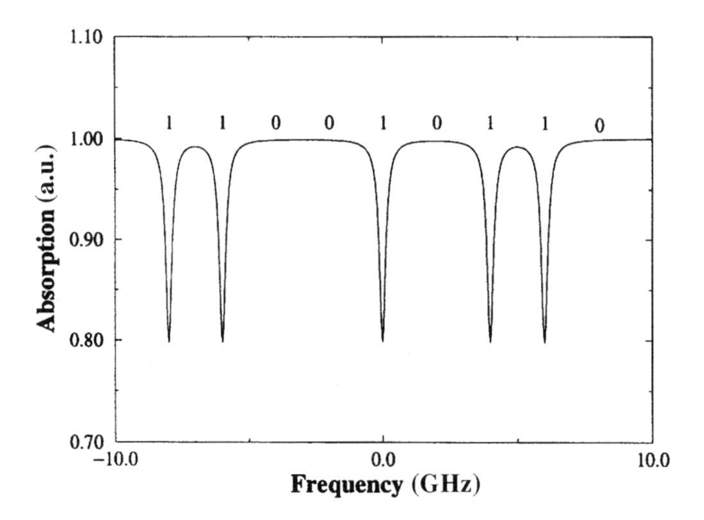

Figure 14.13. Sequence of spectral holes labeled as a bit pattern.

2. The temperature stability of spectral holes is not sufficient: The most efficient burning mechanism is photochemical hole burning. However, at higher temperatures the rate of the backreaction increases strongly. Therefore, to our current knowledge, deep spectral holes cannot be burned photochemically above 100 K [81].
3. In order to use the full spectral bandwidth, for example, by doping the polymer with different dyes, narrow-frequency ($\Delta\nu \approx 10\,\text{MHz}$) diode lasers are required that allow scanning over more than 100 nm without mode-hopping.

Therefore, an optical memory based on the hole-burning technique is limited to temperatures below 100 K and to low reading and recording speeds (the latter as long as no suitable diode lasers are commercially available). Especially the need for cryogenic temperatures renders the system uncompetetive in comparison to holographic polymers, which provide wavelength and angle multiplexing at room temperature [82].

ACKNOWLEDGMENTS

The research of S. Z. and D. H. presented in this section has been supported by the Deutsche Forschungsgemeinschaft (SFB279) and by the Volkswagen-Stiftung (AZ I/70526). Our experiments benefited from stimulating discussions and cooperative efforts with Yu. G. Vainer and R. I. Personov (Russian Academy of Science) and P. Neu and R. J. Silbey (MIT). We are indebted to L. Kador and B. M. Kharlamov for valuable comments on this manuscript.

REFERENCES

1. W. Reese, *J. Macromol. Sci.–Chem.* **A3**(7), 1257 (1969).
2. R. B. Stephens, G. S. Cieloszyk, and G. L. Salinger, *Phys. Lett.* **38A**(3), 215 (1972).
3. G. Federle and S. Hunklinger, *J. Physique* **C9**, *Tome 43*, C9-505 (1982).
4. L. Piche, R. Maynard, S. Hunklinger, and J. Jäckle, *Phys. Rev. Lett.* **32**, 1426 (1974).
5. A. Nittke et al, *J. Low Temp. Phys.* **98**, 517 (1995).
6. R. C. Zeller and R. O. Pohl, *Phys. Rev. B* **4**(6), 2029 (1971).
7. P. W. Anderson, B. I. Halperin, and C. M. Varma, *Philos. Mag.* **25**, 1 (1972).
8. W. A. Philips, *J. Low Temp. Phys.* **7**, 351 (1972).
9. H. Maier, B. M. Kharlamov, and D. Haarer, *Phys. Rev. Lett.* **76**(12), 2085 (1996).
10. S. J. Zilker and D. Haarer, *Chem. Phys.* **220**, 167 (1997).
11. P. Neu, R. J. Silbey, S. J. Zilker, and D. Haarer, *Phys. Rev. B* **56**, 11571 (1997).

12. M. Berg, C. A. Walsh, L. R. Narasimhan, K. A. Littau, and M. D. Fayer, *J. Chem. Phys.* **88**(3), 1564 (1988).
13. D. W. Pack, L. R. Narasimhan, and M. D. Fayer, *J. Chem. Phys.* **92**(7), 4125 (1990).
14. T. Giering and D. Haarer, *J. Lumin.* **66&67**, 299 (1996).
15. T. Giering and D. Haarer, *Chem. Phys. Lett.* **261**, 677 (1996).
16. S. Jahn, K.-P. Müller, and D. Haarer, *J. Opt. Soc. Am. B* **9**(6), 925 (1992).
17. T. Giering, P. Geissinger, L. Kador, and D. Haarer, *J. Lumin.* **64**, 245 (1995).
18. U. Buchenau, M. Prager, N. Nücker, A. J. Dianoux, N. Ahmad, and W. A. Phillips, *Phys. Rev. B* **34**(8), 5665 (1986).
19. K. Barth and W. Richter, *J. Lumin.* **64**, 63 (1995).
20. D. Dab, A. Heuer, and R. J. Silbey, *J. Lumin.* **64**, 95 (1995).
21. D. Haarer and R. Silbey, *Phys. Today* **5**, 58 (1990).
22. J. L. Skinner and W. E. Moerner, *J. Phys. Chem.* **100**(31), 13251 (1996).
23. L. R. Narasimhan, K. A. Littau, D. W. Pack, Y. S. Bai, A. Elschner, and M. D. Fayer, *Chem. Rev.* **90**(3), 439 (1990).
24. L. Kador, *Phys. Stat. Sol. (B)* **189**, 11 (1995).
25. B. M. Kharlamov, D. Haarer, and S. Jahn, *Opt. Spectr.* **46**(2), 302 (1994).
26. F. T. H. den Hartog et al, *J. Lumin.* **66&67**, 1 (1996).
27. H. C. Meijers and D. A. Wiersma, *Phys. Rev. Lett.* **68**(3), 381 (1992).
28. H. C. Meijers and D. A. Wiersma, *J. Chem. Phys.* **101**(8), 6927 (1994).
29. G. Schulte, W. Grond, D. Haarer, and R. Silbey, *J. Phys. Chem.* **88**(2), 679 (1988).
30. J. R. Klauder and P. W. Anderson, *Phys. Rev.* **125**(3), 912 (1962).
31. J. L. Black and B. I. Halperin, *Phys. Rev. B* **16**(6), 2879 (1977).
32. T. L. Reinecke, *Solid State Commun.* **32**, 1103 (1979).
33. S. Hunklinger and M. Schmidt, *Z. Phys. B* **54**, 93 (1984).
34. R. J. Silbey, J. M. A. Koedijk, and S. Völker, *J. Chem. Phys.* **105**(3), 901 (1996).
35. G. W. Gray and P. A. Windsor, *Mol. Cryst. Liq. Cryst.* **26**, 305 (1974).
36. S. J. Zilker. Ph.D. thesis, Universität Bayreuth, 1997.
37. H. Fidder, S. De Boer, and D. A. Wiersma, *Chem. Phys.* **139**, 317 (1989).
38. J. M. A. Koedijk, R. Wannemacher, R. J. Silbey, and S. Völker, *J. Phys. Chem.* **100**(51), 19945 (1996).
39. W. Breinl, J. Friedrich, and D. Haarer, *J. Chem. Phys.* **81**(9), 3915 (1984).
40. G. Hannig, H. Maier, D. Haarer, and B. M. Kharlamov, *Mol. Cryst. Liq. Cryst.* **291**, 11 (1996).
41. A. L. Burin and Yu. Kagan, *JETP* **79**, 347 (1994).
42. A. L. Burin and Yu. Kagan, *JETP* **80**, 761 (1995).
43. S. N. Coppersmith, *Phys. Rev. Lett.* **67**, 2315 (1991).
44. A. Heuer and P. Neu, *J. Chem. Phys.* **107**(20), 8686 (1997).
45. H. Frauenfelder, S. G. Sligar, and P. G. Wolynes, *Science* **254**, 1598 (1991).

46. J. Gafert, H. Pschierer, and J. Friedrich, *Phys. Rev. Lett.* **74**, 3704 (1995).
47. H. Maier, R. Wunderlich, D. Haarer, B. M. Kharlamov, and S. G. Kulikov, *Phys. Rev. Lett.* **74**(26), 5252 (1995).
48. X. Liu et al, *Phys. Rev. Lett.* **78**(23), 4418 (1997).
49. B. J. Baer, R. A. Cromwell, and E. L. Chronister, *Chem. Phys. Lett.* **237**, 380 (1995).
50. A. Elschner, L. R. Narasimhan, and M. D. Fayer, *Chem. Phys. Lett.* **171**, 19 (1990).
51. R. Jankowiak and G. J. Small, *Phys. Rev. B* **47**(22), 14805 (1993).
52. S. J. Zilker, D. Haarer, Yu.G. Vainer, A. V. Deev, M. A. Kol'chenko, and R. I. Personov, *Mol. Cryst. Liq. Cryst.* **314**, 143 (1998).
53. B. Jackson and R. Silbey, *Chem. Phys. Lett.* **99**(4), 331 (1983).
54. M. Maier, *Appl. Phys. B* **41**, 73 (1986).
55. V. M. Agranovich, V. K. Ivanov, R. I. Personov, and N. V. Rasumova, *Phys. Lett. A* **118**(5), 239 (1986).
56. L. Kador, S. Jahn, D. Haarer, and R. J. Silbey, *Phys. Rev. B* **41**, 12215 (1991).
57. R. B. Altmann, D. Haarer, N. I. Ulitsky, and R. I. Personov, *J. Lumin.* **56**, 135 (1993).
58. S. Ullmann. Masters Thesis, Universität Bayreuth, 1996.
59. H. M. Graf, O. Zobel, A.J. East, and D. Haarer, *J. Appl. Phys.* **75**, 3335 (1994).
60. L. Kador, R. I. Personov, W. Richter, Th. Sesselmann, and D. Haarer, *Polymer J.* **19**, 61 (1987).
61. W. Richter, G. Schulte, and D. Haarer, *Opt. Comm.* **51**, 412 (1984).
62. Th. Sesselmann, W. Richter, D. Haarer, and H. Morawitz, *Phys. Rev. B* **36**, 760 (1987).
63. W. E. Henke, W. Yu, H. L. Selzle, E. Q. Schlag, D. Wutz, and S. H. Lin, *Chem. Phys.* **97**, 205 (1985).
64. B. B. Laird and J. L. Skinner, *J. Chem. Phys.* **90**, 3274 (1989).
65. W. E. Moerner and L. Kador, *Phys. Rev. Lett.* **62**, 2535 (1989).
66. M. Orrit and J. Bernard *Phys. Rev. Lett.* **65**, 2716 (1990).
67. W. P. Ambrose, T. Basché, and W. E. Moerner, *J. Chem. Phys.* **95**, 7150 (1991).
68. M. Orrit, J. Bernard, A. Zumbusch, and R. I. Personov, *Chem. Phys. Lett.* **196**, 595 (1992).
69. U. P. Wild, F. Güttler, M. Pirotta, and A. Renn, *Chem. Phys. Lett.* **193**, 451 (1992).
70. M. Croci, H.-J. Müschenborn, F. Güttler, M. Pirotta, and A. Renn, *Chem. Phys. Lett.* **212**, 71 (1993).
71. A. Müller, W. Richter, and L. Kador, *Chem. Phys. Lett.* **241**, 547 (1995).
72. A. Müller, W. Richter, and L. Kador, *Mol. Cryst. Liq. Cryst.* **283**, 185 (1996).
73. B. Kozankiewicz, J. Bernard, and M. Orrit, *J. Chem. Phys.* **101**(11), 9377 (1994).
74. R. Kettner, J. Tittel, Th. Basché, and C. Bräuchle, *J. Phys. Chem.* **98**(27), 6671 (1994).
75. Yu. G. Vainer, T. V. Plakhotnik, and R. I. Personov, *Chem. Phys.* **209**, 101 (1996).
76. S. J. Zilker, D. Haarer, and Yu. G. Vainer, *Chem. Phys. Lett.* **273**, 232 (1997).
77. J. Tittel, R. Kettner, Th. Basché, C. Bräuchle, H. Quante, and K. Müllen, *J. Lumin.* **64**, 1 (1995).

78. S. J. Zilker, D. Haarer, Yu. G. Vainer, and R. I. Personov, *J. Lumin.*, **76/77**, 157 (1998).
79. L. A. Rebane, A. A. Gorokhovskii, and J. V. Kikas, *Appl. Phys. B* **29**, 235 (1982).
80. D. Haarer, *Mol. Cryst. Liq. Cryst.* **236**, 541 (1993).
81. R. Ao, L. Kümmerl, and D. Haarer, *Adv. Mat.* **7**, 495 (1995).
82. Th. Bieringer, R. Wuttke, D. Haarer, U. Gener, and J. Rübner, *Macromol. Chem. Phys.* **196**, 1375 (1995).

CHAPTER

15

SPECTROSCOPY OF SINGLE MOLECULES IN SOLID MATRICES AT CRYOGENIC TEMPERATURES

PHILIPPE TAMARAT, FEDOR JELEZKO, BRAHIM LOUNIS, AND MICHEL ORRIT

Centre de Physique Moléculaire Optique et Hertzienne, CNRS et Université Bordeaux I, 33405 Talence Cedex, France

15.1. INTRODUCTION

At the turn of last century, much indirect proof for the existence of atoms and molecules had accumulated owing to a compelling body of evidence. Yet the direct observation of molecules remained elusive for the best part of the twentieth century, until the advent of scanning probe microscopes in the early 1980s. The breakthrough of the scanning tunnelling microscope (STM), making direct investigations of atoms and molecules routine in many laboratories, lifted a psychological barrier and opened the door for further experiments in micro-physics. The subject of this chapter is the optical detection and study of single molecules in condensed matter [1]. Early attempts, such as by J. Perrin [2], to observe fluctuations in the fluorescence intensity from dye molecules in black soap films directly were doomed for the lack of convenient detectors. Although the equipment needed to detect single molecules optically (photomultiplier tubes and microscopes) was already available in the 1950s and 1960s, such experiments were only tried and achieved in the late 1980s, in the wake of the STM revolution. The first experiments on single-molecules (or ions, color centers, etc.) gave noisy signals [3] or were performed on large biomolecules containing some tens of chromophores [4]. A modulated absorption technique applied to a molecular crystal yielded the first single molecule signals [5], but the signal-to-noise ratio improved dramatically when the powerful fluorescence technique was applied to the same system, yielding sharp spectral lines from single molecules which were reported in 1990 [6]. Independently, single molecules were detected in a liquid solution by the fluorescence bursts they emitted when crossing the exciting laser beam [7]. A few years later, it became possible to image single molecules on surfaces or in thin films at room temperature, either with a scanning near-field

Shpol'skii Spectroscopy and Other Site-Selection Methods, Edited by Cees Gooijer, Freek Ariese, and Johannes W. Hofstraat.
ISBN 0-471-24508-9

optical microscope (SNOM) [8], or with conventional far-field confocal microscopes [9]. The latter kind of microscopes are much easier to operate than SNOMs and are already in use in many laboratories, particularly for the imaging of biological samples. Therefore, they confer a considerable potential to single molecule methods in basic and applied chemistry and biology.

The number of molecules dealt with in most areas of science (not to mention everyday life) is huge. One might thus ask: Why investigate single molecules? There are two ways to answer this question. Single molecules (here again, *molecule* means any small assembly of atoms) are our only doorway into the world at the nanometer level, for example, as parts of the tiny machines envisioned in molecular electronics. The ever-shrinking size of electronic devices will make single-"molecule" investigations inescapable in a near future. But, from a more fundamental point of view, each single-molecule is also a representative of the large ensembles we use in conventional experiments. The many averagings implied in most of these experiments are entirely removed by single-molecule measurements. They highlight inhomogeneities and deviations from the average, and make detailed statistical studies possible, as scanning probe microscopies have forcefully demonstrated in materials science. One very important example of inhomogeneities are time-dependent fluctuations, which are made directly observable by the selection of individual systems. Many of these fluctuations could never be observed as such with conventional methods of time-resolved spectroscopy since these require the initial synchronization of all subsystems. Photoinduced processes can be initiated by a laser pulse, which provides the synchronization, but other properties, like protein folding, for instance, cannot be light driven. However, the observation of a single protein molecule could directly reveal folding as it spontaneously evolves with time. The optical selection of single "molecules" is a general method that starts to be applied to other systems than colored organic molecules: Nanoparticles [10, 11], quantum dots [12, 13], ions, color centers [14] have all been studied individually in recent years and months. Many of the recent advances in the field of single molecules have been achieved at room temperature. Strikingly similar blinking behaviors have been found for different systems [15], although the blinking mechanisms are probably different. The photobleaching processes that limit the amount of fluorescence emitted by single molecules are beginning to be investigated, the mobility and colocalization of labeled lipids and proteins are being investigated in membranes [16], and possible applications to DNA sequencing or to immunological assays are being explored (see, e.g., Dovichi and Chen [17]). In the context of this volume, we focus on single-molecule experiments at low temperatures. The rigid environment in solid matrices brings in new features with respect to room-temperature studies. In many cases, photobleaching disappears in cryogenic conditions, because the diffusion of small molecules like oxygen and water is suppressed and because the solid cage conserves the molecular geometry. More significantly, for some well-chosen host–guest couples the Shpol'skii effect leads to well-defined spectroscopic sites and to narrow zero-phonon lines. Such sharp lines are extremely sensitive to all kinds of perturbations from the environment or from outside, and open a wide field for fine spectroscopic experiments.

This chapter is set up as follows: We give a brief description of the principles and experimental methods of single-molecule spectroscopy in Section 15.2. In Section 15.3, we discuss the physics and chemistry of the host–guest systems known so far to give single-molecule signals. Different kinds of experiments on single molecules will be illustrated by some examples in Section 15.4, and the conclusion and an outlook are presented in Section 15.5.

15.2. SINGLE-MOLECULE SPECTROSCOPY AT LOW TEMPERATURES

15.2.1. Principles

When trying to isolate single molecules optically, the volume and concentration of the excited sample must be sufficiently small. This spatial selection can be performed by tightly focusing a laser beam in a three-dimensional sample, although one often uses surfaces and thin films (e.g., obtained by spin coating) to improve the selectivity in one spatial dimension. In the excited volume, the molecules at the beam focus will contribute most significantly to the total signal. The experiments at room temperature rely on spatial selection only, which leads to excited volumes of the order of one cubic micrometer (for a bulk sample) and to concentrations of guest molecules smaller than 10^{-10} M. Cryogenic conditions open up the possibility for another selection scheme, where molecules are selected by the frequency of the exciting laser. For well-chosen host–guest systems, the broad, inhomogenous, optical line is the superposition of the many sharp, homogeneous, zero-phonon lines (ZPL [18]) of each single molecule in the sample. When the number of molecules is small enough, the narrow lines are spectrally resolved from each other. Single molecules can be isolated in the spectrum by scanning a narrow exciting source across the inhomogeneous band [19]. In addition to the convenience of studying single molecules in a frequency spectrum, the narrow ZPL has another, essential advantage: Since a significant part of the single molecule's oscillator strength is concentrated in a narrow frequency range, a gigantic absorption cross-section appears at resonance. Its maximum value is about five orders of magnitude larger than at room temperature!

The advantages of the ZPL for the study of single molecules are therefore decisive for fine spectroscopy: Narrow lines can be detected easily; they can probe perturbations from the surrounding solid or perturbations applied externally with extreme sensitivity. However, the ZPL phenomenon is far from being observable in all host–guest systems. The ZPL is discussed in detail in other chapters of this volume (see the chapters by Personov, Renge, and Völker). The ZPL arises from the direct transition between the ground and the excited states of the guest molecule, without creation or destruction of any vibrational quantum (here we do not distinguish between lattice phonons, localized phonons, and intramolecular vibrations, so that our ZPL is actually that of the pure electronic component only). At sufficiently low temperatures, the strength of the ZPL is proportional to the quantum overlap between the ground vibrational states of the ground and the excited electronic states. It can be shown that this overlap is a product of factors $\exp(-S_i/\hbar\omega_i)$, where S_i is the relaxation energy of mode number i, that is, the energy of the nuclear configuration of the ground (excited) state in the excited (ground)-state potential, and $\hbar\omega_i$ is the vibrational quantum of mode i [18]. For example, the usual dye molecules present strong charge rearrangements between ground and excited states, while being rather flexible. Therefore, the argument of the exponential is often larger than 10 or 100, making the ZPLs of many dyes hopelessly weak. To present a strong ZPL, a guest should not see its insertion geometry change too much upon excitation (small S), and its vibration frequency ω in the insertion site should be as high as possible. These two requirements are somewhat contradictory, since a good fit of the guest in the site will in general lead to a high libration frequency, but also to a strong interaction with the host and therefore to a high S. This explains that, to this day, finding a host–guest system with a strong ZPL remains largely a matter of trial and error. The combination of simulations of the molecular packing with quantum-chemical calculations will certainly improve our understanding of the strength of ZPL's in various host–guest systems. In general, though, a good fit of the guest in the vacancy left by one or a few host molecules in the host lattice will ensure the presence of a ZPL, since low-frequency librations will be absent.

Beyond the presence of a ZPL, an efficient spectroscopy of single molecules requires other conditions. Since guest molecules are detected in practice via their fluorescence, they must present a fairly strong oscillator strength (thus only strong singlet–singlet transitions need be considered) and a high fluorescence yield. Fortunately, such is the case for the S_0–S_1 transitions of many planar aromatic hydrocarbons [20], which also give strong ZPL's in Shpol'skii matrices and other molecular crystals. A very important feature is the maximum emission rate of the guest. The bottleneck in the excitation–emission cycle is usually the triplet manifold. In order to keep the triplet population as low as possible, the intersystem crossing rate from the excited singlet must be weak and the triplet lifetime short.

Finally, the detailed observation of a single molecule demands the detection of many photons (at least several thousands, which in turn requires at least a few millions of excitation–emission cycles). Both the guest molecule and the matrix must therefore be as stable as possible. Crystals will be good candidates as matrices, but some polymers are also stable enough (see section 15.3). Photo-induced changes in the guest or host lead to persistent spectral hole-burning ([21, 22]; see also the chapters by Personov, Völker, Small, Zilker, and Haarer). Since such processes will prevent further excitation of the guest molecule, it will be lost after a short irradiation time. Host–guest systems that present efficient hole burning will not be suited to single-molecule spectroscopy. Photochemically reactive guests are therefore eliminated, as well as the important class of the hydrogen-bonded hosts, where proton mobility is very high.

15.2.2. Experimental Methods

Once a convenient host–guest system has been found, the detection of single molecules is very easy, provided the two following conditions on the experimental setup are fulfilled:

1. The sample must be dilute enough, and the excited spot small enough, to isolate single molecules in a fluorescence excitation spectrum.
2. The collection and detection of fluorescence must be very efficient; that is, the emission must be collected over a wide solid angle.

These two conditions apply for room temperature as well as for cryogenic experiments. Typical detection efficiencies are of a few thousandths for most cryogenic setups and of a few percent for room-temperature setups using microscope objectives to collect fluorescence and avalanche photodiodes as detectors.

The excited volume is an important parameter in single-molecule experiments, because (together with the purity of the host and the concentration of the guest) it determines the background of the fluorescence signal. The signal-to-background ratio is inversely proportional to the excited volume, since all the illuminated host and guest molecules, except the single guest molecule considered for the signal, add to the background. The smallest sample volume is of a few cubic micrometers in the best case of a diffraction-limited spot. This is difficult to achieve at liquid helium temperatures, because usual optics are designed to work at ambient conditions, but some microscope objectives keep a reasonable optical quality at low temperatures. In some cases, the excited volume may be as large as a thousand cubic micrometers, while the signal-to-background ratio is still much higher than unity for intense molecules such as terrylene.

Various optical designs fulfilling the above two conditions have been developed in different groups. The two main elements of the setup are the selecting element, which defines the excited volume, and the collecting element, which gathers fluorescence. The selecting elements can either focus an exciting laser beam onto a small spot (lens, microscope objective, mirror, . . .) or just diaphragm the laser beam (cleaved single-mode fiber, pinhole, tapered fiber, . . .). The collecting element must have a large numerical aperture (parabolic or elliptic mirror, aspheric lens, microscope objective, index gradient lens, . . .). In principle, any combination of one selecting and one collecting element gives a possible design for single-molecule isolation, including combinations where the same element both selects and collects, and which we call "confocal" (note that in a strict sense the term *confocal* applies to all these combinations, since the focal points of the two elements must coincide). In practice, only a few of these combinations have been implemented as yet. We discuss here the advantages and drawbacks of the most widely used designs, which are schematically represented in Figure 15.1.

1. The fiber-paraboloid [6]: A single-mode optical fiber carries the sample, which is excited by light propagating in the fiber core. The fiber end is crudely placed at the focus of a collecting parabolic mirror, which reflects the fluorescence toward the detector. This very rugged design does not require any adjustment in the cryostat. The excited spot cannot be moved across the sample, but this has the benefit of preventing any drift of the spot

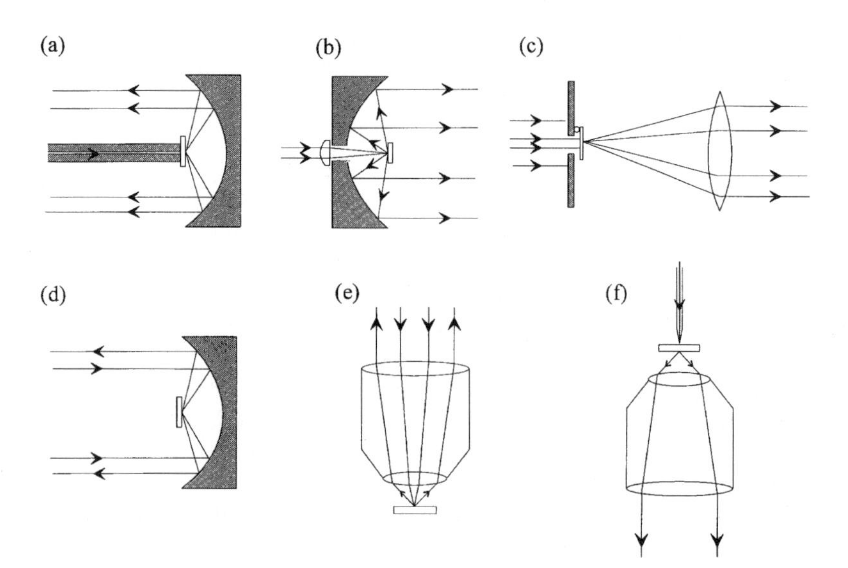

Figure 15.1. Schematic representation of the main optical designs used for single-molecule detection, imaging, and spectroscopy: (a) fiber-paraboloid; (b) lens-paraboloid; (c) pinhole-objective lens; (d) confocal paraboloid; (e) confocal microscope; (f) scanning near-field microscope.

due to experimental factors (temperature changes, helium bath fluctuations, etc.).

2. The lens-paraboloid [23]: The sample is excited at the focal spot of a small lens, which has to be adjusted with respect to the sample to accomodate for temperature changes upon cooling. The collector is again a paraboloid. This design allows the laser spot to scan the sample. It is also easy to vary the polarization continuously.
3. The pinhole-lens [24]: The sample is fixed on a screen and excited through a pinhole. The fluorescence light is collected by a microscope objective. Here, scanning is impossible, but polarization can be changed easily.
4. The confocal microscope [25, 26]: This well-known type of microscope is used in many areas of science and is well adapted to detect single molecules, since the focal spot is diffraction limited and can be imaged very efficiently on the small, sensitive area of an avalanche photodiode. Cryogenic versions of the confocal microscope [27] are more difficult to build and to operate because of space requirements and because of the poorer quality of optics and scanning elements at low temperatures, but they also give good results.
5. The scanning near-field optical microscope (SNOM) [8]: The SNOM is a combination of a tapered and metal-coated optical fiber for selection with a microscope objective for collection. The tapered fiber provides subwavelength resolution. A topographic image is obtained with each optical image, so that a correlation of single-molecule signals with the topographic structures of the sample becomes possible. However, the SNOM is difficult to build and to operate. One of the main problems is the reproducibility and lifetime of the tip. There are no cryogenic versions available commercially, and very few of them work effectively in research groups.

In addition to the optical design, single-molecule spectroscopy requires the well-known equipment used for site-selective spectroscopy and spectral hole-burning. The most essential and expensive parts are a tunable single-frequency laser, a variable-temperature cryostat providing superfluid helium temperatures by pumping on a helium bath, and a sensitive photon-counting detection. The laser must be chosen according to the spectral range of the guest molecules. Many single-molecule experiments are done in the red part of the spectrum, with tunable dye lasers operated with rhodamine dyes. Three types of detectors can be used: photomultipliers (PMT) or avalanche photodiodes (APD) [28] can measure instantaneous fluorescence intensities and fluorescence excitation spectra, while charge-coupled devices (CCD) [26, 29] are often used in spectrographs to record fluorescence spectra. Very useful time-resolved information can be obtained from time-correlated single photon counting [30] and from autocorrelation functions of the fluorescence intensity. These are obtained with an electronic correlator, the logarithmic versions of which are particularly convenient [31].

A single-molecule experiment usually starts as follows: Once all optical adjustments have been completed, one tries to get a fluorescence excitation signal by exciting in the middle of the inhomogeneous band, where the probability to find molecules is highest. Depending on the guest concentration, the signal arises from single-molecules (less than one on average) or from many molecules (statistical fine structure, [32]). In the latter case, one has to shift the laser frequency to the wings of the inhomogeneous band to decrease the number of molecules. With a broad frequency scan (usually 30 GHz), single-molecules appear as sharp spikes. One has to zoom in on each spike for further experiments (e.g., saturation studies), where the single-molecule line is recorded for various laser powers.

15.3. HOST–GUEST SYSTEMS FOR SINGLE-MOLECULE SPECTROSCOPY

In Section 15.2, we have given the most important features of guest molecules for single-molecule spectroscopy (SMS). They must be rigid and have good fluorescence yields and favorable triplet photophysical parameters. All the known molecules giving rise to sharp single-molecule lines at low temperatures belong to, or derive from, the class of polycyclic aromatic hydrocarbons (PAH; see Nakhimovsky et al., [20]). There are to this day about ten guest molecules that are known to yield such sharp lines in suitable solid matrices. We can order them according to the mean absorption wavelength of their first singlet–singlet transition: diphenyloctatetraene (DPOT, 445 nm; [33]), perylene (Pr, 448 nm; [34]), peryleneamidinimide (PAI, 570 nm; [35]), terrylene (Tr, 574 nm; [36]) and tetra-*tert*-butyl-terrylene (TBT, 575 nm; [11]), benzo-di-phenanthro-bisanthene (BDPB, 580 nm; [37]), dinaphthopyrene (DNP, 580 nm; [38]), dibenzanthanthrene (DBATT, 589 nm; [39]), pentacene (Pc, 593 nm; [6]), an unknown impurity of polyethylene (X, 602 nm; [40]), terrylene-diimide (TDI, 650 nm; [41]), dibenzoterrylene (DBT, 750 nm; [28]). The hosts in which these compounds were studied as single-molecules are molecular crystals of aromatics (para-terphenyl, naphthalene, benzophenone, biphenyl, anthracene, ...), molecular crystals of linear *n*-alkanes, known as Shpol'skii matrices (from octane to hexadecane), and various polymers: polyethylene (PE, [34]), polyvinylbutyral, polymethylmethacrylate and polystyrene [42], polyisobutylene [43]. The chemical structures of host and guest molecules are presented in Figure 15.2.

The various combinations of guests and hosts for which single-molecule experiments have been performed or tried are presented in Table 15.1. As one can see immediately, the range of wavelengths that has been most extensively investigated so far is the red part of the spectrum (570–620 nm), because guest

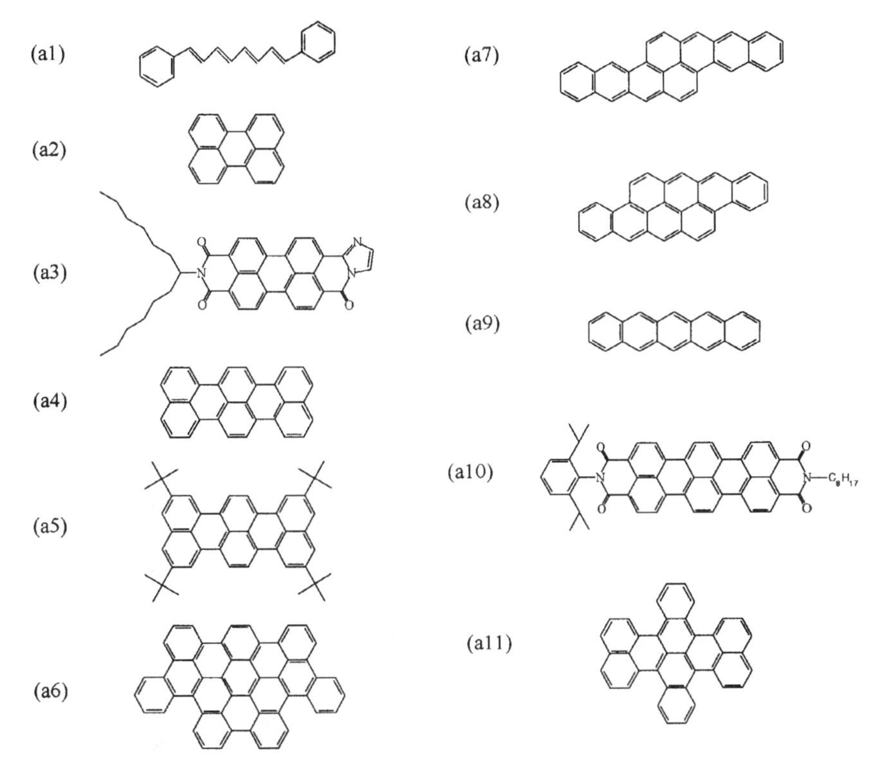

Figure 15.2. Chemical structures of the guest and host molecules used in single-molecule studies at cryogenic temperatures, with the abbreviations as used in the text and in the table. (a1) diphenyl-octatetraene DPOT; (a2) perylene Pr; (a3) substituted perylene-amidinimide PAI; (a4) terrylene Tr; (a5) tetra-*tert*-butyl terrylene TBT; (a6) benzo-diphenanthro-bisanthene BDPB; (a7) dinaphthopyrene DNP; (a8) dibenzanthanthrene DBATT; (a9) pentacene Pc; (a10) substituted terrylenediimide TDI; (a11) dibenzoterrylene DBT. (b1) para-terphenyl pTP; (b2) naphthalene N; (b3) anthracene Ac; (b4) benzophenone BPh; (b5) biphenyl BP; (b6) durene D; (c1) *n*-octane C8; (c2) *n*-nonane C9; (c3) *n*-decane C10; (c4) *n*-dodecane C12; (c5) *n*-tetradecane C14; (c6) *n*-hexadecane C_{16}; (d1) polyethylene PE; (d2) polyisobutylene PIB; (d3) polyvinylbutyral PVB; (d4) polymethylmethacrylate PMMA; (d5) polystyrene PS.

molecules absorbing in this range can be excited with convenient single-frequency dye lasers. The reference numbers in Table 15.1 indicate the systems where single-molecule detection was achieved. Negative results are indicated by crosses (X). Since these are usually unpublished, we indicated only the negative results from our group. The most widely studied guest molecule is probably terrylene (or its variant TBT). Tr keeps its high fluorescence yield and its high saturation intensity (weak bottleneck effect from the triplet) in many matrices; many other hosts would probably work for Tr and have yet to be tried. Although

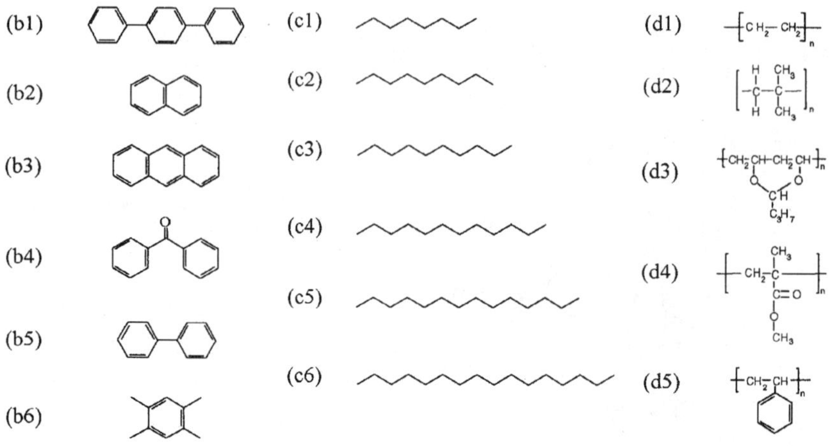

Figure 15.2. (*continued*)

pentacene was the first molecule detected by fluorescence excitation, it is not widely studied because its saturation intensity in *p*-terphenyl is weak, and because it becomes even weaker in other hosts such as naphthalene [44], presumably because the ISC yield to the triplet manifold is enhanced by molecular distortions. On the other hand, we see in Table 15.1 that the most popular host so far is hexadecane (C_{16}), for practical reasons: C_{16} forms Shpol'skii matrices, is liquid at room temperature, and evaporates only slowly, which makes it particularly easy to manipulate. In the following, we describe each class of systems with a few typical examples, considering Shpol'skii matrices separately from other molecular crystals.

15.3.1. Molecular Crystals

When a perfect crystal is cooled down to low temperatures, its periodic structure determines the optical absorption spectra of guest molecules. The only degrees of freedom that may affect optical line shapes are the librations of host and guest molecules in the lattice (acoustic phonons are weakly coupled and have a low density of states at low frequencies). Since the libration correlation time is very short (of the order of $h/k_B T$, i.e., a few ps), the line shapes are broadened by optical dephasing, and the temperature dependency of the broadening is proportional to the number of phonons present; that is, it follows an activated law at low temperature. One very good example of such a behavior is DBATT in a naphthalene (N) crystal [45]. The optical absorption line shows a single narrow site, broadened inhomogeneously by crystal defects (see Figure 15.3). These defects do not lead to any significant additional dynamics of the host. The lines of

Table 15.1. Host–Guest Systems for SMS.

Various host–guest combinations for which narrow lines from single guest molecules have been observed. The guest molecules are ordered by their average absorption wavelength, the hosts by their structure (molecular crystals, Shpol'skii matrices, polymers). The crosses indicate negative results (no single-molecule lines) and the numbers indicate the corresponding references.

Guest	DPOT	Pr	PAI	Tr	TBT	BDBP	DNP	DBATT	Pc	X	TDI	DBT
λ(nm)	445	448	570	574	575	580	580	589	593	602	650	750
Hosts												
pTP				[44]					[6]			
N		X						[28]	[76]			[28]
Ac		X		[42]								
BPh				[75]								
BP		[50]										
D		[38]						[38]				
C8		[38]										
C9		[74]										
C12				[53]				[39]				
C14	[33]			[53]		[37]						
C16		[38]		[51]			[38]	[39]		[40]	[41]	
PE		[34]	[35]	[36]	[43]				X	[40]	[41]	[38]
PIB				[38]	[43]							
PVB				[42]								
PMMA				[42]								
PS				[42]								

single DBATT molecules are shown in Figure 15.3. They are very stable. Neither spontaneous nor photoinduced spectral jumps could be observed in tens of molecules. Their homogeneous width is lifetime limited below 4 K, then starts to increase at higher temperature with an activation energy of about $40\,\mathrm{cm}^{-1}$, characteristic of librations of DBATT in the crystal lattice. The spread of linewidths that was found for certain samples with a free surface was attributed to changes in radiative lifetimes due to orientation-dependent refractive index corrections at the surface [45]. Such host–guest systems with very stable single molecule lines are ideal for nonlinear and quantum–optical measurements, where long accumulation times and high–frequency stabilities of the source and molecule are needed. The ideal behavior of DBATT/N can only be expected for simple crystals such as naphthalene, which present no phase transitions, and if the guest molecule is well embedded in the lattice. Therefore, well-defined and narrow sites in fluorescence or absorption spectra will be good indications that single-molecule lines may be stable. Systems with broad inhomogeneous lines, even when they present ZPLs, are more likely to show jumping single-molecule lines. Molecules outside the main sites will have a tendency to show richer dynamics than those within the sites.

A very important example of this rule is pentacene in a *p*-terphenyl crystal, Pc/pTP. This system presents four well-defined optical sites, corresponding to substitutions of the four pTP molecules in the triclinic unit cell by Pc molecules. Only sites O_1 and O_2 present single-molecule lines, for the triplet yield of the two other sites are unfavorable to SMS. These lines are stable in the center of the inhomogeneous distribution, but become more and more agitated as one goes farther into the wings [23]. The jumping lines are thought to arise from those Pc molecules that lie close to the walls between domains of different symmetries in the crystal [46]. The flips of the central phenyl rings of pTP molecules in a wall cause frequency shifts of the lines of nearby Pc molecules. The central phenyl ring of a pTP molecule has two equilibrium positions tilted in symmetric ways with respect to the external phenyl rings. The central ring angle can be seen as a spin-$\frac{1}{2}$ moment, and the crystal as an Ising antiferromagnet undergoing an order–disorder transition at a temperature of about 190 K. This transition leads to a lowering of the crystal symmetry, from monoclinic to triclinic, and to a doubling of the unit cell. At domain walls, a flip of a central ring does not require much energy. A domain wall can therefore be modeled as a two-dimensional lattice of flipping two-level systems, whose uncorrelated jumps cause a random walk of single-molecule lines along the frequency axis. Note that the jumps are spontaneous here; that is, they do not depend on the excitation intensity [23].

The case of terrylene in the same crystal, Tr/pTP, is quite different [44]. In two of the four sites at low temperature, the single-molecule lines are photostable, while they undergo large photoinduced jumps in the other two sites (X_1 and X_4). The photoinduced jumps of single-molecules in site X_1 were investigated [47]

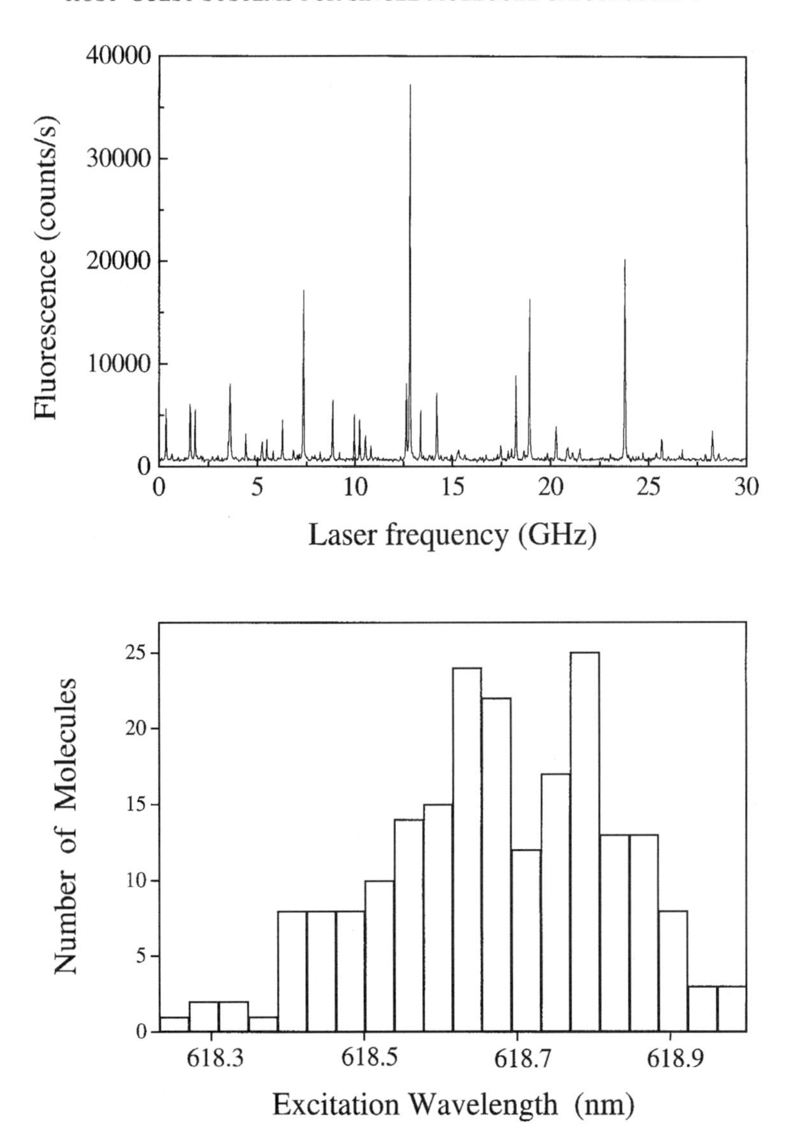

Figure 15.3. Example of a 30 GHz (1 cm^{-1}) scan of the exciting laser frequency, showing the sharp single-molecule lines of DBATT in a naphthalene crystal. The histogram displays the distribution of the lines versus the wavelength.

and found to be reversible. This system could therefore be the basis of a single-molecule optical switch. In high-quality crystals, the jump amplitude, 843 GHz, is constant. Its large value shows that the motion involves the guest molecule itself and/or its first solvent shell. The problem of the conformational changes leading to such large jumps has to be investigated further by molecular-mechanical simulations.

Perylene in biphenyl, Pr/BP, is a superficially similar system. Since Pr/BP is a shorter analogue of Tr/pTP, one would expect to find again spontaneous and photoinduced spectral jumps. However, there is a fundamental difference between BP and pTP crystals. Although the phenyl rings of BP are also tilted with respect to each other, and, like pTP, BP molecules are flat on average at room temperature, the phenyl ring of BP explores the two symmetric positions of the double well during its vibration at room temperature. Upon cooling, the vibration mode in the double well becomes soft, and the phenyl ring localizes in one of the wells. The interactions between neighboring BP molecules lead to an incommensurate phase that persists down to liquid helium temperatures. Fluctuations of the phase of the distortion wave lead to elementary excitations of the lattice called *phasons* [48, 49]. Like acoustic phonons, the frequency of phasons depends linearly on the wave vector for small wave vectors. However, unlike acoustic phonons, phasons involve rotations of the phenyl rings of BP molecules and are therefore strongly coupled to the optical line of guest molecules. Since the correlation time of phasons must be very short, the fluctuations of the optical transition frequency must lead to dephasing rather than to spectral jumps. The coupling of single Pr molecules to phasons in BP can explain why all SM lines are much broader than the lifetime limit, as was recently reported [50].

15.3.2. Shpol'skii Matrices

The crystals of linear *n*-alkanes are commonly used matrices for all kinds of aromatic molecules. They are called Shpol'skii matrices and often give well-defined sites with narrow inhomogeneous widths (usually a few cm^{-1}, [20]). Several Shpol'skii systems present single-molecule lines (see Table 15.1). Here, we describe terrylene in *n*-hexadecane (Tr/C_{16}; [51]) as a generic example. The bulk absorption and fluorescence spectrum of Tr/C_{16} is more complex than those discussed in Section 15.3.1 because the Shpol'skii site appears on a broad absorption background, attributed to ill-inserted molecules [52, 53]. An example of single-molecule lines of Tr/C_{16} is shown in Figure 15.4 (from [51]). Some single-molecule lines are quite stable, but a significant fraction of them undergo spontaneous and photoinduced spectral jumps. The fraction of jumping molecules is larger for molecules out of the site. The lines of single-molecules in Tr/C_{16} present a whole distribution of linewidths, with a cutoff for the lifetime-limited width, 42 MHz. Broadening beyond the natural linewidth is explained by dynamical fluctuations in the surrounding matrix, which lead to dephasing or to spectral diffusion, depending on the correlation time of the fluctuations (see, e.g., Brown and Orrit [54]). In amorphous systems, special low-energy excitations known as "two-level systems" (TLS; see Section 15.3.3) are thought to be the origin of these fluctuations. Because of the wide spread of TLS characteristic times, spectral diffusion is thought to be the dominant broadening mechanism (the characteristic times lie between nanoseconds and seconds, while the characteristic times for dephasing are shorter than a nanosecond). The spontaneous spectral jumps of single Tr molecules [42, 51] and saturation studies [53] prove the existence of spectral diffusion in Tr/C_{16}. The temperature dependence of single-molecule linewidths also shows the heterogeneity of the matrix at a nanometric scale [53]. The existence of such complex dynamics in what, after all, is a molecular crystal may be somewhat surprising. Shpol'skii matrices are highly polycrystalline (or snowlike) and contain a high concentration of defects. As in amorphous systems, the complex dynamics of these defects can give rise to spectral diffusion and dephasing. The insertion of Tr in alkanes of varying lengths was studied for bulk systems [52] and for single-molecules [53]. The complexity and the width of the spectrum increase with alkane length beyond *n*-decane, which appears to be the best host for terrylene.

15.3.3. Polymers

Because they are easily prepared and because of their good optical qualities, polymers are often used as matrices for persistent spectral hole-burning. Polyethylene was one of the first matrices for single-molecule spectroscopy [34]. Single terrylene molecules were studied in polyethylene (Tr/PE; [36]) and in polyisobutylene (Tr/PIB; [43]), and found to present very similar features, although PE is semicrystalline and PIB a completely amorphous polymer. In polymer matrices, no single-molecule line is completely stable. Even at a temperature of 1.7 K, spontaneous and photoinduced jumps occur. Studies of the optical saturation and of the variation of line shape with temperature are difficult because the jumps remove the molecular line from the scanning range of the laser. Many of the features and behaviors observed on single-molecule lines in polymers can be understood within the standard TLS model [55]. This model assumes that certain small regions of an amorphous material (a glass, polymer, etc.) present two nearly degenerate energy states (within thermal energy at the experiment's temperature) and flip from one configuration to the other. The flip arises from a tunneling event assisted by acoustical phonons. The model explains the diversity of individual molecules, which have different distributions of TLS in their environments. The widths of SM lines are broadly distributed in polymers, even more so than in hexadecane, as illustrated by the examples of Figure 15.5. Some molecules are mainly coupled to only one TLS and are found to jump

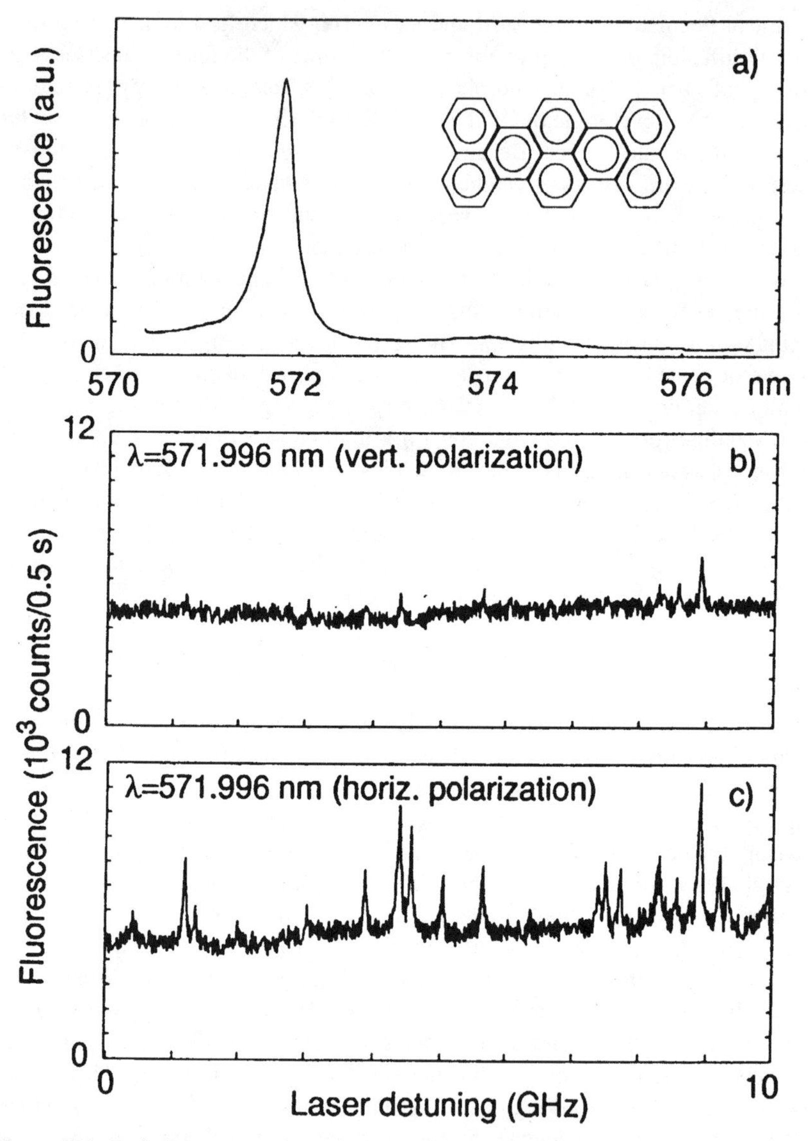

Figure 15.4. Excitation spectra of terrylene in *n*-hexadecane at 1.7 K, with the inhomogeneous distribution (a), and spectra recorded for two orthogonal excitation polarizations (b) and (c) with a fiber-paraboloid setup, showing polarized single Tr molecule lines (from Moerner et al. [51]).

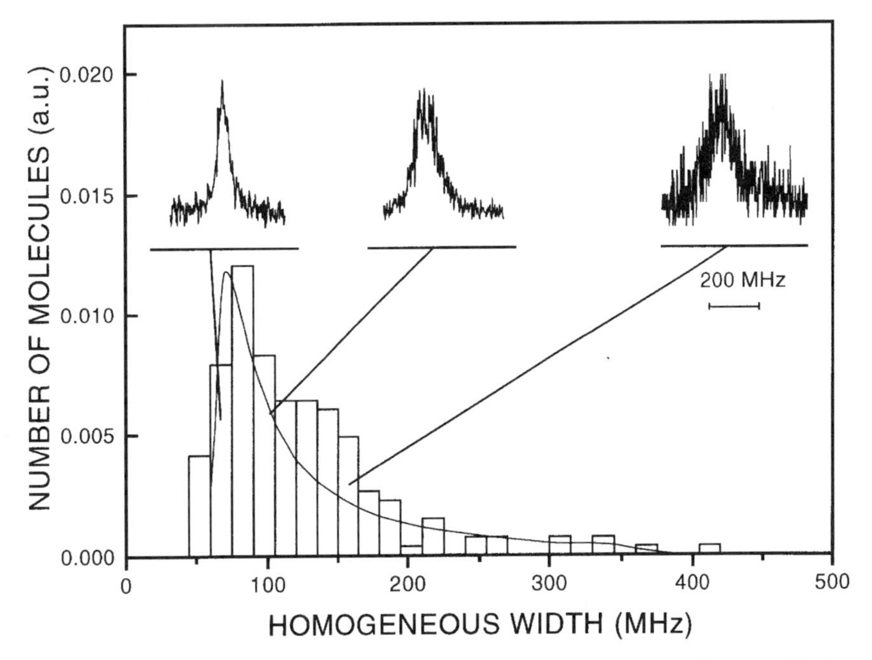

Figure 15.5. Distribution of single-molecule linewidths for terrylene in polyethylene at 1.8 K. Three examples of lines with different widths are presented (from Orrit et al. 1996 [77]).

between two frequencies, but most molecules present more complex behaviors. The characteristic flipping time of a TLS can be measured directly if the jumps are slow enough (slower than a few seconds), by measuring the single-molecule frequency as a function of time. For shorter times (down to microseconds) the fluorescence intensity correlation function [56] gives the flipping time directly. Other methods for the analysis of the spectral diffusion of single-molecules will be discussed in Section 15.4. The behavior of single-molecules becomes more and more complex when one goes from simple molecular crystals to Shpol'skii matrices and to polymers. In other polymers like PVB, PMMA, PS [42], the single Tr lines are even broader and weaker. The rate of spectral jumping was so high in PMMA and PS that detailed studies were very difficult.

15.4. SOME EXAMPLES OF SMS EXPERIMENTS

The purpose of this section is to give a few examples of spectroscopic experiments that can be done on single-molecules. The experiments fall into three classes:

1. Studies of the molecule itself; even in cases where molecules are strictly identical, some experiments are even easier to do with single-molecules than with ensembles!
2. Studies of the interaction between light and matter: A single-molecule can be seen in many respects as a simple electronic two-level system. Its interactions with laser fields can be captured by simple theoretical models. Such simple systems help test nonlinear and quantum optics in condensed matter.
3. Studies of dynamical fluctuations in the solid surrounding a single-molecule. Such fluctuations at low temperature can arise from various degrees of freedom: concerted movements of several nuclei, protons, electrons, etc.

15.4.1. Molecular Physics

One of the easiest experiments to do on a single-molecule line is to change the exciting laser intensity. When increasing the laser power, one observes optical saturation, that is, broadening of the line and saturation of the signal at its maximum [23]. In most guest molecules, the triplet manifold plays an important role in this saturation [57]. Because the triplet lifetime is long, a strong irradiation of the molecule leads to a buildup of the triplet population, and therefore to saturation. Another way of considering the phenomenon is to look at the instantaneous state of the molecule. At the beginning of the experiment, the molecule undergoes excitation–emission cycles between its two singlet states. Then an intersystem crossing transition to the triplet manifold occurs, and the molecule is trapped in the dark triplet state for the whole triplet lifetime. Eventually, the molecule goes back to the ground singlet state, and resumes fluorescence. The fluorescence photons are therefore emitted in bursts or bunches. This phenomenon appears very clearly on the autocorrelation function of the fluorescence intensity [6] or directly on the fluorescence signal as a function of time [58]. If there were a single triplet state, one would observe a single exponential component in the correlation function. Since the triplet manifold consists of three different sublevels, each with its own spin polarization, population, and decay rates, there should be three exponential components, although only two are usually observed [44, 59, 60]. Now, since the populations of the different triplet sublevels differ, a resonant microwave can transfer population from one sublevel to another, thereby changing the average dark time and the average fluorescence rate. Such an experiment is called *optically detected magnetic resonance* [61, 62], and enables a fine spectroscopy of the triplet manifold. It must be noted that, even if the average intensity is not significantly modified by the microwave, it is possible to detect the resonance by studying the correlation function. Figure 15.6 shows how the fluorescence correlation function changes under microwave irradiation for two different microwave powers. Equivalently, the bright and dark fluorescence intervals can be recorded as functions of the microwave frequency and nicely reveal the magnetic resonance, as was recently demonstrated [63]. The ODMR method has been pushed further to detect the nuclear spin states of a single-molecule that possesses only a few protons or ^{13}C atoms [64], and even to achieve nuclear magnetic resonance of a single proton [65]!

15.4.2 Nonlinear and Quantum Optics

Some single-molecules are simple electronic two-level systems that can be used as test benches for laser–matter interactions, that is, of the field of nonlinear and quantum optics. The requirements on a host–guest system for this purpose are as follows. The single-molecule lines must be very stable and the triplet effects should be as small as possible, because they will severely limit the magnitude of the observed effects [66]. In addition, the lines should be fairly narrow. Since a

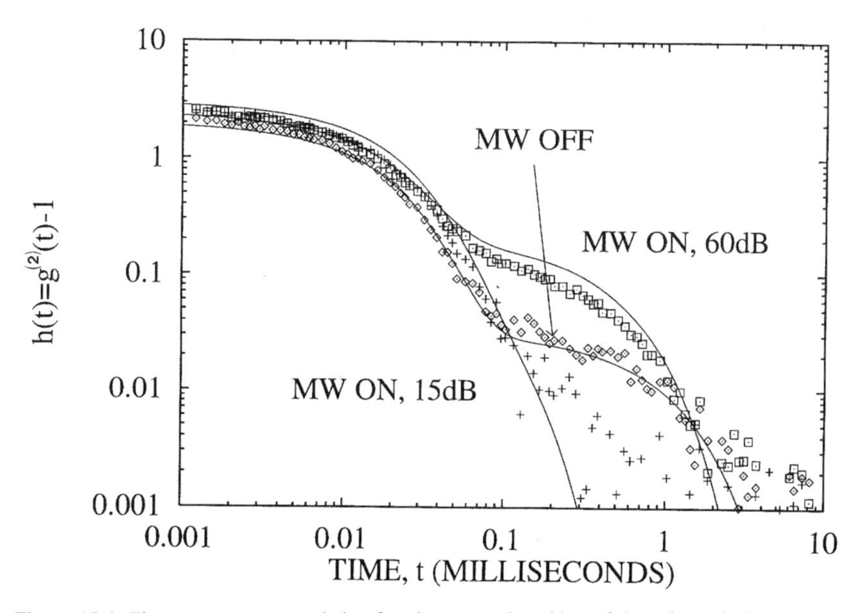

Figure 15.6. Fluorescence autocorrelation function versus logarithm of time, for a single pentacene molecule in a *p*-terphenyl crystal at 1.8 K. The functions are plotted without microwave irradiation, and for two different microwave (MW) powers. For low MW power, the long-lived component increases, but it disappears at high MW power because of coherent driving between X and Z sublevels (from Brown et al. 1994 [59]).

high-frequency stability is required, small frequency intervals are easier to scan with a single frequency-stabilized laser and acousto-optical modulators [67]. Convenient systems for such measurements are pentacene and terrylene in *p*-terphenyl, and dibenzanthanthrene in naphthalene.

The simplest nonlinear effect on an optical two-level system is its saturation, due to the finite lifetime of the excited state. Just as for the three-level system discussed in Section 15.4.1, the optical line broadens and the fluorescence signal saturates when the laser power increases. The antibunching of fluorescence photons is a quantum-mechanical effect that appears because a single-molecule emits only one photon at a time. Immediately after one fluorescence photon has been observed, the probability of observing a second photon is nil. The measurement of the first photon has projected the molecule into its ground state, from which a second photon cannot be emitted. The experiment runs like a fluorescence autocorrelation measurement (see Section 15.4.1), but on a shorter time scale, with a setup of the Hanbury–Brown–Twiss type [68, 69] in order to avoid distortions from dead time and afterpulses of the detector. In addition to the antibunching dip at short times, the correlation function shows Rabi oscillations of the Bloch vector in the laser field.

A simple way of monitoring nonlinear effects in the interaction of a molecule with a strong laser beam (pump beam) is to observe the response to a weak probe beam at a different frequency. The single-molecule is driven by two laser fields with different frequencies. In single-molecule experiments, pump and probe beams are superposed and sent onto the sample. Therefore, it is impossible to separate pump and probe because of the aperture of the beams and because of scattering by sample defects. The only way to measure the combined effect of pump and probe is to compare it to the effect of the pump alone. Under the strong pump illumination, the molecular levels are shifted and new resonances appear. The shift of the levels (called light shift) leads to a shift of the molecular transition, which can be measured as a function of the pump intensity and frequency [66]. The variation of the shift with pump detuning is shown in Figure 15.7, and is seen to deviate from a simple perturbation formula, which would be represented by a straight line. For a weak probe, the pump–molecule system is called a dressed molecule, in analogy with the dressed atom [70]. It displays three resonances: the light-shifted molecular transition, the Rayleigh or two-wave mixing structure at the pump frequency, and the hyper-Raman resonance corresponding to the absorption of two pump photons and to the stimulated emission of one probe photon. The hyper-Raman structure can be detected by an enhanced fluorescence as the probe is scanned across resonance [67]. When both the pump and the probe beams are strong, more complex, multiphoton structures appear due to a transient motion of the Bloch vector of the molecule in the beating of the pump and the probe beams. All these effects and structures are in very good agreement with a description of the molecule by optical Bloch

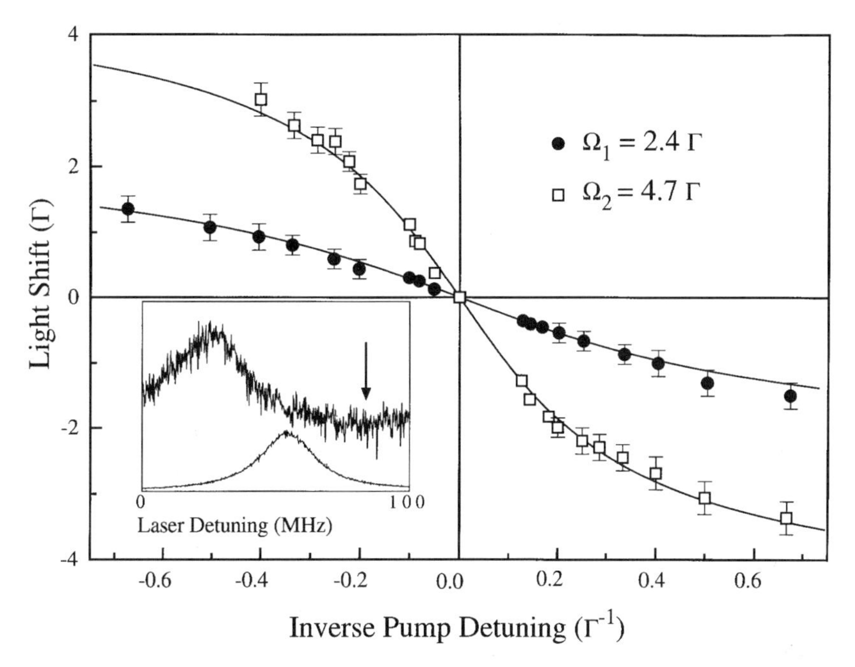

Figure 15.7. Shift of the molecular electronic transition by a near-resonant pump laser wave (light shift, or ac-Stark effect), as a function of the inverse detuning between pump and molecular transition. Deviations from a straight line appear for small detuning. The unperturbed and shifted molecular lines are shown in the inset (from Lounis et al. 1997 [67]).

equations. This shows that single-molecules are very well described by a simple two-level picture [45].

More recently, direct transitions have been observed between the dressed molecular states of a single-molecule in a laser beam [71]. This transition, called Rabi transition, is observed when an RF field is added to a strong, nearly resonant laser field. At weak laser power, the simultaneous absorption of a laser photon and the absorption/emission of an RF photon can be seen as the absorption of a frequency-modulated single-molecule line. The strength of the sidebands in the absorption spectrum is given by Bessel functions of the modulation index, the ratio of the frequency excursion under the RF field to the RF frequency. At high laser power, the RF and laser fields compete with each other to dress the molecule, and the spectra are more complex. However, for weak RF power, the RF transition occurs when the RF frequency exactly matches the frequency difference between dressed molecular levels, that is, the generalized Rabi frequency. Again, the observed spectra are in very good agreement with calculations from optical Bloch equations. These experiments show that it is possible to modulate a single-molecule at high frequency, and that its Bloch vector can be manipulated in further experiments.

15.4.3. Matrix Dynamics and Intermolecular Interactions

A single-molecule can be considered as a very small probe of its surroundings, much like the tip of an STM or AFM. Unfortunately, in present experiments, the single-molecule is frozen in the solid matrix and cannot be scanned to image its neighborhood. Nevertheless, the time-dependent information it relays about its environment is very useful, as the following examples will show.

As was mentioned in Section 15.3, matrix fluctuations will lead to dephasing of the optical transition if the correlation time of the fluctuations is short (shorter than T_2, the transverse coherence time) and to spectral diffusion if this correlation time is long. Dephasing will appear as a broadening of the lines, spectral diffusion as jumps of the lines between different frequencies. Spectral diffusion events can be either spontaneous (assisted by thermal fluctuations) or photo-induced (assisted by the optical excitation of the guest molecule). Studies of the spectral diffusion of single molecules were first applied to the dynamics of disordered matrices. As discussed in Sections 15.3.2 and 15.3.3, many observations, such as the shape and widths of the single-molecule lines or the statistical distribution of their linewidths can be understood within the standard TLS model. Figure 15.8 presents an example of a molecule coupled to a single TLS model, where a correlation function proves that the two peaks in the spectrum originate from the same molecule. By recording the correlation function at a different temperature, the jump rate of the TLS can be studied as a function of temperature, which gives indications about the flipping mechanism of the TLS [56]. It is also possible to deduce valuable information from many repeated scans of the frequency axis, though the accessible time scale of the spectral jumps lies in the range of seconds to hundreds of seconds. From each scan, it is in principle possible to locate the maximum frequency of each single-molecule. By plotting this frequency as a function of time, one obtains a frequency trajectory [23]. However, complex behaviors often make it impossible to assign a given spectral structure to a single-molecule, particularly when jumps are fast and when several molecules have lines in the investigated interval. In this case, the full spectral information as a function of frequency and time (which we call a spectral track) gives more direct and more reliable information. An example of a spectral track of a molecule coupled to three independent TLSs is shown in Figure 15.9. The study of spectral tracks may help to check in a detailed manner whether all single-molecule observations conform to the standard two-level model. Behaviors such as frequency drifts, changes in the rates of flipping of some TLS, three-state or multistate jumping [51] show that the standard TLS model does not capture all the possible microscopic situations in an amorphous system. A further question

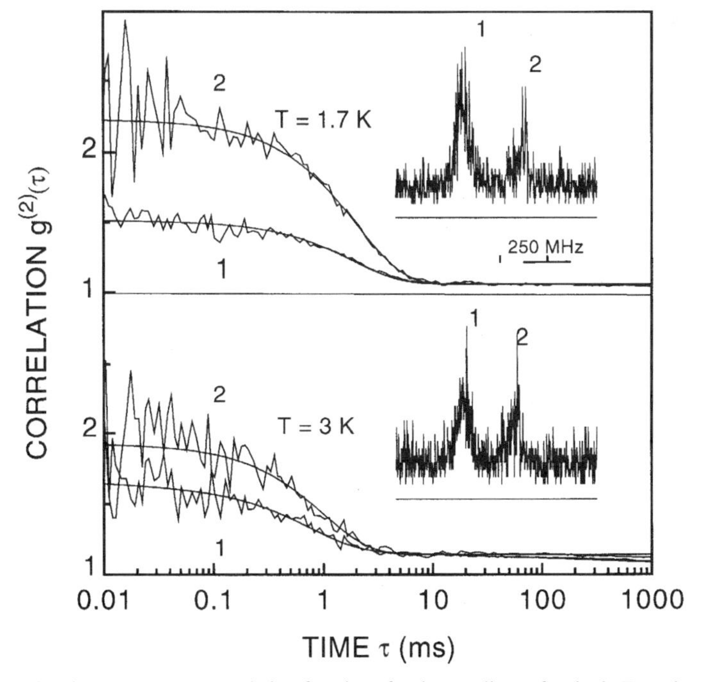

Figure 15.8. Fluorescence autocorrelation functions for the two lines of a single Tr molecule in PE, recorded at two different temperatures. The spectra are presented as insets. All data fit very well with a simple model, where the molecule is coupled to a single nearby two-level system of the polymer (from Orrit et al. 1996 [77]).

that must be addressed in a statistical way is the origin, spontaneous or photoinduced, of the spectral diffusion and spectral jumps of single-molecules.

Two-level systems are just an example of the dynamical fluctuations present in solids. Other slow processes, such as the tunneling of methyl groups [72], could be investigated in the same way. As we have seen in Section 15.3 on the example of Pr/BP, single-molecule lines can also be broadened by low-energy excitations with short correlation times, like phonons and phasons. The different widths of different molecules could show how local conditions affect the coupling to low-energy excitations.

Finally, it would be very interesting to introduce controlled two-level systems in the environment of single-molecules. A first experiment in this direction was performed recently [73]. Triphenylene molecules were introduced in an *n*-hexadecane matrix, along with probe terrylene molecules. Upon excitation, the triphenylene molecules underwent intersystem crossing into their triplet states, then into their ground singlet states. The single terrylene lines underwent spectral

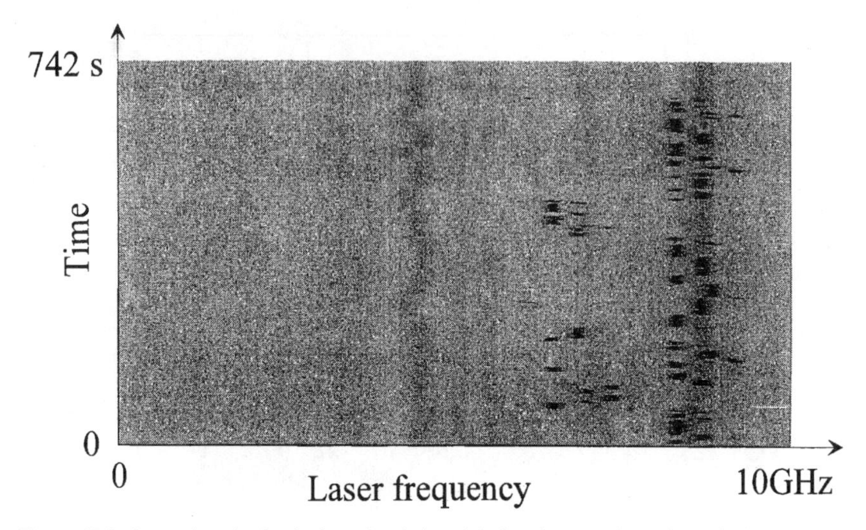

Figure 15.9. Spectral track of a single-molecule in polyisobutylene at 1.8 K. The excitation spectrum is plotted as a density of gray along the horizontal axis, and time along the vertical axis. The spectral jumps and eight different spectral positions of the molecule can be interpreted easily by the coupling of the single-molecule to three independent two-level systems in its vicinity.

jumps, taking on different spectral positions during the lifetime of the triphenylene triplet (about 20 s). A statistical analysis of these shifts will give information about the interaction between triphenylene and terrylene and about the changes of the molecular shape in the triplet state. Other kinds of molecules with metastable states could be used as controllable two-state systems, such as photochromes, spin-transition complexes, or charge-transfer molecules.

15.5. CONCLUSION AND OUTLOOK

Not only can single-molecules now be detected in solids at low temperatures with a good signal-to-noise ratio, but it has been found that many different measurements can be done on a single-molecule, as witnessed by the many experiments undertaken in this field in the past eight years. Of course, the requirements of a narrow zero-phonon line, of a high fluorescence rate and of a high stability severely restrict the number of useful host–guest systems. However, the accuracy and sensitivity of cryogenic experiments are so high that very valuable information can be gained, even on a limited number of model systems. Although single-molecule methods provide knowledge primarily about the photodynamics of the molecules themselves, they also tell a lot about light–matter interactions, and they can be used to gain information about many dynamical aspects of the solid state.

Similar methods are applied to inorganic systems, nanoparticles, and quantum dots in solid-state physics. Some experiments such as optical saturation, which would be nearly impossible on ensembles, are straightforward with single-molecules. One striking example is that of optical saturation, which is an elementary measurement on a single-molecule, but which is much more complex on an inhomogeneous band consisting of many molecules. The data thus obtained are free from any average, which makes them much more reliable, and easy to compare to theoretical models. However, since each single-molecule is an individual, reproducible statistics must be recovered by measuring large numbers of molecules, then plotting histograms and statistical correlation plots between different molecular properties. This aim will only be reached through the parallel measurement of many molecules, for example, by imaging larger samples and automatically treating the data. Among the characteristic advantages of single-molecule methods, the time-resolved investigation of dynamical processes is particularly important. Such information is impossible to obtain with ensembles, because, even with short light pulses, one cannot generally synchronize an ensemble of systems (for instance, it would be impossible to synchronize an ensemble of many two-level systems in a glass with a light pulse). This feature is also particularly interesting for biosystems at room temperature, for example, for the investigation of protein dynamics.

The field of applications for single-molecule methods at room temperature is very broad and promising in biology. Single-molecule studies will complement new microscopic techniques such as confocal, two-photon and second-harmonic generation, and scanning near-field microscopies, by providing the last step toward nanometer objects. Nevertheless, cryogenic studies of single quantum systems, among them single-molecules, in well-defined conditions will remain very active. The future field of molecular electronics, or the wide realm of the interaction of light with small structures in condensed matter, will greatly benefit from the use of single-molecules, nanoparticles, and quantum dots. Even if applications of such nanometric devices are still several years off, the investigation of single quantum systems is a very useful tool for fundamental science, which will provide us with deeper insight into physics and chemistry at a nanometer scale.

REFERENCES

1. Th. Basché, W. E. Moerner, M. Orrit, and U. P. Wild, eds., *Single Molecule Optical Detection, Imaging and Spectroscopy*, VCH, Weinheim, 1997.
2. J. Perrin, *Notice sur les travaux scientifiques*, Privat, Toulouse, 1923, p. 45.
3. R. Lange, W. Grill, and W. Martienssen, *Europhys. Lett.* **6**, 499 (1988).
4. D. C. Nguyen, R. A. Keller, J. H. Jett, and J. C. Martin, *Anal. Chem.* **59**, 2158 (1987).

5. W. E. Moerner, and L. Kador, *Phys. Rev. Lett.* **62**, 2535 (1989).
6. M. Orrit, and J. Bernard, *Phys. Rev. Lett.* **65**, 2716 (1990).
7. E. B. Shera, N. K. Seitzinger, L. Davis, R. A. Keller, and S. A. Soper, *Chem. Phys. Lett.* **174**, 553 (1990).
8. E. Betzig, and R. J. Chichester, *Science* **262**, 1422 (1993).
9. J. K. Trautman and J. J. Macklin, *Chem. Phys.* **205**, 221 (1996).
10. M. Nirmal, B. O. Dabbousi, M. G. Bawendi, J. J. Macklin, J. K. Trautman, T. D. Harris, and L. E. Brus, *Nature* **383**, 802 (1996).
11. J. Tittel, R. Kettner, Th. Basché, C. Bräuchle, H. Quante, and K. Müllen, *J. Lumin.* **64**, 1 (1994).
12. J.-Y. Marzin, J.-M. Gérard, A. Izraël, D. Barrier, and G. Bastard, *Phys. Rev. Lett.* **73**, 716 (1994).
13. D. Gammon, S. W. Brown, E. S. Snow, T. A. Kennedy, D. S. Katzer, and D. Park, *Science* **277**, 85(1997).
14. A. Gruber A. Dräbenstedt, C. Tietz, L. Fleury, J. Wrachtrup, and C. von Borczyskowski, *Science* **276**, 2012 (1997).
15. Th. Basché, *J. Lumin.* **76&77**, 263 (1998).
16. G. J. Schütz, H. Schindler, and Th. Schmidt, *Biophys. J.* **73**, 1 (1997),.
17. N. J. Dovichi, and D. D. Chen in Th. Basché, W.E. Moerner, M. Orrit, and U. P. Wild, eds., *Single Molecule Optical Detection, Imaging and Spectroscopy*, VCH, Weinheim, 1997, Chapter 3.
18. K. K. Rebane, *Impurity Spectra of Solids*, Plenum Press, New York, 1970.
19. H. Talon, L. Fleury, J. Bernard, and M. Orrit, *J. Opt. Soc. Am. B* **9**, 825 (1992).
20. L. Nakhimovsky I. M. Lamotte, and J. Joussot-Dubien, *Handbook of Low Temperature Electronic Spectra of Polycyclic Aromatic Hydrocarbons*, Physical Sciences Data 40, Elsevier, 1989.
21. R.I. Personov, in V. M. Agranovich, and R. M. Hochstrasser, eds., *Spectroscopy and Excitation Dynamics of Condensed Molecular Systems*, North Holland, Amsterdam, 1983, Chapter 10.
22. W. E. Moerner, ed., *Persistent Spectral Hole-Burning: Science and Applications*, Springer, Berlin, 1988.
23. W. P. Ambrose, Th. Basché, and W. E. Moerner, *J. Chem. Phys.* **95**, 7150 (1991).
24. U. P. Wild, F. Güttler, M. Pirotta, and A. Renn, *Chem. Phys. Lett.* **193**, 451 (1992).
25. S. Nie, D. T. Chiu, and R. N. Zare, *Science* **266**, 1018 (1994).
26. L. Fleury, Ph. Tamarat, B. Lounis, J. Bernard, and M. Orrit, *Chem. Phys. Lett.* **236**, 87 (1995).
27. J. Tittel, W. Göhde, F. Koberling, Th. Basché, A. Kornowski, H. Weller, and A. Eychmüller, *J. Phys. Chem. B* **101**, 3013 (1997).
28. F. Jelezko, B. Lounis, and M. Orrit, *J. Phys. Chem.* **100**, 13892 (1996).
29. P. Tchénio, A. B. Myers, and W. E. Moerner, *Chem. Phys. Lett.* **213**, 325 (1993).

30. M. Pirotta, F. Güttler, H. Gygax, A. Renn, J. Sepiol, and U. P. Wild, *Chem. Phys. Lett.* **208**, 379 (1993).
31. L. Fleury, A. Zumbusch, M. Orrit, R. Brown, and J. Bernard, *J. Lumin.* **56**, 15 (1993).
32. W. E. Moerner, and T. P. Carter, *Phys. Rev. Lett.* **59**, 2705 (1987).
33. T. Plakhotnik, D. Walser, M. Pirotta, A. Renn, and U. P. Wild, *Science* **271**, 1703 (1996).
34. Th. Basché., and W. E. Moerner, *Nature* **355**, 335 (1992).
35. R. Kettner, J. Tittel, S. Mais, and Th. Basché, oral contribution at the colloquium "Single Molecule Spectroscopy: New Systems and Methods," held in Ascona, Switzerland, March 1996.
36. M. Orrit, J. Bernard, A. Zumbusch, and R. I. Personov, *Chem. Phys. Lett.* **196**, 595 (1992); **199**, 408 (1992).
37. M. Vacha, and T. Tani, *J. Phys. Chem. A* **101**, 5027 (1997).
38. Ph. Tamarat, F. Jelezko, B. Lounis, and M. Orrit, unpublished results (1998).
39. A.-M. Boiron, B. Lounis, and M. Orrit, *J. Chem. Phys.* **105**, 3969 (1996).
40. L. Fleury, Ph. Tamarat, B. Kozankiewicz, M. Orrit, R. Lapouyade, and J. Bernard, *Mol. Cryst. Liq. Cryst.* **283**, 81 (1996).
41. S. Mais, J. Tittel, Th. Basché, C. Bräuchle, W. Göhde, H. Fuchs, G. Müller, and K. Müllen, *J. Phys. Chem. A* **101**, 8435 (1997).
42. B. Kozankiewicz, J. Bernard, and M. Orrit, *J. Chem. Phys.* **101**, 9377 (1994).
43. R. Kettner, J. Tittel, Th. Basché, and C. Bräuchle, *J. Phys. Chem.* **98**, 6671 (1994).
44. S. Kummer, Th. Basché, and C. Bräuchle, *Chem. Phys. Lett.* **229**, 309 (1994).
45. F. Jelezko, B. Lounis, and M. Orrit, *J. Chem. Phys.* **107**, 1692 (1997).
46. Ph. D. Reilly, and J. L. Skinner, *J. Chem. Phys.* **102**, 1540 (1995).
47. F. Kulzer, S. Kummer, R. Matzke, C. Bräuchle, and Th. Basché, *Nature* **387**, 688 (1997).
48. H. Cailleau, F. Moussa, C. M. E. Zeyen, and J. Bouillot, *Sol. State Comm.* **33**, 407 (1980).
49. J. Etrillard, J. C. Lasjaunias, B. Toudic, and H. Cailleau, *Europhys. Lett.* **38**, 347 (1997).
50. P. J. Walla, F. Jelezko, Ph. Tamarat, B. Lounis, and M. Orrit, *Chem. Phys.* **233**, 117 (1998)
51. W. E. Moerner, T. Plakhotnik, T. Irngartinger, M. Croci, V. Palm, and U. P. Wild, *J. Phys. Chem.* **98**, 7382 (1994).
52. K. Palewska, J. Lipinski, J. Sworakowski, J. Sepiol, H. Gygax, E. C. Meister, and U. P. Wild, *J. Phys. Chem.* **99**, 16835 (1995).
53. M. Vacha, Y. Liu, H. Nakatsuka, and T. Tani, *J. Chem. Phys.* **106**, 8324 (1997).
54. R. Brown, and M. Orrit, in T. Basché, W. E. Moerner, M. Orrit, and U. P. Wild, eds., *Single Molecule Optical Detection, Imaging and Spectroscopy*, VCH, Weinheim, 1997, Chapter 1.4.
55. W. A. Phillips, *Amorphous Solids: Low Temperature Properties*, Springer, Berlin, 1981.

56. A. Zumbusch, L. Fleury, R. Brown, J. Bernard, and M. Orrit, *Phys. Rev. Lett.* **70**, 3584 (1993).
57. H. de Vries, and D. A. Wiersma, *J. Chem. Phys.* **70**, 5807 (1979).
58. Th. Basché, S. Kummer, and Ch. Bräuchle, *Nature* **373**, 132 (1995).
59. R. Brown, J. Wrachtrup, M. Orrit, J. Bernard, and C. von Borczyskowski, *J. Chem. Phys.* **100**, 7182 (1994).
60. M. Vogel, A. Gruber, J. Wrachtrup, and C. von Borczyskowski, *J. Phys. Chem.* **99**, 14915 (1995).
61. J. Köhler, J. A. J. M. Disselhorst, M. C. J. M. Donckers, E. J. J. Groenen, J. Schmidt, and W. E. Moerner, *Nature* **363**, 242 (1993).
62. J. Wrachtrup, C. von Borczyskowski, J. Bernard, M. Orrit, and R. Brown, *Nature* **363**, 244 (1993).
63. A. C. J. Brouwer, E. J. J. Groenen, and J. Schmidt, *Phys. Rev. Lett*, **80**, 3944 (1998).
64. J. Köhler, A. C. J. Brouwer, E. J. J. Groenen, and J. Schmidt, *Chem. Phys. Lett.* **228**, 47 (1994).
65. J. Wrachtrup., A. Gruber, L. Fleury, and C. von Borczyskowski, *Chem. Phys. Lett.* **267**, 179 (1997).
66. Ph. Tamarat, B. Lounis, J. Bernard, M. Orrit, S. Kummer, R. Kettner, S. Mais, and Th. Basché, *Phys. Rev. Lett.* **75**, 1514 (1995).
67. B. Lounis, F. Jelezko, and M. Orrit, *Phys. Rev. Lett.* **78**, 3673 (1997).
68. Th. Basché, W. E. Moerner, M. Orrit, and H. Talon, *Phys. Rev. Lett.* **69**, 1516 (1992).
69. S. Kummer., S. Mais, and Th. Basché, *J. Phys. Chem.* **99**, 17078 (1995).
70. C. Cohen-Tannoudji, J. Dupont-Roc, and G. Grynberg, *Atom-Photon Interactions*, Wiley, New York, 1992.
71. Ch. Brunel, B. Lounis, Ph. Tamarat, and M. Orrit, *Phys. Rev. Lett*, **81**, 2679 (1998).
72. M. Johnson, K. Orth, J. Friedrich, and H. P. Trommsdorff, *J. Chem. Phys.* **105**, 9762 (1996).
73. H. Bach, A. Renn, and U. P. Wild, *Chem. Phys. Lett.* **266** 317 (1997).
74. M. Pirotta, A. Renn, M. H. V. Werts, and U. P. Wild, *Chem. Phys. Lett.* **250**, 576 (1996).
75. M. Croci, V. Palm, and U. P. Wild, *Mol. Cryst. Liq. Cryst.* **283**, 137 (1996).
76. S. Kummer, C. Bräuchle, and Th. Basché, *Mol. Cryst. Liq. Cryst.* **283**, 255 (1996).
77. M. Orrit, J. Bernard, R. Brown, and B. Lounis, in E. Wolf, ed., *Progress in Optics XXXV*, Elsevier, 1996, p. 61.

CHAPTER

16

HIGH-RESOLUTION LUMINESCENCE EXCITATION–EMISSION MATRICES

JOHANNES W. HOFSTRAAT

Philips Research, Department of Polymers and Organic Chemistry, Prof. Holstlaan 4, NL-5656 AA Eindhoven, The Netherlands, and University of Amsterdam, Institute of Molecular Chemistry, Molecular Photonics Group, Nieuwe Achtergracht 129, NL-1018 WS Amsterdam, The Netherlands

URS P. WILD

Swiss Federal Institute of Technology, Physical Chemistry Laboratory, ETH-Zentrum, CH-8092 Zürich, Switzerland

16.1. INTRODUCTION

Conventional, room-temperature, electronic absorption and emission spectra of molecules yield broad and structureless bands. Hence such spectra do not give much detailed information on these compounds, so that they are of little use for their identification and accurate quantification in analytical chemistry, or for the investigation of their structure in spectroscopic studies. In particular, for polycyclic aromatic compounds (PACs), an important class of environmental contaminants, it is necessary to be able to perform the determinations with a high degree of specificity: strongly related structures—with very similar electronic spectra at room temperature—may show completely different toxicological effects. For many studies dealing with structure, interaction, and dynamics of molecules in relevant environments (e.g., in biochemical or physical studies), the use of techniques based on electronic spectra is limited due to the lacking specificity of the information. The large bandwidths of the transitions (several hundreds or even thousands of cm^{-1}) preclude the observation of vibrational fine structure, which would be able to reveal subtle structural changes of the molecules. Moreover, they make the detection of spectral shifts—indicative of changes in interaction of the molecular electronic states with their immediate surroundings—much more difficult.

The reason for this lack of spectral resolution lies in a number of processes that lead to the broadening of the spectral bands in electronic spectra. For

Shpol'skii Spectroscopy and Other Site-Selection Methods, Edited by Cees Gooijer, Freek Ariese, and Johannes W. Hofstraat.
ISBN 0-471-24508-9

molecules in solution the main cause for spectral broadening is in the rapid Brownian motion of the chromophores—even on the limited time scale of the fluorescence process—which leads to an averaging of the energies of the vibronic ground and excited states. By cooling the solution or by increasing the viscosity of the matrix (e.g., by embedding the chromophores in a polymer host), the motions can be slowed down, but even for molecules in cryogenic solution generally broad bands are observed. The reason is that in most matrices the chromophores experience a variety of different environments, all leading to different energies of the vibronic ground and excited states and hence to statistically broadened band shapes in the spectrum. Only by homogenization of the matrix, that is, by providing a well-defined solvent cage for the chromophore, narrow bands may be observed. An example of such a well-defined matrix is the crystalline phase that is provided by solidified *n*-alkanes, commonly referred to as the Shpol'skii matrix. Details on the structure and peculiarities of Shpol'skii matrices can be found in the chapters by Renge and Wild (Chapter 2) and by Lamotte (Chapter 3) in this volume. A general introduction on high-resolution spectroscopy can be found in the chapter by Hofstraat et al., also published in the Wiley Chemical Analysis series [1].

The use of low-temperature luminescence methods enables the measurement of highly resolved spectra allowing for specific identification. Via the so-called Shpol'skii method, named after its Russian inventor, Evgenii Shpol'skii, vibrationally resolved spectra are obtained with fingerprint specificity like infrared or Raman spectra, but obtained with the same level of sensitivity as fluorescence spectra. Particularly for the determination of the generally strongly fluorescent PACs, such methods have proven to be very useful. In contrast to laser-site-selection (or fluorescence-line-narrowing) techniques, which are also frequently applied and will be extensively discussed below, where selective *optical* excitation results in the formation of a well-defined subset of excited molecules, the vibrationally resolved fluorescence spectra in Shpol'skii spectroscopy are solely attributable to a homogenization of the local structure of the host material. Therefore, both absorption (excitation) and emission spectra show narrow lines. The highly resolved character of the excitation spectrum has been exploited for further selectivity and sensitivity enhancement in "laser-excited Shpol'skii spectroscopy" [2, 3].

The ultimate specificity is obtained by considering both the excitation and the emission spectra at the same time. This approach is taken when the excitation–emission matrix (EEM) is considered. The EEM, also referred to as total luminescence spectrum (TLS) or two-dimensional spectrum (TDS), comprises all information from both the excitation and the emission spectra of the fluorescent molecules.

The EEM also serves other purposes. Since an overall view is obtained of both the excitation and emission properties of the luminophores in a sample, straightforward and easily interpretable information is obtained on all effects that are related to energy selection, that is, effects observed in the luminescence spectra as a result of the energy used to bring the system into the excited state. An obvious example is in fluorescence line narrowing, where just the fact that an energetically well-defined excitation source is applied—which forms a narrow subset of excited chromophores, or "isochromat"—is the basis of the high-resolution emission spectrum obtained. Other examples are edge-excitation effects (in general leading to red shifts of the observed spectral bands), energy transfer or solvent reorientation effects, and spectral features related to the multisite structure of crystalline (e.g., Shpol'skii) spectra.

In this chapter first the principles and general applications of EEM will be described; subsequently several instrumental and experimental aspects will be discussed. The main part of the chapter will be devoted to a thorough evaluation of data obtained for low-temperature high-resolution spectra, obtained in both crystalline (Shpol'skii) and amorphous matrices (fluorescence-line-narrowing or site-selection spectroscopy).

16.2. EXCITATION–EMISSION MATRICES

Commonly, when luminescence spectra are recorded, either the emission spectrum is measured at a fixed excitation wavelength, or, alternatively, the excitation spectrum is obtained by monitoring a fixed emission wavelength. However, the emission and the excitation spectrum contain complementary information. In the EEM the full information present in luminescence spectra is exploited. It is obtained by measuring luminescence intensities for all kinds of combinations of luminescence emission and excitation frequencies in a certain interval:

$$I_{\text{lum}} = I_{\text{lum}}(\nu_{\text{ex}}, \nu_{\text{em}}) \tag{16.1}$$

The two- (or rather three-, since obviously the intensity is the third variable) dimensional information usually is represented in isointensity plots, in which points of equal intensity are connected in a graph with the excitation and emission wavelength or frequency as axes. Sometimes also pseudo-3D plots are applied, in which the spectrum is viewed as a "mountain landscape" from a viewing angle that may be chosen so that the spectral details are most visible.

EEMs may just be used to derive the maximum of information from a sample, that is, for analytical purposes. Since the excitation and emission wavelengths may be scanned over a wide range, comprehensive information is obtained about the luminescent entities present. As such, EEMs have been applied in both environmental and pharmaceutical analysis. The work discussed in the bulk of the literature on this subject is mainly directed toward analytical applications and on systems measured at room temperature, that is, under conditions not suitable for the generation of highly resolved luminescence spectra. Therefore, no thorough discussion of these papers will be given in this chapter. General reviews have been published by Warner et al. [4] and Ndou and Warner [5], and with specific focus to analysis of PACs by Kershaw and Fetzer [6]. In biological applications EEMs appear to be very useful in the characterization of phytoplankton [7]. A particular problem in the interpretation of EEMs lies in the wealth of information they contain, which renders them not as easily interpretable as conventional luminescence spectra. Subtle changes in band shapes, which may indicate the presence of several compounds with similar photophysical properties, are not easily detected in the isointensity plots. Significant effort therefore has been invested in the development of chemometrical tools for the interpretation of the data. A recent example is the work of Roch [8], who reports the successful use of rank annihilation factor analysis to detect PACs in petroleum products. Various other methods of factor analysis have been reported as well, among others by Neal [9, 10] and by Wang et al. [11]. Principal component analysis was applied for the classification of algae species by Henrion et al. [12]. Since the spectral characteristics reflect the pigment composition of algae, which forms the basis for their classification (for instance: "green" or "blue-green" algae), such classification is a more straightforward task than the identification of individual environmental contaminants. All these approaches have in common that they try to extract from the EEM the number of compounds present in the sample and their individual spectral features.

An additional dimension has been introduced in the measurement of EEMs by McGown and co-workers: By applying phase-resolved detection of the fluorescence generated at different modulation frequencies not only the excitation and emission properties, but also the fluorescence lifetime of the analytes could be used for characterization (Millican and McGown [13]). The so-called PREEMs (Phase-resolved EEMs) were applied for the study of humic substances, very important, but complicated, environmental constituents, as reported by, for instance, Mobed et al. [14].

Not many publications are found in which EEMs are used for photophysical studies. Since the main potential of EEMs, in our opinion, lies in their use for the study of energy-selection effects (i.e., in the study of excitation-dependent emission spectra), this observation is all the more surprising. One reason may be that the measurement of such effects would require sophisticated, not commercially available, equipment, since the effects are relatively small in liquid solutions. An example is the so-called edge-excitation red shift, observed when molecules are excited in the red side of the S_1 0–0 transition of their absorption spectrum. It mainly appears in polar solutions under conditions where the time needed for solvent reorientation is long compared to the lifetime of the excited state of the solute (Itoh and Azumi [15]). Obviously, the observed effects are strongly viscosity, and therefore also temperature, dependent (Itoh and Azumi [16]). Therefore, for samples in cryogenic solution the excitation dependency of emission spectra may be impressive. Since no major, Brownian, movements occur during the time scale of the luminescence process, energy-selection effects may be significant: the isochromat may be selected across a wide wave-number range, in particular, when the inhomogeneous broadening of the spectral bands is large in comparison to the homogeneous bandwidth of the individual chromophores, as is the case in most noncrystalline solids at low temperatures.

In this chapter, focus will be directed to the use of EEMs in low-temperature matrices in particular, since under such conditions highly resolved spectra may be obtained.

Depending on the nature of the solvent, roughly three types of EEMs can be distinguished, as schematically depicted in Figure 16.1 in the form of isointensity plots [17]. The EEM in Figure 16.1a is obtained for a fluorescent molecule in a liquid solution or in a completely amorphous matrix under nonselective, broad-band, excitation. For a compound obeying Kasha's rule and that emits from a completely relaxed state the EEM can be represented as the simple product of the emission and excitation spectra obtained under conventional conditions:

$$I(\nu_{ex}, \nu_{em}) = kS_{em}(\nu_{em})S_{ex}(\nu_{ex}) \tag{16.2}$$

where $S_{em}(\nu_{em})$ and $S_{ex}(\nu_{ex})$ represent the one-dimensional (1D) emission and excitation spectra, respectively. The 1D spectra are depicted along the sides of the EEM plot. Under these conditions the bandwidth of the spectra of the individual chromophores is of the same order of magnitude as the total bandwidth of the

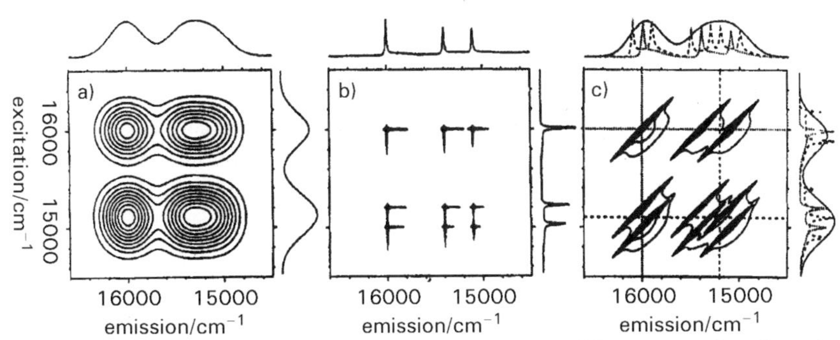

Figure 16.1. Schematic representation of the EEM of a dye in three different types of matrices: (a) a room-temperature spectrum for a liquid solution of the dye; (b), (c) FLN spectra for a crystal and for an amorphous solid doped with the same dye. The solid lines above and at the right-hand side of the EEMs represent the integrated 1D spectra of the EEM. The dotted lines correspond to simple "single-site" spectra (first excitation/emission region), the dashed lines represent "two-site spectra (second excitation–emission region). (Reprinted from [17], with permission from the Society of Applied Spectroscopy.)

ensemble, and no excitation-energy-dependent features are noted in the EEMs. In Figure 16.1b the EEM is shown for a molecule in a crystalline matrix, for example, a Shpol'skii-type matrix. The lines are narrow and exhibit a weak shoulder to the low-energy side in emission and to the high-energy side in excitation, resulting from coupling of the electronic transitions to the matrix vibrations (electron–phonon coupling). Again, the EEM can be represented by Eq. 16.2; the corresponding 1D spectra are shown at the side of the EEM. In this case the homogeneous bandwidth of the individual chromophores and the bandwidth of the ensemble are similar, as in the first type of EEM, but now they are both so narrow that vibrational features may be noted in the excitation and in the emission spectrum. The third type of EEM, shown in Figure 16.1c, is obtained when selective excitation is applied of molecules in an amorphous, cryogenic medium. The excitation wavelength determines the location and the shape of the—narrow-line—emission spectrum. Now no simple relation can be made between the conventional 1D spectra (which in fact consist of the summed contributions of many narrow line spectra) and the EEM, and Eq. 16.2 cannot be applied. On the basis of the EEM both the energy-selection process and the inhomogeneous broadening process (and hence the interaction between the chromophores and their direct surroundings) can be studied. EEMs therefore offer unique possibilities for the investigation of amorphous materials at cryogenic temperatures.

The application of low-temperature EEMs has been reported mainly by Wild and co-workers, both in combination with Shpol'skii matrices and with selective excitation approaches. A review paper on—in particular, low-temperature—total luminescence spectroscopy has been published by Kallir et al. [18].

16.3. EXPERIMENTAL ASPECTS

In principle, EEMs can be recorded with standard luminescence spectrometers, provided they are equipped with advanced data acquisition and data processing software. At present, many of the more expensive luminescence spectrometers have the possibility to acquire EEMs. Several authors have published inexpensive interfaces to enable the acquisition of EEMs on conventional spectrofluorometers (Rangnekar and Oldham [19]; Munoz de la Pena et al. [20]). A major drawback of the measurement of EEMs with scanning monochromators is the significant amount of time needed to acquire the full EEM: The matrix in fact is built up by performing a large number of repetitive scans. The amount of time needed can be significantly reduced by making use of parallel detectors, such as diode arrays or CCDs [21]). Recently, a laser-based fluorescence EEM instrument has been reported for field measurements. The system utilizes a Raman shifter in combination with the third or fourth harmonic of a Nd:YAG laser to scan the excitation wavelength and a diode-array detector for parallel detection of the emitted light [22, 23].

Conventional fluorometers are not suited for the measurement of vibrationally resolved, high-resolution EEMs, since they lack resolving power and/or sensitivity. To obtain a resolution on the order of $10\,cm^{-1}$, which is a minimum requirement for a vibrationally resolved spectrum, long-path-length monochromators with inherently low throughput have to be applied. Most EEMs reported in this chapter were recorded with a high-resolution, computer-controlled spectrometer, based on the original instrument reported by Vo-Dinh and Wild in 1973 [24], though significantly updated since then. A diagram of the system is shown in Figure 16.2. For excitation the output of a 2.5-kW high-pressure xenon lamp is passed first through a water filter and then through a SPEX 0.85-m double monochromator. The sample is contained in an Oxford Instruments CF204

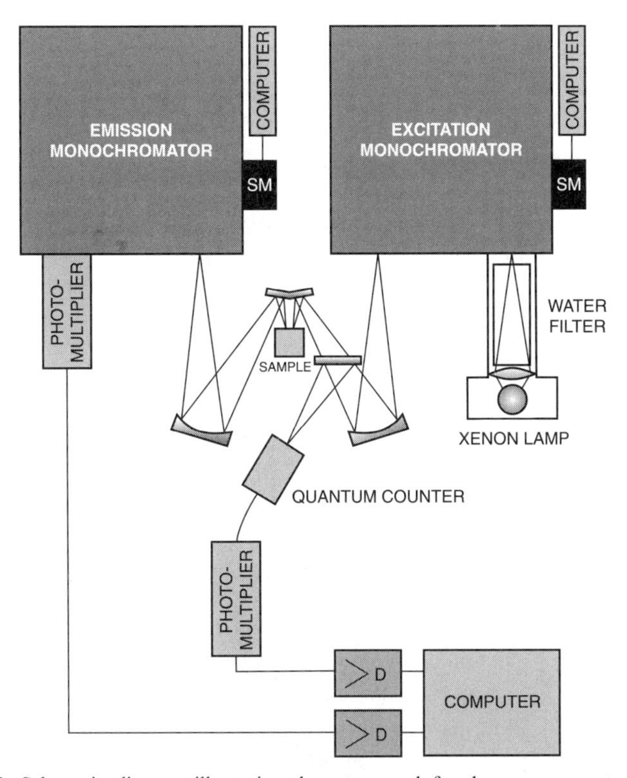

Figure 16.2. Schematic diagram illustrating the setup used for the measurement of the EEMs discussed in this chapter. The emission and excitation monochromator are identical, so that fully equivalent emission and excitation spectra may be obtained. SM, stepping motor; the whole experiment is fully computer controlled. The quantum counter corrects for the fluctuations in the intensity of the excitation light. (Reprinted from [18], with permission of Acta Physica Polonica.)

helium flow-through cryostat, which can be operated in the 3–300 K temperature range, or in a home-built liquid helium cryostat, operating at 4.2 K or lower (under reduced pressure). Front-face illumination and detection are applied. For detection, a SPEX 0.85-m double monochromator is also used for separation of the fluorescence light, followed by a RCA 31034 photomultiplier tube as single photon detector. The experiment is controlled by a modified DEC PDP11/73A computer that runs the monochromators and acquires the data points at equidistant points in wave numbers. For high-resolution EEMs it is crucial to record the spectra in energy rather than in wavelength units, to be able to evaluate the vibronic structure of the chromophores. The spectra are corrected for the wavelength dependence of the excitation light intensity using a concentrated solution of oxazine-1-perchlorate in ethanol as a quantum counter (in the 350–690 nm interval; see also Hofstraat and Latuhihin [25]). The emission light of the quantum counter solution is guided through an optical fiber to a Hamamatsu R928 photomultiplier tube, operated in single-photon-counting mode. The measurement of the quantum counter is also controlled by the computer. The resolution of the low-temperature spectra that have been obtained is mostly determined by the instrumental bandwidths of the excitation and emission monochromators, which for most experiments is 0.2–0.3 nm. For high-resolution spectroscopy this is a relatively large bandwidth, but it is required due to the low throughput of the high-resolution monochromators. Typical high-resolution EEMs of about 30,000 data points are recorded in about 12 h, with a 1 s sampling time per point. The recording time can be substantially reduced by application of parallel detection, which can be applied in combination with high-resolution luminescence spectroscopy, as has been demonstrated by Hofstraat et al. [26].

16.4. EEMS AND FLUORESCENCE-LINE-NARROWING SPECTROSCOPY

16.4.1. Fluorescence-Line-Narrowing Spectroscopy

In fluorescence-line-narrowing (FLN) spectroscopy, narrow-band excitation is employed to select a subset of molecules that are promoted into the excited state: Only those molecules that have an energy difference between ground and excited state that coincides with the narrow-banded excitation energy will be excited. If the energy-selected subset remains intact in the excited state, the emission spectrum also will show narrow lines. This condition first of all requires that there be a strong energetic correlation between the state into which the molecule has been excited and the state from which the molecule emits. In general, therefore, no narrow phosphorescence spectrum is observed unless the excitation was directly into the T_1–S_0 transition (an exception has been reported by Suter et al. [27] for some 1-indanones). A second requirement is that the sample is kept at very low temperatures, in order to prevent appreciable solvent reorientation on the time scale of the excited-state lifetime. Under such circumstances the observed transitions may be described in terms of zero-phonon lines and phonon wings (Rebane [28], see also Chapter 8). Furthermore, interactions between the guest molecules and the host matrix should not be too strong, so that the narrow zero-phonon lines dominate the spectrum. Finally, charge-transfer processes (both intra- and intermolecularly) or energy-transfer processes (mostly only between guest molecules, since the host matrix in general is transparent at the wavelengths employed) must be absent, since they will alter the originally selected isochromat.

The FLN technique has as a main advantage over other low-temperature high-resolution techniques that it can be applied to many types of molecules, both polar and apolar and even ionic, in a variety of matrices, as detailed elsewhere in this volume (Chapters 8, 10, 11). Application of FLN has been reported for, among others, PACs, polar (hydroxy-substituted) metabolites of PACs, large molecules of biological interest like chlorophylls and porphyrins, ionic dye molecules, etc. As matrix materials glasses, TLC plates, polymers, filter paper, and membranes have been employed. Even for molecules in very complicated systems as proteins (see Chapter 12) and for molecules adducted to DNA (see Chapter 11), FLN spectra have been obtained. The FLN technique hence enables one to obtain very specific information on the structure of molecules and interactions between molecules and their surroundings under a variety of experimental circumstances.

16.4.2. EEMs in Fluorescence-Line-Narrowing Spectroscopy

16.4.2.1. Introduction

By making use of EEMs, a lot of additional information can be obtained, in particular on the excitation wavelength dependence of the FLN spectra. The excitation wavenumber is crucial for the appearance of the FLN spectrum that is obtained, since it determines the actual set of molecules that is brought into the excited state. The narrow distribution of energies is able to excite only those molecules with a transition energy that matches the photon energy of the exciting light, according to the "energy-selection" principle (see also Chapter 8 by Jankowiak in this volume). In many publications reference is also made to the term *site selection* when describing the FLN technique. It should be noted that the term *site* does not have a physical meaning, since many molecules that occupy different microenvironments may have the same transition energy. Several publications have appeared that discuss the influence of the excitation wavenumber on 1D-FLN spectra [29–33]. Three excitation regions are discerned: purely electronic excitation to the S_1 state, excitation to the lower vibronic levels of the

S_1 state, and, finally, excitation to higher vibronic levels of the S_1 state or to higher electronic states.

In the first excitation region the excitation source probes the inhomogeneously broadened 0–0 region of the guest molecule (see Fig. 16.3a), at the red end of the absorption spectrum. In this region the energy of the exciting source is just sufficient to excite one subset of molecules, with the laser excitation energy also being the origin of the FLN emission spectrum: Each maximum observed in the spectrum corresponds to a specific transition $S_1(v' = 0) \rightarrow S_0(v'')$, with v'' representing a vibrational state in the ground state S_0. In the FLN emission spectrum only vibronic emissions can be observed since the 0–0 transition coincides with the intense scatter of the excitation light; only when time-gated detection of the EEM is employed, the 0–0 transition can be observed [34]. When

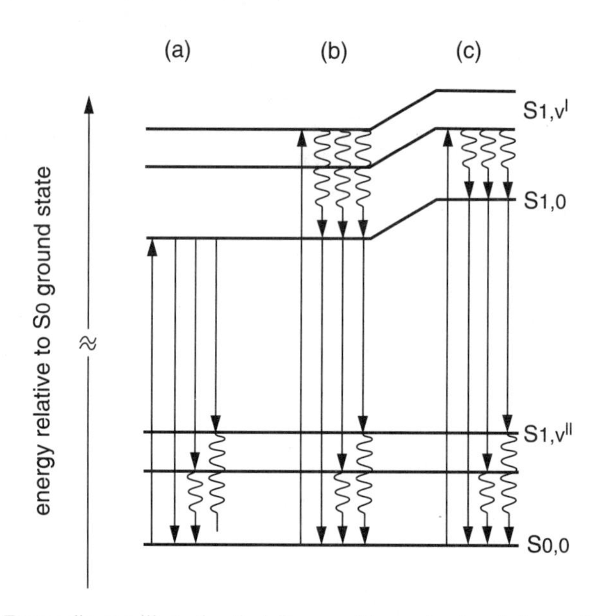

Figure 16.3. Energy diagram illustrating the influence of the mode of excitation on the appearance of the FLN spectrum; (a, b) show a molecule at the red end of the inhomogenous absorption band. Excitation with (a) low energy leads to a "single-site" spectrum (cf. dotted emission spectrum of Fig. 16.1c and the emission spectrum shown in Fig. 16.6a). Higher energy excitation into vibrationally excited S_1 states (b, c) leads to a multisite emission spectrum, if the inhomogeneous broadening is large in comparison to the difference in vibrational energy between adjacent bands (see also the dashed lines in Fig. 16.1c and the emission spectrum shown in Fig. 16.6b). (c) shows a molecule with a higher 0–0 transition energy than the molecule depicted in (a) and (b). Excitation of the molecule with the low 0–0 transition energy into a higher $S_1(v'_2 = 1)$ state as depicted in (b), may lead to a simultaneous excitation of the molecule with a higher-energy 0–0 transition into a lower vibrational level $S_1(v' = 1)$, as shown in (c)! (Reprinted from [17], with permission from the Society of Applied Spectroscopy.)

the excitation wavenumber is changed, the positions of the lines in the FLN spectrum change accordingly: The spectrum as a whole shifts linearly with the excitation energy (see Fig. 16.1c).

In the second region the excitation energy has been increased, so that now more than one subset of molecules, with the same $S_1(v') \leftarrow S_0(v'' = 0)$ transition energy, but each with different $S_1(v' = 0) \rightarrow S_0(v'' = 0)$ transition energies, is created. In general, the second region starts when the lower vibronic region of the molecular S_1 state is probed (see Fig. 16.3b). Here, the excited molecules first lose their excess energy radiationlessly in proceeding to the vibrational ground state of the excited electronic state. During this process the energy-selected population remains preserved, the initial and the final excited-state being coupled by an energetically well-defined vibrational quantum. The FLN spectrum shows its 0–0 transition at an energetic distance of exactly the excited state vibrational quantum away from the excitation wavenumber. Again, upon changing the excitation wavenumber, the complete FLN spectrum shifts by the same amount. As more than one isochromat has been formed in the $S_1(v' = 0)$ state, the total emission spectrum is composed of contributions of the individual emission spectra of all isochromats that have been excited. The total emission spectrum therefore shows a multiplet structure with the individual sets of spectra displaced with respect to each other by the difference in vibrational energies of the corresponding excited states. The larger the inhomogeneous broadening, the larger the density of the vibronic bands in the probed region. Upon excitation into the lower S_1 vibronic region, two situations can be discerned. When excitation energies up to about 400–500 cm^{-1} above the 0–0 transition are employed, in general a region is reached where the vibronic density of states is low. Hence in most cases only one or two inhomogeneously broadened bands are reached, and a simple FLN spectrum emerges. Note that when the inhomogeneous broadening is appreciable, it is also possible that at the same time the 0–0 transition and a vibronic band are probed. In the region up to about 2000 cm^{-1} above the 0–0 transition region the density of the vibronic states is much higher than for the lower energies. In this region it is very likely that several inhomogeneously broadened vibronic bands are probed with the same excitation energy. Then a complicated multiplet structure is observed in the spectrum, with all the 0–0 lines separated from the excitation wave number by the energy of their corresponding excited-state vibration. Besides, it is noted that in some analytical studies these vibrations especially are very important to distinguish between compounds of very similar structure (see Chapter 11 by Ariese and Jankowiak and Chapter 10 by Gooijer and Kok in this volume).

In the third region the excitation energy matches highly energetic S_1 vibronic transitions and/or higher excited electronic states. Here the density of vibrational states becomes high and at the same time the natural, homogeneous, linewidth becomes large due to the very short lifetime of the probed transitions. As a consequence no highly resolved FLN spectra are obtained for excitation in this region.

16.4.2.2. Room-Temperature EEM of Oxazine-4 in PVB Film

Figure 16.4 shows the EEMs obtained for the ionic dye oxazine-4-perchlorate (further abbreviated to oxazine-4) in polyvinylbutyral (PVB) polymer film (see [17] for further details). The spectrum on the left shows that obtained at room temperature. The spectrum, as expected, does not exhibit vibrational fine structure as a result of strong spectral broadening. The 0–0 transition, with excitation and emission maxima at 16040 and 15860 cm^{-1}, respectively, is partially invisible due to instrumental suppression of the stray light. In the EEM the stray light suppression results in a diagonal stripe across the figure, in which no isointensity lines are drawn. The region in the EEM to the left and above the blocked stripe is due to anti-Stokes fluorescence from thermally populated excited-state levels. Cuts taken at various positions along the emission axis and along the excitation axis show the same spectral features, as illustrated by the two cuts indicated in Figure 16.5. In Figure 16.5a the 1D-excitation spectrum recorded at an emission wavelength of 15900 cm^{-1} is shown, with two maxima, the 0–0 transition at 16040 cm^{-1} and another maximum at 16580 cm^{-1}, and, furthermore, a prominent shoulder at 17300 cm^{-1}. In Figure 16.5b the 1D-emission spectrum obtained for excitation at 17000 cm^{-1} is depicted; it only shows one maximum, the 0–0 transition at 15860 cm^{-1}. The EEM clearly shows that the room-temperature emission spectrum of oxazine-4 indeed is independent of the excitation process, and hence stems from a fully equilibrated state. The room-temperature EEM therefore can be described by Eq. 16.2, to a good approximation.

16.4.2.3. Low-Temperature EEM of Oxazine-4 in PVB Film

FLN Emission Spectra. On the right-hand side of Figure 16.4 the spectrum of oxazine-4 in PVB recorded at 8 K under selective excitation conditions is shown. The FLN spectra, being induced by selective excitation of molecules in low-temperature matrices, show strong dependence on the excitation wavenumber. It is clear that now no symmetric and simple EEM is obtained. The figure is dominated in the upper part by slanting, ellipsoidal features that run parallel to the blocked stripe resulting from the stray light rejection procedure. Because of the low temperature at which the EEM was recorded, no emission is observed in the anti-Stokes region. The origin of the slanting features was discussed above (cf. Fig. 16.1c).

When the HREEM displayed in Figure 16.4b is considered, the general dependence of the FLN spectrum on excitation wave number can be easily verified. If the excitation wavenumber is smaller than 16,000 cm^{-1}, only the 0–0

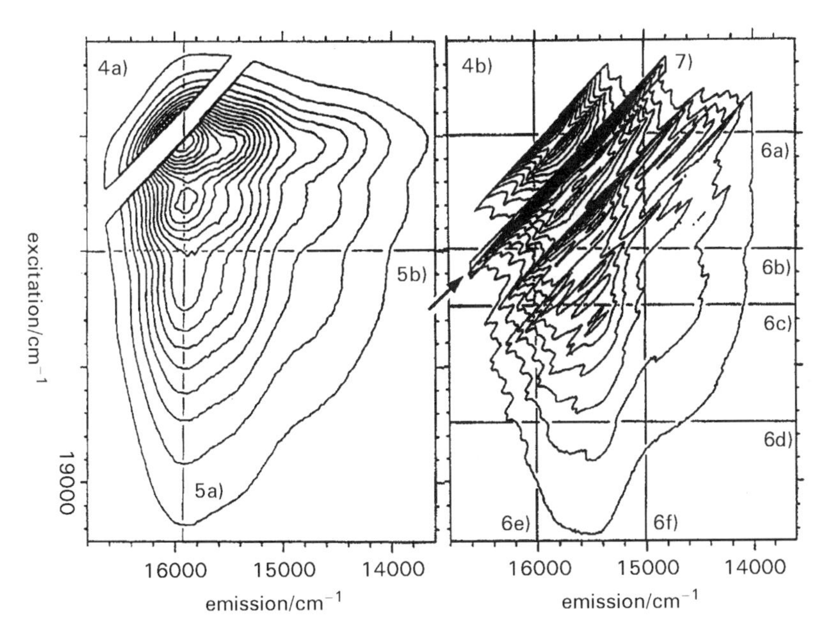

Figure 16.4. Contour plots of the EEM of oxazine-4-perchlorate in polyvinylbutyral film, obtained at room temperature (a) and at 8 K (b). The resolution was 30 cm^{-1} for the room-temperature EEM, and 20 cm^{-1} for the low-temperature EEM and was limited by the slits used for recording the EEM. The straight lines in the spectra indicate the positions of the cuts represented in Figures 16.5 and 16.6. (b) The parallelogram is also indicated (see arrow), as shown in more detail in Figure 16.7. (Reprinted from [17], with permission from the Society of Applied Spectroscopy.)

transition region of oxazine-4 is excited, and a "single-site" spectrum is obtained. When the excitation wavenumber is increased, the structure of the FLN spectra becomes more complicated, since more vibronic levels are involved in the energy-selection process. Above 18,000 cm^{-1} the ellipsoids become blurred, indicating that the FLN effect is not obtained any more, because of the many overlapping vibronic absorption bands in resonance with the excitation wavenumber.

The observations are supported by taking cuts through the EEM, both in the emission and in the excitation direction. The cuts are displayed in Figure 16.6. In Figure 16.6a–d emission spectra are shown obtained for increasing excitation wavenumbers. Figure 16.6a shows a simple, "single-site" emission spectrum, the excitation wavenumber coinciding with the 0–0 transition, at 16,000 cm^{-1}. The main peak observed in the spectrum belongs to a groundstate vibronic emission at about 585 cm^{-1} from the origin. Figure 16.6b shows the emission spectrum obtained for an excitation wavenumber of 17,000 cm^{-1}. The complicated spectrum is dominated by two peaks at 16,424 and 15,840 cm^{-1}. The first peak, at 576 cm^{-1} below the excitation wavenumber, can be attributed to the

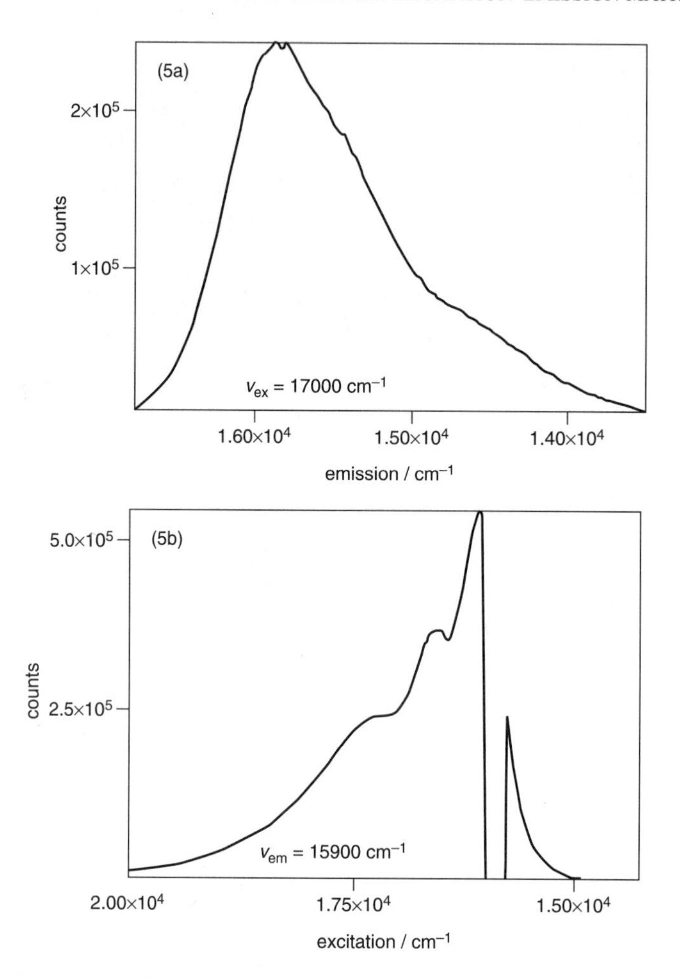

Figure 16.5. Cuts through the EEM, of Figure 16.4a, recorded for oxazine-4 in polyvinylbutyral at room temperature: (a) The excitation spectrum moniored at 15,900 cm^{-1}. (b) The emission spectrum recorded with an excitation wavenumber of 17,000 cm^{-1}. The positions of the two 1D scans are indicated in Figure 16.4a by straight lines. (Reprinted from [17], with permission from the Society of Applied Spectroscopy.)

0–0 emission derived from selective excitation in the region of a strong vibronic excited-state mode with the same wave number. The second peak, situated at 1160 cm^{-1} below the excitation wave number, is assigned to a superposition of two emitting subsets. The molecules of the first subset are excited to the 576-cm^{-1} vibrational mode in S_1 and emit, after relaxation to the vibrational ground state of S_1, to the 585-cm^{-1} vibrational mode of S_0. The molecules of the second subset, which emit in the same region, are due to excitation into the 1157-cm^{-1} vibrational mode of S_1 (the first overtone of the 576-cm^{-1} mode), with a corresponding emission to the vibrational ground state of S_0. The emission spectrum at an excitation wave number of 17,500 cm^{-1} (Fig. 16.6c) is even more complex than the spectrum shown in Figure 16.6b. A loss of fine structure and a further shift in intensity to the red side of the emission spectrum are observed. The two most prominent bands, at 15,975 and 15,894 cm^{-1}, are assigned to 0–0 emissions of subsets of molecules that have been excited into the 1519 and 1602-cm^{-1} vibronic modes of S_1, respectively. Finally, the emission spectrum of Figure 16.6d, obtained for excitation at 18,500 cm^{-1}, shows an almost complete loss of resolution. The only sharp feature, observed at 16,311 cm^{-1}, can be attributed to the 0–0 emission due to selective excitation in the 2178-cm^{-1} mode of S_1 (i.e., a combination band of the 576 and 1602-cm^{-1} excited-state vibrational modes).

FLN Excitation Spectra. Also two narrow-line excitation spectra are depicted in Figure 16.6. The excitation spectra show vibrational resolution just like the emission spectra, provided that only a very narrow spectral bandpass is used to monitor the emission light. Similar observations are made for the monitoring wavenumber dependency of the excitation spectra, as have been presented above for the emission FLN spectra. For excitation spectra, the more the monitoring wavenumber is shifted to the blue, the simpler the excitation spectrum that is obtained. Figure 16.6e shows the excitation spectrum that is obtained for a monitoring wavenumber of 16,000 cm^{-1}. In this case the most intense transition that is observed besides the 0–0 excitation corresponds to the 0-576 cm^{-1} excitation to S_1. The phonon wing of the $S_1(v' = 0) \leftarrow S_0(v'' = 0)$ excitation can just be discerned at the red end of the spectrum for excitation wavenumbers below 16,500 cm^{-1}. The excitation spectrum monitored at 15,000 cm^{-1} (Fig. 16.6f) shows emission to different vibrational states of S_0, so that a more complicated spectrum is obtained. The main peak in this spectrum is at 15,591 cm^{-1} and is due to a 0–0 excitation whose vibronic emission band at 591 cm^{-1} is monitored. As the emission wavenumber is smaller than any 0–0 transition of the $S_1 \leftarrow S_0$ excitation in the inhomogeneous distribution, no 0–0 emission is observed, and the excitation region below 15,591 cm^{-1} shows only dark counts of the photomultiplier. A further red shift of the monitoring wavenumber results in more vibronic emission transitions that are observed simultaneously, resulting in a loss of resolution of the excitation spectra, similar to what was discussed above for the emission FLN spectra.

Figures 16.6e,f clearly demonstrate that high-resolution FLN excitation spectra can be obtained by using energy-selective detection of the emitted light. FLN excitation spectra have as an important advantage over FLN emission spectra that no detrimental hole-burning effects occur. Hole burning causes a decrease of the narrow lines in the FLN emission spectrum under prolonged laser

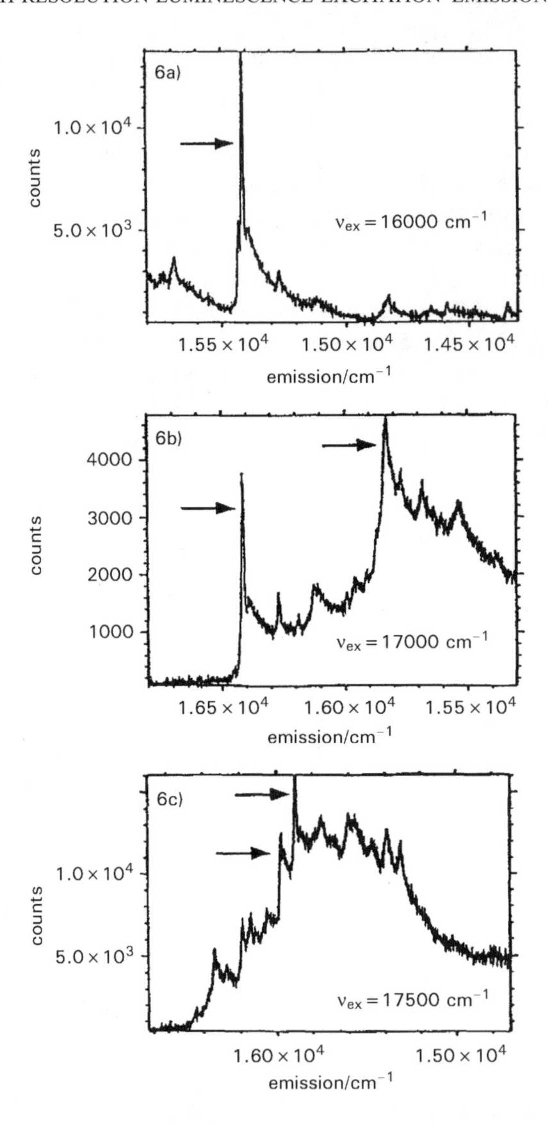

Figure 16.6. One-dimensional FLN emission (a–d) and excitation (e,f) spectra of oxazine-4 in polyvinylbutyral recorded at about 8 K. The spectral resolution was 6 cm^{-1}. (a–d) Emission spectra obtained for excitation wave numbers of 16,000, 17,000, 17,500 and 18,500 cm^{-1}, respectively. (e,f) Excitation spectra obtained while monitoring at an emission wavenumber of 16,000 and 15,000 cm^{-1}, respectively. The arrows in the spectra point to peaks discussed in the text. The positions of the 1D scans is visualized in Figure 16.4a by straight lines. (Reprinted from [17], with permission from the Society of Applied Spectroscopy.)

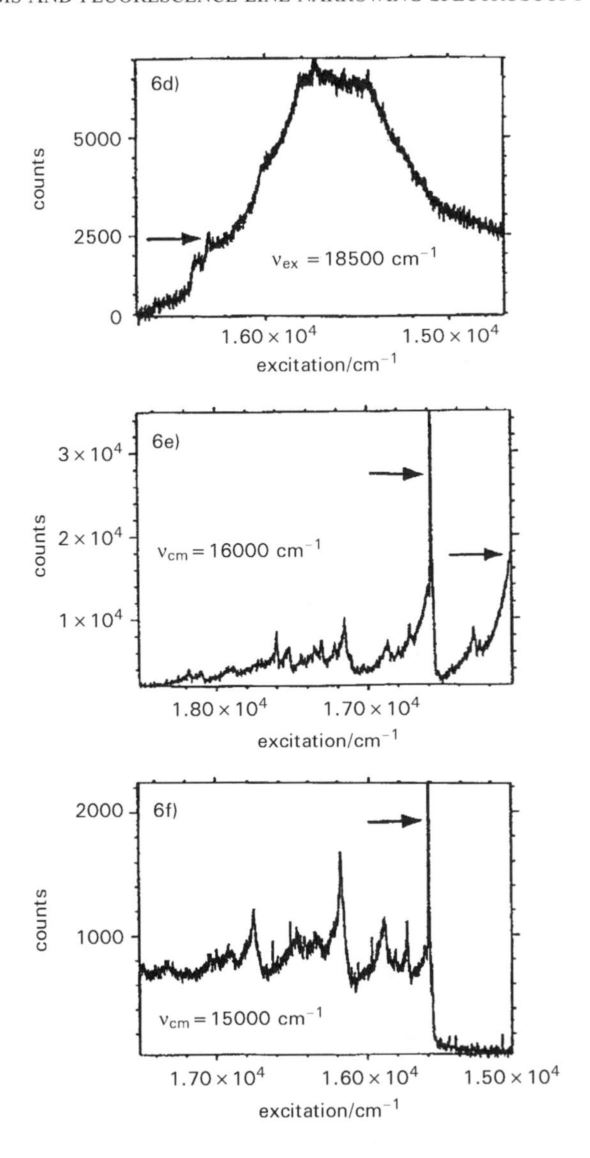

Figure 16.6. (*continued*)

irradiation. If excitation spectra are recorded, the excitation wavenumber changes continuously while scanning, so that hole-burning effects will be absent, or at least strongly reduced, provided that the rate of hole burning is slower than the scan speed.

For oxazine-4 in PVB the interpretation of the FLN spectra is complicated due to the strong inhomogeneous broadening of the spectral bands. For the 0–0 transition an inhomogeneous bandwidth of 550 cm^{-1} was estimated by measuring the full width at half-maximum (FWHM) of a synchronous cut in the EEM, where the excitation wave number of each data point is given by $\sigma_{ex} = \sigma_{em} + 100\,cm^{-1}$ (i.e., just outside the range blocked in the stray light rejection procedure). As a result of the broad inhomogeneous distribution, the FLN spectrum shows an intricate structure of narrow bands due to vibronic emission bands and vibronic excitation bands, which is difficult to unravel when only one-dimensional data are available. An additional complicating factor is that, for the compound concerned, the energies of the vibrational modes in the ground state and in the first excited state are very similar. By examination of the HREEM, however, the vibronic structure of both ground and excited state can be easily interpreted. Also, on the basis of the HREEM excitation and emission, wavenumbers can be identified for which simple, "single-site" emission and excitation spectra are obtained. This is in general the case when large emission wavenumbers (for excitation spectra) or small excitation wavenumbers (for emission spectra) are applied. On the basis of the energy-selection process this can be easily understood: small excitation wavenumbers probe the lower part of the energy distribution, so that the chances of exciting molecules into their vibronic levels are minimized. Similarly, for large emission wavenumbers only the upper part of the energy distribution is monitored.

In the HREEM the vibronic structure of the oxazine-4 molecule can be evaluated across the inhomogeneous band, that is, insight can be obtained into the interaction between the molecule and the matrix for a wide range of isochromats. Figure 16.7 shows a 2D-synchronous scan obtained by taking a number of consecutive scans of the excitation monochromator in the range 615–550 cm^{-1} above the emission wavenumber across the full wavenumber range examined in the HREEM shown in Figure 16.4b. The scanned region is indicated as a parallelogram in Figure 16.4b, but it has been transformed to a rectangular HREEM in Figure 16.7, for convenience. The wavenumber range that is observed comprises the main vibronic features of the emission and excitation spectrum, at about 585 and 576 cm^{-1}, respectively. According to Miller et al. [35], the strong band is attributed to in-plane skeletal deformation of the three-ring system corresponding to the phenoxazine skeleton, γ(CCC) and γ(CNC).

From Figure 16.7 it is clear that the vibrational frequencies decrease with increasing excitation energy. Emission wavenumbers below 15,600 cm^{-1} in the 2D-synchronous scan depicted in Figure 16.7 correspond to excitation wavenumbers below 16,215 cm^{-1}. Such an excitation energy is so low that the molecules can only be excited to the vibrational ground state of S_1. The features observed for emission wavenumbers below 15,600 cm^{-1} therefore can only be due to vibronic effects in the ground state. In this region the most intense vibronic

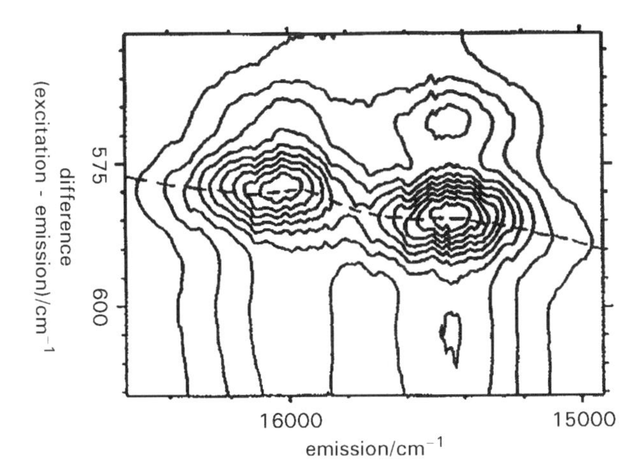

Figure 16.7. 2D-synchronous scan obtained for oxazine-4 in PVB at 6 K. The resolution in the excitation direction was 6 cm^{-1}, determined by the monochromator bandpass. The resolution in the emission direction was 8 cm^{-1}, determined by the monochromator bandpass and the step width. On the horizontal axis of the spectrum the emission wavenumbers are indicated. On the vertical axis the wavenumber difference between excitation and emission is given. The EEM was recorded by scanning the excitation monochromator between 615 and 550 cm^{-1} above the monitoring wavenumber of each subscan. The dashed line connects the maxima of these excitation spectra. The 2D-synchronous scan is indicated in Figure 16.4b as a parallelogram (see arrow in EEM). This parallelogram was distorted to the rectangle depicted here. The excitation–emission difference is enlarged by a factor of 15 as compared to the emission wavenumber scale. (Reprinted from [17], with permission from the Society of Applied Spectroscopy.)

band undergoes a wavenumber shift from 591 cm^{-1}, when emission is detected at 14,800 cm^{-1}, to 585-cm^{-1} for detection at 15,300 cm^{-1}. The vibrational mode remains at this energy up to an emission wavenumber of about 15,600 cm^{-1}. Then a new shift occurs, now due to the fact that the region is reached where both 0–0 and vibronic excitation may take place. The band shifts to 580 cm^{-1} at an emission energy of about 16,100 cm^{-1}. The 580-cm^{-1} mode represents an excited-state vibrational mode (now excitation is in the vibronic part of the excited state, and 0–0 emission is observed). When the excitation wavenumber is shifted further into the blue, the vibrational energy also shifts, to about 576 cm^{-1}, at an emission wave number of 16,600 cm^{-1}.

The observed shift of the ground-state 0–$\sim$585 cm^{-1} and the excited state 0–$\sim$578 cm^{-1} bands with excitation energy is interesting, since it implies that the vibrational energy of these modes in the ground state and in the excited state is not equal for all isochromats. Both in the ground and in the excited state a 5-cm^{-1} decrease is noted in the vibrational energy when the excitation wavenumber is blue shifted across the inhomogeneously broadened absorption band. The energy changes must be due to the selective excitation and detection of molecules with specific interactions with the matrix. The selective excitation of the oxazine-4 molecules in PVB therefore apparently creates an ensemble of molecules not only with a well-defined energy difference between ground and excited state, but also with a well-defined interaction with the environment. It is not so surprising that the 0–$\sim$580 cm^{-1} bands in oxazine-4 are so strongly influenced by the interaction with the matrix. The C–N–C skeletal deformation of the phenoxazine ring is influenced by interaction with polar, positively charged, molecules and in particular by H-bond formation (see, for instance, Van den Berg and Völker [36] for a detailed study of the related molecule resorufin). The observed effects are somewhat related to the well-known edge-excitation red shifts, where a red shift of the emission spectrum is noted upon excitation in the low-energy edge of the 0–0 absorption band of a chromophore. The edge-excitation effects are due to the fact that the low excitation energy results in the selection of a specific set of molecules with strong interaction with the (in general polar) matrix, and hence with a strong solvatation energy.

16.4.2.4. Solvent Relaxation of Oxazine-4 in 2-MTHF Glass

Solvent relaxation effects are particularly strong in the spectra of molecules with different dipole moments in the ground and excited states. Such molecules show a significant Stokes shift between the maxima of the absorption and emission band when they are dissolved in a polar solvent of low viscosity. The instantaneous change in the dipole moment upon excitation brings the solute–solvent system in a nonequilibrium state, which relaxes by rearrangement of the solvent shell, to optimize the interaction according to the new situation. The rearrangement can be followed by time-resolved fluorescence, provided that the solvent reorientation may take place during the lifetime of the excited state. Obviously, the viscosity of the solvent plays an important role in the reorientation process. In the glass-forming solvent 2-methyltetrahydrofuran, the relaxation time may be varied over many orders of magnitude by choosing the appropriate temperature. The effects of solvent reorientation may be observed in the EEM as well, as is obvious from Figure 16.8, displaying the EEMs of oxazine-4 in 2-MTHF. Contrary to the PVB matrix, discussed in the previous section, 2-MTHF is a liquid with low viscosity at room temperature. Upon cooling obviously the viscosity increases.

In Figure 16.8 EEMs are shown obtained at 135 and 117 K, respectively. Although the temperature difference is rather small, the two EEMs are totally different in appearance. Figure 16.8a ($T = 135$ K) shows no correlation between excitation and emission, just like the room-temperature spectrum of oxazine-4 in PVB. In Figure 16.8b, on the contrary, a strong correlation between excitation and emission is observed, indeed showing FLN features. Energy-selection features are found after $S_1(v' = 0) \leftarrow S_0(v'' = 0)$ excitation for both the $S_1(v' = 0) \rightarrow$

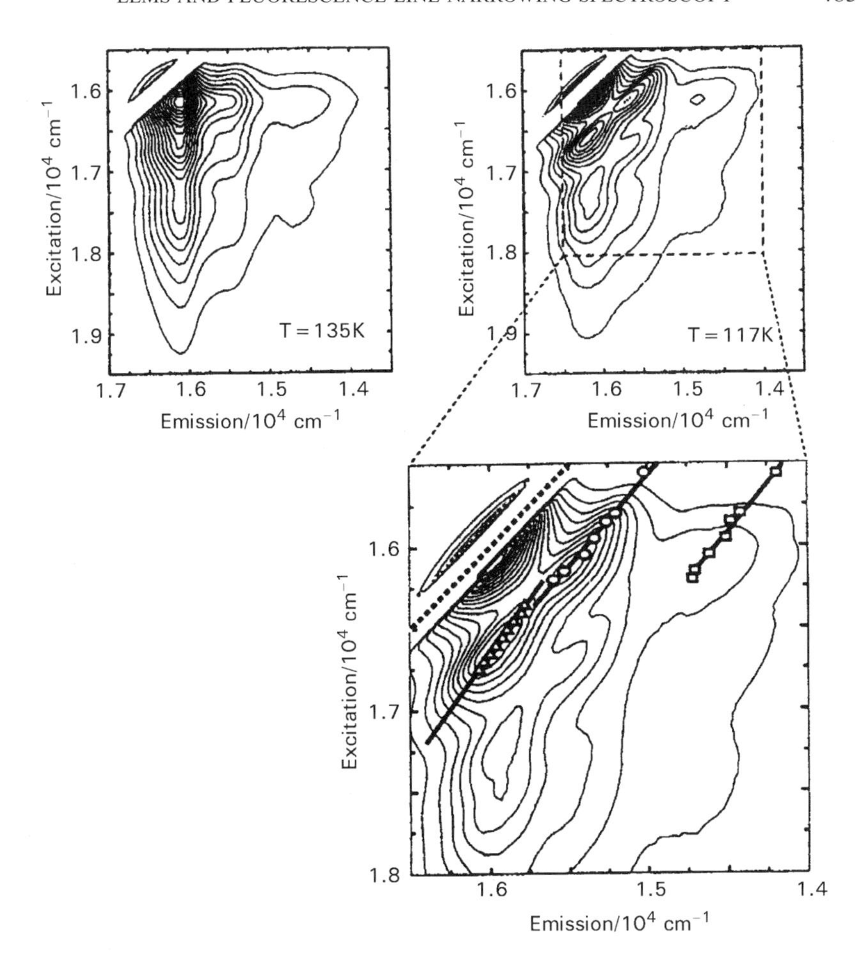

Figure 16.8. EEMs of oxazine-4 in 2-methyltetrahydrofurane at (a) $T = 135\,\text{K}$ (emission after full relaxation in the excited state) and (b) $T = 117\,\text{K}$ (energy selection is observed). For (b) a detailed view of the spectrum is also given to illustrate that the emission from $S_1 v' = 0)$ shows a strong dependence on excitation energy. The lines drawn through the maxima of the $S_1(v' = 0) \rightarrow S_0(v'' = 1)$ emission, indicated as circles, and the maxima of the $S_1(v' = 0) \rightarrow S_0(v'' = 2)$ emission, indicated as rectangles, have the same slope ($m = 1$) as the stray-light line, the dashed line in the figure. It is clear that the slope through the maxima of the $S_1 v' = 0) \rightarrow S_0(v'' = 0)$ emission (indicated as triangles), obtained after excitation into the first vibrational level of the excited electronic state, that is, $S_1(v' = 1) \leftarrow S_0(v'' = 0)$, has significantly increased ($m = 1.4$). After $S_1(v' = 1) \leftarrow S_0(v'' = 0)$ excitation, the dependence on the excitation energy is nearly smeared out. The observations indicate that excess vibrational energy accelerates the relaxation process. (From Ref. 37, with permission of Elsevier Science Publishers.)

$S_0(v'' = 0)$ emission (only partially visible due to the stray light rejection) and for the $S_1(v' = 0) \rightarrow S_0(v'' = 1)$ and $S_1(v' = 0) \rightarrow S_0(v'' = 2)$ emissions. In Figure 16.8b the emission maxima of the three emission processes are indicated in more detail: $S_1(v' = 0) \rightarrow S_0(v'' = 0)$ emission is indicated with a dashed line, the $S_1(v' = 0) \rightarrow S_0(v'' = 1)$ and $S_1(v' = 0) \rightarrow S_0(v'' = 2)$ emissions with circles and rectangles, respectively. The lines drawn through the emission maxima show a direct dependence on the excitation energy; that is, the slopes of the lines are $m = 1$. In the figure the $S_1(v' = 0) \rightarrow S_0(v'' = 0)$ emission maxima after $S_1(v' = 1) \leftarrow S_0(v'' = 0)$ excitation are also shown, which also show energy-selection features, but with a significantly higher slope ($m = 1.4$). The change in slope may be explained by the rapid conversion of vibrational energy into thermal energy (it is known that the dissipation of vibrational energy proceeds with a time constant of 10 ps, according to Scherer et al. [38]). As a result of the local heating, the relaxation process is accelerated, so that the excitation dependence of the emission maximum is less pronounced. The contours of the emission following excitation in the second vibronic level of S_1 suggest a further increase of the slope of the energy-selected emission maxima, indicating even more complete relaxation prior to emission. Further evidence for this interpretation of the effects observed in the EEMs was found in time-resolved fluorescence studies [37].

16.5. EEMS AND HOLE-BURNING SPECTROSCOPY

Persistent spectral hole burning is a phenomenon that was observed for the first time in organic glasses by Personov and co-workers (Kharlamov et al. [39, 40]), and independently by Rebane and co-workers (Gorokhovskii et al. [41]). Later work, mainly by the group of Small [42, 43], has demonstrated that the holes—the permanent reduction of fluorescence intensity due to intense, prolonged, laser irradiation—can be caused by two processes. The first process is called photochemical hole burning, and is due to selective photochemistry, leading to permanent reduction of the FLN intensity, since the photoselected chromophores are converted chemically and are—after relaxation to the ground state—no longer excited at the laser wavelength. The second process is referred to as nonphotochemical hole burning; it is attributed to changes in the solvent cage surrounding the photoselected chromophores when in the excited state. Once relaxed to the ground state the chromophores exhibit a small change of the energy gap between electronic ground and first excited state, so that they are no longer in resonance with the narrow laser excitation line. Nonphotochemical hole burning appears to be a rather general phenomenon for molecules in glasses. Since the hole burning results from a energy-selective process—only the excited molecules (i.e., the molecules with an energy difference between ground and excited state that exactly matches the excitation energy), may undergo hole burning—in principle, very narrow holes will be obtained. The hole width primarily is determined by the laser bandwidth and by the natural linewidth of the probed molecules. Hole-burning experiments can therefore be used to study very fast relaxation phenomena (see the chapter by Rätsep and Small in this volume, in which rapid photosynthetic processes are studied by means of hole-burning spectroscopy).

In FLN spectra hole burning leads to a gradual reduction of the intensity of the narrow zero-phonon lines when the sample is irradiated for a longer period of time [44, 45]. It is generally an unwanted phenomenon in FLN spectra, although hole burning may also be used in an advantageous way, to remove strong phonon wings, which may be subtracted because they do not undergo as strong hole burning as the narrow lines [26] (see also Chapter 8).

In HREEMs effects of persistent spectral hole burning may also be noted. To demonstrate what effects hole burning may have, the HREEM of oxazine-4 doped in a polyvinylbutyral film was investigated before and after irradiation with an argon-ion laser pumped dye laser, filled with Rhodamine 6G (power 5 mW, bandwidth about 1 cm^{-1}) at 16,109 cm^{-1}. The holes were examined in different ways: First of all a synchronous scan was made in the 0–0 region, at a wavenumber difference of about 590 cm^{-1}, that is, following the main vibronic emission as discussed above for Figure 16.7. The excitation monochromator was scanned from 15,360 to 17,000 cm^{-1}; the synchronous scan is shown in Figure 16.9a. The synchronous scan shows two main bands corresponding to excitation in the $S_1(v' = 0) \leftarrow S_0(v'' = 0)$ region (maximum at about 15,930 cm^{-1}) and in the $S_1(v' = 1) \leftarrow S_0(v'' = 0)$ region (maximum at about 16,530 cm^{-1}, i.e., about 600 cm^{-1} above the 0–0 transition), respectively. In both peaks a hole is discerned, slightly above the maximum wave number, at 16,109 and 16,694 cm^{-1}, corresponding to the 0–0 hole and the 0–585 (vibronic) hole, respectively. The two holes are shown in detail in Figures 16.9b and c.

In Figure 16.10 a 2D-synchronous scan is depicted, made by scanning the excitation monochromator in the region from 630 to 530 cm^{-1} above the emission wavenumber, which was scanned from 15,550 to 15,450 cm^{-1}. In other words, the excitation monochromator was scanned in the area of the $S_1(v' = 0) \leftarrow S_0(v'' = 0)$ transition, and the main vibronic emission band was observed. The HREEM in Figure 16.10a shows the region prior to burning. The narrow region of the zero-phonon line is clearly observed on the basis of the isointensity lines. The excitation maximum is about 590 cm^{-1} above the emission wavelength, reflecting the energy of the ground-state vibronic mode in this wave-number region. The HREEM in Figure 16.10b shows the same region after burning. The hole is clearly observed at an excitation wave number of 16,109 cm^{-1}, that is, the laser wavenumber used. Not only is the $S_1(v' = 0) \leftarrow S_0(v'' = 0)$ hole observed for the 0–593 cm^{-1} vibronic emission, but also another hole is visible that is shifted over about 17 cm^{-1}, corresponding to a 0–570 cm^{-1} vibronic mode in emission (see weak signal in Figure 16.7). At limited resolution only one vibronic

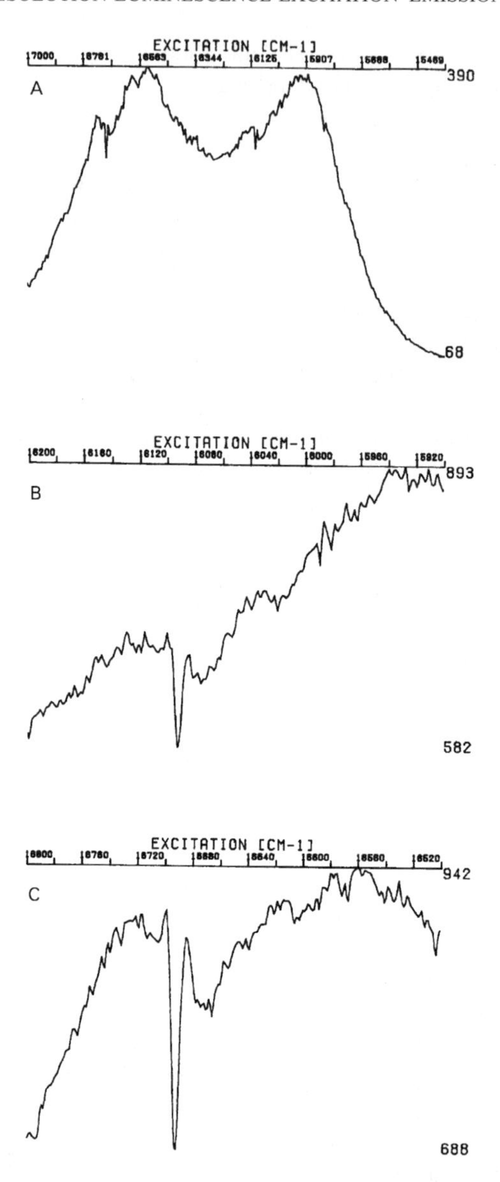

Figure 16.9. Synchronous scan of oxazine-4 in polyvinylbutyral, recorded at about 8 K. The spectra were obtained by simultaneously scanning the excitation and emission monochromator at a constant wave-number difference of 585 cm^{-1}. (a) Two contours are observed in the spectrum, representing excitation in the 0–0 region (excitation maximum at about 15,900 cm^{-1}) and excitation in the vibronic region (excitation maximum at about 16,600 cm^{-1}). Two holes are observed: The first hole is found at an excitation energy of 16,109 cm^{-1}, exactly coinciding with the energy of the laser lines that burned the hole, that is, the 0–0 hole observed in a vibronic emission band [shown in detail in (b)]. The second hole is observed at 16,694 cm^{-1}, where the hole is observed while exciting in the vibronic band at an energy of 585 cm^{-1} above the 0–0 transition [depicted in detail in (c)].

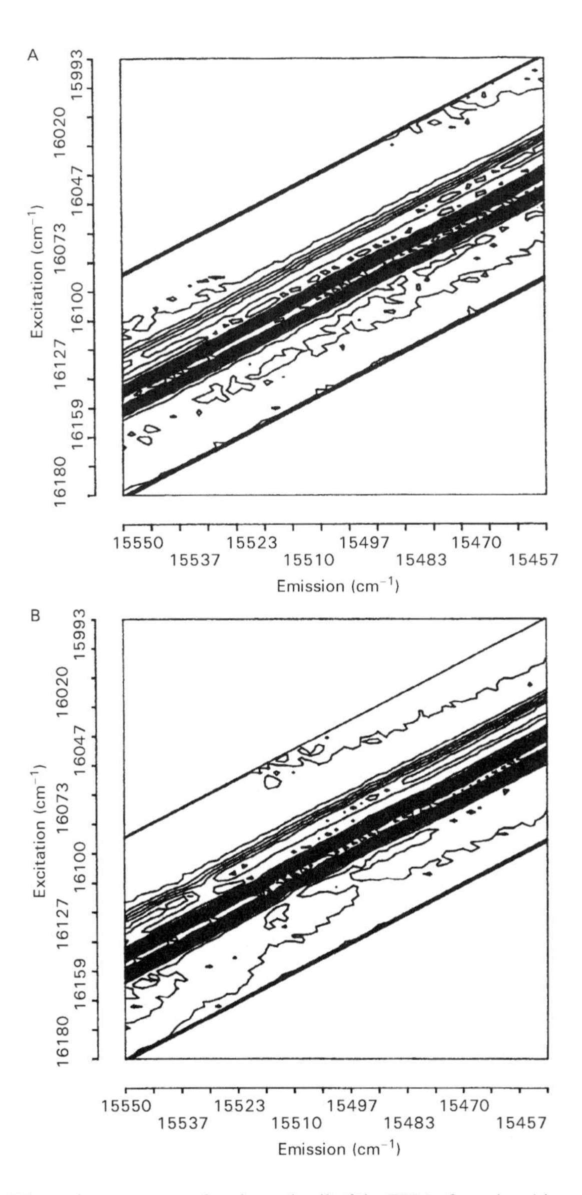

Figure 16.10. 2D-synchronous scans showing a detail of the EEM of oxazine-4 in polyvinylbutyral at 8 K, measured at a resolution of 2 cm^{-1}. Excitation scans were taken at excitation energies of between 630 and 530 cm^{-1} above each emission wavenumber. (a) The 2D synchronous scan as obtained before hole burning. (b) The 2D-synchronous scan obtained after burning with a dye laser at an energy of 16,109 cm^{-1}. The hole is clearly observed in the spectrum, with its maximal decrease in intensity for excitation indeed at the used burning energy and an emission wave number of 15,516 cm^{-1}, that is, at an energy of 593 cm^{-1} below the excitation wave number. (c) A pseudo-3D image of the hole observed in this region; in this figure both the hole in the major band at a vibronic energy of 593 cm^{-1} and that in the minor band at a vibronic energy of 570 cm^{-1} are observed.

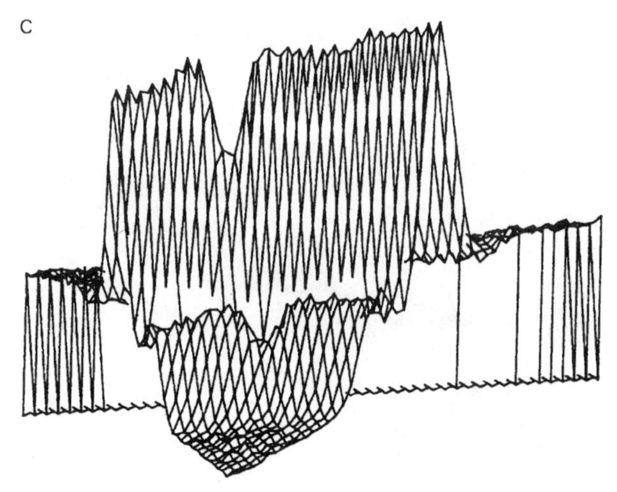

Figure 16.10. (*continued*)

band at intermediate wave-number position is observed. The HREEM shown in Figure 16.10 was recorded with a very narrow bandpass, corresponding to a spectral resolution of 2 cm^{-1}, so that the weak vibronic transition can clearly be observed. The 570-cm^{-1} ground state vibrational mode is also due to the phenoxazine γ(CCC) and γ(CNC), as discussed above, but, in analogy to the study of Van den Berg and Völker on resorufin in ethanol [36], the 593-cm^{-1} mode should be attributed to the H-bonded oxazine-4 and the 570-cm^{-1} mode to free oxazine-4 molecules. The PVB matrix does contain a significant amount of hydroxyl groups (only on the order of 65% of the OH groups in PVB is acetylated). The two holes are clearly visible in the pseudo-3D plot shown in Figure 16.10c.

The width of the hole is about 3 cm^{-1}, which means that it is fully determined by the instrumental resolution (bandpass 2 cm^{-1}, scan step size 1 cm^{-1}). The limited resolution immediately indicates the weakness of HREEM for hole-burning spectroscopy—unless short-lived excited states are considered: the full potential of hole burning can only be employed when very narrow, lifetime-limited holes can also be studied.

16.6. EEMS AND SHPOL'SKII SPECTROSCOPY

16.6.1. The Origin of the Shpol'skii Effect

Since the discovery of the Shpol'skii effect, a number of papers have appeared discussing its origin. Chapter 3 by Lamotte in this volume contains a thorough discussion of this subject; so only some general remarks will be made here. A major parameter that determines whether a highly resolved Shpol'skii spectrum is obtained or not lies in the combination of the *n*-alkane solvent and the solute. A thorough discussion of matrix effects has been reported by Pfister [46] and by Dekkers et al. [47], who both come to the conclusion that the solute should exactly match the dimensions of the solvent molecules to obtain a good fit and hence a narrow-line Shpol'skii spectrum (the "key-and-hole" rule). Still for a number of molecules—even in matching *n*-alkanes—apart from narrow lines broad bands are also observed in the Shpol'skii spectra. Some examples are acenaphthene in *n*-pentane and pyrene in *n*-heptane. The sample preparation and, in particular, the rate of freezing, or, more precisely, the solidification rate of the sample (see Chapter 3), appears to be crucial to obtain narrow-line Shpol'skii spectra for such "difficult" systems [48, 49]. The hypothesis put forward by these authors is that the narrow lines are due to isolated molecules that occupy substitutional sites in the (micro)crystalline matrix, and that the broad bands are due to isolated molecules in intercrystalline regions, subject to an amorphous environment. Since the solubility of the PACs in the crystalline *n*-alkanes is limited, slow freezing of the matrix leads to segregation; the segregated molecules are forced to the intercrystalline regions.

Support for this hypothesis can be found directly in the HREEM. Figure 16.11 shows a HREEM for a dilute solution of pyrene in *n*-heptane (10^{-6} M). Two types of molecules are recognized in the spectrum. Predominantly, molecules in substitutional sites are observed, which give narrow bands, in a "gridlike" pattern, as discussed above for the molecules in a crystalline environment (cf. Fig. 16.1b). Both the emission and the excitation spectrum are dominated by the narrow bands. Due to the stray light rejection procedure, the 0–0 excitation–emission peak, at 26,875 cm^{-1}, is absent from the HREEM. However, at the same time features are observed that clearly display an energy selection type of behavior. Their emission wavenumber varies with the excitation wavenumber, in exactly the same way as described for the FLN spectra discussed above. Such behavior is typical for molecules in a high-viscosity, glassy environment. These spectral features are therefore attributed to molecules occupying amorphous, intercrystalline sites. For the system pyrene in *n*-heptane, prepared by rapid cooling (the sample was immersed in liquid helium), the molecules in the crystalline sites represent the vast majority, as is particularly clear from the pseudo-3D spectrum shown in Figure 16.11b.

16.6.2. Site Structure in Shpol'skii Spectra

In many instances the peaks in Shpol'skii spectra appear to be split; that is, the same bands are present as doublets or triplets, sometimes even multiplets. The

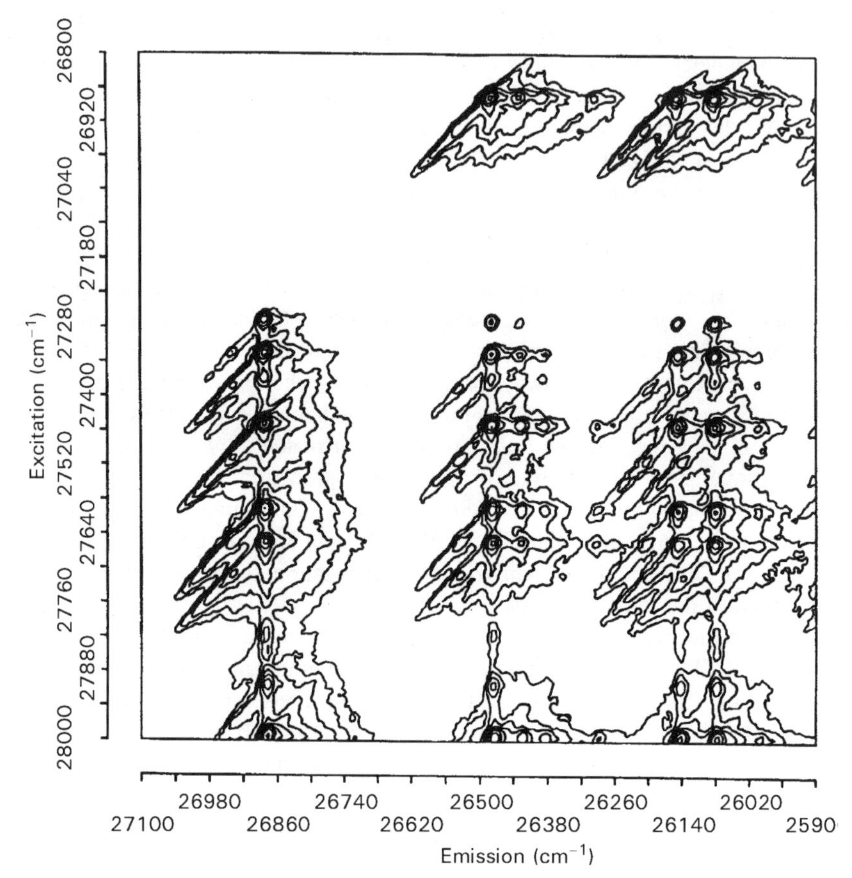

Figure 16.11. High-resolution fluorescence EEM obtained for pyrene in *n*-heptane at 10 K, (a) shown as an isointensity plot and (b) as a pseudo-3D plot. Particularly in the isointensity plot it is clearly observed that the pyrene molecules in the *n*-heptane matrix may occupy crystalline sites, characterized by sharp, needlelike bands and weak phonon side bands, or amorphous, glassy sites, characterized by the typical slanting features, which show energy-selection effects, analogous to what is observed in FLN spectra.

multiplet structure is attributed to the occurrence in the matrix of sets of molecules that occupy different, well-defined substitutional sites, so that each set shows a distinct narrow-line spectrum, which has been shifted depending on the interaction with the matrix.

Interestingly, the interaction energies of the molecules in the excited singlet state (as observed in the fluorescence emission spectrum) and in the excited triplet state (observed in the phosphorescence spectrum) do not necessarily have to be the same. Figure 16.12 shows a detail of the HREEM of chrysene in *n*-hexane,

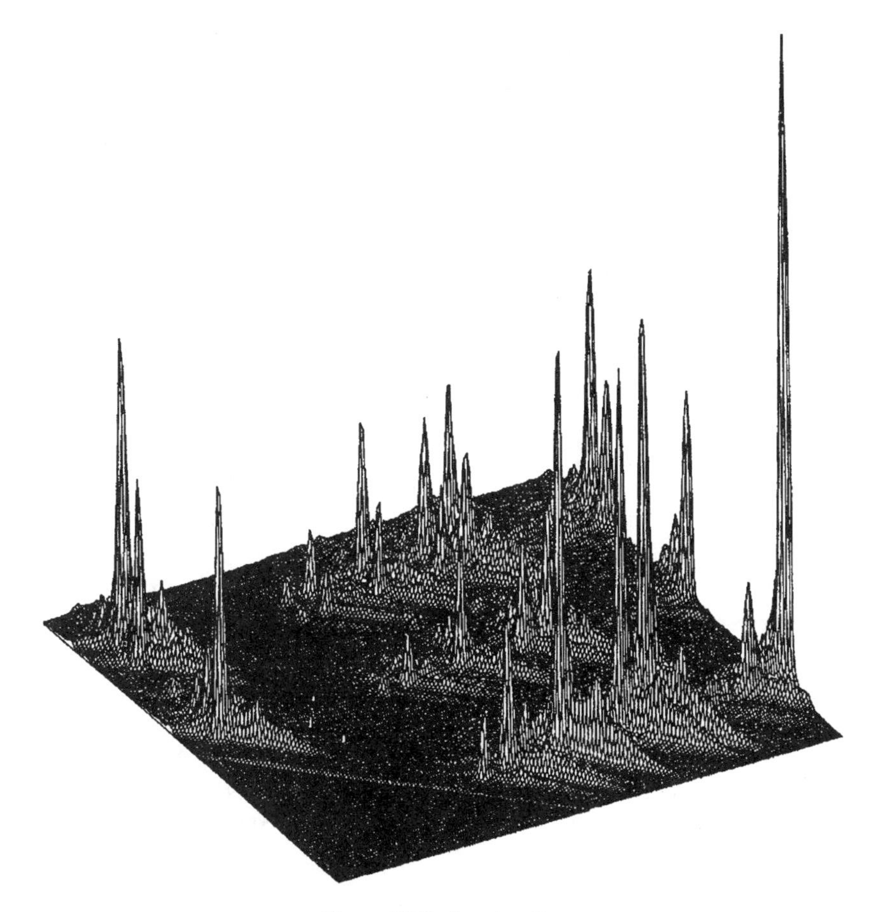

Figure 16.11. (*continued*)

focusing on the region of the phosphorescence spectrum where excitation takes place in the $S_1(v' = 0) \leftarrow S_0(v'' = 0)$ region and emission is observed in the $T_1(v' = 0) \rightarrow S_0(v'' = 0)$ region. Both regions (the former region observed in the excitation spectrum, the latter in the emission spectrum) show a triplet structure, but the order of the transitions in the two states is different. Obviously, there is a one-to-one match between the molecules that give rise to a particular site in the S_1 state and a corresponding site in the T_1 state.

On the basis of the HREEM the multiplet structure can be analyzed. The main transition is observed at an excitation wave number of 27,710 cm^{-1}; the corresponding 0–0 emission from the triplet state occurs at 20,068 cm^{-1}. The

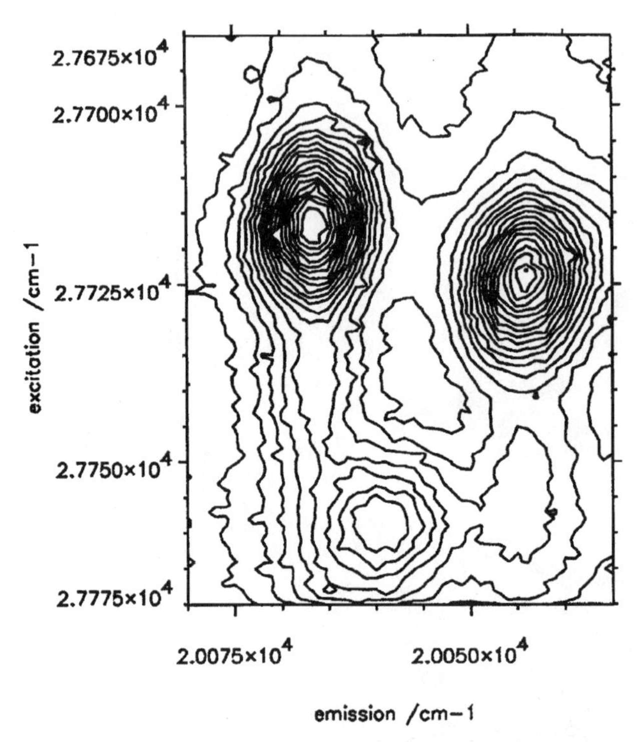

Figure 16.12. EEM obtained for the phosphorescence of chrysene in *n*-hexane at 10 K. Only the 0–0 region is shown as obtained for excitation in the electronic singlet 0–0 region and emission in the electronic triplet 0–0 region. On the basis of the HREEM the relation between the multiplet structure in the singlet and the triplet state can easily be made.

second intense transition is found in excitation at 27,722 cm^{-1} and shows phosphorescence emission at 20,045 cm^{-1}. The third, minor transition is excited at 27,754 cm^{-1} and emits at 20,060 cm^{-1}. Clearly, the order of the multiplets in the singlet manifold is different from that in the triplet manifold, which in fact is not surprising since the electron distribution in the triplet state may be much different from that in the excited singlet state. The complicated multiplet structure of chrysene in *n*-heptane (displaying triplets, separated by no more than 20 cm^{-1} or 0.6 nm in the phosphorescence spectrum) can be straightforwardly interpreted with the aid of the HREEM. Alternatively, only by analysis of a large number of highly resolved conventional 1D spectra, could the multiplet structure have been unraveled.

16.6.3. Sphol'skii EEMs and Synchronous Scans in Environmental Analysis

16.6.3.1. Introduction

The application of low-temperature EEMs for the determination of PAHs in environmental samples has been reported by Hofstraat and Wild [50]. On the basis of EEMs on the one hand, very selective determinations of selected PAHs can be achieved, but on the other hand simplified detection schemes can be devised. Since the complete 2D fluorescence spectrum of the mixture of PACs present in a particular sample is observed in the EEM, optimum conditions for synchronous scan fluorescence measurements (i.e., the optimum wavelength or wave-number difference between excitation and emission monochromators) can be chosen at a glance. Particularly, low-temperature constant-energy synchronous scans (CESSs) are useful for detection of PACs and PAC derivatives, since one may focus on structural characteristics, based on the selective detection of characteristic vibrations. Both approaches will be discussed below.

16.6.3.2. Excitation–Emission Matrices

To demonstrate the applicability of EEMs for analysis of PAHs, in view of the time-consuming data acquisition we will focus on the analysis of the two priority pollutants benzo[*a*]pyrene and benzo[*k*]fluoranthene. The former is known as a highly carcinogenic compound, in contrast to the latter [51]. The two PAHs exhibit fluorescence in the same wavelength region, around 400 nm. On the basis of their conventional, broadband, spectral characteristics they can hardly be distinguished, but their Shpol'skii spectra provide sufficient resolution to identify and quantify them independently. The S_1–S_0 0–0 transition of benzo[*a*]pyrene in *n*-octane is found at 403.0 nm, that of benzo[*k*]fluoranthene at 403.5 nm. Major vibronic emission bands are observed at 408.5 nm for benzo[*a*]pyrene and at 412.8 nm for benzo[*k*]fluoranthene.

Experiments were done on the National Research Council of Canada standard reference material HS-6, originating from a polluted harbor in Nova Scotia. The materials have been freeze dried, sieved over 125-μm mesh, and homogenized. PAH concentrations have been determined by the NRCC Atlantic Research Laboratory in Halifax, Canada; typical concentrations for the priority pollutants benzo[*a*]pyrene and benzo[*k*]fluoranthene are 2.2 ($\pm$0.4) and 1.43 ($\pm$0.15) μg/g sediment, respectively. About 2 g of dry sediment was extracted in a Soxhlet apparatus with acetone/hexane 1:3 (during 4 h at 80°C). The crude extracts with PAHs at 10^{-6}–10^{-7} M concentrations were carefully transferred to a solution in the same volume of *n*-octane for Shpol'skii spectroscopy.

Figure 16.13 shows the EEM obtained by scanning the emision monochromator from 24,500 to 23,875 cm^{-1}, and the excitation monochromator from 24870 to 24570 cm^{-1}. The spectrum indeed shows the features as predicted in Figure 16.1. Major Shpol'skii bands are observed at an excitation wavenumber of 24,818 cm^{-1} and an emission wavenumber of 24,480 cm^{-1} (peak a: benzo[*a*]-pyrene, vibronic band at 338 cm^{-1}), and at an excitation wavenumber of 24,782 cm^{-1} and emission wave numbers 24,230 and 23,980 cm^{-1} (peaks labeled b: benzo[*k*]fluoranthene, vibronic bands at 552 and 802 cm^{-1}, respectively). In addition, some weaker bands are observed, due to other vibronic transitions. The two PAHs are clearly distinguishable due to the highly specific combination of narrow excitation and emission bands. In addition to the typical "spikes" obtained for the molecules embedded in the crystalline (Shpol'skii) matrix, some additional features are noted. The weak features extending from the

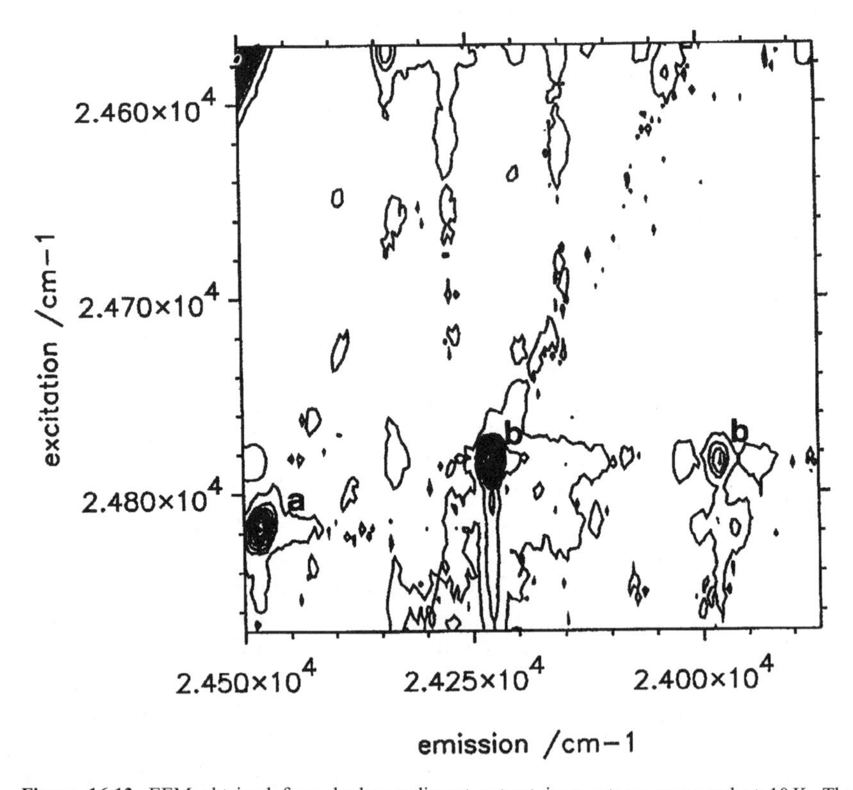

Figure 16.13. EEM obtained for a harbor sediment extract in *n*-octane, measured at 10 K. The emission monochromator was scanned from 24,500 to 23,875 cm^{-1} , the excitation monochromator from 24,870 to 24,570 cm^{-1}. Main peaks in this HREEM are due to (a) benzo[*a*]pyrene and (b) benzo[*k*]fluoranthene. (Reprinted from [50], with permission from Plenum Publishing Corp.)

spikes along the excitation and emission axes are due to electron–phonon coupling, coupling between electronic transitions of the PAH guest molecules with the collective matrix vibrations or phonons. For benzo[*a*]pyrene and benzo[*k*]fluoranthene in *n*-octane the electron–phonon coupling is weak. In addition, some weak slanting bands are observed, which also comprise the spikes. These bands are due to the selective excitation of the PAH molecules, which occupy noncrystalline amorphous sites, as discussed above

The interpretation given above can be further illustrated when the EEM is cut at the excitation wave numbers 24,817.3 and 24,783.4 cm^{-1}, respectively, the 0–0 transitions of benzo[*a*]pyrene and benzo[*k*]fluoranthene, as shown in Figure 16.14. The main peaks described above are easily identified. From the spectra the spectral bandwidth can be estimated. For both PAHs it amounts to approximately 20 cm^{-1} full width at half-maximum (FWHM), which is clearly dominated by the instrumental conditions. The instrumental bandwidth is 12 cm^{-1} and the step size 5 cm^{-1} for the spectra depicted in Figures 16.13 and 16.14. The relatively large bandwidth and step size are required to keep the total spectral acquisition time within reasonable limits. The EEM in Figure 16.13 is composed of almost 10,000 points, which have been measured with a dwell time of 1 s. In combination with the time needed to scan the monochromators, the total time needed to obtain the EEM is about 6 h.

16.6.3.3. Constant-Energy Synchronous Fluorescence

An alternative approach is to record a constant-energy synchronous fluorescence (CESF) scan of the sample. Synchronous fluorometry is a well known method for selective measurement of PAHs; in the most commonly applied approach a constant wavelength difference is used between the excitation and emission monochromators, which are scanned in tandem. The main advantage of the CESF scan over convential synchronous fluorescence methods is that the constant wavenumber difference reflects the vibrational structure of the analyte. The wavenumber spectrum reflects the energy levels of the molecules. The CESF can therefore be applied to monitor a particular structural feature of the analytes, by choosing a wavenumber difference that represents a vibrational mode. Some examples are $\Delta\nu = 1600\,cm^{-1}$ for aromatic compounds, $\Delta\nu = 1700\,cm^{-1}$ for compounds containing a carbonyl group, or $\Delta\nu = 1200\,cm^{-1}$ for compounds containing C–F bonds. Constant-energy synchronous fluorimetry has been pioneered by the group of Winefordner [52, 53].

In the EEM it was observed that a characteristic frequency for a ground-state vibrational mode of benzo[*k*]fluoranthene was 552 cm^{-1}. Therefore, a CESF scan was made with a constant energy difference of 550 cm^{-1} between the excitation and emission wavenumbers. The excitation wavenumber was scanned from

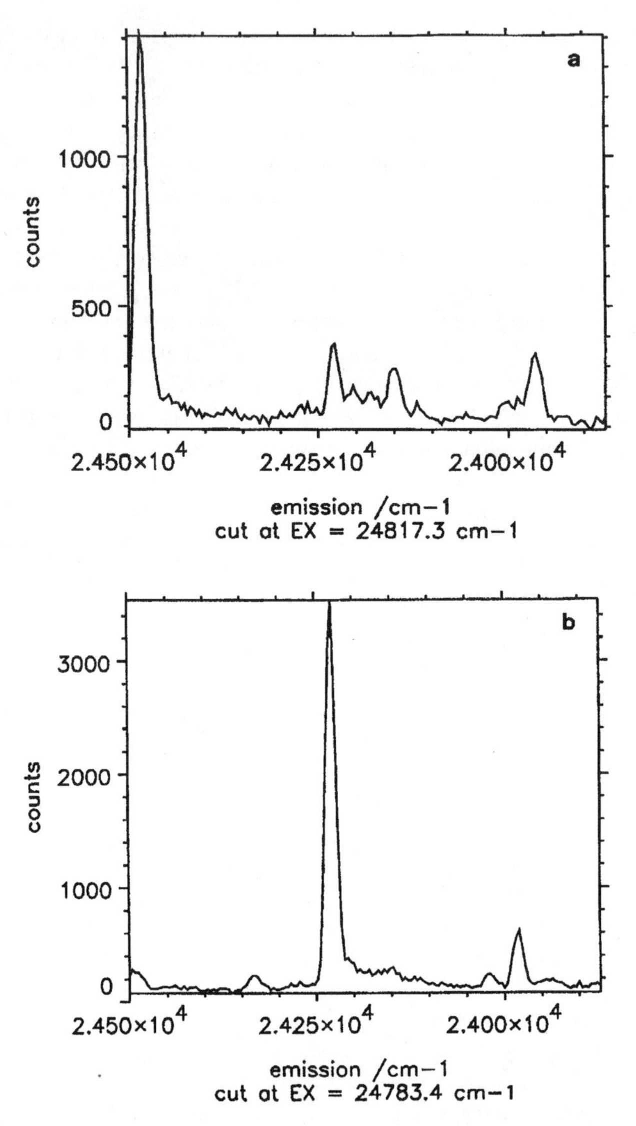

Figure 16.14. Cuts from the EEM shown in Fig. 16.13 a excitation wavenumbers: (a) 24,817.3 cm^{-1} (benzo[*a*]pyrene 0–0 transition) and 24,783.4 cm^{-1} (benzo[*k*]fluoranthene. 0–0 transition). (Reprinted from [50], with permission from Plenum Publishing Corp.)

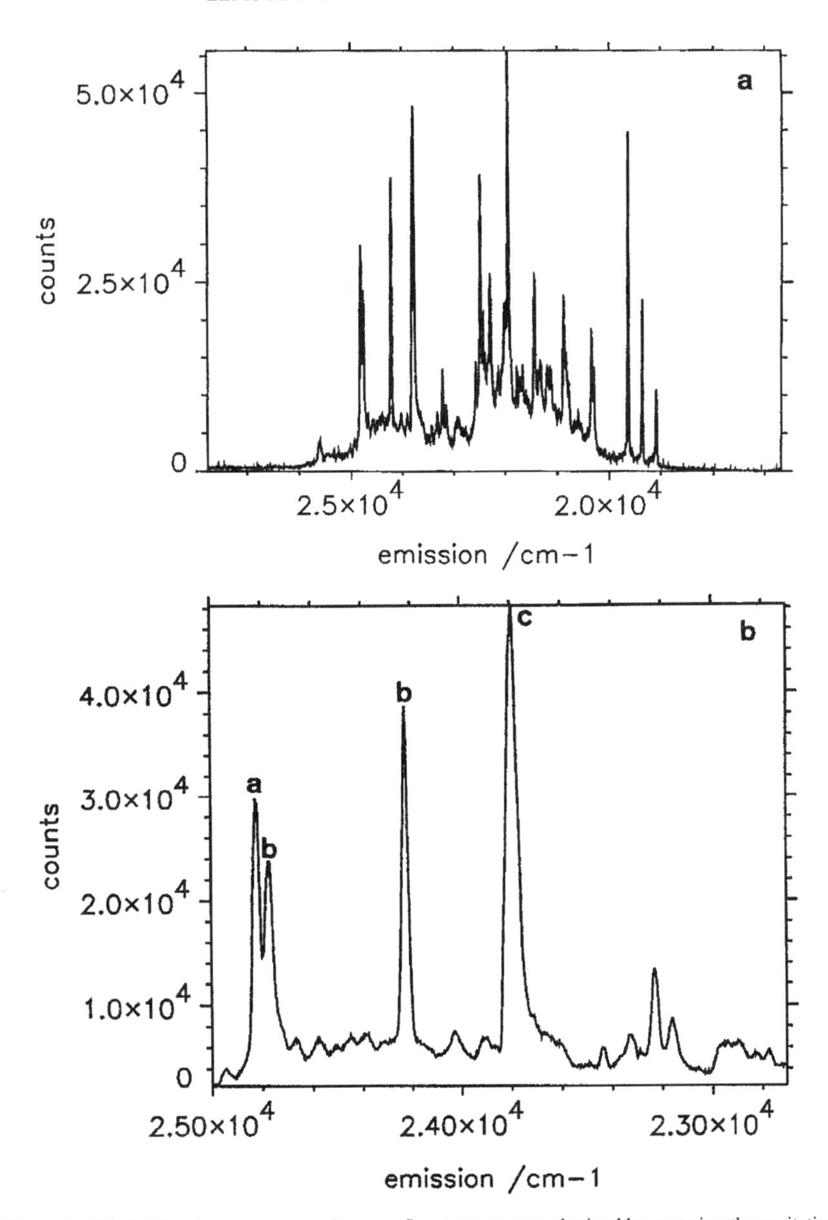

Figure 16.15. (a) Constant-energy synchronous fluorescence scan obtained by scanning the excitation monochromator from 28,000 to 16,000 cm^{-1} at a constant energy difference of 550 cm^{-1} between the excitation and emission monochromator. In the detail of the total spectrum depicted in (b) three major constituents of the sample can be identified: benzo[*a*]pyrene (peak indicated by the letter a in the figure), benzo[*k*]fluoranthene (two peaks, indicated by the letter b) and benzo[*ghi*]fluoranthene (indicated by the letter c). (Reprinted from [50], with permission from Plenum Publishing Corp.)

28,000 cm^{-1} (357 nm) to 16,000 cm^{-1} (625 nm). Variations in the excitation intensity were corrected with the oxazine-1 quantum counter (see the experimental section). The CESF obtained is shown in Figure 16.15. A large number of well-resolved peaks are observed across this wide wave-number range. Let us focus on the 25,000–23,000 cm^{-1} region, where the resonances of benzo[*a*]pyrene and benzo[*k*]fluoranthene are expected. Four major peaks are noted. The first peak (excitation at 25,374 cm^{-1}, emission at 24,824 cm^{-1}) corresponds to excitation in an excited-state vibronic band of benzo[*a*]pyrene and subsequent 0–0 emission. The second peak (excitation at 25,327 cm^{-1}, emission at 24,777 cm^{-1}) corresponds analogously to vibronic excitation and 0–0 emission for benzo[*k*]fluoranthene. The third peak (excitation at 24,778 cm^{-1}, emission at 24,228 cm^{-1}) can be attributed to 0–0 excitation followed by vibronic emission of benzo[*k*]fluoranthene; benzo[a]pyrene does not have a significant transition corresponding to a 550 cm^{-1} vibrational mode in the ground state, and is therefore not observed in this part of the scan. The strongest band in the CESF is found for excitation at 24,344 cm^{-1} and emission at 23,794 cm^{-1}. This peak is due to benzo[*g,h,i*]fluoranthene; a vibronic excitation and the 0–0 emission of this compound exactly match the observed excitation and emission wavenumbers. The NRCC data sheet of the SRM unfortunately only mentions the EPA priority pollutants, to which benzo[*g,h,i*]fluoranthene does not belong. Figure 16.15 shows a number of additional bands, which may obviously be identified as well.

The interpretation of a CESF scan with a wavenumber difference of 550 cm^{-1} is fairly simple, since not many transitions are found in the highly resolved fluorescence spectra that represent such a relatively small difference in energy. The number of allowed vibronic transitions with such a low vibrational energy is low in PAHs. Most bands therefore can be attributed to either vibronic excitation and 0–0 emission, or to 0–0 excitation and vibronic emission. When larger energy differences are applied, other combinations of excitation and emission wavenumbers also become possible, such as vibronic excitation in combination with vibronic emission. Fortunately, in general the oscillator strength of 0–0 transitions is significantly larger than that of vibronic transitions, so that the most intense bands can be easily attributed. The use of EEMs obviously is very helpful for the interpretation of the CESF scans, because the signals can be comprehensively evaluated. Obviously, the advantage of CESF over HREEM is that a significantly shorter acquisition time suffices to obtain the desired information.

16.7. CONCLUSIONS

HREEMs can be used to reveal vibronic features of luminescence spectra, both in the ground and in the excited state, via the emission and excitation spectra, respectively. In particular, HREEMs are useful in spectra that show clear energy-selection effects. The most prominent energy-selection effects appear in the fluorescence-line-narrowing spectra, which are obtained as a result of (narrow-bandwidth) energy selection. Not only are the vibrationally resolved spectra obtained, but information on the inhomogeneous bands is also acquired at the same time. The study of matrix interactions, of solvent reorientation and solvent dynamics, and of interactions between the energy-selected chromophores with other chromophores is greatly facilitated by the HREEMs. Even hole-burning effects can be observed in HREEMs, although in the present examples the usefulness of the approach is still strongly limited by the limited resolution of the spectra.

HREEMs of molecules in Shpol'skii matrices can be applied to obtain highly specific vibrationally resolved fingerprints, which are useful to identify trace contaminants, for instance, in environmental samples. Even molecules with similar spectral characteristics can be distinguished on the basis of such spectra, as has been demonstrated by the example of a harbor sediment sample containing benzo[*a*]pyrene and benzo[*k*]fluoranthene, the former being a notorious carcinogenic compound. A comprehensive set of data is obtained with combinations of excitation and emission wavenumbers, which allows for the screening of many PAHs, including all priority pollutants, in one analysis step.

The acquisition of HREEMs is time consuming, unless special instrumental setups are applied, which are composed of parallel excitation and detection hardware (e.g., distributed excitation using prisms and parallel detection with an intensified CCD detector). However, such hardware is not easily prepared so that high-resolution data can be obtained; at least a resolution on the order of 1 nm is required to get vibrational resolution in fluorescence spectra. New developments like optical parametric oscillators, intense laser sources with a very broad tuning curve of adequate linewidth for high resolution spectroscopy, will enable one to obtain laser-excited HREEMs, which couple the specificity of the high-resolution excitation–emission matrices to the high sensitivity afforded by the high-intensity excitation source.

An alternative way to get fingerprint data over a wide wavelength region in a single scan is by synchronous fluorescence spectroscopy. High-resolution CESF can be applied to obtain vibrationally resolved spectra for many PAHs as well, at a significantly shorter acquisition time than HREEMs. When a relatively small energy difference ($800\,cm^{-1}$) is applied between excitation and emission wavenumber, simple CESF spectra are obtained, which allow for rapid and specific screening of PAHs in environmental samples. Larger energy differences lead to more complex spectra, with spectral features showing up due to various combinations of vibronic excitation and vibronic emission of the same molecule. For the interpretation of such complex spectra it is useful to have HREEM data available, which allow the straightforward identification of peaks derived from one species. Once such identifications have been made, obviously CESF can be applied without further need of HREEM "background information".

REFERENCES

1. J. W. Hofstraat, N. H. Velthorst, and C. Gooijer, in S. G. Schulman, ed., *Methods in Luminescence Spectroscopy—Methods and Applications, Part 2*, John Wiley & Sons, New York, 1988, p. 283.
2. A. P. D'Silva and V. A. Fassel, *Anal. Chem.* **56**, 985A (1984).
3. J. W. Hofstraat, W. J. M. van Zeijl, F. Ariese, J. W. G. Mastenbroek, N. H. Velthorst, and C. Gooijer, *Mar. Chem.* **33**, 301 (1991).
4. I. M. Warner, G. Patonay, and M. P. Thomas, *Anal. Chem.* **57**, 463A (1985).
5. T. T. Ndou and I. M. Warner, *Chem. Rev.* **91**, 493 (1991) .
6. J. R. Kershaw and J. C. Fetzer, *Polycycl. Aromat. Compd.* **7**, 253 (1995).
7. P. B. Oldham, E. J. Zillioux, and I. M. Warner, *J. Marine Res.* **43**, 893 (1985).
8. Th. Roch, *Anal. Chim. Acta* **356**, 61 (1997).
9. S. L. Neal, *Appl. Spectrosc.* **47**, 1161 (1993).
10. S. L. Neal, *J. Chemom.* **8**, 245 (1994).
11. J.-H. Wang, J.-H. Jiang, J.-F. Xiong, Y. Li, Y.-Z. Liang, and R.-Q. Yu. *J. Chemom.* **12**, 95 (1998).
12. R. Henrion, G. Henrion, M. Bohme, and H. Behrendt, *Fres. J. Anal. Chem.* **357**, 522 (1997).
13. D. W Millican and L. B. McGown, *Anal. Chem.* **61**, 580 (1989).
14. J. J. Mobed, S. L. Hemmingsen, J. L. Autry, and L. B. McGown, *Environ. Sci. Technol.* **30**, 3061 (1996).
15. K. Itoh and T. Azumi, *J. Chem. Phys.* **62**, 3431 (1975).
16. K. Itoh and T. Azumi, *J. Chem. Phys.* **65**, 2550 (1976).
17. J. W. Hofstraat, R. Locher, and U. P. Wild, *Appl. Spectrosc.* **44**, 1317 (1990).
18. A. J. Kallir, G. W. Suter, S. E. Bucher, E. Meister, M. Lüönd, and U. P. Wild, *Acta Phys. Polonica* **A71**, 755 (1987).
19. V. M. Rangnekar and P. B. Oldham, *Anal. Instrum.* **19**, 125 (1990).
20. A. Munoz de la Pena, J. A. Murillo, J. M. Vega, and F. Baringo, *Comput. Chem.* **12**, 213 (1988).
21. Y. Talmi, D. C. Baker, J. R. Jadamec, and W. A. Saner, *Anal. Chem.* **50**, 936A (1978).
22. S. J. Hart, Y.-M. Chen, J. E. Kenny, B. K. Lin, and T. W. Best, *Field Anal. Chem. Technol.* **1**, 343 (1993).
23. T. A. Taylor, G. B. Jarvis, H. Xu, A. C. Bevilaqua, and J. E. Kenny, *Anal. Instrum.* **21**, 141 (1993).
24. T. Vo-Dinh and U. P. Wild, *Appl. Opt.* **12**, 1286 (1973).
25. J. W. Hofstraat and M. J. Latuhihin, *Appl. Spectrosc.* **48**, 436 (1994).
26. J. W. Hofstraat, M. Engelsma, J. H. de Roo, C. Gooijer, and N. H. Velthorst, *Appl. Spectrosc.* **41**, 625 (1987).

27. G. W. Suter, U. P. Wild,, and A. R. Holzwarth, *Chem. Phys.* **102**, 205 (1986).
28. L. Rebane, A. A. Gorokhovskii, and J. Kikas, *Appl. Phys. B* **29**, 235 (1982).
29. E. I. Al'shits, R. I. Personov, A. M. Pyndyk, and V. I. Stogov, *Opt. Spectrosc. (USSR)* **39**, 156 (1975).
30. K. Cunningham, J. M. Morris, J. Fünfschilling, and D. F. Williams, *Chem. Phys. Lett.* **32**, 581 (1975).
31. M. N. Sapozhnikov and V. I. Alekseev, *Chem. Phys. Lett.* **87**, 487 (1982).
32. M. N. Sapozhnikov and V. I. Alekseev, *Chem. Phys. Lett.* **107**, 265 (1984).
33. R. I. Personov in V. M. Agranovich, and R. M. Hochstrasser, eds., *Spectroscopy and Excitation Dynamics of Condensed Molecular Systems*, North-Holland, Amsterdam, 1983, Chap. 10, p. 556.
34. E. Meister, Thesis ETH, Zürich, Switzerland, Nr. 8724 (1988).
35. S. K. Miller, A. Baiker, M. Meier, and A. Wokaun, *J. Chem. Soc., Faraday Trans. I* **80**, 1305 (1984).
36. R. Van den Berg and S. Völker, *Chem. Phys.* **128**, 257 (1988).
37. E. Görlach, H. Gygax, P. Lubini, and U. P. Wild, *Chem. Phys.* **194**, 185 (1995).
38. P. O. J. Scherer, A. Seilmeier, and W. Kaiser, *J. Chem. Phys.* **83**, 3948 (1985).
39. B. M. Kharlamov, R. I. Personov, and L. A. Bykovskaya, *Opt. Commun.* **12**, 191 (1974).
40. B. M. Kharlamov, R. I. Personov, and L. A. Bykovskaya, *Opt. Spectrosc.* **39**, 137 (1975).
41. A. A. Gorokhovskii, R. K. Kaarli, and L. A. Rebane, *JETP Lett.* **20**, 216 (1974).
42. J. M. Hayes, K. P. Stout, and G. J. Small, *J. Chem. Phys.* **74**, 4266 (1981).
43. G. J. Small in V. M. Agranovich, and R. M. Hochstrasser, eds., *Spectroscopy and Excitation Dynamics of Condensed Molecular Systems*, North-Holland, Amsterdam, 1983, Chap. 9, p. 515.
44. J. W. Hofstraat, A. J. Schenkeveld, M. Engelsma, C. Gooijer, and N. H. Velthorst, *Spectrochim. Acta* **45A**, 139 (1989).
45. J. W. Hofstraat, M. Engelsma, C. Gooijer, and N. H. Velthorst, *Spectrochim. Acta* **45A**, 491 (1989).
46. C. Pfister, *Chem. Phys.* **2**, 171 (1973).
47. J. J. Dekkers, G. Ph. Hoornweg, C. MacLean, and N. H. Velthorst, *J. Mol. Spectrosc.* **68**, 56 (1977).
48. J. Rima, L. Nakhimovsky, M. Lamotte, and J. Joussot-Dubien, *J. Phys. Chem.* **83**, 4302 (1984).
49. J. W. Hofstraat, I. L. Freriks, M. E. J. de Vreeze, C. Gooijer, and N. H. Velthorst, *J. Phys. Chem.* **9** 3, 184 (1989).
50. J. W. Hofstraat and U. P. Wild, *J. Fluoresc.*, **8**, 319 (1998).

51. W. Karcher, R. J. Fordham, J. J. Dubois, P. G. J. M. Glaude, and J. A. M. Ligthart, eds., *Spectral Atlas of Polycyclic Aromatic Compounds, Vol. 1*, Reidel/Kluwer, Dordrecht, The Netherlands, 1983.
52. E. L. Inman and J. D. Winefordner, *Anal. Chem.* **54**, 2018 (1982).
53. L. A. Files, B. T. Jones, S. Hanamura, and J. D. Winefordner, *Anal. Chem.* **58**, 1440 (1986).

CHAPTER

17

SITE-SELECTIVE LASER SPECTROSCOPY OF DEFECTS IN INORGANIC MATERIALS

KEITH M. MURDOCH AND JOHN. C. WRIGHT

Department of Chemistry, University of Wisconsin, 1101 University Ave, Madison, WI 53706, U.S.A.

17.1. INTRODUCTION

Dye lasers and parametric oscillators are intense tunable light sources with narrow spectral bandwidths. They are ideal for selectively exciting optically active species in samples that have two or more sites with distinguishable spectral features. High-resolution monochromators and optical detectors can selectively monitor the fluorescence emitted by such sites. Together these are the principal components in site-selective laser experiments. Single-site excitation spectra are obtained by selectively monitoring a fluorescence feature of a particular site and scanning the laser through the absorption region of interest. Similarly, single-site fluorescence spectra are obtained by selectively exciting an absorption transition of a site and scanning the monochromator through an emission region.

Most tunable laser sources operate in the near-ultraviolet, visible, or near-infrared regions, where they can access the electronic and vibronic transitions of atoms or molecules. Sites of chemically distinct species in a sample will have different electronic states, and their transitions can generally be distinguished spectroscopically. Multiple sites of a particular species, with different ligands or coordination parameters, can be spectrally discriminated because their states are perturbed by interactions with the local environment. Sites with very different local environments will have distinct electronic transitions. Smaller variations in local strain produce subsites that appear collectively as inhomogeneously broadened transitions. Lasers and monochromators with narrow spectral bandwidths can select groups of subsites within inhomogeneously broadened transitions to produce line-narrowed spectra. These spectra are used to determine the distribution of subsites created by a strain-broadening mechanism.

When the absorption or emission lines of different sites overlap, transitions from all of these sites will appear in the spectra. However, transitions belonging to a common site will always appear with the same relative intensities, regardless of the wavelength being monitored or excited. The polarization of spectra arises

Shpol'skii Spectroscopy and Other Site-Selection Methods, Edited by Cees Gooijer, Freek Ariese, and Johannes W. Hofstraat.
ISBN 0-471-24508-9

from the symmetry properties of a species in a particular site and can be utilized experimentally to achieve additional site selectivity. Polarization experiments can also be interpreted using group theory to determine site symmetries and identify the observed transitions. Often ligand interactions will modify the excited-state lifetimes of a species, either by changing radiative transition probabilities or through energy-transfer processes that relax the species nonradiatively. The changes in dynamics are exploited in experiments that achieve additional site selectivity by using pulsed excitation and time-gated detection of the resulting fluorescence.

Energy upconversion occurs when species are optically pumped at one wavelength and emit at shorter wavelengths. Upconversion can be achieved by two mechanisms. Sequential excitation involves just one species, which is excited sequentially by a ground-state absorption and an excited-state absorption. Energy-transfer upconversion involves an energy transfer between two excited species, where a donor relaxes and transfers its excitation to an acceptor that emits from a higher level. Transfer is only possible when the donor and acceptor species are spatially close. Energy-transfer upconversion can thus be used to improve the selectivity between sites that are isolated and sites that are constituents of clusters. This capability is particularly valuable for detecting cluster sites in samples dominated by isolated sites.

Upconversion mechanisms are also used to excite species to higher energies than would otherwise be possible with available laser sources. Two-photon absorption involves the simultaneous absorption of two photons whose combined energy matches a transition of the optically active species. It is another method for accessing higher energy states. Since the quantum and symmetry selection rules are different for one- and two-photon transitions, two-photon spectroscopy can also improve site selectivity, particularly in samples with intrinsic high-symmetry sites.

Although different types of inorganic materials have been investigated using site-selective laser spectroscopy, the literature is dominated by studies of crystals and glasses containing rare-earth ions. The valence 4*f* electrons of a lanthanide interact weakly with their local environment due to shielding by the filled 5*s* and 5*p* orbitals. Local crystal fields weakly split the 4*f* orbital, and sharp line transitions are observed between the resulting crystal-field levels. An analogous situation exists for the 5*f* actinides. For this reason rare-earth ions are particularly amenable to site-selective techniques.

A considerable volume of work has been published on rare-earth-doped alkaline- earth halide crystals, particularly the fluorites. The fluorite crystal structure is a lattice of F^- cubes, with the cations (Ca^{2+}, Sr^{2+}, or Ba^{2+}) occupying alternate cube centers. Rare-earth ions (R^{3+}) are usually trivalent and substitute for the divalent cations. Charge compensation is therefore required to maintain neutrality and is provided by the inclusion of additional interstitial fluoride ions (F_i^-). Many different ionic configurations are adopted including cubic symmetry R^{3+} sites that are remote from any F_i^-, single R^{3+} associated with single neighboring F_i^-, and clusters of R^{3+} associated with multiple F_i^-. Site-selective experiments have shown that the relative populations of these centers are determined by the R^{3+} concentration in a sample, its thermal history, and any pressure treatments. Additional R^{3+} centers are produced by chemical modifications involving the substitution of lattice cations or anions for other species. As a consequence of their great variety of R^{3+} sites, these crystals are spectroscopically rich. They will be the subject for many of the examples presented here.

17.2. REVIEW OF SITE-SELECTIVE SPECTROSCOPY APPLIED TO INORGANIC MATERIALS

Much research has been reported on the application of high-resolution laser spectroscopy to the study of defects in inorganic materials. This is a short overview of the most important techniques and the different types of inorganic systems that have been investigated.

Spectral hole burning includes a broad range of laser saturation experiments. These techniques [1–4] and the systems to which they have been applied [5, 6] have been reviewed in detail by other authors. Generally, a narrow-bandwidth laser selectively excites a subset of sites within an inhomogeneously broadened absorption transition, depleting their ground-state population. The same laser, or another probe laser, is then scanned across the transition to record the difference in absorption between the subsites that were pumped and those that were not pumped. Usually the bleaching is transient and is only observed during the short time that the selected species remain electronically excited. Very narrow bandwidth lasers can be used to deplete specific hyperfine and superhyperfine levels of metal ions in crystals at cryogenic temperatures. These narrow holes can persist for many seconds and are deleted by spin relaxation between the hyperfine levels. Persistent or permanent hole burning occurs when the excitation laser causes photophysical or photochemical changes to the selected subsites. Laser-selective excitation spectroscopy is a very sensitive way to observe the resulting spectral holes. Often spectral holes are accompanied by antiholes. Antiholes are the absorption features associated with subsites that have been selectively excited and have subsequently returned to the ground electronic state. These subsites have modified environments or populate different hyperfine states after hole burning.

Metal impurities in crystals and glasses, particularly rare-earth ions, are sensitive probes of their local environment. They have been studied extensively by spectral hole burning [1, 5]. Holes as narrow as a few kHz have been observed using an ultra-high-resolution laser [7]. Hyperfine and superhyperfine structure [8–10] is best observed in the well-ordered environment of a crystal. Persistent room-temperature spectral hole burning has been observed in Eu^{2+} [11] and Sm^{2+} [12] doped materials due to photoionization and subsequent trapping of the ionized electron. Persistent spectral hole burning is possible in glasses when the host ions that surround a metal impurity can adopt many different configurations with similar energies. Laser excitation induces the physical transformations between configurations. This type of hole burning is observed for a variety of glasses and dopants [13, 14]. Much of the work on dynamical processes in glasses [15], including spectral diffusion, has been performed on organic systems [16, 17]. Persistent spectral hole burning has been observed for defect centers in diamond [18, 19] and has been observed at room temperature in neutron-irradiated diamond [20]. Color centers have also been investigated using spectral hole burning [21].

High-density optical information storage and optical data processing are the principal applications envisioned for hole burning. Much work has been done towards developing and characterizing holographic storage and retrieval methods based on hole-burning systems [22]. These can be integrated with coherent time-domain techniques [23], such as stimulated photon-echo generation [24], to yield high-speed high-density data storage. A significant advance has been the development of photon-gated hole burning, initially using a sequential mechanism for the photoionization of Sm^{2+} ions [25]. Hole burning by a narrow selective laser is gated by a second laser. The resulting holes can be selectively probed without being significantly degraded by the probing beam. Spectral hole-burning systems with very narrow homogeneous linewidths could be used as frequency standards [7]. Persistent spectral hole-burning systems are being investigated as potential frequency references for laser stabilization [26].

Fluorescence line-narrowing techniques are applied to systems with metastable electronic states and inhomogeneously broadened transitions. They require a correlation between the effects of inhomogeneous broadening on the absorption and fluorescence transitions. A narrow-bandwidth laser is used to excite subsites within an inhomogeneously broadened absorption transition. A high-resolution monochromator is used to analyze the resulting fluorescence. Both excitation and fluorescence spectra can be obtained by fluorescence line narrowing. The technique has been applied to investigate transition elements [27, 28] and rare-earth ions [13, 29, 30] in crystals and glasses. Time-resolved fluorescence line narrowing probes the transfer of energy between subsites within inhomogeneously broadened transitions [31, 32].

Coherent spectroscopic techniques are used to investigate optical dephasing and coherence loss processes. Photon echoes are generated to measure the dephasing times of metal ions in solids. Very sharp optical resonances (less than 1000 kHz wide), with correspondingly long dephasing times, have been measured in rare-earth doped crystals [33, 34]. Optical free induced decay is also applied to measure coherence losses [1, 35, 36]. A narrow-bandwidth laser prepares a subset of sites in a coherent superposition of states. Homodyne or heterodyne detection is used to monitor the evolution of this coherence. Single-molecule excitation is an alternative way to probe defects in the strained environment of a crystal or glass [37].

17.3. LASER SPECTROSCOPIC STUDIES OF RARE-EARTH-DOPED FLUORITES

17.3.1. Defect Chemistry at Low Rare-Earth Concentrations

At low R^{3+} concentrations—less than ca. 0.05 mol %—isolated R^{3+} sites prevail in all fluorite crystals. Optical Zeeman spectra show that the dominant R^{3+} site in CaF_2 has tetragonal C_{4v} symmetry [38]. It has a charge-compensating F_i^- occupying one of the nearest-neighbor (NN) interstitial sites (Fig. 17.1a). A much smaller concentration of trigonal C_{3v} symmetry sites is also found in CaF_2 crystals doped with the smaller R^{3+}, from Tb^{3+} [39] through Yb^{3+} [40]. ESR experiments show that the preferred R^{3+} site symmetry in SrF_2 changes across the rare-earth series [41]. While the tetragonal site is dominant at the La^{3+} end, trigonal site concentrations increase towards the Lu^{3+} end, becoming dominant around Dy^{3+} and Ho^{3+}. This trend suggests that R^{3+} ion size is an important site determinant, as these two ions have approximately the same ionic radius as Sr^{2+}. Laser spectroscopy shows that while the tetragonal Ho^{3+} site is still dominant in $SrF_2:Ho^{3+}$ [42], only trigonal Er^{3+} sites were identified in $SrF_2:Er^{3+}$ crystals [43, 44]. In $SrF_2:Er^{3+}$ the F_i^- occupy a next-nearest-neighbor (NNN) interstitial site (Fig. 17.1b). This type of trigonal site has also been identified in $SrF_2:Eu^{3+}$ [45, 46] and $SrF_2:Dy^{3+}$ [46, 47]. The trigonal site found in $SrF_2:Ho^{3+}$ is spectroscopically distinct, and its specific ionic configuration has not yet been established [48]. This other type of trigonal site was also found as a minor center in both $CaF_2:Er^{3+}$ [49] and $CaF_2:Ho^{3+}$ [42]. Trigonal R^{3+} sites are dominant in BaF_2, as expected for the larger Ba^{2+} cation. The trigonal site in $BaF_2:Er^{3+}$ was spectroscopically very similar to the NNN compensated site of $SrF_2:Er^{3+}$ [50].

Cubic symmetry R^{3+} sites are difficult to observe optically because intraconfigurational electric-dipole transitions are parity forbidden in high-symmetry environments. Their magnetic-dipole transitions can be seen weakly in time-gated spectra that discriminate against lower symmetry sites with shorter excited-state lifetimes [51]. The cubic Er^{3+} site concentrations were measured by EPR in $CaF_2:(0.01\%)Er^{3+}$ [52] and $CaF_2:(0.1\%)Er^{3+}$ [38] and compared to the concentrations of the two other isolated Er^{3+} sites. At both Er^{3+} concentrations the tetragonal site concentration is only slightly greater than the cubic site concentration, although both are an order of magnitude greater than the trigonal site concentration. In contrast, for $CaF_2:(0.01\%)Eu^{3+}$ and $CaF_2:(0.08\%)Eu^{3+}$

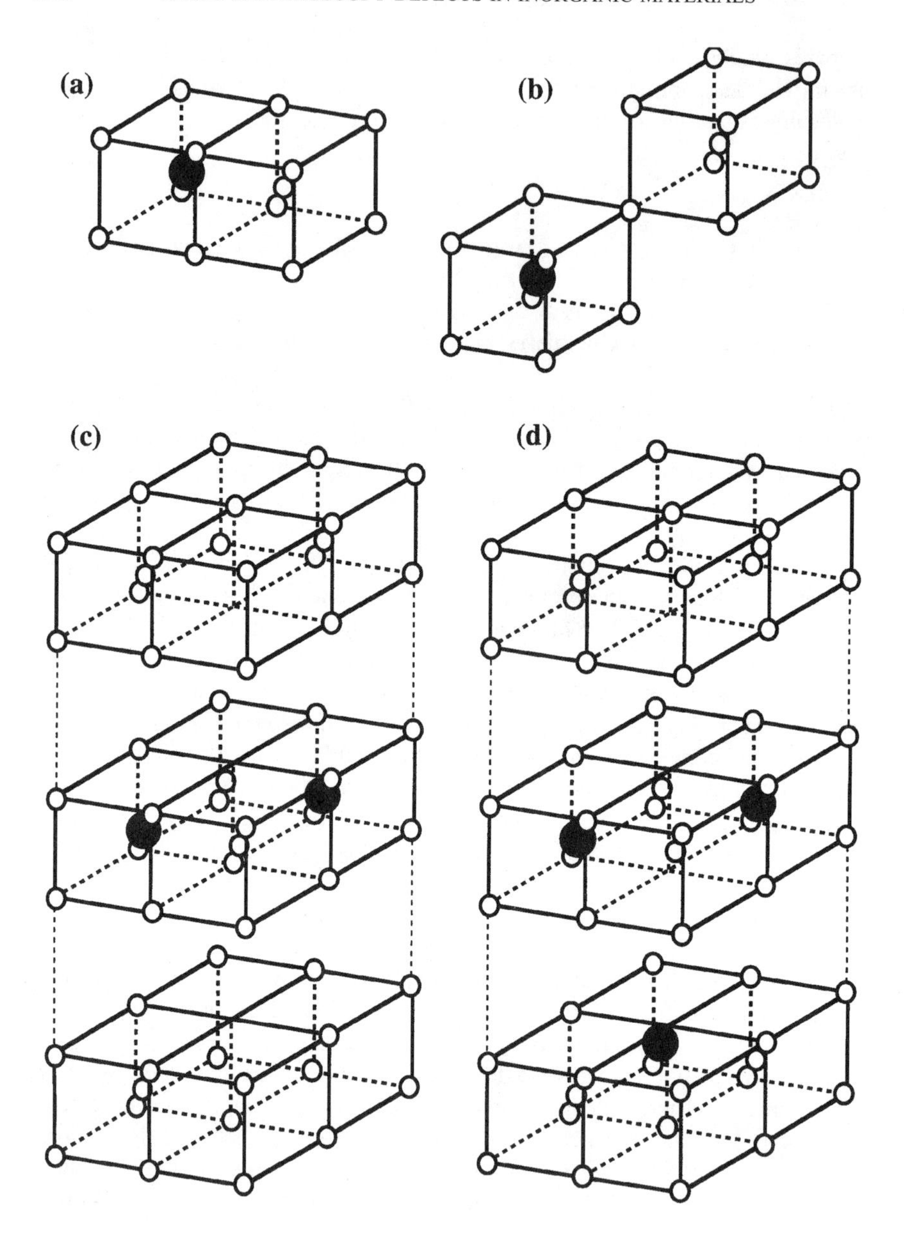

Figure 17.1. The ionic structures of important R^{3+} centers found in doped fluorite crystals. R^{3+} appear as filled spheres and F^{-} as open spheres. (a) the tetragonal (C_{4v}) symmetry site of an isolated R^{3+}, labeled the A site for Eu^{3+} and Er^{3+}. (b) a trigonal (C_{3v}) symmetry site of an isolated R^{3+}, labeled the J site in $SrF_2:Eu^{3+}$ and $SrF_2:Er^{3+}$. (c) a dimer R^{3+} cluster, labeled the R cluster in $CaF_2:Eu^{3+}$. (d) a trimer R^{3+} cluster, labeled the Q cluster in $CaF_2:Eu^{3+}$.

the cubic site concentration is more than five times the tetragonal site concentration. All of the site preferences described for low dopant concentration crystals are consistent with the results of model calculations that include pair potentials, ionic polarization, and ion size effects [53, 54].

Site-selective laser spectra have been reported for Pr^{3+} [55, 56], Nd^{3+} [57], Eu^{3+} [58], Tb^{3+} [59], Ho^{3+} [42, 60], Er^{3+} [43, 49], and Tm^{3+} [61] at low concentrations in CaF_2 and SrF_2. Representative spectra are presented for a $SrF_2:(0.05\%)Tb^{3+}$ crystal in Figure 17.2. A broadband excitation spectrum showing the $^7F_6 \rightarrow {}^5D_4$ absorption transitions appears in Figure 17.2(a). It was created by scanning a dye laser over this absorption region while monitoring the integrated fluorescence from a broad 30 nm range centered on the $^5D_4 \rightarrow {}^7F_5$ fluorescence maximum. This spectrum is not site selective and shows all the optically active Tb^{3+} sites. It is effectively an absorption spectrum, as the radiative quantum efficiencies of all these Tb^{3+} sites are approximately 100% [59]. Figures 17.2(b) and (c) are site selective and show the trigonal and tetragonal symmetry Tb^{3+} sites. These spectra were obtained by monitoring the specific $^5D_4 \rightarrow {}^7F_5$ transitions given in the figure caption. Good selectivity was achieved, as the fluorescence transitions of these two sites are very distinct. A minor site was also present in this crystal at low concentrations (Fig. 17.2d), and its origin is discussed in Section 17.3.4.

17.3.2. Defect Chemistry at High Rare-Earth Concentrations

At R^{3+} concentrations higher than ca. 0.05 mol % other sites become prominent. Many of these are sites in clusters of R^{3+}, whose relative populations exhibit a strong dependence on the R^{3+} concentration in the sample. Cluster sites have been studied in $CaF_2:Pr^{3+}$ [55], $SrF_2:Pr^{3+}$ [62], $CaF_2:Eu^{3+}$ [58], $SrF_2:Eu^{3+}$ [45], $CaF_2:Er^{3+}$ [49, 52], $SrF_2:Er^{3+}$ [43], $BaF_2:Er^{3+}$ [50], and $CaF_2:Ho^{3+}$ [60]. Other halide salts studied using these techniques include $CdF_2:Eu^{3+}$ [63], $CdF_2:Er^{3+}$ [64], $PbF_2:Eu^{3+}$ [65], $PbF_2:Er^{3+}$ [66], and $SrCl_2:Eu^{3+}$ [67, 68]. Figure 17.3(a) is a broadband excitation spectrum showing the $^4I_{15/2} \rightarrow {}^4F_{5/2}$ absorption transitions of the Er^{3+} sites in a $CaF_2:(0.2\%)Er^{3+}$ crystal. Five distinct types of site were identified in the site-selective spectra (Figs. 17.3b–f), obtained by monitoring the specific $^4S_{3/2} \rightarrow {}^4I_{15/2}$ transitions given in the figure caption. Sites A and B are most prominent in crystals with lower Er^{3+} concentrations, Site C appears at higher concentrations, while the D(1) and D(2) bands increase dramatically with concentration and are dominant in the 0.2 mol % Er^{3+} crystal [69]. Site A is the isolated Er^{3+} site of tetragonal symmetry, and Site B is an isolated Er^{3+} site of trigonal symmetry. These assignments are consistent with measurements of Zeeman splittings in their $^4I_{15/2}(1) \rightarrow {}^4S_{3/2}(2)$ and $^4F_{5/2}(1)$ transitions [38]. Site C is similar spectrally to Site B, but exhibits more than the $(2J+1)/2$ crystal-field levels expected for

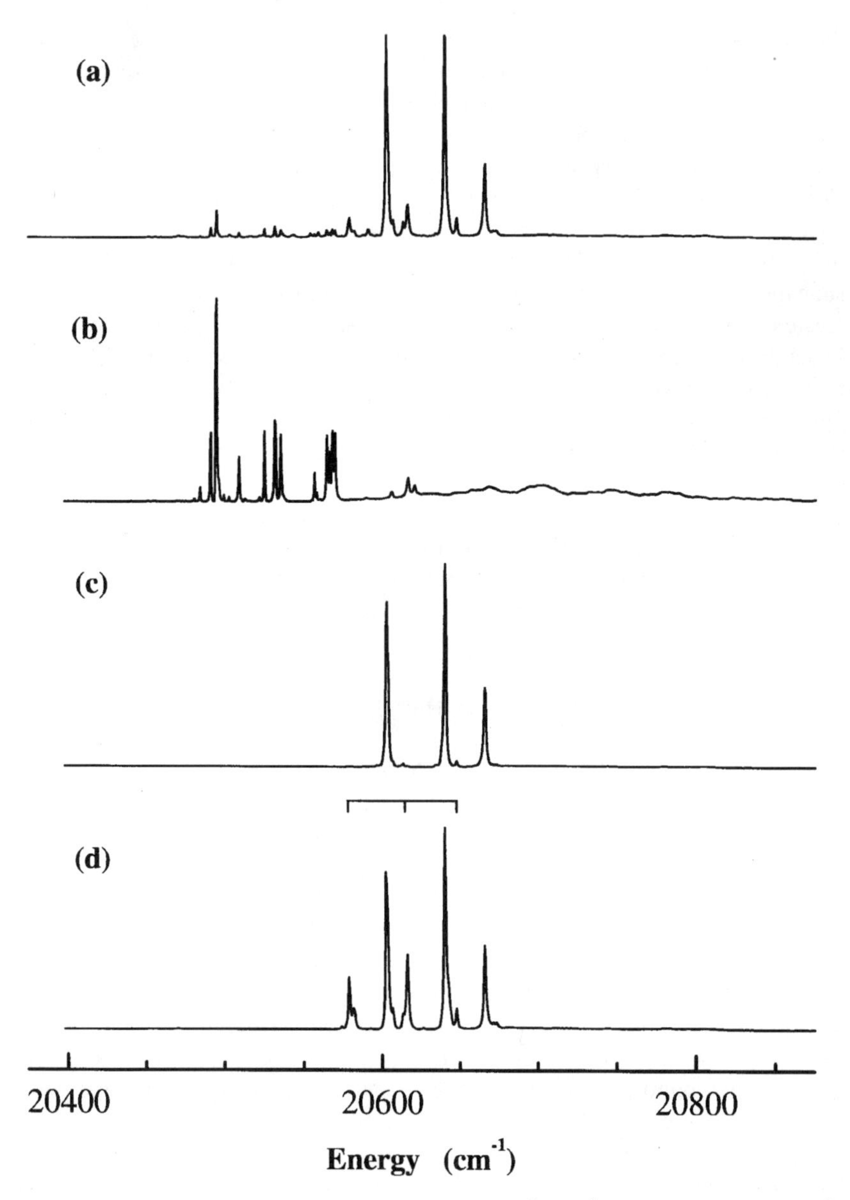

Figure 17.2. (a) Broadband excitation spectrum showing the $^7F_6 \rightarrow ^5D_4$ transitions of all the Tb^{3+} sites in a $SrF_2 : 0.05\% Tb^{3+}$ crystal at 10 K. It was recorded by scanning a dye laser through the absorption region while monitoring the total $^5D_4 \rightarrow ^7F_5$ fluorescence centered at 545 nm. Site-selective excitation spectra were recorded by monitoring fluorescence at specific energies using a double monochromator; (b) a trigonal symmetry site (monitoring 18,459 cm^{-1}); (c) the tetragonal symmetry site (18,494 cm^{-1}); and (d) a tetragonal symmetry mixed site (18,490 cm^{-1}).

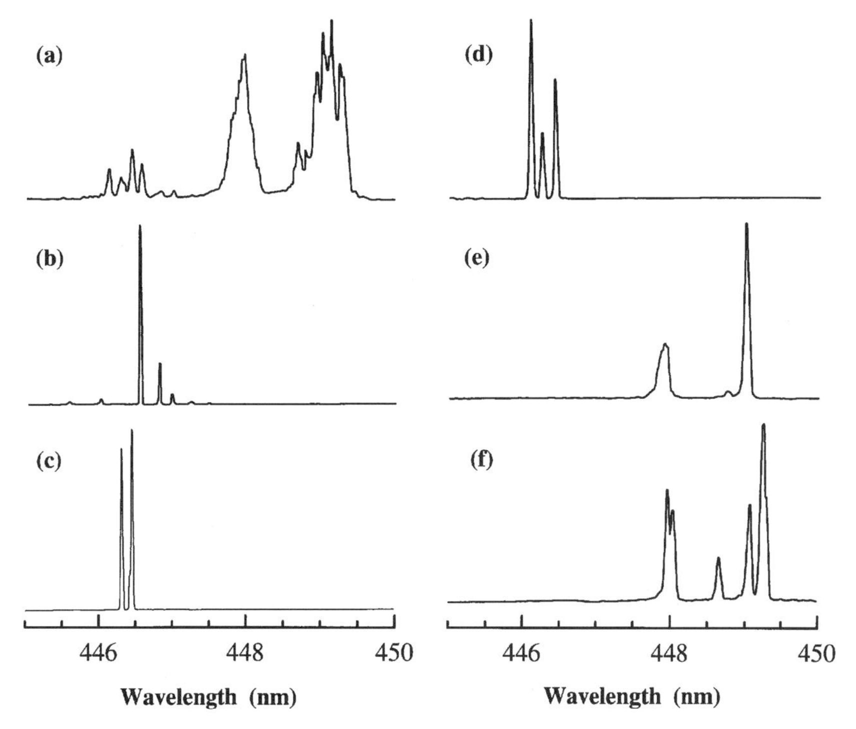

Figure 17.3. (a) Broadband excitation spectrum showing the $^4I_{15/2} \rightarrow {}^4F_{5/2}$ transitions of all the Er^{3+} sites in a $CaF_2 : 0.2\%Er^{3+}$ crystal at 10 K, while monitoring the $^4S_{3/2} \rightarrow {}^4I_{15/2}$ fluorescence. Site-selective excitation spectra of (b) the tetragonal site A (835.4 nm), (c) the trigonal site B (835.8 nm), (d) the cluster site C, (e) the cluster site D(1a) (843.1 nm), and (f) the cluster site D(2a) (654.4 nm). (From Fig. 1 of Ref. 49.)

each multiplet of a single R^{3+}. This multiplicity suggests that it is a cluster with at least two inequivalent Er^{3+}. The D(1) and D(2) absorption bands actually comprise many transitions associated with different sites having similar crystal fields. Although these transitions overlap considerably, they could still be correlated by carefully stepping the monochromator through a fluorescence band and recording a series of excitation spectra. Those lines that appear in repeated spectra with the same relative intensities are associated with the same site. This characterization is demonstrated in Figure 17.4 for a D(1) band in a $CaF_2 : (0.2\%)Er^{3+}$ crystal, for which eleven separate sites were identified, labeled D(1a) through D(1k). Four sites were identified in the D(2) bands by the same method and labeled D(2a) through D(2d). The multiplets of each D(2) site have more than $(2J + 1)/2$ levels; therefore, a D(2) cluster must contain at least two inequivalent Er^{3+}. In contrast, the D(1) sites never have more than $(2J + 1)/2$ levels. Therefore, a D(1) cluster comprises equivalent Er^{3+}.

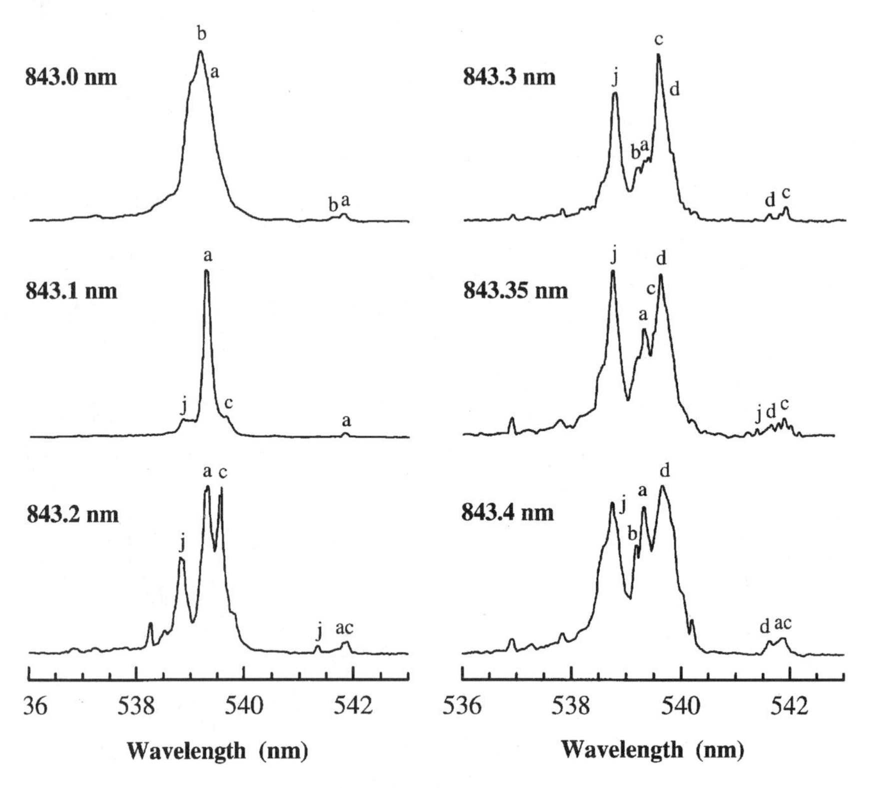

Figure 17.4. Site-selective excitation spectra showing the labeled Er^{3+} sites of the D1 cluster in a $CaF_2:0.2\%Er^{3+}$ crystal. They were recorded while monitoring the fluorescence at the wavelengths indicated. (From Fig. 14 of Ref. 49.)

The observation of efficient upconversion or cross relaxation is also evidence for cluster formation, as both processes depend on efficient energy transfer between R^{3+}. ${}^4S_{3/2} \rightarrow {}^4I_{13/2}$ and ${}^4I_{15/2}$ upconversion fluorescence was observed for the C, D(1), and D(2) sites while selectively pumping their respective ${}^4I_{15/2} \rightarrow {}^4F_{9/2}$ transitions [49]. Neither the A nor B sites exhibit such upconversion, confirming their assignments as isolated Er^{3+} sites. A complicated mechanism was required to account for the fluorescence transients of the C, D(1), and D(2) sites, involving competition between multiphonon relaxation and phonon-assisted energy-transfer processes [70]. Coupled rate equations were derived to describe this relaxation mechanism, and their solutions are the time-varying populations of the excited states. These solutions fit well to the experimental fluorescence transients, confirming this model for the relaxation dynamics of Er^{3+} clusters. No evidence was found for energy transfer between the ions of different Er^{3+} clusters.

Mixed dopant crystals can be useful for identifying R^{3+} cluster sites. Quite different fluorescence transients were measured for Er^{3+} in the C, D(1), and D(2) sites of a mixed $CaF_2:(0.01\%)Er^{3+}:(0.2\%)Yb^{3+}$ crystal, due to efficient cross relaxation from the $^4I_{11/2}$ multiplet of Er^{3+} to the $^2F_{5/2}$ multiplet of Yb^{3+} [70]. The presence of Yb^{3+} at these concentrations also slightly shifted the spectral lines of the C, D(1), and D(2) sites, while the A and B sites remained unperturbed. These results support the previous Er^{3+} site assignments and are also consistent with cluster identifications from dielectric relaxation measurements [71].

Site-selective spectra of clusters in the mixed crystal were used to determine the correspondence between these Er^{3+} cluster sites and the Yb^{3+} sites found in other Yb^{3+} doped crystals [72]. For example, excitation spectra obtained by scanning the laser through an absorption region of Yb^{3+}, while selectively monitoring upconversion fluorescence from a known Er^{3+} site established which specific Yb^{3+} excitation transitions are associated with that site. The Yb^{3+} equivalents of the C, D(1), and D(2) sites were identified in this way. Transitions belonging to Tb^{3+} cluster sites were then identified from site-selective spectra of mixed $CaF_2:Tb^{3+}:Yb^{3+}$ crystals. Energy upconversion from the $^2F_{5/2}$ multiplet of Yb^{3+} to the $^5D_4$ multiplet of Tb^{3+} requires a cooperative mechanism, which is only possible in clusters with at least two Yb^{3+} and one Tb^{3+}. Upconversion fluorescence was observed from the Tb^{3+} equivalents of the D(1) and D(2) sites; so these clusters must comprise more than two R^{3+}, while the C cluster is probably a dimer. Dielectric relaxation measurements support this dimer assignment for the C site [73]. Selective spectra of mixed $CaF_2:Eu^{3+}:Er^{3+}$ [58] and $CaF_2:Tb^{3+}:Yb^{3+}$ [72] crystals show that some of the sites found in Eu^{3+}- and Tb^{3+}-doped crystals are not equivalent to any of the sites reported for Er^{3+} [49]. This result shows again that ion size effects are an important determinant of defect chemistry.

Lifetime measurements were used to identify Gd^{3+} cluster sites in mixed $SrF_2:Gd^{3+}:Ce^{3+}$ crystals [74]. While the Gd^{3+} concentration was fixed, increasing the concentration of Ce^{3+} ions resulted in a disproportionate increase in the number of Gd^{3+} cluster sites. EPR measurements on mixed $SrF_2:Gd^{3+}:Ce^{3+}$ crystals found that the cubic Gd^{3+} site population unexpectedly increased faster than both the tetragonal and trigonal symmetry isolated Gd^{3+} sites as the total R^{3+} concentration was increased from 0.1 to 1 mol % [75]. Similarly, EPR measurements over the Er^{3+} concentration range 0.001 through 0.2 mol % in $CaF_2:Er^{3+}$ showed that the ratio of cubic to A sites was constant at lower concentrations, but increased at higher concentrations [52]. The same anomalous behavior was also exhibited by $CaF_2:Eu^{3+}$ crystals doped with 0.014 and 0.084 mol % Eu^{3+} [58], where the concentration of both sites could be measured optically. These results cannot be accounted for by defect models that assume simple statistical distributions of R^{3+} ions. Measurements of the relative populations of all the R^{3+} sites over a broad range of R^{3+} concentrations are necessary to understand their defect chemistry.

The concentrations of all the noncubic Er^{3+} sites in $CaF_2:Er^{3+}$ were determined by laser-selective spectroscopy for Er^{3+} concentrations between 0.0001 and 1.0 mol % [69]. This study was possible because relaxation mechanisms had already been established for each site [70]. To determine site concentrations from excitation spectra, the absorption cross sections, the quantum efficiency for populating the emitting level, the fluorescence branching ratio of the monitored transition, and the ratio of radiative to nonradiative relaxation must be known for each site. Absorption cross sections were deduced from the branching ratio for fluorescence from the terminal to the initial state of the excitation transition and the radiative lifetime of this terminal state. The final site concentrations are presented in Figure 17.5(a). For Er^{3+} concentrations up to 0.05 mol %, all the sites increased, although the Er^{3+} cluster sites increased much more quickly than the isolated Er^{3+} sites. Above 0.05 mol % the A and B sites actually decreased with Er^{3+} concentration, while the D(1) and D(2) cluster sites continued to increase. Above 0.5 mol % the C site concentration also started to decline. X-ray studies have shown that a hexamer cuboctahedron center with six R^{3+} and twelve F_i^- is dominant at high R^{3+} concentrations [76]. Since optical spectra of a $CaF_2:(10\%)Er^{3+}$ crystal show that D(2a) is the only site present, the D(2) cluster is identified as the cuboctahedron [52]. Calculations show that this center should be preferred for R^{3+} heavier than Gd^{3+}, while the lighter ions will adopt dimer clusters [77], a conclusion supported by dielectric relaxation experiments [78].

An important factor determining the defect chemistry of these R^{3+} sites may be that the cations lose their mobility in the lattice at much higher temperatures than the F_i^- [69]. R^{3+} cluster sites start to form when a crystal is annealed below ca. 1200 K [79] and the R^{3+} distribution becomes fixed at some lower temperature. The F_i^- ions are still mobile and become distributed among the various R^{3+} centers. Scavenging of excess F_i^- by the cluster sites would create negatively charged clusters and explain the excess of cubic sites at high R^{3+} concentrations. Competition for F_i^- results in the dissociation of some isolated R^{3+} ions from their neighboring F_i^-, which creates cubic R^{3+} sites without local charge compensation. It has been suggested that there is a covalent attraction between the F_i^- ions [80]. This interaction would explain the scavenging of additional F_i^- by R^{3+} clusters, as a cluster with equal numbers of R^{3+} and F_i^- would not be locally charge compensated. However, very similar defect behavior was observed in $SrCl_2:Eu^{3+}$ crystals [68]. F^- are smaller and much more electronegative than Cl^-. If covalent interactions were important, there would be significant differences between the defect behavior of fluorides and chlorides. The model of F_i^- scavenging by clusters would account for all the results discussed so far and could

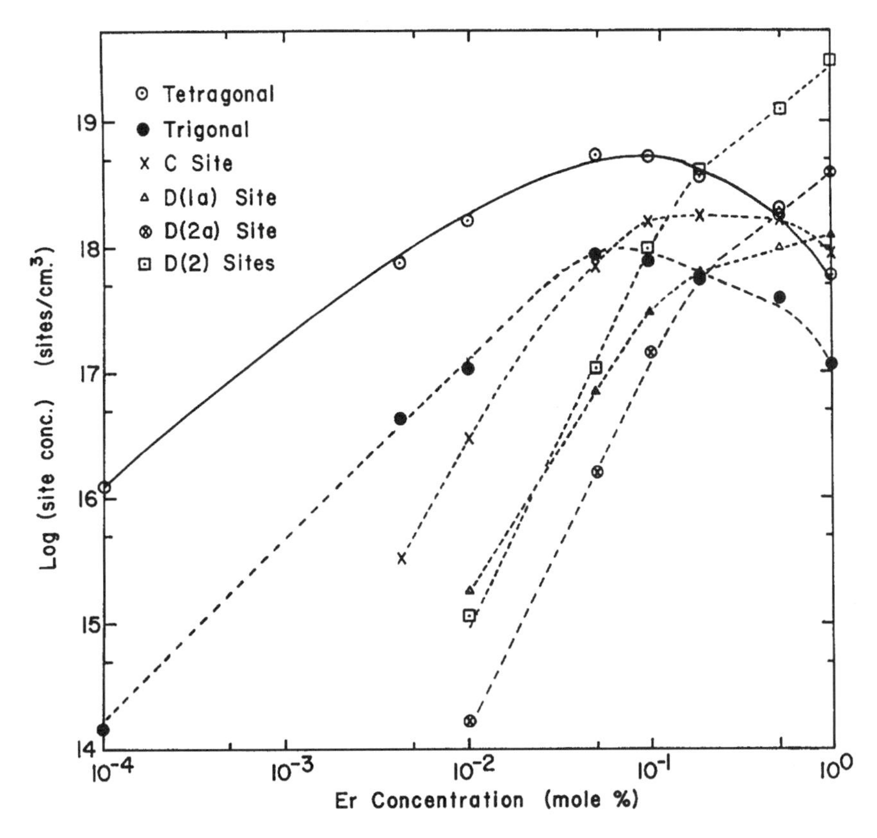

Figure 17.5. (a) Absolute Er^{3+} site concentrations in $CaF_2:Er^{3+}$ crystals for different Er^{3+} concentrations. The D(2) data include all the D(2) sites and a very small contribution from the D(1) site. (From Fig. 3 of Ref. 69)

be tested by measuring the effects of annealing at different temperatures on the relative populations of the R^{3+} sites.

17.3.3. Changes in Defect Chemistry Due to Physical Modifications

Nonselective experiments on CaF_2 doped with moderate concentrations of Er^{3+} have shown that Er^{3+} clusters begin to disappear when the crystals are annealed at temperatures higher than 1200 K and then quenched [79]. Dielectric relaxation studies of annealed $CaF_2:Er^{3+}$ crystals confirmed earlier assignments of the C site to a dimer and the D(2a) site to a larger cluster [71]. The absolute concentrations of all the Er^{3+} sites in crystals of $CaF_2:(0.01)Er^{3+}$ and $CaF_2:(0.1\%)Er^{3+}$ were measured for annealing temperatures between 800 and

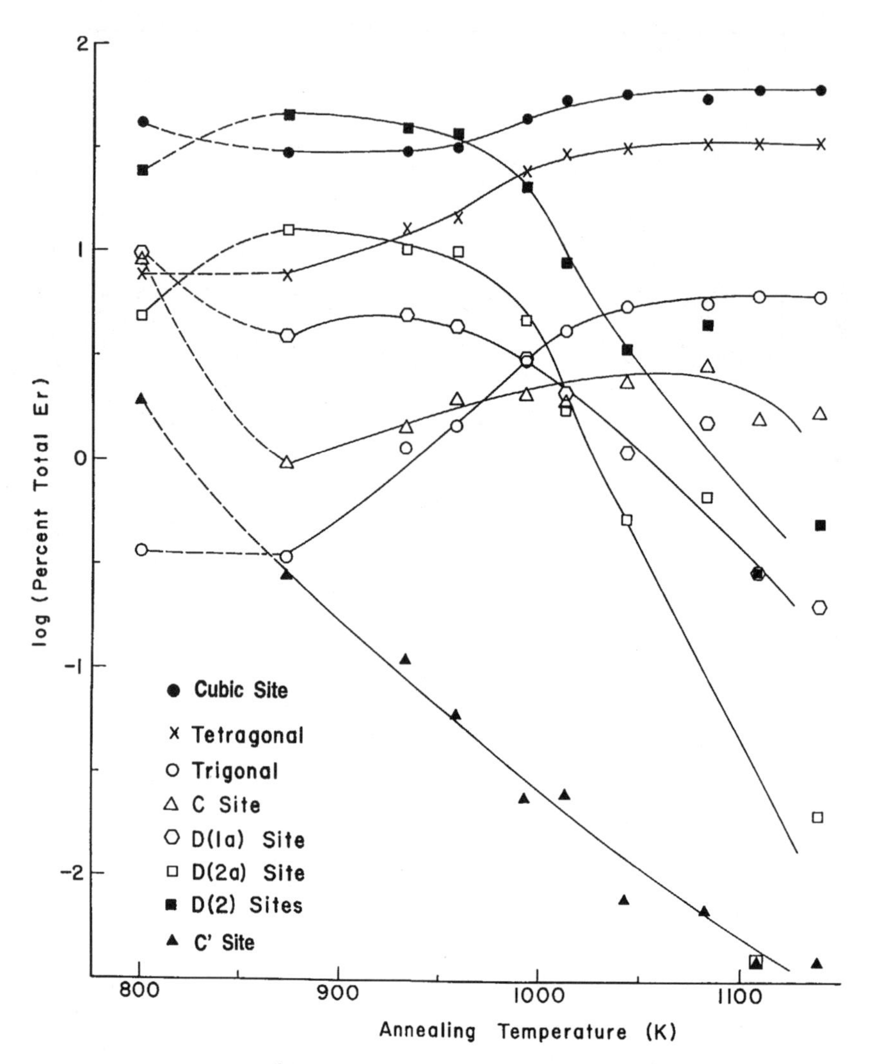

Figure 17.5. (b) Relative Er^{3+} site concentrations in a $CaF_2:0.1\%Er^{3+}$ crystal for different annealing temperatures. The D(2) data include all the D(2) sites. (From Fig. 4 of Ref. 52)

1140 K [52]. Crystals were annealed at each temperature for at least 68 h, until the relative site concentrations reached a steady state, and then quenched. Although they were quenched by immersion in ice water, the cooling time is still a few seconds, which is longer than the migration times characteristic of the cations or F_i^-. Crystals annealed at less than 850 K never attained a thermodynamic equilibrium, and their site concentrations exhibited some dependence on their

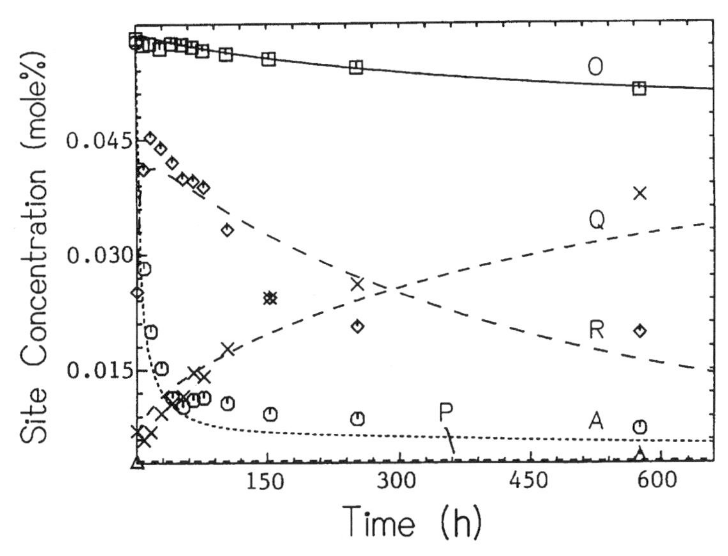

Figure 17.5. (c) Absolute Eu^{3+} site concentrations in a $CaF_2:0.19\%Eu^{3+}$ crystal that was first quenched from 1200 K and then annealed at 835 K for the times indicated. (From Fig. 4 of Ref. 81)

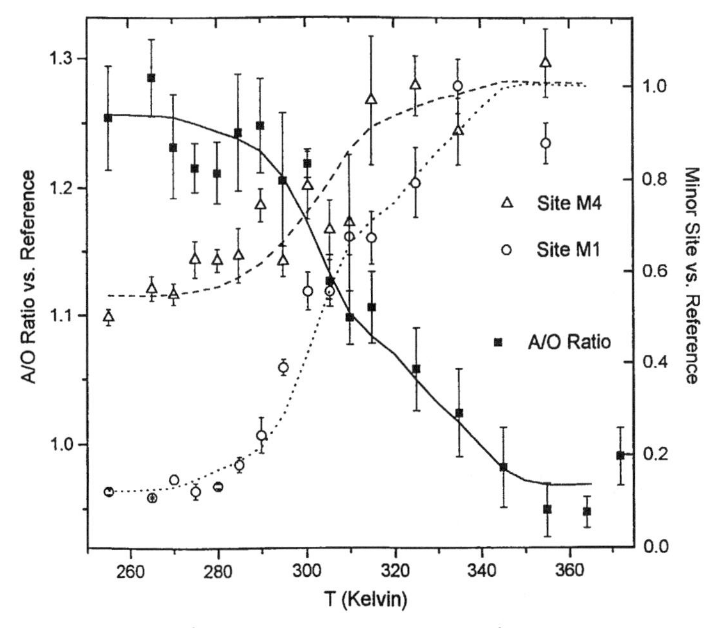

Figure 17.5. (d) Relative Er^{3+} site concentrations in a $CaF_2:Eu^{3+}$ crystal that was treated at high pressure and then annealed at the different temperatures indicated. (From Fig. 4 of Ref. 85; reprinted with permission from Elsevier Science—NL, Amsterdam.)

thermal history. Spectra for 0.1 mol% Er^{3+} are shown in Figure 17.6, while the site concentration changes are summarized in Figure 17.5(b). Generally there was an increase in the concentration of the three isolated Er^{3+} sites and a decrease in the Er^{3+} cluster sites as the annealing temperature was increased. This change is consistent with a model in which Er^{3+} clusters form at lower temperatures and isolated sites are created from their dissociation. The ratio of A to cubic sites increases with annealing temperature as cluster sites break up and return their additional F_i^- to isolated unassociated Er^{3+} sites. At the lowest annealing temperatures, between 800 and 900 K, the D(2) cluster population actually increases with annealing temperature while the C cluster population decreases. This result suggests that D(2) clusters scavenge F_i^- from C clusters, which break up as the crystal is heated through this temperature range. It confirms that both the C and D(2) clusters have an excess negative charge. Two anomalies appear

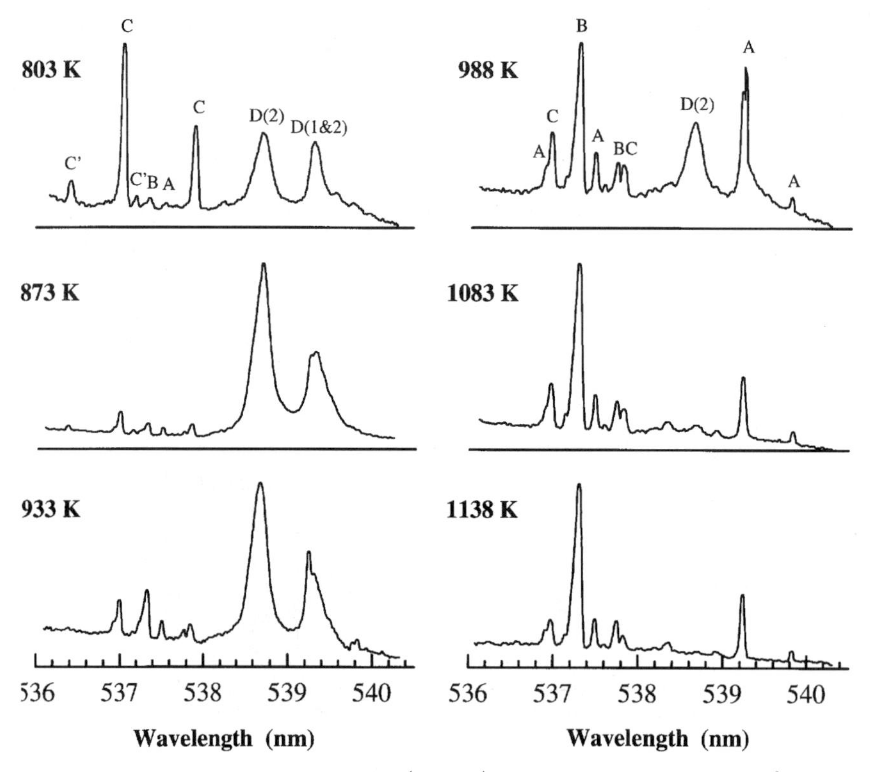

Figure 17.6. Absorption spectra showing the ${}^4I_{15/2} \rightarrow {}^4S_{3/2}$ transitions of a $CaF_2:0.1\%Er^{3+}$ crystal that has been annealed at the different temperatures indicated. (From Fig. 1 of Ref. 123; reprinted with permission from Elsevier Science—NL, Amsterdam.)

when comparing the relative populations of isolated Er^{3+} sites in crystals with different Er^{3+} concentrations that have been annealed at higher temperatures. The ratios of A to cubic sites and A to B sites both decrease with Er^{3+} concentration [52]. These results cannot be accounted for by cluster scavenging, as the clusters dissociate at lower temperatures and cannot affect any high temperature F_i^- equilibrium.

Very similar results were obtained for Eu^{3+} doped CaF_2 crystals [58], with the two Eu^{3+} clusters (labeled R and Q) breaking up at higher annealing temperatures to create isolated Eu^{3+} sites. While the concentration of A sites increased, the cubic site concentration remained almost unchanged, although the dissociation of the clusters should release additional F_i^-. These F_i^- should be attracted to the unassociated Eu^{3+}. In crystals that were quenched from a common high temperature to eliminate clusters, it was found that the ratio of tetragonal to cubic sites was smaller at higher Eu^{3+} concentrations. Both of these results are incompatible with competing equilibrium models that only consider the scavenging of additional F_i^- by clusters.

Relaxation from nonequilibrium site distributions was studied in detail for $CaF_2:Eu^{3+}$ [81]. A nonequilibrium distribution can be created by quenching crystals from high temperatures. Subsequent annealing at a lower temperature slowly restores the equilibrium distribution appropriate for that lower temperature. This process of relaxation is monitored by measuring the absolute site concentrations after different annealing times. The results are shown in Figure 17.5(c) for a $CaF_2:(0.19\%)Eu^{3+}$ crystal that was quenched from ca. 1200 K and then annealed at 835 K for the times indicated. Initially the R site concentration increased and the A site decreased. Then there was a more gradual decline in the A site population. Over hundreds of hours the R site population decreased, accompanied by a commensurate increase in the Q-site population. However, there was no appreciable change in the Eu^{3+} cubic site population during 600 h of annealing. Apparently the formation of R clusters occurs via the aggregation of A sites. These R clusters can then become constituents of Q clusters. This dynamical behavior shows that Eu^{3+} retain their lattice mobility down to 835 K or less. Both of the clusters must scavenge the more mobile F_i^-, as there would otherwise be a decrease in the cubic site population over time. Various kinetic models were tried to establish the specific ionic configurations and dynamical behavior of these centers. The only model that fitted all the data obtained over a range of Eu^{3+} concentrations had an R dimer (comprising two Eu^{3+}, two displaced lattice F^- in interstitial sites, and three F_i^-) and a Q trimer (comprising three R^{3+}, two displaced lattice F^-, and four F_i^-). An R cluster is formed by the union of two A sites and an additional F_i^-. A Q cluster arises when an R cluster combines with another A site. Their ionic configurations are shown in Figures 17.1(c) and (d). These are consistent with the earlier observation that the R cluster comprises inequivalent Eu^{3+} while the Q cluster has equivalent

Eu^{3+} [58]. Both of these structures are predicted to be stable in CaF_2 [77, 82]. The combination of two Q clusters and subsequent scavenging of additional F_i^- would create the hexamer cluster observed for Er^{3+} and smaller R^{3+}. Hexamers with different numbers of F_i^- would account for the various D(2) sites observed in $CaF_2:Er^{3+}$ crystals, as these additional F_i^- change the symmetry of particular Eu^{3+} sites within the cluster.

Quite different rate constants were found for A site ↔ cubic site interconverson while fitting the relaxation kinetics model for different Eu^{3+} concentrations in CaF_2 [81]. The association between Eu^{3+} and F_i^- was weaker at higher Eu^{3+} concentrations. This difference is consistent with the smaller A-to-cubic site ratios observed at high Eu^{3+} concentrations in quenched crystals [58]. In addition, a Eu^{3+} B site was found in mixed dopant CaF_2 crystals with co-dopants smaller than Tb^{3+} [83], even though no trigonal Eu^{3+} site was found in pure $CaF_2:Eu^{3+}$ [58]. All these results are clear evidence for nonideality contributions to the site dynamics. These contributions are equivalent to the Debye–Hückel effects observed in solutions [84]. Coulombic interactions between R^{3+} centers and their surrounding defects will cause nonrandom site distributions that shield charged species and reduce their activity. However, Coulombic interactions are not sufficient to explain all the observed behavior [52]. Strain-mediated interactions will also contribute to the nonrandom distribution of sites, enhancing the screening of charged species. One effect of such nonideality contributions may be to change the attractive interaction between an isolated R^{3+} and its associated F_i^-. This would explain the anomalously high populations of cubic R^{3+} sites found at high R^{3+} concentrations and at high annealing temperatures. It is also likely that R^{3+} cluster to reduce the significant strain caused by incorporating them into the lattice.

F^- are mobile almost down to the 273 K quenching temperature of ice water; so thermal treatments are not sufficient to create truly nonequilibrium site distributions. Pressure treatments are an alternative way to create nonequilibrium distributions. The relaxation dynamics of Eu^{3+} sites in CaF_2 crystals was investigated following hydrostatic compression to above 30 kbar in a diamond-anvil cell [85, 86]. Pressure alters the relative free energies of defect structures, and F_i^- will migrate between R^{3+} centers to minimize strain energy. However, pressure also reduced F_i^- mobility; so the cell was heated to at least 670 K during compression. The new F_i^- distribution was preserved by cooling the cell down to 170 K prior to releasing the pressure. An ambient pressure equilibrium was then restored by annealing the crystal for a set time at successively higher temperatures. This relaxation was monitored spectroscopically after each cycle of heating and cooling. Figure 17.7 shows the relative intensities of sites in $CaF_2:(0.19\%)Eu^{3+}$ before and after pressure treatment. These changes were less dramatic than those produced by thermal annealing in Figures 17.5(b) and (c). Only the A site, the cubic site, and three minor sites (M1, M4, and N) were affected significantly by pressure treatment. The cubic site intensity decreased by 19%, while the A site increased by 5%. Allowing for their relative oscillator strengths, these changes corresponded to comparable changes in absolute site concentration. Their recovery to an ambient pressure distribution is plotted in Figure 17.5(d) for different annealing temperatures. At these temperatures only the F_i^- will be mobile. The ratio of A sites to cubic sites ("O" sites in the figure) decreased at annealing temperatures above 290 K and was fully restored at 355 K. All three minor sites decreased in intensity after pressure treatment and were restored by annealing. These sites probably have association energies similar to the A site and compete for F_i^-. The M sites were attributed to Eu^{3+} sites in perturbed R clusters, while the N site was identified as an isolated Eu^{3+} site [86]. Other centers, such as the clusters Q and R, were unaffected by the pressure treatment and must have higher association energies.

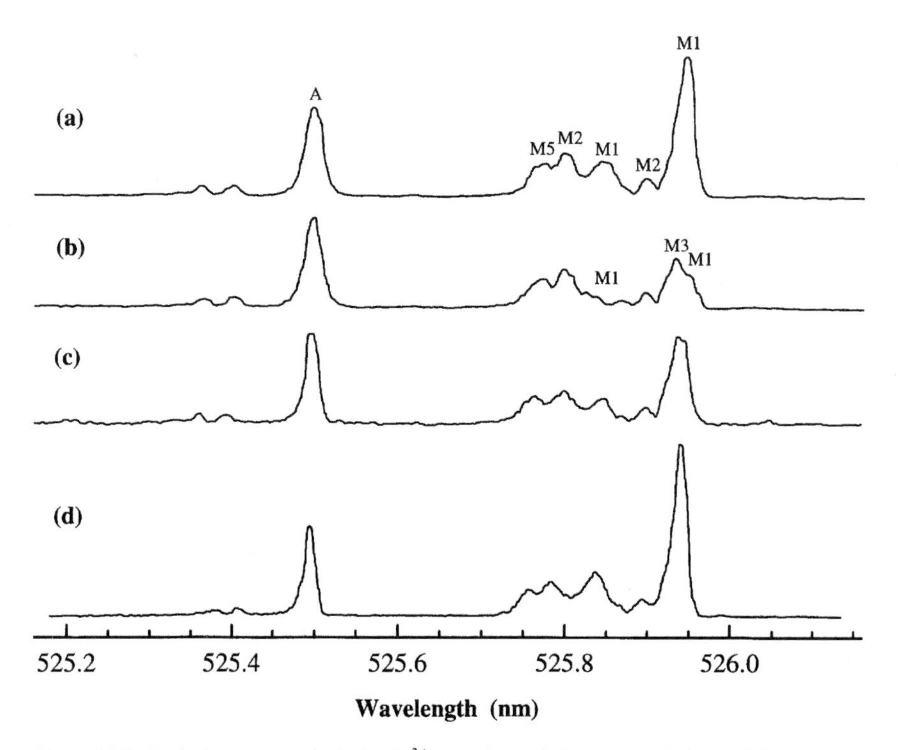

Figure 17.7. Excitation spectra of a $CaF_2 : Eu^{3+}$ crystal recorded at 12 K (a) before and (b) after high-pressure treatment. Spectra were then recorded (c) after annealing for 30 min at 305 K and (d) after annealing for another 30 min at 335 K. These were recorded while selectively monitoring the $^5D_0 \rightarrow {}^7F_1$ fluorescence from these minor sites. (From Fig. 3 of Ref. 85; reprinted with permission from Elsevier Science—NL, Amsterdam.)

The changes that were observed have been fitted to a model of F_i^- diffusion between R^{3+} centers with slightly different F_i^- association [85, 86]. The dissociation of any particular center must be characterized by a single binding energy, as the sharp lines in site-selective spectra indicate relatively homogeneous environments. However, the site relaxation kinetics could only be fitted by including at least four Arrhenius energies, over the range 1.04 through 1.37 eV. Each Arrhenius energy will comprise a binding and a diffusion energy, so the results suggests that at least four different diffusion constants apply to F_i^- migration between the A site and the other F_i^- traps. The binding energy of the Eu^{3+} and F_i^- in an A site was measured previously in conductivity experiments as 0.68 eV [87]; so the four diffusional energies were calculated as 0.70, 0.84, 0.94, and 1.03 eV. The cluster sites M1 and M2 were identified with the lower diffusional energies. Thus the lattice region around one of these clusters has both higher strain and increased F_i^- mobility, a manifestation of the same strain-mediated interdefect interactions that promote R^{3+} clustering and cause nonideality effects in the F_i^- distributions. The inverse relationship between strain and F_i^- mobility was confirmed by the observation that no site changes were observed at pressures greater than 20 kbar at 670 K [86]. This observation shows that F_i^- lose their mobility completely at higher pressures.

The existence of cubic R^{3+} sites in these rare-earth-doped crystals indicates the presence of F^- traps. At the high pressures created by hydrostatic compression F^- are forced out of some traps and become associated with isolated R^{3+} to reduce the overall strain energy. The F^- traps could be other R^{3+} centers or lattice defects unrelated to the R^{3+}.

The most important features of the model to explain the defect chemistry of R^{3+} centers in fluorite crystals can now be summarized. Each R^{3+} admitted into the lattice is charge compensated by an F_i^-, although the two ions may not be immediately associated. The presence of R^{3+} defects in the lattice causes considerable strain, which is minimized by defect ion clustering. R^{3+} can diffuse through CaF_2 at temperatures above 800 K. For annealing temperatures below 870 K the site distribution is determined by the enthalpy component of the free energy and R^{3+} clusters are dominant. Above 1070 K isolated R^{3+} sites are preferred. F_i^- remain mobile down to room temperature and diffuse between the various R^{3+} centers to minimize the strain energy. Their association energy is a function of entropy, strain, and Coulombic forces and is generally higher for clusters than for isolated R^{3+}. Each R^{3+} can associate with a F_i^-, and each cluster can trap additional F_i^-; so the number of potential F_i^- trap sites exceeds the number of F_i^-. R^{3+} centers and other defects thus compete for F_i^-. Electronegative clusters of up to six R^{3+} become stable in crystals annealed at lower temperatures and with high R^{3+} concentrations, while many isolated R^{3+} are not compensated locally producing a substantial population of cubic symmetry R^{3+} sites.

17.3.4. Changes in Defect Chemistry Due to Chemical Modifications

Figure 17.2(d) is a spectrum of a mixed Tb^{3+} site in $SrF_2:Tb^{3+}$, formed from a tetragonal symmetry Tb^{3+} site by substitution of an on-axis NN Sr^{2+} by a Ca^{2+}. On-axis substitution preserves the tetragonal symmetry and explains the similarity of the mixed and parent site spectra. The mixed site appears as a satellite of the main tetragonal symmetry Tb^{3+} site. Their transitions are so close in energy that the dominant parent site could not be excluded in this nominally site-selective spectrum. Trace quantities of CaF_2 are known to contaminate the SrF_2 starting material used to grow these crystals. Such mixed sites have been studied extensively in Pr^{3+} [88], Er^{3+} [89], and Ho^{3+} [42] doped mixed crystals of the type $Ca_{1-x}Sr_xF_2$. Their transitions always appear as satellites to those of their parent R^{3+} sites. Nine mixed sites were found for Pr^{3+} and polarization measurements were used to determine their symmetries. Sites of C_{4v}, C_s, and C_1 symmetry were found, indicating that both on- and off-axis substitutions occur around the tetragonal symmetry parent. The lower-symmetry sites could also be identified spectroscopically from crystal-field splittings of the γ_5 doublet states of the C_{4v} symmetry parent site. Similar results were obtained for mixed Ho^{3+}-doped crystals. Such mixed sites may explain some of the unidentified isolated R^{3+} sites found in CaF_2 and SrF_2 crystals doped with other R^{3+}.

Anion substitutions also create new R^{3+} sites. Fluorites react with H_2O or O_2 at high temperatures, with each O^{2-} ion substituting for a lattice F^- and creating an additional F^- vacancy. O^{2-} also produce a charge-compensated isolated R^{3+} site, labeled G1, with a divalent O^{2-} replacing one of the eight lattice F^- surrounding the trivalent R^{3+} [40]. The defect chemistry of oxygen-compensated R^{3+} sites in fluorite crystals has been investigated both experimentally [90, 91] and theoretically [92].

Reduction of the fluorites by contact with hot aluminum in an atmosphere of hydrogen introduces H^- ions into the crystal lattice. These H^- replace F^- in regular lattice sites of cubic symmetry [93]. Near a single charge-compensated R^{3+}, an H^- ion will substitute preferentially for its associated F_i^- ion, without changing the R^{3+} site symmetry [94]. At higher hydrogen concentrations, multihydrogen R^{3+} centers appear as H^- begin to replace lattice F^- around the R^{3+}. These substitutions will break the axial symmetry of the parent site. The symmetry of a specific R^{3+} site depends on the number of H^- and their lattice positions. The same sites are observed for all three isotopes of hydrogen, with differences in their spectra arising from small isotopic shifts in the crystal-field levels and their quite different local-mode vibrational energies. The terms *hydrogenation* and *hydrogenic* will be taken to include all three hydrogen isotopes.

Multihydrogenic ion R^{3+} centers have been studied in CaF_2 and SrF_2 crystals doped with low concentrations of Pr^{3+} [95, 96], Nd^{3+} [57], Tb^{3+} [59], Ho^{3+}

[42], Er^{3+} [44, 97], and Tm^{3+} [61]. Each R^{3+} is associated with a family of multihydrogen centers, because of the number of different anion substitutions that are possible. The concentration of multihydrogenic R^{3+} centers increases over a period of months if a hydrogenated crystal is stored at room temperature [95]. This result suggests that the hydrogenic ions are mobile and become associated with R^{3+} to reduce the overall strain energy. In all the multihydrogenic R^{3+} centers whose ionic configurations could be determined, it was found that the substitutional hydrogenic ions occupy only those lattice sites between the R^{3+} and its charge-compensating H^-.

The five multihydrogenic centers of Pr^{3+}, CS(1) through CS(5), are the most studied. Site-selective spectra of their ${}^3H_4 \rightarrow {}^3P_0$ excitation transitions are shown in Figure 17.8. The parent tetragonal center has a doublet $|{}^3H_4\Gamma_5\gamma_5>$ ground state, so splitting of this transition for the centers CS(1) through CS(4) indicates Pr^{3+} site symmetries that are lower than C_{4v}. A ${}^3H_4 \rightarrow {}^1I_6$ transition also appears in these spectra. This intermultiplet transition is spin forbidden, but gains intensity due to high-order interactions facilitated by the proximity of the ${}^1I_6(\gamma_1)$ and ${}^3P_0(\gamma_1)$ levels. Centers of C_{2v}, C_s, or C_1 symmetry can be created by anion substitution around a Pr^{3+}. Polarization measurements were used to determine the specific Pr^{3+} site symmetry of each CS center [96]. This was possible because the symmetries of many crystal-field states of the C_{4v} parent center had been identified previously [56]. Group theory predicts differences in polarization behavior between Pr^{3+} sites where the C_4 axial symmetry is broken by a component in a parallel (010) plane and those where there is a component in a diagonal (110) plane. The CS(1), CS(3), and CS(4) centers were found to have diagonal symmetry planes consistent with C_s or C_{2v} symmetry, while the CS(2) center has a parallel symmetry plane consistent with C_{2v} symmetry. The CS(5) center retains the C_{4v} symmetry of the parent center.

Excited-state lifetime measurements are a useful indicator of the number of hydrogenic ions about a R^{3+}. Hydrogenic ions in F^- sites have large local-mode vibrational energies. The fundamental excitation is $965\,cm^{-1}$ for an H^- and $694\,cm^{-1}$ for a D^- in the cubic symmetry lattice site of CaF_2 [93]. Low symmetry splittings are observed in the vibrations of hydrogenic ions in R^{3+} centers: two levels for the interstitial hydrogenic ion of a C_{4v} symmetry center [94] and three levels for the substitutional sites of multihydrogenic centers [98]. These local modes couple to the R^{3+} electronic states. The coupling is evident from the observation of hydrogenic vibronic transitions in optical R^{3+} spectra [94], the quenching of R^{3+} lifetimes through nonradiative relaxation [99], and the splittings in H_i^- vibrational modes due to electron–phonon coupling with the R^{3+} [100]. The isotopic differences in the vibrational energies dramatically affect the lifetimes of those R^{3+} states that are quenched by multiphonon relaxation through the hydrogenic local modes. Higher-order relaxation mechanisms are required for heavier isotopes, because of their smaller local mode energies; so the

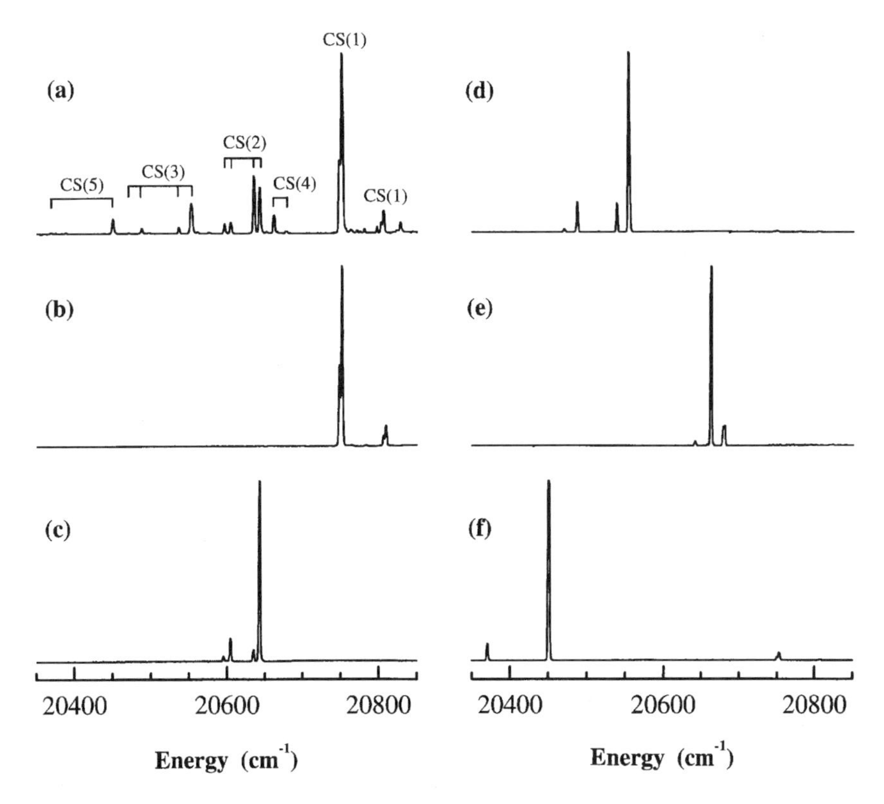

Figure 17.8. (a) A broadband excitation spectrum showing the $^3H_4 \rightarrow ^3P_0$ transitions of all the multihydrogenic Pr^{3+} CS sites in a deuterated $SrF_2:(0.05\%)Pr^{3+}$ crystal at 10 K. It was recorded while monitoring the $^1D_2 \rightarrow ^3H_4$ fluorescence centered at approximately 615 nm. Site-selective spectra were recorded by monitoring the fluorescence at specific energies using a double monochromator: (b) the CS(1) site, (c) the CS(2) site, (d) the CS(3) site, (e) the CS(4) site, and (f) the CS(5) site.

quenching is less efficient. Each hydrogenic ion in an R^{3+} center contributes to the nonradiative relaxation. As the substitutional hydrogenic sites are nearly equivalent, the total contribution of the substitutional hydrogenic ions to the nonradiative relaxation rate is proportional to their number. Lifetime measurements show that the CS(1) center has one substitutional hydrogenic ion, the CS(2) and CS(4) centers have two each, the CS(3) center has three, and the CS(5) center has four [101]. These results are consistent with earlier models for some of these Pr^{3+} centers [95]. Their ionic configurations are illustrated in Figure 17.9.

Multihydrogenic R^{3+} centers are noteworthy in that they exhibit permanent fluorescence bleaching at 10 K [99]. Bleaching is essentially spectral hole-burning where the laser bandwidth is comparable to the linewidth of the

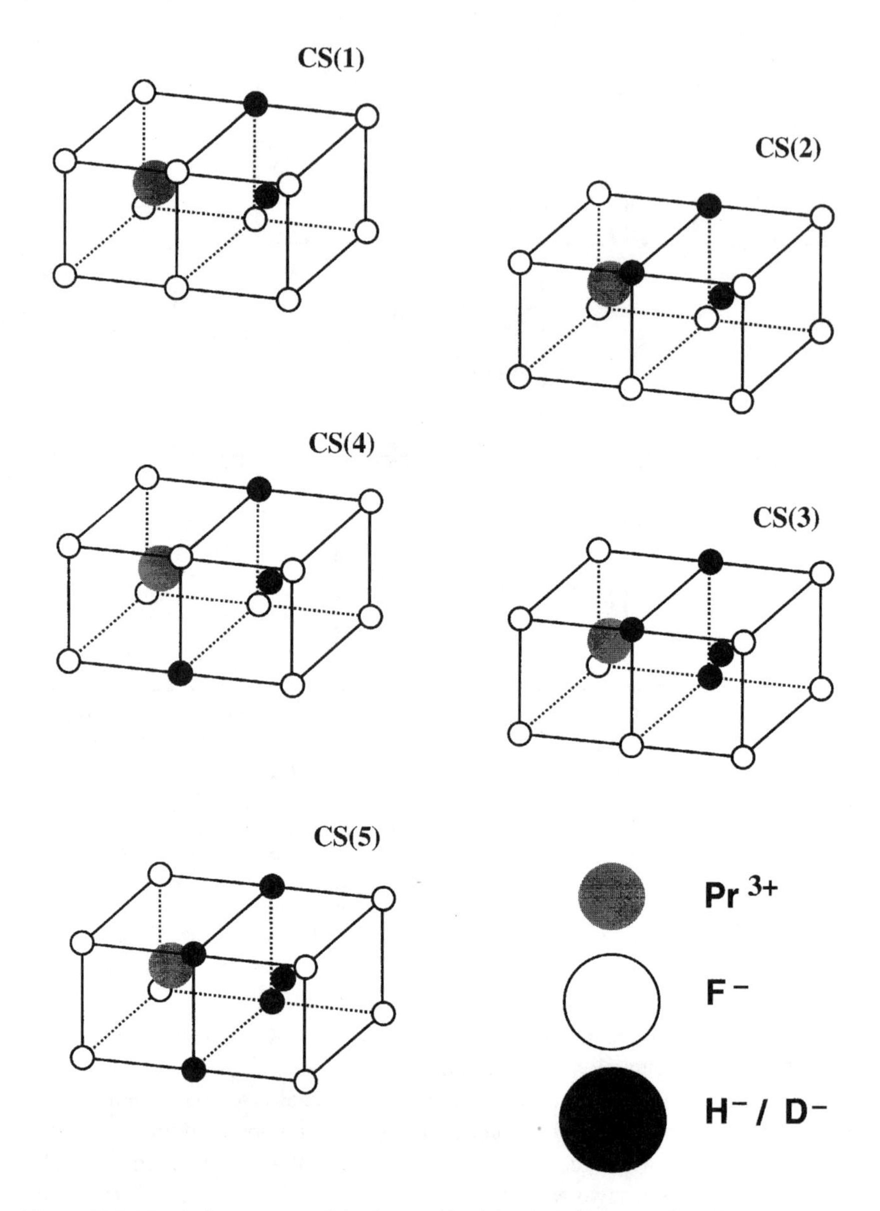

Figure 17.9. The ionic structures of the five multihydrogenic Pr^{3+} centers identified in CaF_2 and SrF_2. The ions in the key are drawn to the same scale as the fluorite lattice.

inhomogeneously broadened absorption transition. Spectral hole-burning within absorption transitions of the CS(1) through CS(4) centers has also been observed using narrow-bandwidth lasers [102]. Two types of bleaching are observed [103]. *Reorientational bleaching* requires a linearly polarized laser beam and an oriented crystal. Fluorescence intensity that is lost during selective excitation in one polarization can be recovered by irradiation in an orthogonal polarization. *Photoproduct formation bleaching* produces new photoproduct centers with distinct excitation energies. Bleaching occurs in either polarization, with no fluorescence recovery on switching the laser polarization. Selective excitation of these new photoproduct centers restores the fluorescence intensity of the original centers.

A variety of bleaching behaviors is exhibited by the Pr^{3+} CS centers in ⟨100⟩ oriented crystals [95]. This is demonstrated in Figure 17.10, where each plot shows the fluorescence intensity recorded during selective excitation of a CS center. The laser beam propagated along one of the ⟨100⟩ crystallographic axes. Its polarization was periodically switched between two orthogonal planes, labeled the vertical and horizontal planes in the figure, to monitor the effects of polarized bleaching. The CS(1) center demonstrates fully reversible reorientational bleaching. Indefinite cycles of fluorescence bleaching and recovery are possible by switching the laser polarization (Fig. 17.10a). The CS(2) center exhibits partially reversible reorientational bleaching, with the intensity of the recovered fluorescence decreasing over successive cycles (Fig. 17.10b). Bleaching is accompanied by the creation of a new photoproduct center, labeled CS(2)*. The CS(3), CS(4), and CS(5) centers exhibit photoproduct formation bleaching. The polarization bleaching curve in Fig. 17.10(e) for the CS(4) center is typical for these centers. The fluorescence observed in each excitation polarization bleaches independently. New photoproduct centers are created, labeled CS(3)*, CS(4)*, and CS(5)*, respectively. Bleaching these photoproduct centers restores their original parent centers.

Fluorescence bleaching arises from a photophysical process [103]. R^{3+} excited by a laser can relax nonradiatively by transferring energy to the local-mode vibrations of neighboring hydrogenic ions. Excited hydrogenic ions can then migrate to different lattice positions by an interstitialcy noncollinear mechanism. This mechanism involves a substitutional hydrogenic ion moving to a vacant neighboring interstitial position and being replaced by the original charge-compensating interstitial hydrogenic ion. Some centers are reoriented to produce equivalent centers. This accounts for reorientational bleaching, since some of the orientations are not excited by the linearly polarized laser beam. However, these changes can be reversed by switching the laser beam to an orthogonal polarization. In other cases distinct centers are created that are also rotated with respect to the original centers. These centers have different absorption energies and can be reverted back to their original ionic configurations by selective excitation. Figure

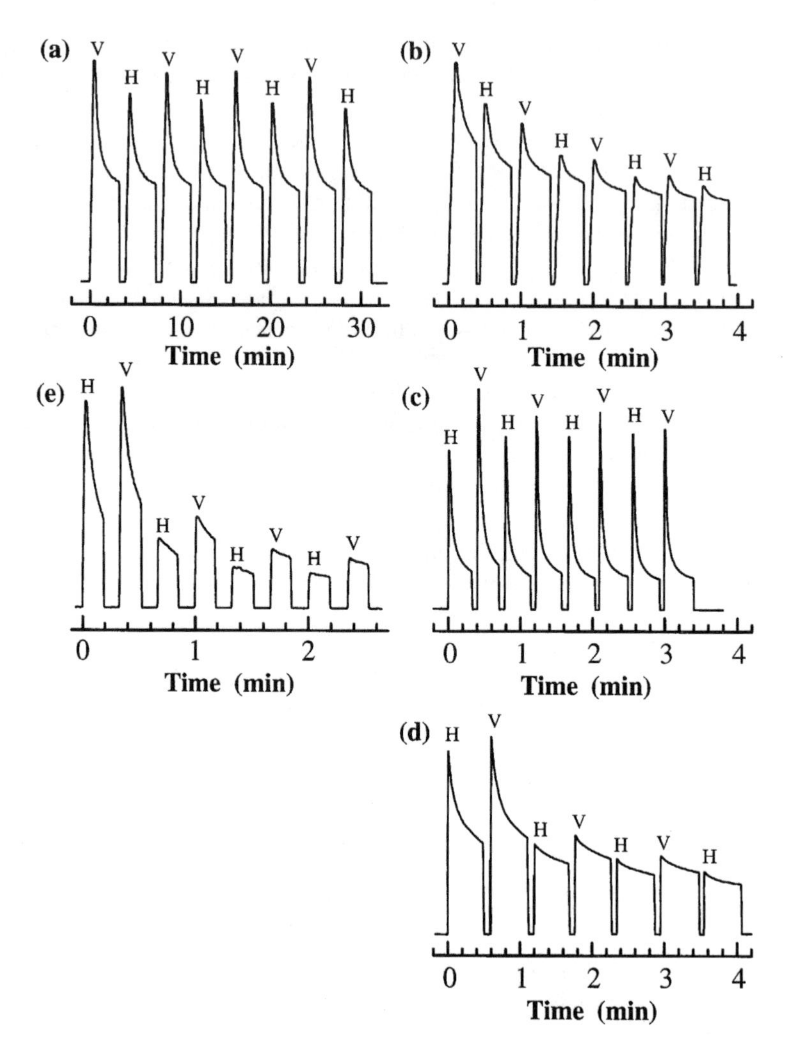

Figure 17.10. Polarized fluorescence bleaching curves for the multihydrogenic Pr^{3+} centers in a deuterated SrF_2 crystal. "V" and "H" denote vertically or horizontally polarized laser excitation. (a) For the CS(1) center exciting the ${}^3H_4(1) \rightarrow {}^1D_2(1)$ transition while monitoring the broadband ${}^1D_2 \rightarrow {}^3H_4$ fluorescence. (b) For the CS(2) center exciting the ${}^3H_4(1) \rightarrow {}^1D_2(1)$ transition while monitoring the broadband ${}^1D_2 \rightarrow {}^3H_4$ fluorescence. (c) For the CS(2) center exciting the ${}^3H_4(1) \rightarrow {}^3P_0$ transition while monitoring the X polarized ${}^1D_2(1) \rightarrow {}^3H_4(2)$ fluorescence. (d) For the CS(2) center exciting the ${}^3H_4(2) \rightarrow {}^3P_0$ transition while monitoring the X polarized ${}^1D_2(1) \rightarrow {}^3H_4(1)$ fluorescence. (e) For the CS(4) center exciting the ${}^3H_4(1) \rightarrow {}^1D_2(1)$ transition while monitoring the broadband ${}^1D_2 \rightarrow {}^3H_4$ fluorescence. (From Fig. 3 of Ref. 95 and Figs. 10 and 11 of Ref. 96.)

17.11 illustrates the specific mechanisms proposed to account for the bleaching behavior of all the Pr^{3+} CS centers [95, 96]. In each case bleaching rotates the CS center by 90°.

Similar reorientation of tetragonal Gd^{3+} centers was detected when they were thermally excited in EPR and dielectric loss experiments [104]. F^-, H^-, and D^- charge-compensated centers exhibited this behavior. Calculations showed that the proposed interstitialcy collinear mechanism has the smallest possible activation energy [54]. No bleaching of the tetragonal single-hydrogenic ion centers has been observed [42, 44, 59, 95, 97]. The local-mode energy of the heavier F^- is too low for efficient excitation by energy transfer from the R^{3+}. Even when a lattice F^- is excited in this way it is apparent that it does not have sufficient vibrational energy to migrate between sites and thus initiate bleaching.

Bleaching rates were found to be a factor of 10 lower for D^- than for H^- varieties of the Pr^{3+} CS centers [95]. Again this difference reflects the relative efficiencies for exciting the local mode vibrations of the different hydrogen isotopes. The CS* photoproduct centers were found to bleach much more rapidly than their original CS centers under identical excitation conditions.

All the possible orientations of a multihydrogenic center are equally populated in an unexposed crystal. Bleaching changes this distribution of centers by preferentially populating those center orientations and/or photoproduct centers that are not resonant with the polarized laser beam. A nonequilibrium distribution is created that has different polarization behavior. These polarization changes can be predicted for specific bleaching mechanisms using the symmetry selection rules for electric-dipole transitions. Polarized bleaching experiments were conducted for all the Pr^{3+} CS centers, and these confirmed the proposed models in Figure 17.11 [96].

The CS(2) center is particularly interesting because it undergoes both reorientational and photoproduct formation bleaching. There are six possible orientations of a C_s symmetry center with parallel symmetry planes in a fluorite lattice. Some of these orientations will be selectively excited when pumping a transition of a particular symmetry. Fluorescence will then be detected from those orientations selectively monitored using a monochromator and a polarization analyzer. In Figure 17.10(c), excitation and fluorescence transitions were chosen to observe an ensemble of CS(2) centers reorienting between two orthogonal orientations that were alternately excited. Fluorescence intensity that was bleached in one polarization was recovered by switching the laser polarization. Very little decrease in the recovered intensity was observed after the first two cycles. This result suggests that reorientational bleaching is more probable than photoproduct formation bleaching for the CS(2) center. In Figure 17.10(d) different excitation and fluorescence transitions were chosen so that reoriented CS(2) centers were still resonant with the laser beam and no fluorescence intensity was lost by reorientational bleaching. However, photoproduct formation

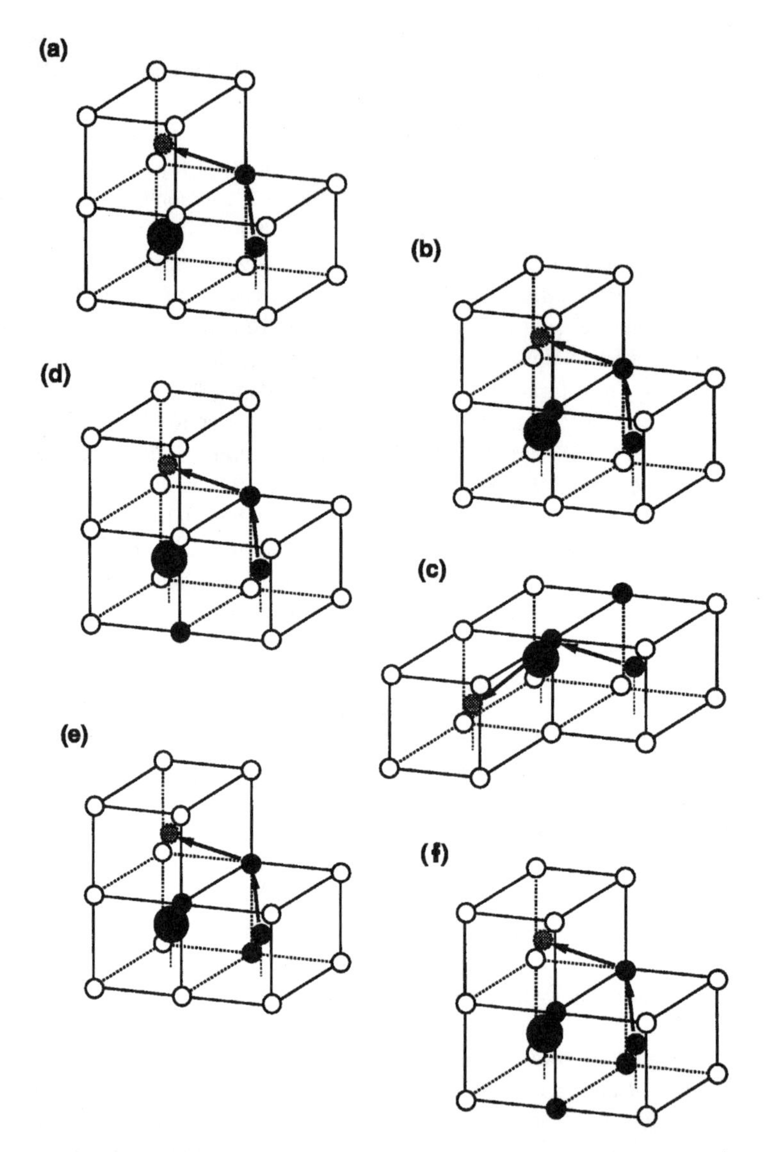

Figure 17.11. The proposed mechanisms for the bleaching of the five multihydrogenic Pr^{3+} centers: (a) reorientational bleaching of the CS(1) center, (b) reorientational bleaching of the CS(2) center, (c) photoproduct formation bleaching of the CS(2) center, (d) photoproduct formation bleaching of the CS(4) center, (e) photoproduct formation bleaching of the CS(3) center, and (f) photoproduct formation bleaching of the CS(5) center. (From Fig. 9 of Ref. 96.)

bleaching still occurs. Different CS(2) orientations were excited in each laser polarization; so two ensembles of CS(2) centers bleached independently to create CS(2)* photoproduct centers. The polarized bleaching curves in Figures 17.10(c) and (d) show separately the effects of reorientational and photoproduct formation bleaching on the CS(2) fluorescence. Selective excitation in just one polarization, "V" or "H" in Figure 10(d), creates only CS(2)* centers rotated by 90° from the excited CS(2) centers. This preferential orientation was confirmed in polarized spectra of the CS(2)* center recorded with a low-power laser beam following such polarization selective bleaching of the CS(2) center.

Bleaching effects are permanent while the crystal is maintained at low temperatures, but can be reversed by heating the crystal to above 120 K [101]. Thermal excitation restores the equilibrium distribution of centers and the photoproduct centers revert back to their original center configurations. The specific temperatures for re-equilibration were measured for all the Pr^{3+} CS centers in SrF_2 and CaF_2 [96, 101]. In each case a lower temperature was found in CaF_2 than in SrF_2, as the smaller CaF_2 lattice produces stronger coupling between the Pr^{3+} and hydrogenic ions. For a given CS center this temperature was only weakly dependent on the hydrogen isotope [101], and the re-equilibration process was therefore independent of the mass of the mobile ion. This result eliminates tunneling processes, which have an explicit mass dependence, and shows that bleaching and re-equilibration are barrier crossing processes [105]. In the proposed model a vibrationally excited center can cross the barrier in the double-well potential associated with the initial and final center configurations. For two equivalent orientations of a center that undergoes reorientational bleaching, an almost symmetric double-well potential is appropriate. Very small isotopic differences were found in the recovery temperatures, and these can be accounted for by the different zero-point energies of the corresponding double-well potentials. A double-well potential for a center and its discrete photoproduct is necessarily asymmetric. The re-equilibration temperature results for the CS(2) center and the absence of any CS* centers at 10 K in crystals cooled from room temperature establish that the CS center configurations have much lower energy minima than the CS* configurations. The multidimensional potential surface in configuration space representing possible rearrangements of a multihydrogenic center is very complicated, particularly in the case of the CS(2) center, which can adopt either equivalent CS(2) or inequivalent CS(2)* ionic configurations.

The measured thermal re-equilibration temperatures were used to determine the approximate barrier energies [96, 105]. For example, the barrier for a D^- CS(2) center reorientation in SrF_2 was calculated to be approximately 2340 cm^{-1}, while the barrier for a D^- CS(2)* center reverting to a D^- CS(2) center was approximately 1900 cm^{-1}. These values can be compared to the activation energy of approximately 3200 cm^{-1} measured for the reorientation of Gd^{3+} tetragonal centers in CaF_2 by EPR line broadening [104]. It is noteworthy that the CS reverting temperatures were almost independent of the hydrogen isotope, even though the bleaching rates of the H^- and D^- centers were quite different. This observation emphasizes the point that the bleaching rate is determined by both the rate at which these hydrogenic ions are excited and the size of the potential barrier to be surmounted.

These examples have demonstrated the great variety of defect centers that can be created by doping fluorites with other anions and cations. Monovalent anions and divalent cations can substitute into regular lattice sites without charge compensation. However, their inclusion still causes strain, which can be minimized by preferential substitution in sites neighboring other defects, particularly R^{3+}. Other ions are charge compensated, often by the inclusion of F_i^- or F^--vacancies, which also prefer to associate with defects. Hydrogenation produces families of hydrogenic R^{3+} centers through the association of R^{3+} and hydrogenic ions. These ions are strongly coupled by the electron–phonon interaction. New centers and nonequilibrium distributions are created by the migration of the associated hydrogenic ions following energy transfer from the optically excited R^{3+}. Laser site-selective experiments have probed how vibrationally excited hydrogenic R^{3+} centers can cross the energy barriers between different possible ionic configurations.

17.4. TRACE ANALYSIS OF METALS BY SELECTIVE EXCITATION OF AN ASSOCIATED FLUORESCENT PROBE

The principles governing the defect chemistry of rare-earth-doped fluorites are now well understood and this knowledge can be applied to other problems in chemistry. One application that has been demonstrated is the identification and measurement of ultratrace quantities of metals [106], particularly nonfluorescent metal ions such as La^{3+}, Ce^{3+}, Gd^{3+}, Lu^{3+}, Y^{3+}, Sc^{3+}, Th^{4+}, and Zr^{4+} [107, 108]. Although these analyte ions cannot be observed directly, they can associate with a fluorescent probe ion in a cluster center, perturbing the crystal field at the probe ion site. The crystal-field levels of a probe ion associated with a particular analyte ion are unique and can be selectively excited so that the analyte ion can be detected at very low concentrations.

Initially all the analyte and probe ions must be incorporated into a fluorite lattice. This has been achieved by precipitating microcrystals of CaF_2 from aqueous solutions containing the sample to be analyzed [109]. Coprecipitation provides preconcentration and separation, as well as an ordered environment for efficient fluorescence. The Ca^{2+} host ion and R^{3+} probe ions were dissolved as nitrates and mixed with a measured amount of the sample. NH_4F was added to precipitate most of the calcium in solution. Experiments showed that all the R^{3+}

ions were incorporated into the CaF_2 coprecipitate [106]. After being filtered and dried the coprecipitate was heated above 300°C for approximately 3 h and allowed to cool. Ignition incorporated oxygen into the lattice producing O^{2-} charge-compensated R^{3+} sites and the subsequent annealing caused these sites to cluster. The presence of the trivalent R^{3+} increases the rate of oxygen incorporation and the O^{2-} displace any charge-compensating F_i^-. The addition of lithium and potassium to the starting solution as nitrate salts increases the number of F^- vacancies in the CaF_2 lattice [91] and so improves the anion and cation mobility [110, 111].

At R^{3+} concentrations of approximately 0.02 mol % and ignition temperatures above 500°C, the dominant oxygen-compensated sites are the single R^{3+} site G1 and a dimer R^{3+} site [91]. At lower temperatures some F_i^- compensated sites persist and the cations do not attain equilibrium. Above 550°C three multiple oxygen sites appear in crystals containing the lighter R^{3+}. These sites are labeled G2, G3, and G4, and comprise a single R^{3+}, a F^- vacancy, and two, three, and four O^{2-}, respectively [91]. They are derived from the G1 site by the successive substitution of O^{2-} for pairs of nearest-neighbor lattice F^-. Annealing over a period of at least an hour enhances the G1 center at the expense of these multiple oxygen sites. The G1 site concentration was found to be insensitive to changes in the analyte ion concentration when this is much lower than the probe ion concentration [107]. Thus the intensity of the G1 fluorescence can be used as an internal reference to account for changes in the optical alignment and positioning of individual samples. The relative intensity of the mixed dimer site is a measure of the analyte ion concentration in the crystal. The quantity of analyte ions in the original sample can then be calculated stoichiometrically.

A number of difficulties must be considered when using this technique. When the analyte concentration is very small, fluorescence from the dominant probe–probe dimers may overwhelm that from the mixed probe–analyte dimers. This interference can be overcome by using a probe ion that has self- quenching levels, such as Er^{3+}, and time gating the detector to measure fluorescence from longer lifetime states of the mixed site [107]. Nd^{3+} can be used as the probe for the analyte Yb^{3+}, which itself quenches Er^{3+} fluorescence [108].

Another problem is chemical interferences that arise when the analyte affects the formation of O^{2-} compensated sites or the formation of R^{3+} clusters or even creates separate phases of the precipitates. Larger R^{3+} have higher R^{3+}–F^- association energies, while lighter R^{3+} have higher R^{3+}–O^{2-} association energies; so probe and analyte ions compete for the available O^{2-}. The interference effects of numerous other anions and cations have also been investigated [91, 106]. It was found that the types of R^{3+} clusters formed are governed by the defect chemistry of the highest concentration R^{3+} [91], which is usually the probe ion, with larger clusters arising in the presence of smaller R^{3+}.

17.5. FLUORESCENCE LINE-NARROWING SPECTROSCOPY IN RARE-EARTH-DOPED MATERIALS

In the applications described above the sites being discriminated generally had different ionic configurations; so their electronic transitions were quite distinct. However, all hosts will contain defects, particularly any R^{3+} dopants, which alter the environment about each site. Thus what appears spectroscopically as a single site is in many cases an ensemble of subsites, with the same ionic configuration locally, surrounded by a random distribution of defects. Fluorescence line narrowing (FLN) can be used to resolve close-lying inhomogeneously broadened transitions or to determine the distribution of subsites within particular inhomogeneously broadened transitions.

Cm^{3+} ions have a $5f^7$ electronic configuration and a $^8S_{7/2}$ ground state. Although a pure S state should not be split by a crystal field, the $^8S_{7/2}$ ground state of Cm^{3+} was split in all the hosts that have been investigated. These splittings ranged from $2\,cm^{-1}$ in $LaCl_3$ [112] to $36\,cm^{-1}$ in ThO_2 [113]. They arise from the large spin-orbit coupling of Cm^{3+} that mixes higher $J = \frac{7}{2}$ terms into the ground-state wavefunction. As for Eu^{2+} and Gd^{3+}, the lanthanide analogues of Cm^{3+} that have $4f^7$ ground states, several mechanisms contribute to split these states [114]. It is difficult to quantify these mechanisms a priori, so it is important to obtain precise measurements of the splittings and to identify the resulting levels experimentally, particularly in the case of Cm^{3+}, where this effect is most pronounced. Although EPR has been useful for measuring ground-state splittings in the lanthanides, it has not worked well for the larger splittings of Cm^{3+}. Laser spectroscopy has been successfully applied to this task, using FLN to resolve these splittings when they are obfuscated by inhomogeneous broadening. All the isotopes of Cm^{3+} are radioactive; so as well as the broadening inherent in doped crystals, radiation damage causes additional strain broadening over time.

Figure 17.12 shows the $^6D_{7/2} \leftrightarrow {}^8S_{7/2}$ transitions of $^{248}Cm^{3+}$ in $LuPO_4$ [115]. This isotope is comparatively long lived, with a half-life of 340,000 years. The broadband fluorescence spectrum in Figure 17.12(a) shows that these transitions had inhomogeneously broadened FWHM line widths of $2.3\,cm^{-1}$ in this sample. The four crystal-field levels of the ground state are barely resolved. In the line-narrowed fluorescence spectrum of Figure 17.12(b), a subset of sites was selectively excited by the laser. The observed fluorescence linewidths were reduced to the FWHM linewidth of the laser, which was approximately $0.8\,cm^{-1}$, and the ground-state levels are now clearly resolved. This spectrum demonstrates an artifact that is often observed in line-narrowing spectroscopy. Additional satellite lines appear because the laser excited subsites from the intended absorption transition and also from the other inhomogeneously broadened transitions that overlap it spectrally. These satellites appear at common

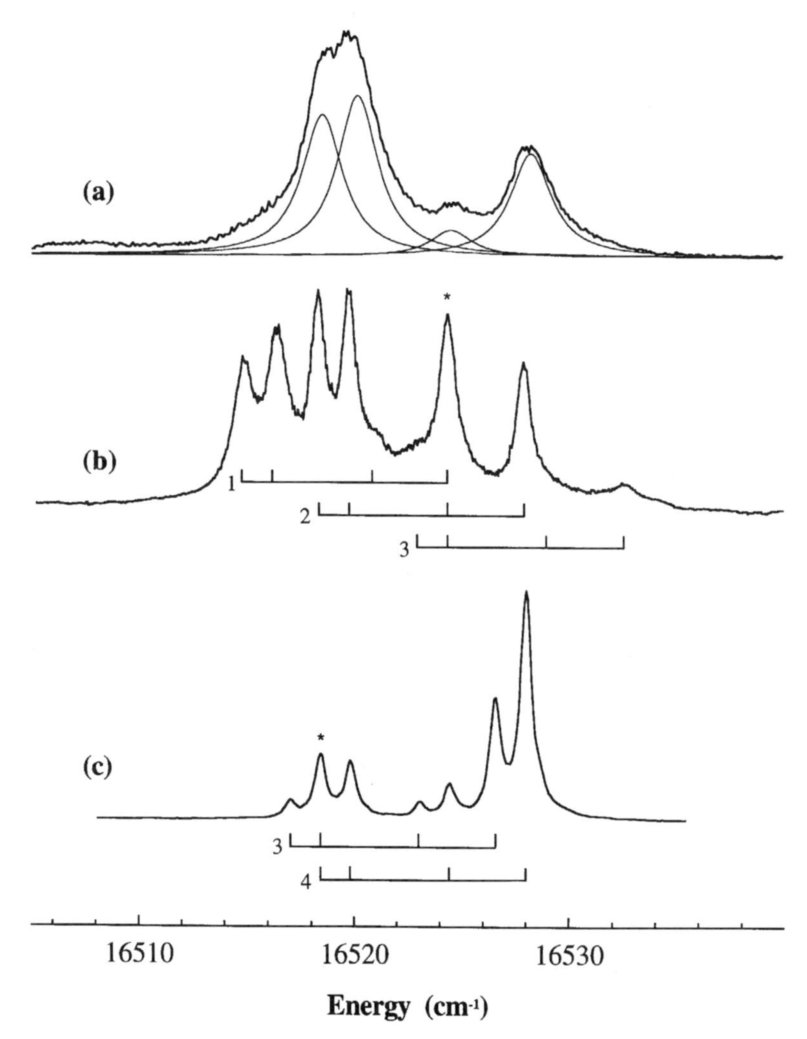

Figure 17.12. (a) The inhomogeneously broadened ${}^6D_{7/2} \rightarrow {}^8S_{7/2}$ fluorescence transitions of Cm^{3+} diluted in a crystal of $LuPO_4$. They have been fitted to Lorentzian functions with linewidths of $2.3\,cm^{-1}$. This spectrum was recorded during broadband excitation at 400 nm. (b) A fluorescence spectrum obtained by selective excitation of subsites in the ${}^8S_{7/2}(2) \leftrightarrow {}^6D_{7/2}(1)$ transition (*) at $16{,}524.5\,cm^{-1}$. The numbers n indicate the originating levels ${}^8S_{7/2}(n)$ of inhomogeneously broadened excitation transitions resonant with the pump laser. (c) An excitation spectrum obtained by selectively monitoring subsites in the $6D_{7/2}(1) \leftrightarrow {}^8S_{7/2}(4)$ transition (*) at $16{,}518.5\,cm^{-1}$. The numbers n indicate the terminating levels ${}^8S_{7/2}(n)$ of inhomogeneously broadened transitions seen by the monochromator. (From Fig. 1 of Ref. 115.)

displacements from the principal lines, equal to one of the crystal-field splittings of the ground state. As one of the satellites is always coincident with the transition being monitored (indicated by a "*" in the figure), this transition appears to be enhanced compared to the others. Satellite lines appear in both fluorescence and excitation spectra (Fig. 17.12b and c, respectively). Their relative intensities can be minimized by selecting a strong transition that is distant from any other strong transitions, in this case the $^8S_{7/2}(1) \leftrightarrow ^6D_{7/2}(1)$ transition at 16,528.0 cm^{-1}. FLN has been used to measure the ground-state splittings of Cm^{3+} in $LuPO_4$ [115], $LaCl_3$ [112, 116], and Cs_2NaYCl_6 [117]. FLN spectroscopy has also been performed on $LuPO_4$ crystals containing the $^{244}Cm^{3+}$ iosotope, which has a half-life of only 18 years, and generates considerably more structural disorder [118].

The weak splitting of the Cm^{3+} ground state can also degrade absorption or broadband laser excitation measurements of excited crystal-field levels. In hosts with weak crystal fields, all the $^8S_{7/2}$ levels are populated at liquid helium temperature, and absorption lines are broadened because they contain contributions from each of these levels. The broadening can be mitigated by sequential excitation, with the first laser exciting the metastable $^6D_{7/2}(1)$ level at 16,528 cm^{-1} and the second laser exciting higher-lying levels by excited-state absorption. The first laser pulse selects a subset of sites within one of the $^8S_{7/2} \rightarrow ^6D_{7/2}$ transitions. Only these subsites can be excited from the isolated $^6D_{7/2}(1)$ level by the second laser pulse and detected as anti-Stokes fluorescence by the monochromator. This technique has been demonstrated for Cm^{3+} in $LuPO_4$ [115] and $LaCl_3$ [116]. It also extends the range of excitation energies available to tunable lasers and has been used to measure Cm^{3+} crystal-field levels up to 40,000 cm^{-1} in $LuPO_4$ [115], $LaCl_3$ [116], Cs_2NaYCl_6 [117], and $CsCdBr_3$ [119]. Two-photon absorption has also been applied to obtain excitation spectra of the high-lying levels of the actinides U^{6+} [120] and Cm^{3+} [121]. However, the small oscillator strengths of two-photon transitions has proved limiting in studies of the trans-uranic elements, which are very radioactive and can only be included in optical hosts at low concentrations.

17.6. CONCLUSIONS

The applications of laser site-selective spectroscopy that have been presented in this chapter have all concerned processes in rare-earth-doped systems that could not have been understood without utilizing this powerful experimental technique. Experiments on doped fluorite crystals have investigated their defect chemistry. It has been shown that this is largely determined by the overall strain caused by the inclusion of these defects. At low temperatures, different defects associate and adopt ionic configurations that minimize the strain energy. At higher temperatures, they dissociate, and isolated defect centers are preferred. The particular ionic configurations that arise are determined by the mobilities of the defect species and the thermal history of the crystal. One application of this association of defects is the detection and measurement of trace quantities of metals. Fluorescence line-narrowing techniques have added utility to the general method of laser site-selective spectroscopy. Future experiments seek to achieve site selectivity with molecular species through the polarization enhancements caused by vibrational resonances in nonlinear four-wave mixing spectroscopy [122].

REFERENCES

1. R. M. Macfarlane and R. M. Shelby, in A. A. Kaplyanskii and R. M. Macfarlane, eds., *Modern Problems in Condensed Matter Sciences*, Vol. 21, 1987, p. 51.
2. K. K. Rebane and L. A. Rebane, in W. E. Moerner, ed., *Topics in Current Physics*, Springer Verlag, New York, Vol. 44, 1988, p. 17.
3. S. Völker, *Ann. Rev. Phys. Chem.* **40**, 499 (1989).
4. R. M. Macfarlane, in B. Di Bartolo, ed., *NATO ASI Series*, Plenum Press, New York, Series B: Vol. 339, 1994, p. 151.
5. R. M. Macfarlane and R. M. Shelby, in W. E. Moerner, ed., *Topics in Current Physics*, Springer Verlag, New York, Vol. 44, 1988, p. 127.
6. K. Holliday and U. P. Wild, in S. G. Schulman, ed., *Chemical Analysis*, John Wiley & Sons, New York, Vol. 77, 1993, p. 149.
7. N. B. Manson, M. J. Sellars, P. T. H. Fisk, and R. S. Meltzer, *J. Lumin.* **64**, 19 (1995).
8. R. M. Macfarlane, R. M. Shelby, and D. P. Burum, *Opt. Lett.* **6**, 593 (1981).
9. N. B. Manson, Z. Hasan, P. T. H. Fisk, R. J. Reeves, G. D. Jones, and R. M. Macfarlane, *J. Phys.: Condens. Matter* **4**, 5591 (1992).
10. J. P. D. Martin, T. Boonyarith, N. B. Manson, M. Mujaji, and G. D. Jones, *J. Phys.: Condens. Matter* **5**, 1333 (1995).
11. D. M. Boye, R. M. Macfarlane, Y. Sun, and R. S. Meltzer, *Phys. Rev. B* **54**, 6263 (1996).
12. K. Holliday, C. Wei, M. Croci, and U. P. Wild, *J. Lumin.* **53**, 227 (1992).
13. R. M. Macfarlane and R. M. Shelby, *J. Lumin.* **36**, 179 (1987).
14. T. Schmidt, R. M. Macfarlane, and S. Völker, *Phys. Rev. B* **50**, 15707 (1994).
15. R. Wannemacher, J. M. A. Koedijk, and S. Völker, *J. Lumin.* **60 & 61**, 437 (1994).
16. P. J. van der Zaag, J. P. Galaup, and S. Völker, *Chem. Phys. Lett.* **174**, 467 (1990).
17. F. T. H. den Hartog, M. P. Bakker, R. J. Silbey, and S. Völker, *Chem. Phys. Lett.* **297**, 314 (1998).
18. R. T. Harley, M. J. Henderson, and R. M. Macfarlane, *J. Phys. C: Solid State Phys.* **17**, L233 (1984).

19. D. Redman, S. Brown, and S. C. Rand, *J. Opt. Soc. Am. B* **9**, 768 (1992).
20. R. Bauer, A. Osvet, I. Sildos, and U. Bogner, *J. Lumin.* **56** 57 (1993).
21. B. Henderson and K. P. O'Donnell, in W. M. Yen, ed., *Topics in Applied Physics*, Springer Verlag, New York, Vol. 65, 1989, p. 151.
22. A. J. Meixner, A. Renn, and U. P. Wild, *J. Chem. Phys.* **91**, 6728 (1989); A. Renn, A. J. Meixner, and U. P. Wild, *J. Chem. Phys.* **92**, 2748 (1990); A. Renn, A. J. Meixner, and U. P. Wild, *J. Chem. Phys.* **93**, 2299 (1990); K. Holliday, C. Wei, A. J. Meixner, and U. P. Wild, *J. Lumin.* **48 & 49**, 329 (1991); S. Bernet, B. Kohler, A. Rebane, A. Renn, and U. P. Wild, *J. Opt. Soc. Am. B* **9**, 987 (1992).
23. W. R. Babbitt and T. W. Mossberg, *Opt. Commun.* **65**, 185 (1988).
24. X. A. Shen, A. D. Nguyen, J. W. Perry, D. L. Huestis, and R. Kachru, *Science* **278**, 96 (1997).
25. A. Winnacker, R. M. Shelby, and R. M. Macfarlane, *Opt. Lett.* **10**, 350 (1985).
26. P. B. Sellin, N. M. Strickland, J. L. Carlsten, and R. L. Cone, *Opt. Lett.* **24**, 1038 (1999).
27. A. Szabo, *Phys. Rev. Lett.* **25**, 924 (1970).
28. M. Grinberg, P. I. Macfarlane, B. Henderson, and K. Holliday, *Phys. Rev. B* **52**, 3917 (1995).
29. R. M. Macfarlane, *J. Lumin.* **45**, 1 (1990).
30. P. M. Selzer, in W. M. Yen and P. M. Selzer, eds., *Topics in Applied Physics*, Springer Verlag, New York, Vol. 49, 1986, p. 115.
31. W. M. Yen and P. M. Selzer, in W. M. Yen and P. M. Selzer, eds., *Topics in Applied Physics*, Springer Verlag, New York, Vol. 49, 1986, p. 141.
32. G. P. Morgan and W. M. Yen, in W. M. Yen, ed., *Topics in Applied Physics*, Springer Verlag, New York, Vol. 65, 1989, p. 78.
33. R. W. Equall, Y. Sun, R. L. Cone, and R. M. Macfarlane, *Phys. Rev. Lett.* **72**, 2179 (1994).
34. R. M. Macfarlane, T. L. Harris, Y. Sun, R. L. Cone, and R. W. Equall, *Opt. Lett.* **22**, 871 (1997).
35. R. M. Shelby, C. S. Yannoni, and R. M. Macfarlane, *Phys. Rev. Lett.* **41**, 1739 (1978).
36. R. M. Shelby, A. C. Tropper, R. T. Harley, and R. M. Macfarlane, *Opt. Lett.* **8**, 304 (1983).
37. H. Talon, L. Fleury, J. Bernard, and M. Orrit, *J. Opt. Soc. Am. B* **9**, 825 (1992).
38. C. W. Rector, B. C. Pandey, and H. W. Moos, *J. Chem. Phys.* **45**, 171 (1966).
39. P. A. Forrester and C. F. Hempstead, *Phys. Rev.* **126**, 923 (1962).
40. U. Ranon and A. Yaniv, *Phys. Lett.* **9**, 17 (1964).
41. M. R. Brown, K. G. Roots, J. M. Williams, W. A. Shand, C. Groter, and H. F. Kay, *J. Chem. Phys.* **50**, 891 (1969).
42. M. Mujaji, G. D. Jones, and R. W. G. Syme, *Phys. Rev. B* **46**, 14398 (1992).
43. M. D. Kurz and J. C. Wright, *J. Lumin.* **15**, 169 (1977).

44. N. J. Cockroft, G. D. Jones, and R. W. G. Syme, *J. Chem. Phys.* **92**, 2166 (1990).
45. J. P. Jouart, C. Bissieux, G. Mary, and M. Egee, *J. Phys. C: Solid State Phys.* **18**, 1539 (1985).
46. K. Lesniak and F. S. Richardson, *J. Phys.: Condens. Matter* **4**, 1743 (1992).
47. M. V. Eremin, R. K. Luks, and A. L. Stolov, *Sov. Phys.—Solid State* **12**, 2820 (1971).
48. H. K. Welsh, *J. Phys. C: Solid State Phys.* **18**, 5637 (1985).
49. D. R. Tallant and J. C. Wright, *J. Chem. Phys.* **63**, 2074 (1975).
50. M. P. Miller and J. C. Wright, *J. Chem. Phys.* **68**, 1548 (1978).
51. D. S. Moore, Ph.D. Thesis, University of Wisconsin, Madison (1980) [University Microfilms International, Ann Arbor, Michigan].
52. D. S. Moore and J. C. Wright, *J. Chem. Phys.* **74**, 1626 (1981).
53. C. R. A. Catlow, *J. Phys. C: Solid State Phys.* **9**, 1845 (1976).
54. J. Corish, C. R. A. Catlow, P. W. M. Jacobs, and S. H. Ong, *Phys. Rev. B* **25**, 6425 (1982).
55. B. M. Tissue and J. C. Wright, *Phys. Rev. B* **36**, 9781 (1987).
56. R. J. Reeves, G. D. Jones, and R. W. G. Syme, *Phys. Rev. B* **46**, 5939 (1992).
57. T. P. J. Han, G. D. Jones, and R. W. G. Syme, *Phys. Rev. B* **47**, 14706 (1993).
58. R. J. Hamers, J. R. Wietfeldt, and J. C. Wright, *J. Chem. Phys.* **77**, 683 (1982).
59. K. M. Murdoch, G. D. Jones, and R. W. G. Syme, *Phys. Rev. B* **56**, 1254 (1997).
60. M. B. Seelbinder and J. C. Wright, *Phys. Rev. B* **20**, 4308 (1979).
61. N. M. Strickland and G. D. Jones, *Phys. Rev. B* **56**, 10916 (1997).
62. C. D. Cleven, S. H. Lee, and J. C. Wright, *Phys. Rev. B* **44**, 23 (1991).
63. S. Mho and J. C. Wright, *J. Chem. Phys.* **77**, 1183 (1982).
64. S. Mho and J. C. Wright, *J. Chem. Phys.* **81**, 1421 (1984).
65. F. J. Weesner, J. C. Wright, and J. J. Fontanella, *Phys. Rev. B* **33**, 1372 (1986).
66. S. Mho and J. C. Wright, *J. Chem. Phys.* **79**, 3962 (1983).
67. J. R. Wietfeldt and J. C. Wright, *J. Chem. Phys.* **83**, 4210 (1985).
68. J. R. Wietfeldt and J. C. Wright, *J. Chem. Phys.* **86**, 400 (1987).
69. D. R. Tallant, D. S. Moore, and J. C. Wright, *J. Chem. Phys.* **67**, 2897 (1977).
70. D. R. Tallant, M. P. Miller, and J. C. Wright, *J. Chem. Phys.* **65**, 510 (1976).
71. J. J. Fontanella, D. J. Treacy, and C. G. Andeen, *J. Chem. Phys.* **72**, 2235 (1980).
72. M. B. Seelbinder and J. C. Wright, *J. Chem. Phys.* **75**, 5070 (1981).
73. C. G. Andeen, G. E. Matthews, M. K. Smith, and J. J. Fontanella, *Phys. Rev. B* **19**, 5293 (1979).
74. P. P. Yaney, D. M. Schaeffer, and J. L. Wolf, *Phys. Rev. B* **11**, 2460 (1975).
75. G. K. Miner, T. P. Graham, and G. T. Johnston, *J. Chem. Phys.* **57**, 1263 (1972).
76. D. J. M. Bevan, J. Strahle, and O. J. Greis, *J. Solid State Chem.* **44**, 75 (1982).
77. P. J. Bendall, C. R. A. Catlow, J. Corish, and P. W. M. Jacobs, *J. Solid State Chem.* **51**,

159 (1984).

78. C. G. Andeen, J. J. Fontanella, M. C. Wintersgill, P. J. Welcher, R. J. Kimble, and G. E. Matthews, *J. Phys. C: Solid State Phys.* **14**, 3557 (1981).
79. J. B. Fenn, J. C. Wright, and F. K. Fong, *J. Chem. Phys.* **59**, 5591 (1973).
80. C. R. A. Catlow, *J. Phys. C: Solid State Phys.* **9**, 1859 (1976).
81. K. M. Cirillo-Penn and J. C. Wright, *Phys. Rev. B* **41**, 10799 (1990).
82. C. R. A. Catlow, A. V. Chadwick, G. N. Greaves, and L. M. Moroney, *Nature* **312**, 601 (1984).
83. S. K. Batygov, Y. K. Voronko, L. S. Gaigerova, and V. S. Federov, *Opt. Spectrosc.* **35**, 505 (1973) [*Opt. Spektrosk.* **35**, 868 (1973)].
84. A. R. Allnatt and P. S. Yuen, *J. Phys. C: Solid State Phys.* **8**, 2199 (1975).
85. A. O. Wright, L. R. Olsen, and J. C. Wright, *Chem. Phys. Lett.* **244**, 395 (1995).
86. L. R. Olsen, A. O. Wright, and J. C. Wright, *Phys. Rev. B* **53**, 14135 (1996).
87. W. Bollman and R. Reimann, *Phys. Stat. Sol. (a)* **16**, 187 (1973).
88. Y. L. Khong, G. D. Jones, and R. W. G. Syme, *Phys. Rev. B* **48**, 672 (1993).
89. I. B. Aizenberg, M. S. Orlov, and A L. Stolov, *Opt. Spectrosc.* **38**, 660 (1975). [*Opt. Spectrosk.* **38**, 1144 (1975)].
90. K. Muto and K. Awazu, *J. Phys. Chem. Solids* **29**, 1269 (1968).
91. M. V. Johnston and J. C. Wright, *J. Phys. Chem.* **85**, 3064 (1981).
92. C. R. A. Catlow, *J. Phys. Chem. Solids* **38**, 1131 (1977).
93. R. J. Elliott, W. Hayes, G. D. Jones, H. F. MacDonald, and C. T. Sennett, *Proc. Roy. Soc. (London)* **A289**, 1 (1965).
94. G. D. Jones, S. Peled, S. Rosenwaks, and S. Yatsiv, *Phys. Rev.* **183**, 353 (1969).
95. R. J. Reeves, G. D. Jones, and R. W. G. Syme, *Phys. Rev. B* **40**, 6475 (1989).
96. K. M. Murdoch and G. D. Jones, *Phys. Rev. B* **58**, 12020 (1998).
97. N. J. Cockroft, D. Thompson, G. D. Jones, and R. W. G. Syme, *J. Chem. Phys.* **86**, 521 (1987).
98. A. Edgar, C. A. Freeth, and G. D. Jones, *Phys. Rev. B* **15**, 5023 (1977).
99. R. J. Reeves, G. D. Jones, N. J. Cockroft, T. P. J. Han, and R. W. G. Syme, *J. Lumin.* **38**, 198 (1987).
100. I. T. Jacobs, G. D. Jones, K. Zdánský, and R. Satten, *Phys. Rev. B* **3**, 2888 (1971).
101. R. J. Reeves, K. M. Murdoch, and G. D. Jones, *J. Lumin.* **66 & 67**, 136 (1996).
102. R. J. Reeves and R. M. Macfarlane, *J. Opt. Soc. Am. B* **9**, 763 (1992).
103. N. J. Cockroft, T. P. J. Han, R. J. Reeves, G. D. Jones, and R. W. G. Syme, *Opt. Lett.* **12**, 36 (1987).
104. A. Edgar and H. K. Welsh, *J. Phys. C: Solid State Phys.* **8**, L336 (1975).
105. T. Attenberger, U. Bogner, G. D. Jones, and K. M. Murdoch, *J. Phys. Chem. Solids* **58**, 1513 (1997).
106. F. J. Gustafson and J. C. Wright, *Anal. Chem.* **51**, 1762 (1979).
107. M. V. Johnston and J. C. Wright, *Anal. Chem.* **51**, 1774 (1979).

108. M. V. Johnston and J. C. Wright, *Anal. Chem.* **53**, 1054 (1981).
109. F. J. Gustafson and J. C. Wright, *Anal. Chem.* **49**, 1680 (1977).
110. F. K. Fong, *Prog. Solid State Chem.* **3**, 135 (1967).
111. C. R. A. Catlow, M. J. Norgett, and T. A. Ross, *J. Phys. C: Solid State Phys.* **10**, 1627 (1977).
112. G. K. Liu, J. V. Beitz, and J. Huang, *J. Chem. Phys.* **99**, 3304 (1993).
113. P. Thouvenot, S. Hubert, and N. M. Edelstein, *Phys. Rev. B* **50**, 9715 (1994).
114. D. J. Newman and W. Urban, *Adv. Phys.* **24**, 793 (1975).
115. K. M. Murdoch, N. M. Edelstein, L. A. Boatner, and M. M. Abraham, *J. Chem. Phys.* **105**, 2539 (1996).
116. M. Illemassene, K. M. Murdoch, N. M. Edelstein, and J. C. Krupa, *J. Lumin.* **75**, 77 (1997).
117. K. M. Murdoch, R. Cavellec, E Simoni, M. Karbowiak, S. Hubert, M. Illemassene, and N. M. Edelstein, *J. Chem. Phys.* **108**, 6353 (1998).
118. G. K. Liu, S. T. Li, V. V. Zhorin, C. K. Loong, M. M. Abraham, and L. A. Boatner, *J. Chem. Phys.* **109**, 6800 (1998).
119. M. Illemassene, N. M. Edelstein, K. M. Murdoch, M. Karbowiak, R. Cavellec, and S. Hubert, in preparation.
120. T. J. Barker, R. G. Denning, and J. R. G. Thorne, *Inorg. Chem.* **26**, 1721 (1987).
121. K. M. Murdoch, A. D. Nguyen, N. M. Edelstein, S. Hubert, and J. C. Gâcon, *Phys. Rev. B* **56**, 3038 (1997).
122. J. C. Wright, P. C. Chen, J. P. Hamilton, A. Zilian, and M. J. LaBuda, *Appl. Spectrosc.* **51**, 949 (1997).
123. D. S. Moore and J. C. Wright, *Chem. Phys. Lett.* **66**, 173 (1979).

INDEX